2013 IEEE Custom Integrated Circuits Conference

(CICC 2013)

San Jose, California, USA
22-25 September 2013

Pages 481-972

IEEE Catalog Number: CFP13CIC-POD
ISBN: 978-1-4673-6145-3

Copyright © 2013 by the Institute of Electrical and Electronic Engineers, Inc
All Rights Reserved

Copyright and Reprint Permissions: Abstracting is permitted with credit to the source. Libraries are permitted to photocopy beyond the limit of U.S. copyright law for private use of patrons those articles in this volume that carry a code at the bottom of the first page, provided the per-copy fee indicated in the code is paid through Copyright Clearance Center, 222 Rosewood Drive, Danvers, MA 01923.

For other copying, reprint or republication permission, write to IEEE Copyrights Manager, IEEE Service Center, 445 Hoes Lane, Piscataway, NJ 08854. All rights reserved.

***This publication is a representation of what appears in the IEEE Digital Libraries. Some format issues inherent in the e-media version may also appear in this print version.**

IEEE Catalog Number: CFP13CIC-POD
ISBN 13:   978-1-4673-6145-3
ISSN: 0886-5930

Additional Copies of This Publication Are Available From:

Curran Associates, Inc
57 Morehouse Lane
Red Hook, NY 12571 USA
Phone: (845) 758-0400
Fax: (845) 758-2633
E-mail: curran@proceedings.com
Web: www.proceedings.com

Table of Contents

Technical Sessions
Monday, September 23 – Wednesday, September 25

Educational Sessions
Tuesday, September 24 and Wednesday, September 25

Session 1 – Plenary Session

Monday, 9/23/2013, 8:15 am
Oak Ballroom
8:15 am
Welcome and Opening Remarks
Awards Presentations
Keynote Speaker Introduction
Aurangzeb Khan, Altia Systems

Keynote Presentation
Digital Analog Design
Mark Horowitz, Stanford University

The past 30 years have seen an enormous growth in the power and sophistication of digital design tools, while progress in analog tools has been much more modest. Digital tools use many abstractions to allow them to validate implementations match the functional models, and the composition of cells matches the composition of the functional models. While there are many reasons why this is more difficult for analog circuits, it can be done. To prove this point, this talk presents how to leverage the fact that the result surface of analog designs are smooth to create ways to formally validate analog models to instances, define analog fault models, and even efficiently explore the effect of process variations.

Session 2 -- Microsystems for Biomedical and Sensing Applications

Monday, 9/23, 10:00 am
Oak Ballroom
Session Chair: Christophe Antoine, Analog Devices
Session Co-Chair: Stephen O'Driscoll, University of California, Davis

10:00 am   **Introduction**

Biomedical and sensing applications present new challenges for IC and system designers. In this session, five papers representing advances in these areas are presented.

10:05 am   **A Broadband Biosensor Interface IC for Miniaturized Dielectric Spectroscopy from**
2-1   **MHz to GHz,** *M. Bakhshiani, M. A. Suster, and P. Mohseni, Case Western Reserve University*   **1**

This paper describes a broadband biosensor interface IC as part of a miniaturized measurement platform for MHz-to-GHz dielectric spectroscopy. Developed in 0.35µm 2P/4M RF CMOS, the IC measures the frequency-dependent S21 magnitude and phase of a microfabricated microfluidic dielectric sensor, when the sensor is loaded with a solution-under-test (SUT).

10:30 am
2-2

Capacitive Proximity Communication with Distributed Alignment Sensing for Origami Biomedical Implants, *M. Loh, A. Emami-Neyestanak, California Institute of Technology*   **5**

Origami implant design is a 3D integration technique which addresses size and cost constraints in biomedical implants. A capacitive proximity interconnect that enables this technique is presented. It embeds an alignment sensor that measures link quality directly and simplifies adaptation to alignment. The sensor and transceiver share functional blocks, saving power and area. Data rates 10-60 Mbps are achieved over 4-12µm parylene-C, with efficiencies up to0.180 pJ/bit.

10:55 am
2-3

An Active Rectifier/Regulator Combo Circuit for Powering Biomedical Implants, *E. Lee, Alfred Mann Foundation*   **9**

A circuit that combines an active rectifier and a linear regulator, which will be referred to as a rectulator, is proposed in this paper. The main transistor in the rectulator is used for both the rectification of the AC input (VAC) and the regulation of the DC output (VO) to reduce the overall voltage drop between VAC and VO. The rectulator has a high power efficiency (nP) and very good amplitude modulation (am) rejection on VAC against AC field strength variations. A full-wave rectulator implemented in a 0.18µm CMOS process showed an AM rejection > 45dB for an AM frequency < 2kHz. A nP of 90.7% was measured for an input frequency of 5MHz, VO = 2.5V and an output power of 10.5mW.

11:20 am
2-4

65nW CMOS Temperature Sensor for Ultra-Low Power Microsystems, *Seokhyeon Jeong, Jae-yoon Sim*, David Blaauw, Dennis Sylvester, University of Michigan, *Pohang University of Science and Technology*   **13**

A temperature sensor using a novel process-invariant temperature sensing element and voltage to current converter is proposed for battery-operated ultra-low power micro systems. Measurements from a 180nm CMOS test chip show power consumption of 65nW with an inaccuracy of +1.3°C/-1.4°C over a temperature range of 0°C to 100°C after 2-point calibration.

11:45 am
2-5

Design and Characterization of Electronic Sensing System for a 13 x 13 Biomechanical Ground Reaction Sensor Array, *Q. Guo, M. A. Suster*, R. Surapaneni, C. H. Mastrangelo and D. J. Young, The University of Utah, *Case Western Reserve University*   **17**

This paper presents the design and characterization of an electronic sensing system interfaced with a high-density flexible biomechanical ground reaction sensor array. The prototype system can measure real-time ground force, shear strain and sole deformation associated with a human bipedal locomotion, thus providing zero-velocity correction to an inertial measurement unit to improve navigation accuracy.

Session 3 -- High-Speed Wireline Timing Recovery and PLLs

Monday, 9/23, 10:00 am
Fir Ballroom
Session Chair: Samuel Palermo, Texas A&M University
Session Co-Chair: Kimo Tam, Analog Devices

10:00 am   **Introduction**

This session presents various design techniques for ADC-based and burst-made timing recovery systems and PLLs with peaking-free transfer functions and low-area utilization.

10:05 am
3-1

Design Metrics for Blind ADC-Based Wireline Receivers (Invited), *Ali Sheikholeslami and Hirotaka Tamura*, University of Toronto and *Fujitsu Labs*   **21**

This paper compares blind ADC-based receivers against binary and phase-tracking ADC-based receivers in terms of their design complexity and cost, and derives equations that relate the required ADC resolution (ENOB) to channel loss and to the characteristics of the FFE/DFE that follow the ADC.

10:55 am
3-2

A 10Gbps, 1.24pJ/bit, Burst-Mode Clock and Data Recovery with Jitter Suppression, *Ming-Chiuan Su, Wei-Zen Chen, Pei-Si Wu*, Yu-Hsian Chen*, Chao-Cheng Lee*, Shyh-Jye Jou, National Chiao Tung University, *Realtek Corp.*   **29**

A 10Gbps, 1/5 rate burst mode CDR is reconfigurable between data gating and phase tracking modes to achieve instantaneous phase-locking with low jitter suppression. Incorporated with 1:5 demultiplexer, it achieves a high energy efficiency of 1.24pJ/bit. The prototype chip is fabricated in UMC 55nm CMOS technology.

11:20 am
3-3

A 9.2-GHz Digital Phase-Locked Loop with Peaking-Free Transfer Function, *Sigang Ryu, Hwanseok Yeo, Yoontaek Lee, Seuk Son, Jaeha Kim, Seoul National University*   **33**

This paper describes a digital phase-locked loop (PLL) that realizes a peaking-free jitter transfer. That is, the PLL's second-order transfer function does not have a closed-loop zero. Such a PLL does not exhibit overshoots in the phase step response and achieves fast settling. Unlike the previously-reported peaking-free PLLs, the proposed PLL implements the peaking-free loop filter directly in digital domain without requiring additional components. A time-to-digital converter (TDC) is implemented as, a set of three binary phase-frequency detectors that over sample the timing error with time-varying offsets, achieving a linear TDC gain and PLL bandwidth insensitive to the jitter condition. And a 9.2-GHz digitally-controlled LC oscillator (DCO) with transformer-based tuning realizes a predictable DCO gain set by a ratio between two digitally-controlled currents. The prototype 9.2-GHz-output digital PLL fabricated in a 65nm CMOS demonstrates a fast settling time of 1.58-µs with 700-kHz bandwidth. The PLL has a 3.477-psrms divided clock jitter and -120dBc/Hz phase noise at 10-MHz offset while dissipating 63.9-mW at a 1.2-V supply.

11:45 am
3-4

A Sub-200 fs RMS Jitter Capacitor Multiplier Loop Filter-Based PLL in 28 nm CMOS for High-Speed Serial Communication Applications, *Burak Çatl, Ali Nazemi, Tamer Ali, Siavash Fallahi, Yang Liu, Jaehyup Kim, Mohammed Abdul–Latif, Mahmoud Reza Ahmadi, Hassan Maarefi, Afshin Momtaz, and Namik Kocaman, Broadcom Corporation*   **37**

An 8.0 GHz to 12.2 GHz PLL with a capacitor multiplier-based active loop filter is designed in a 28 nm digital CMOS process. A passive loop filter-based version of the PLL is also implemented for comparison. While the PLL area is comparable to that of digital PLLs, the PLL performance is as good as that of an analog PLL that employs a passive loop filter. The capacitor multiplier-based active loop filter PLL has a jitter performance of 198 fs(rms), while its passive loop filter-based counterpart shows a jitter performance of 195 fs (rms). The PLL occupies 0.093 mm^2 and consumes 15.5 mA at 1.0V.

Session 4 -- RF Building Blocks

Monday, 9/23, 10:00 am
Pine Ballroom
Session Chair: Andrea Mazzanti, University of Padova
Session Co-Chair: Earl McCune

10:00 am **Introduction**

Moving beyond traditional technologies this session presents digital PLL nonlinearity cancellation, integrated tunable duplexer, smartphone T/R switch, and a programmable broadband integrated phase shifter.

10:05 am **Nonlinearity Cancellation in Digital PLLs (Invited),** *S. Levantino, C. Samori, Politecnico*
4-1 *di Milano* **41**

The spur level in digital fractional-N PLLs is often bounded by TDC resolution and linearity. Methods for mitigating TDC nonlinearity tend to increase phase-noise level. By contrast, PLL architectures based on digital-to-time converters enable nonlinearity cancellation and spur reduction with no added noise at lower design complexity and power consumption.

10:55 am **Hybrid Transformer-Based Tunable Integrated Duplexer with Antenna Impedance**
4-2 **Tracking Loop,** *S. Abdelhalem, P. Gudem*, L. Larson**, University of California at San Diego, *Qualcomm Inc., **Brown University* **49**

Electrical balance between the antenna and the balance network impedances is crucial for achieving high isolation in a hybrid transformer duplexer. In this paper, an auto calibration loop for tuning a novel integrated balance network to track the antenna impedance variations is introduced. It achieves an isolation of more than 50 dB in the transmit and receive bands, with an antenna VSWR within 2:1, and between 1.7 and 2.2 GHz. The duplexer, along with a cascaded direct-conversion receiver, achieves a noise figure of 5.3 dB, a conversion gain of 45 dB and consumes 34 mA. The insertion loss in the transmit path was less than 3.8 dB. Implemented in a 65-nm CMOS process, the chip occupies an active area of 2.2 mm^2.

11:20 am **A Smartphone SP10T T/R Switch in 180nm SOI CMOS with 8kV ESD Protection by** **53**
4-3 **Co-Design,** *X.S. Wang, X. Wang*, F. Lu**, L. Wang**, R. Ma**, Z. Dong**, L. Sun, A. Wang**, C.P. Yue, D. Wang***, A. Joseph***, University of California, Santa Barbara., *OmniVision Technologies., **University of California, Riverside., ***IBM Microelectronics*

This paper reports the first 8kV+ ESD-protected SP10T transmit/receive (T/R)antenna switch for quad-band (0.85/0.9/1.8/1.9GHz) GSM and multiple WCDMA smartphones fabricated in an 180nm SOI CMOS. A novel physics-based switch-ESD co-design methodology is applied to ensure full-chip optimization for a SP10T test chip and its ESD protection circuit simultaneously.

11:45 am **A Lumped Component Programmable Delay Element for Ultra-Wideband** **57**
4-4 **Beamforming,** *Naga Rajesh, Shanthi Pavan, Indian Institute of Technology Madras*

We introduce a ladder filter based programmable time delay element for beamforming in Ultra-Wideband (UWB) systems. When compared to conventional methods based on the tapped delay line architecture, our technique achieves lower power dissipation, better area efficiency, and finer delay and gain resolution more efficiently. The proposed architecture is more scalable, has better parasitic absorption capability and highly programmable with delay and gain resolution dependent only on transconductor resolution. A prototype delay line designed for the 3.1-10.6 GHz UWB range achieves a delay range of 80 ps with 0.5 ps resolution and a gain range of -30 dB to +10 dB with 0.15 dB step. Fabricated in a 0.25 μm SiGe BiCMOS process, the delay element occupies an active area of 1 mm^2 and consumes 47 mW from a 2.5 V supply. A four antenna beamforming system using the delay element can achieve scanning range of +/-51deg with resolution of 0.86 deg for antenna spacing of 10 mm.

Session 5 -- Beyond 14nm Technology Circuit Interaction

Monday, 9/23, 10:00 am
Cedar Ballroom
Session Chair: Rajiv Joshi, IBM TJ Watson Research Center
Session Co-Chair: Ramnath Venkatraman, LSI Corp.

10:00 am **Introduction**

This session covers technology-circuit interaction needs beyond 14nm including advanced device and lithographic considerations. Multiple patterning and FinFETs add significant complexity. This session further covers efficient methodology and solutions needed for product metrics applied to a myriad of design-technology choices.

10:05 am **The Past, Present, and Future of Design-Technology Co-Optimization (Invited)**, *G.*
5-1 *Yeric, B. Cline, S. Sinha, D. Pietromonaco, V. Chandra, and R. Aitken, ARM* **61**

Design-Technology Co-Optimization (DTCO) has evolved into a multi-faceted, multi-lateral co-optimization below 20nm, where double patterning and FinFETs create significant complexities. Effective DTCO now involves end product metrics applied to a myriad of technology choices. A future of even more complex lithography, devices, and reliability will drive continued evolution in DTCO.

10:55 am **From 2D-Planar to 3D-Non_Planar Device Architecture: A Scalable Path Forward?**
5-2 **(Invited)**, *G. Shahidi, IBM T.J. Watson Research Center* **69**

11:45 am **Foundations for Scaling to 7nm and Beyond (Invited)**, *R. Schenker, V. Singh, Intel*
5-3 *Corporation* **77**

Session 6 – Forum Session: 20 nm Design Challenges

Monday, 9/23, 1:30 pm
Oak Ballroom
Organizers – Pavan Hanumolu, Kimo Tam, Manoj Sachdev
Moderator – Manoj Sachdev

1:30 **Introduction**

1:35 pm **Analog/wireline design in an increasingly digital process,** *Matt Straayar, Maxim*
6-1
Analog design in mostly digital process is a challenge. What are issues of embedding sensitive analog circuits surrounded by noisy digital transistors in 20 nm? **N/A**

2:00 pm **Tools/rules driving designers vs. designers driving tools?,** *Ravi Subramanian, Berkeley*

6-2   *Design Automation*   **N/A**

Design automation is the key to enhance designer productivity. Are we relying too heavily on tools in scaled geometries? Are design rules & tools are too restrictive in 20 nm?

2:25 pm   **DFM Issues for 20 nm Analog,** *Stacy Ho, Mediatek*   **N/A**
6-3

The emergence of finFETs is an important evolutionary step for transistor scaling? How one can exploit its benefits for analog circuits while ensuring yield and reliability? What for the Design For Manufacturing (DFM) challenges that we should be mindful of?

2:50 pm   **Analog design in 20 nm - putting it all together,** *Madhukar Reddy, Maxilinear*   **N/A**
6-4

Successful implementation of analog circuit in Systems on Chip (SoC) requires experience, expertise. How one should execute design, exploit tools while ensuring good yield and reliability in 20 nm?

Session 7 -- Power Management

Monday, 9/23, 1:30 pm
Fir Ballroom
Session Chair: Christoph Sandner, Infineon
Session Co-Chair: Raj Amirtharajah, UC Davis

1:30 pm   **Introduction**

Effective power management requires innovative techniques to minimize system cost while maximizing efficiency. This session covers a wide array of advances in power management spanning battery interfaces, high voltage converters, LED lighting, digital control, and energy harvesting.

1:35 pm   **BIF--Battery Interface Standard for Mobile Devices (Invited),** *W. Furtner, S.*
7-1   *Schaecher, M. Littow*, L. Cimaz*, and P. Leinonen**, Infineon Technologies AG,*
          *ST-Ericsson and **Nokia   **81**

The MIPI® Alliance Battery Interface (BIF) is the first comprehensive battery communication interface standard for mobile devices. MIPI BIF is a robust, scalable and cost-effective single-wire communication interface between the mobile terminal and smart or low cost batteries. It is suited for removable batteries as well as for embedded batteries.

2:00 pm   **A 40V 10W 93%-Efficiency Current-Accuracy-Enhanced Dimmable LED Driver with**
7-2   **Adaptive Timing Difference Compensation for Solid-State Lighting Applications,** *D.*
          Park and H. Lee, University of Texas at Dallas   **89**

This paper describes a floating buck dimmable LED driver for solid-state lighting applications. Adaptive timing difference compensation is proposed to enable the driver to achieve high accuracy of the average LED current, fast settling time, and high-frequency operation over a wide range of input voltages and number of LED loads. The power efficiency of the proposed LED driver is benefited from the capabilities of using synchronous rectification and having no sensing resistor in the power stage. The synchronous rectification under high input supply voltage is enabled by a proposed high-speed and low-power gate driver with pseudo-digital level shifters. Implemented in a 0.35μm 50V CMOS process, the proposed 40V LED driver can operate at 1MHz and achieve 93% peak power efficiency when driving up to 10 series-connected LEDs. It has only 2.8% current error from the average LED current of 345mA and settles within 8.5μs under different line and load variations. The performances of the proposed driver significantly outperform all state-of-the-art counterparts.

2:25 pm
7-3

A Stackable Switched-Capacitor DC/DC Converter IC for LED Drivers with 90% Efficiency, *Chengrui Le, Mitchell Kline*, Daniel L. Gerber*, Seth R. Sanders*, Peter R. Kinget, Columbia University, *University of California, Berkeley* **93**

A stackable switched-capacitor DC-DC converter IC for a hybrid-SC-resonant LED driver is presented. The IC can handle a range of input voltages through chip-stacking in the voltage domain. The tested driver delivers 17.6W with 90% peak efficiency and maintains >85% efficiency over a rectified voltage range from 160VDC to 180VDC.

2:50 pm
7-4

A 100V Gate Driver with Sub-Nanosecond-Delay Capacitive-Coupled Level Shifting and Dynamic Timing Control for ZVS-Based Synchronous Power Converters, *Z. Liu and H. Lee, The University of Texas at Dallas* **97**

A high-voltage high-speed gate driver to enable synchronous rectifiers with zero-voltage-switching (ZVS) operation is presented in this paper. A capacitive-coupled level-shifter (CCLS) is developed to achieve negligible propagation delay and static current consumption. With only 1 off-chip capacitor, the proposed gate driver possesses strong driving capability and requires no external floating supply for the high-side driving. A dynamic timing control is also proposed not only to enable ZVS operation in the converter for minimizing the capacitive switching loss, but also to eliminate the converter short-circuit power loss. Implemented in a 0.5μm HV CMOS process, the proposed CCLS of the gate driver can shift up a 5V signal to the 100V DC rail with sub-nanosecond delay, improving the FoM by at least 29 times compared with that of state-of-the-art counterparts. The dynamic dead-time control properly enables ZVS operation in a synchronous buck converter under different input voltages (30V to 100V). The power losses of the high-voltage buck converter are thus greatly reduced under different load currents, achieving a maximum power efficiency improvement of 11.5%.

3:15 pm   **BREAK**

3:30 pm
7-5

A Compact 120-MHz 1.8V/1.2V Dual-Output DC-DC Converter With Digital Control, *S. Arora, D.K. Su, B. A. Wooley, Stanford University* **101**

A dual-output cascaded dc-dc converter for embedded applications uses a programmable switching frequency up to 120 MHz, output stage segmentation, and cascoding to achieve high power efficiency with small output inductors. A fast digital constant-off-time controller provides the suppression of cross-regulation among multiple output voltages.

3:55 pm
7-6

A Monolithic Digitally Controlled Ripple-Based DC-DC Converter with Digital Inductor Current Sensor, *Man Pun Chan, Philip K.T. Mok, The Hong Kong University of Science and Technology* **105**

A ripple-based digital controller is presented which has incorporated a digital inductor current sensor that does not require extra ADCs and occupies a small chip. Both the digital sensor and controller are fully synthesizable with the total chip area of 0.048 mm^2 in 0.13-μm digital CMOS process.

4:20 pm

A Fully Integrated Battery-Connected Switched-Capacitor 4:1 Voltage Regulator with 70% Peak Efficiency Using Bottom-Plate Charge Recycling, *T. Tong, X. Zhang,*

7-7   *W. Kim, D. Brooks, G.-Y. Wei, Harvard University*   **109**

This work presents a switched-capacitor (SC) DC-DC voltage regulator that converts a 3.7V battery voltage down to ~0.8V in order to power the 'brain' SoC of a flapping-wing microrobotic bee. A cascade of two 2:1 SC converters offers high efficiency for a 4:1 conversion ratio. A charge recycling technique reduces the flying capacitor's bottom-plate parasitic loss by 50% and overall conversion efficiency reaches 70%. The output droop is less than 10% of the nominal output voltage for a worst-case 47mA load step.

4:45 pm   **A 110nA Synchronous Boost Regulator With Autonomous Bias Gating for Energy**
7-8   **Harvesting,** *Khondker Z. Ahmed and Saibal Mukhopadhyay, Georgia Institute of Technology* **113**

An autonomously bias gated synchronous boost regulator consuming 110nA at 1V is demonstrated in 130nm CMOS. The IC generates regulated 1V output from 30mV input, starts up autonomously (battery-less) at 265mV, and regulates output ranging from 0.78V-3.3V. The peak efficiency is 83% with 10µA and 85% with 10mA load.

5:10 pm   **A Power Sensor with 80ns Response Time for Power Management in**
7-9   **Microprocessors,** *S. Bhagavatula, B. Jung, Purdue University*   **117**

A real-time, on-chip power sensor that estimates load currents and on-chip temperatures concurrently is presented. It occupies an area of 0.11mm X 0.09mm in 130n mCMOS technology. With a simplified 1-point calibration and a response time of 80ns, it shows improvements in input dynamic range by 10 X, response time by 6 X and sensitivity by 3 X over previous such sensors. A current reference with a measured temperature coefficient 91ppm/C (-20Cto 120C) is presented. This reference is used for online calibration of the power sensor to enable greater tolerance to PVT variations and aging effects.

Session 8 -- AMS Verification in Advanced Technologies

Monday, 9/23, 1:30 pm
Pine Ballroom
Session Chair: Hidetoshi Onodera, Kyoto University
Session Co-Chair: Yu (Kevin) Cao, Arizona State University

1:30 pm   **Introduction**

Verification of AMS circuits is increasingly challenging. This session presents novel AMS simulation and emulation techniques, and advanced reliability and performance issues with technology scaling.

1:35 pm   **Discretization and Discrimination Methods for Design, Verification, and Testing of**
8-1   **Analog/Mixed-Signal Circuits (Invited),** *J. Kim, J. Lee, D.-G. Song, T. Kim, K.-H. Kim, S. Jung, S. Youn, Seoul National University*   **121**

This paper describes how the difficult problems of designing, verifying, and testing analog circuits in presence of variability can be converted to easier ones by discretizing the search spaces or discriminating one case from another. For instance, discretizing the continuous design space of analog circuits enables the use of an efficient, predictive global circuit optimizer. Also, discretizing the initial condition space of a circuit enables one to establish its global convergence property over the entire space by exploring only a small number of samples. Lastly, discriminating the test responses of the circuit with and without a fault in consideration of the underlying statistical distribution provides a formal guide on how to quantify the fault coverage of analog/mixed-signal circuit tests. It is noteworthy that it is variability that introduces the cross-correlations in the performance metrics, convergence behaviors, and test responses between two nearby candidates in consideration and therefore enables the use of discretization and discrimination methods listed in this paper. The proposed methods are demonstrated on the practical examples of sizing an operational amplifier, verifying the correct start-up of a coupled ring oscillator, and composing a test suite for screening faults in a digitally-controlled phase interpolator circuit.

2:25 pm
8-2

Indirect Performance Sensing for On-Chip Analog Self-Healing via Bayesian Model Fusion, *S. Sun, F. Wang, S. Yaldiz, X. Li, L. Pileggi, A. Natarajan*, M. Ferriss*, J. Plouchart*, B. Sadhu*, B. Parker*, A. Valdes-Garcia*, M. Sanduleanu*, J. Tierno*, D. Friedman*, Carnegie Mellon University, *IBM T.J. Watson Research Center* **129**

On-chip analog self-healing requires low-cost sensors to accurately measure performance metrics. In this paper, we propose a novel approach of indirect performance sensing based upon Bayesian model fusion to facilitate inexpensive-yet-accurate on-chip measurement. A 25GHz differential Colpitts VCO is used to validate the proposed indirect performance sensing and self-healing methodology.

2:50 pm
8-3

Fast FPGA Emulation of Background-Calibrated SAR ADC with Internal Redundancy Dithering, *G. Wang, Y. Chiu, University of Texas at Dallas* **133**

A custom FPGA emulation platform for the verification of a slowly adapted, background calibration technique for successive-approximation-register (SAR) analog-to-digital converter (ADC) is demonstrated in an Altera DE4 board. The internal redundancy of a sub-binary SAR is exploited for the identification of ten leading bit weights in a 14.5-bit SAR ADC using pseudorandom bit sequence (PRBS) injection with background correlation. Experimental results reveal that the FPGA emulation achieves a 3000× speedup for the same simulation executed on a general-purpose microprocessor.

3:15 pm

BREAK

3:30 pm
8-4

Circuit Reliability Simulation Using TMI^2 (Invited), *Min-Chie Jeng, Cheng Hsiao, Ke-Wei Su, Chung-Kai Lin, Taiwan Semiconductor Manufacturing Company* **137**

This paper reviews existing circuit aging simulation approaches with focus on TMI^2.

4:20 pm
8-5

Scalable Behavior Modeling for 3D Field-Programmable ESD Protection Structures, *L. Wang, X. Wang*, Z. T. Shi**, R. Ma, C. Zhang, Z. Dong, F. Lu, H. Zhao** and A. Wang, University of California, Riverside, *Omnivision Technologies, **Marvell Semiconductor* **144**

This paper reports new accurate and scalable behavioral modeling for novel 3D field-programmable ESD protection circuits, which enables post-Si on-chip ESD protection design simulation. New field-programmable ESD protection devices were fabricated in CMOS-compatible processes. The behavior models were developed from ESD testing results and verified in SPICE circuit simulation.

4:45 pm
8-6

Quasi-3D method: Time-efficient TCAD and Mixed-Mode Simulations on finFET Technologies, *G. Hellings, S-H Chen, D. Linten, M. Scholz, G. Groeseneken, imec,* **148**

The Quasi-3D allows to drastically speed up TCAD and mixed-mode simulations of finFET technologies, by solving on well-chosen 2D finFET cross sections. The method accurately reproduces important transistor metrics requiring only 1/20th of the simulation time.

5:10 pm
8-7

Gate Stack Resistance and Limits to CMOS Logic Performance, *R. A. Wachnik, S. Lee, L. H. Pan, N. Lu, H. Li, R. Bingert **, M. Randall, S. Springer, C. Putnam, IBM Corporation, **ST Microelectronics*   **152**

Measured data from five generations of CMOS technology including polysilicon and High-K metal gate stacks shows a trend of increasing gate resistance. The data are analyzed to determine horizontal and vertical components in terms of scalable model parameters. Gate resistance affects performance of a 20nmreplacement gate technology.

Session 9 -- Wireless Transceivers

Monday, 9/23, 1:30 pm
Cedar Ballroom
Session Chair: Julian Tham, Broadcom
Session Co-Chair: Jonathan Borremans, imec

1:30 pm

Introduction

This session presents papers on advances in wide-band receiver designs and a calibrated software-defined radio. It also presents WLAN, GPS and ultra low-power receivers and transceivers.

1:35 pm
9-1

IIP2 and HR Calibration for an 8-Phase Harmonic Recombination Receiver in 28nm, *B. van Liempd, J. Borremans, S. Cha*, E. Martens, H. Suys, J. Craninckx, imec vzw, *Renesas Electronics Corporation*   **156**

Fully integrated CMOS receivers achieve high linearity and low noise due to harmonic recombination, but suffer from limited IIP2 and harmonic rejection due to mismatch and inaccuracies. This paper presents an 8-phase harmonic recombination receiver with independent IIP2, HR3 and HR5 calibration techniques. Calibrated >80dBm IIP2, >70dB HR3 and >75dB HR5 are measured.

2:00 pm
9-2

Advances in the Design of Wideband Receivers (Invited), *D. Murphy, M. Mikhemar, A. Mirzaei, H. Darabi, Broadcom Corporation*   **160**

To be practical, wideband receivers must tolerate large out-of-band blockers, which can desensitize the receiver through gain compression or reciprocal mixing with LO phase noise. This paper reviews how a new noise-cancelling receiver architecture – that utilizes 3 important circuit innovations – mitigates gain compression without compromising noise figure. While the architecture is still susceptible to reciprocal mixing, it is shown how a recently proposed reciprocal mixing cancelling technique (if incorporated into the receiver) can eliminate the need for a dramatic rise in LOGEN current.

2:50 pm	**An Asymmetric Dual-Channel Reconfigurable Receiver for GNSS in 180nm CMOS,**
9-3	*Nan Qi, Baoyong Chi, Yang Xu, Zhou Chen, Jun Xie, Yang Xu, Zheng Song, Zhihua Wang, Tsinghua University* **168**

A fully integrated dual-channel reconfigurable receiver supporting all the GNSS signals is presented. The two channels share the frequency synthesizer and RF front-end circuits, but employ separate asymmetric IF strips to support simultaneous dual-constellation reception. The 2nd IF strip can be configured to a dual-conversion mode to lower the sampling rate.

3:15 pm	**BREAK**

3:30 pm	**A 5-GHz 11.6-mW CMOS Receiver for IEEE 802.11a Applications,** *A. Homayoun, B.*
9-4	*Razavi, University of California, Los Angeles* **172**

A direct-conversion receiver employs a 1-to-6 transformer as a low-noise amplifier along with passive mixers and noninvasive baseband filters. Realized in 65-nm CMOS technology, the receiver provides an average noise figure of 5.3dB and a sensitivity of -70 dBm at a data rate of 54 Mb/s. The prototype draws 11.6 mW from a 1-V supply and occupies an active area of 0.18 mm^2.

3:55 pm	**An Adaptive Predistorter for Wireless LAN RFSoC with embedded PA and T/R switch**
9-5	**in 55nm CMOS,** *K. Muhammad, M.-C. Chen, K.-H. Wang, K.-P. Ma, Y.-L. Hiseh, W.-S. Hsu, Y.-Y. Fu, M.-C. Lee, S.-Y. Hsiao, C.-M. Hung, MStar Semiconductor Inc.* **176**

We present an adaptive predistortion system for a WLAN transceiver in 55nm CMOS. The forward DSP path utilizes complex gain predistortion while the APD module in the feedback path computes AMAM and AMPM coefficients by comparing ideal transmit signal with the distorted signal from the receiver. This module operates with various calibration signals generated on-chip in addition to TX data. Measurement results show improvement of EVM by 1dB with the proposed approach. Improvement of P1dB of more than 3dB was obtained using fully automatic processing. The total solution utilizes 120k gates.

4:20 pm	**A 116nW Multi-Band Wake-Up Receiver with 31-bit Correlator and Interference**
9-6	**Rejection,** *S. Oh, N. Roberts*, D. Wentzloff*, Samsung, *University of Michigan* **180**

This paper presents a 116nW wake-up radio complete with crystal reference, interference compensation, and baseband processing, such that a selectable 31-bit code is required to toggle a wake-up signal. The baseband processor detects interferers and dynamically adjusts the receiver's sensitivity, mitigating the jamming problem to previous energy-detection wake-up radios.

4:45 pm	**A 11µW Sub-pJ/bit Reconfigurable Transceiver for mm-Sized Wireless Implants,** *A.*
9-7	*Yakovlev, J. Jang, D. Pivonka, A. Poon, Stanford University* **184**

A wirelessly powered 11µW transceiver for mm-sized wireless implants supporting TDMA has been designed and demonstrated through 35mm of porcine heart. The communication links have configurability for operation in diverse biological environments. The forward link achieves 4-20Mbps at 0.3pJ/bit, and the reverse link achieves 0.7-2Mbps at 0.7pJ/bit.

Session 10 – Panel Session

"Can biomedical electronic startups make money??"

Monday, 9/23, 3:30 pm
Oak Ballroom
Session Chair: Pedram Mohseni, Case Western Reserve University
Moderator: John McNeill, Worcester Polytechnic Institute

-Straight semiconductor startups are marginally viable
-Green/solar startups took the money and ran
-Will bioelectronics fare any better?

Panelists   Arjang Hassibi, UT Austin & Insilixa

Patrick Chiang, Fudan University

Chris Raanes, Viewray

Monday Poster Session

Monday, 9/23, 5:00 pm – 7:00 pm
Donner/ Siskiyou/ Cascade Ballrooms

M-1   **An All-Digital Time Difference Hold-and-Replication Circuit utilizing a Dual Pulse Ring Oscillator,** *Tetsuya Iizuka, Teruki Someya, Toru Nakura, Kunihiro Asada, University of Tokyo*   **188**

This paper presents a time-domain analog signal hold-and-replication circuit which holds an input time interval of two signal transitions and replicates it any number of times.65nm CMOS implementation accepts 100ps to 1.2ns time interval while occupying40x60μm^2 area. A TDC resolution enhancement application is also demonstrated in this paper.

M-2   **A 15-Bit Binary-Weighted Current-Steering DAC with Ordered Element Matching,** *T. Zeng, K. Townsend, J. Duan*, D. Chen, Iowa State University, *Broadcom Corporation*   **192**

This paper introduces a 15-bit binary-weighted current-steering DAC in a 130nm CMOS technology. The core area is less than 0.42mm^2, among which the MSB area is well within 0.021mm^2. Measurement results have shown that the DNL and INL can be reduced from 9.85LSB and 17.41LSB to 0.34LSB and 0.77LSB, respectively.

M-3   **A 500 MS/s 76dB SNDR Continuous Time Delta Sigma Modulator with 10MHz Signal Bandwidth in 0.18μm CMOS,** *Rune Kaald, Bjørnar Hernes*, Christian Holdø*, Frode Telstø*, Ivar Løkken*, Norwegian University of Science and Technology, *Hittite Microwave Norway*   **196**

A 5th order continuous-time delta sigma modulator is designed in 0.18um CMOS. At a sampling rate of 500MHz it achieves 76dB SNDR over a 10MHz bandwidth consuming 58mW. 5th order noise shaping is realized with 4 op amp based RC integrators and a VCO realizing an integrator and a 4 bit quantizer. A THD of-82.3dBc is achieved without calibration of feedback DACs. We address two problems related to VCO quantizers which have local feedback to also work as integrators with a high-speed excess loop delay compensation using capacitive summation and a method for reducing the switching activity of the output codes.

M-4   **A 0.1-3GHz Cell-Based Fractional-N All Digital Phase-Locked Loop Using ΔΣ Noise-Shaped Phase Detector,** *Yao-Chia Liu, Wei-Zen Chen, , Mao-Hsuan Chou*, Tsung-Hsien Tsai*, Yen-Wei Lee, Min-Shueh Yuan, National Chiao Tung University and *TSMC*   **200**

A 0.1-3 GHz, cell-based, fractional-N ADPLL with $\Delta\Sigma$ noise-shaped phase detector is presented. By dithering the reference phase and quantization phase error through an additional feedback path, linear phase detection and zero stabilization are accomplished without resort to sophisticated time to digital converter (TDC). The measured rms jitter from a 3GHz carrier is 1.9 ps with a multiplication factor of 60. Implemented in TSMC 40nm general purpose superb CMOS technology, the chip size is 280µm x 240µm. Keywords: TDC, $\Delta\Sigma$ phase detector, fractional-N ADPLL

M-5 **A Direct-Battery Hookup, Fully Integrated Stereo Headphone Module with 82 mW Output Power and 110 dB PSRR,** *Khaled Abdelfattah, Sherif Galal, Iuri Mehr, Alex Jianzhong Chen, Ahmet Tekin*, Xicheng Jiang, Todd L. Brooks, Broadcom Corp.,*Semtech* **204**

A complete stereo ground-referenced headphone module that supports direct battery hookup is integrated on a 40 nm mobile baseband SoC. Several techniques were employed to guarantee the reliability of the module circuitry under high output swing and limited safe operating regions in this low-voltage technology. Additional techniques to reduce area and enable low-cost integration were also employed. The module delivers 3.24 Vpp to a 16Ω load (82 mW) and achieves 100 dB dynamic range (DR), 110 dB PSRR, and 84 dB THD+N with an area of 0.675mm^2 on the SoC.

M-6 **A Fast-Locking Digital DLL with a High Resolution Time-to-Digital Converter,** *Dandan Zhang, Hai-gang Yang, Zhujia Chen, Wei Li, Zhihong Huang, Lijiang Gao, Wenrui Zhu, Institute of Electronics, Chinese Academy of Sciences* **208**

A fast-locking DLL is presented in this paper. By adopting a novel high resolution TDC, the total locking time is reduced to 8 clock cycles and shortened by 80% to 94.6% compared to previous closed-loop architectures .The measured RMS and p-p jitters are 2.3ps and 10ps respectively.

M-7 **A Stochastic Sampling Time-to-Digital Converter with Tunable 180-770fs Resolution, INL less than 0.6LSB, and Selectable Dynamic Range Offset,** *J. Tandon, T. Yamagichi, S. Komatsu, K. Asada, VDEC-D2T, University of Tokyo* **212**

We introduce a stochastic time-to-digital converter (TDC) that has 180-770fs tunable resolution, less than 0.6LSB INL, and selectable dynamic range offset .Previous arbiter-based TDCs have fine resolution but small dynamic range which is difficult to calibrate. Our approach uses comparators as decision elements to precisely control dynamic range offset.

M-8 **A 50 µW/Ch Artifacts-Insensitive Neural Recorder Using Frequency-Shaping Technique,** *J. Xu, Z. Yang, National University of Singapore* **216**

This paper presents a frequency-shaping (FS) neural recording interface that can inherently reject electrode offset, 5-10 times increase input impedance,4.5-bit extend system dynamic range, and provide much more tolerance to motion artifacts and 50/60 Hz power noise interferences. It is supposed to be more suitable for long-term brain-machine-interface (BMI) experiments.

M-9 **A Bipolar >40-V Driver in 45-nm SOI CMOS Technology,** *Yousr Ismail, Chang-Jin Kim, Chih-Kong Ken Yang, University of California, Los Angeles* **220**

A novel, switched-capacitor output driver combining both voltage-conversion and pulse-drive is introduced. The driver is implemented in 45-nm SOI CMOS technology and uses only process-compliant devices. It achieves a maximum output drive of 44 V and has a 36 KΩ output resistance while consuming 28 mA from a 1.5-V supply.

M-10   **High-Sensitivity Photodetection Sensor Front-End, Detecting Organophosphourous Compounds for Food Safety,** *L. Wan, Y. Qin, P. Chiang*, G. Chen, R. Liu, Z. Hong, Fudan University and *Oregon State University*   **224**

A high-sensitivity, high-dynamic range photo detection sensor front-end is presented, suitable for low-cost hand-held food safety systems. This sensor-on-a-chip for detecting organophosphorus compounds incorporates a non-chip deep N-well photo detector, pulse width modulation, and a folded reference. Measurement results show an input optical power dynamic range of 71dB, a sensitivity of $3.6nW/cm^2$.

M-11   **A 16-Channel, 359 μW, Parallel Neural Recording System Using Walsh-Hadamard Coding,** *Vahid Majidzadeh, Alexandre Schmid and Yusuf Leblebici, Swiss Institute of Technology (EPFL)*   **228**

Application of an algebraic coding to a multichannel parallel neural recording system is presented. The Walsh-Hadamard coding enables back-end hardware sharing between recording channels, using a linear and orthogonal superposition of the analog inputs. Moreover, this technique preserves the temporal information of the channels in contrast to the conventional architectures which use time-multiplexed ADC. In the proposed architecture a single ADC operates on a superposed signal, thereby the dynamic range of the ADC is effectively shared between channels benefiting from the sparsity characteristics of the channels. A 16-channel parallel recording system is implemented as a proof of concept. The system is implemented in a 0.18 μm CMOS technology and occupies $1.99 \ mm^2$ of silicon area. The input-referred noise of a single channel integrated from 10 Hz to 100 kHz equals to $4.1 \ \mu V_{rms}$, and the effective power consumption of each channel is measured at 22.4 μW from a 1.2 V power supply, which results in a system level NEF of 5.6.

M-12   **Analysis of Deviation from Pelgrom Scaling Law in V_{th} Variability of Pocket-implanted MOSFET,** *K. Sakakibara, Y. Miura, T. Kumamoto, S. Tanimoto, Renesas Electronics Corporation*   **232**

Deviation from Pelgrom scaling law in threshold voltage variability of pocket-implanted MOSFET is attributed to an increasing behavior of offset-voltage variability in weak and moderate inversion regions. This increasing behavior can be completely removed by using both-side ring gate structure. This means that the deviation is caused by subthreshold hump.

M-13   **Low Power ARM® Cortex™-M0 CPU and SRAM Using Deeply Depleted Channel (DDC) Transistors with VDD Scaling and Body Bias,** *V. Agrawal, N. Kepler, D. Kidd, G. Krishnan, S. Leshner, T. Bakishev, D. Zhao, P. Ranade, R. Roy, M. Wojko, L. Clark, R. Rogenmoser, M. Hori*, T. Ema*, S. Moriwaki*, T. Tsuruta*, T. Yamada*, J. Mitani*, and S. Wakayama*, SuVolta Inc. and *Fujitsu Semiconductor Ltd.*   **236**

130-D Knowles Drive, Deeply Depleted Channel™ (DDC) technology demonstrates more than 50% power reduction for ARM® Cortex™-M0 CPU cores and SRAMs at matched performance via VDD scaling and body biasing. DDC technology also demonstrates 35% speed improvement at matched power, improved SNM, 150mV 8Mb SRAM VDDmin improvement, and 5x SRAM retention leakage reduction.

M-14   **Highly Efficient CMOS Rectifier Assisted by Symmetric and Voltage-Boost PV-Cell Structures for Synergistic Ambient Energy Harvesting,** *K. Kotani, Tohoku University*   **240**

A highly efficient CMOS RF rectifier assisted by symmetric PV cells was developed as an example of the synergistic ambient energy harvesting concept. Output-voltage-boosted PV cell structures were also developed to improve the efficiency of this rectifier. Under typical indoor lighting conditions, 4x PCE than a conventional rectifier was achieved.

M-15   **A Slew-Rate Based Process Monitor and Bi-directional Body Bias Circuit for Adaptive Body Biasing in SoC Applications,** *S. Lee, E. Boling, A. Kuo, R. Rogenmoser, SuVolta, Inc.* **244**

A process monitor based on slew-rate measurement has been applied to a body bias control system to detect the process corners and adjust the body bias voltage to meet the power and performance requirements for SoCs. A 55nm testchip includes a new pulse extender and a bi-directional body bias circuit.

M-16   **Comparison of Modeling Approaches through Hierarchical Behavioral Modeling of a GNSS Receiver Front-end,** *Z. Chen, Y.Wang, J. Driesen*, F. Garzia**, S. Koehler**, F. Henkel*, R. Wunderlich, S. Heinen, IAS RWTH, *IMST, **IIS Fraunhofer*   **248**

This paper analyzes and compares the mixed-signal modeling approaches (conservative, timed data flow, and event-driven, as well as the base band modeling approach) through the hierarchical behavioral modeling of a GNSS (Global Navigation Satellite System) receiver front-end. Based on the result of the comparison, one hierarchical modeling flow is finally derived comprising multi-modeling approaches with reduced manual modeling effort.

M-17   **Pulse Amplification Based Dynamic Synchronizers with Metastability Measurement using Capacitance De-rating,** *B. Giridhar, M. Fojtik, D. Fick, D. Sylvester, D. Blaauw, University of Michigan*   **252**

We present dynamic buffer based synchronizers where only pulses (rather than stable intermediate voltages) cause metastability. This unique feature is exploited by amplifying such pulses to improve MTBF by $>10^6$x over jamb latches and double flip-flops at 2GHz in 65nm CMOS. A new on-chip metastability measurement method is also proposed.

M-18   **FireBird: PowerPC e200 Based SoC for High Temperature Operation,** *Radisav Cojbasic, Omer Cogal, Pascal Meinerzhagen, Christian Senning, Conor Slater, Thomas Maeder, Andreas Burg, Yusuf Leblebici, Ecole Polytechnique Federale de Lausanne (EPFL)*   **256**

This work presents FireBird, the first PowerPC based SoC for reliable operation beyond 200C. This paper proposes to customize a PowerPC e200 based SoC by using a dynamically reconfigurable clock frequency, exhaustive clock gating, and electromigration-resistant power supply rings. The custom testing procedure showed the expected maximum operating frequency reduction from 38MHz at room-temperature to 30MHz at 200C. The maximum power dissipation at 3.3V supply voltage was 1.2W and the idle state static leakage current was 3.4mA. Silicon measurements proved that this design outperforms PowerPC based SoCs available in the high-temperature microcontrollers market which are not operational at temperatures above 125C.

M-19   **A 1/10000 Lower Error Rate Achievable SSD Controller with Message-Passing Error Correcting Code Architecture and Parity Area Combined Scheme,** *K. Li, M. Ito, A. Esumi, Siglead, Inc.*   **260**

A new Error Correcting Code (ECC) solution to improve the reliability of NAND is proposed. Implemented in SSD controller IC, it is confirmed that more than1/10000 lower error rate, and 1.7x longer endurance of SSD can be achieved. This solution consists of a Message-Passing ECC architecture and a Parity Area Combined ECC scheme.

M-20 **45pW ESD Clamp Circuit for Ultra-Low Power Applications,** *Yen-Po Chen, Yoonmyung Lee, Jae-Yoon Sim*, Massimo Alioto**, David Blaauw, and Dennis Sylvester, University of Michigan, *Pohang University of Science and Technology, **University of Siena* **263**

Novel ultralow-leakage ESD power clamp designs are proposed and implemented in 0.18μm CMOS. Limiting both subthreshold leakage and GIDL, the proposed designs consume 43pW at 25°C and 119nW at 125°C with 4500V HBM level and 400V MM level protection, marking an 18-139× leakage reduction over conventional ESD clamps.

M-21 **A 1.14mW 750kb/s FM-UWB Transmitter with 8-FSK Subcarrier Modulation,** *F. Chen, Y. Li, D. Lin, H. Zhuo, W. Rhee, J. Kim*, D. Kim*, and Z. Wang, Tsinghua University and *Samsung Advanced Institute of Technology* **267**

A noninvasive energy-efficient FM-UWB transmitter is implemented in 65nm CMOS for stereo hearing aid. 8-FSK subcarrier modulation is employed to triple data rate by a fast-settling PLL. The FM-UWB signal is generated by an FLL-assisted ring VCO and a class AB PA. The 3.5-4GHz 750kb/s transmitter consumes 1.14mW,achieving 1.5nJ/bit.

M-22 **A 2.4 GHz Energy-Efficient 18-Mbps FSK Transmitter in 0.18 μm CMOS,** *Jingjing Chen, Weiyang Liu, Peng Feng, Haiyong Wang, and Nanjian Wu, Institute of Semiconductors, Chinese Academy of Sciences* **271**

This paper presents a 2.4 GHz energy-efficient phase locked loop (PLL)-based transmitter (TX) integrated in 0.18-μm CMOS technology. By using Twin-VCO transmission scheme, the data rate of the transmitter is free of loop bandwidth of PLL with stable carrier frequency. Measured results show that The TX achieves an energy efficiency less than 0.64 nJ/bit at a data rate of 18 Mbps.

M-23 **A 60GHz, Linear, Direct Down-Conversion Mixer with mm-Wave Tunability in 32nm CMOS SOI,** *M.A.T. Sanduleanu, A. Valdes-Garcia, Y. Liu, B. Parker, S. Shlafman*, B. Sheinman*, D. Elad*, S. Reynolds, D. Friedman, IBM T.J. Watson Research Center and *IBM Haifa R&D* **275**

The gain/linearity trade-off is exploited to achieve the best linearity performance of a mm-Wave down-conversion system. The achieved linearity (IIP3) for the whole down-conversion chain is better than 11.06dBm for 5.8dB gain at 60GHz. The down-converter occupies 1.38mm^2 in 32nm CMOS SOI and consumes 19.2mW from 1V supply.

M-24 **A Fully Integrated Highly Linear Receiver with Automatic IP$_2$ Calibration Schemes for Multi-Standard Applications,** *A. Borna, Y. Wang*, C. Hull*, H. Wang**, A. Niknejad, UC Berkeley, *Intel Corp., ** Georgia Institute of Technology* **279**

This paper presents an entire receiver chain with fully integrated self-calibration circuitries for suppressing the 2nd-order intermodulation distortions in Homodyne receivers for multi-standard applications. All the potential sources for IM2 generation are identified and tackled independently by architectural and calibration techniques, which results in a robust IP$_2$

enhancement.

Session 11 -- Power and Heterogeneous Technology Circuit

Tuesday, 9/24, 9:00 am
Oak Ballroom
Session Chair: Takamaro Kikkawa, Hiroshima University
Session Co-Chair: Philippe Jansen, Texas Instruments

9:00 am   **Introduction**

This session focuses on technology circuit interactions for heterogeneous 3D stacked-silicon FPGA, advanced high-voltage GaN electronics and future generation nanowire transistors.

9:05 am   **40V MESFETs Fabricated on 32nm SOI CMOS,** *W. Lepkowski, S. Wilk, J. Kam, T.*
11-1   *Thornton, Arizona State University*   **283**

40V N-channel MESFETs fabricated at a commercial 32nm SOI CMOS foundry without changing any of the process flow or including additional mask steps. Current drives of 110mA/mm with peak cut-off frequency of 30.5GHz and maximum oscillation frequency of 34.5GHz were observed.

9:30 am   **Recent Advances in GaN Power Electronics (Invited),** *Karim Boutros, Rongming Chu,*
11-2   *Brian Hughes,HRL Laboratories, LLC*   **287**

Gallium Nitride power devices are poised to replace silicon-based MOSFETs in power switching applications having weight and volume constraints, while simultaneously needing a high overall efficiency. With its projected 100x performance advantage over silicon, GaN is a game changing technology for energy-efficient power electronics. This paper reviews the advantages of GaN material and devices, the performance of these devices in power circuits, and the potential applications for this technology.

9:55 am   **Prospective for Nanowire Transistors (Invited),** *J.P. Colinge, S. Dhong, Taiwan*
11-3   *Semiconductor Manufacturing Company*   **291**

The multigate nanowire FET architecture allows for ultimate short-channel control and push Moore's law down to sub-5nm gate lengths. This paper reviews nanowire transistor device physics as well as circuit prospects in the fields of CMOS logic, memory, analog, RF and integrated sensor applications.

10:45 am   **BREAK**

11:05 am   **Interconnect and Package design of a Heterogeneous Stacked-Silicon FPGA**
11-4   **(Invited),** *E. Wu, K. Abugharbieh, B. Banijamali, S. Ramalingam, P. Wu, C. Wyland, Xilinx,*
    Inc.   **299**

Session 12 -- Wireline Transmitter and Receiver Design Techniques

Tuesday, 9/24, 9:00 am
Fir Ballroom
Session Chair: Dennis Fischette, AMD
Session Co-Chair: Jaeha Kim, Seoul National University

9:00am   **Introduction**

This session presents state of the art wireline transmitter and receiver circuits, including advanced low-power equalization and impedance-matching techniques.

9:05 am
12-1

A 5Gb/s 3.2mW/Gb/s 28dB Loss-compensating Pulse-Width Modulated Voltage-Mode Transmitter, *S. Saxena, R. K. Nandwana, and P. K. Hanumolu, Oregon State University*   **307**

A voltage mode transmitter employs pulse width modulation (PWM) based equalization of NRZ input data at 5Gb/s and compensates 28dB channel loss at 2.5GHz. Fabricated in a 90nm CMOS process, the proposed transmitter achieves a horizontal eye opening of 0.3UI with BER<10^{-12} and consumes only 16mW power of which 2.5mW is consumed by the digital PLL.

9:30 am
12-2

Current-Steering Pre-Emphasis Transmitter with Continuously Tuned Line Terminations for Optimum Impedance Match and Maximum Signal Drive Range, *Gerrit W. den Besten, Harold G. Hanson, Ranjeet K. Gupta, NXP Semiconductors*   **311**

A configurable 24-segment current-steering transmitter with linear continuously-tuned active line terminations is presented. A linearized resistor-MOSFET termination topology is proposed for accurate output levels (σ=1%) and good impedance matching (σ=2%) enabling larger drive levels by better supply utilization. The concept is implemented in 0.16 μm CMOS for 1-6Gbps.

9:55 am
12-3

Design Techniques for CMOS Backplane Transceivers Approaching 30-Gb/s Data Rates (Invited), *J. Bulzacchelli, IBM T.J. Watson Research Center*   **315**

This paper highlights design techniques for extending backplane transceiver data rates by describing a 28-Gb/s prototype implemented in 32-nm SOI CMOS and featuring a source-series terminated driver with 4-tap FFE, a two-stage peaking amplifier with active feedback, and a 15-tap DFE. Equalization is demonstrated over a 35-dB loss channel.

10:45 am   **BREAK**

11:05 am
12-4

Design Considerations for Low-Power Receiver Front-End in High-Speed Data Links (Invited), *S. Shekhar, J. E. Jaussi, F. O'Mahony, M. Mansuri, B. Casper, Intel Corporation*   **323**

This paper presents different design considerations for the receiver front-end (RXFE) in low-power, high-speed data links. Specifications for the RXFE are defined and explained in detail, including their impact on the overall link performance. Based on these specifications, low-power RXFE topologies are then analyzed to illustrate the design and performance tradeoffs. Techniques to properly characterize and measure the RXFE specifications are also provided, supplemented with measurement results from three different low-power links operating at 10Gb/s, 16Gb/s and 20Gb/s.

11:55 am
12-5

An 8mW Frequency Detector for 10Gb/s Half-Rate CDR using Clock Phase Selection, *M.S. Jalali, R. Shivnaraine, A. Sheikholeslami, M. Kibune*, H. Tamura*, University of Toronto, *Fujitsu Laboratories Limited*   **331**

A half-rate single-loop CDR with a new frequency detection scheme is introduced. The proposed frequency detector selects between the clock phases (I and Q) to reduce cycle slipping, hence improving lock time and capture range. This frequency detector, implemented within a 10Gb/s CDR in Fujitsu 65nm CMOS, consumes only 8mW, but improves the capture range by up to 3.6 . The measured capture range with the FD is from 8.675Gb/s to 11Gb/s.

Session 13 -- mm-Wave Circuits and Systems

Tuesday, 9/24, 9:00 am
Pine Ballroom
Session Chair: John Rogers, Carleton University
Session Co-Chair: Howard Luong, Hong Kong University of Science & Tech.

9:00 am   **Introduction**

This session will present the latest advances in mm-Wave (>30 GHz) circuits and systems, including transceivers, power amplifiers, synthesizers, VCOs and dividers.

9:05 am
13-1
    A 60 GHz Linear Wideband Power Amplifier using Cascode Neutralization in 28 nm CMOS, *Siva V Thyagarajan, Ali M Niknejad, Christopher D Hull*, University of California Berkeley, *Intel Corporation*   **335**

This paper presents the design of a 60GHz linear wideband power amplifier (PA) in deeply scaled 28nm CMOS technology. The PA utilizes cascode drain-source neutralization to improve stability and low-k transformer techniques to achieve high bandwidth. The PA delivers a saturated output power of 16.5dBm with a peak PAE of 12.6% and achieves a bandwidth of 11GHz with a peak gain of 24.4dB.

9:30 am
13-2
    Compact High-Power 60 GHz Power Amplifier in 65 nm CMOS, *Payam M. Farahabadi, Kambiz Moez, University of Alberta*   **339**

This paper presents a compact 60 GHz power amplifier utilizing a novel 4-waymulti-conductor power combiner. Fabricated in 65 nm CMOS process, the measured gain of the 0.19 mm^2 power amplifier is 18.8 dB at 60 GHz. A maximum saturated output power of 18.3 dBm is measured with the 15.9% peak power added efficiency.

9:55 am
13-3
    CMOS Low-Power Transceivers for 60GHz Multi Gbit/s Communications (Invited), *V. Vidojkovic, V. Szortyka, K. Khalaf,G. Mangraviti, B. Parvais, K. Vaesen, S. Brebels, A. Spagnolo, M. Libois, J. Long*, K. Raczkowski, P. Raghavan, A. Bourdoux, L. Min, C. Soens, V. Giannini, P. Wambacq, imec, *Delft University of Technology*   **343**

The availability of 9GHz bandwidth around 60GHz in combination with simple modulations schemes, low-cost radio ICs and small antenna size allows for multi Gbit/s wireless communications. In this article the potential of 60GHz wireless communications is evaluated from system, application and user point of view. Further, design challenges for 60GHz CMOS transceivers are identified. State-of-the-art designs show that short-range high-data rate radio links based on CMOS ICs can be made, potentially helped with beamforming.

10:45 am   **BREAK**

11:05 am
13-4
    A CMOS 21-48GHz Fractional-N Synthesizer Employing Ultra-Wideband Injection-Locked Frequency Multipliers, *A. Li, S. Zheng, J. Yin, X. Luo*, H. Luong, The Hong Kong University of Science and Technology, *Huawei Technologies Co. Ltd.*   **351**

Higher-order LC tanks are proposed to widen the locking range of mm-Wave injection-locked frequency multipliers. Employing a chain of such multipliers, a wideband fractional-N frequency synthesizer is demonstrated in 65nm CMOS. An output tuning range from 20.6-48.2GHz (80.2%) is measured with phase noise<-107dBc/Hz at 1MHz offset while consuming 148mW.

11:30 am 13-5	**A 75.7GHz to 102GHz Rotary-traveling-wave VCO by Tunable Composite Right /Left Hand T-line,** *Shunli Ma, Wei Fei*, Hao Yu*, Junyan Ren, Fudan University, *Nanyang Technological University* **355**

With the use of tunable composite-right/left-hand transmission line, this paper provides a wide frequency-tuning-range mechanism for Mobius-ring rotary-traveling-wave VCO in millimeter-wave region. Measurement results show 29.5% tuning range with center frequency at 89.3GHz, and phase noise from-100.08dBc/Hz to -98.7dBc/Hz with 10MHz offset, demonstrating state-of-art FOMT of -177.78dBc/Hz.

11:55 am 13-6	**Transformer-Based Dual-Band VCO and ILFD for Wide-Band mm-Wave LO Generation,** *Yue Chao, Howard C. Luong, Hong Kong University of Science and Technology* **359**

This paper presents wide-band transformer-based mm-wave dual-band VCO and ILFD. Based on two novel design techniques, the circuit is designed and fabricated in TSMC 65nm CMOS process and measures a state-of-art performance.

Educational Session 1

Tuesday, 9/24, 9:00 am
Cedar Ballroom
Session Chair: Foster Dai, Auburn University
Session Co-Chair: Earl McCune, RF Communications Consulting

E-1 9:00 am – 10:30 am	**Concurrent Design of ESD Protection and Integrated Circuits for Optimization and Prediction,** *Albert Wang, University of California, Riverside* **363**

As semiconductor technologies continuously advance into nano nodes, while integrated circuits (IC) become faster and more complex, on-chip ESD protection design quickly emerges as a huge IC design barrier nowadays. Major ESD design challenges include the followings: First, how to conduct simulation-based quantitative design to achieve ESD protection design optimization and prediction? Second, how to minimize ESD-induced parasitic effects that affect IC performance. Third, how to perform co-design of ESD protection and IC to achieve ESD protection and core circuit design optimization simultaneously. This lecture discusses critical aspects and techniques in practical ESD protection designs, including a mixed-mode ESD simulation-design method for design prediction, accurate RF ESD design characterization, complex ESD-IC interactions, ESD+IC co-design for whole-chip design optimization, etc. Real-world design examples will be presented.

10:30 am   **BREAK**

Educational Session 2

Tuesday, 9/24, 11:00 am
Cedar Ballroom

Session Chair: Eric Naviasky, Cadence
Session Co-Chair: Gerrit den Besten, NXP Semiconductors

E-2
11:00 am –
12:30 pm

Characterization of Matching, Variability, and Low-Frequency Noise for Mixed-Signal Technologies, *Hans Tuinhout, NXP*   **397**

Parametric mismatches and low frequency noise are major performance limiters as well as notorious causes for redesigns of high performance mixed-signal (HPMS) circuits and systems. Consequently, it is extremely important to measure, analyze, interpret, model and document these effects for mixed-signal technologies.

Part one of this educational lecture discusses parametric mismatch benchmarks and variability characterization challenges for active and passive devices. Part two focuses on low frequency noise, in particular on the emerging challenge in this field, namely variability of 1/f noise and associated Random Telegraph Noise.

These topics will be exemplified with results from (Bi)CMOS technologies, ranging from the current HPMS cash cow technologies (140 to 250 nm minimum dimensions), up to more advanced 40 nm node devices which can be seen as the stepping stone to some of the ultimate challenges of sub 10 nm devices that will mark the end of the CMOS shrink roadmap.

Luncheon Keynote

Tuesday, Sept. 24, 12:20 – 1:50 pm
Sierra Ballroom
Tickets for the luncheon are for sale at the Registration Desk

Connecting with the Emerging Nervous System of Ubiquitous Sensing, *presented by Joseph Paradiso, MIT Media Laboratory*   **N/A**

Embedded sensors are touching every phase of our lives as they diffuse into the objects and environments around us. We'll exhibit a "phase change" within a few years, however, once this sensor information becomes networked and available to applications running outside of each device's domain that will be at least as profound as the web was to computers. Accordingly, this talk will overview the broad theme of interfacing humans to the ubiquitous electronic "nervous system" that sensor networks will soon extend across things, places, and people. I'll illustrate this through two avenues of research - one looking at a new kind of digital "omniscience" (e.g., building different kinds of browsers for sensor network data) and the other looking at buildings & tools as "prosthetic" extensions of humans (e.g., making HVAC systems an extension of your sense of comfort), drawing from many projects that are running in my group at the MIT Media Lab.

Session 14 – Panel Session

"Do You Need to Plug In to Get Your Fill of Bits?"
Tuesday, 9/24, 2:00 pm
Oak Ballroom
Moderator: Sam Palermo, Texas A&M University

Applications such as video streaming and data sharing/backup are driving demand for increased device-to-device data transfer bandwidth for in/adjacent-room communication on the order of 10-20m. While the demand for higher data rates is growing, consumers are also becoming accustomed to and beginning to expect broadband wireless connectivity for their devices. To replace electrical cable-based links, wireless links will need to demonstrate competitive or superior data rates, energy efficiency, reliability, cost, and in some applications, latency. On the other hand, optical interconnects offer the flexibility of traditional electrical cable-based systems at potentially higher data rates and lower power and latency. However, the reliability and cost of optical cable systems are open issues. This panel aims to answer the question: "Do You Need to Plug In to Get Your Fill of Bits?" In other words, can future wireless systems support the 10+Gb/s data rates that future systems will demand? If not, what is best approach? Traditional electrical or emerging optical cable solutions?

Panelists: Elad Alon, University of California - Berkeley
                   Marc Loinaz, Broadcom
                   Payam Heydari, University of California - Irvine
                   Tirdad Sowlati, Broadcom
                   Drew Alduino, Intel

Session 15 - Forum Session: Electrical and Photonics I/O Test and Debug

Tuesday, 9/24, 2:00 pm
Fir Ballroom
Session Chair: Mike Li, Altera
Session Co-Chair: Takahiro Yamaguchi, Advantest Laboratories

2:00 pm   **Introduction**

This session addresses test and debug challenges associated with multi-Gbits/s to more than 30 Gbits/s I/Os built with electrical and photonic ICs.

2:05 pm   **The Future of High Speed Electrical and Photonics IO Testing: Facing Complex**
15-1   **Challenges,** *Salem Abdennadher, Intel Corporation*   **N/A**

Where is High Speed IO manufacturing Test heading? As technology continues to scale to increase system bandwidth, decrease power dissipation, die area and system cost, the challenges associated with test seem to expand exponentially. There has been a rise in defect occurrences as Serial Electrical IO interfaces instances and test complexity rise. In addition, introduction of optical IO's in main stream products is introducing unprecedented challenges. Technology process variation and process uncertainty is also affecting the performance of these circuits.

Intel Test community wonder if the current technologies and strategies are adequate in the short term and what they should focus on now to deal with issues that are surfacing in the 2013-2015 timeframe. Current 22nm analog test coverage issues will persist to be an issue with 14nm process and beyond or even get worse. With Signal Headroom becoming too small to design analog circuit with sufficient signal integrity, HVM tests need to screen not only for manufacturing defects but also for design marginality and process uncertainty.

In this talk will present new HVM test techniques to meet the ever increasing test complexity challenges. Such as developing methodologies to test optical interfaces depending on their level of integration (Hybrid or Full). Providing new innovative approaches in areas such as: Complete No Touch Testing (NTT) methodology and systematic defect capture in Electrical High Speed IO circuits.

2:40 pm   **Testability Improvement for 12.8 GB/s Wide IO DRAM Controller by Small Area**
15-2   **Pre-bonding TSV Tests and a 1 GHz Sampled Fully Digital Noise Monitor,** *T. Nomura,*
R. Mori, M. Ito*, K. Takayanagi, T. Ochiai, K. Fukuoka, K. Otsuga, K. Nii, S. Morita, T.

*Hashimoto, T. Kida, J. Yamada, H. Tanaka, Renesas Electronics Corporation, *Renesas Micro Systems Corporation* **453**

A Wide IO DRAM controller chip was designed, and fabricated with Through Silicon Via (TSV) technology. The memory interface consists of 1200 TSVs including 512 bit data signals, which introduces new challenges in testability. To address these challenges, testing schemes by dedicated circuitry are proposed. TSV test circuitry is implemented in the micro-IOs placed in between the fine pitch TSV arrays, which can detect and reject TSV defects prior to stacking process. Another circuitry is for monitoring power noise, where we are aware of 512 bit Data simultaneously switching noise. We also introduced a impedance optimization scheme associated with the noise monitoring circuitry, where Vmin was improved for 30mV by appropriate optimization. We achieved 12.8GB/s operation, while IO power was reduced by 89% compared to LPDDR3.

3:15 pm
15-3
Design Verification and Testing of High Speed Silicon Photonics Links, *Brian Welch, Luxtera, T. Nomura, R. Mori, M. Ito*, K. Takayanagi, T. Ochiai, K. Fukuoka, K. Otsuga, K. Nii, S. Morita, T. Hashimoto, T. Kida, J. Yamada, H. Tanaka, Renesas Electronics Corporation, *Renesas Micro Systems Corporation* **N/A**

This paper looks at the verification and test techniques that can be deployed in silicon photonics solutions, and how it mirrors those that are used in conventional CMOS design.

3:50 pm **BREAK**

4:05 pm
15-4
CMOS Photonics: Product Test and Debug Challenges, *David Piede, Cisco Systems* **N/A**

CMOS photonics is a new and exciting technology that offers potential improvements in cost, power, integration, and size over current photonic and electronic technologies. As with any new technology, there are new challenges associated with productization. We will focus on the test and debug challenges associated with known-good-die (KGD), and with the heterous integration of electronics and CMOS photonics in a package format.

4:40 pm **Panel Discussion and Q&A**

Session 16 -- Nyquist Rate A/D Converters

Tuesday, 9/24, 2:00 pm
Pine Ballroom
Session Chair: Mohammad Ranjbar, Cirrus Logic
Session Co-Chair: John McNeill, Worcester Polytechnic Insittute

2:00 pm **Introduction**

A/D converters are key building blocks in many electronic systems. Their applications range from low-speed, low-power sensors to high-speed wireless or wireline communication systems. Papers in this session cover a range of speeds form 250 kSps to 12.8 GSps, and resolutions from 5 to 12 bits.

2:05 pm
16-1
A 12.8GS/s Time-Interleaved SAR ADC with 25GHz 3dB ERBW and 4.6b ENOB, *Y. Duan, E. Alon, UC Berkeley* **457**

This paper presents a 12.8GS/s 32-way hierarchically time-interleaved SAR ADC with 4.6-bit

ENOB in 65nm CMOS. The prototype utilizes multi-stage sampling and a cascode sampler circuit to enable greater than 25GHz 3dB effective resolution bandwidth (ERBW). We further employ a pseudo-differential SAR ADC to save power and area. The core circuit occupies only 0.23mm^{2}and consumes a total of 162mW from dual 1.2V/1.1V supplies.

2:30 pm
16-2
A 10GS/s 6b Time-Interleaved ADC with Partially Active Flash sub-ADCs, *Xiaochen Yang, Robert Payne*, Jin Liu, University of Texas at Dallas, *Texas Instruments* **461**

A 10GS/s 6b time-interleaved ADC in 65nm CMOS is presented. A partially-active flash sub-ADC is proposed to improve the power efficiency, a source-follower based boot-strap T&H circuit reduces input kickback and improve ADC bandwidth, and timing skew is corrected with duty-cycle calibration. The measurement shows a FOM of 197fJ/conv-step.

2:55 pm
16-3
An 8-Bit 4-GS/s 120-mW CMOS ADC, *Hegong Wei, Peng Zhang*, Bibhu Datta Sahoo**, and Behzad Razavi, University of California-Los Angeles, *Tsinghua University, **Amrita University* **465**

A four-channel time-interleaved pipelined ADC employs a new timing calibration technique to suppress mismatch-induced spurs and achieve a Nyquist-rate SNDR of 44.4 dB. Designed in 65-nm CMOS technology, the ADC draws 120 mW, providing an FOM of 219 fJ per conversion step.

3:20 pm
16-4
A 7.1-mW 1-GS/s ADC with 48-dB SNDR at Nyquist Rate, *S. Hashemi, B. Razavi, University of California, Los Angeles* **469**

A two-stage pipelined ADC employs a double-sampling residue amplifier, two interleaved precharged DACs, and a new calibration scheme to correct for residue gain error, offset, and nonlinearity. Realized in 65-nm CMOS technology and sampling at 1 GHz, the prototype exhibits an FOM of 25 fJ/conversion-step while drawing 7.1 mW from a 1-V supply.

3:45 pm
BREAK

4:00 pm
16-5
A 0.55 V 7-bit 160 MS/s Interpolated Pipeline ADC Using Dynamic Amplifiers, *J. Lin, D. Paik, S. Lee, M. Miyahara, A. Matsuzawa, Tokyo Institute of Technology* **473**

This paper presents a 0.55 V, 7-bit, 160 MS/s pipeline ADC using dynamic amplifiers. In this ADC, dynamic amplifiers with a common-mode detection technique are used as residual amplifiers to increase its robustness against supply voltage lowering and to remove the unnecessary static power consumption achieving clock-scalability in power performance.

4:25 pm
16-6
A 95-MS/s 11-bit 1.36-mW Asynchronous SAR ADC with Embedded Passive Gain in 65nm CMOS, *Jae-Won Nam, David Chiong, Mike Shuo-Wei Chen, University of Southern California* **477**

An asynchronous SAR ADC with embedded passive gain is fabricated in 65nm CMOS. The prototype ADC demonstrates a peak SNDR of 63.1dB and SFDR of 75.2dB at 95MS/s. It dissipates 1.36mW at 1.1V supply and achieves the lowest FoM among the recently published ADCs with similar specification (>10 ENOB, >10MS/s).

4:50 pm
16-7
A 24μW 12b 1MS/s 68.3dB SNDR SAR ADC with Two-Step Decision DAC Switching, *Yung-Hui Chung, Meng-Hsuan Wu, Hung-Sung Li, MediaTek, Inc.* **481**

A 12-bit SAR ADC employs a new DAC switching technique for improving the ADC linearity and tolerating the DAC settling errors. At 1MS/s, it consumes 24uW from a 0.9V supply. At the Nyquist-rate input, measured SNDR and SFDR are 68.3dB and 82dB, respectively. It achieves a FoM of 11.7fJ/conversion-step.

5:15 pm
16-8
A 3.3fJ/conversion-step 250kS/s 10b SAR ADC Using Optimized Vote Allocation, *M. Ahmadi, W. Namgoong, University of Texas at Dallas* **485**

A 10b SAR ADC that supports a flexible differential input swing from 0.4V to 1V is presented. The proposed ADC employs a non-binary architecture along with a majority vote comparison using optimized vote allocation. The prototype achieves ENOB ranging from 7.1b to 9.1b and FOM from 3.3 to 6.8fJ/conversion step.

Educational Session 3

Tuesday, 9/24, 2:00 pm
Cedar Ballroom
Session Chair: Christoph Sandner, Infineon Technologies Austria AG
Session Co-Chair: Hoi Lee, University of Texas at Dallas

E-3
2:00 pm – 3:30 pm
Single-Inductor-Multiple-Output DC-DC Converter Design, *Philip Mok, Hong Kong University of Science and Technology* **489**

Multiple well-regulated power supplies are essential for reducing power consumption and isolating the coupling noise between different functional blocks in VLSI design. With the increasing number of functional blocks in SoC applications, the need for a cost and efficiency effective solution of multiple power supplies is growing. Single-Inductor-Multiple-Output (SIMO) switching regulator, which provides several output voltages with only one inductor, is one of promising solutions and becomes a hot topic in DC-DC converter design due to the cost and volume reduction. However, with one inductor shared by all the outputs to accumulate and transfer energy from the input, cross regulation easily appears at outputs when a change in the inductive energy is induced by a load transient at one output. These unwanted voltage variations affect the performance or even the function of the loading devices. This talk will discuss the operation principle of SIMO switching converters and their design issues. To minimize cross regulation of a SIMO regulator, several control techniques will be presented and their pros and cons will be discussed.

3:30 pm **BREAK**

Educational Session 4

Tuesday, 9/24, 4:00 pm
Cedar Ballroom
Session Chair: Foster Dai, Auburn University
Session Co-Chair: Jonathan Borremans, IMEC

E-4
4:00 pm – 5:30 pm
Advanced Digital Phase-Locked Loops, *Salvatore Levantino, Politecnico di Milano* **524**

This tutorial will introduce the fundamentals of digital phased-locked loops for wireless applications. After reviewing the basic architectures, the tutorial will analyze the mechanisms of generation of limit cycles, which manifest themselves as spurious tones in the output spectrum even when synthesizing integer-N channels. Then, loop-parameter settings and design strategies for spur elimination and phase-noise optimization will be derived. Next, we will move to the fractional-N case, in which quantization and nonlinearity add new sources of spur tones, and we will review the different design techniques which helps mitigate such impairments. Finally, examples of state-of-the-art implementations of frequency synthesizers and direct-FM modulators based on digital PLLs will be discussed.

Tuesday Poster Session

Tuesday, 9/24, 5:00 pm – 7:00 pm

T-1 **All-Digital 90° Phase-Shift DLL with a Dithering Jitter Suppression Scheme,** *D. H. Jung, K. Ryu, J. H. Park, W. Lee*, S. O. Jung, Yonsei University, *Samsung Electronics* **572**

We propose a 90° phase-shift digital delay-locked loop (DLL) with a new dithering jitter suppression scheme. Delay-line control code dithering is effectively suppressed by comparing the distribution of the input and the output clock jitter. And the phase shift and duty cycle correction accuracy are enhanced by MDLL based phase-shift structure.

T-2 **A 1Gb/s Reconfigurable Pulse Compression Radar Signal Processor in 90nm CMOS,** *Jun Li, Hirohito Mukai*, Mehmet Parlak, Michiaki Matsuo*, James F. Buckwalter, University of California San Diego and *Panasonic Corporation* **576**

This paper presents a reconfigurable analog signal processing circuit for pulse compression radar. Adapting bandwidth for the range of the target is proposed for radar systems. The baseband signal processor includes a high-speed correlator/integrator, a 4-bit flash analog-to-digital converter (ADC) and a multi-range delay lock loop (DLL). The DLL generates multi-phase clock to align the template signal with the received signal. The circuit is fabricated in90-nm CMOS and can be configured to work from 50Mb/s to 1Gb/s with Barker codes. An SNR of 8.5dB is demonstrated for 1Gb/s. The total power consumption is 33mW at 1Gb/s.

T-3 **A 5GS/s 4-bit Time-Based Single-Channel CMOS ADC for Radio Astronomy,** *A. Macpherson, J. Haslett, L. Belostotski, University of Calgary* **580**

A 4-bit 5GS/s 65nm time-based analog-to-digital converter (ADC) targeting the next-generation Square Kilometre Array (SKA) is presented. This ADC is composed of an analog voltage-to-time converter (VTC) front end and a digital time-to-digital converter (TDC) back end. The two components can be physically separated to minimize the impact of digital noise from the ADC on high-gain, high-sensitivity receiver chains common in radio telescopes.

T-4 **A 6b 800MS/s 3.62mW Nyquist AC-coupled VCO-Based ADC in 65nm CMOS,** *P. K. Sharma, M. S-W. Chen, University of Southern California* **584**

A 6-bit 800MS/s Nyquist VCO-Based ADC is proposed. The ADC utilizes an analog differentiator, replacing the conventional digital differentiator to avoid quantization noise shaping and achieve Nyquist operation with embedded DC rejection, first order anti-aliasing filtering and improved VCO linearity without calibration. The ADC achieves peak SNDR of 34dB with over 400MHz input bandwidth and occupies an active area of 0.015mm2 while consuming 3.65mW.

T-5 **A Fully-Digital Beat-Frequency Based ADC Achieving 39dB SNDR for a 1.6mV$_{pp}$**

Input Signal, *Bongjin Kim, Weichao Xu, Chris H. Kim, University of Minnesota* **588**

A fully-digital VCO-based ADC employing a beat frequency detection scheme is demonstrated in 65nm. The proposed design is highly effective in measuring extremely small changes in the VCO frequency within a short sampling time. Direct A-to-D conversion of a 1.6mVpp differential input signal with 39dB SNDR was experimentally verified.

T-6 **A 4–15-GHz Ring Oscillator based Injection-Locked Frequency Multiplier with Built-in Harmonic Generation,** *J. Xu, J. Hu, B. Ciftcioglu, H. Wu, University of Rochester* **592**

This paper presents a new inductorless injection-locked frequency multiplier(ILFM) designed to achieve wide locking range and low power dissipation. The ILFM integrates harmonic generation in each stage and realizes multiphase injection simultaneously. A multiply-by-2 ILFM prototype is implemented and demonstrated the wide locking range of the proposed ILFM.

T-7 **WITHDRAWN**

T-8 **Power Management Circuits for a 15-µA, Implantable Pressure Sensor,** *Steve Majerus* and Steven L. Garverick, Case Western Reserve University, *APT Rehabilitation Research and Development Center* **596**

An ASIC for wireless bladder pressure sensing incorporates power-management circuitry, limiting active time of instrumentation circuitry and minimizing telemetry rate. Measured results with prerecorded bladder signals indicate that5% of acquired samples merit transmission, resulting in an average telemetry rate of 1.5 Hz and total IC current of 12.8 µA.

T-9 **A Novel Voltage-Programmed Pixel Circuit with V_T-Shift Compensation for AMOLED Displays,** *M. Yang, N. Papadopoulos, C-H. Lee, W.S. Wong, M. Sachdev, University of Waterloo* **600**

A novel voltage-programmed pixel circuit using hydrogenated amorphous silicon(a-Si:H) thin-film transistors (TFTs) for active-matrix organic light-emitting diode (AMOLED) displays is proposed. The threshold voltage shift (ΔVT) of the drive TFT due to electrical stress is compensated by the change of gate-to-source voltage (ΔVGS) generated by the ΔVT-dependent charge transfer from the drive TFT to a TFT-based Metal-Insulator-Semiconductor (MIS)capacitor. Another MIS capacitor is used to improve OLED drive currents. Measurement results verify the effectiveness and speed of the proposed pixel circuit.

T-10 **Design for Manufacturing Layout Analyses Correlate Layout to Physico-Chemical Yield Loss Mechanisms,** *C.P. Tan, C. Zhou, Y. Tian, C. Liu, H.-M. Lam, J. Zhang, M. Lu, GLOBALFOUNDRIES Singapore Pte. Ltd.* **604**

We introduce a case-based learning workflow in the foundry for managing layout weakpoints and implementing layout analyses checks. In this work, we describe case studies that demonstrate how layout analyses can be used to detect layout weakpoints and correlate them to actual physico-chemical mechanisms behind defects observed on silicon.

T-11 **A Split-Foundry Asynchronous FPGA,** *B. Hill, R. Karmazin, C. Ortega, J. Tse, R. Manohar, Cornell University* **608**

We present the first published measurements of a complex digital integrated circuit fabricated in both standard and split-foundry processes. Our1.3-million-transistor asynchronous FPGA operates at over 300MHz in 130nm. We discuss the challenges inherent in split design and our automated layout tools that address them.

T-12   **A 40-nm 8T SRAM with Selective Source Line Control of Read Bitlines and Address Preset Structure,** *S. Yoshimoto, S. Miyano*, M. Takamiya**, H. Shinohara*, H. Kawaguchi*, and M. Yoshimoto, Kobe University, *Semiconductor Technology Academic Research Center, and **University of Tokyo*   **612**

This paper presents a 40-nm 8T SRAM in which bit lines are partially discharged by a selective source line control (SSLC) for low-power operation. The proposed SSLC scheme reduces a read bit line voltage swing in an unselected column with af loating source line (SL) of dedicated read ports.

T-13   **AOT-Controlled Dual-Mode AVP Buck Reglator with AEAF Mechanism,** *Hsin-Lun Li, Chia-Cheng Pao, Bo-Ming Chen, Chien-Hung Tsai, National Cheng Kung University*   **616**

A novel adaptive voltage positioning (AVP) buck regulator using adaptive on-time (AOT) control targeted for applications with low-ESR output capacitors is proposed. In this work, AOT control is adapted to keep the system's switching frequency quasi-fixed or independent of the input supply voltage and the AVP mechanism is realized without the need to use conventional error amplifier compensator or extra current-sensing circuit. For ensuring the system's switching frequency not entering the range of acoustic frequency at light load, an AEAF (avoid entering acoustic frequency) circuit is also proposed. For comparison purpose, the implemented buck regulator can be set to operate under AVP or non-AVP mode. This work has been fabricated and verified with a standard 0.18µm CMOS technology. Experimental results show excellent transient recovery time of 4µs (under AVP mode), ±0.11% switching frequency variation (for the specified input voltage range), and 91% peak conversion efficiency.

T-14   **Switched-Capacitor Filter based Type-III Compensation for switched-mode Buck Converters,** *G. Bawa, A.Q. Huang, North Carolina State University*   **620**

A switched-capacitor filter based Type-III compensation for regulation of Buck converters is presented. Compared to the all-analog filter, the proposed compensator can be fully-integrated onto the die, resulting in reduced footprint and cost. The filter time-constants also scale linearly with the converter's switching time-period, resulting in increased programmability and ease-of-use.

T-15   **Estimation of Passive Mixer Output Bandwidth Using Switched-Capacitor Techniques,** *Essam S. Atalla, Frank Zhang*, Abdellatif Bellaouar*, Poras T, Balsara, The University of Texas at Dallas and *NVIDIA Corp.*   **624**

Passive mixers have become an essential component of SAW-less receivers. It is well known that the passive mixer behaves as a switched-capacitor circuit (SC)but to the authors knowledge, there is no reported analysis of the mixer impedance that truly accounts for the SC behavior. In this paper, we present for the first time a closed form of the passive mixer output impedance based on SC techniques. We prove that the fundamental lower limit of the mixer impedance is proportional to the well-known switched capacitor resistor$1/\left(f_{LO}C\right)$ and different from the previously reported mixer switch ON resistance. We also explain that the equation is useful in estimating output bandwidth of passive mixer based front-ends with general LNA load impedance. We finally show that our bandwidth estimation matches measured results of two receiver front-ends.

T-16   **How to Reduce Power in 3D IC Designs: A Case Study with OpenSPARC T2 Core,** *Moongon Jung, Taigon Song, Yang Wan, Young-Joon Lee, Debabrata Mohapatra*, Hong Wang*, Greg Taylor*, Devang Jariwala*, Vijay Pitchumani*, Patrick Morrow*, Clair Webb*, Paul Fischer*, and Sung Kyu Lim, Georgia Institute of Technology, *Intel Corporation*   **628**

The power benefit of 3D ICs is demonstrated with an OpenSPARC T2 core. Four design techniques are explored to optimize power in 3D ICs: 3D floor planning, intra-block metal layer usage control, dual-Vth, and FUB folding. With aforementioned methods, the total power saving of 21.2% is achieved against the 2D counterpart.

T-17   **A General-purpose Vision Processor with 160x80 Pixel-Parallel SIMD Processor Array,** *A. Lopich, P. Dudek, University of Manchester*   **632**

In this paper we present a vision processor, which incorporates a 160×80 SIMD array of pixel-processors. The processor operates with a 100MHz clock and 1.8Vsupply. The device provides 640 GOPS (binary) and 23 GOPS (greyscale) consuming 0.5 W. The chip occupies 50mm^2 and is fabricated in a standard 0.18 μm CMOS process. The I/O interface supports 200 M Pixels/s (greyscale), 1.6 G Pixels/s (binary) and 40 M Pixels/s (address-event readout) data rate, and PE-parallel image sensing mode for embedded high-speed vision applications. Experimental results indicate that the performance of the presented chip approaches the efficiency of recently reported application-specific vision processors, while providing full programmability and thus being adjustable to a wide range of applications.

T-18   **A Programmable Analog Frequency-Locked Loop for VCO Characterization and Test with 8 ppm Resolution,** *S. Aouini, J.F. Bousquet, N. Ben-Hamida, L. Jakober, J. Wolczanski, C. Kurowski, Ciena Corporation*   **636**

We present a digitally controlled analog frequency-locked loop for VCO characterization and test. The scheme allows a frequency tuning better than 8ppm. The AFLL comprises a 17-bit frequency counter, a sigma-delta modulator used for dithering the correction signal, a charge-pump and capacitance used as integrator and a VCO.

T-19   **Detection of Early-Life Failures in High-K Metal-Gate Transistors and Ultra Low-K Inter-Metal Dielectrics,** *Y.M. Kim, J. Seomun*, H.-O. Kim*, K.-T. Do*, J.Y. Choi*, K.S. Kim*, M. Sauer**, B. Becker**, S. Mitra, Stanford University, *Samsung Electronics, **University of Freiburg*   **640**

We derive signatures for early-life failures (ELF) in 28nm high-K/metal-gate transistors and ultra low-K inter-metal dielectrics. We also demonstrate that the derived ELF signatures can be successfully detected using a clock control technique, activated during periodic on-line self-test and diagnostics in robust systems, without requiring expensive concurrent error detection.

T-20   **A Fully Differential Ultra-Compact Broadband Transformer Based Quadrature Generation Scheme,** *Jong Seok Park, Shouhei Kousai*, and Hua Wang, Georgia Institute of Technology, *Toshiba Corporation*   **644**

This paper presents a fully differential ultra-compact broadband transformer-based quadrature generation scheme implemented within only one inductor-footprint. A 5GHz design in a 65nm CMOS only occupies 260µm-by-260µmarea and achieves 0.82dB insertion loss (5GHz) with 3.8° maximum phase error and ±0.5dB amplitude mismatch within 13% bandwidth (4.75GHz to 5.41GHz).

T-21 **A -173 dBc/Hz @ 1 MHz offset Colpitts Oscillator using AlN Contour-Mode MEMS Resonator,** *Jabeom Koo, Augusto Tazzoli*, Jeronimo Segovai-Fernandez*, Gianluca Piazza*, Brian Otis, University of Washington, *Carnegie Mellon University* **648**

A differential Colpitts oscillator using AlN MEMS CMR designed in 0.13 µm CMOS is presented in this work. The oscillator operates at 1.16 GHz, with a total power consumption of 4.2 mW at 1 V supply. It achieves a phase noise of -143.6dBc/Hz, -173.3 dBc/Hz at 100 kHz and 1 MHz offset frequency respectively with a figure of merit (FOM) of 228.3 dB. Current-based temperature compensation was employed to reduce oscillator drift across temperature.

T-22 **An Ultra-Broadband Compact Mm-Wave Butler Matrix in CMOS for Array-Based MIMO Systems,** *J. Park, T. Chi, H. Wang, Georgia Institute of Technology* **652**

This paper presents an ultra-broadband compact mm-wave Butler Matrix utilizing new transformer-based swapped-port couplers. It is implemented as a 4×4 Butler Matrix at 63GHz in a 65nm CMOS process with 0.335×0.215mm2, and it achieves9.8GHz bandwidth, 2.77dB insertion loss, and better-than 17dB array peak-to-null-ratio over 57GHz and 67GHz.

T-23 **A 1.2 pJ/b 6.4 Gb/s 8+1-Lane Forwarded-Clock Receiver with PVT-Variation-Tolerant All-Digital Clock and Data Recovery in 28nm CMOS,** *Shuai Chen, Hao Li*, Liqiong Yang, Zongren Yang, Weiwu Hu and Patrick Yin Chiang*, Chinese Academy of Science, *Oregon State University* **656**

This paper presents an energy/area-efficient forwarded-clock receiver fabricated in 28 nm CMOS process. The receiver consists of 8 data lanes plus one forwarded clock lane, and adopts a novel all-digital clock and data recovery (CDR) based on delay-locked loop (DLL). The proposed all-digital DLL-based CDR uses the calibration and the update techniques to achieves a robust PVT-variation tolerance as well as a low power/area consumption. The measurement results show that our receiver can work at a data rate of 6.4 Gb/s with BER<10e-12 and consume only 7.5 mW per lane under 0.85 V power supply. The receiver core merely occupies an area of 0.02 mm*mm per lane.

T-24 **A True 4-Cycle Lock Reference-Less All-Digital Burst-Mode CDR Utilizing Coarse-Fine Phase Generator with Embedded TDC,** *Tetsuya Iizuka, Satoshi Miura*, Yohei Ishizone*, Yoshimichi Murakami*, Kunihiro Asada, University of Tokyo, *THine Electronics, Inc.* **660**

This paper presents a reference-less all-digital burst-mode CDR using a coarse-fine phase generator with embedded TDC. It achieves true 4-cycle lock without warm-ups, and eliminates dynamic power consumption in a stand-by state. Fabricated in 65nm CMOS, this CDR operates from 1.40 to 2.06Gb/s and consumes9.6mW at 2.06Gb/s with 80x80µm^2.

Session 17 -- Variation and Analog Modeling

Wednesday, 9/25, 9:00 am
Oak Ballroom
Session Chair: Trent McConaghy, Solido Design
Session Co-Chair: Brian Chen, Agilent

9:00 am   **Introduction**

This session discusses modeling process variation in analog and SRAM circuits, as well as analog thermal noise and distortion modeling.

9:05 am
17-1
Thermal Noise Modeling of Nano-scale MOSFETs for Mixed-signal and RF Applications (Invited), *Chih-Hung Chen, David Chen, Ryan Lee, Peiming Lei, and Daniel Wan, United Microelectronics Corporation*   **664**

This paper presents the thermal noise in nano-scale MOSFETs – from measurement, characterization, modeling, and potential technology enhancement for future low power, mixed-signal, and radio-frequency (RF) applications. Experimental data from five CMOS technology nodes, namely 180 nm, 130 nm, 90nm, 65 nm, and 40 nm nodes are presented and discussed.

9:55 am
17-2
A Model-Agnostic Technique for Simulating Per-Element Distortion Contributions, *Nagendra Krishnapura, Rakshitdatta K. S., Indian Institute of Technology*   **672**

The nonlinearity of an element can be altered while maintaining the operating point and first-order terms by appropriately combining two instances of the nonlinear element with complementary scaling factors for incremental voltages above the operating points. Per-element distortion contributions in a circuit can then be determined by altering the nonlinear terms by known factors and simulating the output distortion in each case. This technique can be used in a standard circuit simulator with the appropriate nonlinear device models butr equires no knowledge of the device model details on the part of the circuit designer. The technique is demonstrated by applying it to a common source amplifier with a nonlinear load and a two stage fully differential opamp.

10:20 am
17-3
Corner Models: Inaccurate at Best and it Only Gets Worst ..., *Colin C. McAndrew, Ik-Sung Lim, Brandt Braswell, and Doug Garrity, Freescale Semiconductor*   **676**

Corner (best- and worst-case) models have been a mainstay of integrated circuit design for decades. Obviously they can be effective, especially for digital CMOS design. However, there are significant inaccuracies that arise when digital CMOS corner models are used for analog circuits, or any types of circuits or measures of circuit performance they were not targeted for. This paper details what corner models can and cannot do, and shows their inadequacies for analog CMOS circuits.

10:45 am   **BREAK**

11:05 am
17-4
Energy Centric Model of SRAM Write Operation for Improved Energy and Error Rates, *Swaroop Ghosh, University of South Florida*   **680**

We propose an energy centric model of SRAM write operation. The model provides useful insight about energy and write error rate. We employ the proposed model for evaluating write assist mechanisms and their potential in reducing intrinsic memory error rates. We also employ it for optimizing energy of memories.

11:30 am
17-5
SRAM Read Current Variability and its Dependence on Transistor Statistics, *Sriramkumar Venugopalan, Vivek Joshi*, Luis Zamudio*, Matthias Goldbach*, Gert Burbach*, Ralf VanBentum*, Sriram Balasubramanian*, University of California, Berkeley, *GLOBALFOUNDRIES*   **684**

Our study breaks down the dependence of SRAM read current (Iread) variability(σIread) into constituting pass-gate (PG) and pull down (PD) NMOS transistor variability. We report a bottoms-up model for σIread including feedback in stacked transistors and discuss its implications on SRAM performance.

11:55 am
17-6

Mismatch Characterization of Small Metal Fringe Capacitors, *V. Tripathi, B. Murmann, Stanford University* **688**

This paper describes a test structure and measurements results pertaining to the characterization of single-layer, lateral-field, 0.45-fF and 1.2-fF unit metal capacitors in a 32-nm SOI CMOS process. The measurement-inferred average standard deviations for these capacitances are 1.2% and 0.8%, respectively, confirming variance scaling according to Pelgrom's matching law.

Session 18 -- Energy Efficient SoC Design

Wednesday, 9/25, 9:00 am
Fir Ballroom
Session Chair: Visvesh Sathe, Advanced Micro Devices
Session Co-Chair: Arif Rahman, Altera Corporation

9:00 am

Introduction

Processors, accelerators and on-chip clocking implementations for energy-efficient SoC design.

9:05 am
18-1

A 1GHz Hardware Loop-Accelerator with Razor-based Dynamic Adaptation for Energy-Efficient Operation, *S. Das, G. Dasika, K. Shivashankar and D. Bull, ARM Ltd.* **692**

We describe the implementation and silicon measurement results from a Razor-based hardware loop-accelerator (RZLA), implementing the Sobeledge-detection algorithm. We demonstrate robust operation with a large Dynamic Voltage Scaling (DVS) range achieved using 50% of the clock-period for timing-speculation. At 1GHz operating frequency, Razor DVS enables34% energy saving on a per-device basis and 33% overall on the entire batch of devices.

9:30 am
18-2

Energy Efficient Recognition and Mining Processor using Scalable Effort Design, *Vinay Chippa, Hrishikesh Jayakumar, Debabrata Mohapatra, Kaushik Roy, Anand Raghunathan, Purdue University,* **696**

A Recognition and Mining(RM) processor, that exploits the inherent application resilience using scalable effort design is implemented in TSMC-65nm technology. Measurements demonstrate energy savings of 1.2-2.3X with no quality-loss, and2X-20X with modest quality reduction due to cross-layer optimization of algorithm, architecture and circuit level scaling mechanisms.

9:55 am
18-3

An Energy-Efficient Coarse-Grained Dynamically Reconfigurable Fabric for Multiple-Standard Video Decoding Applications, *L. Liu, C. Deng, D. Wang, M. Zhu, S. Yin, P. Cao*, S. Wei,Tsinghua University,*Southeast University* **700**

We introduce a coarse-grained dynamically reconfigurable fabric consisting of16x16 Processing Elements. Line-Switched Mesh Connect routing and Hierarchical Configuration Context organization scheme are proposed. Measured results show that the fabric has great advantage in energy efficiency compared with the state-of-art designs when processing video decoding and some other computation-intensive applications.

10:20 am
18-4

SURFEX: A 57fps 1080P resolution 220mW Silicon Implementation for Simplified Speeded-Up Robust Feature with 65nm Process, *L. Liu, W. Zhang, C. Deng, S. Yin, S. Cai, S. Wei, Tsinghua University*   **704**

Speeded-Up Robust Feature algorithm is optimized for silicon implementation and a 57fps 1080P 220mW ASIC is presented. Methods including orientation assignment& descriptor extraction reorganization, memory accesses improvement and etc. are introduced. Experimental results show proposed architecture has great advantages in performance and power consumption compared with the state-of-art designs.

10:45 am   **BREAK**

11:05 am
18-5

Supply-Noise Resilient Adaptive Clocking for Battery-Powered Aerial Microrobotic System-on-Chip in 40nm CMOS, *Xuan Zhang, Tao Tong, David Brooks, Gu-Yeon Wei, Harvard University*   **708**

A battery-powered aerial microrobotic System-on-Chip (SoC) has stringent weight and power budgets, which requires fully-integrated solutions for both clock generation and voltage regulation. Supply-noise resilience is important yet challenging for such SoC systems due to a non-constant battery discharge profile and load current variability. This paper proposes an adaptive-frequency clocking scheme that can tolerate supply noise and improve performance when implemented with an integrated voltage regulator (IVR). Measurements from a `brain' SoC, implemented in 40nm CMOS, demonstrate 2x performance improvement with adaptive-frequency clocking over conventional fixed-frequency clocking. Combining adaptive-frequency clocking with open-loop IVR extends error-free operation to a wider battery voltage range (2.8 to 3.8V) with higher average performance.

11:30 am
18-6

Distributed clock generator for synchronous SoC using ADPLL network, *E. Zianbetov, D. Galayko, F. Anceau, M. Javidan, C. Shan, O. Billoint**, A. Korniienko*, E. Colinet**, G. Scorletti*, J. M. Akre, J. Juillard***, UPMC LIP6 Lab, *Ampere lab, **CEA-LETI, ***Supelec* **712**

This paper presents a novel architecture of on-chip clock generation employing a network of oscillators synchronized by the distributed all-digital PLLs(ADPLLs). The implemented prototype has 16 clocking domains operating synchronously in a frequency range of 1.1-2.4 GHz. The synchronization error between the neighboring clock domains is less than 60 ps. The fully digital architecture of the generation offers flexibility and efficient synchronization control suitable for use in synchronous SoCs.

11:55 am
18-7

A 920MHz Quad-core Cryptography Processor Accelerating Parallel Task Processing of Public-key Algorithms, *Shuai Wang, Jun Han, Yang Li, Yifan Bo, Xiaoyang Zeng, Fudan University*   **716**

The wireless access point (AP) devices of the next generation requires to implement the high-complexity public-key ciphers efficiently on programmable processors. Therefore, this paper presents a quad-core processor that accelerates public-key computations by enabling high-speed parallel task processing.

Session 19 -- Analog Techniques I

Wednesday, 9/25, 9:00 am
Pine Ballroom
Session Chair: Don Thelen, ON Semiconductor
Session Co-Chair: Jerry (Xicheng) Jiang, Broadcom

9:00 am   **Introduction**

This session showcases a collection of analog techniques that enable high-performance analog design.

9:05 am   **Parallel Gain Enhancement Technique for Switched-Capacitor Circuits,** *Hariprasath*
19-1   *Venkatram, Benjamin Hershberg, Taehwan Oh, Manideep Gande, Kazuki Sobue*, Koichi Hamashita*, Un-Ku Moon, Oregon State University, *Asahi Kasei Microdevices*   **720**

This paper presents a unified classification model for gain enhancement techniques used in the design of high performance amplifiers. A parallel gain enhancement technique is proposed for switched capacitor circuits which combine the best features of the existing gain enhancement techniques found in continuous-time and discrete-time amplifiers. This technique utilizes two dependent closed loop amplifiers to enhance the open loop DC gain of the main amplifier. This replicated parallel gain enhancement (RPGE) technique enables a very high DC gain amplifier with an improved harmonic distortion performance. A proof of concept pipeline ADC in a 0.18 μm CMOS process using RPGE technique achieves 75 dB SNDR, 91 dB SFDR, -87 dB THD at 20 MS/s. The measured 13 bit DNL and INL is +0.75/-0.36 and +0.88/-0.92 LSB respectively. The ADC operates from a supply voltage of 1.3 V, consumes 5.9 mW, occupies 3.06 sq. mm and achieves a figure of merit of 65 fJ/CS.

9:30 am   **Sampling Circuits That Break the kT/C Thermal Noise Limit (Invited),** *R. Kapusta, H.*
19-2   *Zhu, C. Lyden, Analog Devices, Inc.*   **724**

This paper presents techniques that prove the kT/C limit of sampled thermal noise is, in fact, not a limit at all. A first feedback technique is demonstrated to reduce thermal noise power by nearly 50%, and a second active noise cancellation technique achieves better than 70% noise power reduction.

10:20 am   **Blind Background Calibration of Harmonic Distortion Based on Selective Sampling,**
19-3   *Manideep Gande, Ho-Young Lee, Hariprasath Venkatram, Jon Guerber, Un-Ku Moon, Oregon State University*   **730**

This paper proposes a blind calibration algorithm for suppressing harmonic distortion in analog to digital converters (ADCs). The proposed algorithm does not need any external calibration signal and is first of its kind. The proposed algorithm relies on the properties of downsampling and orthogonality of sinusoidal signals to estimate the harmonic distortion coefficients. The algorithm can be operated in both foreground and background modes to remove even and odd harmonics simultaneously. The algorithm is demonstrated on a first-order ring oscillator based delta sigma ADC, whose performance is harmonic distortion limited. Built in 0.13 μm, the algorithm improves the SNDR of the ADC by 39dB while improving SFDR by 45 dB.

10:45 am   **BREAK**

11:05 am   **CMOS Millimeter Wave Phase Shifter Based on Tunable Transmission Lines,** *Wayne*
19-4   *H. Woods, Alberto Valdes-Garcia*, Hanyi Ding, Jay Rascoe, IBM Semiconductor Research and*

*Development, *IBM T.J. Watson Research Center* **734**

Design and measurements are presented of a new type of phase shifter, fabricated in a 32 nm SOI technology and operating at 60 GHz, which consists of novel tunable t-line sections that use FET switches to control L and C separately to minimize Z0 variation while changing delay.

11:30 am
19-5

Charge Steering: A Low-Power Design Paradigm (Invited), *Behzad Razavi, University of California, Los Angeles* **738**

Discrete-time charge-steering circuits consume less power than their continuous-time current-steering counterparts even at high speeds. This advantage can be exploited in the design of semi-analog circuits such as latches, demultiplexers, and CDR circuits as well as mixed-mode systems such as ADCs. Employing charge steering in 65-nm CMOS technology, a 25-Gb/sCDR/deserializer consumes 5 mW and a 10-bit 800-MHz pipelined ADC draws 19 mW.

Educational Session 5

Wednesday, 9/25, 9:00 am
Cedar Ballroom
Session Chair: Gordon Roberts, McGill University
Session Co-Chair: Takamaro Kikkawa, Hiroshima University

E-5
9:00 am –
10:30 am

Design for Nanoscale Patterning, *Puneet Gupta, UCLA* **746**

This tutorial explains how layout and circuit design interact with lithography choices. Lithography technology is rapidly evolving and has started to impose unusual restrictions on design layout. The tutorial will give a brief introduction to current and upcoming lithography technologies. We especially focus on multi-patterning technologies such as LELE double patterning and SADP. We will discuss design enablement of multi-patterning technologies, especially in context of cell-based digital designs. Models for electrical impact of lithography imperfections such as polysilicon/active rounding and overlay errors will be outlined. We will also briefly explore role of design in lithography technology development.

10:30 am   **BREAK**

Educational Session 6

Wednesday, 9/25, 11:00 am
Cedar Ballroom
Session Chair: Howard Luong, Hong Kong University of Science & Technology
Session Co-Chair: Earl McCune, RF Communications Consulting

E-6
11:00 am –
12:30 pm

Low Power Chip & System Design for Biomedical Applications, *Brian Otis, Google, Inc.* **772**

Advances in chip and system design will help define the next generation of wireless sensors for biomedical applications. This talk will investigate system and circuit design techniques for body-worn systems, implantable chips, and wireless sensors. These areas present unique challenges at the interface between the IC and the body that cannot be solved by technology scaling alone. Traditional circuit blocks, architectures, and even assembly techniques need to be questioned. Several future applications will demand thin-film realization and biocompatibility of complex systems.   RFID-like techniques are highly useful for many of these emerging biomedical applications. High-Q RF resonators are useful for minimizing power and size of low power radios. We'll discuss a few examples of the above.

Session 20 -- Advanced Memory Topics

Wednesday, 9/25, 1:30 pm
Oak Ballroom
Session Chair: Koji Nii, Renesas
Session Co-Chair: Toshiaki Kirihata, IBM

1:30 pm	**Introduction**

This session covers scaling challenges, latest advances, and future trends on spin-torque, MRAM, NAND, and logic-compatible flash, TCAM, and 6T/8T SRAM.

1:30 pm	**Introduction**

This session covers scaling challenges, latest advances, and future trends on spin-torque, MRAM, NAND, and logic-compatible flash, TCAM, and 6T/8T SRAM.

1:35 pm 20-1	**ST-MRAM Fundamentals, Challenges, and Applications (Invited),** *T. Andre, S.M. Alam, D. Gogl, C.K. Subramanian, H. Lin, W. Meadows, X. Zhang, N.D. Rizzo, J. Janesky, D. Houssameddine, J.M. Slaughter, Everspin Technologies* **799**

MRAM technology emerged from research and development into volume production within the last decade in the form of Toggle MRAM. Spin-Torque MRAM has reached the level of customer sampling, offering higher density and bandwidth. This paper describes the devices, fundamental circuit challenges, and applications of this evolving MTJ based memory.

2:25 pm 20-2	**Scaling Challenges of NAND Flash Memory and Hybrid Memory System with Storage Class Memory & NAND flash memory (Invited),** *Ken Takeuchi, Chuo University* **807**

SSDs and emerging storage class non-volatile memories such as PCRAM, ReRAM and MRAM have enabled innovations in various nano-scale VLSI memory systems for personal computers, smart phones, tablets and enterprise servers. This paper discusses the scaling challenges of 2D and 3D NAND flash memory and then provides a state-of-the-art hybrid memory solution with storage class memory and NAND flash memory for the big data solid-state storage system.

3:15 pm	**BREAK**

3:30 pm 20-3	**A 28nm High Density 1R/1W 8T-SRAM Macro with Screening Circuitry against Read Disturb Failure,** *M. Yabuuchi, H. Fujiwara, Y. Tsukamoto. M. Tanaka, K. Nii, Renesas Electronics Corporation* **813**

We developed a high density 1R/1W SRAM macro based on 8T-SRAM with a novel scheme for Design for Testing. To achieve a smaller Macro area, a differential sense amplifier is introduced to read the data, where the reference voltage for reading 0/1 data is generated by unselected cell array. In addition, we proposed a screening test circuit for read disturb operation.

3:55 pm	**A HKMG 28nm 1GHz Fully-Pipelined Tile-able 1MB Embedded SRAM IP with**

20-4 **1.39mm^2 per MB,** *M. Z. Kuo, O. Takahashi, P. L. Yang, C. C. Lin, M.J. Wang, P.W. Wang, S. H. Dhong, Taiwan Semiconductor Manufacturing Company* **817**

A fully-pipelined tile-able 1MB SRAM IP with a 0.127µm2 cell in a HKMG 28nmbulk technology has an area of 1.39mm2/MB with 79.2% array efficiency. It operates with 2-cycle latency up to 1GHz. The no-repair hardware has a circuit limited yield of 99.92 and 53% at 100 and 850MHz, respectively with 0.75V VDD.A Data Retention Voltage of 0.42V has been measured.

4:20 pm **A Bit-by-Bit Re-Writable Eflash in a Generic Logic Process for Moderate-Density**
20-5 **Embedded Non-Volatile Memory Applications,** *Seung-Hwan Song, Ki Chul Chun, Chris H. Kim, University of Minnesota* **821**

A bit-by-bit re-writable embedded flash memory is demonstrated in a generic65nm logic process for moderate-density embedded non-volatile memory applications. The proposed 6T embedded flash memory cell improves the overall cell endurance by eliminating redundant program/erase cycles without disturbing cells in the unselected wordlines. A multi-story high voltage switch utilizes four boosted supply levels generated by a compact voltage doubler based on-chip negative charge pump.

4:45 pm **Tail-Bit Tracking Circuit with Degraded VGS Bit-Cell Mimic Array for a 50%**
20-6 **Search-Time and 200mV Vmin Improvement in a Ternary Content Addressable**
Memory, *Igor Arsovski, Travis Hebig, John Goss, Paul Grzymkowski, Josh Patch , IBM* **825**

A memory sense-timing circuit uses VGS degradation to emulate the behavior of weak memory tail-bits, improving Tail-Bit Tracking (TBT) across process, voltage and temperature. The TBT circuit generates timing for a TCAM search operation reducing Vmin by 200mV, and improving sense-time by 50%. Implemented in 32nm HKMG SOI process the 2Kx640b TCAM achieves 0.60V and 1G search/sec.

Session 21 -- Oversampled ADC's

Wednesday, 9/25, 1:30 pm
Fir Ballroom
Session Chair: Eric Naviasky, Cadence
Session Co-Chair: Hasnain Lakdawala, Qualcomm

1:30 pm **Introduction**

This session has 4 papers on noise shaping ADC's. The first two papers take advantage of the noise shaping 1~ inherent in a VCO. The last two offer novel solutions to feedback DAC imperfections.

1:35 pm **A 1.8mW 2MHz-BW 66.5dB-SNDR Delta-Sigma ADC Using VCO-Based Integrators**
21-1 **with Intrinsic CLA,** *Kyoungtae Lee, Yeonam Yoon, Nan Sun, The University of Texas at Austin* **829**

This paper presents a scaling-friendly continuous-time closed-loop VCO-based Delta-Sigma ADC. It uses the VCO as both quantizer and integrator, and thus, obviates the need for power-hungry scaling-unfriendly OTAs and precision comparators. It arranges two VCOs in a pseudo-differential manner, which cancels out even-order distortions. More importantly, it brings an intrinsic CLA capability that automatically addresses DAC mismatches. The prototype ADC in 130nm CMOS occupies a small area of 0.03mm^2 and achieves 66.5dB

SNDR over2MHz BW while sampling at 300MHz and consuming 1.8mW under a 1.2V power supply. It can also operate with a low analog supply of 0.7V and achieves 65.8dB SNDR while consuming 1.1mW. The corresponding FOMs for the two cases are 0.25pJ/step and 0.17pJ/step, respectively.

2:00 pm
21-2

A 50MHz bandwidth, 10-b ENOB, 8.2mW VCO-based ADC enabled by filtered-dithering based linearization, Abhishek Ghosh, Sudhakar Pamarti, University of California, Los Angeles   **833**

A dithering technique for linearization of VCO-based ADCs is proposed. The proposed technique conditions the signal to the VCO input to appear as whitenoise thereby eliminating spurious signal content arising out of the VCO nonlinearity. The technique, thus obviates the need for power-hungry digital calibration techniques or expensive front-end loop-filters. A prototype implementation (in 65nm CMOS) based on the technique achieves 10-b ENOB in digitizing signals with 50MHz bandwidth consuming 8:2mW at an FoM of 90fJ/conv.step.

2:25 pm
21-3

A Reconfigurable Delta-Sigma Modulator With Up To 100 MHz Bandwidth Using Flash Reference Shuffling, T. Caldwell, D. Alldred, Z. Li, Analog Devices, Inc   **837**

A reconfigurable 65 nm continuous-time low-pass delta-sigma modulator operates with a sampling frequency from 491 MHz to 1536 MHz, a signal bandwidth from 10MHz to 100 MHz, and a dynamic range of 75.4 dB to 62.8 dB, respectively. Reference shuffling in the flash ADC is used to improve the linearity of the flash and DAC, while also increasing the highest sampling rate and bandwidth of the modulator.

2:50 pm
21-4

A 10-MHz Bandwidth 70-dB SNDR 640MS/s Continuous-Time Sigma-Delta ADC Using Gm-C Filter with Nonlinear Feedback DAC Calibration, J. Huang, S. Yang, J. Yuan, Hong Kong University of Science and Technology   **841**

Traditionally, wide-band (>10MHz) continuous-time sigma-delta ADCs with Gm-C filters have poor linearity. This paper introduces a novel on-chip calibration scheme to compensate the Gm-cell's nonlinearity. Measurements of the 640MS/s CTSD modulator show the best SNDR and power efficiency among Gm-C-based modulators. The FOM is also comparable to active-RC-based modulators.

3:15 pm   **BREAK**

Session 22 - AMS System Simulation Techniques

Wednesday, 9/25, 1:30 pm
Pine Ballroom
Session Chair: Larry Nagel, Omega Enterprises
Session Co-Chair: Colin McAndrew, Freescale

1:30 pm   **Introduction**

This session presents new, innovative, and efficient techniques that extend the state of the art for analog mixed-signal (AMS) system-level simulations.

1:35 pm
22-1

Algorithmic Nonlinear Macromodeling: Challenges, Solutions and Applications in Analog/Mixed-Signal Validation (Invited), C. Gu, Intel Corporation   **845**

Analog/Mixed-Signal validation at the system level is becoming increasingly important as more electrical bugs are caused by the interaction among various circuit blocks. While hand-crafted behavioral models and linear models are still most widely used among designers, there is an increasing need for automatic behavioral modeling tools which capture low-level nonlinear behaviors in the circuit. This paper discusses challenges and difficulties of algorithmic nonlinear macromodeling, and reviews a series of recently developed techniques. In particular, we study the behavioral modeling problem from the perspective of projection in the state space defined by voltages and currents. We review a few nonlinear macromodeling techniques from the projection perspective, and demonstrate the model accuracy and computational efficiency compared to transistor-level models and linear models.

| 2:25 pm 22-2 | **Event-Driven Simulation of Volterra Series Models in System Verilog,** *J-E. Jang, S-J. Yang, J. Kim, Seoul National University* **853** |

This paper presents a true event-driven methodology to simulate weakly-nonlinear analog circuits in System Verilog. We express a continuous-time signal as a linear combination of basis functions and reformulate a Volterra series model into a set of linear differential equations with an explicit notion on initial conditions. Two circuit examples showed 300~1000× speed-up compared to SPICE with the same accuracy.

| 2:50 pm 22-3 | **A Verilog Piecewise-Linear Analog Behavior Model for Mixed-Signal Validation,** *S. Liao, M. Horowitz, Stanford University* **857** |

Mixed-signal validation requires simulating the entire chip through a large number of test vectors, which makes pin-accurate and fast Verilog functional models of analog circuits with reasonable fidelity valuable. We describe an extensible approach to creating these models, by mapping continuous signals into discrete events and avoiding explicit time integration.

| 3:15 pm | **BREAK** |

| 3:30 pm 22-4 | **Advancements in High-Speed Link Modeling and Simulation (Invited),** *Mike Peng Li, Masashi Shimanouchi, Hsinho Wu, Altera Corporation* **861** |

This paper starts with reviewing the status of techniques and methods used in recent high-speed link simulation and modeling for signaling, integrated circuits, board circuits, and associated limitations and challenges, and then discusses new advancements that can overcome them.

| 4:20 pm 22-5 | **Structure-Aware High-Dimensional Performance Modeling for Analog and Mixed-Signal Circuits,** *S. Sun, X. Li, C. Gu*, Carnegie Mellon University, *Intel Corp* **869** |

Efficient high-dimensional performance modeling of nanoscale analog and mixed signal (AMS) circuits is challenging. In this paper, we propose a novel structure-aware modeling technique to accurately solve the model coefficients by exploiting the underlying structure of AMS circuits, and hence dramatically reduce the number of sampling points for performance modeling.

Session 23 -- Analog Techniques II

Wednesday, 9/25, 3:25 pm
Fir Ballroom
Session Chair: Hasnain Lakdawala, Qualcomm

Session Co-Chair: Eric Naviasky, Cadence

3:25 pm   **Introduction**

The session includes topics in time to digital converters, fast locking PLL's and double sampling sigma delta ADC's.

3:30 pm
23-1
A 148fs$_{rms}$Integrated Noise 4MHz Bandwidth All-Digital Second-Order ΔΣ Time-to-Digital Converter Using Gated Switched-Ring Oscillator, *W. Yu, K.S. Kim, S. Cho, KAIST*   **873**

This paper presents an all-digital second-order ΔΣ time-to-digital converter(TDC) by using switched-ring oscillator (SRO) and gated switched-ring oscillator (GSRO). Unlike conventional multi-stage noise-shaping (MASH) TDC using the SRO, the proposed TDC does not require complex calibration to compensate for the error from frequency difference between the SROs. The prototype TDC achieves 148fsrms integrated noise and 80.4dB dynamic range in4MHz signal bandwidth at 400MS/s while consuming 6.55mW in a 65nm CMOS process.

3:55 pm
23-2
A 0.84ps-LSB 2.47mW Time-to-Digital Converter Using a Charge Pump and a SAR-ADC, *Zule Xu, Seungjong Lee, Masaya Miyahara, and Akira Matsuzawa, Tokyo Institute of Technology*   **877**

We propose a 0.84ps-LSB, 2.47mW, 0.06mm² time-to-digital converter (TDC) using a charge pump and a SAR-ADC in 65nm CMOS. Sub-pico second time resolution is attainable by quantizing the time in charge domain. Low power consumption and small area are also feasible by using the SAR-ADC.

4:20 pm
23-3
A double-sampling cross noise-coupled Sigma Delta modulator with a reduced amount of opamps, *M. De Bock, P. Rombouts, UGhent*   **881**

A second order double-sampling cross noise-coupled split-path Sigma Delta modulator is presented. The implementation of the noise-coupling is incorporated into the second integrator using a novel delaying feed-forward circuit. A prototype integrated in a 130nm CMOS technology achieves 77.8dBdynamic range and 71.4dB SNDR over a 5MHz bandwidth.

4:45 pm
23-4
A Novel OTA-Based Fast Lock PLL, *Mezyad Amourah, Sandeep Krishnegowda, Morgan Whately, Cypress Semiconductor*   **885**

A novel fast lock scheme for phase-locked loops (PLLs). The proposed scheme uses a simple operational transconductance amplifier (OTA) to achieve significant reduction in PLL lock acquisition time without affecting PLL noise performance. The new fast lock schemes were implemented in multi-port SRAM chip to provide frequencies from 400MHz to 1.6GHz, The chip was fabricated using 65nm CMOS process. Silicon measurements across corner lots show significant reduction in PLL lock time, by a factor of 6.5X, over device operating conditions.

Educational Session 7

Wednesday, 9/25, 1:30 pm
Cedar Ballroom
Session Chair: Mike Li, Altera
Session Co-Chair: Gerrit den Besten, NXP Semiconductors

E-7	**Trends, Possibilities and Limitations of Photonic Integrated Circuits and Lasers,**
1:30 pm – 3:00 pm	*John E. Bowers, University of California, Santa Barbara* **889**

A number of important breakthroughs in the past decade have focused attention on Si as a photonic platform. We review here recent progress in this field, focusing on efforts to make lasers, amplifiers, modulators and photodetectors on or in silicon. We also describe progress in silicon photonic integrated circuits. The impact active silicon photonic integrated circuits could have on interconnects, telecommunications and on silicon electronics is reviewed.

3:00 pm **BREAK**

Educational Session 8

Wednesday, 9/25, 3:30 pm
Cedar Ballroom
Session Chair: Jerry Jiang, Broadcom
Session Co-Chair: Eric Naviasky, Cadence

E-8	**A/D Converter Circuit and Architecture Design for High-Speed Data Communication,**
3:30 pm – 5:00 pm	*Boris Murmann, Stanford University* **934**

A number of important breakthroughs in the past decade have focused attention on Si as a photonic platform. We review here recent progress in this field, focusing on efforts to make lasers, amplifiers, modulators and photodetectors on or in silicon. We also describe progress in silicon photonic integrated circuits. The impact active silicon photonic integrated circuits could have on interconnects, telecommunications and on silicon electronics is reviewed.

A 24μW 12b 1MS/s 68.3dB SNDR SAR ADC with Two-Step Decision DAC Switching

Yung-Hui Chung[1,2], Meng-Hsuan Wu[1], Hung-Sung Li[1]

[1]MediaTek Inc., Hsin-Chu, Taiwan

[2] National Taiwan University of Science and Technology, Taipei, Taiwan

Abstract—**A 12-bit SAR ADC is presented with a FoM of 11.7fJ/conversion-step. The ADC employs a two-step decision DAC switching technique for improving the ADC's linearity and switching energy using smaller input capacitance. It effectively eliminates the largest capacitor-DAC middle-code transition error. The proposed switching scheme can also tolerate DAC settling errors without requiring redundant capacitors. The ADC core occupies an area of 0.079mm² in a 0.11μm CMOS process. At 1MS/s, it consumes 24μW from a 0.9V supply. Measured DNL and INL are 0.29LSB and 0.55LSB respectively. Measured SNDR and SFDR are 68.3dB and 82dB respectively. Measured ENOB is 11.05b at the Nyquist-rate input.**

I. INTRODUCTION

For energy-efficient and low voltage applications, such as wireless sensor network, biomedical sensors and portable instruments, power-efficient ADCs play a crucial role in enabling autonomous operation. As CMOS process advances, successive approximation register (SAR) ADCs have drawn more and more attention due to their unique advantages for operating at low voltage, with excellent conversion power efficiency and small footprint.

Considering high-resolution SAR ADCs, DAC linearity is among the main hurdles for getting high-accuracy outputs. A 12b SAR ADC was reported in [1] using a complex digital calibration process to achieve good linearity. Reference [2] proposed a 10b variable window SAR ADC in which an extra coarse sub-ADC was introduced for improving linearity. Moreover, DAC settling time is another bottleneck for SAR ADCs conversion efficiency. Digital error correction scheme [3] uses redundant capacitors and conversion cycles to tolerate DAC settling error with simple digital logics. However, it incurs extra input capacitance and reduced dynamic range. An addition-only digital error correction [4] was reported without extra redundant capacitors. But its error tolerance range is quite restrained. This paper presents a 12b SAR ADC that employs a Two-Step Decision (TSD) DAC switching technique for both improving linearity and compensating DAC settling error without needing redundant capacitors. It achieves 12b linearity with smaller input capacitance and therefore lowers the power consumption for input and reference buffers.

II. TSD DAC SWITCHING TECHNIQUE

In SAR ADCs, capacitive DAC linearity is typically dominated by the capacitor mismatch. Illustrated in Fig. 1a, a conventional binary DAC has a characteristic big middle-

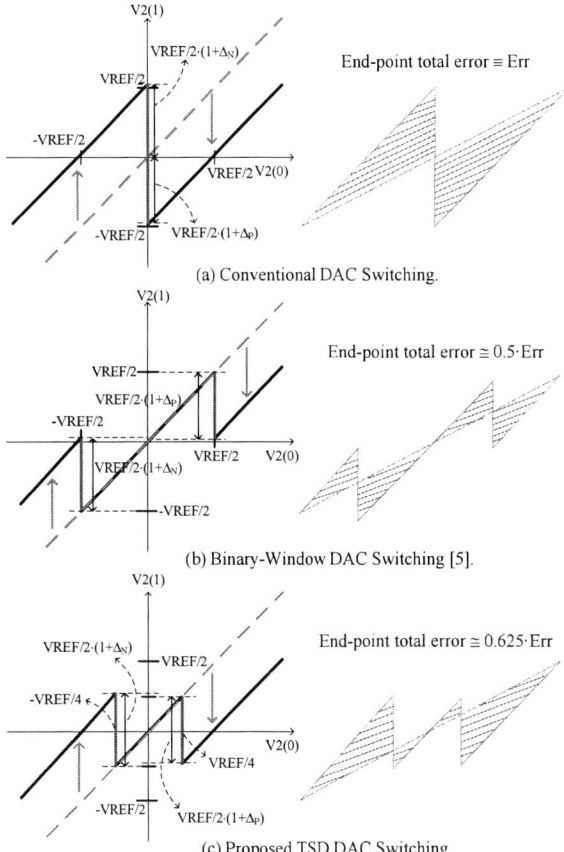

(a) Conventional DAC Switching.

(b) Binary-Window DAC Switching [5].

(c) Proposed TSD DAC Switching.

Fig. 1. Comparison of MSB DAC switching errors due to capacitor mismatches.

code switching glitch when the MSB capacitor is switched in or out [3]. Here V2 is the differential comparator input voltage. V2(0) (x-axis) represents the initial input voltage seen by the comparator. V2(1) (y-axis) is the comparator input after the MSB capacitor is switched. The input range is ±VREF. When there is a capacitor mismatch, the resultant voltage on the DAC will not be exactly VREF/2, rather they become VREF/2×(1+Δ_P) and VREF/2×(1+Δ_N) for the plus and minus terminals respectively. The maximum combined error for the zero input would be VREF/2×($|\Delta_P|+|\Delta_N|$). To improve the resultant INL by half, the capacitance must be 4X larger. The power consumption of input and reference buffers is also 4X to maintain the same operation speed.

The major INL error results from the accumulation of one half of the total capacitances vs. that of the other half. If the MSB capacitor switching can be avoided for middle input, the amount of total accumulated INL error can be reduced. The idea, as shown in Fig. 1b, is to shift the first switching to +1/2 or -1/2 of the full input range so INL error is improved [5]. Although the INL error (area by red dotted-line) in Fig. 1b is only half of the amount in Fig. 1a, the DAC settling error cannot be properly tolerated to speed up the ADC operation and achieve good conversion efficiency. Therefore, both DAC linearity and settling error tolerance must be put into consideration. In this work, the MSB transition boundaries are shifted to ±1/4 points of the full input range. As shown in Fig. 1c, the proposed DAC switching scheme removes the middle-code switching error and kind of splitting it into two smaller errors at the ±VREF/4 input. The largest switching error is now shifted to +VREF/4 or -VREF/4. Somewhat similar to a pipelined ADC, the error tolerance is ±VREF/4. But in this work, the scheme tolerates the DAC settling error and comparator noise, rather than the sub-ADC offset.

Unlike [2], the decision boundary is not determined by including two extra comparators and one resistor string. Such an auxiliary circuit will consume static power. In addition, its performance would be degraded if the comparator offset is too much. In this work, the boundary is set by the proposed two-step decision (TSD) DAC switching scheme. Fig. 2a depicts the TSD-SAR ADC operation. Each output bit is obtained by executing two consecutive comparisons with respective DAC switching. Taking the MSB as an example, in the first step the polarity of V2(0) is sensed, $q_{11}=1$ when V2(0)>0 and $q_{11}=0$ when V2(0)<0. Depending on the first comparison result, the (MSB-1) capacitor of either the plus or minus terminal is connected to GND. V2' is then driven to V2(0)-VREF/4 for $q_{11}=1$ or V2(0)+VREF/4 for $q_{11}=0$. In the second step, the comparator is activated again to sense the polarity of V2' and outputs q_{12} where $q_{12}=1$ if V2'>0 and $q_{12}=0$ if V2'<0. Next, the two comparison results (q_{11}, q_{12}) are combined to give the final MSB DAC switching decision (b_{11}, b_{12}), as shown in Fig. 2a. There are three possible outcomes: (1) V2(0)>VREF/4, plus-terminal MSB capacitor is connected to GND and yields V2(1)=V2(0)–VREF/2; (2) V2(0)<–VREF/4, minus-terminal MSB capacitor is connected to GND and yields V2(1)=V2(0)+VREF/2; and (3) V2(0) is bounded within –VREF/4 and VREF/4, plus or minus-terminal (MSB-1) capacitor is switched back thus V2(1) is put back to V2(0). For the rest of the ADC conversion cycles, the same procedure is followed and repeated. There are total 23 evaluation cycles for a 12-bit TSD-SAR ADC. Finally, overall comparison results {(b_{11}, b_{12}), (b_{21}, b_{22})... (b_{91}, b_{92}), (b_{A1}, b_{A2}), (b_{B1}, b_{B2}), b_C} are encoded to generate a 12-bit digital output, similar to a 1.5b/stage pipelined ADC. The comparator input V(2) vs. conversion cycle is plotted in Fig. 2b. Here the no-switching zone refers to the area where there is no DAC switching happening for that particular conversion cycle. The voltage range of the no-switching zone is scaled down by a factor of 2 from one bit to the next. The DAC settling error tolerance range, shown by the dotted red-box, is also plotted. They are VREF/4 for the MSB DAC switching, VREF/8 for (MSB-1) DAC switching etc.

In this work, TSD-SAR can tolerate 50% error for each DAC settling, comparator offset variation and comparator noise during the sub-conversion evaluation, four times

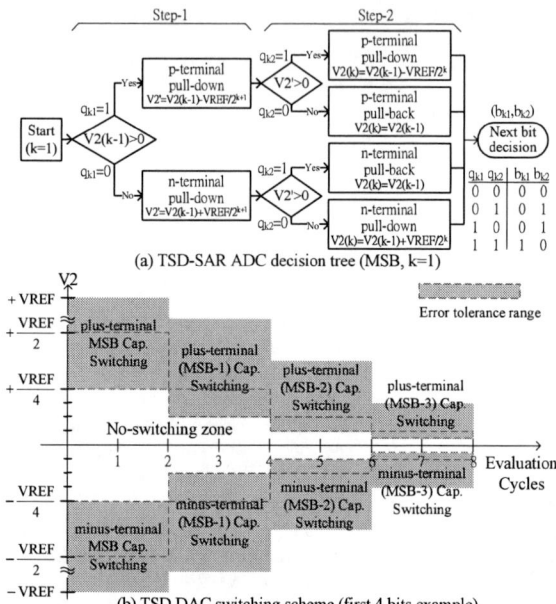

(a) TSD-SAR ADC decision tree (MSB, k=1)

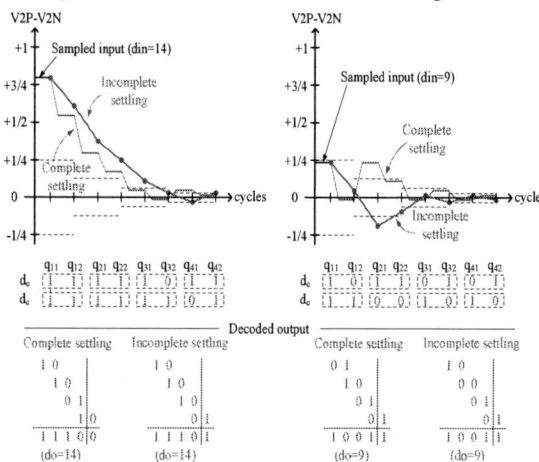

(b) TSD DAC switching scheme (first 4 bits example)

Fig. 2. TSD SAR ADC decision tree and switching scheme.

Fig. 3. Two 4-bit TSD-SAR ADC examples.

bigger error tolerant range than that in [2]. Larger DAC error tolerance effectively shortens the DAC settling time thus extra number of comparison can be compensated. For high-resolution SAR ADC, the DAC settling time is still longer than the comparison time, especially for scaled CMOS technologies. It means that faster CMOS process favors TSD-SAR. In fact, the time reserved for DAC settling is the major contributor to delay when CMOS is scaled. To estimate this waiting time, some extra delay circuits are implemented [6]. With the proposed TSD DAC switching scheme, it becomes easier to implement the delay circuits with less power consumption. For high-speed operation, the TSD DAC switching can be implemented only for the beginning of the first few evaluation cycles to reduce the burden of the additional comparison cycles.

Fig. 3 shows two 4-bit TSD-SAR ADC examples. The left figure is an example of larger input (din is 14). The blue line shows a complete settling case for reference. The red line is for incomplete DAC settling, which yields different evaluation result. After encoding, both final digital outputs are the same. The right figure is another example of a

978-1-4673-6145-3/13 $31.00 © 2013 IEEE

smaller input (din is 9). Due to large settling errors for the first and second DAC switching, the evaluation process is very different from that of the complete settling case. However, with the TSD-SAR encoding, both resultant digital outputs reveal the same values.

III. CIRCUIT IMPLEMENTATION

Fig. 4 shows the schematic of the proposed 12b SAR ADC. It consists of bootstrapped switches, a hybrid-DAC, a low-noise dynamic comparator and a TSD-SAR control unit. Two capacitor arrays (plus- and minus-terminals in the figure) are used for differential operation. VREF and GND represent the top and bottom reference voltages, respectively. In this design, VREF is connected to the ADC power supply. For a top-plate sampling binary-weighted 12-bit SAR ADC using C-DAC, MSB capacitor is $2^{10} \times C_U$, where C_U being the LSB unit capacitor. The single-ended total capacitance would be around 10pF with a 5fF unit capacitor. Such large capacitance incurs excessive driving power for the ADC input and reference buffers. To minimize the input capacitance thus power, the hybrid-DAC uses resistor-DAC (R-DAC) to generate sub-reference voltage (VREF/4) for the last three bits. Taking plus-terminal capacitor as an example, first nine capacitors ($CP_{11} \sim CP_3$, $CP_k = 2^{k-3} \times C_U$) are connected to VREF directly by their respective control switches. The last three capacitors ($CP_2 = 2C_U$, $CP_1 = CP_0 = C_U$) are connected to the sub-reference voltage VREF/4. As shown in Fig. 4, the total single-ended capacitance is thus reduced to 2.58pF, which still less than the total capacitance in [1] without any calibration. Furthermore, this hybrid DAC avoids LSB capacitors that are too small to maintain the matching reliability for mass production. The accuracy of the R-DAC is simple to maintain. The settling time of last three LSB DAC switching can be easily achieved even using a large resistor string (the total resistance is 1MΩ).

A programmable dynamic comparator [6] is adopted for power efficient operation. This comparator has two operation modes, high-speed mode and low noise mode, controlled by LN_mode. As shown in Fig. 4, this ADC uses a clock signal CK with about 50% duty ratio. It helps to avoid using a power-hungry input buffer. Internal clock 'CKc' is generated by an asynchronous clocking scheme. It also helps the conversion speed and removes the need of a ~13X-speed clock signal. First five MSB capacitors are designed with the split-capacitor structure to reduce the comparator input common-mode variations for different samples and ensure reliable comparator function.

At the beginning of each sampling period (CK is '1'), all capacitors are connected to their reference voltages through switches. At the falling edge of CK, the analog input signals, V1P and V1N, are sampled on the capacitor array and the comparator's input nodes, V2P and V2N. At the beginning of hold mode (CK is '0'), the comparator makes the first comparison to yield the comparator output (d_c). If V2P>V2N, d_c is '1'; otherwise, d_c is '0'. After the comparison is finished, a 'valid' signal is then generated to trigger the next comparison. Based on the comparator output d_c, the TSD-SAR control unit decides the switch connections for all capacitors in the plus- and minus-terminals. The TSD-SAR control unit receives comparison result 'd_c' sequentially and encodes them into a 12-bit digital output 'D_o'.

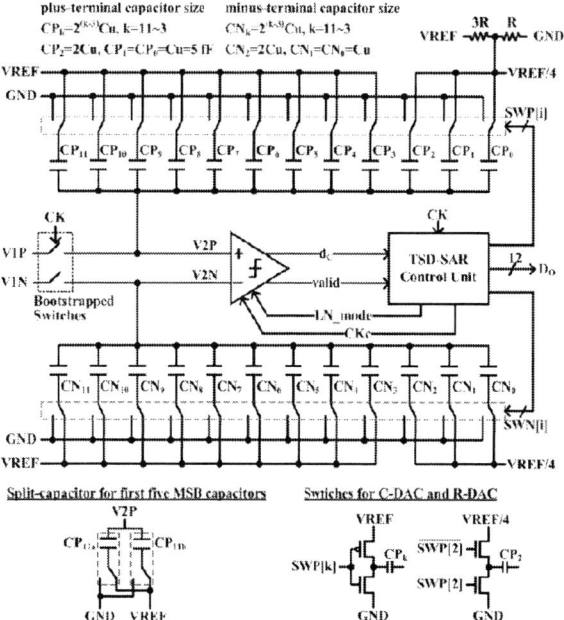

Fig. 4. The proposed ADC architecture.

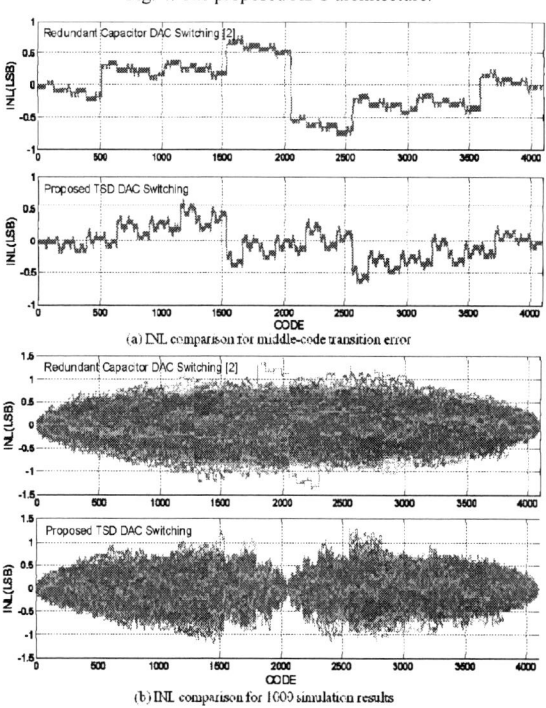

(a) INL comparison for middle-code transition error

(b) INL comparison for 1000 simulation results

Fig. 5. Simulated INL Comparison between the 12-bit redundant-capacitor SAR ADC and the proposed TSD-SAR ADC.

To illustrate the MSB glitch removal, one comparison of mismatch simulation result is shown in Fig. 5a. Top plot is the INL plot using conventional DAC switching scheme and bottom plot is the result using the proposed TSD DAC switching scheme. It is obviously that the MSB glitch is eliminated. In fact, it is shifted to ±1/4 of the full scale input range. The major linearity error is effectively improved. In Fig. 5b, these two 12-bit SAR ADCs are simulated with 1000 Monte-Carlo runs to see the linearity improvement. Capacitor mismatch data is extracted from UMC 0.11um CMOS process. The single-ended total capacitance is

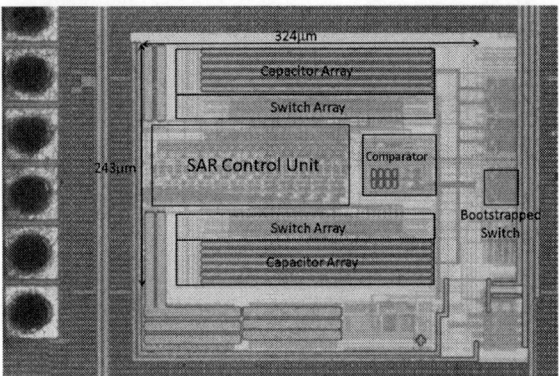

Fig. 6. The ADC chip micrograph.

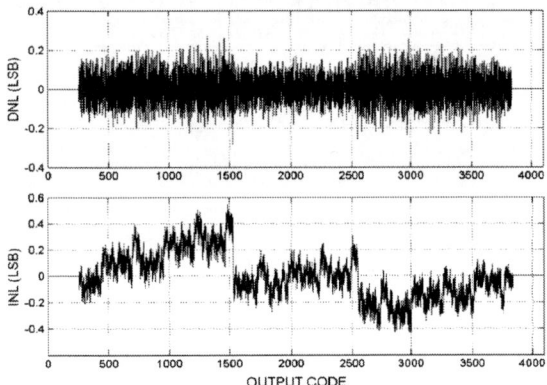

Fig. 7. Measured DNL and INL.

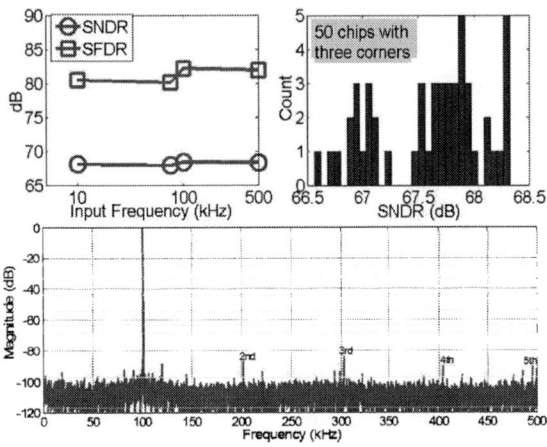

Fig. 8. Measured dynamic performance.

2.56pF. There is a characteristic notch in the TSD-SAR INL plot to improve linearity significantly.

IV. MEASUREMENT RESULT

Fig. 6 shows the ADC chip micrograph. The chip is fabricated in a 0.11μm CMOS technology and occupies 0.079mm². Operating at 1MS/s, it consumes a total current of 26.7μA from a 0.9V supply. As shown in Fig. 7, the maximum DNL and INL are -0.29/+0.28LSB and -0.42/+0.55LSB respectively. As evident in the INL plot, the middle-code glitch is gone. Fig. 8 depicts the measured SNDR and SFDR versus input frequency at 1MS/s. The measured SNDR and SFDR are maintained over 68dB and 80dB up to Nyquist frequency. To evaluate its robustness,

TABLE I. 12-BIT SAR ADC PERFORMANCE COMPARISON.

Specification (Unit)	[1]	[6]	[7]	This work
Supply Voltage (V)	1.2	0.9	1.0	0.9
Power (mW)	3.02	0.0165	0.025	0.024
Sampling rate (MS/s)	22.5	1	0.1	1
SNDR (dB)	71.1	67.3	65.3	68.3
SFDR (dB)	90.3	87	71	82
DNL/INL (LSB)	NA	0.4/0.56	0.66/0.68	0.29/0.55
Input capacitance (pF)	3.6	3.1	NA	2.58
FOM (fJ/conv.-step)	51.3	8.47	165	11.7
ENOB (bit)	11.5	10.92	10.55	11.05
Core Area (mm²)	0.09	0.076	0.63	0.079
Technology (nm)	130	110	180	110

fifty ADC chips from three process corners (30 TT-chips, 10 FF-chips and 10 SS-chips) are verified. The measured SNDR varies from 66.5dB to 68.4dB, which variance is smaller than conventional SAR ADCs to get better yield. The FFT spectrum (4096-pt) is shown in Fig. 6. With a -1dBFS 100kHz sine wave input, it achieves SFDR of 82dB and SNDR of 68.3dB at 1MS/s. The total input capacitance is only 2.58pF. Table I represents the 12-bit SAR ADC comparison result. This ADC has good SNDR performance and smallest input capacitance for 12-bit resolution.

V. CONCLUSION

A 12-bit 1-MS/s SAR ADC fabricated in a 0.11 μm digital CMOS process is presented. The proposed two-step decision (TSD) DAC switching is proven to effectively linearize the SAR ADC performance without beefing up the sampling capacitors, thus achieving both small die area and low power consumption. It also provides large error tolerance for DAC settling, comparator offset variation and comparator noise during evaluation. The designed ADC consumes 24 μW from a 0.9V supply and achieves good power efficiency, with a FoM of 11.7fJ/conversion-step. The ultra low power consumption feature finds itself useful for battery-powered systems.

REFERENCES

[1] Wenbo. Liu, Pingli Huang, and Yun Chiu, "A 12 b 22.5/45 MS/s 3.0 mW 0.059 mm² CMOS SAR ADC achieving over 90 dB SFDR," in *IEEE ISSCC Dig. Tech. Papers*, pp.380–381, Feb. 2010.

[2] Chun-Cheng Liu, et al., "A 1V 11fJ/Conversion-Step 10bit 10MS/s Asynchronous SAR ADC in 0.18μm CMOS," in *IEEE Symposium on VLSI Circuits*, Jun. 2010, pp. 241-242.

[3] Chun-Cheng Liu, et al., "A 10b 100MS/s 1.13mW SAR ADC with binary scaled error compensation," in *IEEE ISSCC Dig. Tech. Papers*, pp.386–387, Feb. 2010.

[4] Sang-Hyun Cho, Chang-Kyo Lee, Jong-Kee Kwon, and Seung-Tak Ryu, "A 550μW 10b 40MSPS SAR ADC with multistep addition-only digital error correction," in *Proc. CICC*, pp.1-4, Sept. 2010.

[5] Yung-Hui Chung, "The Swapping Binary-Window DAC Switching Technique for SAR ADCs," in *Proc. IEEE International Symposium on Circuits and Systems (ISCAS) 2013, in press.

[6] Meng.-Hsuan. Wu, Yung-Hui Chung, and Hung-Sung Li, "A 12-bit 8.47-fJ/Conversion-Step 1-MS/s SAR ADC using Capacitor-Swapping Technique," in *Proc. IEEE A-SSCC*, pp.157–160, Nov. 2012.

[7] N. Verma, and A. Chandrakasan, "A 25 μW 100kS/s 12 bit ADC for wireless micro-sensor applications," in *IEEE ISSCC Dig. Tech. Papers*, pp.222–223, Feb. 2006.

A 3.3fJ/conversion-step 250kS/s 10b SAR ADC Using Optimized Vote Allocation

Muhammad Ahmadi, *Student Member IEEE*, Won Namgoong, *Senior Member IEEE*
Department of Electrical Engineering, University of Texas at Dallas
Email: mxa091220@utdallas.edu, namgoong@utdallas.edu

Abstract—This paper presents a 10b successive approximation register (SAR) analog-to-digital converter (ADC) that operates at 0.5V supply voltage and supports a flexible differential input dynamic range from 0.4V to 1V. The proposed ADC employs a majority vote comparison along with a non-binary architecture to alleviate the effect of comparator noise in scaled input voltage swings. To maximize performance subject to comparator power level constraints, the allocation of votes is optimized for each bit cycle. The prototype, fabricated in 65nm CMOS process, achieves ENOB ranging from 7.1b to 9.1b and FOM from 3.3 to 6.8fJ/conversion step while operating at 250kS/s.

I. INTRODUCTION

Wireless sensor networks monitor environmental conditions such as temperature, pressure, light, and/or sound. The sensor output signals are digitized with an ADC before processing, storage, and/or transmission. In many autonomous sensor nodes that are powered by an energy harvesting source [1], the energy harvester can only provide low output voltage with limited power. The ADC, therefore, needs to dissipate minimal power while operating at reduced supply voltages (V_{DD}). The ADC should also support signals with reduced input dynamic range, since the linear region of the amplifier output preceding the ADC may be a fraction of the supply voltage.

Charge redistribution SAR ADC has been popular in sensor node applications due to its simplicity and low power consumption at medium sampling speed and resolution [2], [3]. Existing works in literature on reduced supply voltage ADCs are based primarily on SAR architecture. A 0.6V SAR ADC using data driven noise reduction technique is reported in [4]. The ADC achieves 9.4 effective number of bits (ENOB) at 40 kS/s conversion rate while consuming $0.072\mu W$ power. A resolution configurable and variable supply voltage (0.4V-to-1V) ADC has been demonstrated in [5]. The ADC achieves a FOM of 22.4 fJ/conversion-step at 0.55V in the 10b mode. In [6], a 0.5V 10b SAR ADC using tri-level comparator and reconfigurable DAC architecture is presented. The achieved ENOB is 7.5b at sampling rate of 1.1MS/s and FOM is 6.3fJ/conversion-step.

The three primary components of a SAR ADC are the digital-to-analog converter (DAC), digital SAR logic, and comparator. By employing a small unit capacitor for the DAC [6] together with digital subthreshold circuits, the power consumption of the DAC and SAR logic can be greatly

This work was supported in part by the National Science Foundation (NSF) under contract ECCS 1002334.

minimized in low frequency applications. Consequently, the comparator has become a major source of power consumption in many low frequency SAR ADCs [4].

When operating at reduced input voltage swings, the primary source of performance degradation is the comparator thermal noise. Low-noise comparator can be used but at the expense of increased power dissipation, since the comparator input-referred thermal noise variance is approximately inversely proportional to the comparator power consumption. To achieve high resolution at scaled supply voltages and/or reduced input dynamic range, we use a non-binary capacitor array so to provide sufficient redundancy in the presence of thermal noise. As decision errors in the last few cycles cannot be corrected, low-noise comparators need to be used. One approach for reducing power is to employ a comparator with two operating modes [7] - a low-power comparison for the non-critical upper bits and a low-noise/high-power comparison for the critical last few cycles. The drawback of this approach is the need to calibrate the offset voltage for each mode of operation. In this work, we propose to perform multiple comparisons to reduce the effective comparator noise before deciding in the critical bit cycles. The number of votes allocated to the last few critical bits are optimized to maximize performance for a given comparator power budget. The prototype, fabricated in 65nm CMOS process, achieves ENOB ranging from 7.1b to 9.1b and FOM from 3.3 to 6.8fJ/conversion-step for 0.4V to 1V differential input dynamic range.

The paper is organized as follows. Section II describes the proposed ADC architecture and majority vote based comparator. Section III formulates the comparator noise requirement in critical bit positions and optimizes the number of votes allocated to each bit step. Finally, measurement results are shown in Section IV and conclusions are drawn in Section V.

II. ADC ARCHITECTURE

Fig. 1 shows the proposed architecture of a 10-bit SAR ADC. The proposed architecture employs a majority vote comparison along with a non-binary redundant capacitor array to alleviate the effects of comparator noise at reduced supply voltages and/or small input dynamic range.

A. Majority Vote Comparison

A majority vote based technique is employed to control the effective comparator noise in each bit cycle. Fig. 2 shows

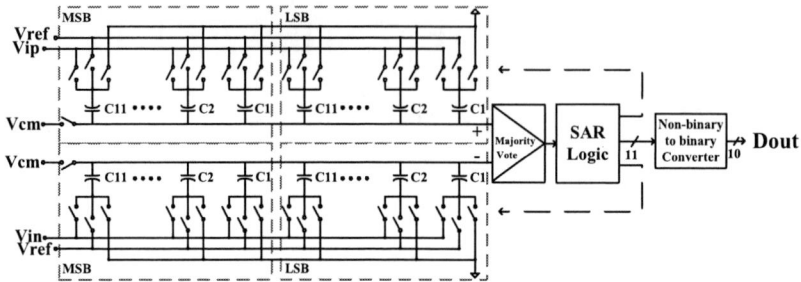

Fig. 1. The proposed SAR ADC block diagram.

a block diagram of the majority vote comparison. When performing majority vote, the comparisons are performed successively. A simple counter and a comparison circuit are used to determine the majority vote. The majority voting circuit can be bypassed and clock-gated whenever there is no need to take multiple votes. If N_i comparisons are performed before deciding, the comparator dynamic power consumption increases by N_i, while the effective input referred noise variance is approximately reduced by N_i. Consequently, performing N_i comparisons is equivalent in effect to employing a comparator that dissipates N_i times more power with attendant input referred noise variance that is approximately N_i times smaller.

An important advantage of voting-based comparator is the flexibility that it provides. For the first several bit cycles where low-noise comparisons are not critical, the proposed scheme allows the comparator to decide based on only one comparison. In the last few critical cycles, the majority vote is enabled and low noise comparison can be performed by taking multiple votes. This approach, therefore, saves comparator power consumption in non-critical bits by operating in high-noise/low-power mode. A potential drawback is the increase in delay due to multiple votes, but as the comparisons can be performed at a rate as fast as 500 MHz at 0.5V supply, the speed overhead is negligible, especially in sensor systems where the sampling frequencies are low.

B. Non-binary Redundant DAC

For an N-bit binary weighted SAR ADC, at most two out of N comparisons are critical because of thermal noise. One of those critical decisions occurs at the last bit. The other critical decision may occur at any of the previous (N-1) comparisons. As simply performing a large number of votes for each bit cycle is not power efficient, a non-binary architecture realized as a split-capacitor array is employed to realize a 10 bit resolution SAR ADC in 11 bit cycles. The capacitor weights are chosen to have enough redundancy in non critical bit cycles. In the prototype ADC, the capacitors which are sized at 512, 250, 128, 64, 32, 16, 8, 6, 4, 2, 1 multiples of unit

capacitance, provide error corrections of up to ±12, ±6, ±6, ±6, ±6, ±6, ±2, 0, 0, 0, 0 LSBs in each cycle, respectively. If any error occurs during the first seven non-critical cycles, it can be corrected assuming decisions in the last four critical cycles are correct. Therefore, a large number of votes are taken in the last four cycles to minimize the probability of a decision error.

III. EFFECTIVE COMPARATOR NOISE OPTIMIZATION

The number of votes allocated to the last four critical bit cycles need not be the same since the signal voltage distribution at the comparator input is different for each of these four bit cycles. Therefore, the optimum number of votes that maximizes performance for a given power constraint can be determined. This section describes the general approach employed to compute the optimum distribution of votes in the last four critical bit cycles.

The mean-squared error (MSE) of the ADC codeword is defined as

$$MSE = E[(C - \hat{C})^2] \tag{1}$$

where $C \in \{0, 1, \ldots, 1023\}$ is the ideal codeword and $\hat{C} \in \{0, 1, \ldots, 1023\}$ is the observed codeword in the presence of comparator noise. In each of the four critical bit cycles, we assume that the comparator decision error is negligible when the comparator input voltage V_{diff} exceeds 2Δ, where Δ represents the LSB voltage, and the comparator noise is in the tolerable range of redundancy for the first seven non-critical bit cycles. The MSE in (1) can then be approximated as

$$MSE \approx \sum_{i=0}^{3} P_{a,i} + 4\sum_{i=1}^{3} P_{b,i} \tag{2}$$

where $P_{a,i}$ and $P_{b,i}$ are the probabilities of a decision error in the ith bit cycle when $|V_{diff}| \le \Delta$ and $\Delta < |V_{diff}| \le 2\Delta$, respectively. $P_{a,i}$ can be shown to be

$$P_{a,i} = \frac{1}{W_i}[Q(\frac{\Delta}{\sigma_i}) - \frac{\sigma_i}{\Delta\sqrt{2\pi}}(1 - e^{\frac{-\Delta^2}{2\sigma_i^2}})] \tag{3}$$

where σ_i^2 is the effective input-referred comparator noise variance at the i-th bit cycle, W_i is the normalized capacitor weights (i.e., $W_3 = 6$, $W_2 = 4$, $W_1 = 2$, $W_0 = 1$) for the ith bit cycle, and the Q-function is defined as

$$Q(x) = \int_x^{\infty} \frac{1}{\sqrt{2\pi}} e^{-t^2/2} dt \tag{4}$$

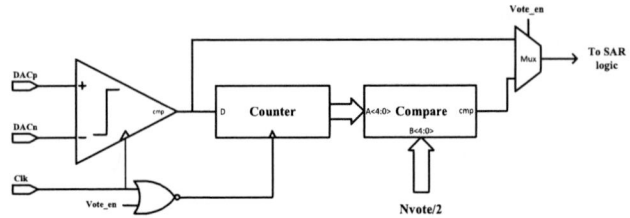

Fig. 2. The proposed majority vote comparison block diagram.

Similarly, $P_{b,i}$ is given by

$$P_{b,i} = \frac{1}{W_i}[2Q(\frac{2\Delta}{\sigma_i}) - Q(\frac{\Delta}{\sigma_i}) + \frac{\sigma_i}{\Delta\sqrt{2\pi}}(e^{\frac{-\Delta^2}{2\sigma_i^2}} - e^{\frac{-2\Delta^2}{\sigma_i^2}})] \quad (5)$$

Using Lagrange multiplier method, the MSE in (2) is minimized subject to the total number of votes available to determine the optimum σ_i values and the corresponding number of votes for each of the bit cycles. Simulation results shows that more than 50% saving in comparator power can be achieved compared to the conventional binary SAR ADC with equally distributed votes.

IV. MEASUREMENT RESULTS

The prototype chip has been designed using 65nm CMOS process. Fig. 3 shows die micrograph of the chip which occupies an active area of $0.072mm^2$. Measured differential non-linearity (DNL) and integral non-linearity (INL) with respect to output code of the ADC are shown in Fig. 4. The maximum DNL and INL are +0.34/-0.36 LSB and +0.45/-0.47 LSB, respectively. The supply voltage is at 0.5V and differential input dynamic range is ±0.5V.

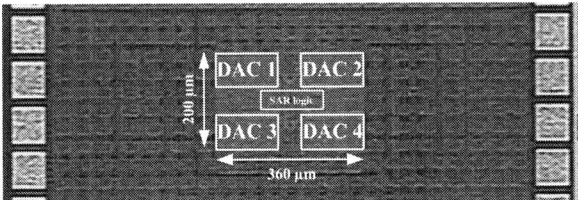

Fig. 3. Die micrograph.

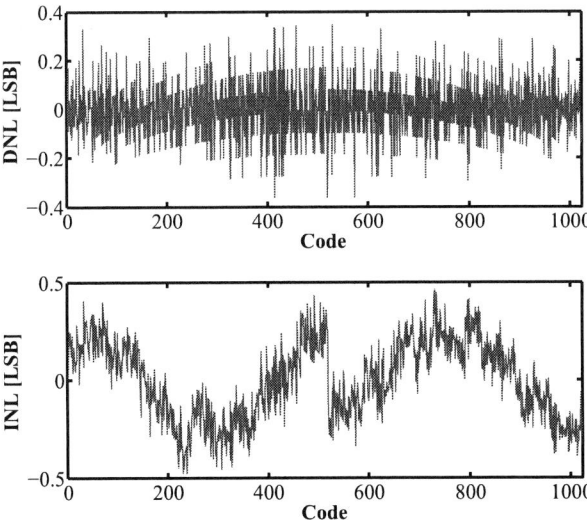

Fig. 4. Linearity measurement.

The measured ENOB and FOM are plotted in Fig. 5 and Fig. 6, respectively, as a function of the DAC reference voltage. The input dynamic range is set to be the same as the DAC reference voltage. The supply voltage is at 0.5V. The measurement is performed for three different total votes: 11, 67, and 127 votes.

Except when 11 votes are taken, which corresponds to one vote per bit cycle, the distribution of votes for the remaining two are determined using the optimization approach in Sec. III. For example, the optimal vote allocation for 67 and 127 total votes when the differential input voltage swing is ±0.5V is given by

$$D_2 = [1 \ 1 \ 1 \ 1 \ 1 \ 1 \ 1 \ 7 \ 11 \ 17 \ 25] \quad (6)$$

$$D_3 = [1 \ 1 \ 1 \ 1 \ 1 \ 1 \ 1 \ 15 \ 21 \ 33 \ 51] \quad (7)$$

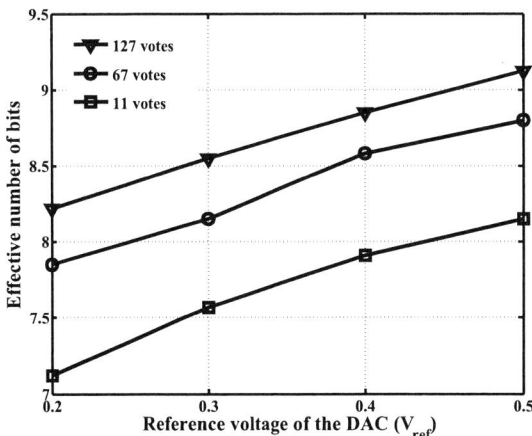

Fig. 5. Measured ENOB as input voltage swing changes.

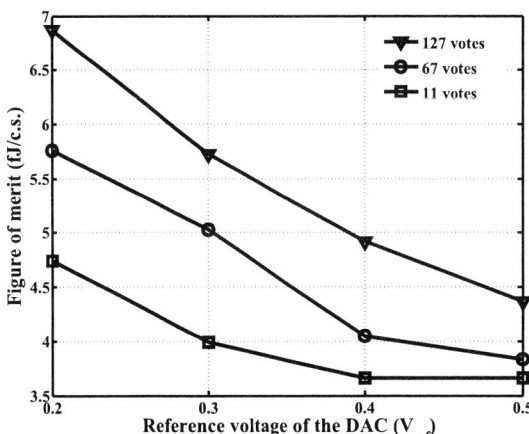

Fig. 6. Measured FOM as input voltage swing changes.

In Fig. 7, the ADC FOM is plotted as a function of ENOB by varying the number of total votes whose distribution has been determined as described in Sec. III. The input voltage swing is ±0.5V. The DAC refence voltage and supply voltage are both at 0.5V. As the total number of votes increases, the ENOB monotonically improves. The FOM, however, also improves briefly then worsens with increasing number of votes as shown in Fig. 7. Compared to when one vote is performed for each bit cycle, which corresponds to ENOB of 8.15b and FOM of 3.7J/conversion-step, the optimum point improves ENOB by 0.3b and the FOM by 0.28fJ/conversion-step.

978-1-4673-6145-3/13 $31.00 © 2013 IEEE

TABLE I
PERFORMANCE COMPARISON

Reference	[5]	[4]	[8]	[6]	This Work		
Technology	$65nm$	$65nm$	$0.18\mu m$	$40nm$	$65nm$		
Area (mm^2)	0.212	0.076	0.125	0.0112	0.072		
Sampling Rate	20kS/s	40kS/s	100kS/s	1.1MS/s	250kS/s		
Resolution (bit)	10	10	10	9	10		
Supply Voltage	0.55V	0.6V	0.6V	0.5V	0.5V	0.5V	0.5V
Differential Input Range	$\pm0.55V$	$\pm0.6V$	$\pm0.6V$	$\pm0.5V$	$\pm0.5V$	$\pm0.5V$	$\pm0.2V$
INL (LSB)	0.57	0.48	0.8	1.1	0.47	0.47	0.47
DNL (LSB)	0.58	0.32	0.7	1.4	0.36	0.36	0.36
ENOB (bit)	8.84	9.42	9.3	7.5	8.45	9.12	8.22
Power (μW)	0.206	0.072	1.32	1.2	0.29	0.60	0.51
FOM (fJ/c-s)	22.4	2.7	21	6.3	3.3	4.3	6.8

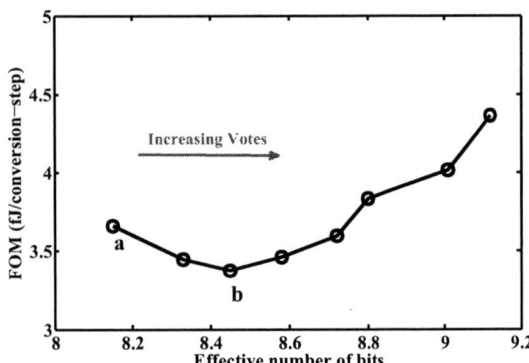

Fig. 7.   Measured FOM versus ENOB at $\pm0.5V$ input swing.

Fig. 8 plots the measured FFT output when a sinusoidal input at 118.4kHz and differential voltage swing of $\pm0.5V$ is sampled at 250kS/s. For this near-Nyquist input, the SNDR and spurious free dynamic range (SFDR) of the ADC at 0.5V supply are 57.14 dB and 64.1 dB, respectively, corresponding to an ENOB of 9.12 bits. A total of 127 votes were performed. Table I summarizes the performance of the proposed ADC and compares with recent low voltage SAR ADCs reported in the literature.

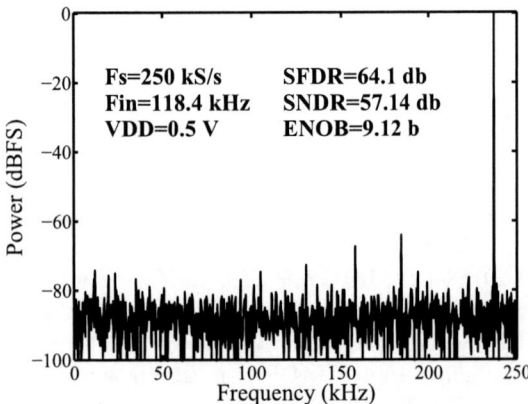

Fig. 8.   Measured output spectrum at 0.5V supply voltage with 1V differential input dynamic range .

V. CONCLUSION

The proposed SAR ADC employs a majority vote comparison along with a non-binary architecture to alleviate the effects of comparator noise when operating at reduced supply voltages. The proposed marjority vote based comparator provides flexibility in controlling the effective comparator noise in each bit cycle. This property enables the ADC to efficiently operate at a wide range of input voltage swings. Furthermore, the optimum distribution of votes to maximize performance for a given power constraint can be determined. To validate the effectiveness of the proposed approach, a 10-bit 250kS/s SAR ADC that supports a flexible differential input dynamic range from 0.4V to 1V has been designed and fabricated in 65nm CMOS process. The ADC achieves ENOB ranging from 7.1b to 9.1b and FOM from 3.3 to 6.8fJ/conversion step.

REFERENCES

[1] B. Calhoun, D. Daly, N. Verma, D. Finchelstein, D. Wentzloff, A. Wang, S.-H. Cho, and A. Chandrakasan, "Design considerations for ultra-low energy wireless microsensor nodes," *Computers, IEEE Transactions on*, vol. 54, no. 6, pp. 727 – 740, jun 2005.

[2] N. Verma and A. Chandrakasan, "An ultra low energy 12-bit rate-resolution scalable SAR ADC for wireless sensor nodes," *Solid-State Circuits, IEEE Journal of*, vol. 42, no. 6, pp. 1196 –1205, june 2007.

[3] R. Ozgun, J. Lin, F. Tejada, P. Pouliquen, and A. Andreou, "A low-power 8-bit sar adc for a qcif image sensor," in *Circuits and Systems (ISCAS), 2011 IEEE International Symposium on*, may 2011, pp. 841 –844.

[4] P. Harpe, E. Cantatore, and A. Van Roermond, "A 2.2/2.7 fJ/conversion-step 10/12b 40kS/s SAR ADC with data-driven noise reduction," in *Solid-State Circuits Conference, 2013. ISSCC 2013. Digest of Technical Papers. IEEE International*, 2013, pp. 270–271.

[5] M. Yip and A. Chandrakasan, "A resolution-reconfigurable 5-to-10b 0.4-to-1V power scalable SAR ADC," in *Solid-State Circuits Conference Digest of Technical Papers (ISSCC), 2011 IEEE International*, feb. 2011, pp. 190 –192.

[6] A. Shikata, R. Sekimoto, T. Kuroda, and H. Ishikuro, "A 0.5 V 1.1 MS/sec 6.3 fJ/conversion-step SAR-ADC with tri-level comparator in 40 nm CMOS," *Solid-State Circuits, IEEE Journal of*, vol. 47, no. 4, pp. 1022 –1030, april 2012.

[7] V. Giannini, P. Nuzzo, V. Chironi, A. Baschirotto, G. Van der Plas, and J. Craninckx, "An 820 μW 9b 40MS/s noise-tolerant dynamic-SAR ADC in 90nm digital CMOS," in *Solid-State Circuits Conference, 2008. ISSCC 2008. Digest of Technical Papers. IEEE International*, feb. 2008, pp. 238 –610.

[8] S.-K. Lee, S.-J. Park, H.-J. Park, and J.-Y. Sim, "A 21 fJ/conversion-step 100 kS/s 10-bit ADC with a low-noise time-domain comparator for low-power sensor interface," *Solid-State Circuits, IEEE Journal of*, vol. 46, no. 3, pp. 651 –659, march 2011.

Single-Inductor-Multiple-Output DC-DC Converter Design

Philip K. T. Mok
Professor
Department of Electronic and Computer Engineering
Hong Kong University of Science and Technology
Clear Water Bay, Hong Kong

September 24, 2013

Outline

- Background of SIMO converters

- Design Issues
  - Conversion Ratios, Efficiency, Driving Capability, Cross-Regulation

- Control Methods
  - Time-Multiplexing Control, Ordered Power Distributive Control, Constant-Charge-Auto-Hopping Control, Vestigial Current Control, Single Feedback-Loop Control, Current-Mode Hysteretic Control

- Summary

Prof. P. Mok , CICC 2013

Needs for DC-DC Converters

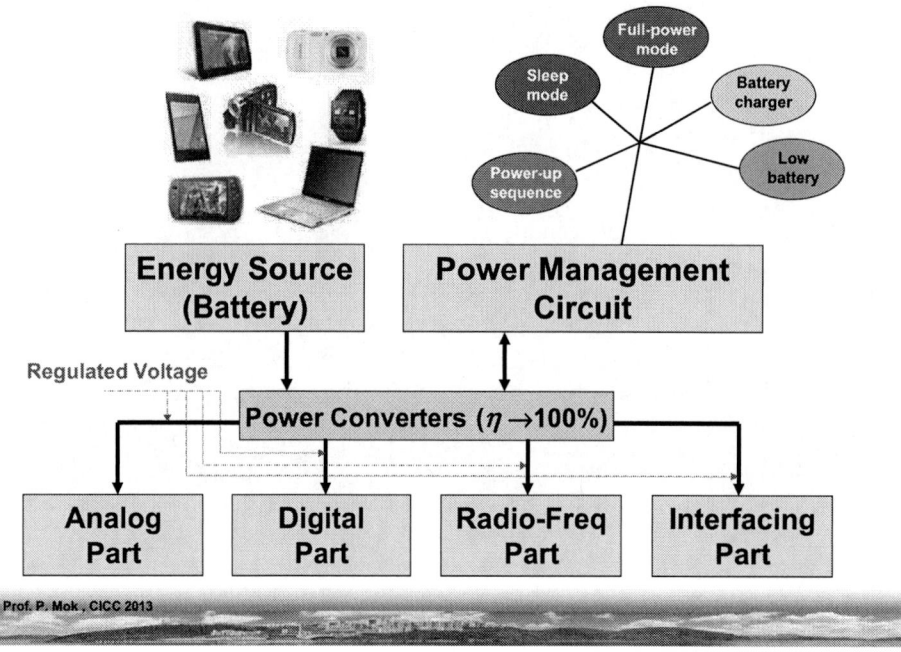

Prof. P. Mok , CICC 2013

Function of Power Converters

Provide a regulated energy source.
1. Low Drop-out Linear Regulator (LDO)
2. Switch-Mode Power Converter
3. Switch-Capacitor Power Converter (Charge Pump)

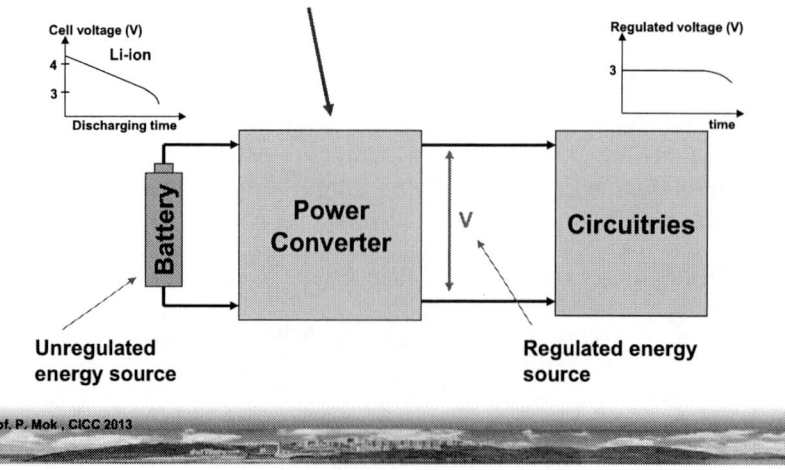

Prof. P. Mok , CICC 2013

Switched-Mode Regulator Basic

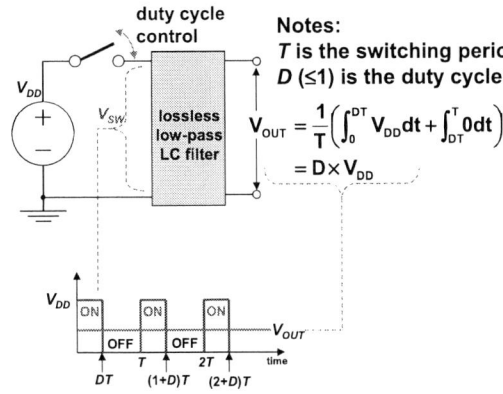

duty cycle control

Notes:

T is the switching period

D (≤ 1) is the duty cycle

$$V_{OUT} = \frac{1}{T}\left(\int_0^{DT} V_{DD}\,dt + \int_{DT}^{T} 0\,dt\right)$$

$$= D \times V_{DD}$$

lossless low-pass LC filter

- V_{OUT} is a function of D and V_{IN}

- Ideal efficiency is 100% with no loss at the switch and averaging circuit

Prof. P. Mok , CICC 2013

Converter Topologies (Synchronous Converters)

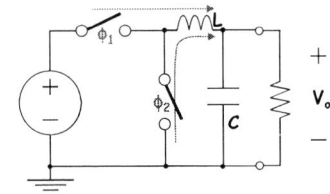

Buck — step-down only

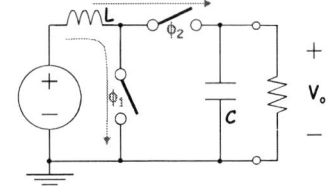

Boost — step-up only

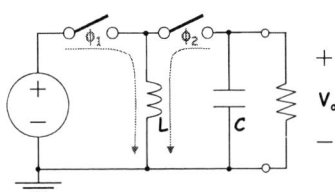

Flyback — step-up and -down

However, V_o is negative

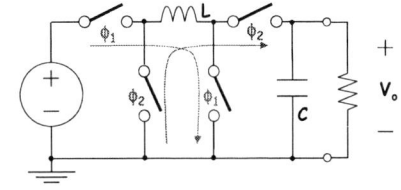

Non-inverting flyback — step-up and -down

V_o is positive but need more switches

Prof. P. Mok , CICC 2013

978-1-4673-6145-3/13 $31.00 © 2013 IEEE 491

Converters for VLSI SoC

- **Increasing numbers of supplies with different voltages and different current consumptions**

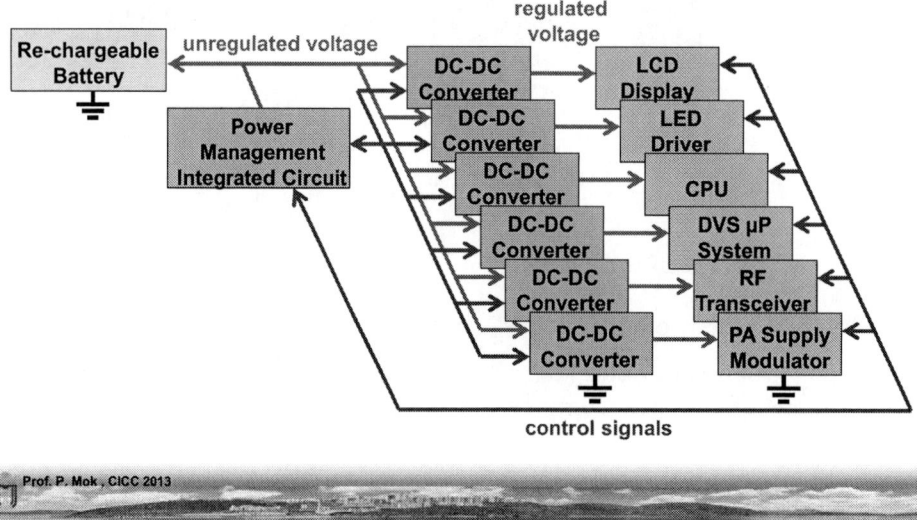

Prof. P. Mok , CICC 2013

Existing Solution for Multiple Supply Voltages

- Existing multiple supply methods

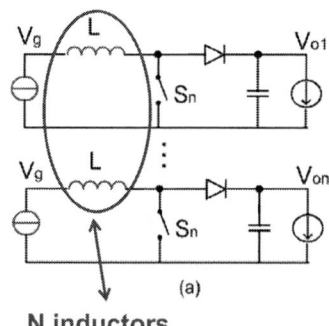

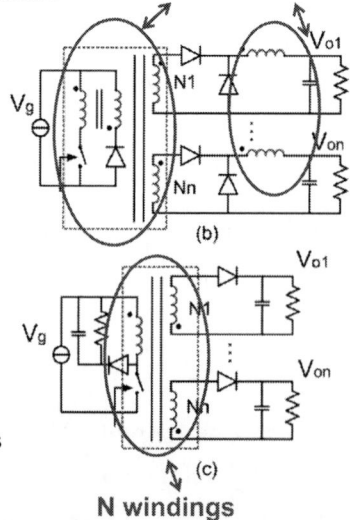

- Non-isolated (a): **N** inductors
- Isolated : (b) **N** inductors and windings
  - (c) **N** windings

Prof. P. Mok , CICC 2013

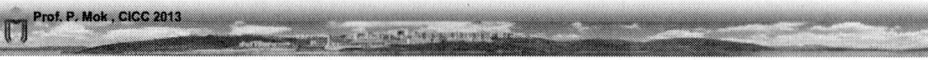

Potential Solution — SIMO

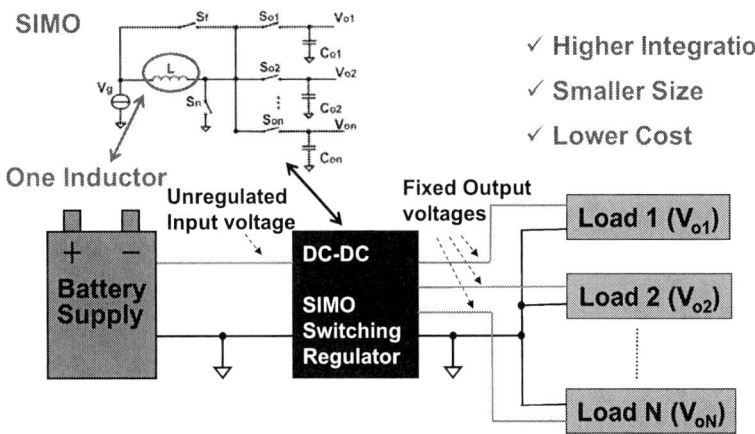

- ✓ Higher Integration
- ✓ Smaller Size
- ✓ Lower Cost

- Conventional: one inductor for one output voltage (SISO)
- SIMO: Single-Inductor-Multiple-Output.
- SIMO: one inductor can provide many outputs

Prof. P. Mok, CICC 2013

Requirements for SIMO

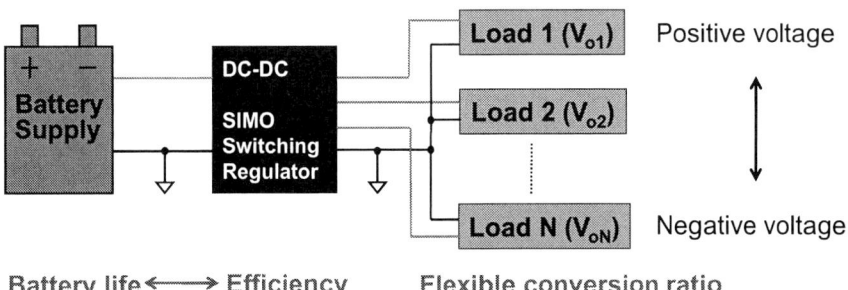

Battery life ⟷ Efficiency   Flexible conversion ratio

- SIMO regulators
  - Multiple-output: area and cost effective methods
  - Each output with flexible conversion ratio
  - More switches, efficiency degradation

Prof. P. Mok, CICC 2013

SISO Standard Topologies

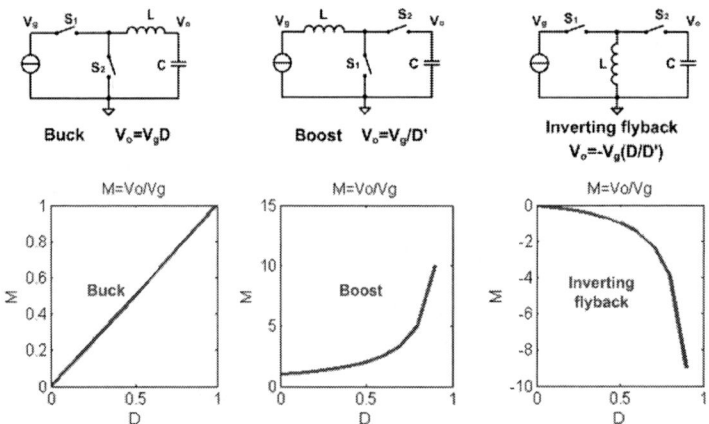

- Basic SISO regulators: L, C, Two switches
- 2-S SISO switching regulators

SISO Topologies with 4 Switches

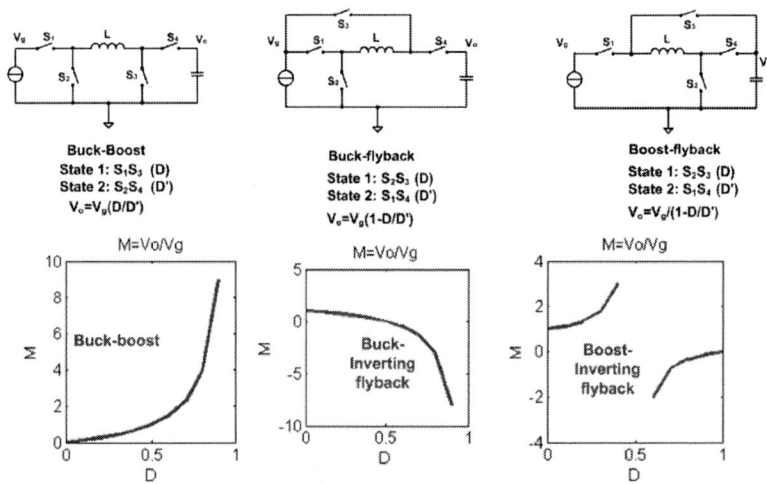

- 4-S SISO switching regulators: L, C, four switches
- Can be configured into corresponding 2-S regulators

SISO Topologies with 5 and 6 Switches

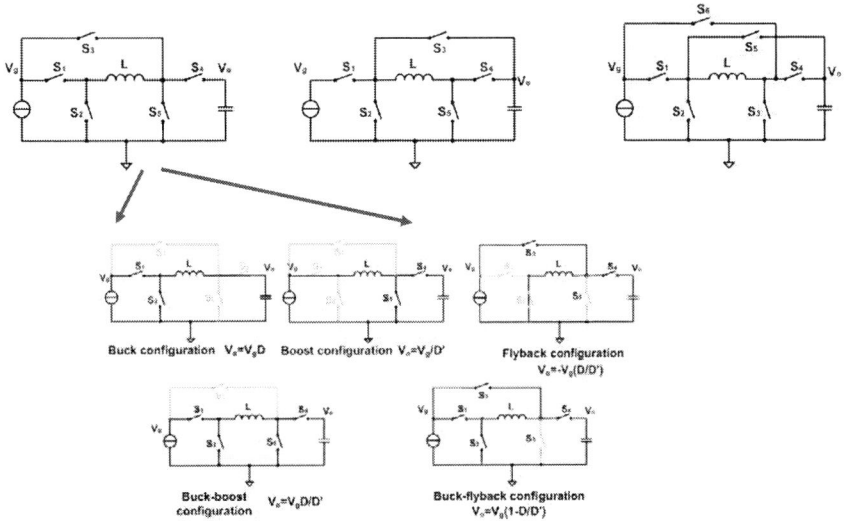

- Can be configured into corresponding 2-S and 4-S regulators

Prof. P. Mok , CICC 2013

Topology Extension to SIMO

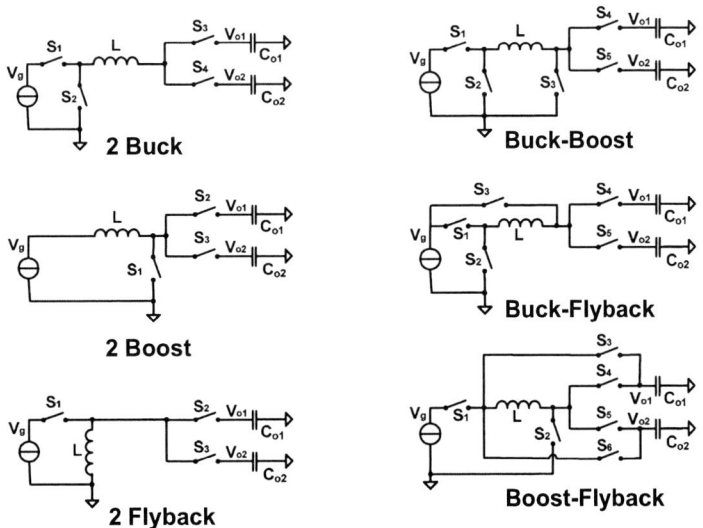

- SIMO corresponding to 2-S and 4-S SISO switching regulators

Prof. P. Mok , CICC 2013

Topology Extension to SIMO

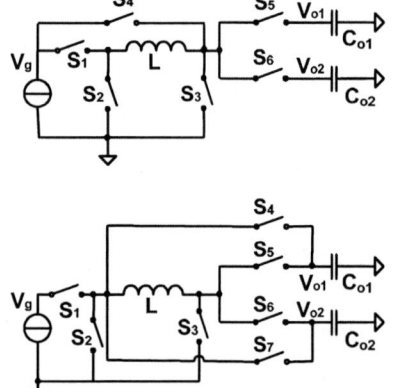

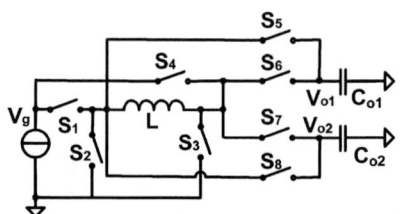

SIMO corresponding to 6-S SISO

SIMO corresponding to 5-S SISO

- Any conversion ratio can be achieved by all these topologies.

Prof. P. Mok , CICC 2013

SIMO Design Issues

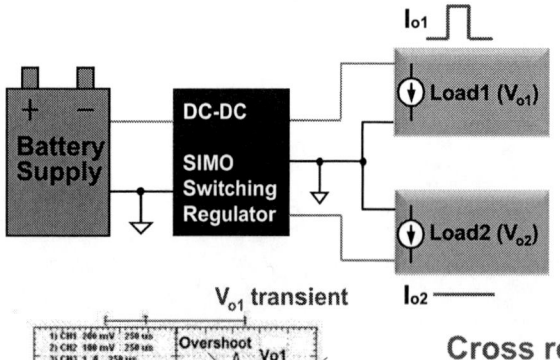

» Efficiency (η)

» Driving Capability (DC)

» Transient Response (TR)

» Cross Regulation (CR)

Cross regulation :

- Any load change at one output will cause voltage variation at the other outputs due to extra or insufficient charge.

Prof. P. Mok , CICC 2013

Cross Regulation

- Cross regulation

  - Degrades system performance.

  - Affects the stability of the regulator

  - Affects the functionality of the loading devices

- Minimize cross regulation

  - Reduce error of the energy transfer

  - Fast correction of the error of the energy transfer

Time-Multiplexing (TM) Control

Use a Single-Inductor-Dual-Output
(SIDO) Boost as an example

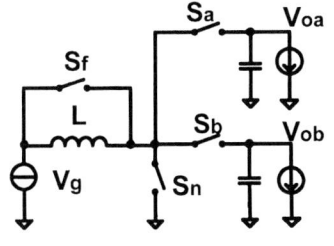

SIDO boost converter

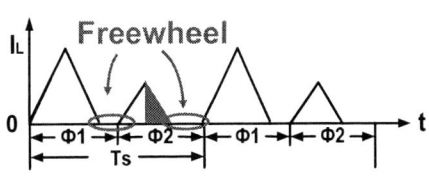

- Operates at Discontinuous
  Conduction Mode (DCM)
- A small freewheel switch is needed
  to eliminate ringing

*D. Ma, et al., JSSC, Jan 2003

Ringing Suppression

When both switches SW_1 and SW_2 are open, L and the parasitic capacitor C_x at V_x form a resonance circuit that leads to severe ringing.

The problem can be solved by adding a small free-wheeling switch to short the inductor when both switches are off.

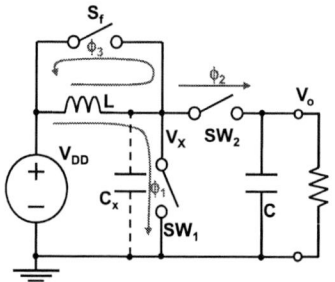

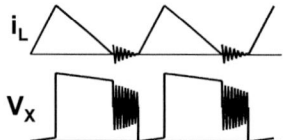

without ringing suppression circuit

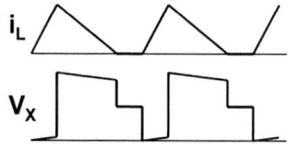

with ringing suppression circuit

Prof. P. Mok, CICC 2013

Time-Multiplexing (TM) Control

- Problem:
  Limited driving capability

- During the transient, if one of the load currents is higher than its limited driving capability, cross regulation occurs

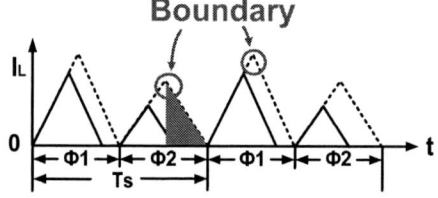

Prof. P. Mok, CICC 2013

978-1-4673-6145-3/13 $31.00 © 2013 IEEE   498

TM: Pseudo-Continuous Conduction Mode (PCCM)

- S_f turns on when the I_L drops to a fixed DC level I_{dc}

- Can improve the driving capability of each output

- A large S_f is needed which will also increase the switching loss

- May not very effective to handle the trade-off between driving capability and power loss

- Difficult to implement unbalance load

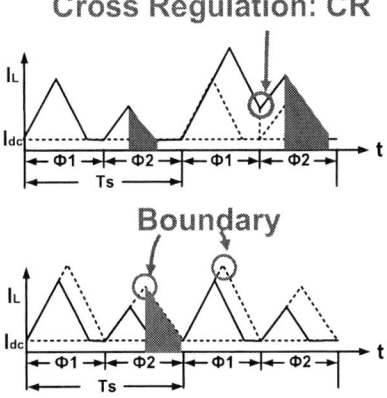

*D. Ma, et al., JSSC, Jun 2003

TM: Adaptive PCCM Control

- Different freewheel currents I_{dc1} & I_{dc2} are used for unbalance load

- Current range is limited and easy to have cross regulation

- Adaptive freewheel current can be used but rather complicated to implement

- A very low resistance freewheel switch is needed

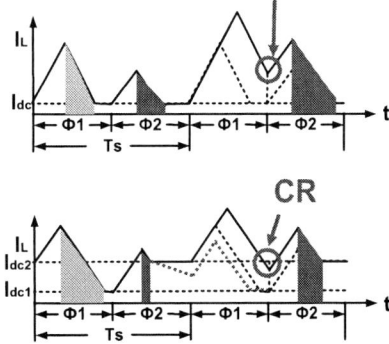

*Y. Zhang, et al., ISCAS, 2011

Ordered Power Distributive Control (OPDC)

- Charged the inductor energy once and transfers to outputs one by one within one period.

- Can operates at DCM (Idc = 0) or PCCM (Idc ≠ 0)

- Also need a free-wheeling switch

- Less number of switching than the TM control

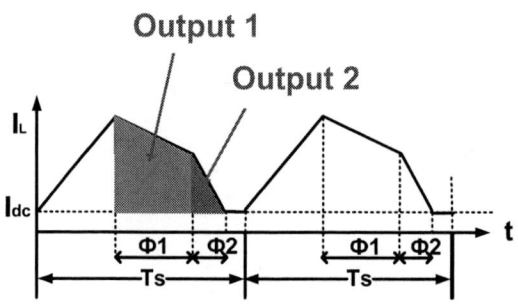

*H.-P. Le, et al., ISSCC, 2007

Ordered Power Distributive Control (OPDC)

- Cross regulation is more serious when load change happens at the former output

- Cross regulation occurs even before reaching the maximum driving capability of the SIDO

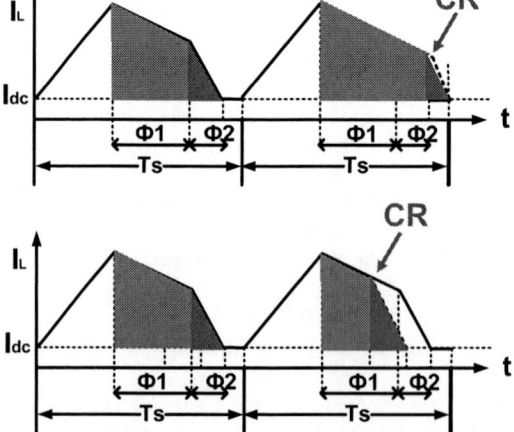

Efficiency Consideration

* OPDC has less switching in one period than TM control
* OPDC has low switching loss as less switching occurs in one period
* OPDC has large conduction loss with high peak inductor current
* Efficiency will be more or less the same with optimized transistor sizing

Prof. P. Mok , CICC 2013

Driving Capability Consideration

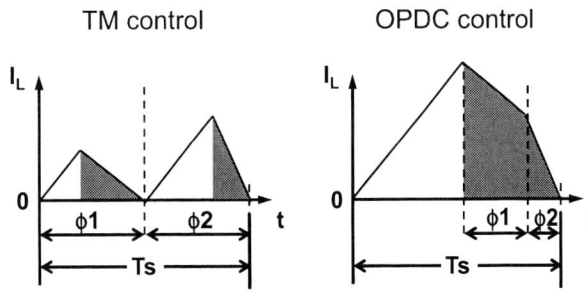

* With the same input and output voltages, OPDC has a higher driving capability
* To increase the driving capability of TM control, increase the switching period Ts or lower the switching frequency
* Same for OPDC control

Prof. P. Mok , CICC 2013

Design Challenge in SIMO (1)

- Cross-regulation

  (the average energy transferred to the unchanged output is constant.)

  - TM
    - Isolation ↔ good
    - Fixed time slot ↔ bad
  - OPDC
    - Fast comparator controlled output ↔ good
    - Charge dependent ↔ bad
  - Both sensitive to load transient step ↔ bad

Prof. P. Mok , CICC 2013

Design Challenge in SIMO (2)

- Driving capability

  (Need a better way to balance the cross regulation and the driving capability.)

  - TM
    - DC level ↔ good
    - Fixed time slot ↔ bad
  - OPDC
    - CCM & DCM operation ↔ good
    - Large current ripple ↔ bad
  - For both, the larger the loading current, the larger the cross regulation ↔ bad

Prof. P. Mok , CICC 2013

Sequential Control

Operation Mode

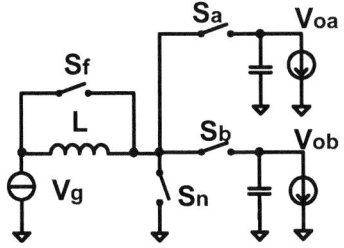

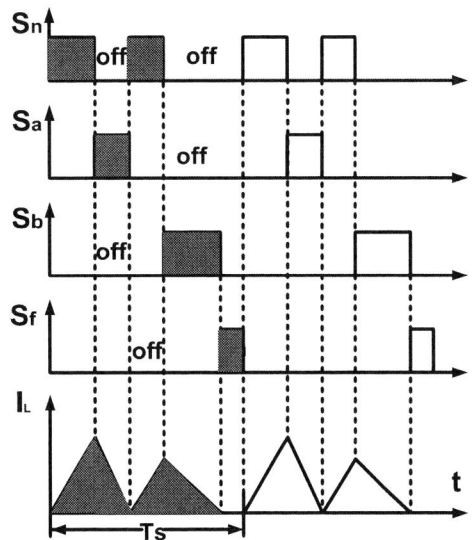

Boundary of DCM & CCM
or DCM operation

*X. Jing, et al., JSSC, 2011

Prof. P. Mok , CICC 2013

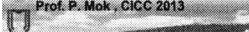

Driving Unbalanced Loads

- Sequential control
  - flexible loading combination

- Limited total driving capability
  - Fixed frequency: BCM

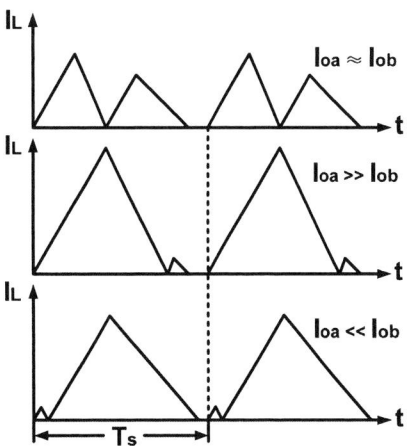

Prof. P. Mok , CICC 2013

Constant-Charge-Auto-Hopping (CCAH)

- Extending driving capability by extending the switching period
- Frequency hopping
  - Multiple fraction of f_D
  - Based on total loads
- Predictable switching noise spectrum

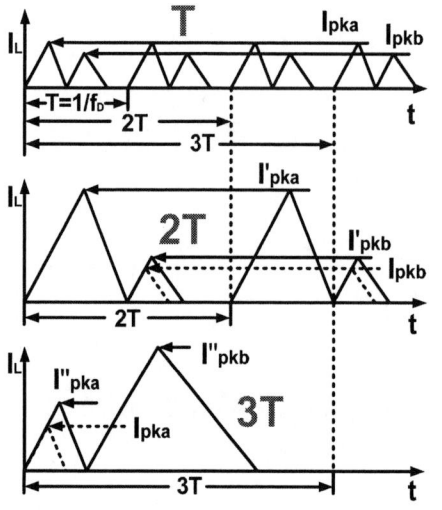

*X. Jing, et al., JSSC, 2011

Constant-Charge-Auto-Hopping (CCAH)

- Deliver a constant average charge per swithing period to the unchanged outputs
- Minimize cross-regulation during frequency hopping
- A faster transient response

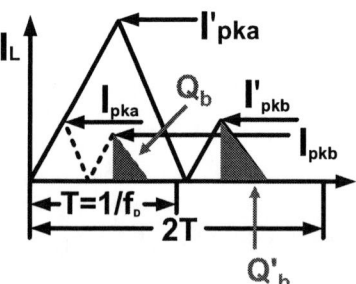

$$I_{ob} = \frac{Q}{T_s} = \frac{Q_b}{T} = \frac{Q'_b}{2T} \Rightarrow Q'_b = 2Q_b$$

$$I'_{pkb} = \sqrt{2}I_{pkb} \qquad I'_{pkb} = \sqrt{N}I_{pkb}$$

Circuit Implementation

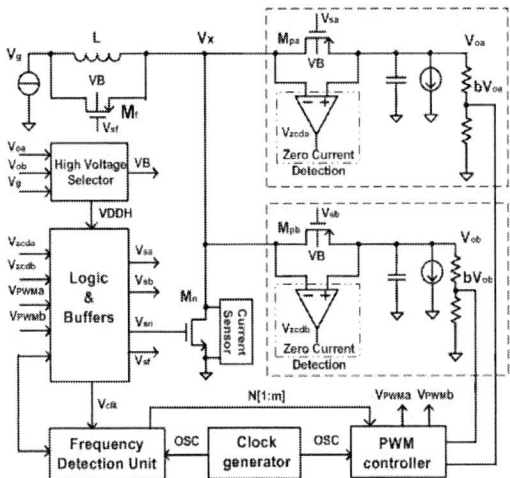
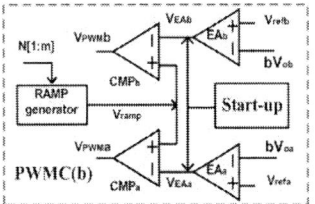

- Use 2 simple voltage-mode PWM feedback loops to control the duty ratios
- Including current sensor and zero current detection

Block Diagram of SIDO Boost Converter

Prof. P. Mok , CICC 2013

Measurement Results (1)

Chip Micrograph

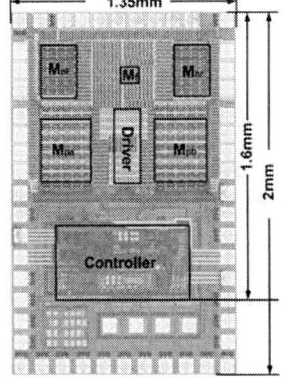

AMS 0.35μm 2P4M CMOS

Steady-state waveforms

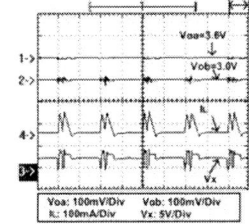

(a) fs=fd, I$_{oa}$ = 2.5mA and I$_{ob}$ = 6mA

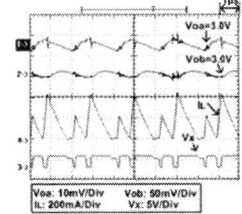

(b) fs=fd/2, I$_{oa}$ = 20mA and I$_{ob}$ = 70mA

Prof. P. Mok , CICC 2013

Measurement Results (2)

Steady-state waveforms

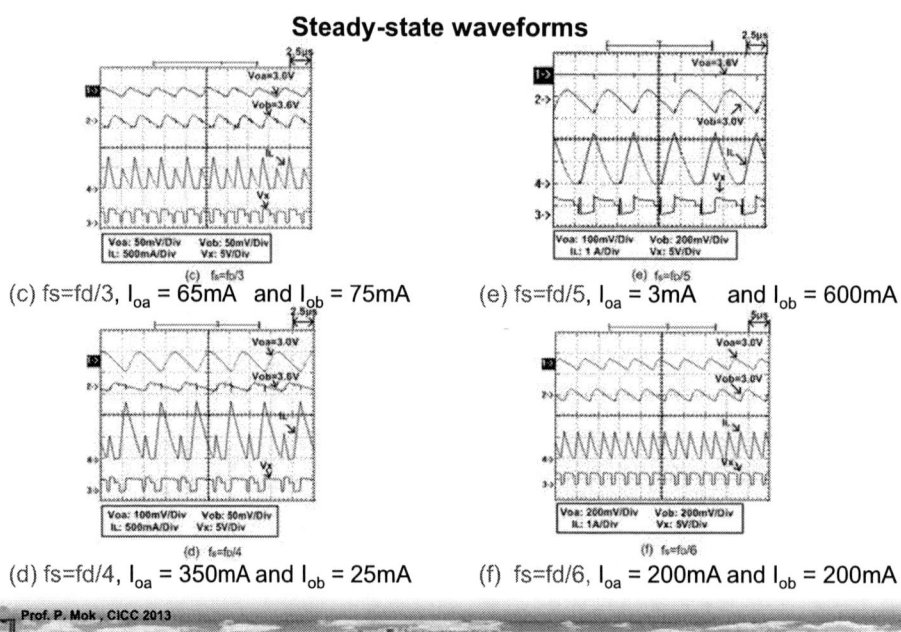

(c) fs=fd/3, I_{oa} = 65mA and I_{ob} = 75mA

(e) fs=fd/5, I_{oa} = 3mA and I_{ob} = 600mA

(d) fs=fd/4, I_{oa} = 350mA and I_{ob} = 25mA

(f) fs=fd/6, I_{oa} = 200mA and I_{ob} = 200mA

Prof. P. Mok , CICC 2013

Measurement Results (3)

Load transient

(a) I_{oa}:20-200mA ; I_{ob}:20mA.

No noticeable CR

(b) I_{oa}:120-240mA ; I_{ob}:20mA.

No noticeable CR

(c) I_{oa}:50-220mA ; I_{ob}:60mA.

No noticeable CR

(d) I_{oa}:100-400mA ; I_{ob}:60mA.

0.033mV/mA CR

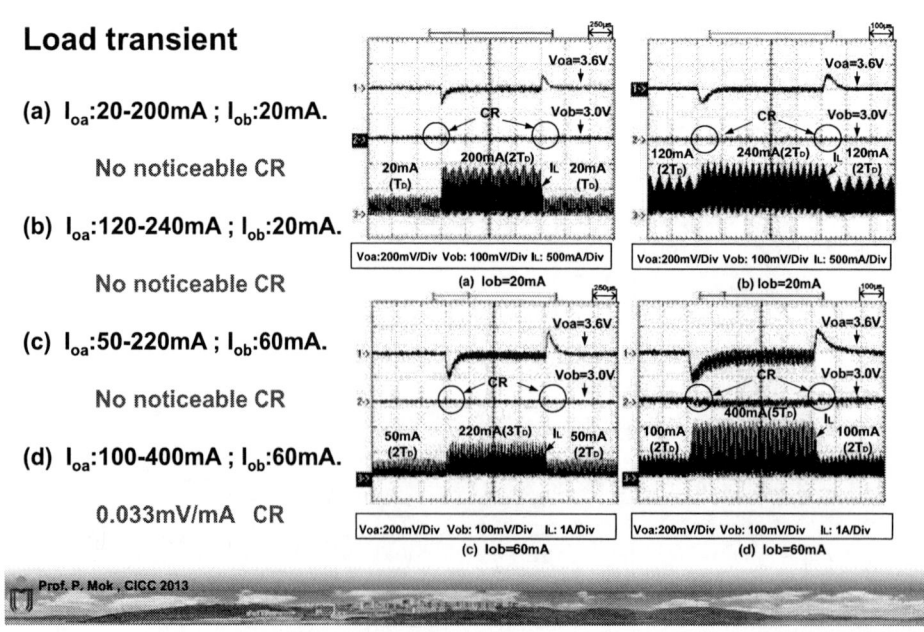

Prof. P. Mok , CICC 2013

Measurement Results (4)

Load transient

(a) I_{ob}:40-120mA ; I_{oa}:20mA.

   No noticeable CR

(b) I_{ob}:40-450mA ; I_{oa}:20mA.

   No noticeable CR

(c) I_{ob}:20-200mA ; I_{oa}:50mA.

   No noticeable CR

(d) I_{ob}:20-300mA ; I_{oa}:50mA.

   0.0714mV/mA CR

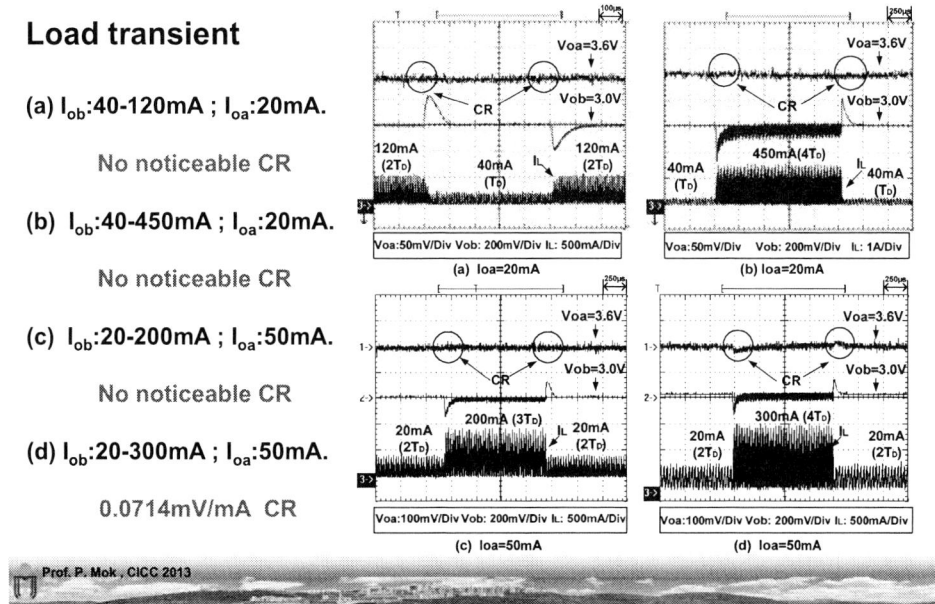

Prof. P. Mok , CICC 2013

Measurement Results (5)

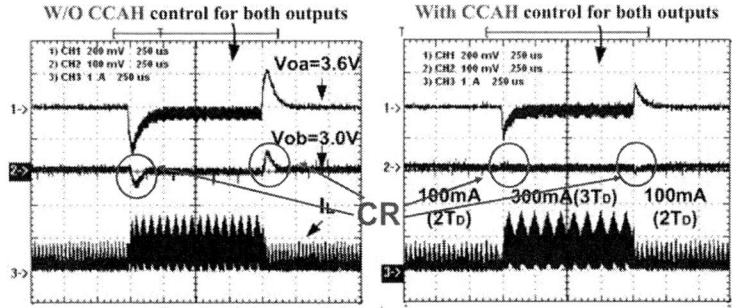

Cross regulation induced by load transient at first output.

- With CCAH control
    - No noticeable cross regulation can be found at the second output
    - Transient response of the first output is improved

Prof. P. Mok , CICC 2013

Measurement Results (6)

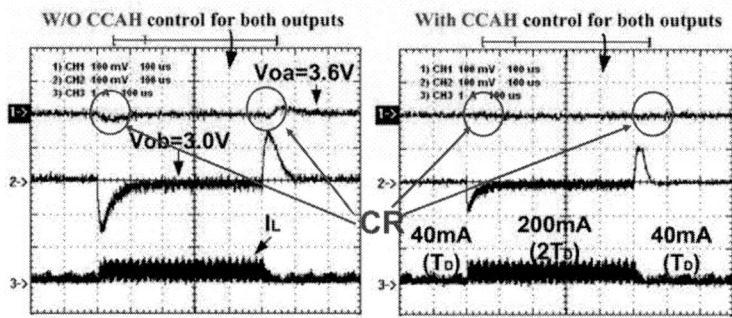

Cross regulation induced by load transient at second output.

- With CCAH control
  - No noticeable cross regulation can be found at the first output
  - Transient response of the second output is improved

Prof. P. Mok , CICC 2013

CCAH Control - Summary

- Advantages:
  - Power handling capability: extended and flexible
  - Cross-regulation: minimized and less sensitive to the load transient step.
  - Transient response: enhanced with inductor current prediction.
  - Noise spectrum: predictable with predefined frequency hopping.

- Drawbacks:
  - DCM operation: the larger the loads, the larger the ripple.
  - PWM operation: Transient response is still limited by the compensation network.

Prof. P. Mok , CICC 2013

OPDC Feedback Control

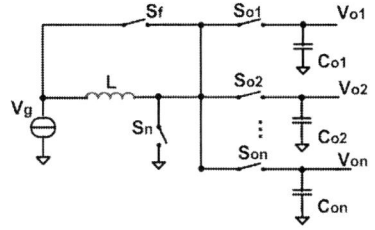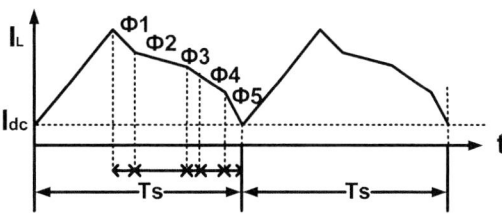

- PWM feedback loops can be used to control the TM SIMO
- Hysteric control can be used for OPDC with only one P-I loop (the last output n)
- All the errors of the preceding bang-bang outputs are transferred and accumulated to the last output

*H.-P. Le, et al., ISSCC, 2007

Prof. P. Mok , CICC 2013

OPDC Implementation

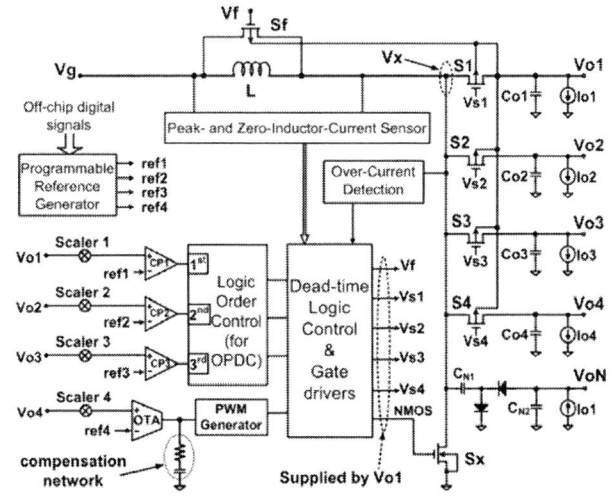

- Vo1 – Vo3 are hysteric control
- Vo4 is PWM control with P-I compensation
- VoN provides a negative output voltage

*H.-P. Le, et al., ISSCC, 2007

Prof. P. Mok , CICC 2013

SIMO with Vestigial Current Control

- Overall energy is controlled to be slightly larger than what all of the outputs need
- One auxiliary output (V_A) is used to take up the excessive energy and is monitored and regulated
- A small vestigial current is used to allow transient load variation in normal operation

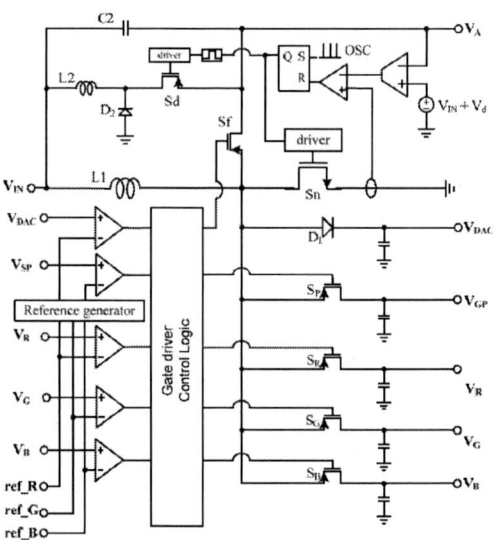

*K.-S. Seol, et al., ISSCC, 2009

Prof. P. Mok , CICC 2013

Ripple Current Reduction

- Reduce the inductor ripple current can reduce the conduction loss
- Smaller peak current can lead to a smaller transistor and thus a smaller switching loss
- Reduce the ripple can improve the efficiency
- Make use of relationship between the input and output voltages and the switching sequence

- Consider a SIDO Buck:

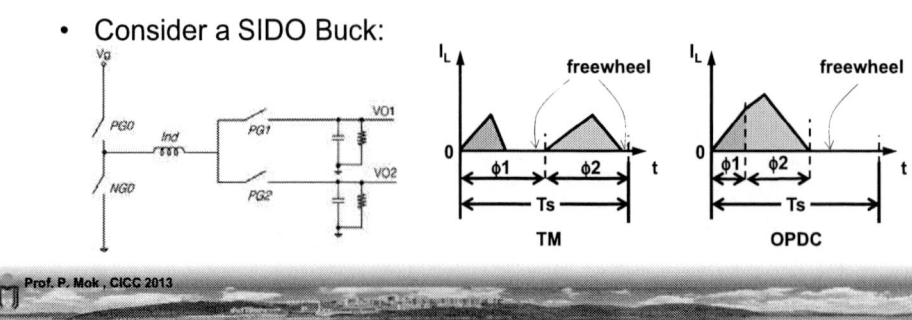

Prof. P. Mok , CICC 2013

OPDC in CCM Operation

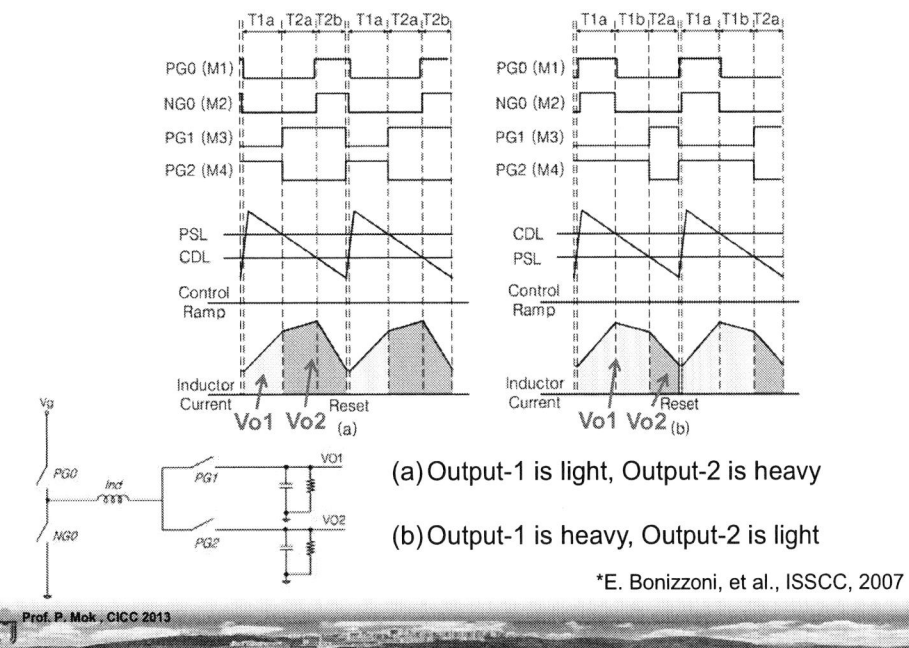

(a) Output-1 is light, Output-2 is heavy

(b) Output-1 is heavy, Output-2 is light

*E. Bonizzoni, et al., ISSCC, 2007

Prof. P. Mok, CICC 2013

Single Feedback-Loop Control SIDO Buck

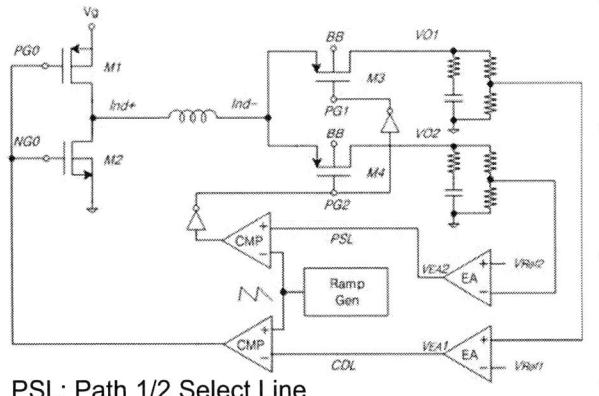

PSL: Path 1/2 Select Line
CDL: Charge/Discharge select Line

- Ordered-Power-Distributive Control Buck converter

- Smart and simple design to achieve very high efficiency

- Two straight-forward voltage feedback loops

- Two outputs are strongly related

*E. Bonizzoni, et al., ISSCC, 2007

Prof. P. Mok, CICC 2013

Low-CR Single Feedback-Loop Control

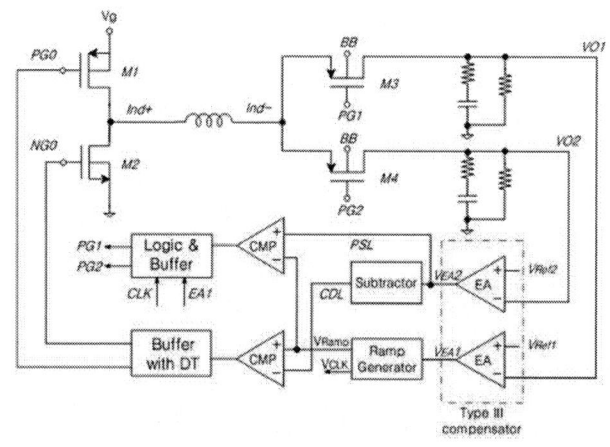

- A Subtractor based on VO2-loop is added to generate both CDL and PSL which always move in the opposite directions

- A DC-Level adjustable ramp generator based on VO1-loop is added to generate the ramp

*C. Huang, et al., ISCAS, 2011

Low-CR Single Feedback-Loop Control

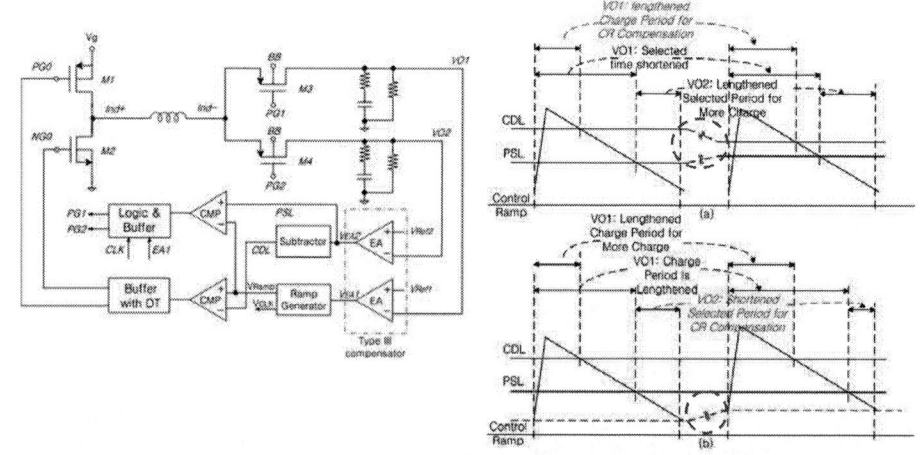

(a) Output-1 is fixed, Output-2 changes from light- to heavy-load;
(b) Output-2 is fixed, Output-1 changes from light- to heavy-load.

Measured Cross-Regulation

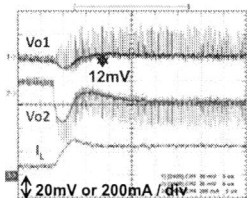

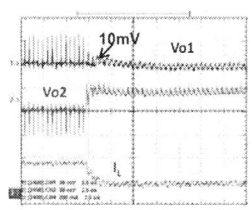

Output-1 fixed at 20mA.
Output-2 changes from 20mA to 200mA.
Cross Regulation: 0.067mV/mA

Output-1 fixed at 20mA.
Output-2 changes from 200mA to 20mA.
Cross Regulation: 0.056mV/mA

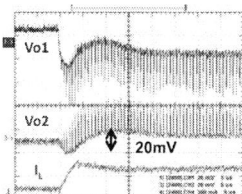

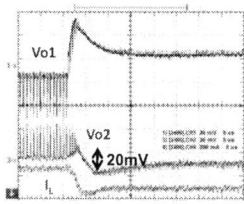

Output-2 fixed at 20mA.
Output-1 changes from 20mA to 200mA.
Cross Regulation: 0.11mV/mA

Output-2 fixed at 20mA.
Output-1 changes from 200mA to 20mA.
Cross Regulation: 0.11mV/mA

Prof. P. Mok , CICC 2013

Fast Transient Consideration

- Fast response
    - PWM with compensation network ↔ bad
    - TM & OPDC limited by PWM controlled stage
    - Hysteretic control ↔ good
- Hysteretic control
    - Voltage-mode, current-mode
    - Variable frequency → noise spectrum
    - Ripple voltage
- Ripple
    - OPDC with large inductor peak current ↔ bad
    - DCM with large driving ability ↔ bad
    - CCM ↔ good
- Cross-regulation
    - CCM ↔ bad
    - Fast response ↔ good

Prof. P. Mok , CICC 2013

Current-Mode Hysteretic Control (CMHC)

- Current-Mode Hysteretic Control
  - General for all types of converter.
  - Better ripple
- Operation
  - DCM – light load
  - CCM – heavy load
- Noise spectrum
  - High frequency synchronization
  - Fixed off-time to limit the maximum frequency

*X. Jing, et al., ISCAS, 2011

Basic Operations

- Operation
  - DCM – light load
  - CCM – heavy load
- 3 operation modes for dual output systems
  - Both light load
    - DCM+DCM
  - One light load, one heavy load
    - DCM+CCM
  - Both heavy load
    - CCM+CCM

Basic Operations

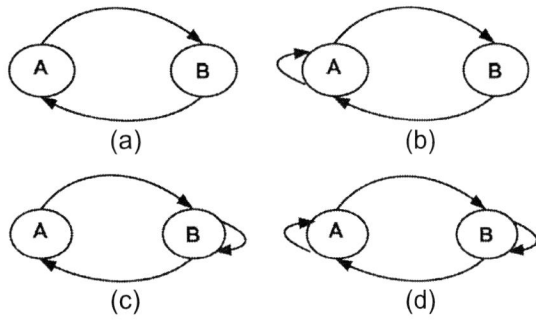

(a) Energy transfers to A then to B then back to A

(b) Energy transfers to A several cycles then to B then back to A.

(c) Energy transfers to A then to B several cycles then back to A.

(d) Energy transfers to A several cycles then to B several cycles then back to A

Prof. P. Mok , CICC 2013

Flexible Operation Mode (1)

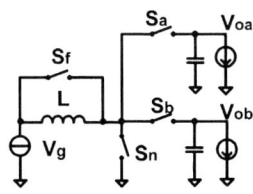

- **DCM+DCM**

- **Charge transfer:**
  - **A + B**
  - **Several A + B**
  - **A + Several B**

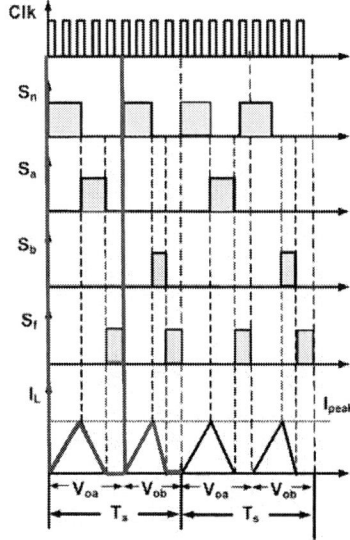

Prof. P. Mok , CICC 2013

Flexible Operation Mode (1)

A + B

Several A + B

A + Several B

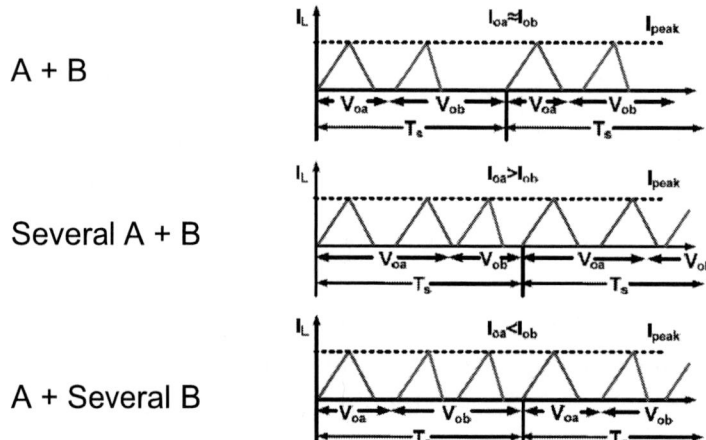

Prof. P. Mok , CICC 2013

Flexible Operation Mode (2)

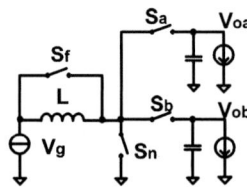

- **DCM+CCM**

- **Charge transfer:**
  - **A + B**
  - **Several A + B**
  - **A + Several B**

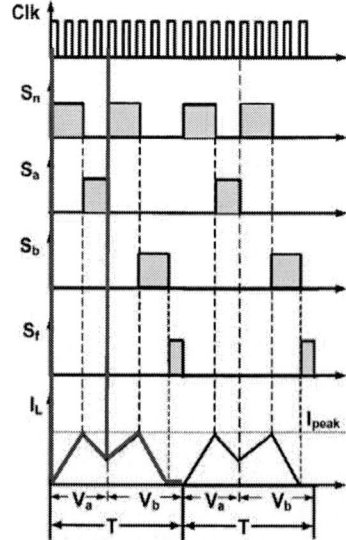

Prof. P. Mok , CICC 2013

978-1-4673-6145-3/13 $31.00 © 2013 IEEE   516

Flexible Operation Mode (2)

A + B

Several A + B

A + Several B

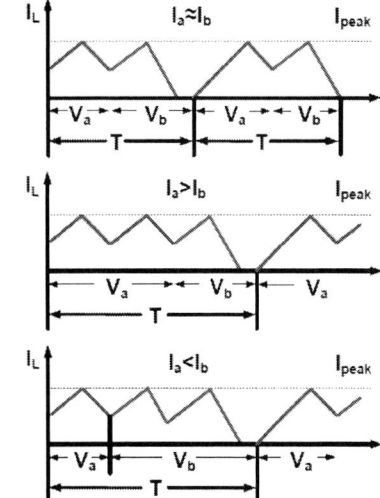

Flexible Operation Mode (3)

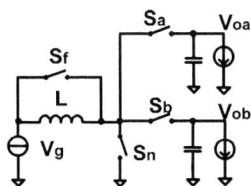

- CCM+CCM

- **Charge transfer:**
  - **A + B**
  - **Several A + B**
  - **A + Several B**
  - **Several A + Several B**

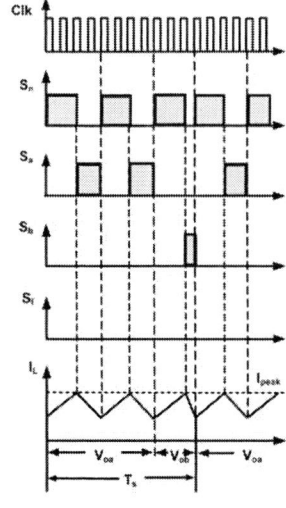

Flexible Operation Mode (3)

A + B

Several A + B

A + Several B

Several A + Several B

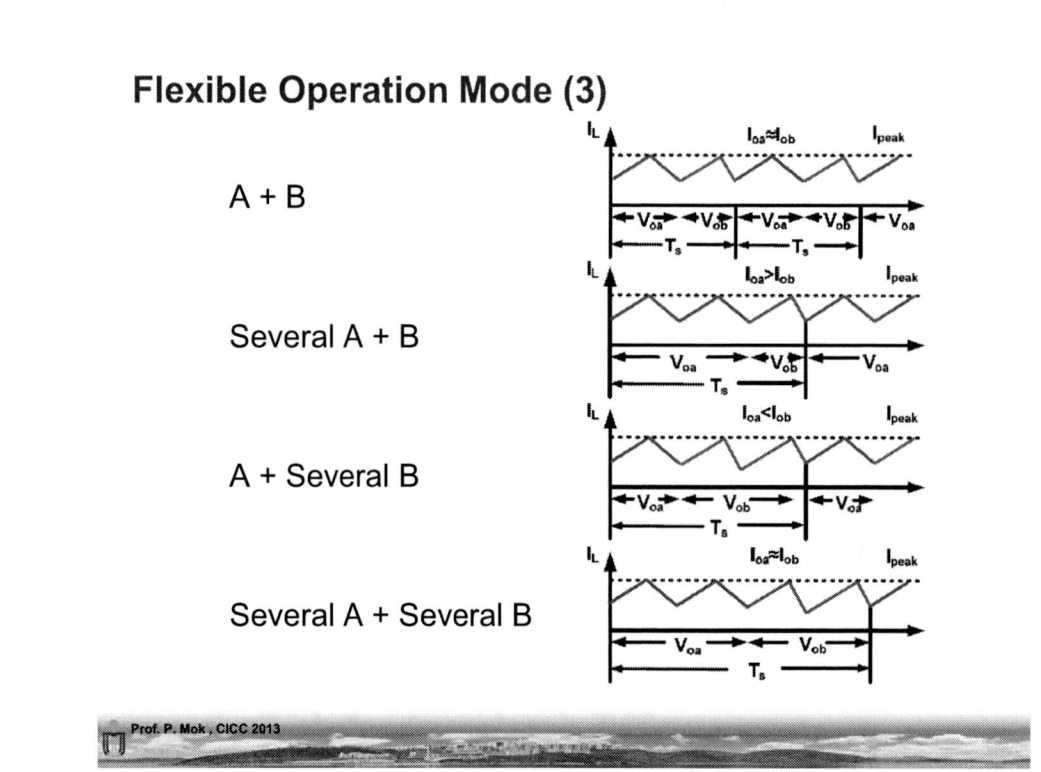

Operation Flow Chart

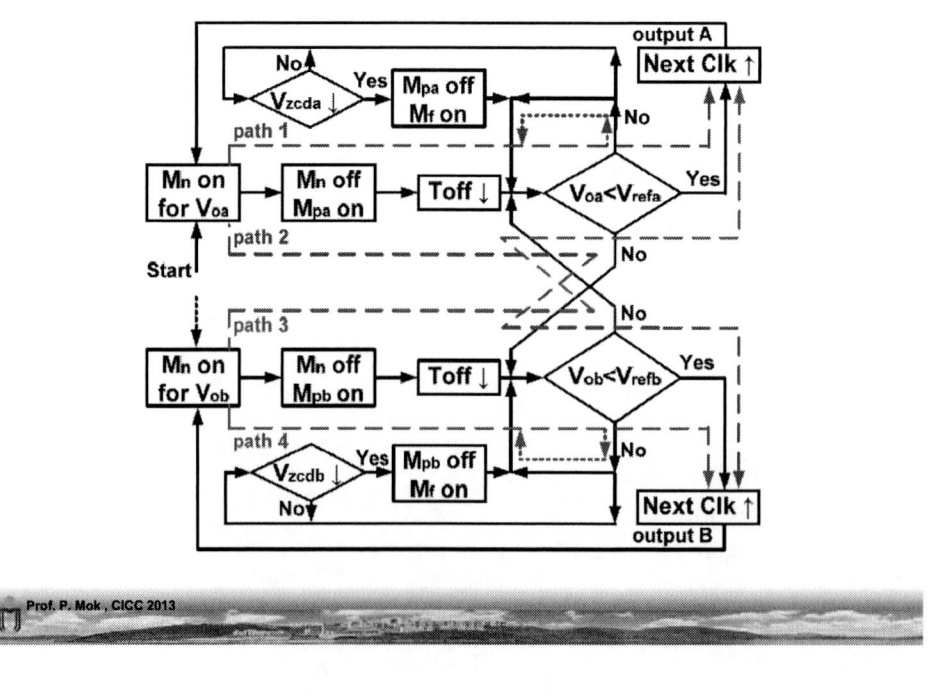

Circuit Implementation

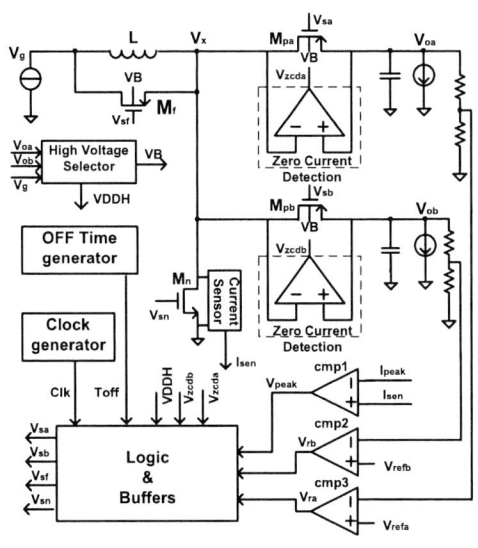

- Very simple implementation

- 1 current-mode hysteretic control loops

- 1 current sensor and 2 zero current detection circuits

- 2 comparators

*X. Jing, et al., ISCAS, 2011

Chip Micrograph

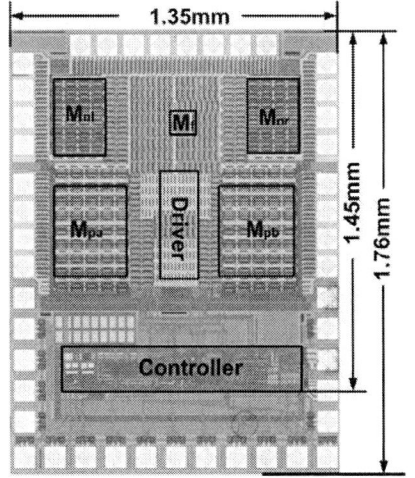

- **Process: 0.35μm 2P4M CMOS**
- **Chip area: 1.35mm × 1.76mm**

Parameter Summary	
process	0.35μmCMOS
V_g	1.8V-2.4V
V_{oa}/V_{ob}	3.0V-3.6V
L	1μH
DCR	0.2Ω
C_{out}	10μF
ESR	0.03Ω
I_{oa}/I_{ob}	0-500mA

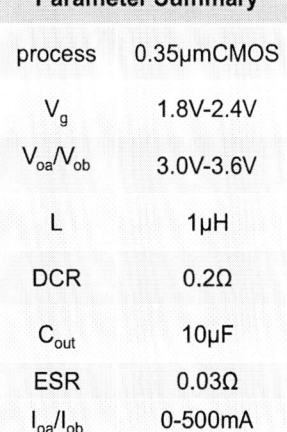

978-1-4673-6145-3/13 $31.00 © 2013 IEEE

Measurements (1)

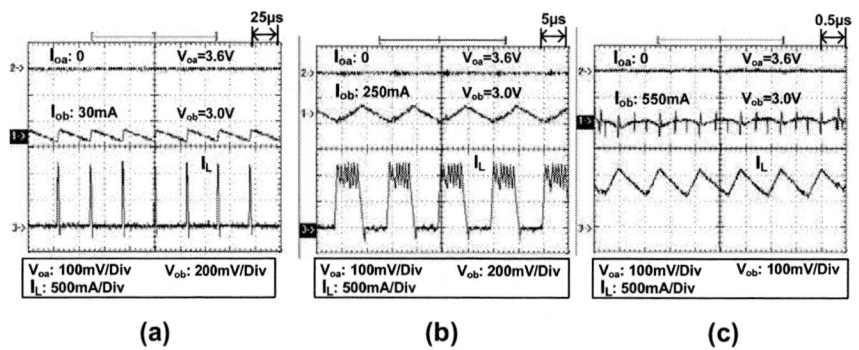

(a)   **(b)**   **(c)**

Steady-state with I_{oa}=0mA :

(a) I_{ob}**=30mA**   **DCM+DCM**

(b) I_{ob}**=250mA**   **DCM+CCM**

(c) I_{ob}**=550mA**   **CCM+CCM**

Measurements (2)

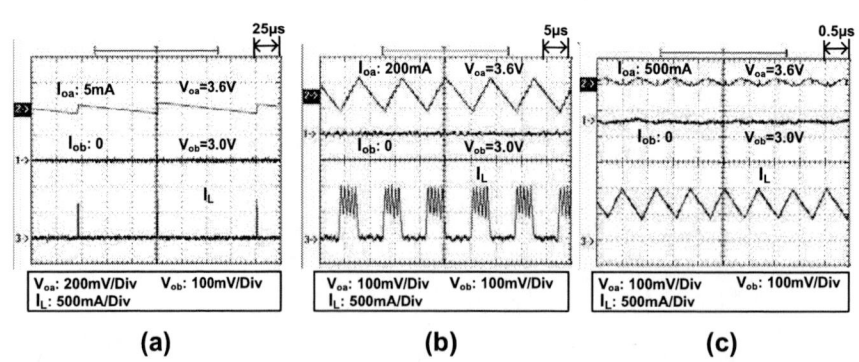

(a)   **(b)**   **(c)**

Steady-state with I_{ob}=0mA :

(a) I_{oa}**=5mA**   **DCM+DCM**

(b) I_{oa}**=200mA**   **DCM+CCM**

(c) I_{oa}**=500mA**   **CCM+CCM**

Measurements (3)

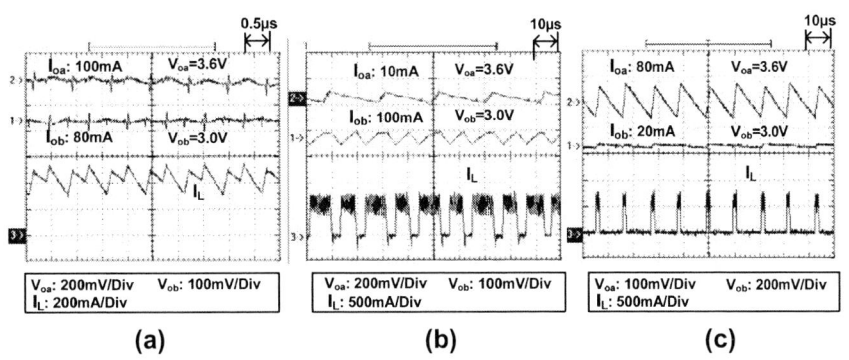

Steady-state with :

(a) I_{oa}=100mA & I_{ob}=80mA   CCM+CCM

(b) I_{oa}=10mA & I_{ob}=100mA   DCM+CCM

(c) I_{oa}=80mA & I_{ob}=20mA   DCM+CCM

Prof. P. Mok , CICC 2013

Measurements (4)

- Load transient at V_{ob}
- I_{ob}=10mA-500mA
- I_{oa}=10mA

- Load transient at V_{oa}
- I_{oa}=10mA-500mA
- I_{ob}=0mA

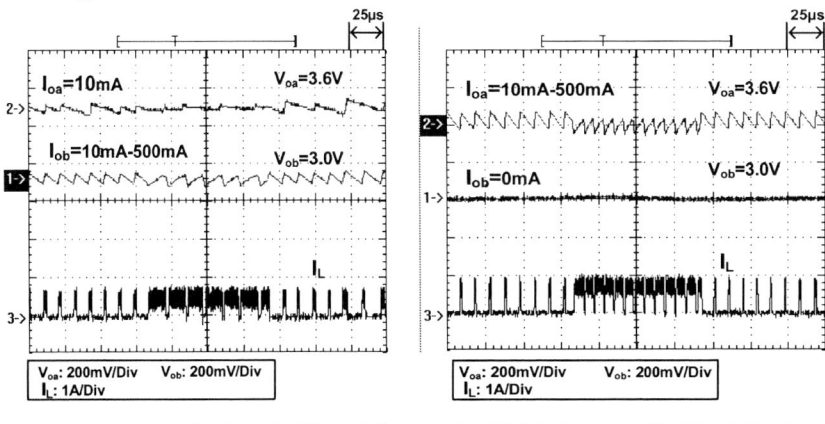

Prof. P. Mok , CICC 2013

Summary (1)

- Conversion Ratios, Efficiency, Driving Capability, Cross-Regulation and Transient Response of SIMO converter has been discussed
- Cross-Regulation is a unique problem in SIMO converter
- Time-Multiplexing (TM) Control and Ordered Power Distributive Control (OPDC) are commonly used control method.
- TM control has limited driving capability while OPDC control has large cross-regulation.
- Several advanced control methods have been introduced to improve the performance of the SIMO converter.

Summary (2)

- Power handling capability extension
  - PCCM: Pseudo CCM operation
  - CCAH: Sequential control with frequency hopping
  - CMHC: CCM operation
- Cross-regulation suppression
  - CCAH: Maintain constant average charge deliveries to the output during transient
  - Vestigial Current Control: move the errors to the last dummy stage
  - CMHC: Fast response to minimize the cross-regulation
- Fast transient response
  - CCAH: PWM with inductor peak current prediction and frequency hopping
  - Vestigial Current Control: Comparator control method
  - CMHC: Hysteretic control with clock synchronization

Summary (3)

- Single-Inductor-Dual-Output (SIDO) converters have been used to demonstrate the ideas. It can be easily extended to the Single-Inductor-Multiple-Output (SIMO) converters

- The same concept can also be extended to Single-Inductor-Multiple-Input-Multiple-Output (SIMIMO)

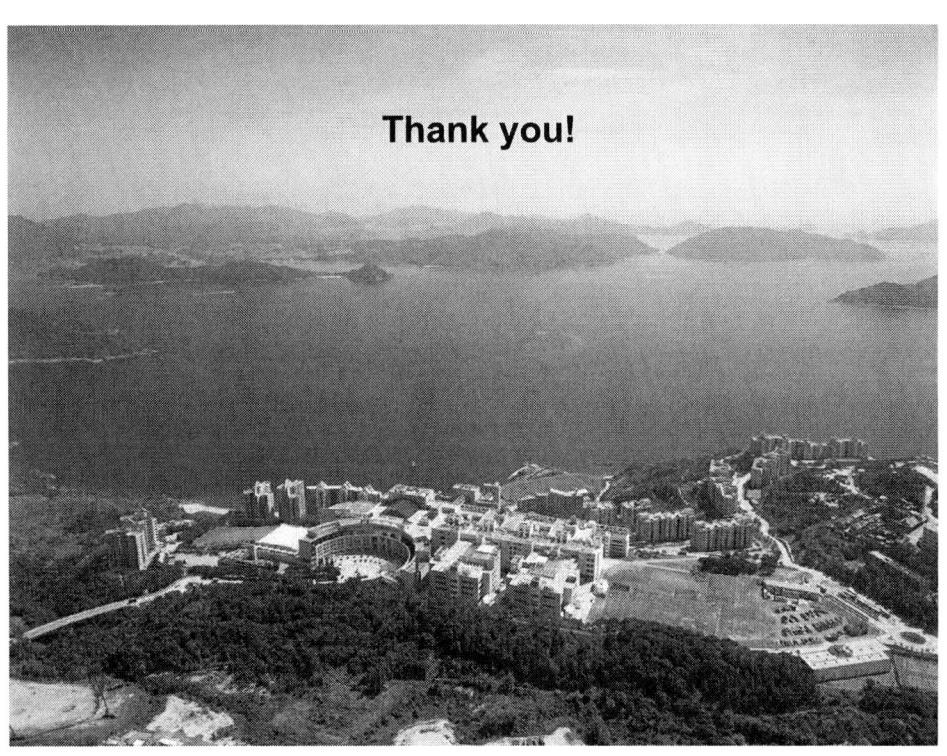

Thank you!

Advanced Digital Phase-Locked Loops

Salvatore Levantino
24 September 2013

S. Levantino © 2013 Politecnico di Milano

Outline of Talk

- **Digital Phase-Locked Loops**
- **Integer-N Synthesis**
- **Fractional-N Synthesis**
- **Direct Frequency Modulation**
- **Practical Implementation**

S. Levantino © 2013 Politecnico di Milano

Analog Charge-Pump PLLs

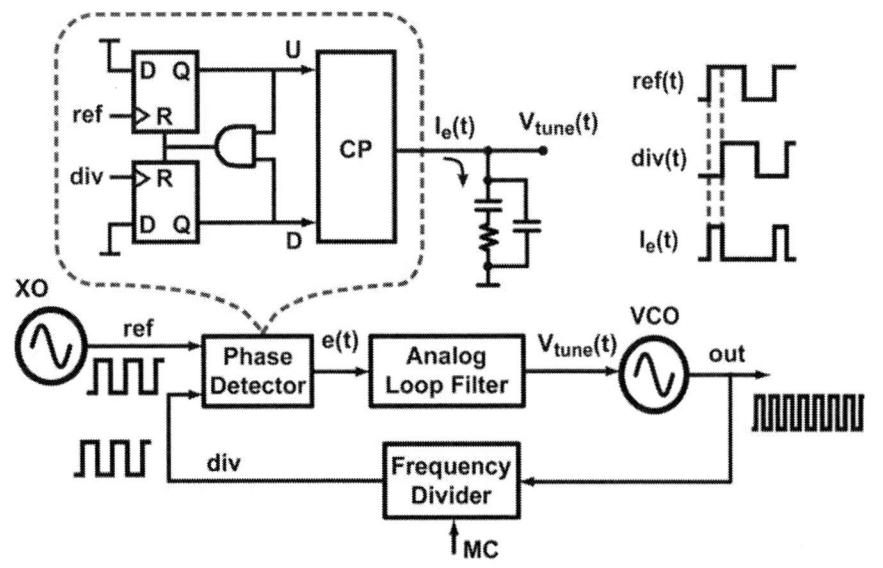

S. Levantino   © 2013 Politecnico di Milano

Motivation to Investigate Digital PLLs

- **LO phase noise limits *wireless applications***
- **Clock jitter limits *digital applications***
- **Analog PLLs do not scale down as process**
- **Spur/noise cancellation techniques are hardly implemented in analog PLLs**

S. Levantino   © 2013 Politecnico di Milano

From Analog to Digital PLL

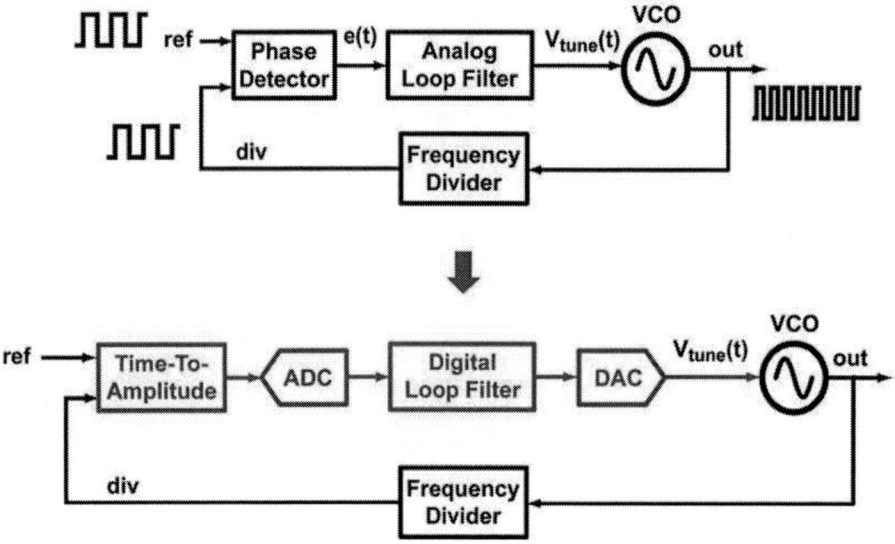

S. Levantino · © 2013 Politecnico di Milano

More-Digital PLL

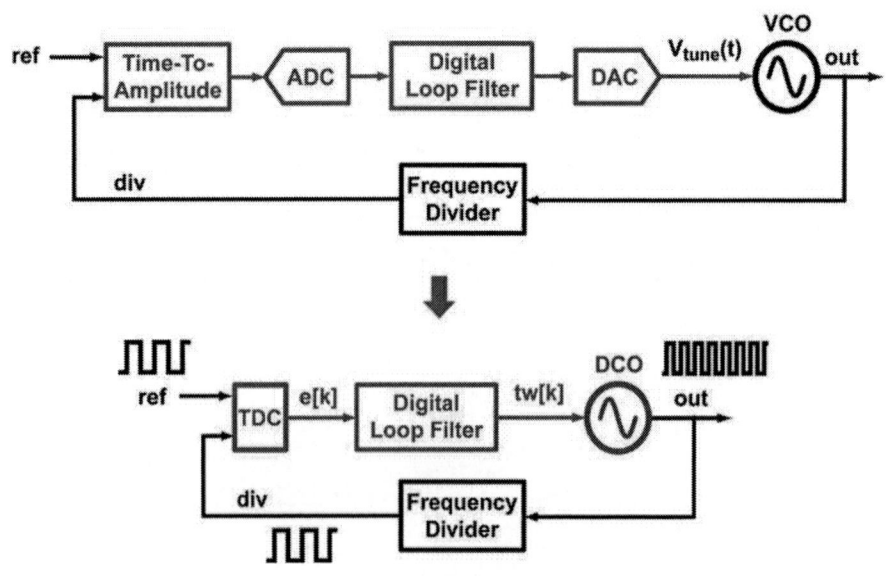

S. Levantino · © 2013 Politecnico di Milano

Time-To-Digital Conversion

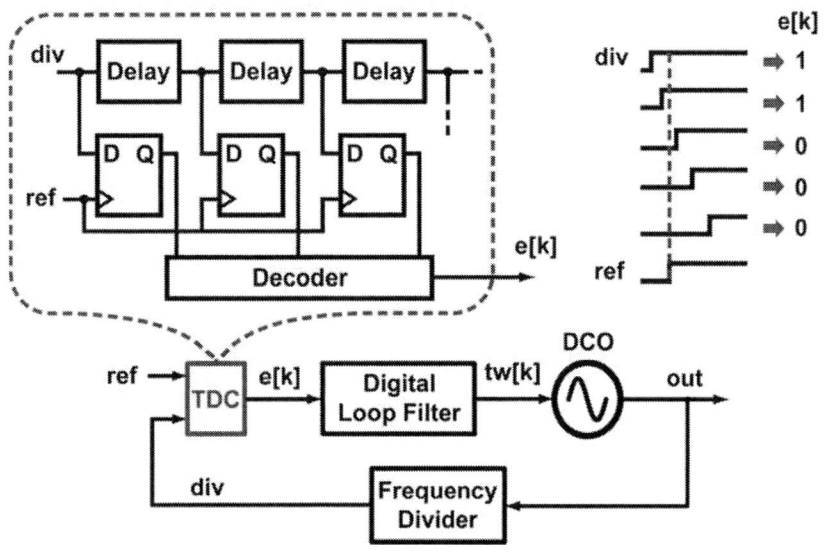

S. Levantino · © 2013 Politecnico di Milano

Digitally-Controlled Oscillator

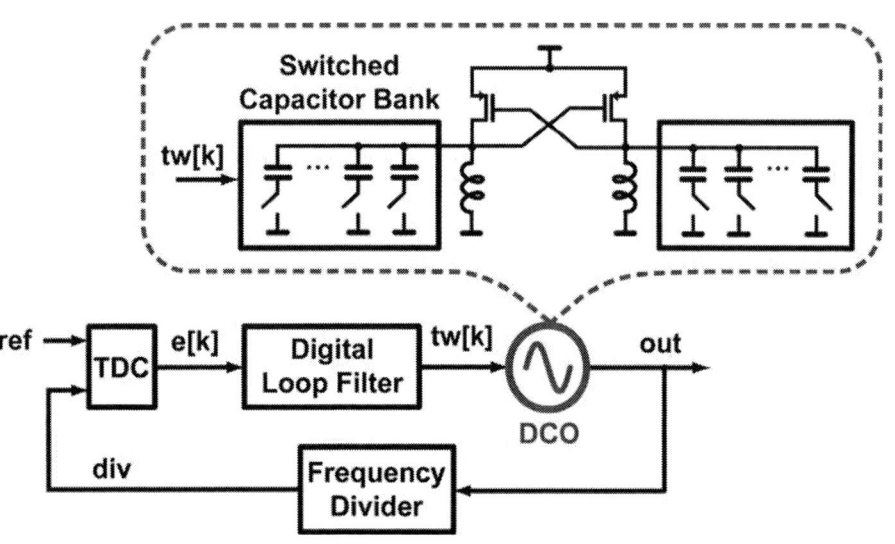

S. Levantino · © 2013 Politecnico di Milano

Digital Loop Filter

Example (FIR filter)

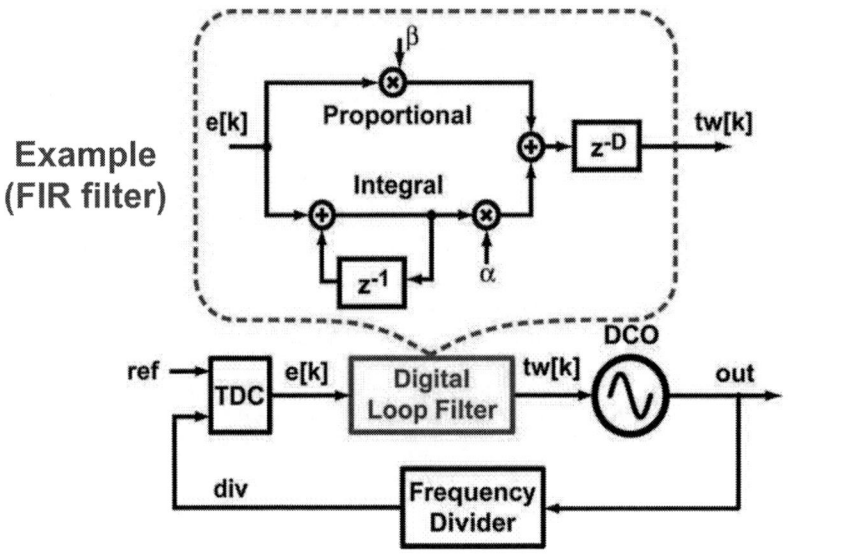

S. Levantino   © 2013 Politecnico di Milano

Integer-N Frequency Synthesis

S. Levantino   © 2013 Politecnico di Milano

TDC and DCO Quantization

Mid-tread Quantizer

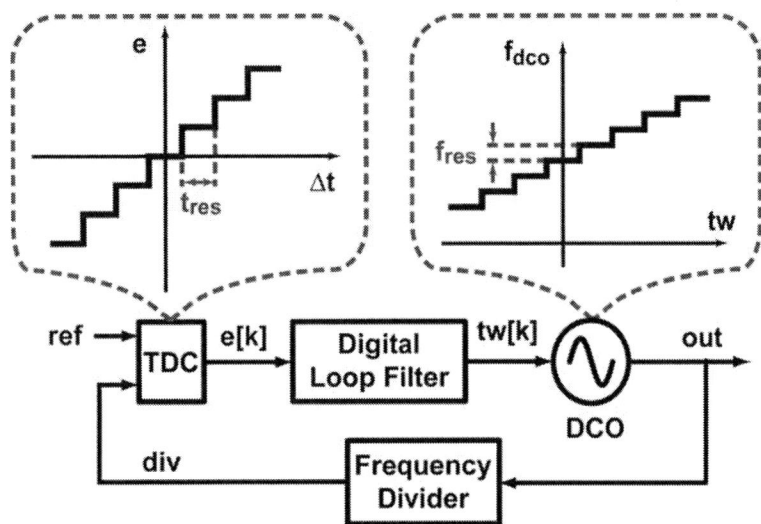

Deriving a Model at Reference Rate

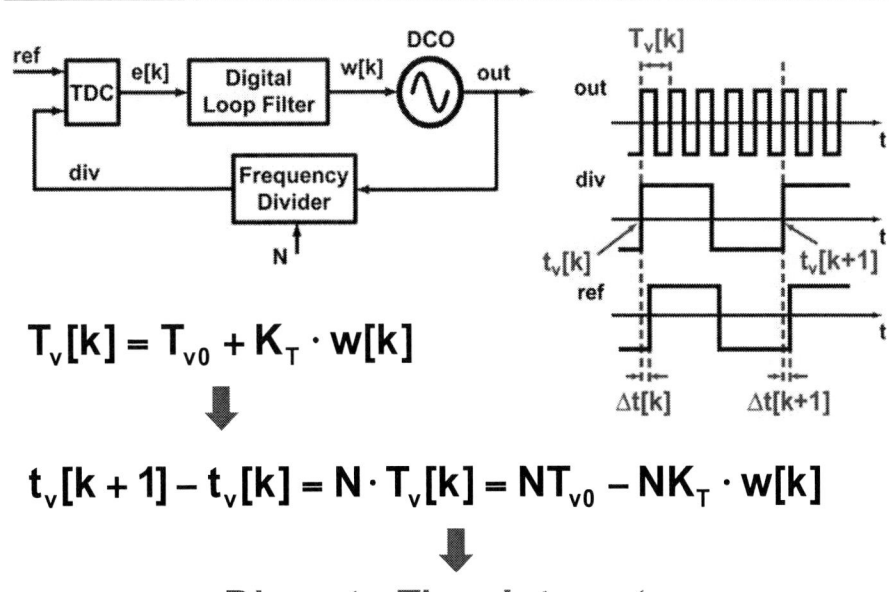

$$T_v[k] = T_{v0} + K_T \cdot w[k]$$

$$t_v[k+1] - t_v[k] = N \cdot T_v[k] = NT_{v0} - NK_T \cdot w[k]$$

Discrete-Time Integrator

DCO and Divider Model at Reference Rate

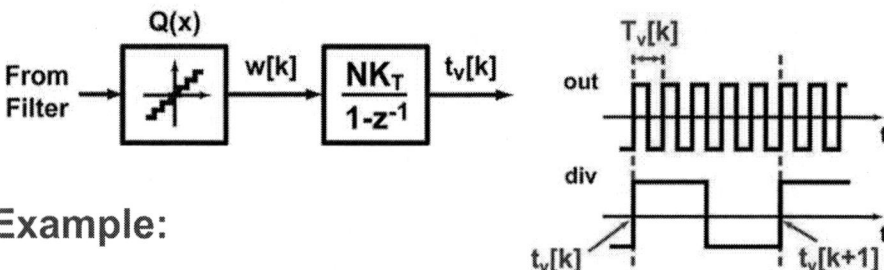

Example:

$$f_{v0} = 3.6 \text{ GHz}$$

$$T_{v0} = f_{v0}^{-1} = 278 \text{ ps}$$

$$f_{res} = 12 \text{ kHz/unit cap}$$

$$K_T = T_{v0}^2 f_{res} = 0.92 \text{ fs/unit cap} \quad \text{(DCO resolution)}$$

S. Levantino   © 2013 Politecnico di Milano

TDC Model

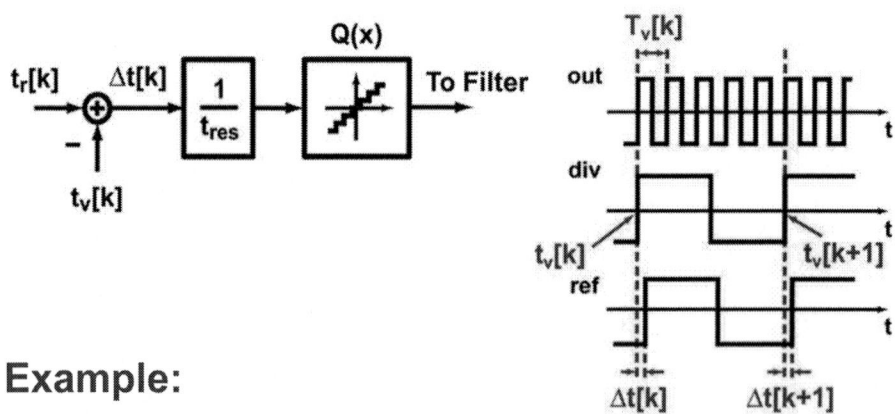

Example:

$$t_{res} = 20 \text{ ps} \quad \text{(TDC resolution)}$$

S. Levantino   © 2013 Politecnico di Milano

DPLL Model

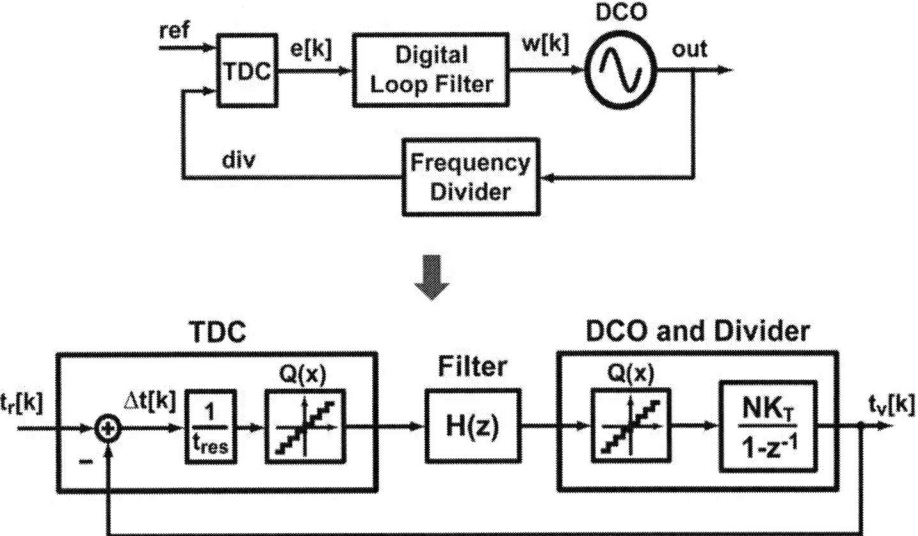

S. Levantino   © 2013 Politecnico di Milano

Conventional Simplification for Analysis

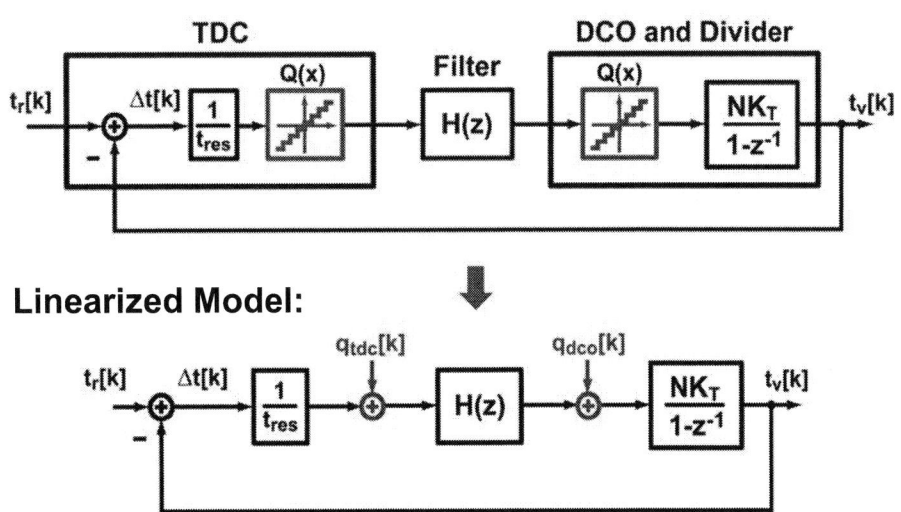

Linearized Model:

Under what condition the system acts as a linear one?

S. Levantino   © 2013 Politecnico di Milano

Loop Filter for a Second-Order DPLL

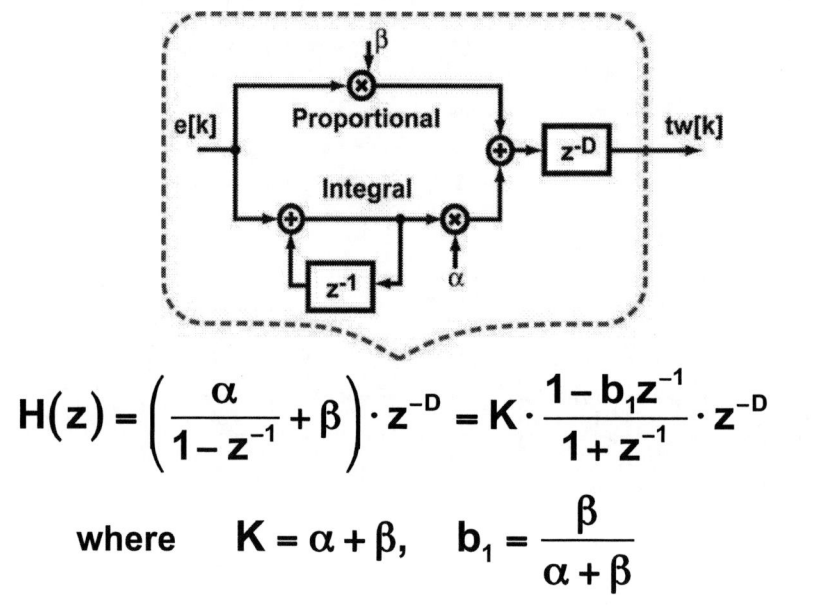

$$H(z) = \left(\frac{\alpha}{1 - z^{-1}} + \beta \right) \cdot z^{-D} = K \cdot \frac{1 - b_1 z^{-1}}{1 + z^{-1}} \cdot z^{-D}$$

where $\quad K = \alpha + \beta, \quad b_1 = \dfrac{\beta}{\alpha + \beta}$

S. Levantino $\qquad$ © 2013 Politecnico di Milano

Second Simplification for Analysis

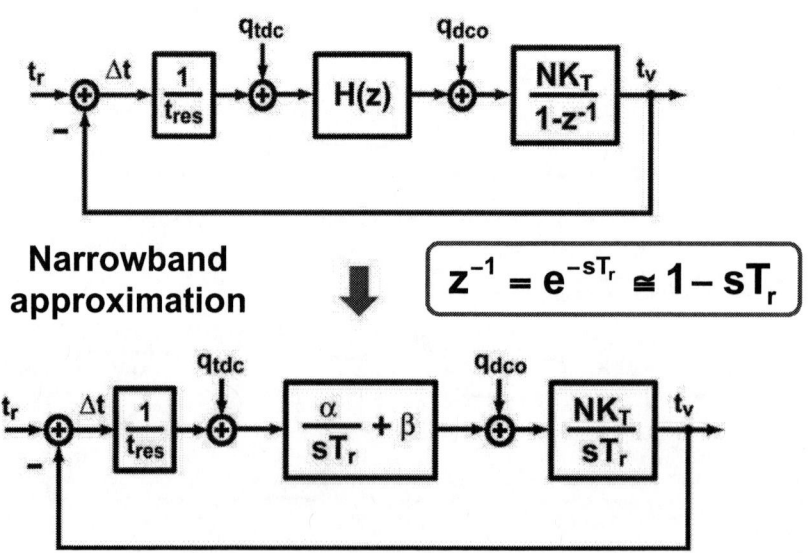

Narrowband approximation $\qquad$ $\boxed{z^{-1} = e^{-sT_r} \cong 1 - sT_r}$

S. Levantino $\qquad$ © 2013 Politecnico di Milano

Loop Bandwidth

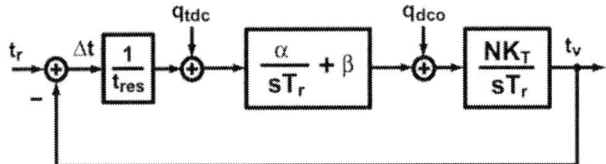

$$G_{loop}(s) = \left(1 + \frac{\alpha}{\beta}\frac{1}{sT_r}\right)\cdot\frac{N\beta K_T}{sT_r t_{res}}$$

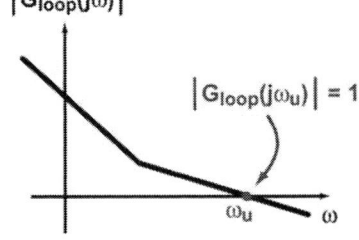

$$\omega_u \cong \frac{N\beta K_T}{T_r t_{res}}, \qquad \omega_z = \frac{\alpha}{\beta}\cdot\frac{1}{T_r}$$

S. Levantino   © 2013 Politecnico di Milano

Narrow Bandwidth Design

Example:

$f_{v0} = 3.6$ GHz

$f_r = 1/T_r = 50$ MHz $\Rightarrow$ N = 72

$K_T = 0.92$ fs/unit cap (DCO resolution)

$t_{res} = 20$ ps (TDC resolution)

$f_u = 100$ kHz, $f_z = 10$ kHz

$$\beta = \frac{\omega_u T_r t_{res}}{N K_T} = 4.5, \quad \alpha = \beta\cdot\omega_z T_r = 5.6\cdot 10^{-3}$$

S. Levantino   © 2013 Politecnico di Milano

Simulated Spectrum (No Random Noise)

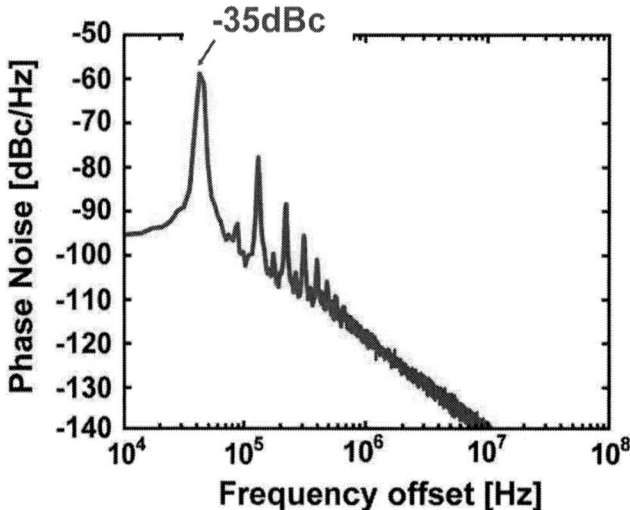

S. Levantino   © 2013 Politecnico di Milano

Adding Random Noise

- **With negligible random noise,** quantization noise $q_{tdc}[k]$ is periodic

- **Adding random noise of same variance,** $q_{tdc}[k]$ becomes a white random process

- **Output in-band noise is approximated by:**

$$\mathcal{L}_{q,tdc} = N^2 \cdot \left(2\pi f_r\right)^2 \cdot \frac{t_{res}^2}{12 f_r} = \frac{2\pi^2}{3} \cdot N^2 t_{res}^2 f_r$$

$$\Downarrow$$

$$\mathcal{L}_{q,tdc} = -92 \text{ dBc/Hz}$$

S. Levantino   © 2013 Politecnico di Milano

Simulation of Second-Order DPLL

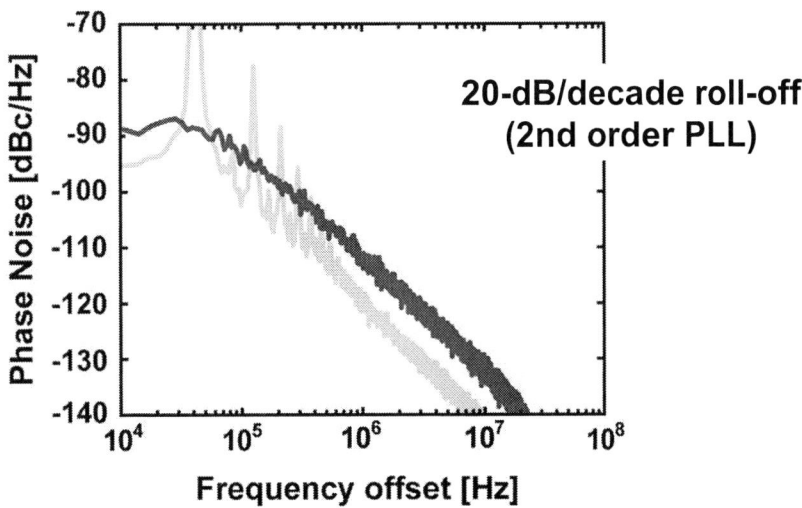

20-dB/decade roll-off
(2nd order PLL)

S. Levantino © 2013 Politecnico di Milano

Wideband Design

(From slide #19)

$$\omega_u \cong \frac{N\beta K_T}{T_r t_{res}}$$

- Wider BW (relative to $1/T_r$) would require either to increase βK_T or reduce t_{res}
- Larger βK_T (DCO frequency granularity) increases truncation noise at DCO input
- Wider BW at same t_{res} (same quantization noise level) may violate phase-noise mask

S. Levantino © 2013 Politecnico di Milano

Wideband Design Example

- **In conventional DPLL, wider bandwidth demands for better TDC resolution**

  Example:

  $$t_{res} = 2 \text{ ps} \qquad \Rightarrow \qquad \mathcal{L}_{q,tdc} = -115 \text{ dBc/Hz}$$

  $$f_u = 1 \text{ MHz}, \quad f_z = 100 \text{ kHz}$$

  $$\beta = \frac{\omega_u T_r t_{res}}{NK_T} = 4.5, \quad \alpha = \beta \cdot \omega_z T_r = 5.6 \cdot 10^{-3}$$

  Same filter coefficients

S. Levantino · © 2013 Politecnico di Milano

Simulation of Wideband Second-Order DPLL

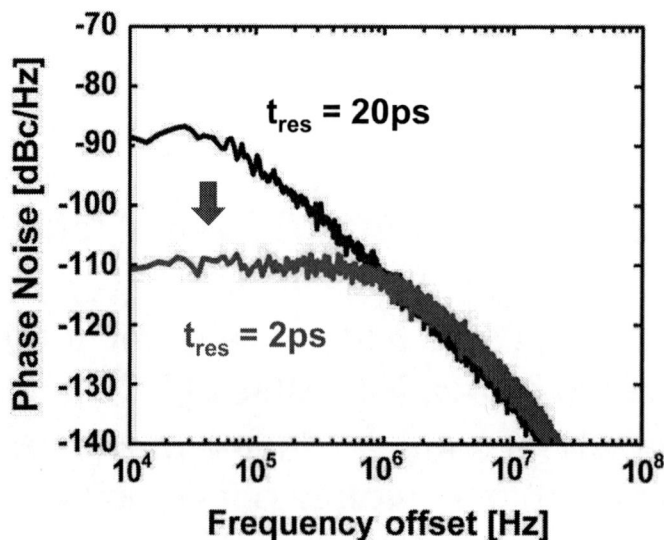

S. Levantino · © 2013 Politecnico di Milano

Conclusion

- In the conventional DPLL scheme, in-band noise is limited by TDC resolution
- Typically, if the system requirements demand for wider PLL bandwidth, TDC resolution must be improved
- In wireless applications, ps or sub-ps TDC resolution may be required

S. Levantino — © 2013 Politecnico di Milano

Integer-N Frequency Synthesis with Bang-Bang TDC

S. Levantino — © 2013 Politecnico di Milano

Mid-rise TDC or Bang-Bang Phase Detect

Mid-rise Quantizer

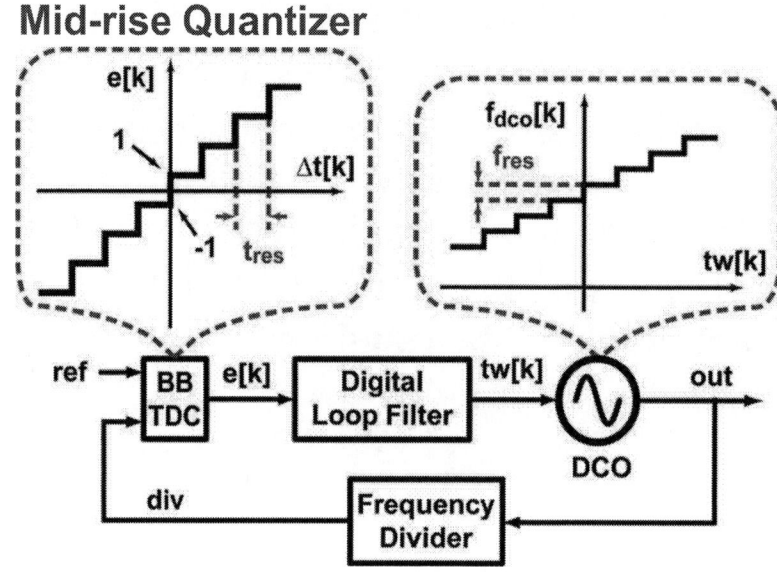

S. Levantino   © 2013 Politecnico di Milano

Model of First-Order DPLL with Quantization

- Proportional filter (gain β) and no latency
- No truncation at DCO input
- No random noise

$$\Delta t[k+1] - \Delta t[k] = T_0 - N\beta K_T \cdot Q\left(\Delta t[k]/t_{res}\right)$$

$$\text{where} \quad T_0 = NT_{v0} - T_r$$

S. Levantino   © 2013 Politecnico di Milano

Behavior of 1st-Order Bang-Bang DPLL

$$\Delta\tau[\kappa+1] - \Delta\tau[\kappa] = \quad -\quad \beta \quad \cdot \quad \left(\Delta\tau[\kappa]/\tau_{\rho\epsilon\sigma}\right)$$

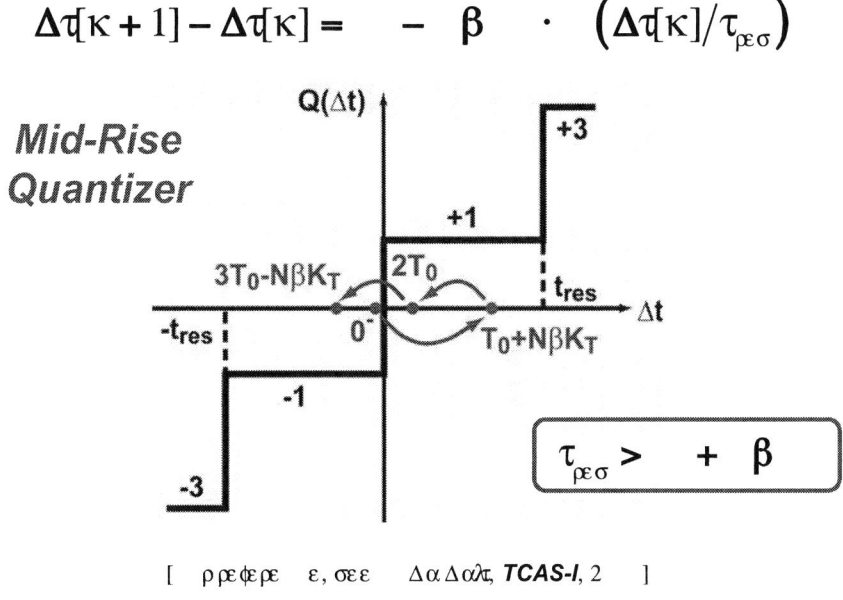

Mid-Rise Quantizer

$$\tau_{\rho\epsilon\sigma} > \quad + \quad \beta$$

[ρρε¢ερε ε, σεε Δα Δαλτ, *TCAS-I*, 2]

1st-Order Bang-Bang DPLL (Continued)

$$\Delta\tau[\kappa+1] - \Delta\tau[\kappa] = \quad -\quad \beta \quad \cdot \quad \left(\Delta\tau[\kappa]/\tau_{\rho\epsilon\sigma}\right)$$

Ιρροατ αλ(/**NβK**)

⬇

p(Δt) is uniform

⬇

$$\sigma_{\Delta\tau} = \frac{\Delta\tau_{\pi\pi}}{\sqrt{12}} = \frac{\beta}{\sqrt{}}$$

$\Delta t_{pp} = 2N\beta K_T$

2nd-Order Bang-Bang DPLL

- PI digital loop filter with latency D
- No truncation at DCO input
- No random noise

Quantization-induced Jitter

$$\sigma_{\Delta t} = \frac{\Delta t_{pp}}{\sqrt{12}} = \frac{(1+D)}{\sqrt{3}} \cdot N\beta K_T$$

[N. Da Dalt, *TCAS-I*, 2005]

S. Levantino   © 2013 Politecnico di Milano

Quantization Noise in Bang-Bang DPLL

(From previous slide)

$$\sigma_{\Delta t} = \frac{(1+D)}{\sqrt{3}} \cdot N\beta K_T$$

- Loop quantization no more limited by t_{res} (TDC), but only by βK_T (DCO granularity)
- Quantization increases as filter latency D (number of clock cycles)

S. Levantino   © 2013 Politecnico di Milano

Example of 2nd-Order Bang-Bang DPLL

Example:

$$f_{v0} = 3.6 \text{ GHz}$$

$$K_T = 0.92 \text{ fs/unit cap}$$

$$f_r = 50 \text{ MHz} \quad \Rightarrow \quad N = 72$$

$$t_{res} = 20 \text{ ps}$$

$$\beta = 4.5, \quad \alpha = 5.6 \cdot 10^{-3}, \quad D = 1$$

$$\sigma_{\Delta t} = \frac{(1+D)}{\sqrt{3}} \cdot N\beta K_T = 344 \text{ fs}_{rms}$$

S. Levantino   © 2013 Politecnico di Milano

Simulation (No Random Noise)

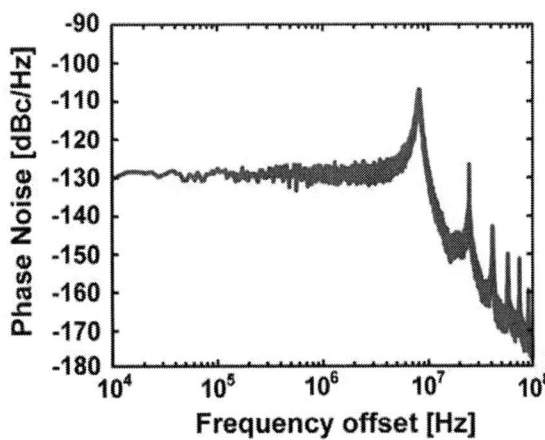

- **The system converges to a limit cycle**
- **Quantization noise is periodic**

S. Levantino   © 2013 Politecnico di Milano

Adding Random Noise

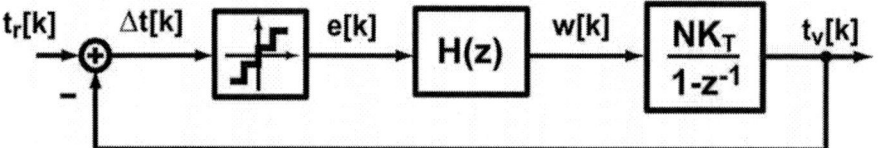

- **Periodicity can be broken only relying on true random noise**
- **If random noise deviation is lower than t_{res}, e[k] will be ±1**
- **The average of e[k] will be proportional to Δt as in a linear phase detector**

S. Levantino   © 2013 Politecnico di Milano

Phase Detect Characteristic is Smoothed

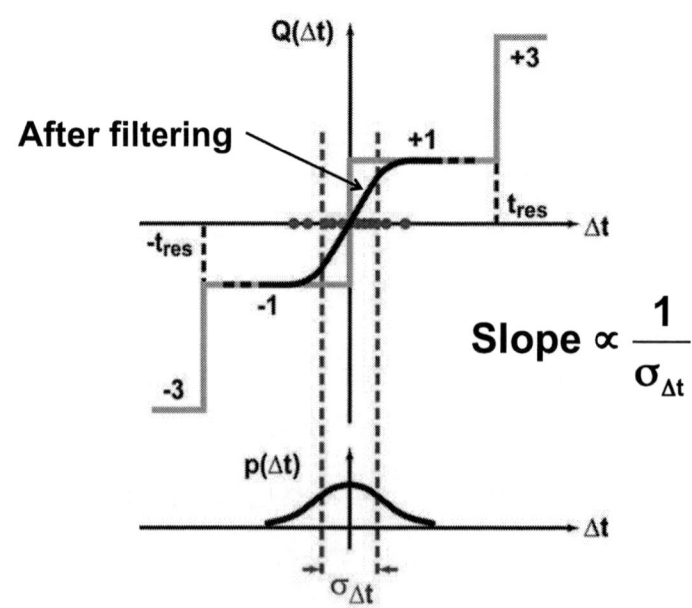

S. Levantino   © 2013 Politecnico di Milano

Model Simplification: Linearization

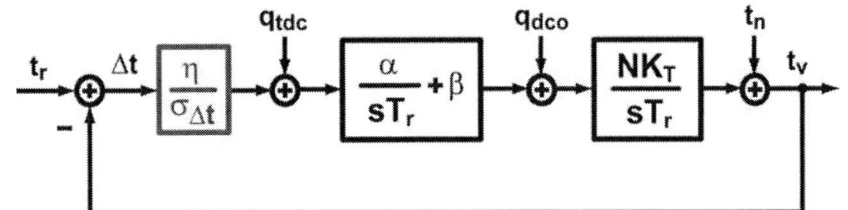

$$K_{pd} = \frac{\eta}{\sigma_{\Delta t}}, \quad \text{where} \quad \eta = \sqrt{\frac{2}{\pi}}$$

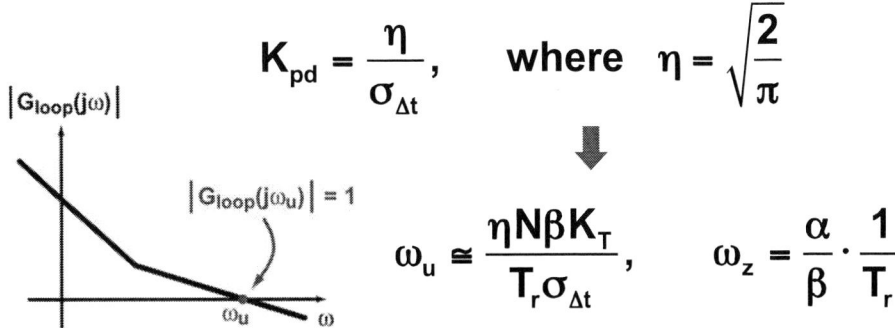

$$\omega_u \cong \frac{\eta N \beta K_T}{T_r \sigma_{\Delta t}}, \qquad \omega_z = \frac{\alpha}{\beta} \cdot \frac{1}{T_r}$$

S. Levantino © 2013 Politecnico di Milano

Computation of Random Noise

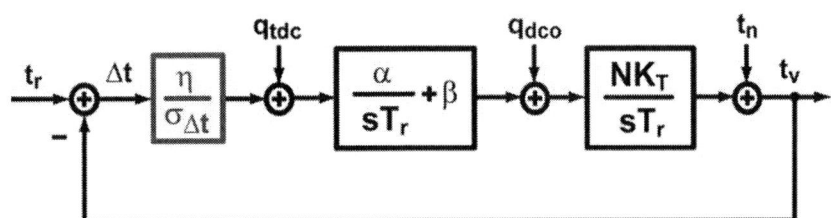

$$\sigma_{\Delta t}^2 = 2 \cdot \int_0^\infty \frac{S_{t_r}(f) + S_{t_n}(f)}{\left|1 + G_{loop}(f)\right|^2} \, df$$

$$\mathcal{L}(f) = \left(2\pi f_{v0}\right)^2 \cdot S_{t_n}(f)$$

S. Levantino © 2013 Politecnico di Milano

Adding White Noise

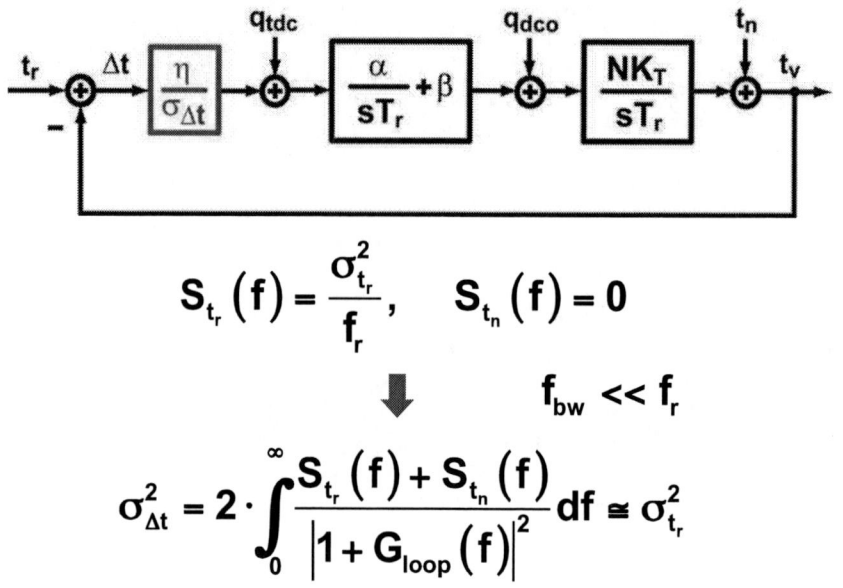

$$S_{t_r}(f) = \frac{\sigma_{t_r}^2}{f_r}, \quad S_{t_n}(f) = 0$$

$$\Downarrow \qquad f_{bw} \ll f_r$$

$$\sigma_{\Delta t}^2 = 2 \cdot \int_0^\infty \frac{S_{t_r}(f) + S_{t_n}(f)}{\left|1 + G_{loop}(f)\right|^2} df \cong \sigma_{t_r}^2$$

S. Levantino   © 2013 Politecnico di Milano

Bang-Bang DPLL with Random Noise

Random-noise-induced Jitter

$$\boxed{\sigma_{\Delta t} \cong \sigma_{t_r}}$$

must be greater or equal to quantization-induced jitter

$$\sigma_{t_r} \geq \frac{(1+D)}{\sqrt{3}} \cdot N\beta K_T \quad \Rightarrow \quad \boxed{\beta K_T \leq \frac{\sqrt{3}}{N(1+D)} \cdot \sigma_{t_r}}$$

[M. Zanuso et al., TCAS-II, 2009]

S. Levantino   © 2013 Politecnico di Milano

Design Example

$$\mathcal{L}(f) = N^2 (2\pi f_r)^2 \cdot \frac{\sigma_{t_r}^2}{f_r} = 4\pi^2 N^2 f_r \sigma_{t_r}^2$$

Example:

$$\mathcal{L}(f) = -118\,dBc/Hz \qquad \text{White reference noise}$$

$$f_{v0} = 3.6\,GHz, \quad \sigma_{t_r} = 344\,fs_{rms}$$

$$\Downarrow \quad D = 1, \ N = 72$$

$$\beta K_T = \frac{\sqrt{3}}{N(1+D)} \cdot \sigma_{t_r} = 4.1\,fs \qquad \begin{array}{c}\text{Maximum}\\ \text{DCO granularity}\end{array}$$

S. Levantino © 2013 Politecnico di Milano

Simulation of 2nd-Order Bang-Bang DPLL

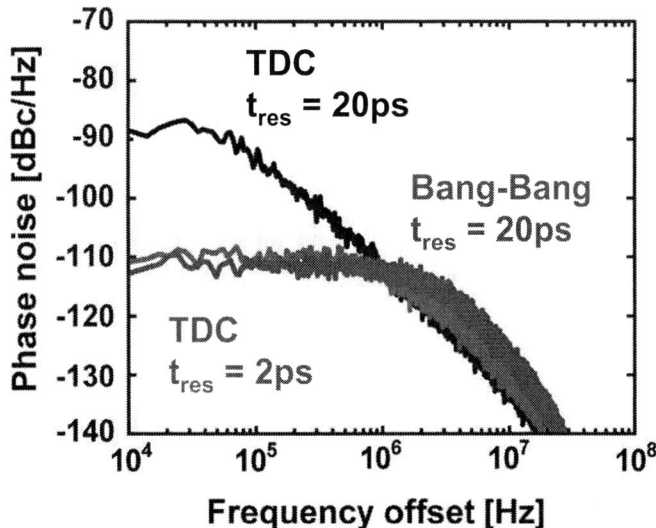

S. Levantino © 2013 Politecnico di Milano

Conclusions

- DPLLs with bang-bang detectors (i.e. coarse TDCs with mid-rise quantization) allows same phase noise performance as those employing high-resolution TDCs

- Thus, at wide loop bandwidths, BBPLLs show much higher efficiency (noise/power trade-off) than conventional DPLLs

- In practice, fine resolution is only required to the DCO, which can be improved leveraging oversampling techniques

S. Levantino   © 2013 Politecnico di Milano

Fractional-N Frequency Synthesis

S. Levantino   © 2013 Politecnico di Milano

ΔΣ Fractional-N Analog PLL

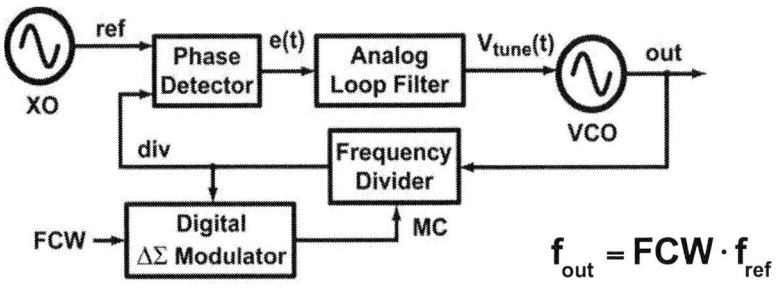

$$f_{out} = FCW \cdot f_{ref}$$

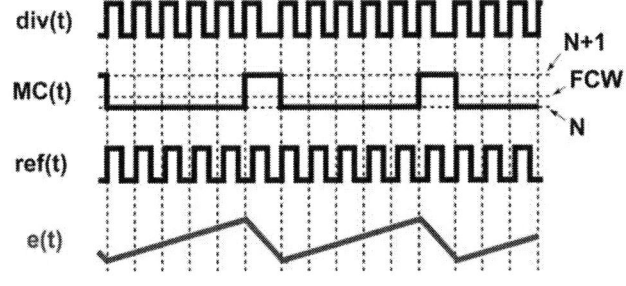

S. Levantino　　　© 2013 Politecnico di Milano

Digiphase Technique to Cancel ΔΣ Noise

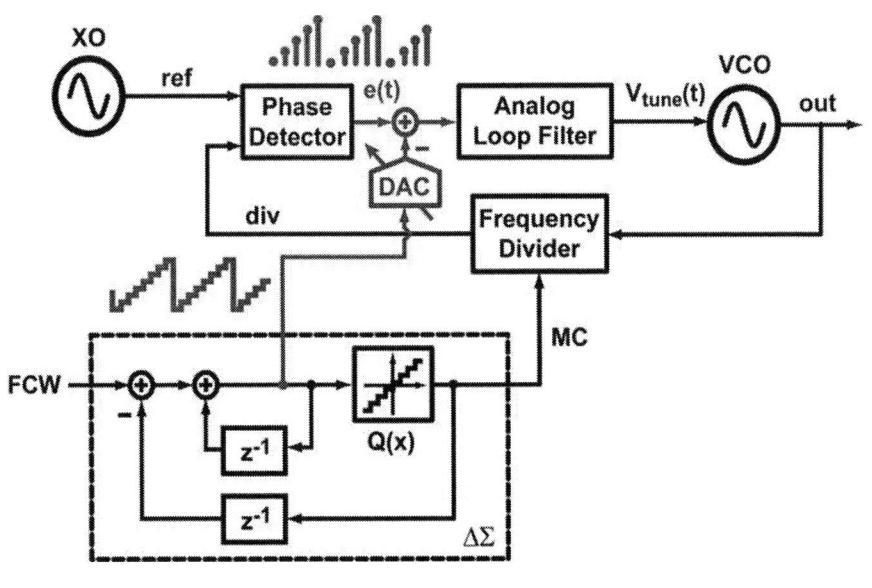

S. Levantino　　　© 2013 Politecnico di Milano

From Analog to Digital Fractional-N PLL

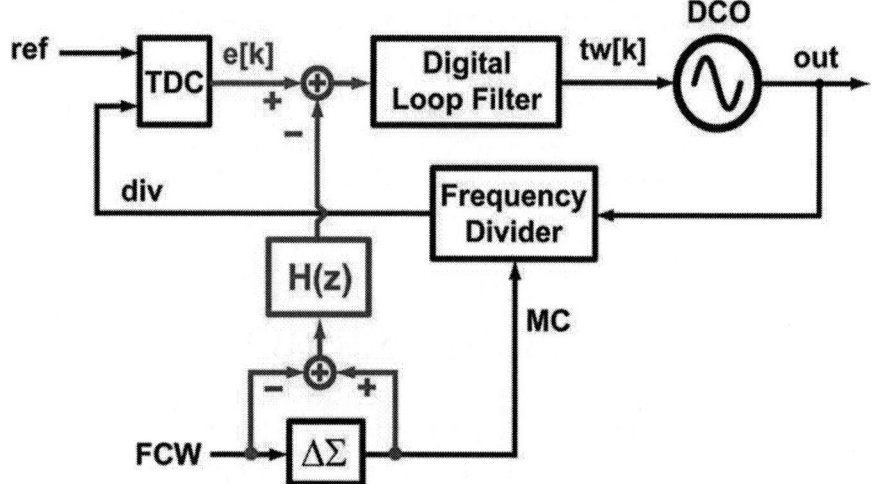

- **How do we build an adaptive filter H(z)?**

S. Levantino   © 2013 Politecnico di Milano

Time-Domain Model

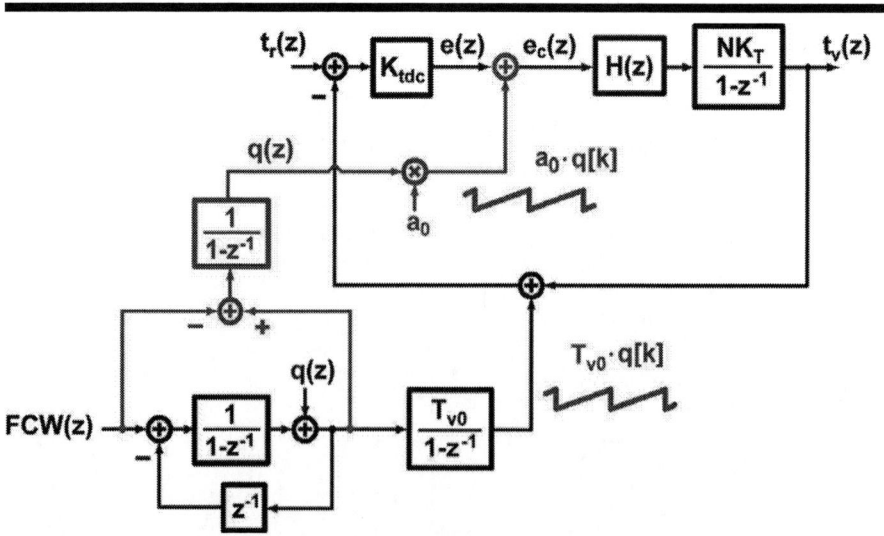

- **$q(z)$ is cancelled out if $a_0 = T_{v0}$**

S. Levantino   © 2013 Politecnico di Milano

Automatic Background Regulation of a_0

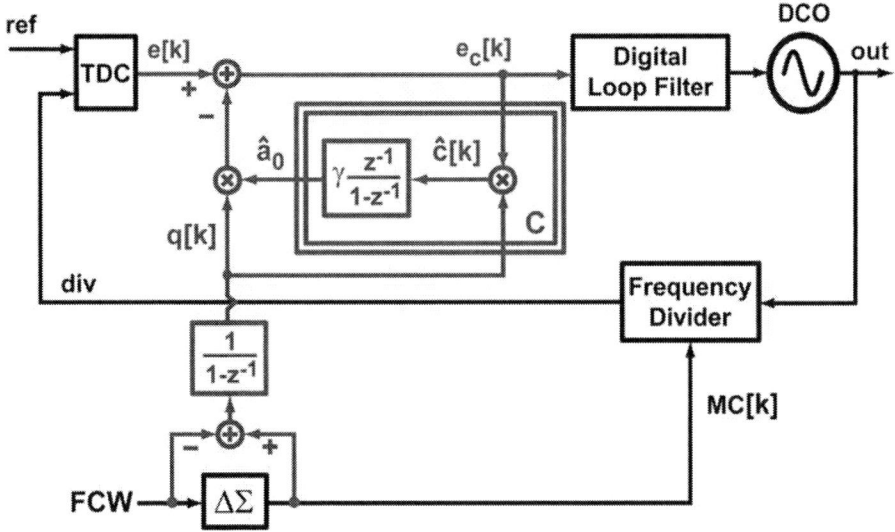

[C.-M. Hsu et al., *ISSCC*, 2008]

Effect of Limited TDC Resolution

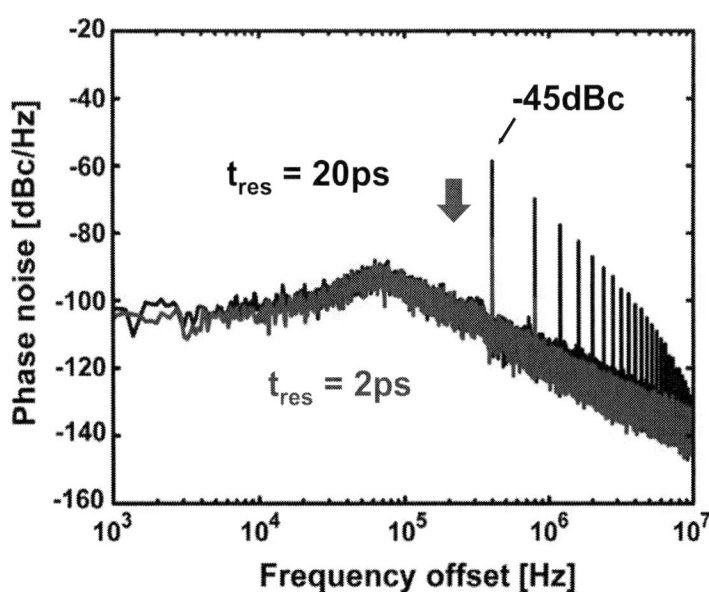

Conclusion

- **Fractional-N DPLLs over their analog counterparts allow simple implementation of** digiphase technique

- **Automatic calibration of cancellation algorithm relies on** digital adaptive filtering

- **However, in the conventional DPLL scheme,** the fractional spur level is limited by TDC resolution

- **In wireless applications,** a 10b-to-12b TDC is typically required

S. Levantino   © 2013 Politecnico di Milano

Fractional-N Frequency Synthesis with Bang-Bang TDC

S. Levantino   © 2013 Politecnico di Milano

Digiphase Technique with Bang-Bang TDC

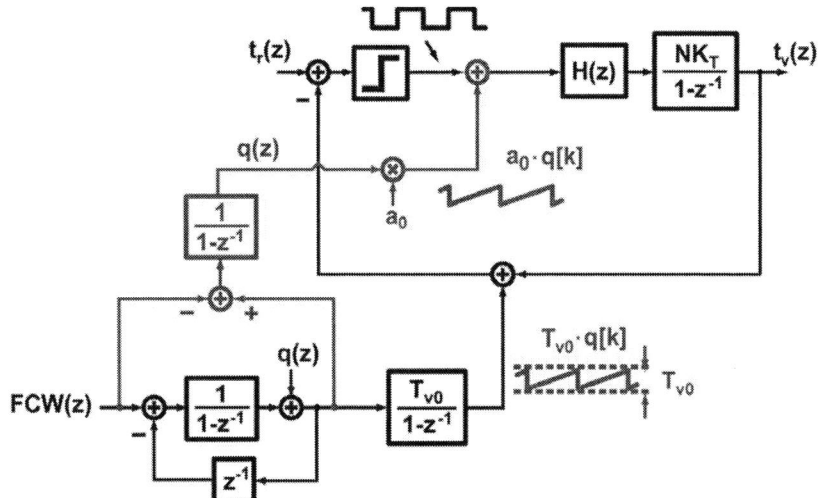

- **ΔΣ quantization error exceeds linear range of the Bang-Bang Phase Detector**

S. Levantino   © 2013 Politecnico di Milano

Solution: Digital-To-Time Converter (DTC)

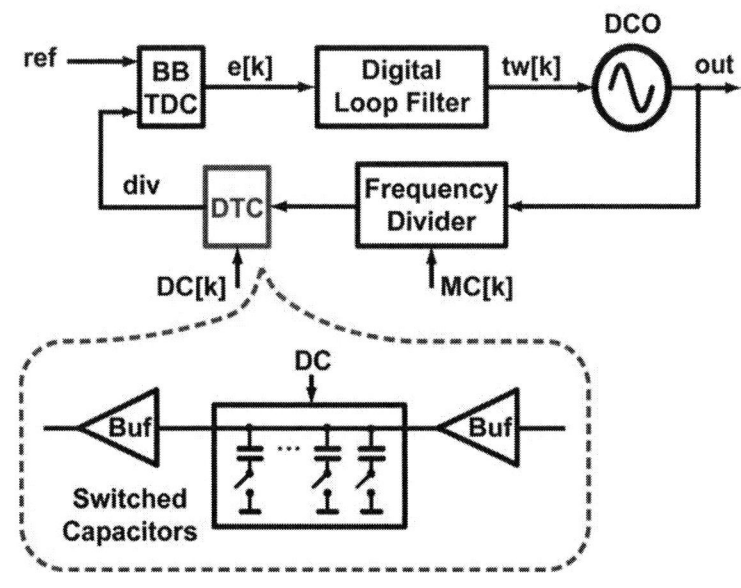

S. Levantino   © 2013 Politecnico di Milano

Bang-Bang Fractional-N DPLL

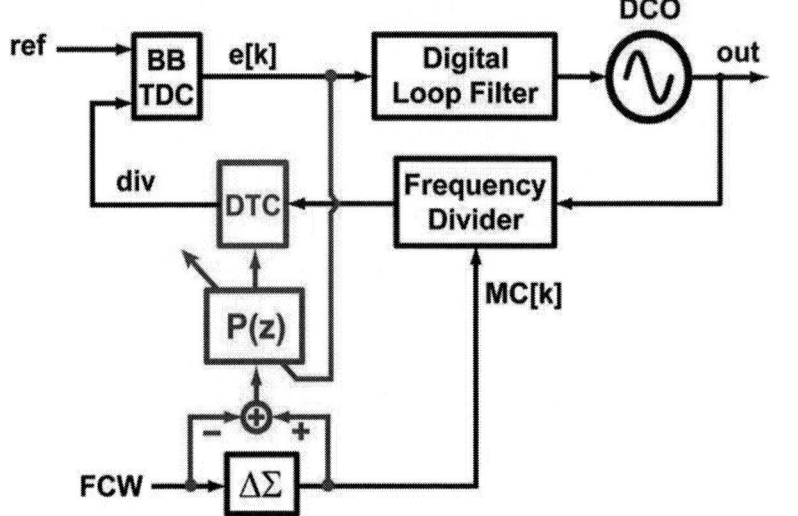

- **ΔΣ noise is cancelled out via DTC**

S. Levantino · © 2013 Politecnico di Milano

Automatic Regulation of DTC Characteristic

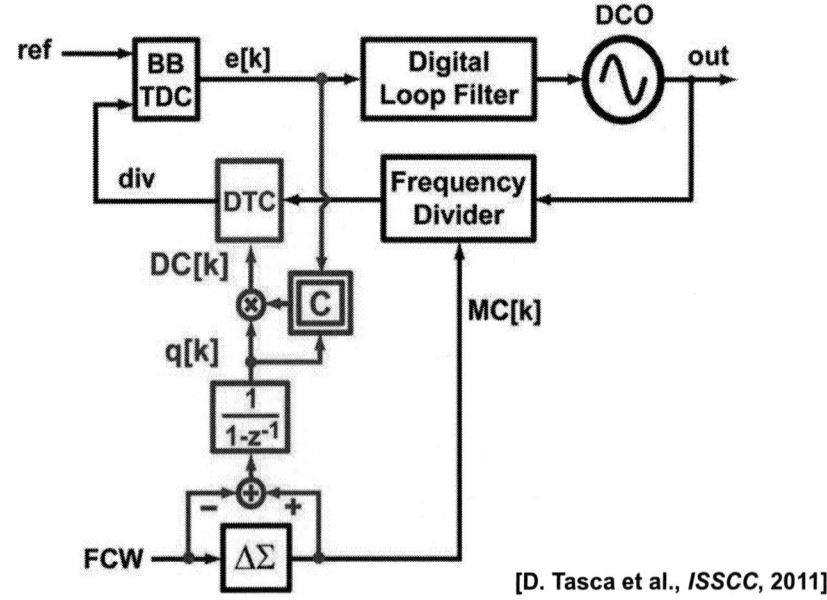

[D. Tasca et al., *ISSCC*, 2011]

S. Levantino · © 2013 Politecnico di Milano

Conclusion

- DPLLs with bang-bang detectors (i.e. coarse TDCs with mid-rise quantization) can be employed even in fractional-N frequency synthesis
- In practice, they require a DTC block with fine resolution, which can be improved leveraging oversampling techniques
- DTCs are easier to be designed than TDCs
- Fractional-N BBPLLs maintain the superior efficiency (noise/power trade-off) over conventional DPLLs

S. Levantino © 2013 Politecnico di Milano

Direct FM Modulation

S. Levantino © 2013 Politecnico di Milano

Direct-FM of a Digital PLL

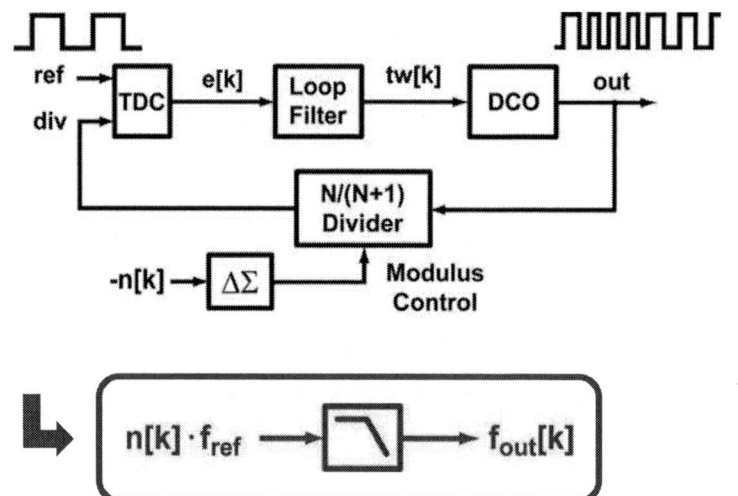

- **Modulation rate is limited by loop BW**

S. Levantino   © 2013 Politecnico di Milano

Two-Path Injection of Modulation Signal

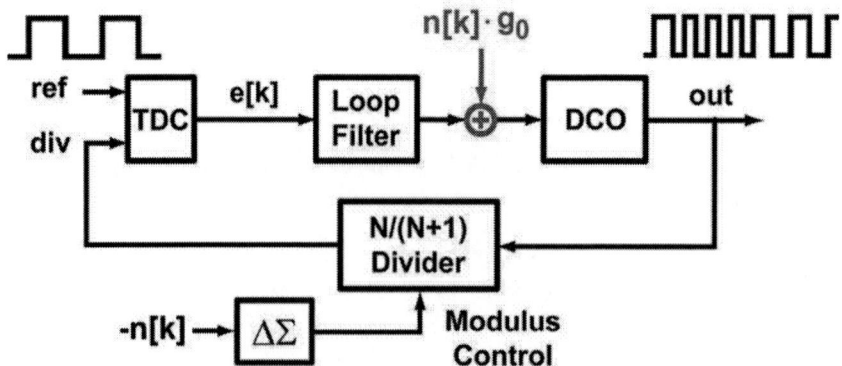

S. Levantino   © 2013 Politecnico di Milano

Output Frequency Expression

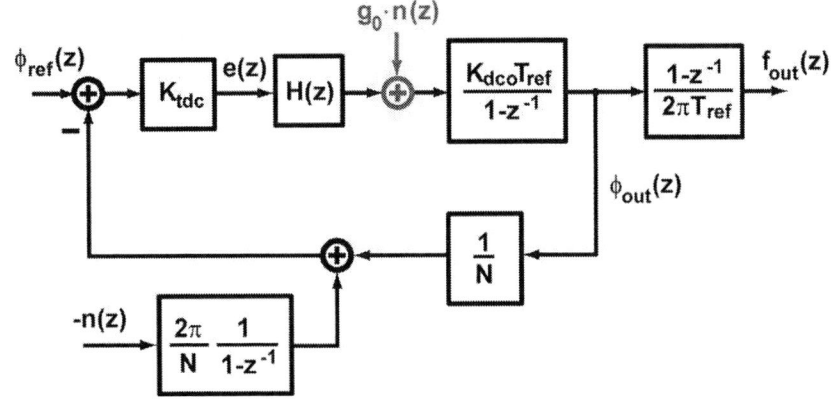

$$f_{out}(z) = n(z) \cdot \left[\frac{g_0 K_{dco}}{2\pi} \cdot \frac{1}{1 + G_{loop}(z)} + f_{ref} \cdot \frac{G_{loop}(z)}{1 + G_{loop}(z)} \right]$$

S. Levantino   © 2013 Politecnico di Milano

Extension of Modulation Bandwidth

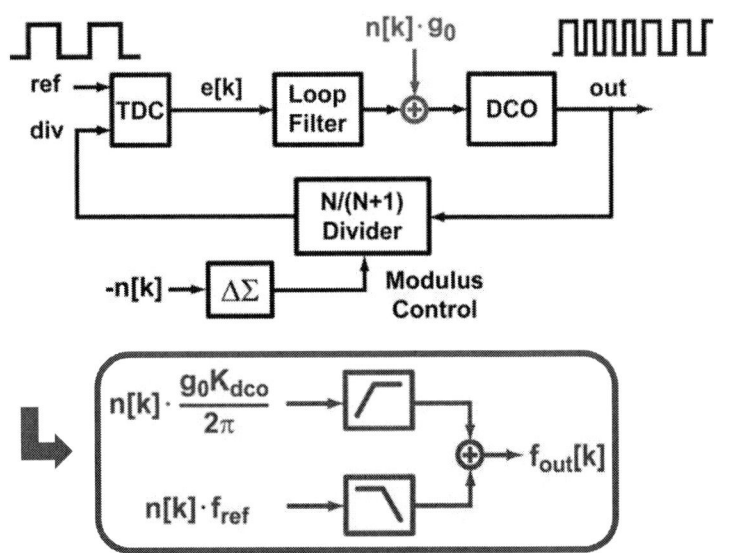

- **All-pass transfer function if $g_0 K_{dco}/2\pi = f_{ref}$**

S. Levantino   © 2013 Politecnico di Milano

Tracking of K_{dco} Variations

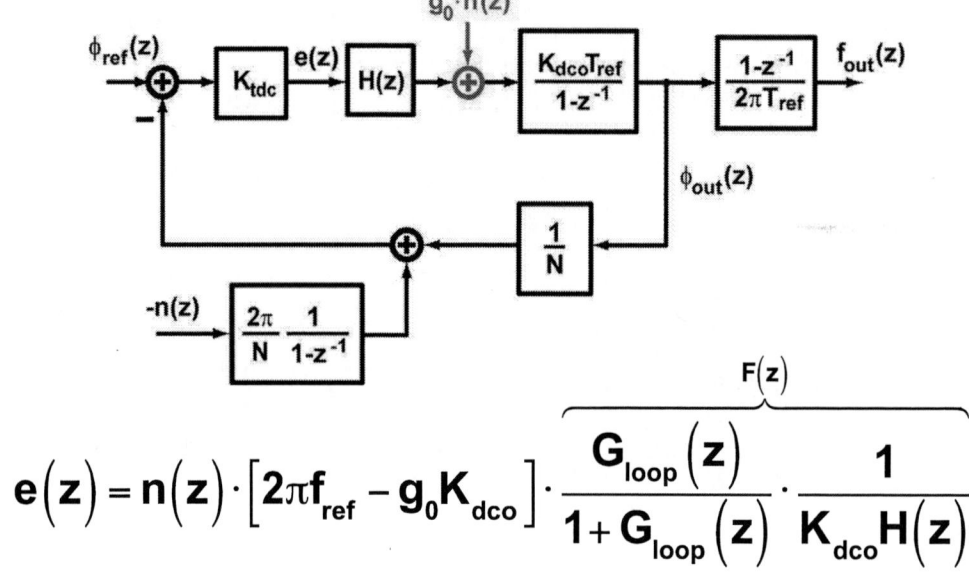

$$e(z) = n(z) \cdot \left[2\pi f_{ref} - g_0 K_{dco} \right] \cdot \overbrace{\frac{G_{loop}(z)}{1 + G_{loop}(z)}}^{F(z)} \cdot \frac{1}{K_{dco} H(z)}$$

Effect of Gain Imbalance

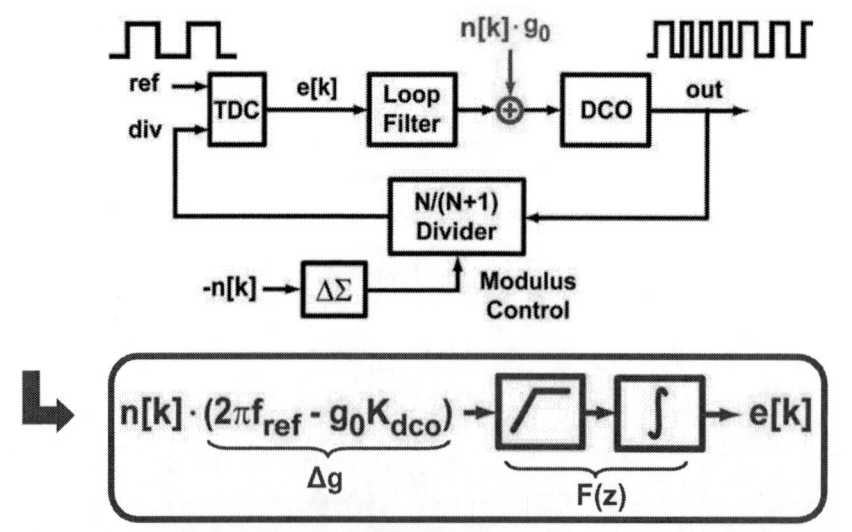

- **Correlation $c[k] = E\{ e[k] \cdot n[k] \}$ is proportional to gain imbalance Δg**

Adaptation of DCO-Path Gain via LMS

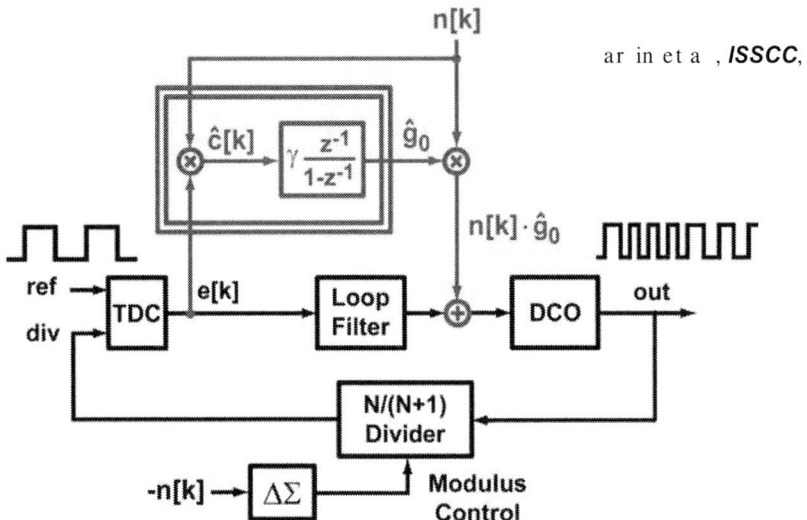

- **LMS loop nulls ĉ[k], thus gain imbalance Δg**

DCO Range Requirements

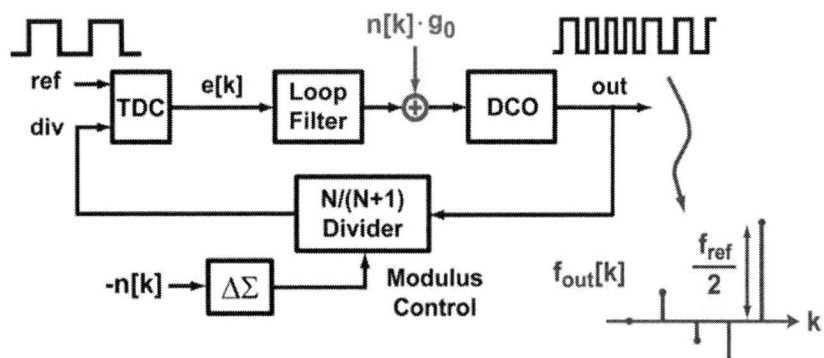

- hase shi t to **π in one ref cycle**
- t ning range in e ess o $_{re}$
- re en y gran arity or er o
- ith inear ontro over - bit

Nonlinearity of DCO w/ Multi-Capacitor-Bank

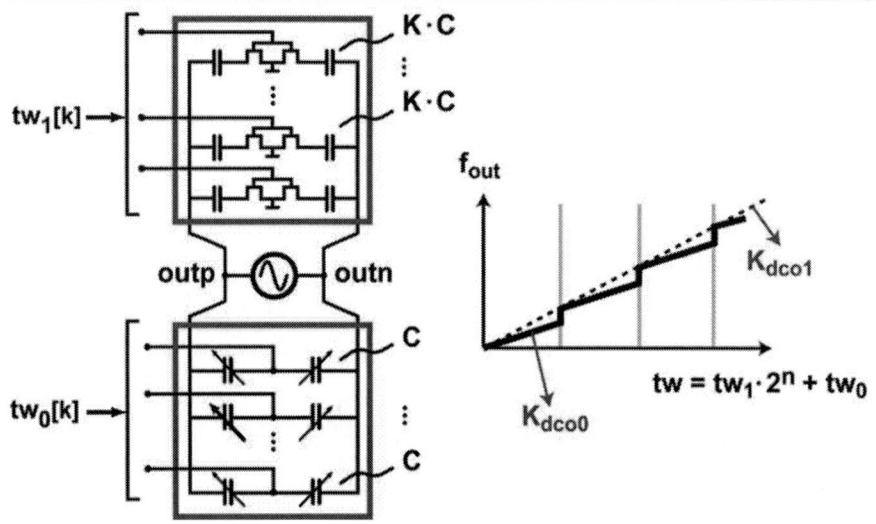

- **Saves area and improves tuning range**
- **Produces *inter-bank* nonlinearity**

S. Levantino   © 2013 Politecnico di Milano

Direct-FM with Multi-Bank DCO

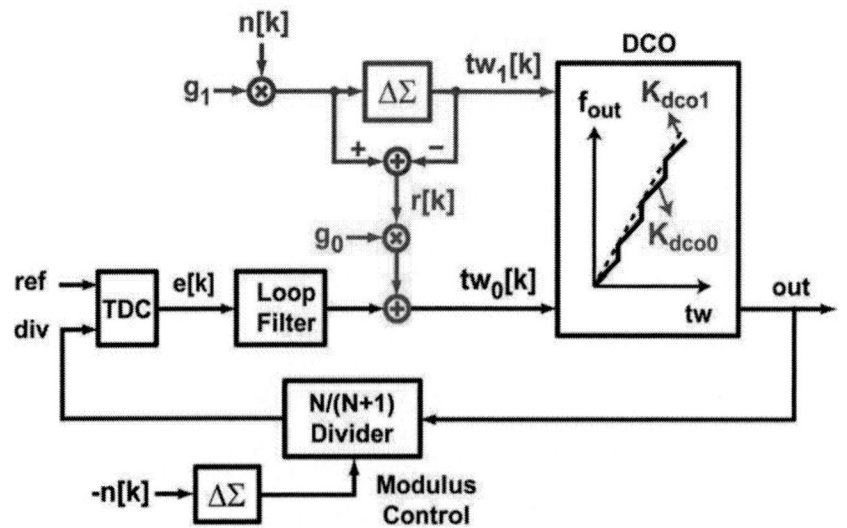

S. Levantino   © 2013 Politecnico di Milano

Model of the Modified System

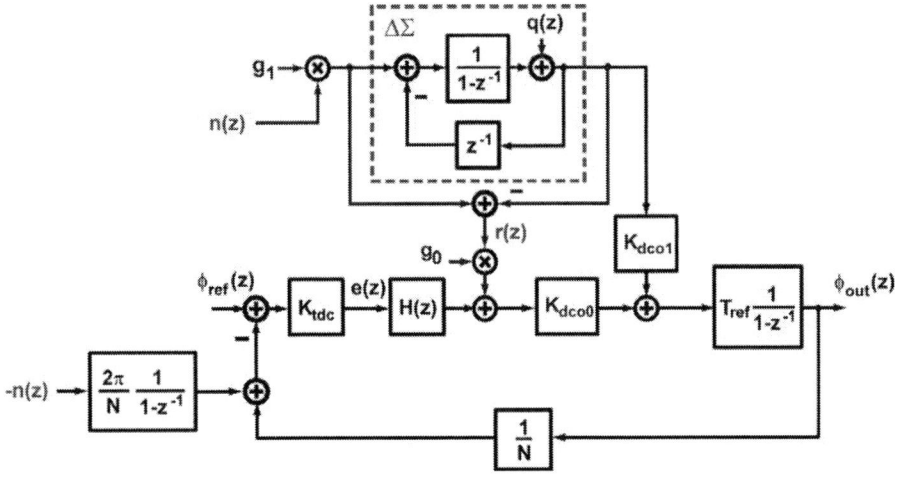

$$e(z) = n(z) \cdot \left[2\pi f_{ref} - g_1 K_{dco1} \right] \cdot F(z)$$
$$+ r(z) \cdot \left[K_{dco1} - g_0 K_{dco0} \right] \cdot F(z)$$

S. Levantino   © 2013 Politecnico di Milano

Direct-FM with Multi-Bank DCO

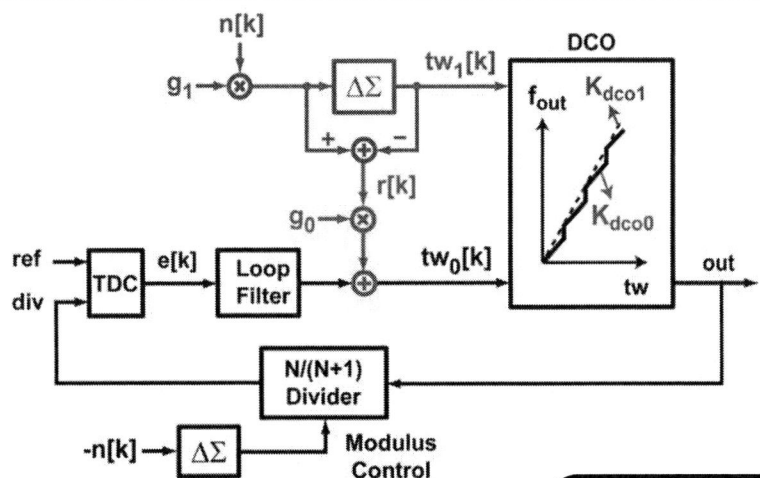

- **All-pass transfer function if:**

$$2\pi f_{ref} = g_1 K_{dco1}$$
$$K_{dco1} = g_0 K_{dco0}$$

S. Levantino   © 2013 Politecnico di Milano

Gains and Correlations

$$e(z) = n(z) \cdot \left[2\pi f_{ref} - g_1 K_{dco1} \right] \cdot F(z)$$
$$+ r(z) \cdot \left[K_{dco1} - g_0 K_{dco0} \right] \cdot F(z)$$

⬇

- **Gain g_1 must be regulated so that it nulls the correlation $c_n[k] = E\{\, e[k] \cdot n[k] \,\}$**
- **Gain g_0 must be regulated so that it nulls the correlation $c_r[k] = E\{\, e[k] \cdot r[k] \,\}$**

S. Levantino   © 2013 Politecnico di Milano

Gain Adaptation of Multi-Bank-DCO

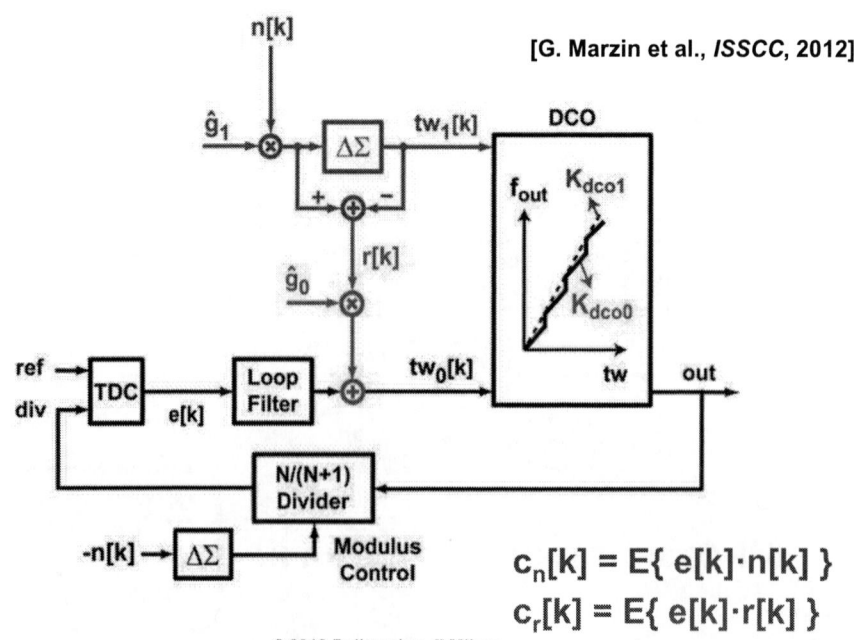

[G. Marzin et al., *ISSCC*, 2012]

$$c_n[k] = E\{\, e[k] \cdot n[k] \,\}$$
$$c_r[k] = E\{\, e[k] \cdot r[k] \,\}$$

S. Levantino   © 2013 Politecnico di Milano

Gain Adaptation of Multi-Bank-DCO (Cont'd)

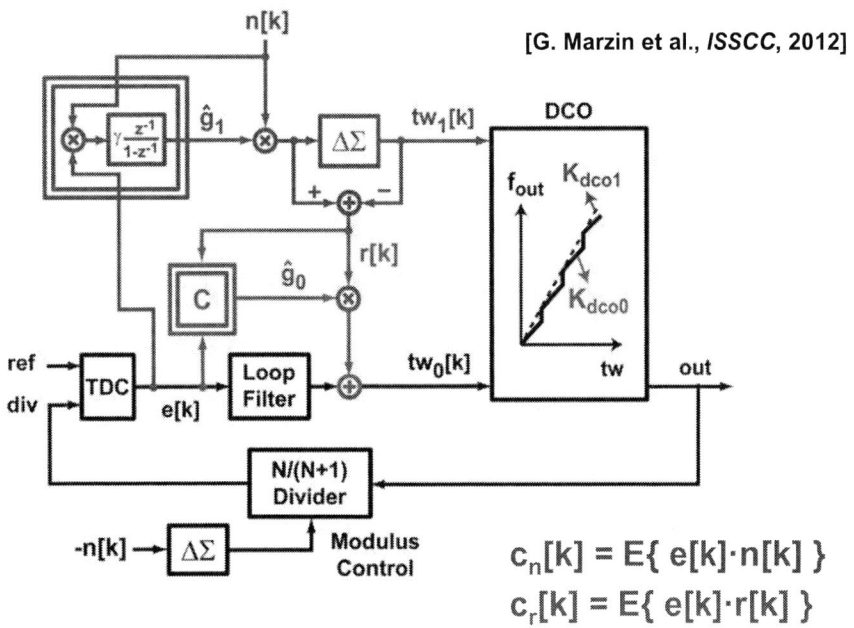

[G. Marzin et al., *ISSCC*, 2012]

$$c_n[k] = E\{ e[k]\cdot n[k] \}$$
$$c_r[k] = E\{ e[k]\cdot r[k] \}$$

Practical Implementation

Implemented Phase Modulator

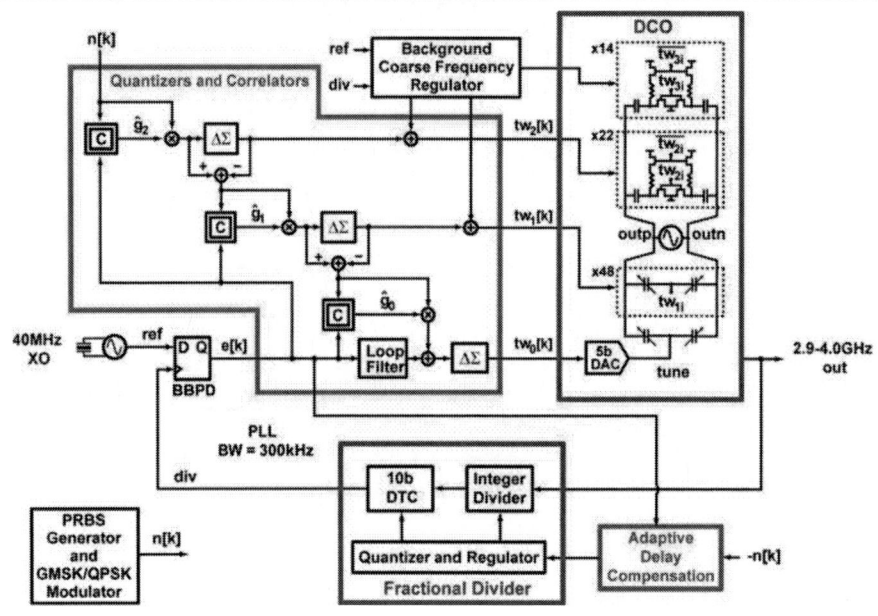

S. Levantino   © 2013 Politecnico di Milano

Practical DTC Schematic

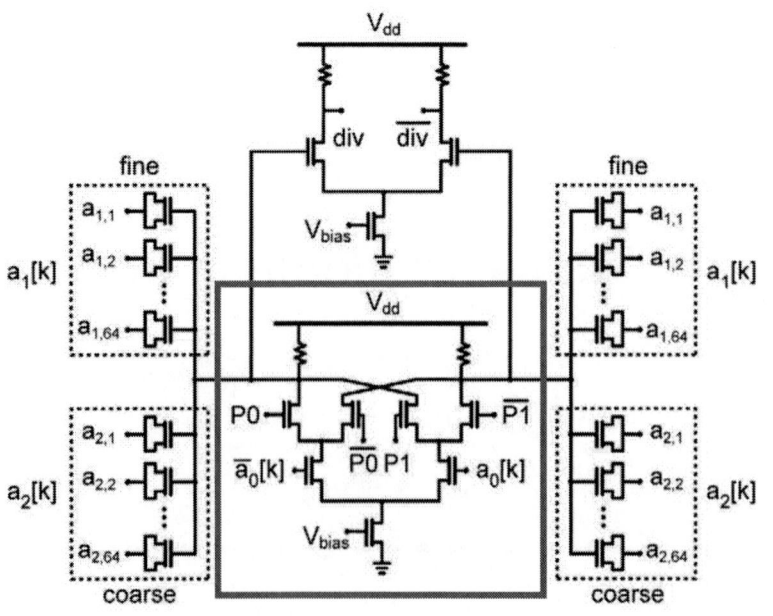

S. Levantino   © 2013 Politecnico di Milano

Practical Bang-Bang-Detector Schematic

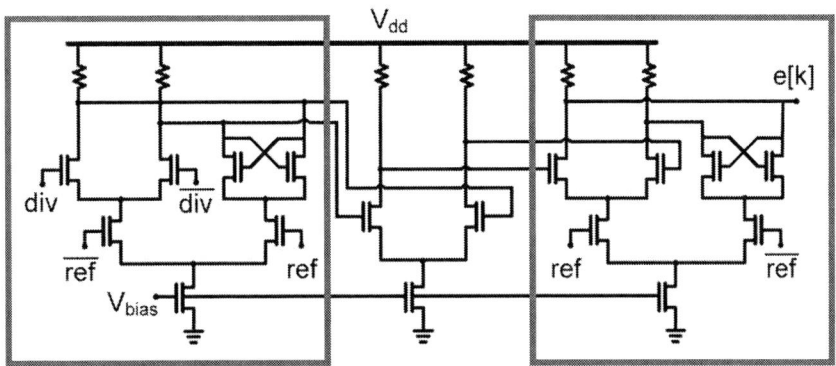

S. Levantino   © 2013 Politecnico di Milano

Die Photograph

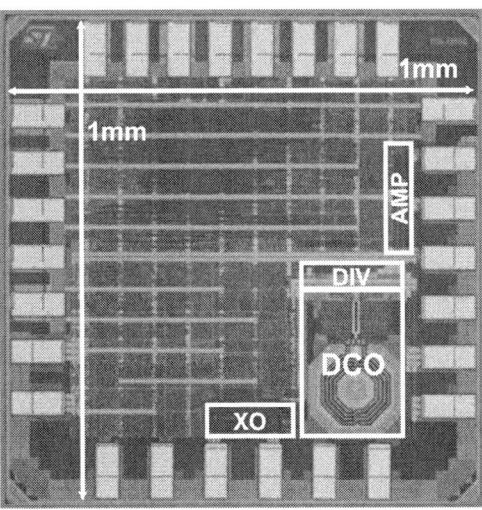

[G. Marzin et al., *ISSCC*, 2012]

- **65nm CMOS**
- **Active area: 0.52mm^2**
- **Chip area: 1.0mm^2**
- **Power: 5mW**
- **Supply voltage: 1.2V**

S. Levantino   © 2013 Politecnico di Milano

Phase Noise of 3.6-GHz Unmodulated Carrier

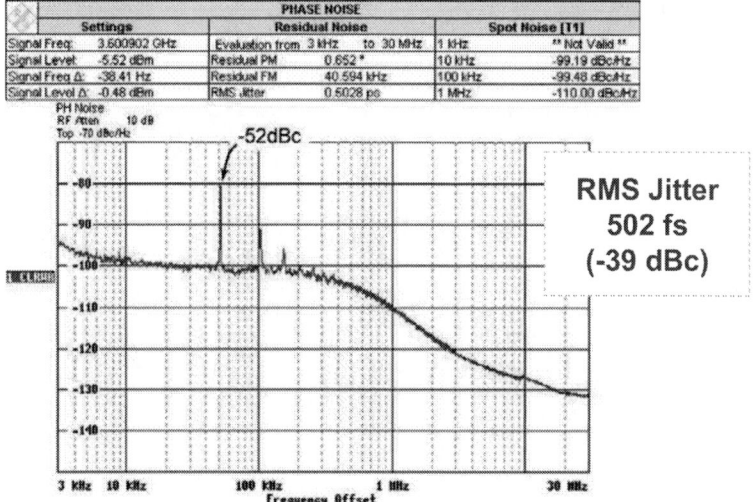

- **Near-integer channel (50 kHz offset)**

S. Levantino © 2013 Politecnico di Milano

LMS Coefficients and Frequency Deviation

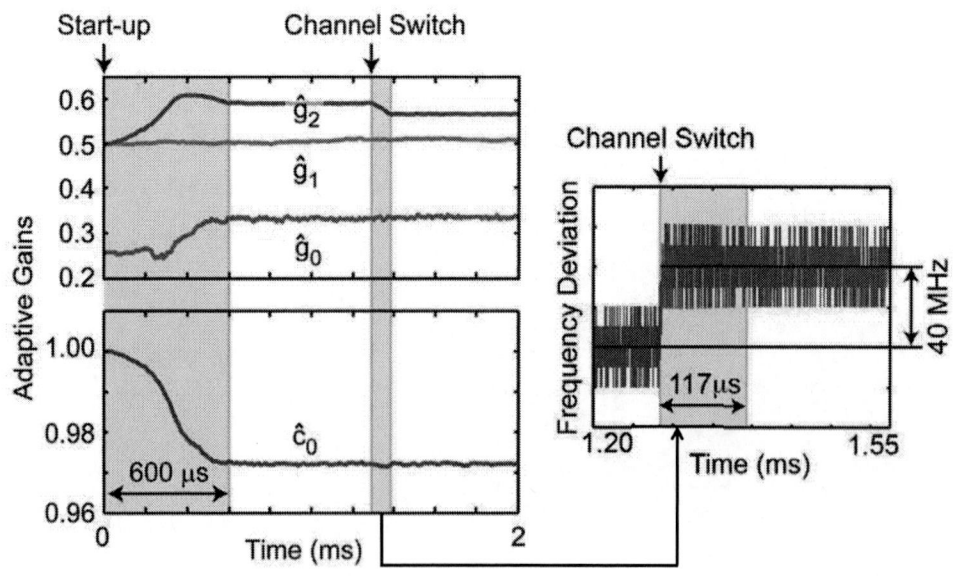

S. Levantino © 2013 Politecnico di Milano

10-Mb/s GMSK Modulation

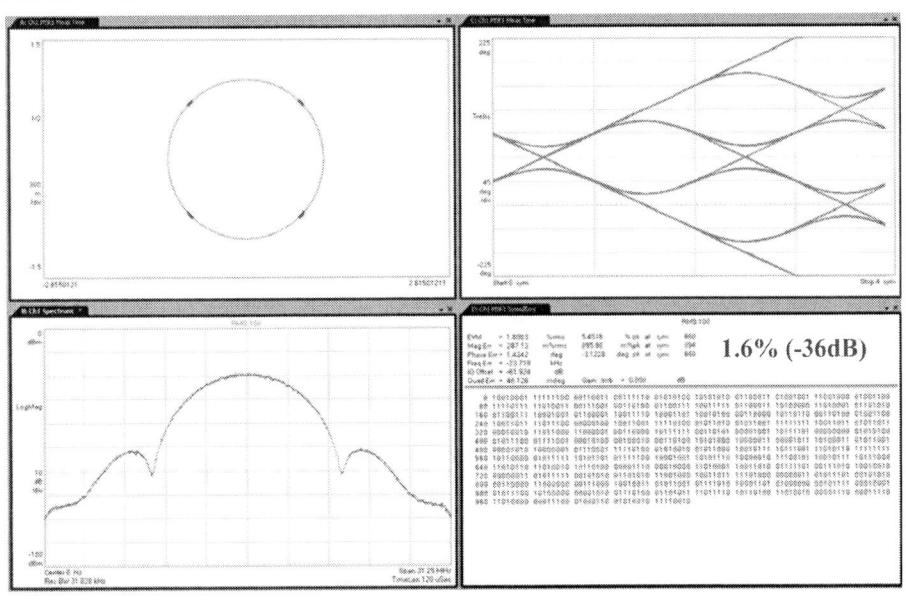

S. Levantino   © 2013 Politecnico di Milano

Measured EVM vs. Bit Rate

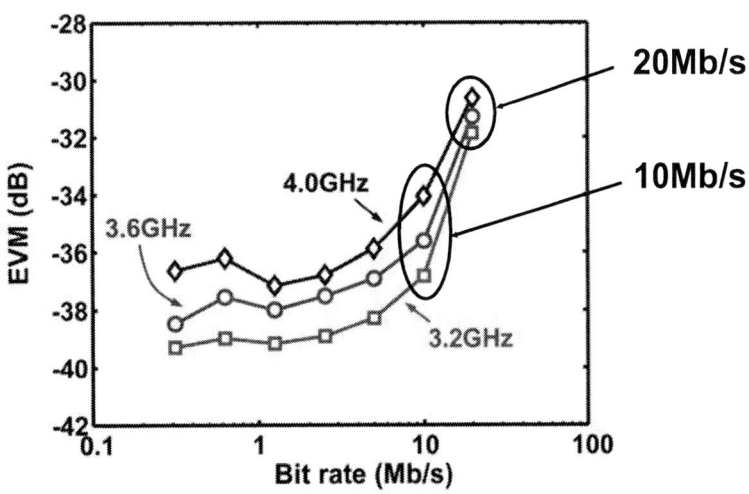

S. Levantino   © 2013 Politecnico di Milano

Out-of-band Emission

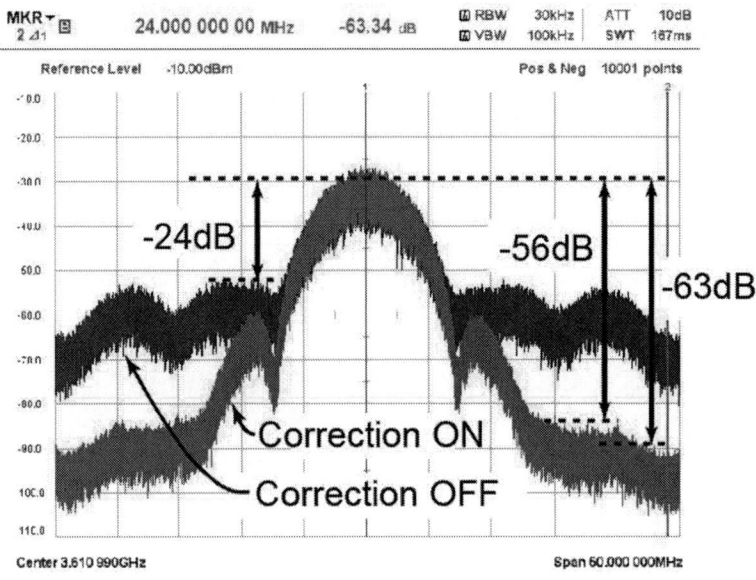

S. Levantino   © 2013 Politecnico di Milano

20-Mb/s QPSK Modulation

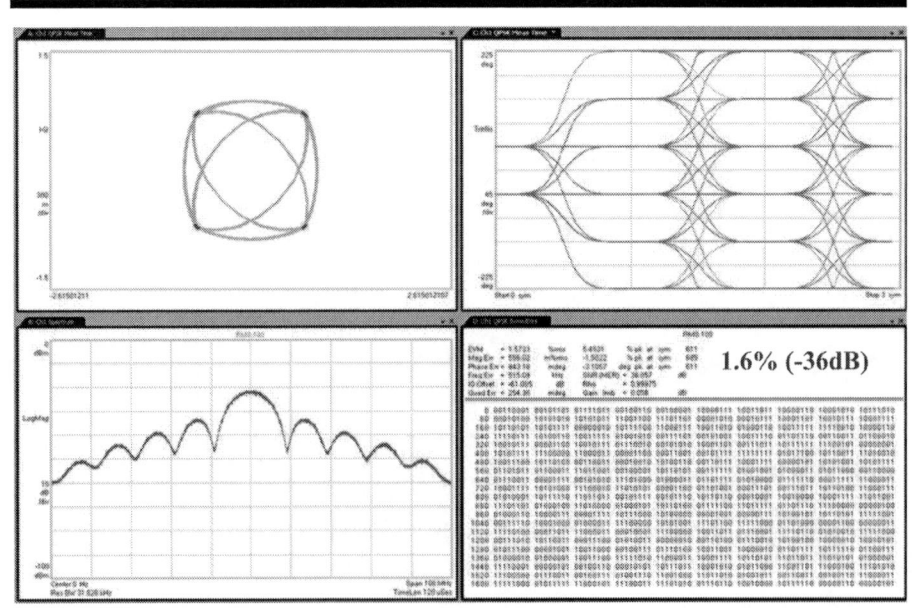

S. Levantino   © 2013 Politecnico di Milano

Measured EVM vs. Bit Rate

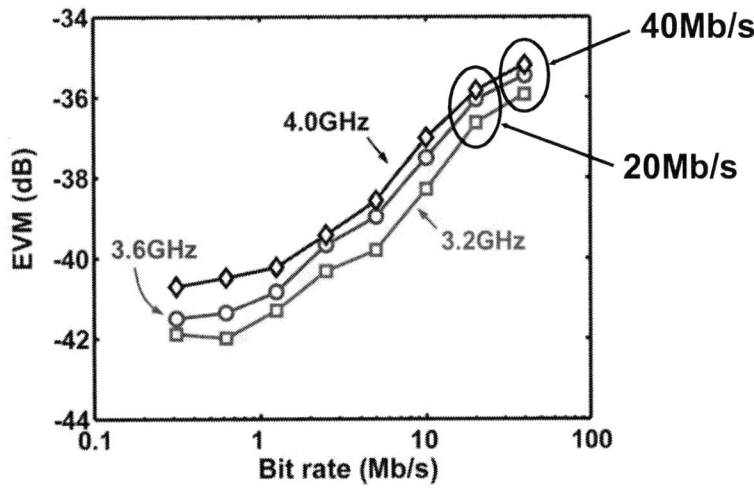

S. Levantino   © 2013 Politecnico di Milano

Performance Summary

Clock Frequency (MHz)	40	
Output (GHz)	2.9-4.0	
Tuning Range	31.9%	
Power Diss. (mW)	5.0	
PLL Bandwidth (kHz)	312	
Phase Noise (dBc)	-39	
In-band Fract. (dBc)	-52 (50 kHz)	
Reference Spur (dBc)	-76 (40 MHz)	
Start-up Settling Time (μs)	600	
Channel Switch Settling Time (μs)	117	
Modulation Type	GMSK	QPSK
Maximum Data Rate (Mb/s)	10	20
Energy/bit (nJ/bit)	0.5	0.25
EVM r.m.s. (dB)	-36	-36
Out-of-band Emission (dBr)	-56	-38
Area occup. (mm²)	0.5	
Process (nm)	65	

S. Levantino   © 2013 Politecnico di Milano

State-of-the-Art Comparison (GMSK)

	P. Su, JSSC11 [25]	H. Shanan, ISSCC09 [31]	This work
Architecture	Phase-Switching	Direct-FM	Direct-FM
Clock Frequency (MHz)	480	52	40
Output (GHz)	2.4	2.4	2.9-4.0
Power Diss. (mW)	62.1	16.0	5.0
Modulation Type	GFSK	GMSK	GMSK
Maximum Data Rate (Mb/s)	40	2	10
Energy/bit (nJ/bit)	1.55	8	0.5
EVM r.m.s. (dB)	-26	-18	-36(*)
Phase Noise (dBc)	N/A	N/A	-39
Out-of-band Emission (dBr)	-49	-57	-56
Area occup. (mm^2)	2.5	1.1	0.5
Process (nm)	180	180	65

(*) EVM equivalent to -39.5dB for a 2.4-GHz carrier

S. Levantino   © 2013 Politecnico di Milano

State-of-the-Art Comparison (QPSK)

	Y.-H. Liu, MTT09 [54]	S. Diao, MTT12 [55]	This work
Architecture	Phase-Switching	Phase-Switching	Direct-FM
Clock Frequency (MHz)	N/A	102	40
Output (GHz)	0.35-0.44	0.915	2.9-4.0
Power Diss. (mW)	3.0	3.5	5.0
Modulation Type	OQPSK	QPSK	QPSK
Maximum Data Rate (Mb/s)	17.5	50	20
Energy/bit (nJ/bit)	0.17	0.07	0.25
EVM r.m.s. (dB)	-23	-24	-36(*)
Phase Noise (dBc)	-45	N/A	-39
Out-of-band Emission (dBr)	N/A	N/A	-38
Area occup. (mm^2)	0.7	0.28	0.5
Process (nm)	180	180	65

(*) EVM equivalent to -55dB for a 0.4-GHz carrier and -44.5dB for 0.9-GHz carrier

S. Levantino   © 2013 Politecnico di Milano

Conclusions

- **Analog PLLs** do not scale down as process and are not amenable to noise-cancellation and other calibration algorithms
- **Digital PLLs** exploit CMOS scaling and allow for simple, accurate implementation of digiphase and two-point modulation
- Typically, **DPLLs require** TDCs with tight resolution **to achieve low phase noise and fractional-spur level, which increase both power consumption and design effort**

S. Levantino   © 2013 Politecnico di Milano

Conclusions (Continued)

- DPLLs with Bang-Bang Detectors (i.e. coarse midrise TDCs) **in combination with a** DTC **allows same phase-noise performance and fractional-spur level at much lower power consumption**
- **Fine resolution is only required to DCO and DTC, which can be both improved leveraging oversampling techniques**
- **Bang-Bang DPLLs achieve** superior noise/ power trade-off **over conventional DPLLs, while reducing design effort**

S. Levantino   © 2013 Politecnico di Milano

References (Integer-N)

- R. C. Walker, "Designing bang-bang PLLs for clock and data recovery in serial data transmission systems," in *Phase Locking in High-Performance Systems*, B. Razavi, Ed. Piscataway, NJ: IEEE Press, 2003.
- Y. Choi et al., "Jitter transfer analysis of tracked oversampling techniques for multigigabit clock-and-data recovery," *TCAS-II*, pp. 775–783, Nov. 2003.
- N. Da Dalt, "Linearized analysis of a digital Bang-Bang PLL and its validity limits applied to jitter transfer and jitter generation," *TCAS-I*, pp. 3663-3675, Nov. 2008.
- A. Rylyakov et al., "Bang-Bang digital PLLs at 11 and 20GHz with sub-200fs integrated jitter for high-speed serial Communication Applications," in *ISSCC* 2009, pp. 94–96.
- P. Madoglio et al., "Quantization Effects in All-Digital Phase-Locked Loops," *TCAS-II*, pp. 1120–1124, Dec. 2007.
- M. Zanuso et al., "Noise Analysis and Minimization in Bang-Bang Digital PLLs," *TCAS-II*, pp. 835-839, Nov. 2009.
- G. Marucci et al., "Analysis and Design of Low-Jitter Digital Bang-Bang Phase-Locked Loops," *TCAS-I*, in press.

References (Fractional-N)

- R. B. Staszewski, et al., "All-digital TX frequency synthesizer and discrete-time receiver for bluetooth radio in 130-nm CMOS," *JSSC*, pp. 2278–2291, Dec. 2004.
- C. Weltin-Wu et al., "Insights Into Wideband Fractional ADPLLs: Modeling and Calibration of Nonlinearity Induced Fractional Spurs," *TCAS-I*, pp. 2259-2268, Sep. 2010.
- C.-M. Hsu et al., "A low-noise wide-BW 3.6-GHz digital $\Delta\Sigma$ fractional-N frequency synthesizer with a noise-shaping time-to-digital converter and quantization noise cancellation," *ISSCC* 2008.
- M. Zanuso et al., "A Wideband 3.6GHz $\Delta\Sigma$ Digital Fractional-N PLL with Phase Interpolation Divider and Digital Spur Cancellation," ISSCC 2010.
- D. Tasca et al., "A 2.9-to-4.0GHz fractional-N digital PLL with Bang-Bang phase detector and 560fsrms integrated jitter at 4.5mW power," in *ISSCC* 2011, pp. 88-90.
- N. Pavlovic, J. Bergevoert, "A 5.3GHz Digital-to-Time-Converter-Based Fractional-N All-Digital PLL," *ISSCC 2011*, pp. 54-55.

References (Modulator)

- M. J. Underhill and R. I. H. Scott "Wideband frequency modulation of frequency synthesisers," *Electronics Letters*, vol. 15, no. 13, pp. 393–394, Jun. 1979.
- M. H. Perrott, M. D. Trott, and C. G. Sodini, "A modeling approach for Δ-Σ fractional-*N* frequency synthesizer allowing straightforward noise analysis," *JSSC*, pp. 1038–1038, Aug. 2002.
- G. Marzin, S. Levantino, C. Samori, and A. Lacaita, "A 20Mb/s phase modulator based on a 3.6GHz digital PLL with -36dB EVM at 5mW power," in *ISSCC* 2012, pp. 342-343.
- P. Su and S. Pamarti, "A 2.4GHz wideband open-loop GFSK transmitter with phase quantization noise cancellation," *JSSC*, pp. 615–626, Mar. 2011.
- H. Shanan, G. Retz, K. Mulvaney, and P. Quinlan, "A 2.4GHz 2Mb/s versatile PLL-based transmitter using digital pre-emphasis and auto calibration in 0.18μm CMOS for WPAN," in *ISSCC* 2009, pp. 420-421.
- Y.-H. Liu, C.-L. Li, and T.-H. Lin, "A 200-pJ/b MUX-based RF transmitter for implantable multichannel neural recording," *MTT*, pp. 2533-2541, Oct. 2009.
- S. Diao, Y. Zheng, Y. Gao, S.-J. Cheng, X. Yuan, M. Je, and C.-H. Heng, "A 50-Mb/s CMOS QPSK/O-QPSK transmitter employing injection locking for direct modulation," *MTT*, pp. 120-130, Jan. 2012.

All-Digital 90° Phase-Shift DLL with a Dithering Jitter Suppression Scheme

Dong-Hoon Jung[1], Kyungho Ryu[1], Jung-Hyun Park[1], Won Lee[2], and Seong-Ook Jung[1]

[1]School of Electrical and Electronic Engineering, Yonsei University, Seoul, Korea
[2]System LSI Division, Samsung Electronics Co., Ltd., Yongin-City, Gyenggi-Do, Korea

Abstract - We propose a 90° phase-shift digital delay-locked loop (DLL) with a new dithering jitter suppression scheme. Delay-line control code dithering is effectively suppressed by comparing the distribution of the input and the output clock jitter. The proposed scheme is analyzed through a stochastic calculation. A test chip is fabricated using a 45-nm CMOS technology, and a 1.95-ps rms and 12.89-ps peak-to-peak jitter are achieved at 800-MHz operating frequency with a 1.1-V supply voltage. The measured power consumption is 1.32 mW at 800 MHz, and the active chip area is 69.9 μm × 49.3 μm.

I. INTRODUCTION

In a double data rate memory interface, a 90° phase-shift DLL with duty cycle correction capability is required to generate a data strobe signal. Conventional 90° phase-shift DLLs [1], [2] divide their delay lines into four equivalent delay-line segments and used their four multi-phase outputs for 90° phase shift generation and duty cycle correction. However, the DLL presented in [1] has a delay-line mismatch problem among its delay-line segments and area overhead due to additional delay lines for duty cycle correction. The DLL presented in [2] proposed a delay-line calibration circuit to solve the mismatch problem but still has the area overhead problem owing to a calibration circuit.

With technology scaling, digital DLLs are becoming more preferable to analog DLLs owing to the increased process variation, supply voltage scaling, and low-power requirement [3]–[6]. Along with this trend, the code dithering of digital DLLs, which enlarges the output jitter and results in poor bit-error rate, is becoming an important issue. To solve the digital code dithering problem, phase detectors (PDs) adopting a hold region [3] have been proposed. However, the PD cannot consider the timing distribution of all input and output edges because delay-line control using a PD compares only one pair of input and output edges. Thus, the PD cannot adaptively respond to the various noise or input clock jitter. As a result, the effectiveness of dithering jitter suppression could be degraded.

In this paper, we propose a high-accuracy all-digital 90° phase-shift DLL with an effective dithering jitter suppression scheme. The proposed dithering jitter suppression scheme with its stochastic analysis and DLL structure are described in section II and III, respectively. Measurement results are discussed in section IV, and finally, we conclude with a brief summary of our proposal in section V.

II. PROPOSED DITHERING JITTER SUPPRESSION TECHNIQUE

The main concept of the proposed dithering jitter suppression scheme is to compare the timing distribution of the input and output clocks. To apply this concept to the proposed DLL, the loop controller shown in Fig. 1 is used to control the delay line. The operation of the proposed loop controller is described as follows.

First, the UP/DN counter counts the number of difference of outputs of PD, PD_{UP} and PD_{DN}, by increasing or decreasing the output of the UP/DN counter, UP/DN_{diff}. Meanwhile, the UP counter counts the total number of sampled PD outputs. After the $N_{sample}[n:0]$ number of PD outputs are sampled, $COMP2_{out}$ is enabled, and the comparator, COMP1, calculates whether the number of PD_{UP} or PD_{DN} is larger than the other by N_{diff}. If the number of PD_{UP} (PD_{DN}) is larger than PD_{DN} (PD_{UP}), then the delay line code is increased (decreased). Otherwise, the delay-line code is maintained. The value of N_{diff} is preset to 4, and the value of $N_{sample}[n:0]$ is preset to 20; however, $N_{sample}[n:0]$ is modified to the optimal value by the optimization feedback loop (OFL), which will be described later in this section.

By calculating the probability of locked state of the proposed and conventional schemes, the effectiveness of the proposed dithering jitter suppression scheme can be proven. The probability of locked state is simply calculated by analyzing the normal distribution of the timing difference between the input and output clocks, T_{diff}. The mean and standard deviation of T_{diff} (μ_{Tdiff}, σ_{Tdiff}) are given in (1) and (2), respectively.

$$\mu_{Tdiff} = \mu_{input} - \mu_{output} \tag{1}$$

$$\sigma_{Tdiff} = \sqrt{\sigma_{input}^2 + \sigma_{output}^2} \tag{2}$$

From (1) and (2), the probabilities for the PD outputs are expressed in (3)–(5).

$$P_{UP} = P(T_{diff} \geq \Delta hold) = 1 - \Phi\left(\frac{\Delta hold - \mu_{Tdiff}}{\sigma_{Tdiff}}\right) \tag{3}$$

$$P_{DN} = P(T_{diff} \leq -\Delta hold) = \Phi\left(\frac{-\Delta hold - \mu_{Tdiff}}{\sigma_{Tdiff}}\right) \tag{4}$$

$$P_{HO} = P(-\Delta hold \leq T_{diff} \leq \Delta hold)$$
$$= \Phi\left(\frac{\Delta hold - \mu_{Tdiff}}{\sigma_{Tdiff}}\right) - \Phi\left(\frac{-\Delta hold - \mu_{Tdiff}}{\sigma_{Tdiff}}\right) \tag{5}$$

In (3)–(5), P_{UP}, P_{DN}, P_{HO}, and $\Delta hold$ are the probabilities for the PD_{UP}, PD_{DN}, and $HOLD$ signals from the PD and the hold region of the PD, respectively. From (3)–(5), the probability of locked state of the proposed loop controller can be derived as (6).

$$P_{LOCK} = \sum_{i=0}^{i=\frac{Nsample}{2}} P_{UP}{}^i P_{DN}{}^i P_{HO}{}^{Nsample-2i} (N_{sample}C_i)[(N_{sample}-i)C_i]$$
$$+ \sum_{k=1}^{N_{diff}} \sum_{i=0}^{\frac{Nsample}{2}} P_{UP}{}^i P_{DN}{}^{i-k} P_{HO}{}^{Nsample-2i+k} \left(_{Nsample}C_i\right)\left[_{(Nsample-i)}C_{(i-k)}\right]$$
$$+ \sum_{k=1}^{N_{diff}} \sum_{i=0}^{\frac{Nsample}{2}} P_{UP}{}^{i-k} P_{DN}{}^i P_{HO}{}^{Nsample-2i+k} \left(_{Nsample}C_i\right)\left[_{(Nsample-i)}C_{(i-k)}\right] \tag{6}$$

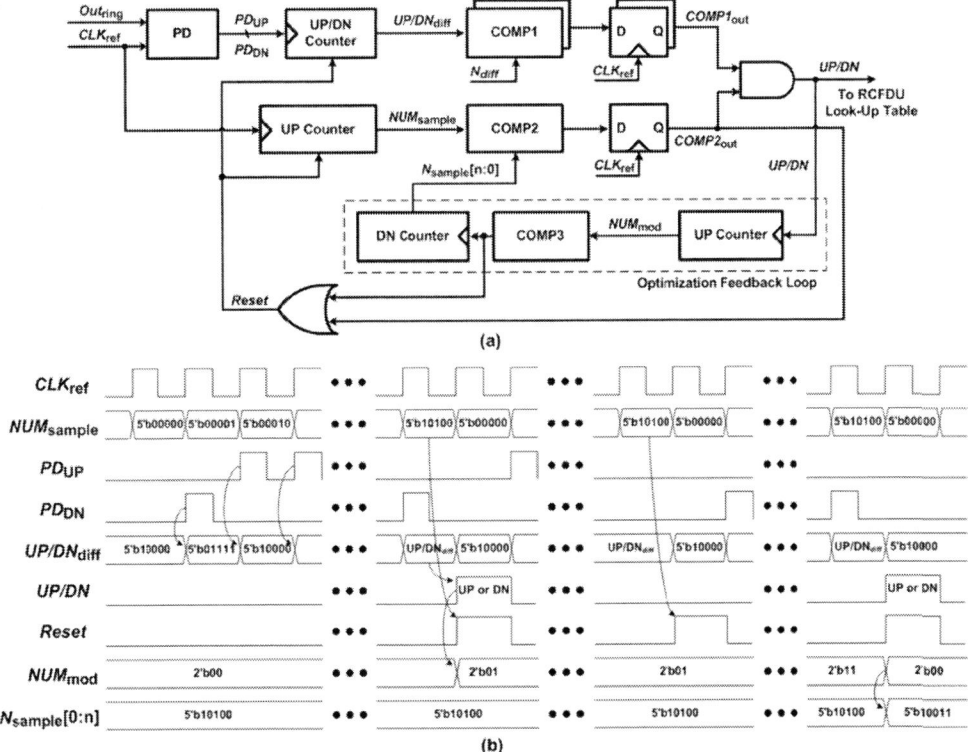

(a)

(b)

Fig.1. (a) Block diagram of the loop controller for dithering jitter suppression and (b) its timing diagram.

MATLAB simulation results for the probability of locked state of the proposed scheme based on (6) and the conventional scheme in [3] are shown in Fig. 2(a). In the simulation, a 4-ps PD hold region, N_{sample} = 11, and a 3-ps σ_{Tdiff} are assumed. Unlike the conventional dithering jitter suppression scheme, the proposed scheme maintains the probability of locked state high inside the lock range, which means that code dithering, is effectively suppressed. Furthermore, the probability of locked state decreases rapidly outside the lock range, which means that the proposed scheme returns to the lock range immediately when the DLL lock is released.

The variations in the noise and input clock jitter are modeled by different σ_{Tdiff}. Fig. 2(b) shows the simulation results of the proposed scheme for various values of σ_{Tdiff}. Fig. 2(b) shows that the optimal $N_{sample}[n:0]$ changes according to σ_{Tdiff}. In addition to the σ_{Tdiff} variation, the hold region of the phase detector also varies owing to the process variation. Because these variations affect the probabilities of the PD outputs in (3)–(5), an adjustment of $N_{sample}[n:0]$ is required to increase the effectiveness of the proposed dithering jitter suppression scheme for various operating conditions. Thus, we adopt an OFL to find the UP counter in the OFL operates to increase NUM_{mod}, which stores the number of delay-line code changes. After a certain amount of delay-line code changes, i.e., more than three times the number of changes in this structure, the OFL decreases $N_{sample}[n:0]$ to increase the probability of locked state and to further suppress code dithering. Through several repetitions of the feedback

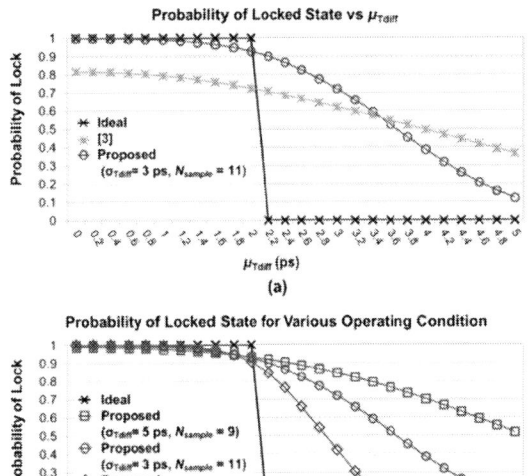

(a)

(b)

Fig. 2. Simulation results for the probability of locked state of: (a) the ideal case, [3], and the proposed scheme and (b) the proposed scheme for the various operating σ_{Tdiff}.

operation, the OFL finds the optimal $N_{sample}[n:0]$. By virtue of the OFL, the proposed dithering jitter suppression scheme is adoptable for various operation conditions.

978-1-4673-6145-3/13 $31.00 © 2013 IEEE   573

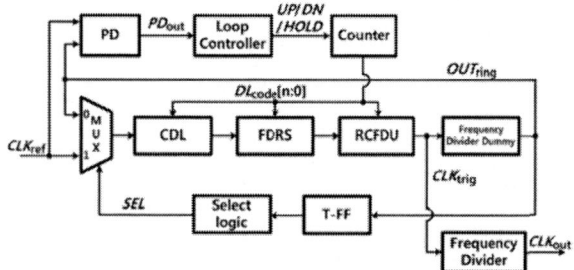

Fig. 3. Block diagram of the proposed 90° phase-shift all-digital DLL

III. PROPOSED DLL STRUCTURE

A. Overall Structure of the Proposed DLL

The overall block diagram of the proposed 90° phase-shift DLL is shown in Fig. 3. The proposed 90° phase-shift DLL is based on the multiplying DLL (MDLL) structure. Instead of dividing the delay line into four segments and using their four multi-phase outputs as in [1], [2], the proposed structure operates a ring oscillator at a frequency twice that of the reference input clock (CLK_{ref}). As a result, the output of the ring oscillator has transitions at the 90°, 180°, 270°, and 360° phases of CLK_{ref}, and a 90° phase-shifted clock is generated by a frequency divider using these edges. The ring oscillator consists of a coarse delay line (CDL), a fine delay range selector (FDRS), a resistance-controlled fine delay line (RCFDU), a frequency divider dummy, and a MUX. The FDRS and RCFDU are designed using the structures proposed in [2]. The proposed structure's delay-line length is required to generate only a quarter of the CLK_{ref} period, because the ring oscillator runs at a frequency twice that of CLK_{ref}. Thus, the delay-line length is reduced to a quarter of the conventional delay line.

Because the edges at 90°, 180°, 270°, and 360° phase timings in the proposed structure are generated by passing a clock through a single ring oscillator, the problem of mismatches between the delay-line segments in [1] does not exist. Thus, the additional calibration circuit required in [2] is also not necessary. In addition, the proposed structure significantly reduces the area required, as only one of the four delay-line segments in [1], [2] is used, and the duty cycle correction can simply be achieved by the frequency divider in Fig. 3, without any additional delay lines.

B. Symmetric Hold Region Phase Detector Design

Fig. 4 shows the PD, and Fig. 5 shows the frequency divider and its dummy. The proposed DLL has a tracking mode, in which DLL searches its control code for a lock, and a hold mode, in which the proposed dithering suppression scheme operates. When the proposed DLL is in the tracking mode, a cross-coupled latch in the proposed PD is formed, and a very small hold region is maintained at less than 2 ps. The small hold region prevents the DLL from a false lock. After the proposed DLL enters the hold mode, a latch connection is released, and the output nodes are charged by feedback through a weak keeper. As a result, the hold region increases to 4 ps–5 ps. In contrast to the PD using a TSPC in [3], the proposed PD has a symmetric hold region for the lead and lag

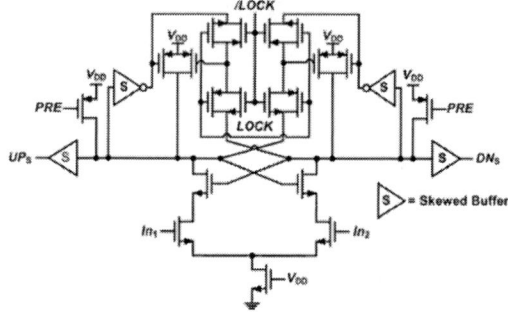

Fig. 4. Proposed symmetric window phase detector.

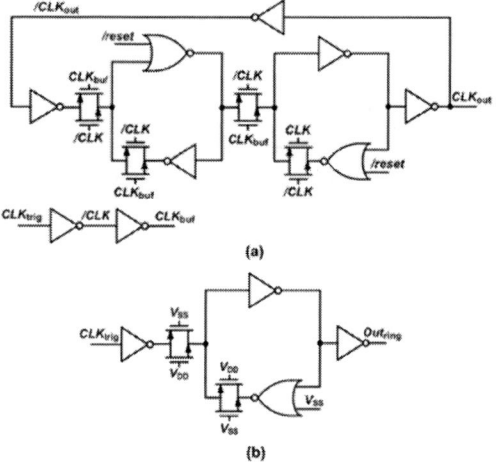

Fig. 5. (a) Frequency divider and (b) its dummy.

cases of input and output and the window size of the proposed PD can be easily controlled by modifying the strength of the feedback keepers. These characteristics result in a smaller offset between the input and output clock distributions, and the efficiency of the dithering jitter suppression scheme can be considerably increased.

C. Frequency Divider and its Dummy Design

A static master-slave flip-flop (MSFF) is used as a frequency divider, as shown in Fig. 5 with its dummy. The main design issue for the frequency divider and its dummy is to match their propagation delays. At the rising edge of the CLK_{trig}, $/CLK_{out}$ propagates through the frequency divider and generates a rising or falling edge of CLK_{out} depending on the value of $/CLK_{out}$. Thus, it is important to match the propagation delays of the frequency divider for both the *HIGH* case of $/CLK_{out}$ (T_{FD_HIGH}) and the *LOW* case of $/CLK_{out}$ (T_{FD_LOW}) to ensure the high duty cycle accuracy of CLK_{out}. On the other hand, the frequency divider dummy has to provide the same amount of delay with T_{FD_HIGH} and T_{FD_LOW} between CLK_{trig} and OUT_{ring} to ensure the output duty cycle and 90° phase-shift accuracy. In addition, it is also important to match the delays of the frequency divider dummy for both the rising edge of CLK_{trig} (T_{FDD_rising}) and the falling edge of CLK_{trig} ($T_{FDD\_falling}$), because the frequency divider dummy has to transfer both rising and falling edges of CLK_{trig}.

978-1-4673-6145-3/13 $31.00 © 2013 IEEE   574

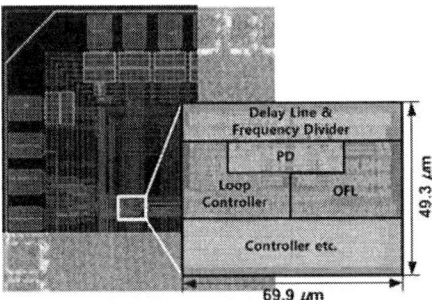

Fig. 6. Microphotograph and core layout of the test chip and performance summary table.

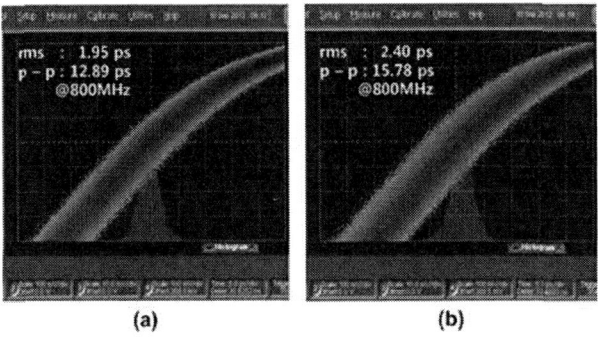

Fig. 7. Measured jitter histogram (a) with the dithering jitter suppression scheme on and (b) with the dithering jitter suppression scheme off.

TABLE I
Jitter enhancement using the proposed dithering jitter suppression scheme at various operating frequencies

	500 MHz	600 MHz	700 MHz	800 MHz
	rms/p-p (ps)	rms/p-p (ps)	rms/p-p (ps)	rms/p-p (ps)
Scheme ON	1.81/13.33	2.38/15.56	2.09/15.22	1.95/12.89
Scheme OFF	1.94/14.44	2.83/19.22	2.21/16.11	2.40/15.78

In summary, T_{FD_HIGH}, T_{FD_LOW}, T_{FDD_rising}, and $T_{FDD\_falling}$ have to be matched for high-accuracy duty cycle correction and phase shift. Even though dynamic FFs have faster propagation delay than the MSFF in Fig. 5(a), the MSFF has the same delay path for *HIGH* and *LOW* data input and thus offers good delay matching between these four delay parameters. Thus, an MSFF for the frequency divider is used instead of a TSPC to enhance the phase-shift and duty-cycle accuracy.

IV. MEASUREMENT RESULTS

A test chip is fabricated using a 45-nm CMOS process. A microphotograph of the test chip with its core layout is shown in Fig. 6. The core area of the proposed DLL is 69.9 μm × 49.3 μm.

Fig. 7 shows the measured jitter of the output clock and its histogram when turning the proposed dithering jitter suppression scheme on and off. When the proposed scheme is turned off, code dithering exists, and the value of the rms jitter increases. However, when the proposed scheme is turned on, the output jitter shows a single Gaussian distribution with a smaller rms and peak-to-peak jitter. Jitter enhancement at each operating frequency is summarized in Table. I.

TABLE II
Performance summary of the proposed structure and comparison with conventional schemes

	[1]	[2]	Proposed
Process	0.13 μm	45 nm	45 nm
Operating frequency	333 MHz – 800 MHz	400 MHz – 800 MHz	400 MHz – 800 MHz
Input duty cycle range	23% – 76% @ 800 MHz	25% – 75% @800 MHz	30 % – 70% @800 MHz
Output duty cycle error	47.8% – 49% @ 800 MHz	-	49.5% – 50.2% @ 800 MHz
Jitter(p-p/rms)	22.2 ps/2.95 ps	20.4 ps/2.48 ps	12.9 ps/1.95 ps
Supply voltage	1.2 V	1.1 V	1.1 V
Active area	0.06 mm²	0.01 mm²	0.0034 mm²
Power	19.2 mW	3.3 mW	1.32 mW

The performance of the proposed structure and comparison with conventional 90° phase-shift DLLs are summarized in Table II. Compared to conventional schemes, the output duty cycle error decreases to less than 0.5% by removing the effects of mismatch using the MDLL structure. In addition, the core area and power consumption decrease by more than 66% and 60%, respectively.

V. CONCLUSION

A 90° phase-shift all-digital DLL with a 50% output duty cycle, dithering jitter suppression technique, and reduced core area is proposed. By adopting an MDLL structure to generate multi-phase clocks for phase-shift and duty cycle correction, the proposed structure successfully removed the mismatch issue in conventional phase-shift DLLs and reduced area and power consumption. In addition, the proposed DLL effectively suppressed the dithering jitter by comparing the timing distribution of the input and output clocks, and its effectiveness is proven through stochastic analysis and calculations.

ACKNOWLEDGMENTS

This work was partly supported by Samsung Electronics. [High-Speed SRAM research & development project].

REFERENCES

[1] J.-H. Bae *et al.*, "An all-digital 90-degree phase-shift DLL with loop embedded DCC for 1.6Gbps DDR interface," in *CICC* 2007, pp. 373–376, Sept. 2007.

[2] Kang. H. C et al., "Process Variation Tolerant All-Digital 90° Phase Shift DLL for DDR3 Interface," in *IEEE Transactions on Circuits and Systems I* : Regular Papers, 2012.

[3] R.-J. Yang et al., "A 2.5 GHz all-digital delay-locked loop in 0.13μm CMOS technology," *IEEE J. Solid-State Circuits*, vol. 42, no.1, pp. 2338–2347, Nov. 2007.

[4] W.-J. Yun, H.-W et al., "A 3.57 Gb/s/pin low jitter all-digital DLL with dual DCC circuit for GDDR3 DRAM in 54-nm CMOS technology," *IEEE Trans. Very Large Scale Integr. Syst.*, vol. 19, no. 9, pp. 1718–1722, Sep. 2011

[5] D. Shin, W. J. et al., "A 0.17–1.4 GHz low-jitter all digital DLL with TDC-based DCC using pulse width detection scheme," in *Proc. Euro. Solid-State Circuits Conf.*, 2008, pp. 82–85.

[6] D. H. Jung, K, Ryu, J. H. Park, S. O. Jung, "A Low-Power and Small-Area All Digital Delay-Locked Loop with Closed-Loop Duty-Cycle Correction," *in Proc. Euro. Solid-State Circuits Conf.*, 2012, pp.181-184.

978-1-4673-6145-3/13 $31.00 © 2013 IEEE

A 1Gb/s Reconfigurable Pulse Compression Radar Signal Processor in 90nm CMOS

Jun Li[1], Hirohito Mukai[2], Mehmet Parlak[1], Michiaki Matsuo[2], and James F. Buckwalter[1], *Member*, *IEEE*,

[1]University of California-San Diego, La Jolla, CA, USA 92093

[2]Panasonic Corporation, Tokyo, Japan

Abstract-This paper presents a reconfigurable analog signal processing circuit for pulse compression radar. Adapting bandwidth for the range of the target is proposed for radar systems. The baseband signal processor includes a high-speed correlator/integrator, a 4-bit flash analog-to-digital converter (ADC) and a multi-range delay lock loop (DLL). The DLL generates multi-phase clock to align the template signal with the received signal. The circuit is fabricated in 90-nm CMOS and can be configured to work from 50Mb/s to 1Gb/s with Barker codes. A sidelobe reduction (SLR) of 15.6dB is demonstrated for 1Gb/s. The total power consumption is 33mW at 1Gb/s.

I. INTRODUCTION

Advances in silicon technology have made possible new sensor applications for consumer markets at millimeter-wave bands that require lower power and cost. As a result, silicon integrated beamforming architectures and phased arrays have been demonstrated for millimeter-wave automotive radar system [1], high-definition content streaming [2], and satellite systems [3]. However, millimeter-wave sensors for target range surveillance could offer low interference due to the spatial selectivity of the beamforming. In general, two main radar systems have been proposed: frequency modulated continuous wave (FMCW) radar [1] and pulse compression radar (PCR) [4-5]. FMCW radar requires two isolated antennas for high receiver sensitivity. For phased array applications, the transmitter and receiver require two separate arrays. Pulse compression radar can be implemented using digital signal modulation and time-division duplexing of the RF between transmit and receive.

In this paper, a baseband signal processing platform for PCR range sensing is proposed. As shown in Figure 1, the direction and range of a target is determined through a combination of the spatial selectivity of the beamformer and analog signal processing. To improve the angle resolution, the overall antenna array size should be increased to improve the antenna directivity. The range resolution of a pulse compression radar is determined by the bandwidth (*BW*) - and consequently - minimum symbol period (chip period) T_s that can be transmitted and received. The range resolution ΔR is

$$\Delta R = \frac{cT_s}{2} = \frac{c}{2BW} \qquad (1)$$

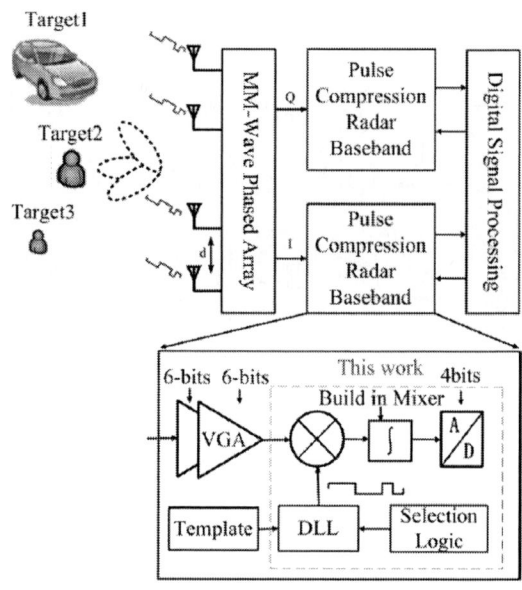

Figure 1. Proposed analog processing based system

where c is the speed of light. Bands available at 60 and 77 GHz offer sufficient bandwidth to allow centimeter-range resolution. For example, 500MHz BW achieves a range resolution of 30 cm from (1).

Wide bandwidth comes at the disadvantage of more integrated noise in the receiver, which reduces the SNR and range of detection. Consequently, a trade-off between range and range-resolution is evident. To improve the range resolution, the signal BW should be increased. To improve the range, the signal BW should be decreased. Therefore, a PCR sensor requires a dedicated baseband signal processing circuit to generate and detect the radar signal with adaptation of the receiver. As shown in Figure 1, the circuit includes a high-speed analog correlator, ADC, and DLL. In this paper, Section II discusses the circuit design of the PCR receiver and Section III presents measurement results.

II. Proposed Baseband Circuits for PCR System

Here, the PCR signal is based on the set of Barker codes with length *N*. A 7-b Barker code is illustrated in Figure 2. Barker codes are particularly well-suited for PCR because of their autocorrelation properties; a peak value of *N* occurs for zero lag and either 0 or 1 for other lags.

978-1-4673-6145-3/13 $31.00 © 2013 IEEE

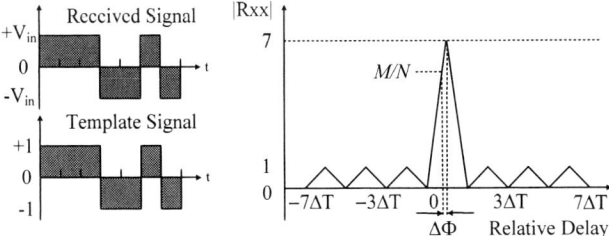

Figure 2. 7-bit barker code correlation

Figure 3 illustrates digital and analog correlation circuitry to detect the Barker code. A digital correlator approach samples the received signal first and performs autocorrelation in digital domain. Since the ADC must operate at the symbol rate, this approach requires a high-resolution, high-speed analog-to-digital converter (ADC) and consumes substantial power to produce the symbol rate correlation. Additionally, a high-speed digital multiplier is also required.

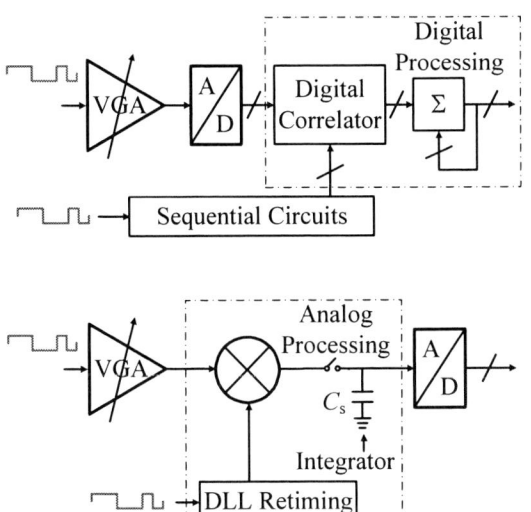

Figure 3. Block diagram of digital correlation (top) and analog correlation (bottom)

The analog correlation approach substantially lowers the power required to correlate the PCR signal. Since the signal correlation occurs in the analog domain, the sampling rate of the ADC is determined by the code rate, which is lower than the symbol rate by N. Analog correlation relies on accurate alignment of the template signal with the received signal. A delay lock loop (DLL) is proposed to provide an 8-phase clock for retiming the template for correlation. Since the power consumption of DLL is low, the analog correlation is preferable for a low power implementation.

Finally, the analog correlation technique limits the reconfiguration for different PCR codes and symbol rates. Our proposed approach can use low code rates for high ranges and increase the symbol rate for nearby targets to get the best range accuracy. This allows control of the range and range resolution of the PCR system. Varying the speed from

50 Mb/s to 1 Gb/s, the minimum range detection can be configured to be 3m and 15cm.

A. Analog Correlator

Figure 4 shows the analog correlator and integrator. Since Friis loss has been compensated by the VGA [5], the input signal swing to the correlator exhibit less dynamic range. Then, the template signal can switch the cascode transistors to perform correlation.

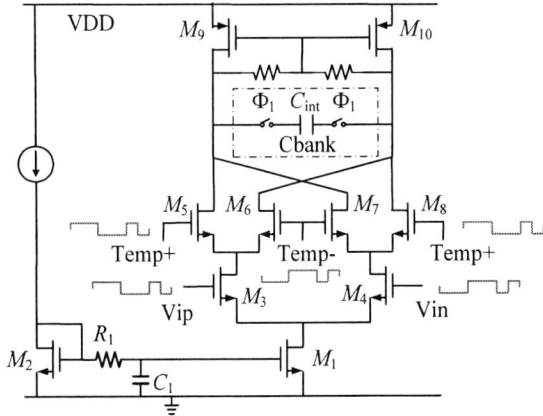

Figure 4. Proposed analog correlator

When the template and received signal have the same polarity, the output current charges the load capacitor C_{int}. When the template and received signal have reversed polarity, load capacitor is discharged. Consequently, if the Barker code template aligns correctly with the received signal, the voltage on the capacitor is charged to its maximum value and represents the desired autocorrelation of the signal. Figure 5 shows the simulation result for the analog correlator when two signals are aligned.

In the 'Reset' mode, the bias of template inputs are shifted to logic high and the output signal will settle to a common mode voltage. Apart from previous work [5], the correlator is controlled digitally to reduce power consumption and system complexity. One problem with the analog integrator occurs when the system is operated over a wide range of symbol rates. At low rates, the voltage that results from the integration increases rapidly because of the long period of the symbol. To compensate, a 2-b capacitor bank is designed for different template speed.

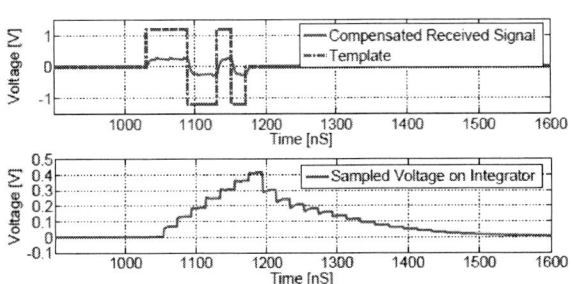

Figure 5. Sampled correlated output

978-1-4673-6145-3/13 $31.00 © 2013 IEEE   577

B. Wide Range Delay Lock Loop (DLL)

Since misalignment between received signal and template signal will degrade the analog correlation, retiming circuitry is needed to fine-tune the phase of template signal. This alignment needs to occur over the entire symbol rate of the system (50Mb/s to 1Gb/s) and a multi-range multiplying wideband DLL is proposed in this paper [6].

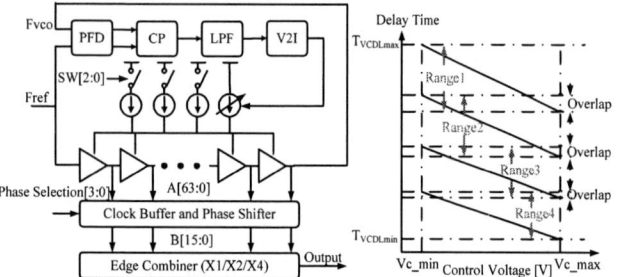

Figure 6. Proposed DLL (left) and tuning range (right).

The wide tuning range DLL schematic is shown in Figure 6. The DLL is based on current-starved delay cells for the voltage controlled delay line (VCDL). A 4-bit thermometer controlled current source controls the delay into 4 discrete values [7]. The minimum delay is controlled with a fine-tuning current source while the 3 other timing steps are digitally controlled to cause an overlapping tuning curve shown in Figure 6. For wide tuning, an edge combiner is provided to allow frequency multiplication by a factor of 2 (X2) and 4 (X4). To generate 8 phases, different combinations are chosen by the digital control bits. These signals retime the template until it aligns with the received signal. Consequently, the DLL can align the template signal with received signal to within 1/8 of the symbol period.

C. A 4-bit Flash Analog to Digital Convertor (ADC)

To sample the autocorrelation of a Barker code ($N \leq 13$), the ADC resolution should be at least 4 bits. A high-speed 4-b Flash ADC illustrated in Figure 7 is designed by utilizing average and interpolation techniques [8]. Interpolation reduces the number of pre-amplifiers from 15 to 6. The output of the integrator in the correlator implements a zero-order hold for the input of the ADC. To reduce glitches, a gray coding encoder between thermometer and binary code is inserted. The ADC can work up to 2GS/s and a level shifter and LVDS transmitter are included for interfacing to a FPGA.

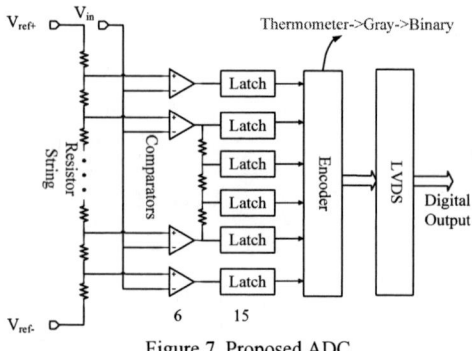

Figure 7. Proposed ADC

III. Measurement Results

The circuit is fabricated in a 90nm digital CMOS process. The chip microphotograph is shown in Figure 8 and has a measured area of 1.3mm². The circuit operates from a 1.2 V supply and consumes 33 mW including 22 mW for the ADC, 8.5 mW for the DLL, and 1.5 mW for the analog correlator.

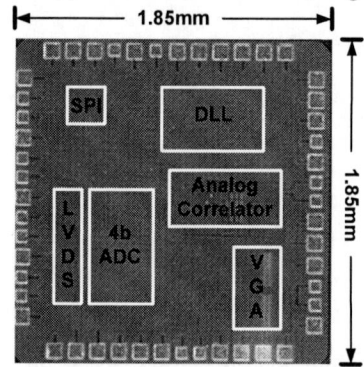

Figure 8. Die photo

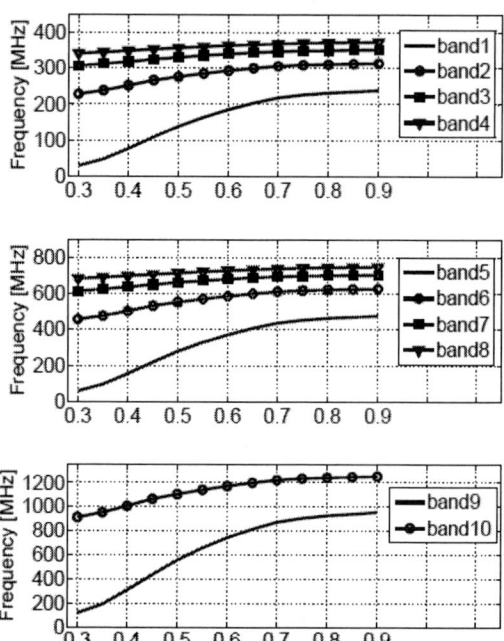

Figure 9. Measured tuning range of the wideband DLL

The DLL performance is demonstrated in Figure 9. In "X1" mode, the DLL output range is 50MHz to 380MHz. In "X2" or "X4" modes, the edge combiner multiplies the frequency to increase the frequency range from 250MHz to 1.2GHz multi-phase signal. The power consumption of the DLL is 8.5mW at 500MHz in the "X2" mode. The RMS and peak-to-peak jitter of the output clock is between 2ps and 13 ps over the entire frequency range. The low jitter does not produce a significant SNR degradation for sampling of the correlated output signals.

For transient testing of the PCR system, a Xilinx ML605 FPGA and Analog Device DAC (AD9738a-FMC-EBZ) board generates the Barker code template signal, a replica of the received signal, and control signals for the chip.

978-1-4673-6145-3/13 $31.00 © 2013 IEEE   578

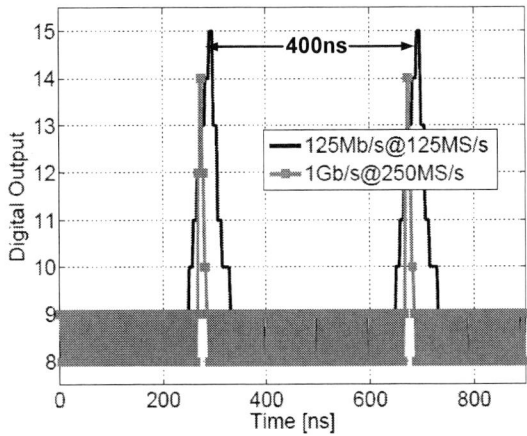

Figure 10. Peak value measurements of 7b barker code

Figure 10 plots the output of the ADC for different DLL phase shifts at different template rates. The code is repeated every 400 ns. The swing of the input signal is ~60 mV. Figure 11 superimposes the peak autocorrelation found for two symbol rates. In both cases, the general shape of the Barker code autocorrelation is evident. However, the peak SNR is degraded by one LSB at the peak measured data rate at 1 Gb/s and arises from memory effects in the correlator. The system SLR degradation is plotted in Figure 12 as a function of DLL phase offset. The SLR degradation is more severe for higher data rates and more tolerant to timing misalignment at lower data rates.

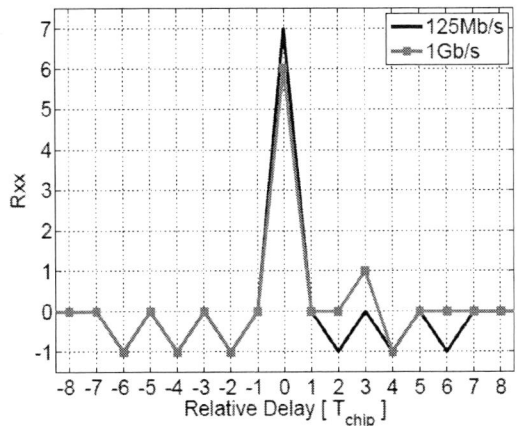

Figure 11. Measurements of 7b barker code

Table I Performance Comparison and Summary

	This Work	[1]	[4]	[5]
Process	90nm CMOS	65nm CMOS	180nm CMOS	90nm CMOS
Radar Type	PCR	FMCW	PCR	PCR
Bandwidth	50MHz ~ 1.2GHz	700MHz	~5MHz	1GHz
Template Code	2/3/5/7 Barker Code	NA	Chirp	2/3/5/7 Barker Code
Active Area	1.3mm²	1.045mm²	5.67mm²	
Power (mW)	COR : 1.5 DLL : 8.5 ADC : 22 Total : 33	Total:243	AWG : 30 COR : 22 ADC : 7.9 Total : 62.6	VGA : 16.8 COR : 4.9 Comp : 0.7 Total : 22.4

Table I compares the performance with previous works. This work demonstrates the lowest power consumption for a PCR signal processing circuit.

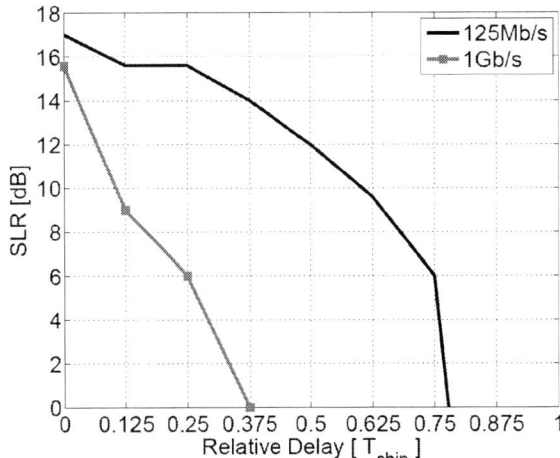

Figure 12. SLR degradation of 7b barker code

Conclusions

This paper presents an analog correlator architecture for the detection of pulse compression radar signals. The signal processing circuit includes an analog correlator, ADC, and DLL and is implemented in 90 nm CMOS. It consumes 33mW at a peak data rate of 1 Gb/s for a 15cm range resolution.

ACKNOWLEDGMENT

The author would like to thank Yan Li, Hao Wang and Jiang Long for the PCB assembling and trouble shooting.

REFERENCES

[1] J. Lee et al., "A Fully-Integrated 77-GHz FMCW Radar Transceiver in 65-nm CMOS Technology," *IEEE JSSC*, vol. 45, pp. 2746-2756, 2010.

[2] J. Kuo et al., "60-GHz four-element phased-array transmit/receive system-in-package using phase compensation techniques in 65-nm flip-chip CMOS process," *IEEE Trans. Microw. Theory Techn.*, vol.60, no. 3, pp. 743–756, Mar. 2012.

[3] S. Y. Kim et al., "A low-power BiCMOS 4-element phased array receiver for 76–84 GHz radars and communication systems," *IEEE J. Solid-State Circuits*, vol. 47, no. 2, pp. 359–367, Feb. 2012.

[4] S. Lee et al., "A CMOS Integrated Analog Pulse Compressor for MIMO Radar Applications," *IEEE Trans. Microw. Theory Techn.*, vol. 58, no. 4, pp. 747-756, Jun. 2010.

[5] M. Parlak, et al.,"Analog Signal Processing for Pulse Compression Radar in 90-nm CMOS," *Microwave Theory and Techniques, IEEE Transactions on*, vol.PP, no.99, pp.1-13. Dec. 2012

[6] Fang Ren Liao et al., "A Programmable Edge-Combining DLL With a Current-Splitting Charge Pump for Spur Suppression," *IEEE Trans. Circuits Syst. II*, vol.57, pp. 946-950, Dec. 2010

[7] C. Kuo et al., "A Multi-Band Delay-Locked Loop with Fast-Locked and Jitter-Bounded Features," *IEEE Asian. Solid-State Circuits Conference*, pp. 441-444, 2008

[8] K. Deguchi et al., "A 6-Bit 3.5GS/s 0.9V 98mW Flash ADC in 90-nm CMOS," *IEEE J. Solid-State Circuits*, vol. 43, no. 10, pp. 2303-2310, 2008.

A 5GS/s 4-bit Time-Based Single-Channel CMOS ADC for Radio Astronomy

Andrew R. Macpherson, James W. Haslett and Leonid Belostotski

RFIC Research Group, Department of Electrical and Computer Engineering

University of Calgary

Calgary, Alberta, Canada

Abstract—A 4-bit 65nm time-based analog-to-digital converter (ADC) targeting the next-generation Square Kilometre Array (SKA) is presented. This ADC is composed of an analog voltage-to-time converter (VTC) front end and a digital time-to-digital converter (TDC) back end. The two components can be physically separated to minimize the impact of digital noise from the ADC on high-gain, high-sensitivity receiver chains common in radio telescopes. At a sampling rate of 5 GS/s the ADC consumes 35 mW from a 1 V supply. After calibration, the ADC achieves a peak SNDR of 22.9 dB, SFDR of 34.0 dB and ENOB of 3.5. At the ERBW of 2100 MHz, SNDR is 18.4 dB, SFDR is 22.3 dB and ENOB is 2.8. The resulting worst-case figure of merit is 1.0 pJ/conversion. This is the highest reported sampling rate for a time-based ADC to date.

I. INTRODUCTION

Time-based processing has received increasing attention in recent years. By encoding signals in pulse delay times rather than voltages, designers can take advantage of the ever-increasing switching speed of CMOS technology, while avoiding the issue of decreased signal-to-noise ratio (SNR) due to supply voltage reduction [1]. One application for time-based processing is high-speed, low-resolution analog-to-digital converters (ADCs) [2,3].

Consider the time-based ADC architecture is shown in Fig. 1. The front-end is a voltage-to-time converter (VTC) that takes in a differential analog voltage signal and a clock. The VTC output consists of two pulse trains. On each clock edge, the VTC samples the input and produces a differential delay between the pulses on its two outputs that is proportional to the input voltage, as shown in Fig. 2. This timing diagram shows three different samples being taken from the differential input voltage, each resulting in a different relative delay Δt between the pulses on the two output channels. The back-end of the ADC includes a time-to-digital converter (TDC) that measures the delay between the two VTC outputs and a thermometer encoder to produce binary outputs. A serial-to-parallel converter (S2P) and a block of digital-to-analog converters (DACs) are also included for calibration purposes, as will be discussed in Section IV.

The Square Kilometre Array (SKA) [4] is a radio telescope project consisting of potentially millions of antennas, composed of phased array feeds and aperture arrays. High-speed ADCs are required for the direct-sampling receivers on each antenna. A single phased array feed antenna may contain hundreds of antenna elements whose signals must be independently amplified and digitized. However, placing an ADC near the antenna is problematic due to digital switching noise interfering with the ultra-low noise, high gain circuits in the analog signal chain. An advantage of the time-based ADC architecture in this work is that the VTCs can be physically separate from the TDCs, with only the two-channel connection

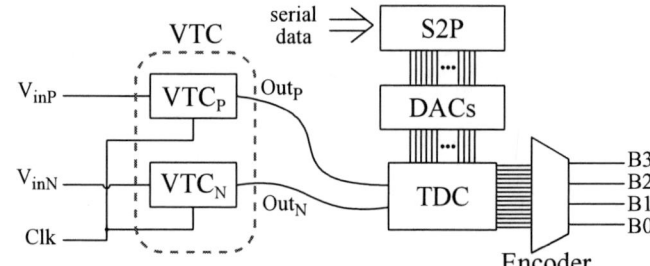

Fig. 1: Time-based ADC architecture.

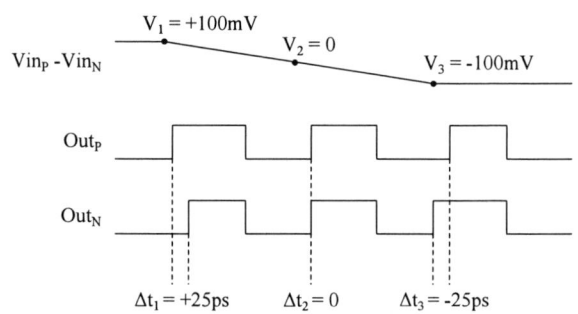

Fig. 2: Timing diagram for VTC operation.

between them. The small, low-power VTCs can be integrated with the analog receiver chains in the antenna feed while the larger TDCs, which produce the bulk of the digital noise, can be located safely away from the feed. In addition, the VTC is designed with a small analog input voltage range, decreasing the gain required in the receiver amplification stages. Finally, the single-channel ADC design being presented is an advantage for radio telescopes because time-interleaving can create frequency spurs that are indistinguishable from astronomical phenomena, leading to uncertainty in observations.

II. VOLTAGE-TO-TIME CONVERTER

The VTC schematic is shown in Fig. 3. This circuit converts an input voltage signal into a pulse delay. In doing so, it also samples the input signal, making an additional sample-and-hold circuit unnecessary. This structure is duplicated to form the positive and negative paths of a pseudo-differential system as shown in Fig. 1.

Referring to Fig. 3, transistors M1-M4 form an inverter with a current-starved ground connection. This block serves to pass falling clock edges with minimal delay, while passing rising clock edges with a delay proportional to the input voltage. The time-varying input is applied to the gate of transistor M3, along with a DC bias voltage. Transistor M4, connected in parallel with M3, is biased with the DC tuning

978-1-4673-6145-3/13 $31.00 © 2013 IEEE

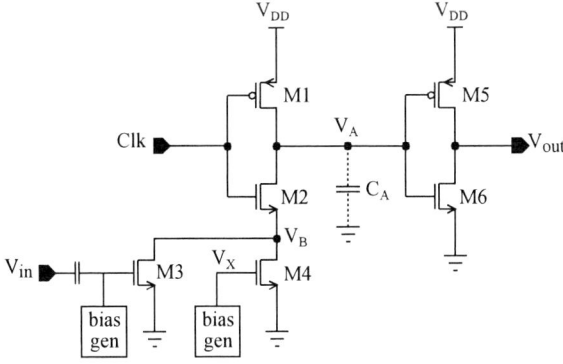

Fig. 3: VTC schematic

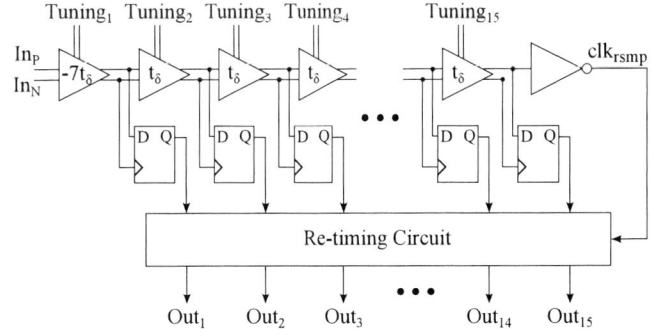

Fig. 5: TDC schematic

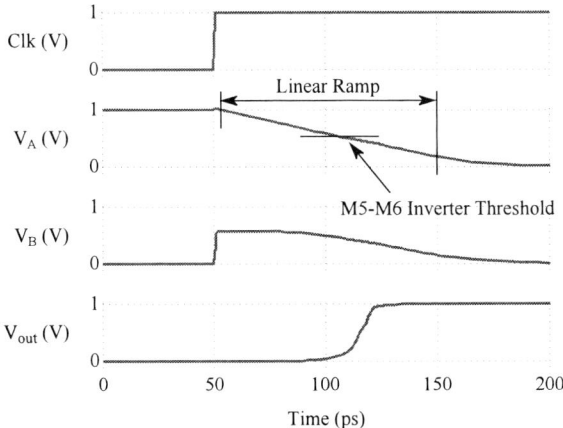

Fig. 4: Simulated waveforms for a single VTC conversion.

voltage V_X. Simulated waveforms for a single VTC clock cycle are shown in Fig. 4. When Clk rises, M3 and M4 are designed to quickly enter saturation to limit the current flow through M2. This results in a voltage ramp on V_A as a constant current is drained from the parasitic capacitance C_A. The current, and thus the slope of the ramp, varies with V_{in}. The second inverter formed by M5-M6 triggers when the ramp reaches its threshold, producing a rising edge on V_{out}, which is then further sharpened with additional inverters.

The VTC is linearized in two ways. First, the pseudo-differential output helps to eliminate even harmonics from the time-based output signal. The second factor involves the bias voltage V_X, which can be tuned to bias the VTC at peak linearity. The gain of the VTC is affected by the input bias voltage V_B as well as V_X, so an iterative design technique is used to select these voltages for peak VTC linearity at the desired gain based on simulations.

III. TIME-TO-DIGITAL CONVERTER

The TDC measures the delay produced by the differential VTC block and converts it to a binary output. It uses a Vernier delay line architecture, as shown in Fig. 5. The designed time resolution of the TDC, t_δ, is 3.1 ps, i.e. the delay span (50ps) divided by 2^N, where N=4 bits. The TDC consists of a chain of 15 differential time delay stages, with the outputs of each stage tapped and sent to a high-speed flip-flop. The flip-flops are designed to exhibit very low metastability using a sense-amplifier architecture based on [5].

The purpose of the delay chain is to produce 15 different delay signals, each one being t_δ apart. At some point along the line, the differential delay of the TDC input signals will change from positive to negative. In other words, a rising edge will occur on the upper (positive) path prior to a rising edge occuring on the lower (negative) path. The location of this transition determines the digital output and is detected by the flip-flops. Each flip-flop acts as a phase detector, producing an output of '1' if the input from the positive path occurs first and an output of '0' otherwise. The end result is a 15-bit thermometer code.

Since the signal takes time to propagate through the delay chain, there is a delay between each flip-flop decision. Over the chain of 15 delay blocks, the total delay can be greater than 200 ps depending on the process corner. In order to synchronize the flip-flop outputs, a re-timing circuit is used. This circuit delays each flip-flop output depending on its position in the chain. For instance, the fifteenth flip-flop output is delayed only minimally, the fourteenth flip-flop output goes through a single delay unit, the thirteenth flip-flop output goes through two delay units, and so on. Each delay unit is designed to match the absolute delay through one delay stage in the TDC chain. Delaying the flip-flop outputs in this manner provides coarse alignment for the outputs, with small variations remaining due to non-idealities such as device mismatches. The final step is to re-clock the outputs using a second set of flip-flops within the re-timing circuit. The re-timing clock (clk_{rsmp}) is generated from the end of the Vernier delay line to ensure that ample setup and hold times are maintained regardless of process corner. A thermometer-to-binary encoder is then used on-chip to produce a 4-bit binary output.

The time delay stages that make up the main TDC delay chain use starved inverters to delay both the rising and falling edges of the signal, thereby maintaining a constant pulse width throughout the chain. The delay of each stage can be adjusted for tuning purposes by changing the bias voltages of the current-starving transistors. The bias voltages are supplied by DACs that are individually programmable using an on-chip serial-to-parallel interface. Tuning the delays allows PVT variations within the delays themselves to be corrected. In addition, the delays can be tuned to compensate for non-linearities, gain errors and offset errors in the VTC output. This is accomplished by adjusting the position of each of the 15 threshold levels in the converter. By way of analogy, it would be equivalent to adjusting each reference voltage from the resistive ladder of a standard flash ADC.

978-1-4673-6145-3/13 $31.00 © 2013 IEEE   581

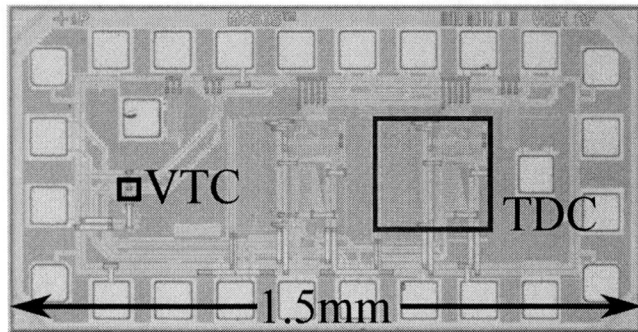

Fig. 6: Die micrograph

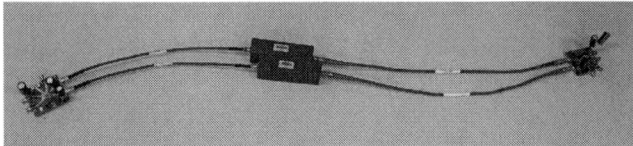

Fig. 7: Measurement setup: VTC and TDC boards connected with coaxial cables. Inverting transformers (Picosecond model 5100) provide a signal inversion required by the TDC due to a design oversight.

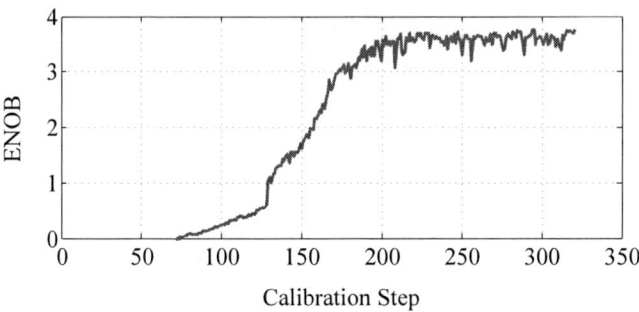

Fig. 8: Measured ADC calibration result.

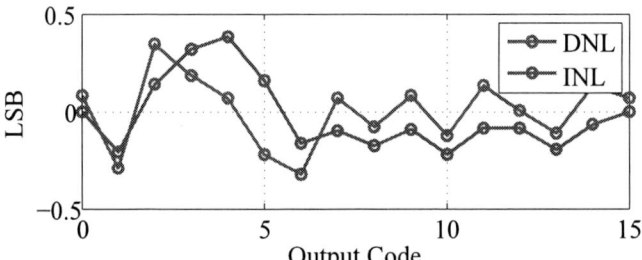

Fig. 9: Measured INL and DNL with DC inputs.

IV. ADC CALIBRATION

The VTC is designed to produce a differential delay of ±25 ps with a differential input of ±100 mV. Simulations show that the delay can vary from ±20 to ±33 ps over process corners. In the event that the actual delay range is different due to PVT variations, the TDC delays are adjusted to compensate during the calibration process.

An algorithm to automatically tune all TDC delays for optimum linearity, gain and offset was developed. Since the main focus of this design was proving the viability of operating the VTC and TDC separately at 5GS/s over a distance and maximum flexibility during testing was desired, full calibration functionality was not implemented on-chip. Instead the algorithm was implemented in MATLAB on a PC attached to the TDC chip, communicating over an on-chip serial interface. During calibration a sinusoidal signal is applied to the VTC input. In each step of the tuning algorithm, a histogram of output values is recorded and compared to the ideal histogram for a digitally sampled sinusoid. Each delay is then either incremented, decremented or left unchanged, depending on whether the corresponding histogram value is too low, too high or within the acceptable range. A new histogram is then recorded, and the cycle is repeated. In this way each delay gets closer to the optimum value with each step until the calibration process is finished.

V. MEASURED DATA

A die micrograph of the fabricated 65nm CMOS ADC containing both a VTC and TDC is shown in Fig. 6. The area of the VTC core is 40x20 μm and the area of the TDC, including the serial-to-parallel converter, DACs and decoder, is 280x270 μm. The ADC was tested with the VTC and TDC each running on their own separate board connected with 1m of coaxial cable, as shown in Fig. 7. In the photo, the boxes in between the VTC and TDC boards are passive inverting transformers

(Picosecond model 5100) needed to correct an inadvertent inversion in the TDC input buffers. The ADC remains functional in spite of a 1.5 dB insertion loss in the transformers as well as cable losses. The output was measured using a 40 GS/s 4-channel real-time oscilloscope. The VTC bias voltages were supplied from off-chip for maximum flexibility. Running at 5 GS/s, the VTC draws 4.1 mA of current and the TDC draws 30.5 mA from a 1 V supply, not including output buffers. The calibration algorithm was used to tune the TDC delays, starting from minimum delays. Fig. 8 shows the performance of the algorithm with a 500 MHz sinusoidal input signal. The calibration only needs to be performed at a single frequency.

With DC inputs, the ADC achieves a maximum DNL and INL of 0.38 and 0.34 LSB respectively, as shown in Fig. 9. For wideband inputs, the measured ERBW is 2100MHz, defined as the frequency range over which SNDR is within 3dB of the low frequency value. At the ERBW frequency, the ADC achieves an SFDR of 22.3 dB, an SNDR of 18.4 dB and an ENOB of 2.8. Maximum DNL and INL are 0.91 and 0.95 LSB respectively. Peak ADC performance occurs with an input of 400 MHz. At this frequency, SNDR is 22.9 dB, SFDR is 34.0 dB, and ENOB is 3.5. Fig. 10 shows an FFT plot of the output spectrum with input frequencies of 400 MHz (peak SNDR) and 2100 MHz (ERBW). The existence of a prominent 2nd harmonic in the output spectrum for the 2100 MHz input but not the 400 MHz input indicates a frequency-dependent mismatch in the pseudo-differential VTCs or the VTC-TDC connections, since even harmonics will ideally be cancelled out by differential operation. Fig. 11 shows the SNDR plotted against frequency.

The Walden figure of merit (FOM) is calculated using the standard formula [6]:

$$FOM = \frac{P}{2^{ENOB} f_s}$$

where P is the power consumption and f_s is the sampling

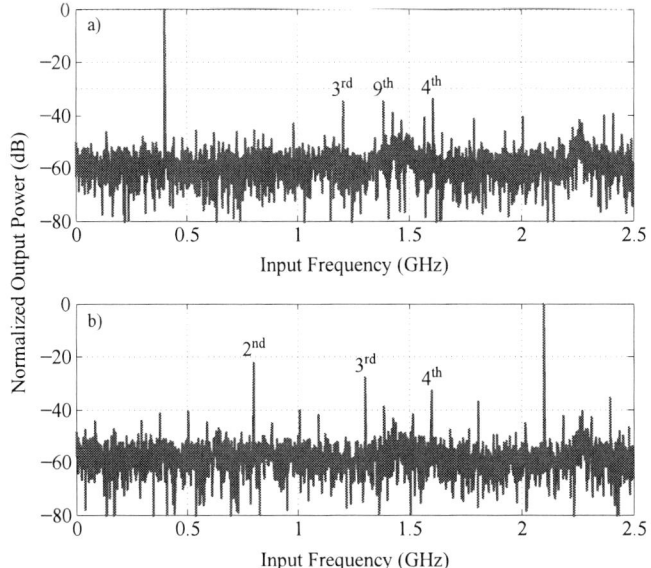

Fig. 10: Measured 5 GS/s FFT plot with significant harmonics indicated, for a) 400 MHz input (peak SNDR) and b) 2100 MHz input (ERBW).

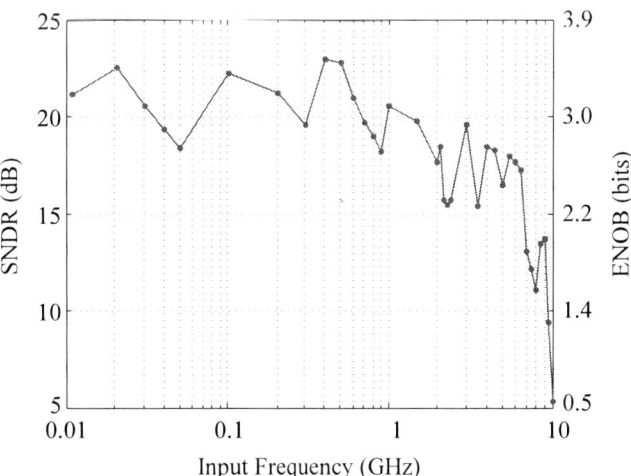

Fig. 11: Measured ADC wideband SNDR and ENOB.

TABLE I: PERFORMANCE COMPARISON

Reference	This Work	[2]	[8]
CMOS Technology	65nm	90nm	65nm
Chip area (mm^2)	0.08	0.04	0.3
Power dissipation (mW)	34.6	12.1	320
Input Range (mV pp)	200 (diff)	100	N/A
Sampling rate (GS/s)	5.0	2.5	5.0
Number of bits	4	3	6
ERBW (MHz)	2100	1320	2500
SNDR Peak/@ERBW (dB)	22.9/18.4	18.6/14.4	34/32
ENOB Peak/@ERBW (bits)	3.5/2.8	2.8/2.1	5.3/5.0
SFDR Peak/@ERBW (dB)	34.0/22.3	N/A	42/41
Max DNL/INL @DC (LSB)	0.34/0.38	0.02/0.04	0.6/0.7
FOM Peak/@ERBW (pJ/conv)	0.62/1.0	0.69/1.1	2.0/1.6

frequency. The 5 GS/s ADC achieves a peak FOM of 0.62 pJ/conversion and an FOM of 1.0 pJ/conversion at the ERBW input frequency. Fig. 12 compares the ERBW FOM to that of other single-channel ADCs with sampling rates of 1 GS/s and above with time-based ADCs highlighted using published data [2,6,7]. The design from JSSC 2011 [3] is shown even though it uses time-interleaving because it is also a time-based ADC. Table I summarizes the measured results and provides a comparison with both a single-channel flash ADC operating at the same frequency [8] and our previous design [2].

VI. CONCLUSIONS

A time-based single-channel 5GS/s 4-bit ADC was presented. The possibility for the VTC and TDC to be physically separated is a unique feature of this time-based architecture. Even when operating over a 1m link, the ADC achieves a comparable FOM to other recent designs with similar sampling rates, most of which use mature architectures. The use of this ADC is being explored for the SKA project.

VII. ACKNOWLEDGEMENT

The authors acknowledge the support of NSERC, Alberta Innovates and CMC Microsystems.

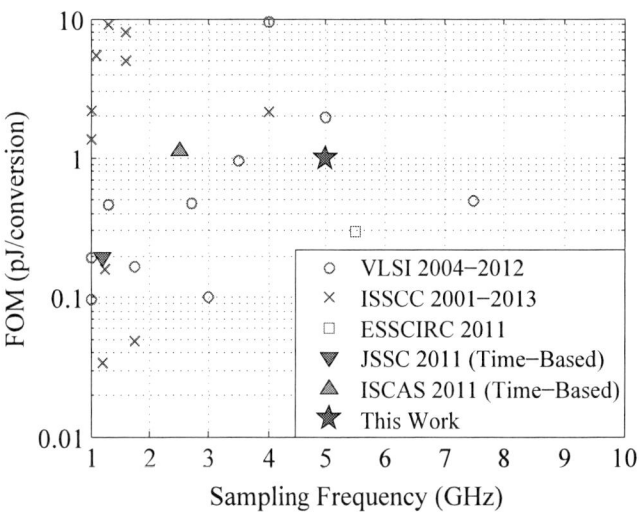

Fig. 12: Figures of merit for published multi-GHz-rate ADCs. All designs are single-channel except [3] from JSSC 2011. The ISCAS2011 design is our previous work [2].

REFERENCES

[1] G. Roberts and M. Ali-Bakhshian, "A Brief Introduction to Time-to-Digital and Digital-to-Time Converters," *Circuits and Systems II: Express Briefs*, vol. 57, no. 3, pp. 153–157, 2010.

[2] A. R. Macpherson, K. A. Townsend, and J. W. Haslett, "A 2.5GS/s 3-bit time-based ADC in 90nm CMOS," *ISCAS*, pp. 9–12, May 2011.

[3] Y. Tousi and E. Afshari, "A Miniature 2 mW 4 bit 1.2 GS/s Delay-Line-Based ADC in 65 nm CMOS," *JSSC*, vol. 46, pp. 2312–2325, Oct. 2011.

[4] P. Dewdney, P. Hall, R. Schilizzi, and T. Lazio, "The Square Kilometre Array," *Proceedings of the IEEE*, vol. 97, no. 8, pp. 1482–1496, 2009.

[5] B. Nikolic, V. G. Oklobdzija, V. Stojanovic, W. Jia, J. K.-S. Chiu, and M. Ming-Tak Leung, "Improved sense-amplifier-based flip-flop: design and measurements," *JSSC*, vol. 35, no. 6, pp. 876–884, 2000.

[6] B. Murmann, "ADC Performance Survey 1997-2012." [Online]. http://www.stanford.edu/ murmann/adcsurvey.html.

[7] W-H. Ma, J. Kao, and M. Papaefthymiou, "A 5.5GS/s 28mW 5-bit flash ADC with resonant clock distribution," in *ESSCIRC*, pp. 155–158, 2011.

[8] M. Choi, J. Lee, J. Lee, and H. Son, "A 6-bit 5-GSample/s Nyquist A/D converter in 65nm CMOS," in *VLSI*, pp. 16–17, 2008.

A 6b 800MS/s 3.62mW Nyquist AC-coupled VCO-Based ADC in 65nm CMOS

Praveen Kumar Sharma, Mike Shuo-Wei Chen
University of Southern California, Los Angeles, CA, USA
Email: praveens@usc.edu, swchen@usc.edu

Abstract **A Nyquist VCO-Based ADC architecture is proposed for AC-coupled systems which are commonly used in high-speed wireline and wireless communications. The proposed ADC utilizes a built-in high pass filter as an analog differentiator, replacing the digital differentiator in conventional oversampling VCO-based ADCs. As a result, it avoids quantization noise shaping and achieves wideband Nyquist operation, first order anti-aliasing filtering and improved VCO linearity without calibration. The ADC prototype achieves peak SNDR of 34dB and SFDR of 50dB with over 400MHz input bandwidth and sampling rate of 800MS/s. It occupies an active area of 0.01mm² and consumes 3.62mW in 65nm CMOS.**
Keywords: analog-to-digital, time-based-quantizer, VCO, all-digital.

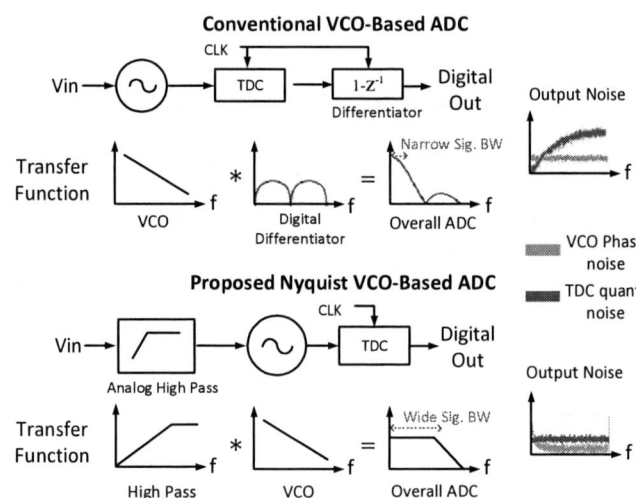

Fig. 1. Comparison of the conventional VCO-Based ADC vs. proposed Nyquist VCO-Based ADC

I. INTRODUCTION

High-speed (GS/s), medium resolution (~6bit) ADCs are in high demand for wideband electronic systems, and many of these systems utilize AC-coupled signal paths, i.e., the communication channel blocks the close-to DC frequency components to mitigate the DC offset, flicker noise impacts, and thermal asperity. For instance, wideband wireless receivers [1-2], hard-drive read channels [3], and various wireline standards [4-5] typically require additional DC blocking and/or high pass filtering in the receiver front end.

On the other hand, the scaled CMOS technology provides increasing intrinsic device speed but lower voltage headroom. This greatly restricts analog designs that are limited by headroom reduction. To make use of the faster speed and finer time resolutions, data converter architectures that utilize time information to quantize analog signals are emerging. They have been limited to either (a) wideband but very low resolution applications because of the non-linear voltage-to-time conversion [6], or (b) high resolution but narrowband applications, mainly due to their time quantization noise shaping property [7-8].

Motivated by the above observations, this paper explores the opportunity to embed the desirable high pass filtering within the ADC by exploiting the unique property of VCO transfer function while performing high-speed sampling. The objective is to minimize the overall implementation complexity and power/area consumption via architectural innovations.

The paper is organized as follows: section II explains the motivation and concepts behind the proposed architecture and

its advantages. Section III explores the transfer function and its dependency on various design parameters. The circuit implementation and design methodology of the critical building blocks is explained in section IV and finally sections V and VI discuss the experimental results and conclusions.

II. PROPOSED ADC ARCHITECTURE

The upper part of Fig. 1 shows a traditional implementation of a VCO-based ADC [7]. A digital differentiator, i.e. $(1-z^{-1})$, is used to calculate the phase difference between consecutive samples, which is approximately the digital representation of the analog input signal. Since the digital differentiator provides first-order noise shaping property, the time-to-digital converter (TDC) quantization noise overwhelms the high frequency spectrum. Moreover, since the signal transfer function of the conventional VCO-based ADC is a Sinc function, it leads to signal attenuation towards high input frequency - as much as 4dB at Nyquist frequency. As a result, the VCO-based ADCs are conventionally operated in oversampling mode [7-8].

To avoid the aforementioned bottlenecks, a wideband Nyquist VCO-based ADC architecture is proposed, as shown in the bottom part of Fig. 1. An analog high-pass filter is inserted prior to the VCO for several reasons. First of all, it provides an analog differentiation function from DC to its 3-dB corner frequency. In conjunction with the continuous time integrator, i.e.

978-1-4673-6145-3/13 $31.00 © 2013 IEEE

the VCO, the signal gain is unity within this band. Beyond the corner frequency, there is a 20dB/decade roll off in the transfer function, and results in the overall first-order anti-aliasing filter response. Secondly, since the digital differentiator is no longer present, a wideband operation is achieved by eliminating quantization noise shaping. Additionally, the analog differentiation attenuates the low frequency signals, reducing the voltage swing at the VCO input, and thus improves the linearity at lower frequencies which is a major limiting factor of the conventional VCO-based ADCs that usually require extra calibration [7-8]. Consequently, this work achieves peak SFDR of 50.1 dB at low input frequencies without any calibration.

III. PROPOSED ADC TRANSFER FUNCTION AND RESOLUTION

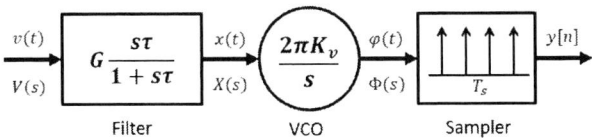

Fig. 2. Frequency-domain model of the proposed ADC

This section derives the theoretically achievable resolution and bandwidth of the proposed ADC as a function of several design parameters. As shown in Fig. 2, the input signal $v(t)$ is sent into a first order analog high pass filter whose response is given by

$$H(s) = G s \tau / (1 + s \tau) \qquad (1)$$

,where τ is the filter time constant and G is the gain of the filter. Since the filter output modulates the VCO frequency, the output phase of the oscillator can be expressed as

$$\Phi(s) = 2\pi K_v \, G \tau V(s) / (1 + s \tau) \qquad (2)$$

,where K_v is the gain of the VCO in hertz/volt. The output phase of the VCO is sampled using a time quantizer and represents the output of the ADC. Therefore (3) shows that the ADC transfer function has a low pass nature with its bandwidth determined by the input filter cutoff frequency.

For determining the resolution of the ADC we will now analyze the output in time domain. The $s\tau$ term in the denominator of (1) can be neglected for input frequencies much lower than the cut off frequency of the filter. Hence the transfer function approximates a differentiator and the output of the filter, $x(t)$, can be written as

$$x(t) = G\tau v'(t) \qquad (3)$$

,where $v'(t)$ is the time derivative of the input signal $v(t)$. The filter output modulates the VCO frequency; therefore, the output phase of the oscillator is

$$
\begin{aligned}
\varphi(t) &= \int_0^t 2\pi K_v \, x(t) dt \\
&= \int_0^t 2\pi K_v \, G\tau v'(t) dt = 2\pi K_v G \tau v(t)
\end{aligned}
\qquad (4)
$$

$\varphi(t)$ is then sampled by the TDC, and the discrete sample can be expressed as

$$Y[n] = G\tau K_v v(nT_s) \qquad (5)$$

Assuming the value of $v(t)$ lies between $-A$ to $+A$, where A is the amplitude of the signal, and the TDC quantizes each time period of the VCO into M steps, each step being $2\pi/M$ radians long; the resolution N of the ADC can be derived as

$$N = log_2(2AG\tau K_v M) \qquad (6)$$

According to (6), the value of τ should be maximized for maximal ADC resolution. However, τ also determines the bandwidth of the ADC as suggested by (2), and the maximum gain of a passive RC filter is 1. Thus, given the designed bandwidth and input amplitude, the ADC resolution solely depends on TDC resolution and VCO gain.

IV. CIRCUIT IMPLEMENTATION

The high-level block diagram of the converter is shown in Fig. 3. It consists of three main sections: the AC-coupled VCO, TDC and the digital processing block. To further enhance the dynamic range and design robustness, the ADC adopts two VCOs to construct a fully differential topology, which is referred as the dual oscillator mode in this paper. Not only can it reject the even order harmonics, but also helps to mitigate the supply noise sensitivity. A simulation shows that a peak supply noise of 100mV can be suppressed by 60dB via this fully differential design. The implementation details of the critical building blocks will be described in the following subsections

A. AC-Coupled VCO

The AC-coupled VCO is composed of a high-pass filter followed by the voltage controlled ring oscillator. The high-pass filter is implemented with the first-order RC circuit comprised of a poly-resistor and Metal-Oxide-Metal (MOM) capacitor. The corner frequency of the high-pass filter is designed around 800MHz to accommodate the wide input bandwidth. To avoid the vulnerability of parasitic capacitance degrading the ADC transfer function (3), the series capacitance of this RC filter is chosen sufficiently large, ~1pf. Moreover, the series resistor is

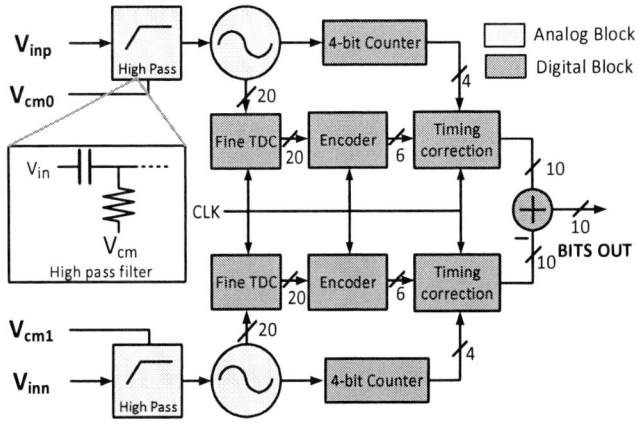

Fig. 3. Top-level block diagram of the proposed ADC architecture

split into on-chip and off chip resistors to facilitate filter bandwidth tuning for testing purpose.

A 5-stage ring oscillator is used as the VCO in the ADC. Each stage consists of a pseudo NMOS inverter with a PMOS load. The gate of the PMOS is controlled by the input voltage which varies its resistance and hence the VCO oscillation frequency. This topology was chosen over regular CMOS inverters because of better linearity over the operating range, and faster switching, shown by lower power delay products, at the cost of higher power consumption due to the static current consumption [9]. The better VCO linearity together with the reduced input voltage swing as a result of the high pass filtering makes this ADC free of linearity calibration, which is typically required in the conventional VCO-based ADC [7-8].

B. Fine Time-to-Digital Conversion

The multi-phase output of the VCO is captured by a fine TDC composed of phase interpolation and latch stage, which serves to quantize the VCO phase information into digital codes. Phase interpolation is implemented in between the delay stages via cross coupled resistor ladders, as shown in Fig. 4. This enables a fine TDC resolution without additional power consumption [8]. Each VCO stage is interpolated four times giving a total of twenty virtual stages which translate to forty phases of the VCO. However, the parasitic capacitances at the internal nodes of the passive RC interpolation result in unequal delay between the phases. This can severely degrade the DNL of the TDC. To alleviate this issue, the resistive ladder used for interpolation (R1-R4) is individually scaled via post layout simulations, such that the delay between each stage is equalized. Another advantage of utilizing a resistor ladder across the delay stage is the improved delay mismatches between the stages due to the better matched resistors. Dummy resistors are also added at the ends of the resistor ladder to improve matching of the layout.

After the phase interpolation, the sense amplifier flip-flops are used to latch the phase information. When the clock is high, the flip flop is in reset mode such that both output nodes are discharged to low. At the falling edge of the clock, the differential current between MP_1 and MP_2 is amplified and regenerated by the positive feedback. Either one of the outputs, i.e. S or R annotated in Fig. 4, will resolve to high, which sets or resets the following NOR-based SR-latch. Note that a PMOS input pair of the sense amplifier was chosen in this prototype because of the low common voltage at the VCO output.

C. Digital Encoding Logic

The output of the fine TDC is converted from cyclic thermometer code to binary representation using a ROM encoder. The outputs of the encoders are discharged during the first half clock period, and then evaluated in the second half. The ROM operation is completely dynamic to reduce power consumption and increase conversion speed.

The coarse TDC is implemented using a 4bit counter to count the VCO cycles. Digital standard cells provided by the

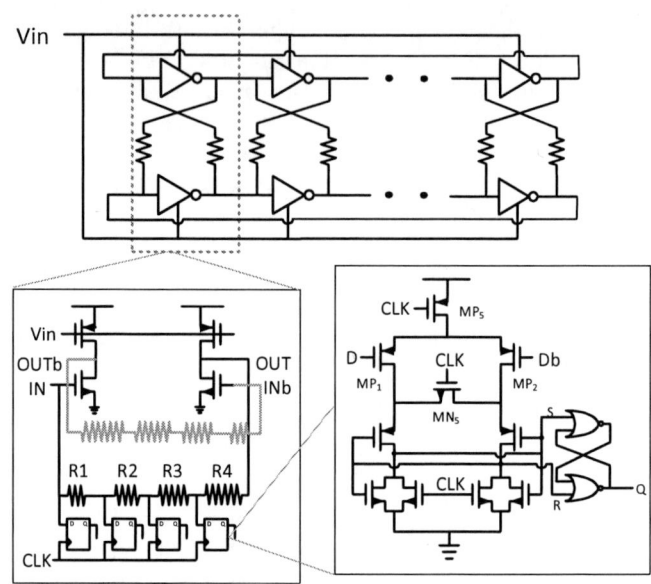

Fig. 4. Proposed VCO topology with fine passive interpolation TDC, and sense amplifier flip-flop used for sampling the VCO phase

foundry were sufficiently fast to implement the high-speed synchronous counter. Since the fine TDC is clocked by the sampling clock, while the coarse TDC is clocked by the faster VCO rising edges, a timing synchronizing logic is required to combine the two results correctly. This is achieved by re-latching the coarse TDC at the next VCO edge arriving after the sampling clock. This ensures that the counter output is always stable while latching.

After both VCOs are sampled, they are subtracted using a custom designed unsigned digital subtracter. Using the overflow bits generated from the coarse TDC, corrections are made to prevent wrapping around errors, which can result in random spikes in the output code. Furthermore, the mantissa of TDC code must be scaled appropriately when combining the two sections for binary base subtraction.

V. MEASUREMENT RESULTS

To properly reconstruct the ADC output waveform, the captured bits are processed as follows: firstly, since AC-coupled system avoids frequency components close to DC, a digital high-pass filter is applied to remove the unwanted signal within this low frequency band, such as DC offset or phase noise from the VCO. In this prototype, the lower cutoff frequency of 400 KHz is designed for the high-pass filter, which sufficiently attenuates the low frequency noise component to avoid SNR degradation. This is analogous to the decimation filter used in a delta-sigma ADC.

Secondly, since two oscillators are used in this prototype for differential operation, the frequency offset between the oscillators manifests as a constant phase ramp added to the output of the ADC. The slope of this ramp is first estimated from the output codes in the foreground by shorting the input to ground, and subtracted from the output to obtain the correct waveform.

For the measurement, the silicon die is directly attached and wire bonded to the PCB board. The peak SNDR achieved is 34.1dB and 30.6dB at 1MHz and 411MHz input frequency respectively. Figure 5 shows the spectrum of the ADC output in the two operation modes and the digital filter used to attenuate the low frequency noise. The single oscillator mode shows the presence of 2nd harmonic tone. This is suppressed in the dual oscillator mode due to the differential topology. The static linearity and dynamic performance of the ADC prototype is plotted in Fig. 6. It achieves DNL of 0.55 LSBs, and INL of 1.4 LSBs. Note that, the higher DNL at code 00 and 39 is due to extra buffer loading at the output of the VCO, which can be improved by adding dummy loads to match the capacitances in the future. The ADC prototype occupies an active area of just $0.01mm^2$, which is smaller than other existing ADC architectures with similar specifications. The ADC dissipates just 3.62mW of total power with the analog section consuming 2.865mW and the digital section consuming 0.755mW. The figure of merit (FOM) of the ADC is 123fJ/conversion-step and 162fJ/conversion-step at low and Nyquist frequency respectively. Fig. 7 shows the die photo of the chip.

VI CONCLUSION

A Nyquist AC-coupled VCO-based ADC is proposed to achieve wide input bandwidth (400MHz), high sampling rate (800MS/s) and yet low complexity. With the low power consumption and inherent AC-coupling property, the ADC is well suited for many wireline and wireless applications. With increasing time resolution in the future scaled CMOS technologies, the proposed ADC architecture is expected to further improve the dynamic range and power efficiency with less silicon area due to its digital intensive implementation.

ACKNOWLEDGEMENTS

The authors would like to thank Dr. Jerry Shiao and National Nano Device Laboratories (NDL), Taiwan for chip fabrication and assistance, and ONR for funding support.

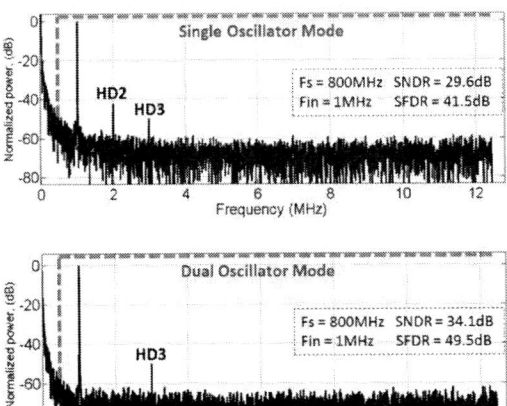

Fig. 5. ADC output spectrum in single oscillator (top) and dual oscillator mode (bottom) operation with output decimated by 32 folds before applying the post-processing digital filter annotated in blue dashed line

REFERENCES

[1] M.S.W. Chen and R. W. Brodersen, "A subsampling radio architecture for ultrawideband communications," *IEEE Trans. on Signal Processing,* vol.55, no.10, pp.5018,5031, Oct. 2007.

[2] B. Afshar, Y. Wang, A.M. Niknejad, "A Robust 24mW 60GHz Receiver in 90nm Standard CMOS," *ISSCC Dig. Tech. Papers,* pp.182,605, 3-7 Feb. 2008.

[3] S. Gopalaswamy, P. McEwen, "Read channel issues in perpendicular magnetic recording," *IEEE Trans. on Magn.,* vol.37, no.4, pp.1929-1931, Jul 2001.

[4] L. Lei, J. M. Wilson, S.E. Mick, J. Xu, L. Zhang, P. D. Franzon, "3Gb/s AC-coupled chip-to-chip communication using a low-swing pulse receiver," *ISSCC Dig. Tech. Papers,* pp.522,614 Vol. 1, 10-10 Feb. 2005.

[5] J. Kim; I. Verbauwhede, M-C.F. Chang, "A 5.6-mW 1-Gb/s/pair pulsed signaling transceiver for a fully AC coupled bus," *IEEE J. Solid-State Circuits,* vol.40, no.6, pp.1331,1340, June 2005.

[6] Y.M. Tousi, E. Afshari, "A Miniature 2 mW 4 bit 1.2 GS/s Delay-Line-Based ADC in 65 nm CMOS," *IEEE J. Solid-State Circuits ,* vol.46, no.10, pp.2312,2325, Oct. 2011.

[7] J. Kim, T. -K. Jang, Y. -G. Yoon and S. Cho, "Analysis and design of voltage-controlled oscillator based analog-to-digital converter," *IEEE Trans. Circuits. Syst. I,* vol.57, no.1, pp.18-30, Jan. 2010.

[8] J. Daniels, W. Dehaene, M. Steyaert, A. Wiesbauer, "A 0.02mm² 65nm CMOS 30MHz BW all-digital differential VCO-based ADC with 64dB SNDR," *VLSI Circuits (VLSIC), 2010 IEEE Symposium on,* pp.155,156, 16-18 June 2010.

[9] V. Beiu, J. Nyathi and S. Aunet, "Sub-Pico Joule Switching High Speed CMOS Circuits are Feasible", *The Second International Conference on Innovations in Information Technology (IIT'05),* September 26-28, 2005.

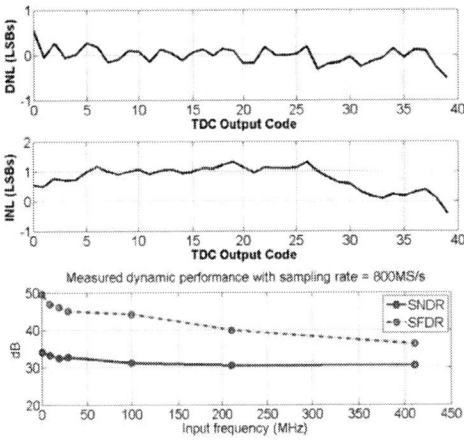

Fig. 6: DNL and INL of the measured fine TDC output (top) and the measured dynamic performance at 800MS/s sampling rate (bottom)

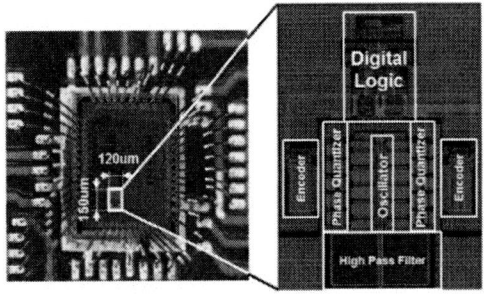

Fig. 7 Die micrograph of the test chip in 65nm CMOS

A Fully-Digital Beat-Frequency Based ADC Achieving 39dB SNDR for a 1.6mV$_{pp}$ Input Signal

Bongjin Kim, Weichao Xu, and Chris H. Kim

University of Minnesota, Minneapolis, MN 55455 USA

Abstract- **A fully-digital VCO-based ADC featuring a novel beat frequency detection scheme is demonstrated in 65nm LP CMOS. The proposed beat frequency based ADC is unique compared to previous VCO-based ADCs in that it is highly effective in measuring extremely small changes (e.g., 0.01%) in the VCO frequency within a short sampling time (e.g., 100 VCO periods). Direct amplifier-less A-to-D conversion of a 1.6mV$_{pp}$ differential input signal with 39dB SNDR and 6.2 ENOB was experimentally verified.**

I. INTRODUCTION

To meet the ever-increasing demands of energy-efficient sensor systems, there has been a great deal of circuit research aiming at developing compact low-power ADCs. For example, recent SAR-ADCs have demonstrated microwatt level power consumption while offering a scalable sampling rate and resolution [1-2]. While ADC remains a key building block for most mixed-signal systems, sensor applications do not always benefit from the improvements in the ADC power and performance because of the Low Noise Amplifier (LNA) and Variable Gain Amplifier (VGA) which are typically required for the signal pre-conditioning. Prior ADC designs however, focus on converting a rail-to-rail analog signal thereby neglecting the overhead of these analog components. Not only does this conventional design paradigm incur a large power and area overhead, it is also susceptible to device noise occurring in the signal amplification and filtering stages. In this work, we present a fully-digital Beat Frequency based ADC (BF-ADC) capable of directly measuring sub-mV input signals without an LNA or VGA. Fig. 1 contrasts the usage scenario of the conventional and proposed ADC.

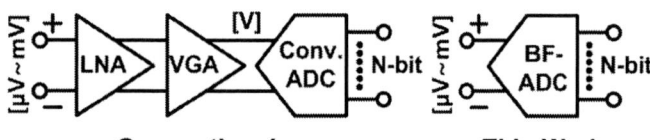

Fig. 1. The focus of this work is on direct acquisition of small input signals without using an LNA or VGA.

II. Beat Frequency (BF) Based ADC

The operating principle of the beat frequency (BF) detection scheme in the context of a VCO-based ADC is illustrated in Fig. 2. The proposed scheme can achieve a high resolution compared to a simple frequency counting method in cases where the frequency difference is extremely small. This is possible by measuring the period of the beat frequency signal which is equivalent to the time it takes for the faster signal to pass, catch up and overtake the slower signal again [3]. To understand better how the BF-ADC scheme works, let's consider a scenario in which the initial difference

between the two VCO frequencies is 1%. This gives an output count of 100 as it takes 100 VCO periods for the slow and fast signal edges to overlap again. Now, suppose the frequency difference becomes 1.01% due to a small change in the input signal. This translates into an output count of 99 as it takes one less period for the fast signal to catch up with the slower one [3]. The same count change from 100 to 99 would have required a larger frequency change of 1% using the linear counting method implying a significantly lower sensing resolution. Note that for the beat frequency detection scheme, the frequency measurement resolution increases exponentially as the two VCO frequencies become closer to each other.

The aforementioned BF detection concept can be readily applied to an analog-to-digital converter as further indicated in Fig. 2. The incoming differential input signals V_N and V_P are first AC-coupled to the supply voltages of two VCOs, respectively. To produce high-resolution beat frequency counter outputs, the DC bias of the negative (or positive) input voltage is set to be slightly higher than that of the positive (or negative) input voltage. This ensures that the VCO biased using the negative (or positive) input voltage is always running faster (or slower) than the VCO biased using the positive input voltage. Using this configuration, the BF counter generates an output count value depending on how close the two VCO frequencies are as illustrated in Fig. 2.

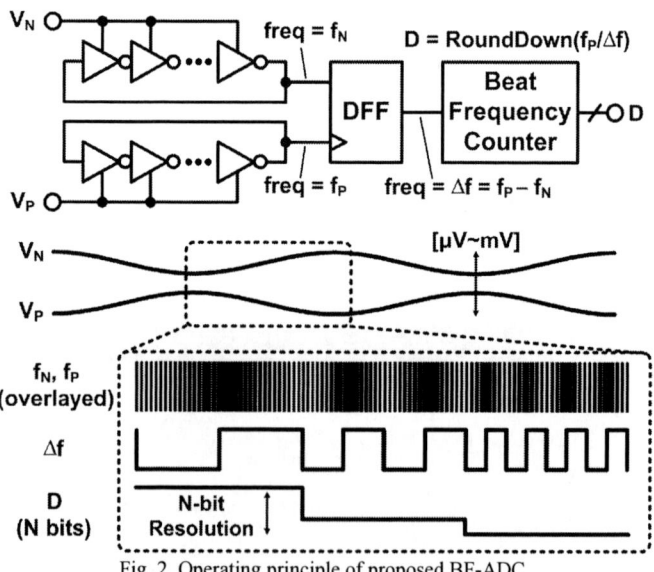

Fig. 2. Operating principle of proposed BF-ADC.

III. Dual Reference BF-ADC

One limitation of a simple BF-ADC scheme described above is that the sensing resolution quickly degrades as the difference between the positive and negative input voltages

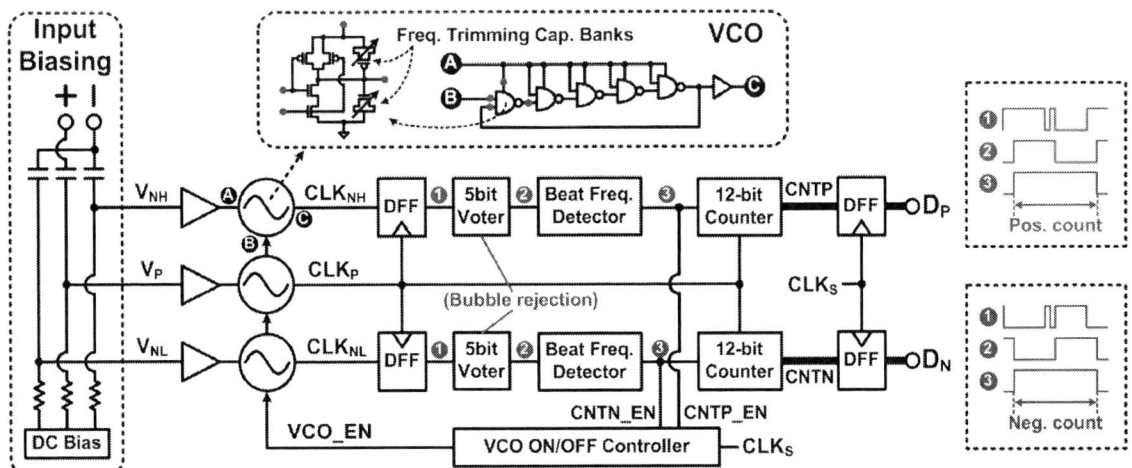

Fig. 3. Dual reference BF-ADC circuit for improved resolution.

becomes larger. To overcome this limitation, we propose a dual reference BF-ADC circuit shown in Fig. 3 where the negative input signal is AC-coupled to the supply voltage of not one but two VCOs with different DC bias levels. Using the two AC-coupled negative input signals as the upper bound and lower bound, we can obtain a high sensing resolution for both positive and negative phases of the differential input signal. The DC bias levels for the two reference VCOs (V_{NH} and V_{NL}) and the main VCO (V_P) in Fig. 3 are set using a simple on-chip voltage bias generator. Each VCO is implemented using 5 static NAND gates with programmable capacitor banks attached to each stage for fine grain frequency trimming. A static D flip-flop and a 5 bit majority voter circuit are used in both the upper and lower paths to generate the beat frequency signal while eliminating any logic bubbles (e.g. lone 1 in a stream of 0's) that may cause logic errors. A 12 bit counter is used to record the number of reference periods corresponding to the period of the beat frequency signal. The output count is then sampled by the main sampling clock CLK_S. The 12 bit positive and negative beat frequency values (D_P and D_N) are used to compute the actual input differential voltage [3]. Fig. 4 shows the timing diagram for a single sampling period of the BF-ADC circuit. To reduce unnecessary switching power, all VCOs are automatically shut off once the sampling is complete. Fig. 5 shows the signal waveforms measured from

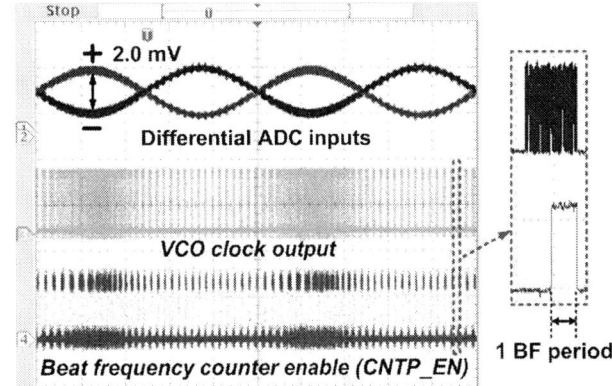

Fig. 5. Waveforms from BF-ADC test chip for a 4 mV$_{pp}$ differential input.

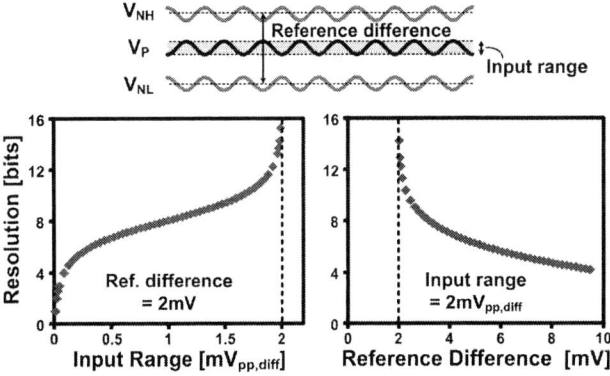

Fig. 6. Simulated resolution as a function of input range and difference between two AC-coupled reference voltages.

the 65nm BF-ADC test chip.

One interesting feature of the proposed BF-ADC is that we can obtain a uniformly high ADC resolution for a range of input signal amplitudes by simply adjusting the reference voltage levels. This eliminates the need for a separate variable gain amplifier reducing the power consumption, area, and complexity of the overall system. This would be particularly attractive for applications such as bio-potential (EEG, ECG, EMG, EOG) acquisition systems which may have to operate across a wide range of input signals with different amplitudes.

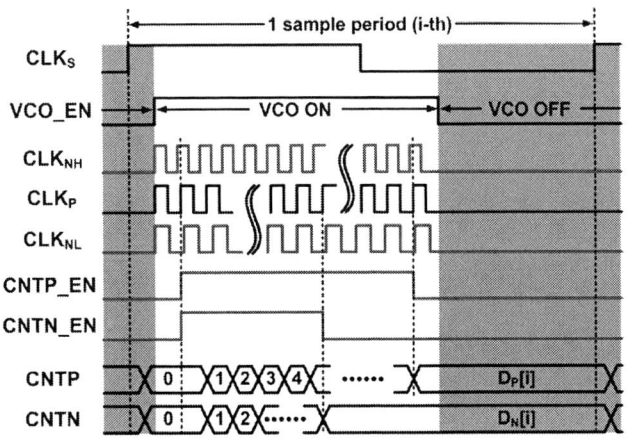

Fig. 4. Timing diagram for one sampling period.

978-1-4673-6145-3/13 $31.00 © 2013 IEEE

Simulation results in Fig. 6 show the BF-ADC resolution as a function of the signal input range and the difference between the upper bound and lower bound reference signals. The resolution increases exponentially as the reference difference approaches the input range, although in practice, the effective resolution, i.e. ENOB, is limited by the noise floor of the input and reference signals as well as the VCO phase noise. Fig. 7 shows the measured BF-ADC resolution from the 65nm test chip illustrating the dependence of the achievable ADC resolution on two key parameters: input range and reference voltage difference. Note that the measured resolution varies with the reference voltage. For instance, the resolution changes from 9bit to 5bit for a 5mV change in the reference voltage difference under a fixed input range of 6mV. Calibration techniques widely used in bio-signal sensing applications can be adopted to compensate for any PVT effects in the reference voltage generator circuit.

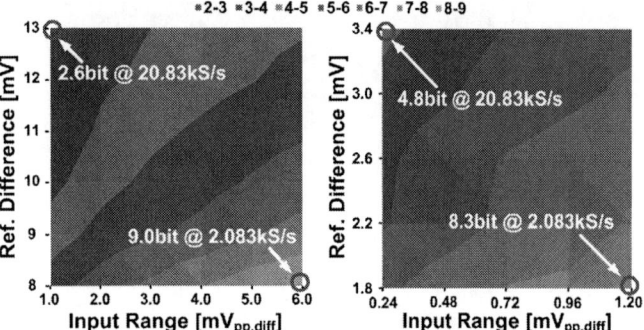

Fig. 7. Measured ADC resolution as a function of input range and reference voltage difference.

IV. Comparison with Conventional VCO-based ADC

VCO-based ADCs have been drawing attention lately owing to their digital-friendly implementation and inherent 1st order noise shaping property. Recent publications have reported high-order CT-$\Delta\Sigma$ loops utilizing a VCO-based ADC as a quantizer achieving very high SNDR (e.g. 78dB [4]). The proposed BF-ADC has the unique property of being able to achieve high resolution for very small input signals when used as a Nyquist rate ADC. Conventional VCO-based ADCs on the other hand take advantage of the noise-shaping property to improve resolution while sampling rail-to-rail input signals.

	Conv. VCO-based ADC	Proposed BF-based ADC
ADC Type	Delta-sigma	Nyquist rate
Main Feature	1st-order noise-shaping	Beat frequency detection
Input Range	Large [V]	Small [μV~mV]
VCO Linearity	Nonlinear K_{VCO} (large range)	Linear K_{VCO} (small range)
Key Circuit Block	VCO + linear counter	VCO + beat freq. counter
Counting Period	Fixed sample period	Variable beat freq. period
Sampling Rate	High speed [MS/s]	Low speed [kS/s]
Reconfigurability	No	Reconfigurable resolution
Applications	Wireless receiver [6, 7]	Sensor applications

Fig. 8. Comparison table of VCO-based ADC versus BF-ADC

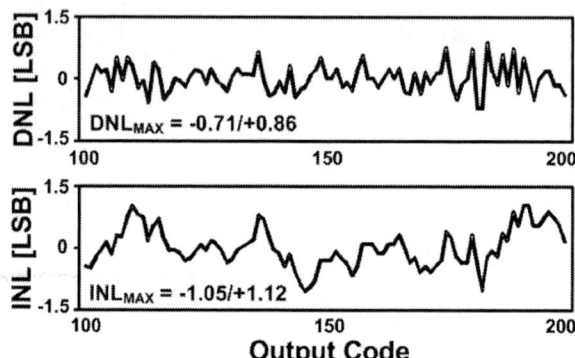

Fig. 9. Normalized DNL and INL from the measured output code.

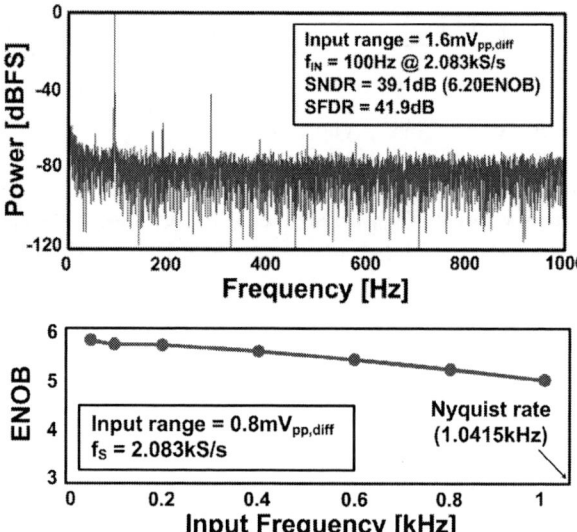

Fig. 10. Measured FFT for a 1.6mV$_{pp}$ AC signal with key ADC specifications (upper). ENOB versus input frequency data (lower).

Another noteworthy difference is that existing VCO-based ADCs have a fixed sampling period while the BF-ADC has a sampling period that is a function of the beat frequency period (i.e. difference between the main and reference frequencies). Finally, as far as linearity is concerned, conventional designs are more susceptible to the VCO's inherent voltage-to-frequency nonlinearity and difficult to achieve high performance unless sophisticated techniques such as the phase-feedback closed-loop [4] or digital calibration [5] are employed. This stems from the rail-to-rail input signal swing requirement of conventional ADCs. In contrast, BF-ADC has a better linearity as it can work for smaller input signals. Simulation results in 65nm show that the variation in K_{VCO} is reduced from ±14% to ±0.1% as the input signal amplitude is reduced from 400mV$_{pp}$ to 1mV$_{pp}$. Fig. 8 compares various features of a conventional VCO-based ADC and the proposed BF-ADC.

V. Test Chip Measurement Results

A BF-ADC test chip was implemented in a 65nm LP process as a proof of concept. The measured DNL and INL were -0.71/+0.86 LSB and -1.05/+1.12 LSB, respectively as shown in Fig. 9. For a sampling rate of 2.083kS/s and an input

	[8] VLSI'07	[4] ISSCC'09	[9] VLSI'12	[1] ISSCC'11	[2] ISSCC'12	This work
ADC Type	Nonlinear Pipelined	VCO-based Delta-Sigma	Pipelined	SAR	SAR	Nonlinear Beat-Freq.
Input Range	Large[V]	Large[V]	Large[V]	Large[V]	Large[V]	Small[mV]
Process / Supply Voltage	0.18μm /1.62V	0.13μm /1.5V	0.18μm /1.3V	65nm /0.4V~1.0V	90nm /1.1V	65nm /0.5V~1.2V
Sampling Rate(f_s) / Power(P)	22MS/s /2.54mW	900MS/s /87mW	30MS/s /2.6mW	20kS/s /206nW	4MS/s /17.44μW	4.17kS/s /0.92μW
Energy Efficiency (P/f_s or $P/(2 \cdot f_{BW})$ for $\Delta\Sigma$)	115.5pJ	2175pJ	86.7pJ	10.3pJ	4.36pJ	220.6pJ
SNDR [dB] @ Input Range	35.6@1V	78.1@2.28V	61.5@2.2V	55.0@1.1V	58.3@1.36V	39.1@1.6mV
ENOB [bit] @ Input Range	2.8@1mV	12.7@2.28V	9.9@2.2V	8.84@1.1V	9.4@1.36V	6.2@1.6mV
Conversion-steps per mV	6.96	2.92	0.43	0.42	0.50	45.95
*FOM$_{1mV}$ [pJ/conv-step]	16.59	744.86	199.64	24.72	8.78	4.80
**Min. Input Amplitude [mV] (Dynamic Range)	0.10 (80dB)	0.28 (78dB)	1.85 (57dB)	1.96 (54dB)	1.65 (58dB)	***0.03 (89dB)
Area [mm²]	0.560	0.450	0.500	0.212	0.047	0.013

*FOM normalized to 1mV (=-60dB), ** Input amplitude at SNDR = 0dB, *** based on measured SNDR of 10dB @0.12mV

Fig. 11. Comparison with previous ADCs.

signal of 1.6mV$_{pp,diff}$, a 39.1dB SNDR (6.2bit ENOB) and a 41.9dB SFDR were achieved (Fig. 10). Fig. 10 also shows the measured ENOB as a function of the input signal frequency, showing a 5.0 to 5.8 ENOB range for a 0.8mV$_{pp}$ differential input signal. Here, the Nyquist rate frequency is 1.0415kHz. Fig. 11 compares the performance of the proposed ADC versus prior ADCs including a nonlinear pipeline ADC [8], a VCO-based ADC [4], a linear pipelined ADC [9], and two state-of-the-art SAR-ADCs [1-2]. BF-ADC shows a FOM (normalized to 1mV$_{pp,diff}$) of 4.80pJ/conv-step, a dynamic range (normalized to 1.2V) of 89dB, and an area of 0.013mm² which is much smaller than the other ADCs. The compact nature and simplicity of the BF-ADC makes it ideally suited for multi-channel bio-signal sensors which require large and power hungry multi-stage amplifiers and signal filters per channel. In contrast, BF-ADC only requires passive signal filters which have a smaller footprint compared to having dedicated amplifier circuits. The test chip microphotograph and feature summary table are given in Fig. 12.

VI. CONCLUSION

In this paper, we have presented a fully-digital VCO-based ADC utilizing the beat-frequency detection scheme for sampling signals with sub-mV amplitudes. To improve the ADC resolution, we propose a dual reference BF-ADC scheme which employs two reference VCOs. One VCO is running at a frequency higher than the maximum frequency of the main VCO while the other VCO has a frequency that is lower than the minimum frequency of the main VCO. This technique helps maintain a consistently high sensing resolution for both positive and negative phases of the input signal. A 65nm test chip demonstrates 6.2 ENOB for a 1.6mV$_{pp}$ input differential signal without any external signal amplification or conditioning circuitry.

REFERENCES

[1] M. Yip, A. P. Chandrakasan, "A Resolution-Reconfigurable 5-to-10b 0.4-to-1V Power Scalable SAR ADC," in *Proc. Int. Solid-State Circuits Conf.*, pp. 190-191, February 2011.

[2] P. Harpe, Y. Zhang, G. Dolmans, K. Philips, H. D. Groot, "A 7-to-10b 0-to-40MS/s Flexible SAR ADC with 6.5-to-16fJ/conversion-step," in *Proc. Int. Solid-State Circuits Conf.*, pp. 472-473, February 2012.

[3] T. Kim, R. Persaud, C. H. Kim, "Silicon Odometer: An On-Chip Reliability Monitor for Measuring Frequency Degradation of Digital Circuits," *in Proc. IEEE Symp. VLSI Circuits*, pp. 122-123, June 2007.

[4] M. Park, M. Perrot, "A 0.13μm CMOS 78dB SNDR 87mW 20MHz BW CT ΔΣ ADC with VCO-Based Integrator and Quantizer," in *Proc. Int. Solid-State Circuits Conf.*, pp. 170-171, February 2009.

[5] J. Daniels, W. Dehaene, M. Steyaert, A. Wiesbauer, "A 0.02mm2 65nm CMOS 30MHz BW All-Digital Differential VCO-based ADC with 64dB SNDR," *in Proc. IEEE Symp. VLSI circuits*, pp. 155-156, June 2010.

[6] J. Kim, W. Yu, H. Yu, S. Cho, "A Digital-Intensive Receiver Front-End Using VCO-Based ADC with an Embedded 2nd-Order Anti-Aliasing Sinc Filter in 90nm CMOS," in *Proc. Int. Solid-State Circuits Conf.*, pp. 176-177, February 2011.

[7] M. Park, M. Perrot, "A 78dB SNDR 87mW 20MHz Bandwidth Continuous-Time ΔΣ ADC with VCO-Based Integrator and Quantizer Implemented in 0.13μm CMOS," *in IEEE Journal of Solid-State Circuits*, pp. 3344-3358, December 2009.

[8] J. Lee, S. Park, J. Kang, J. Seo, J. Anders, M. Flynn, "A 2.5mW 80dB DR 36dB SNDR 22MS/s Logarithmic Pipeline ADC," in *Proc. IEEE Symp. VLSI Circuits*, pp. 194-195, June 2007.

[9] B. Hershberg, S. Weaver, K. Sobue, S. Takeuchi, K. Hamashita, U. Moon, "A 61.5dB SNDR Pipelined ADC Using Simple Highly-Scalable Ring Amplifiers," in *Proc. IEEE Symp. VLSI Circuits*, pp.32-33, June 2012.

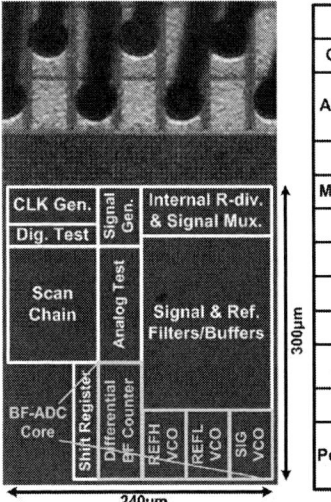

Process		65nm LP CMOS
Operating Voltage		0.5V ~ 1.2V
Area	Test chip	0.072mm²
	BF-ADC core	0.013mm²
Input Range		120μV ~ 6mV
Max. Sampling Rate		20.83kS/s
ADC Resolution		2.6bit ~ 9bit
DNL		-0.71/+0.86 [LSB]
INL		-1.05/+1.12 [LSB]
Dynamic Range		89dB
SNDR @ 1.6mV		39.1dB (ENOB=6.20)
SFDR @ 1.6mV		41.9dB
Power	VCO @ 2MHz	0.86μW
	Digital Power	0.06μW

Fig. 12. Chip microphotograph and feature summary table.

A 4–15-GHz Ring Oscillator based Injection-Locked Frequency Multiplier with Built-in Harmonic Generation

Jie Xu, Jianyun Hu, Berkehan Ciftcioglu, and Hui Wu

Department of Electrical and Computer Engineering, University of Rochester

Email: hui.wu@rochester.edu

Abstract—This paper presents a new injection-locked frequency multiplier (ILFM) designed to achieve wide locking range and low power dissipation. The ILFM is constructed using an injection-locked ring oscillator (ILRO) without any inductors. It is injection-locked at each ILRO stage by an injection signal, which is generated by the harmonic generation integrated in each stage from the input signal, with a progressive phase shift. The multiphase injection locking scheme and built-in harmonic generation improve the injection efficiency, increase the locking range, and reduce the power consumption. A multiply-by-2 ILFM prototype is implemented on a test chip using 130-nm CMOS. The measured locking range is 116% (4-15 GHz) with 0.3-V input amplitude. The ILFM prototype consumes 8 mW at 1.8 V, and occupies a chip area less than 0.01 mm^2.

Index Terms—injection locking, frequency multiplier, ring oscillator, inductorless, locking range

I. INTRODUCTION

Frequency multipliers are important building blocks in frequency synthesis, clock distribution and other RF systems [1], [2]. A conventional design of frequency multipliers uses nonlinear devices to generate harmonics from the input signal, and then selects the desired harmonic frequency by a high-Q filter. These frequency multipliers can operate at up to millimeter-wave frequencies but require large input signal and exhibit poor harmonic suppression. In recent years, injection-locked oscillators are increasingly used as frequency multipliers because of their high operation frequency, good phase noise and low power consumption [3]–[6].

Most of these injection-locked frequency multipliers (ILFM) are based on injection-locked LC oscillators. Typically several harmonic frequencies are generated by a nonlinear circuit, and then the desired harmonic is selected by the injection-locking oscillator [3]–[6]. Because of the bandpass characteristic of the LC tank, these ILFMs can achieve low phase noise and good suppression of undesired harmonics. However, they exhibit relatively narrow locking range due to the resonant LC thank, and require inductors which occupy large chip area. These limitations pose significant challenges in scaled CMOS and prevent their wide adoption in systems-on-chip applications. For example, an injection-locked clocking system [2] uses many ILFMs as local clock generators, which require wide locking range and small chip area.

Inductorless injection-locked ring oscillators (ILRO) present a promising solution to the locking range and chip area problems. They occupy smaller chip area since no inductors are used. They can achieve wider locking range because the effective Q is equal to one, much smaller than the LC oscillators. Similar to the LC-oscillator based case, a ring-oscillator based ILFM can be constructed by generating harmonics using a nonlinear circuit before injection-locking the ring oscillator, as illustrated in Fig. 1-a. In [7], for example, a small duty-cycle pulse is generated, which contains a large number of harmonics, and then used to injection-lock the ring oscillator whose free-running frequency is near the desired harmonic. However, directly injecting a harmonic-rich signal into the ring oscillator may result in many strong harmonic components in the output, especially at low frequency, because the ring oscillator without inductors has lowpass rather than bandpass filtering characteristic. Further, such single-point injection results in a small locking range due to limited phase shift from a single delay stage [8]. Note that simply boosting the injected pulse power may quickly disrupt the ring oscillator operation and cause loss of lock.

Multiphase injection has been proposed to enlarge the locking range of ILROs [8], in which multiple signals with progressive phases are injected into all stages of the ILRO. An ILFM can be constructed by adding a nonlinear harmonic generation circuit, as illustrated in Fig. 1-b. In [9], for example, the input signal frequency is multiplied by 4 through a pulse generator, and then the multiplied signal passes a delay line to generate multiphase signals, which injection-locks the ILRO. Two ILFMs with adjacent locking ranges are combined to achieve 57% locking range, with the aid of additional sensing and control circuits.

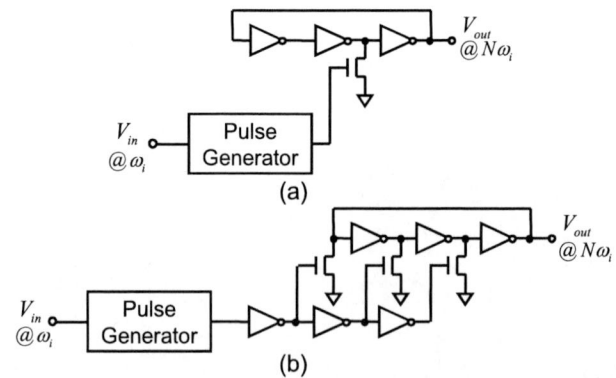

Fig. 1. Prior ring oscillator based ILFM topologies: (a) A single-input ILRO similar to [7]. (b) A multiphase ILRO similar to [9]. Note that frequency multiplication is done by the separate pulse generator.

In most of these earlier designs such as [9], the harmonic generation is carried out first, and then multiphase injection signals are generated and distributed for ILRO. At high frequencies, this scheme causes several problems: first, the multiphase generation circuit now operates at the multiplied frequency, which results in large power consumption. Second, there is larger signal attenuation at the multiplied frequency, leading to imbalance in the multiphase injection signals and limiting the locking range. In this paper, we present a new ILFM design in which multiphase input signals are directly injected into the ILFM without being frequency multiplied. Instead, the ILFM incorporates efficient harmonic generation in each delay stage to generate the desired harmonic. The new design further improves the locking range while reducing the circuit complexity and power consumption.

II. ILFM DESIGN

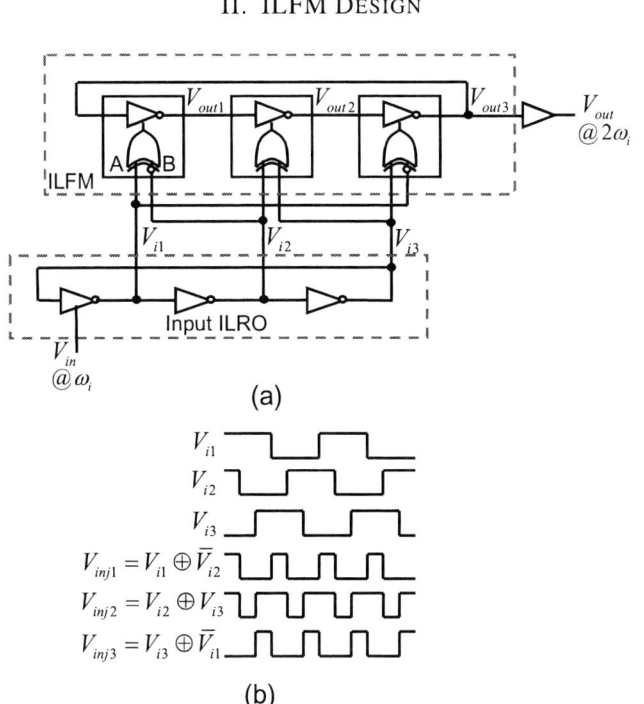

(a)

(b)

Fig. 2. The proposed ILFM based on a multiphase ILRO. (a) Harmonic generation is integrated into each delay stage. The input ILRO generates the multiphase input signals. (b) The multiphase input signals (V_{i1-3}) and the injection signals ((V_{inj1-3})).

Fig. 2-a shows the proposed ILFM based on a 3-stage ILRO. Another ILRO is added at the input to generate the multiphase input signals (V_{i1-3}) for the ILFM. Note that The input ILRO operates at ω_i, and the ILFM at $\omega_{inj} = 2\omega_i$ in this implementation. Note that the input ILRO is not an essential part of the ILFM, and can be replaced with other multiphase generation circuits or removed if multiphase signals already exist.

As discussed in Sec. I, it is more energy-efficient to tightly integrate the harmonic generation with injection locking in an ILFM, especially for ILRO-based ones. In this ILFM, each delay stage incorporates an XOR function to mix the two input signals (e.g. V_{i1} and V_{i2} for Stage I) and generate the injection signal, as shown in Fig. 2-b. Note that V_{i1-3} have 120° phase shift consecutively at ω_i, and the XOR between every two of V_{i1-3} can generate three injection signals still with 120° phase shift in sequence at ω_{inj}. The multiphase injection signals are configured in progressive phase shift to widen the ILFM's locking range [8].

This design has several distinctive advantages as compared to the prior arts: First, the tight integration of the harmonic generation function into the ILFM allows efficient injection locking, and hence increases the locking range and/or reduces the required input RF signal amplitude. Second, saving a separate edge combiner reduces circuit complexity and power consumption. Third, the multiphase generation circuit (the input ILRO in this case) now operates at input frequency (ω_i) instead of the multiplied output frequency, which further reduces the power consumption.

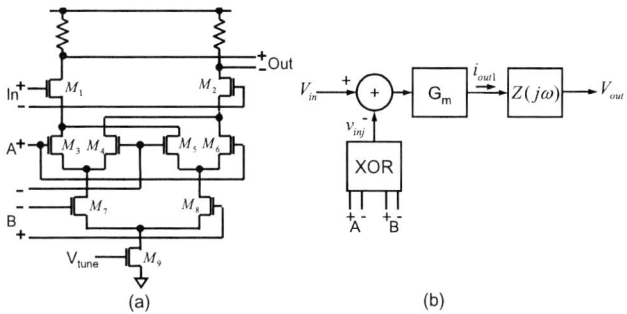

(a)   (b)

Fig. 3. (a) Circuit implementation of each delay stage in the prototype ILFM. (b) The model of the delay stage.

Fig. 3-a shows the delay stage designed for the prototype ILFM in the test chip, with the circuit model shown in Fig. 3-b. The top part is the CML delay cell. The lower part is an XOR gate based on a Gilbert cell. Because the output voltages of the input ILRO have relatively large amplitudes, the transistors in the XOR gate operate in the hard switching mode. Transistor M_9 is added to tune the dc current. Note that the current is reused by the XOR gate and the ILRO cell, which saves power and allows efficient injection similar to [3]. The four transistor stack, however, needs relatively high power supply voltage. To lower Vdd, a folded design can be utilized by replacing one stack of NMOS with PMOS in the XOR gate.

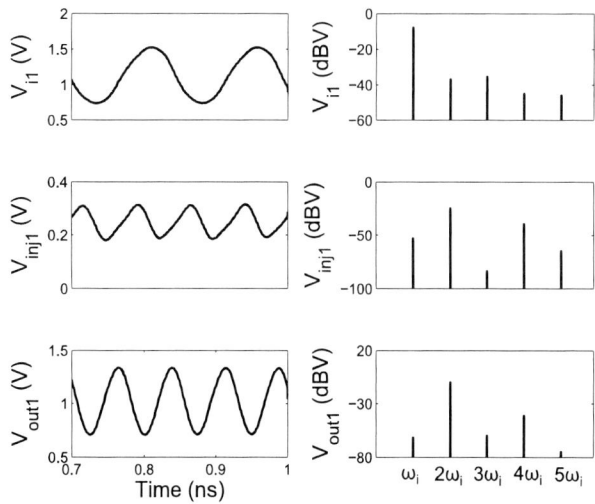

Fig. 4. The simulated waveforms and spectra of the input ILRO output, the injection signal and the ILFM output, respectively. The input signal amplitude is 300 mV and input frequency is near the half of free-running frequency of the ILFM.

To give an example of ILFM operation, Fig. 4 shows the

978-1-4673-6145-3/13 $31.00 © 2013 IEEE

simulated waveforms and spectra of the input ILRO output, the injection signal and the ILFM output, respectively. The input signal has amplitude 0.3 V and frequency near the half of free-running frequency of the ILFM. The output voltages of the input ILRO have mainly fundamental frequency. In the injection signals generated by the XOR operation, the second harmonic has stronger amplitude than other harmonics. After the injection locking and the lowpass filtering of the ILFM, in the outputs of the ILFM, the second harmonic frequency has the dominant function, while the other harmonics are suppressed. There are still fundamental frequency and higher odd harmonics with small amplitudes, because of the mismatch in the XOR gate. Simulation results also show the three output voltages have nearly 120° phase shift among them for both the input ILRO and the ILFM.

III. PROTOTYPE EXPERIMENTAL RESULTS

The new ILFM are designed and fabricated in a 130-nm BiCMOS technology using only CMOS transistors. The chip photo is shown in Fig. 5. The area of ILFM with the input ILRO is only 0.08x0.09 mm^2. To facilitate testing using the single-ended signal source, a single-ended-to-differential converter is added at the input. An output buffer is added for the prototype to drive the 50-Ω load.

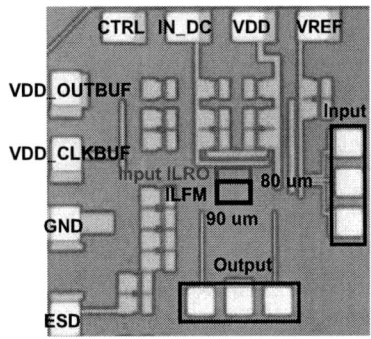

Fig. 5.   Chip photo of the proposed ILFM.

The measured free-running frequency of ILFM are 13.6 GHz, with output power at -25 dBm approximately, which can be improved using larger gain output buffer. Fig. 6 shows the measured output spectra of ILFM when injection-locked by the signal with an amplitude of 0.3 V at different frequencies. With the increase of injection frequency, the output power drops because of the decreased gain of the stage in ILRO and increased attenuation in the signal path.

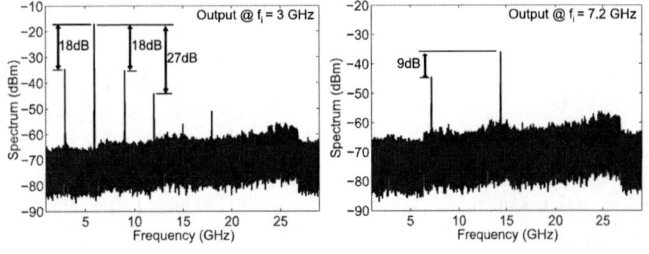

Fig. 6.   Output spectrum of ILFM when injection-locked by the signal with amplitude 0.3 V at different frequencies.

Fig. 7 shows the measured output power of fundamental and second harmonic frequencies of the ILFM within the locking range at different input signal amplitudes. The ILFM has wide locking range 4-15 GHz at input signal amplitude 0.3 V. When the input signal amplitude decreases to 0.15 V, the locking range is still as large as 4.4-14.2 GHz. This is because the input ILRO can convert the relatively small amplitude input to relatively large amplitude multiphase signals for injection-locking the ILFM. The suppression of the fundamental frequency is 16-26 dB in most of the locking range at input signal amplitude 0.3 V.

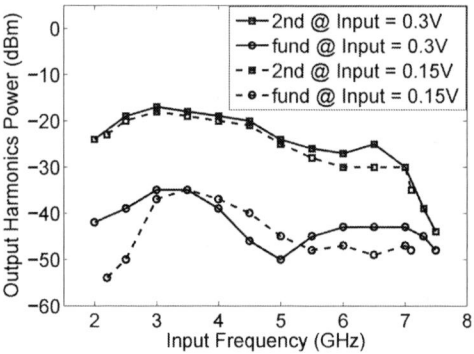

Fig. 7.   Measured output power of fundamental and second harmonic frequencies of the ILFM at input signal amplitude 0.3 V and 0.15 V.

The phase noise of the ILFM is shown in Fig. 8. The phase noise is -67 dBc/Hz @ 1 MHz offset when free running, and -103 dBc/Hz @ 1MHz offset when injection-locked. Larger than 30 dB phase noise suppression can be achieved at 1 MHz offset frequency. When injection-locked, the phase noise of the ILFM is approximately 6 dB higher than the measured input near 1 kHz offset, which is the theoretical degradation value due to multiplication by 2. At higher offset frequency, the phase noise of the ILFM is dominated by ring oscillators themselves, and therefore has larger difference from the measured input phase noise.

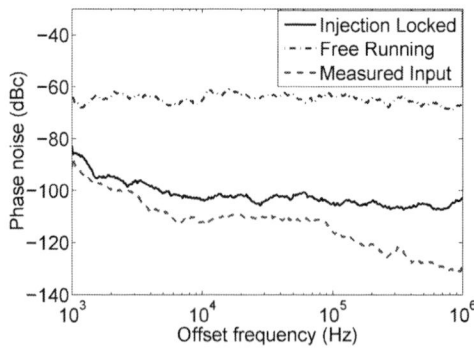

Fig. 8.   Measured phase noise of the ILFM when free running and injection-locked with comparison to the signal source.

The ILFM alone consumes 8.1 mW, and the power consumptions of the ILFM with the input ILRO together is 18 mW at 1.8 V.

Fig. 9 compares the performance of the proposed ILFM with recent published frequency multipliers [3]–[7], [9]–[13]. As can be seen, the ILROs [7], [9] have wider locking range and

978-1-4673-6145-3/13 $31.00 © 2013 IEEE

Ref.	Topology	Process	Multiplier	Output frequency (GHz)	Frequency range (%)	Fundamental suppression (dB)	Chip size without pads (mm²)	Phase noise (dBc/Hz)	P_{dc} (mW)
This work	ILRO	130 nm CMOS	2	4-15	115.8	16-24	0.0072	-103 @ 1 MHz	8.1
[9]	ILRO	65 nm CMOS	4	2.3-4*1	55.7	NA	0.149	NA	60.5
[7]	ILRO	65 nm CMOS	8-14	3.96-7.34*2	60.4	NA	0.0016	-113 @ 1 MHz	12
[5]	LC ILO	90 nm CMOS	3 (exclude 2, 4-8)	27.5-30	8.7	18	0.036	-92.3 @ 100 kHz	4
[6]	LC ILO	130 nm CMOS	2	11-15*3	30.8	45	0.08325	-130 @ 1 MHz	5.2
[4]	LC ILO	180 nm CMOS	3	3915 MHz*3 @ 26.5 GHz	14.8	22.65	0.4554*4	-130 @ 1 MHz	2.95
[3]	LC ILO	180 nm CMOS	2	3.12-3.28	5	42	0.04	-126 @ 1 MHz	2.2
[3]	LC ILO	180 nm CMOS	3	4.68-4.92	5	40	0.04	-124 @ 1 MHz	3.7
[12]	VCO + LC Tripler	180 nm CMOS	3	21.83-27.33	22.4	>18	0.6552	-101 @ 1 MHz	11.1
[13]	VCO + LC doubler	180 nm CMOS	2	21-25.8	20.5	NA	0.5589	-100 @ 1 MHz	15
[10]	Harm. gen.+LC filter	90 nm SOI CMOS	2	26.5-28.5	7.3	11.2	0.1	NA	10
[11]	Harm. gen.+LC filter	800 nm SiGe HBT	2	15.4-18	15.6	25	0.245	NA	22-28
[11]	Harm. gen.+LC filter	800 nm SiGe HBT	2	34.6-37.6	8.3	35	0.35	NA	95-114

*1 Combining 2 ILFMs with adjacent locking ranges *2 Including some unlocked frequency bands *3 Using tuning   varactors   *4 Including pads

Fig. 9.   Performance comparison with other related work.

smaller area than the LC ILOs [3]–[6]. However, LC ILOs has better phase noise and fundamental suppression. The proposed ILFM achieves the widest locking range due to the efficient mixing and multiphase injection in each stage. It also has small area without inductors.

IV. CONCLUSION

In this paper, a new wideband inductorless frequency multiplier based on multiphase ILROs is presented. Integrating harmonic generation into the ILFM significantly reduces circuit complexity and power consumption. A prototype frequency doubler was fabricated in 130-nm CMOS to demonstrate the design concept. The 116% locking range achieved is the largest for ILFMs reported to date. This ILFM design will be highly attractive for applications such as on-chip clock distribution where compact local clock generators with wide frequency range are needed.

ACKNOWLEDGMENT

This work is partially supported by NSF ECCS-0901701 and CNS-1217662. The authors would like to acknowledge MOSIS Education Programs support for the test chip fabrication.

REFERENCES

[1] C.Y. Wu et al. A Phase-Locked Loop With Injection-Locked Frequency Multiplier in 0.18-um CMOS for V-Band Applications. *IEEE Trans. Microwave Theory Tech.*, 57(7):1629–1636, Jul. 2009.

[2] L. Zhang et al. Injection-Locked Clocking: A Low-Power Clock Distribution Scheme for High-Performance Microprocessors. *IEEE Trans. on Very Large Scale Integration Systems*, 9(16):1251–1256, Sep. 2008.

[3] L. Zhang et al. A 1.6-to-3.2/4.8 GHz dual-modulus injection-locked frequency multiplier in 0.18 um digital CMOS. In *IEEE RFIC Symp. Dig. Papers*, pages 427–430, 2008.

[4] M.C. Chen et al. Design and Analysis of CMOS Subharmonic Injection-Locked Frequency Triplers. *IEEE Trans. Microwave Theory Tech.*, 56(8):1869–1878, Aug. 2008.

[5] S.W. Tam et al. Simultaneous sub-harmonic injection-locked mm-wave frequency generators for multi-band communications in CMOS. In *IEEE RFIC Symp. Dig. Papers*, pages 131–134, 2008.

[6] E. Monaco, M. Pozzoni, F. Svelto, and A. Mazzanti. Injection-Locked CMOS Frequency Doublers for μ-Wave and mm-Wave Applications. *IEEE J. Solid-State Circuits*, 45(8):1565–1574, Aug. 2010.

[7] N. Kanemaru et al. A ring-VCO-based injection-locked frequency multiplier using a new pulse generation technique in 65 nm CMOS. In *Proc. International SoC Design Conf.*, pages 32–35, 2011.

[8] J.C. Chien et al. Analysis and Design of Wideband Injection-Locked Ring Oscillators With Multiple-Input Injection. *IEEE J. Solid-State Circuits*, 42(9):1906–1915, Sep. 2007.

[9] D. Dunwell et al. A 2.34GHz injection-locked clock multiplier with 55.7% lock range and 10-ns power-on. In *IEEE Custom Integrated Circuits Conf. Dig. Tech. Papers*, pages 1–4, 2012.

[10] F. Ellinger et al. Ultracompact SOI CMOS Frequency Doubler for Low Power Applications at 26.5-28.5 GHz. *IEEE Microwave and Wireless Components Letters*, 14(2):53–55, Feb. 2004.

[11] J.J. Hung et al. High-power high-efficiency SiGe Ku- and Ka-band balanced frequency doublers. *IEEE Trans. Microwave Theory Tech.*, 53(2):754–761, Feb. 2005.

[12] P.K. Tsai et al. A CMOS Voltage Controlled Oscillator and Frequency Tripler for 22-27 GHz Local Oscillator Generation. *IEEE Microwave and Wireless Components Letters*, 21(9):492–494, Sep. 2011.

[13] G. Bu et al. A 24 GHz Indirect VCO in 0.18 um CMOS Technology. In *Proc. Microw. Integr. Circuits Conf.*, pages 71–74, 2008.

Power Management Circuits for a 15-µA, Implantable Pressure Sensor

Steve Majerus[1,2] and Steven L. Garverick[1]

[1]Dept. of Electrical Engineering and Computer Science
Case Western Reserve University
Cleveland, OH

[2]APT Rehabilitation Research and Development Center
Louis Stokes Cleveland Veterans Affairs Medical Center
Cleveland, OH

Abstract-A custom IC for wireless bladder pressure sensing incorporates power-management circuitry to limit the active time of the instrumentation circuitry and to minimize telemetry rate. Instrumentation circuits are operated with low duty factors in a pipelined manner to generate 100-Hz pressure samples. Telemetry rate is adapted according to sample activity, which is measured using a 2nd-order FIR filter. Measured results indicate that the number of transmitted samples is less than 5% of those acquired, resulting in an average telemetry rate of 1.5 Hz and corresponding current of 0.6 µA. A low-power regulator and clock oscillator set a baseline current draw of 7.5 µA while instrumentation and digital circuits consume an average of 4.7 µA, for a total average consumption of just 12.8 µA.

I. INTRODUCTION

From the time that miniature, subcutaneously-implanted electronics were first demonstrated in chronic applications, clinicians and biomedical designers have been increasingly focused on implanting similar devices within the abdominal cavity. Such devices can provide vital physiological measurements of a number of organs and systems and have been demonstrated acutely, but chronic implantation is difficult due to several complications.

Chronically-implanted systems deep within the abdominal cavity must be small to prevent erosion, migration, or other complications. Furthermore, the distance from the implanted device to an external transceiver can be 20 cm or greater in obese patients. Size-constrained RF-powered sensors have very low efficiency at this distance [1] and passive sensors such as [2] are only suitable for implantation depths less than a few cm.

The deep implantation distance necessitates onboard power storage, which is typically a rechargeable lithium-ion battery in chronically-implanted systems that provide real-time telemetry. In this work, we present a multi-tiered power management scheme for an implantable wireless pressure sensor. This approach achieves low average and peak current consumption drawn during active periods, while still achieving sufficient performance for physiological measurements.

II. POWER MANAGEMENT METHODOLOGY AND DESIGN

A schematic for the pressure sensor IC is shown in Fig. 1a. The dynamic current draw of the instrumentation circuitry is controlled by a power management unit (PMU) which controls power gating. An adaptive transmission engine computes the sensed pressure activity and adjusts the transmitter active

time. Details of the wireless battery recharger were published in [3] and are beyond the scope of this manuscript.

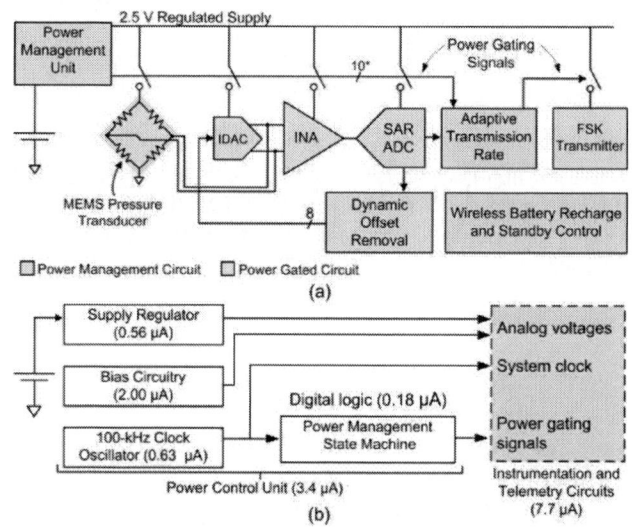

Fig. 1. (a) Block diagram of the wireless pressure sensor IC showing selective power gating switches used to limit power consumption of telemetry and instrumentation circuits. The schematic of (b) shows the individual circuits within the always-running power management unit (after [4]).

The PMU proposed here is a collection of circuits responsible for controlling the overall current draw of the ASIC, as detailed in Fig. 1b. The PMU circuits are always running and set the baseline current draw of the system. The power consumption of these circuits was reduced as much as possible while maintaining acceptable performance and matching. The power consumption of other analog circuits is reduced by a digital finite state machine which turns the instrumentation circuitry on/off in a pipelined manner to minimize the active time for each stage.

A. Low-Quiescent-Current Linear Voltage Regulator

Because the wireless sensor is powered by a secondary cell with fairly large internal resistance, the battery voltage changes significantly with load current, in addition to the slow decrease caused by depletion of stored charge. The PMU uses a simple linear voltage regulator to reduce instrumentation circuit and clock oscillator voltage coefficient requirements. The regulator schematic is presented in Fig. 2.

The regulator topology deviates from the traditional low-drop-out configuration by generating its bandgap reference from the regulated supply, V_{REG}, rather than the unregulated

978-1-4673-6145-3/13 $31.00 © 2013 IEEE

input voltage, V_{BAT}. This configuration provides excellent voltage regulation because the bandgap supply sensitivity is reduced by the loop gain of the regulator, but proper startup and overall stability is more difficult to achieve. A voltage divider formed by matched, long composite transistors M_{F1-F6} provides a feedback factor of ½ to the error amplifier; the resulting output voltage of the regulator is twice V_{BG}.

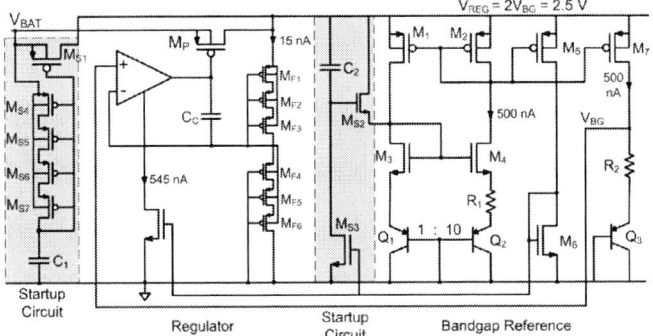

Fig. 2. Low dropout regulator schematic with bandgap reference derived from regulated supply. Startup circuits guarantee proper startup into the stable operating mode in which $V_{REG} = 2V_{BG}$.

This circuit has potentially two undesirable operating points, a zero-current mode and a second mode in which FETs M_{1-7} are in triode, and Q_3 carries a very low current. The AC-coupled startup networks provide pulses of current that overcome these modes, provided that the input supply VBAT has a ramp-up rate greater than 100 V/s [5].

B. Low-Power, 100-kHz Clock Oscillator

The clock source for the PMU state machine is a low-power, current-limited relaxation oscillator, shown in Fig. 3.

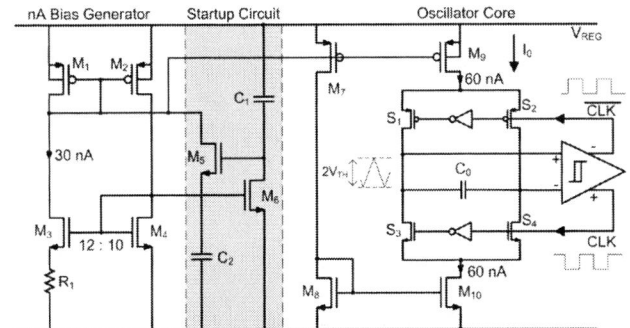

Fig. 3. Low-power, 100-kHz clock oscillator schematic, including nA current reference generator and startup circuit. Total dynamic current draw is 630 nA.

The oscillator symmetrically charges capacitor C_0 at a fixed rate of I_0 through an H-bridge switch configuration. When the differential voltage across C_0 exceeds the comparator hysteresis threshold V_{TH} [6], the capacitor polarity is flipped. The circuit produces a triangle wave of amplitude $\pm V_{TH}$ across C_0 and a corresponding binary clock waveform at the comparator outputs. The oscillation period is given by $2C_0V_{TH}/I_0 + 4T_D$, where T_D is the total loop delay from the oscillator output through C_0. An oscillation frequency of 100 kHz was designed with $C_0 = 3.4$ pF, $V_{TH} = 55$ mV, $I_0 = 60$ nA and T_D of 940 ns.

The charging current I_0 is copied from the nA bias generator

of Fig. 3 [7], which was designed using inversion-coefficient methodology [8] to ensure reasonable current matching. Matched devices M_{1-2}, M_{7-9}, and M_{8-10} were biased in moderate inversion with inversion coefficients of 0.3. Transistors M_3 and M_4 were biased deep into weak inversion with IC = 0.03 to create a 30-nA reference current with $R_1 = 230$ kΩ.

The common-mode level of C_0 is determined by the ratio of small-signal output resistances of M_9 and M_{10}. Since NMOS devices have lower r_0 and increased junction leakage, the comparator uses a PMOS input pair and can tolerate input common-mode levels including 0V. Static differential-mode offsets caused by non-equal charge injection and comparator differential pair offset can change the clock duty cycle from the nominal 50%. Dynamic differential-mode offsets such as the comparator input-referred thermal and $1/f$ noise and the uncorrelated channel noise between M_9 and M_{10} plus dynamic changes in T_D are the primary sources of oscillator jitter.

C. Power management state machine

The PMU state machine controls power consumption through low-duty-cycle operation of analog circuits. This is possible because of the huge speed difference between instrumentation circuitry and the required 100-Hz sampling rate for bladder pressure. The proposed PMU state machine applies power activation to individual circuits, creating a sample pipeline in which sensed information is passed between sequential stages as charge stored on switched capacitors. Stages are switched off as samples are acquired and conveyed to the following stage. The PMU state machine timing diagram illustrated in Fig. 4 shows power gating signals of Fig. 1a.

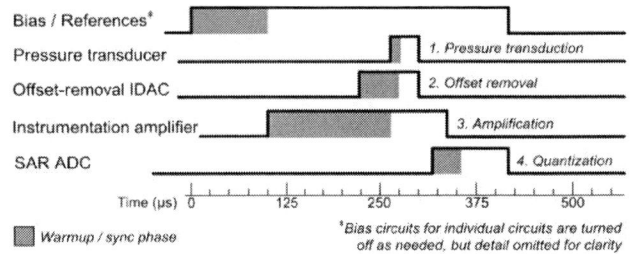

Fig. 4. Timing diagram of PMU signals used during sample acquisition (transmission occurs in the next 450 μs). Circuits are sequentially activated to minimize active time per stage. This figure was modified from that in [4].

Fast-settling elements, such as the piezoresistive pressure transducer and IDAC, have lower duty factors than switched-capacitor and bias circuits, which have longer warmup, step response, and settling limitations. To reduce peak current and RF interference, the FSK transmitter is separately activated for 450 μs after each sample period as determined by the adaptive-rate transmitter described in Section III.

The sample rate of the pressure sensing system is 100 Hz, and the PMU state machine requires just 475 μs to acquire a sample, although some instrumentation circuits are only activated for a small fraction of this time. The time-average current consumption of individual instrumentation circuits is thus reduced to between 1.4 and 6.5 percent. The average current for the pressure sensor IC is dominated by the transmitter when the transmission rate is greater than 25 Hz, as shown in Fig. 5. The PMU, piezoresistive pressure transducer, and digital circuits dominate power consumption at lower rates.

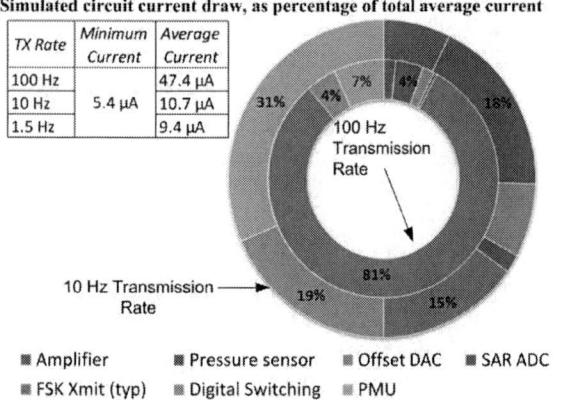

Simulated circuit current draw, as percentage of total average current

TX Rate	Minimum Current	Average Current
100 Hz		47.4 µA
10 Hz	5.4 µA	10.7 µA
1.5 Hz		9.4 µA

■ Amplifier ■ Pressure sensor ■ Offset DAC ■ SAR ADC
■ FSK Xmit (typ) ■ Digital Switching ■ PMU

Fig. 5. Current used by various circuits as percentage of total IC current.

III. ACTIVITY DETECTOR FOR ADAPTIVE TRANSMISSION RATE CONTROL

A sample rate of 100 Hz is required to capture fast transients in bladder pressure, but a fixed, 100-Hz telemetry would account for 81% of the system current. Because bladder contractions are intermittent, significant power savings can be achieved by only transmitting "active" samples. A digital implementation of an activity detector designed specifically for bladder pressure signals is presented in Fig. 6a.

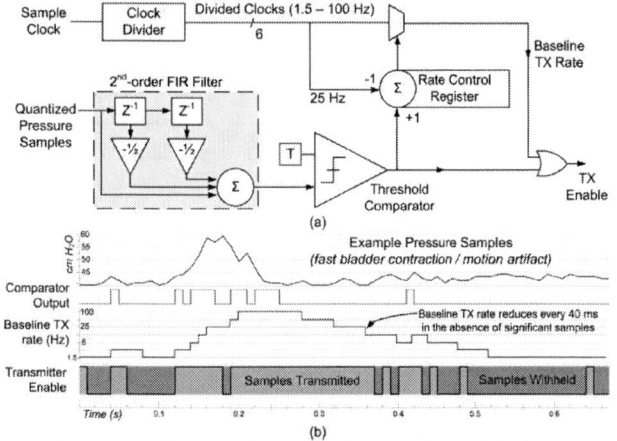

Fig. 6. The adaptive transmission activity detector in (a) determines the appropriate transmission rate for samples. Simulated waveforms in (b) show that samples are transmitted more rapidly during periods of high pressure activity.

The activity detector is based on the first and second differences of the signal [9]. When these differences are combined, the expression becomes that of a 2nd-order FIR filter given by

$$y(x) = f(x) + \alpha f(x-1) + \beta f(x-2). \quad (1)$$

Coefficients α and β can be selected such that $y(x)$ is an indicator of activity. Coefficient values of -½ were chosen to create a high-pass filter with a peaking response at $F_S/4$, unity gain at $F_S/2$ and zero DC gain. The FIR filter output is compared to a threshold by a magnitude comparator; if the sample

is significant enough, it is transmitted.

A rate control register, which sets the baseline transmission rate from 1.5 – 100 Hz, is adjusted based on the level of pressure activity. Samples are transmitted at the baseline rate even if the comparator does not indicate activity. The rate control register is incremented when the magnitude comparator detects activity, and is decremented at a constant rate of 40 ms. Thus, transmission rate remains high for a period of time after activity. Waveforms demonstrating this operation are shown in Fig. 6b for representative, non-voiding bladder contractions.

IV. EXPERIMENTAL RESULTS

The PMU circuits were integrated with instrumentation circuits to produce a pressure sensor IC for wireless bladder pressure monitoring. The 6.25 mm^2 IC was fabricated in a 0.5-µm CMOS process and a die photo is presented in Fig. 7.

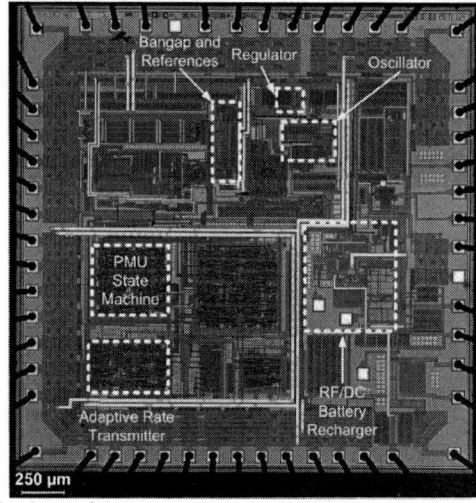

Fig. 7. Die photo of the pressure sensor IC. Power management circuits, including the battery recharger [3], are highlighted and occupy 0.75 mm^2.

A. Dynamic current draw of pressure sensor IC

The current consumption of the pressure sensor IC was measured to verify proper PMU function and to determine leakage currents not modeled through simulation. The IC draws a very low current with intermittent bursts of high peak current, when samples are acquired and transmitted. Dynamic current draw was measured by amplifying the voltage drop across a 10-Ω shunt resistor. An oscilloscope trace of the dynamic current draw is shown in Fig. 8.

The minimum current draw of the IC is set by the PMU DC current and was measured to be 7.5 µA. This was about 40% higher than expected, possibly due to greater digital power consumption and leakage current through bulk junction diodes and gate dielectrics. Peak current of 520 µA occurred during sample transmission. The time-averaged current is a function of transmission rate, but measured to be 12.8 – 50.6 µA for transmission rates between 1.5 and 100 Hz. For the levels of pressure activity observed in human bladders, it is expected that the average transmission rate will be about 3 Hz, in which case the anticipated current draw for the IC is less than 15 µA.

978-1-4673-6145-3/13 $31.00 © 2013 IEEE

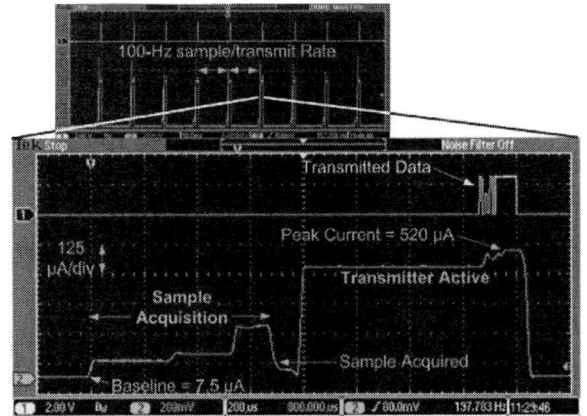

Fig. 8. Measured dynamic current draw of the pressure sensor IC with PMU control. Assuming 100-Hz data transmission, peak current of 520 µA occurs with an average 10% duty factor over a baseline current of 7.5 µA.

B. Adaptive transmission of bladder pressure signals

The performance of the adaptive-rate transmitter was verified by driving the pressure sensor amplifier with a differential voltage source to emulate precise and rapid pressure changes. Pre-recorded voltage waveforms from animal and human trials were used and both compressed and un-compressed ADC samples were captured, as plotted in Fig. 9.

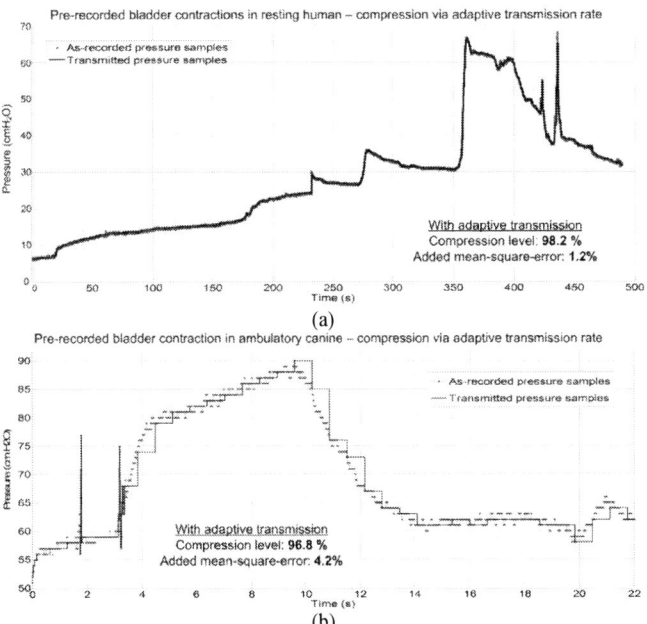

Fig. 9. Measurement of the adaptive-rate transmitter with pre-recorded bladder pressure from (a) a resting human, and (b) an ambulatory canine

The measurement shown in Fig. 9a is based on pre-recorded vesical pressure from a resting human with bladder dysfunction. The data did not contain any high-frequency motion artifacts, and only 1.8% of the recorded samples were transmitted. Transmitter power was reduced accordingly, and the mean-square difference between the two data sets is just 1.2%.

A shorter recording with transient motion artifacts was applied to the system, as shown in Fig. 9b. The adapted transmission rate was only 3.2% of the original sample rate but high-frequency events were transmitted at increased rates as

expected. The system introduced a fair amount of error but the fidelity of the transmitted samples was sufficient to capture the dynamics of the pressure signal.

In these examples, the adaptive-rate transmission yielded dramatic energy savings but added mean-square difference, in part because meaningless noise had been removed. Relative shapes of bladder pressure curves are still apparent and yield important information about bladder dysfunction.

V. CONCLUSION

Power management circuits integrated onto a wireless bladder pressure IC reduced overall and peak current draw by selectively activating circuitry throughout sampling periods. An adaptive-rate transmitter further reduced power consumption by adjusting transmission rate based on the activity level of samples. Measurements of the IC indicate that the PMU was effective at limiting average current draw to 12.8-50.6 µA for transmission rates of 1.5 – 100 Hz. Testing of the adaptive transmission rate with pre-recorded bladder signals confirmed that power savings greater than 95% can be achieved without sacrificing important signal dynamics.

ACKNOWLEDGMENT

The authors would like to thank Dennis Bourbeau, PhD, research investigator at the FES Center and Louis Stokes Cleveland VA Medical Center for providing bladder pressure recordings. This work was supported by the Rehabilitation Research Service of the U.S. Department of Veterans Affairs.

REFERENCES

[1] R.-F. Xue, K.-W. Cheng and M. Je, "High-Efficiency Wireless Power Transfer for Biomedical Implants by Optimal Resonant Load Transformation," *IEEE Transactions on Circuits and Systems I: Regular Papers*, vol. 60, no. 4, pp. 867-874, 2012.

[2] N. Xue, S. Chang and J. Lee, "A SU-8-based microfabricated implantable inductively coupled passive RF wireless intraocular pressure sensor," *in Proceedings of the International Conference of IEEE Engineering in Medicine and Biology Society*, 2011.

[3] M. A. Suster and D. J. Young, "Wireless recharging of battery over large distance for implantable bladder pressure chronic monitoring," *in 16th Intl. Solid-State Sensors, Actuators and Microsys. Conf.*, Beijing, 2011.

[4] S. Majerus, S. Garverick, M. Suster, P. Fletter and M. Damaser, "Wireless, ultra-low-power implantable sensor system for chronic bladder pressure monitoring," *ACM Journal of Emerging Technology*, vol. 8, no. 2, pp. 11.1-11.13, 2012.

[5] Q. Khan, S. Wadhwa and M. Kulbhushan, "Low power startup circuits for voltage and current reference with zero steady state current," *in Proc. of 2003 Intl. Symp. on Low Power Electronics and Design*, Seoul, 2003.

[6] S. Majerus and S. Garverick, "Telemetry platform for deeply implanted biomedical sensors," *in Proc. of the Fifth International Conference on Networked Sensing Systems (INSS 2008)*, Kanazawa, 2008.

[7] Vittoz and J. Fellrath, "CMOS Analog Integrated Circuits Based on Weak Inversion Operation," *IEEE Journal of Solid-State Circuits*, Vol. SC-12, no. 3, pp. 224-231, 1977.

[8] M. Colombo, G. I. Wirth and C. Fayomi, "Design Methodology Using Inversion Coefficient for Low-Voltage Low-Power CMOS Voltage Reference," *in Proceedings of the 23rd Symposium on Integrated Circuits and System Design (SBCCI '10)*, New York, 2010.

[9] R. Rieger and J. T. Taylor, "An Adaptive Sampling System for Sensor Nodes in Body Area Networks," *IEEE Transactions on Neural Systems and Rehabilitation Engineering*, vol. 17, no. 2, pp. 183-189, 2009.

A Novel Voltage-Programmed Pixel Circuit with V_T-Shift Compensation for AMOLED Displays

Maofeng Yang, Nikolas P. Papadopoulos, Czang-Ho Lee, William S. Wong, and Manoj Sachdev

University of Waterloo, Waterloo, Ontario, Canada, N2L 3G1

Email: {m29yang, npapadop, czang-ho.lee, wswong, msachdev}@uwaterloo.ca

Abstract—A novel voltage-programmed pixel circuit using hydrogenated amorphous silicon (a-Si:H) thin-film transistors (TFTs) for active-matrix organic light-emitting diode (AMOLED) displays is proposed. The threshold voltage shift (ΔV_T) of the drive TFT due to electrical stress is compensated by the change of gate-to-source voltage (ΔV_{GS}) generated by the ΔV_T-dependent charge transfer from the drive TFT to a TFT-based Metal-Insulator-Semiconductor (MIS) capacitor. Another MIS capacitor is used to improve OLED drive currents. Measurement results verify the effectiveness and speed of the proposed pixel circuit.

I. INTRODUCTION

A-Si:H TFTs are used to build the backplanes of AMOLED displays for the low fabrication cost and good uniformity. However, the OLED current drop caused by the ΔV_T of the drive TFT under electrical stress reduces pixel lifetime, so several compensation mechanisms have been proposed [1]–[4]. Typical voltage-programmed and current-programmed [1] pixel circuits use two or more TFTs in series with OLED, resulting in higher static power consumption. Besides, conventional V_T-generation methods [1] need complicated control signals, increasing the costs in external driver. The pixel circuit using optical feedback [2] has the photosensor instability problem and the light interference from neighboring pixels [3]. External-driver compensation [4] uses complicated external drivers and has limitations in the resolution of compensation.

In this paper, a voltage-programmed pixel circuit using a novel ΔV_T-compensation mechanism is proposed. The ΔV_T-dependent charge transfer from the drive TFT to a TFT-based MIS capacitor is utilized to generate the ΔV_{GS} of the drive TFT to compensate its ΔV_T. The proposed design uses only one TFT in series with OLED, reducing static power consumption. Control signals are simple, reducing the cost of external driver. The proposed circuit design and operation are presented in Section II. Section III presents analysis and simulation/measurement results. Section IV is the summary.

II. PIXEL CIRCUIT AND OPERATION

The proposed pixel circuit and driving scheme are shown in Fig. 1. V_{prog} is an external programming voltage driver. T_0 is the drive TFT to hold the OLED current in the driving phase ($I_{DS,0}^{driv}$). T_1 is a switch TFT. T_2 and T_3 are the TFTs with drains and sources respectively connected, being used as MIS capacitors. In the driving phase, T_2 provides the compensation for $\Delta V_{T,0}$, and T_3 injects positive charge onto node A to improve the OLED current levels.

A. Programming Phase

TFT gate charge includes two parts: (1) Q_{ch}, the gate charge due to the gate-to-channel capacitance; (2) Q_{ov}, the

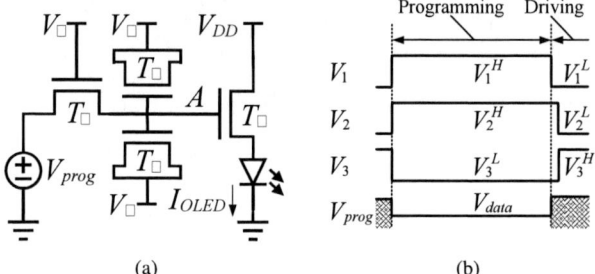

Fig. 1. (a) Schematic and (b) driving scheme of the proposed pixel circuit.

gate charge due to the overlap capacitance between gate and source/drain. In programming phase, T_1 is turned on by V_1^H, T_2 is turned off by V_2^H, and T_3 is turned on by V_3^L. After voltages settle down, T_0 is in saturation mode, T_2 is off, T_1 and T_3 are in triode mode. According to [5], the Q_{ch} of T_0 is

$$Q_{ch,0}^{prog} \approx 2/3 C_i W_0 L_0 \left(V_{data} - V_{OLED}^{prog} - V_{T,0} \right), \quad (1)$$

where C_i is the unit-area channel capacitance, W_0, L_0, and $V_{T,0}$ are the width, length, and threshold voltage of T_0. V_{data} is the data voltage provided by V_{prog} to control $I_{DS,0}^{driv}$. V_{OLED}^{prog} is the set-point OLED voltage and assumed as independent from $\Delta V_{T,0}$ (see Appendix).

B. Driving Phase

In the driving phase, T_1 is turned off by V_1^L, and T_2 is turned on by V_2^L. To improve the OLED drive current, V_3 is switched up to V_3^H to inject positive charge from T_3's gate onto node A. After node voltages settle down,

$$Q_{ch,0}^{driv} \approx 2/3 C_i W_0 L_0 \left(V_{G,0}^{driv} - V_{OLED}^{driv} - V_{T,0} \right), \quad (2)$$

$$Q_{ch,2}^{driv} = C_i W_2 L_2 \left(V_{G,0}^{driv} - V_2^L - V_{T,2} \right), \quad (3)$$

where $V_{G,0}^{driv}$ is the set-point voltage on node A, and V_{OLED}^{driv} is the set-point OLED voltage. Besides, the Q_{ov} of TFTs are

$$Q_{ov,0}^{driv} = C_{ov} W_0 L_{ov} (2 V_{G,0}^{driv} - V_{OLED}^{driv} - V_{DD}), \quad (4)$$

$$Q_{ov \to A,1}^{driv} = -C_{ov} W_1 L_{ov} (V_{G,0}^{driv} - V_1^L), \quad (5)$$

$$Q_{ov,2}^{driv} = 2 C_{ov} W_2 L_{ov} (V_{G,0}^{driv} - V_2^L), \quad (6)$$

$$Q_{ov,3}^{driv} = 2 C_{ov} W_3 L_{ov} (V_{G,0}^{driv} - V_3^H), \quad (7)$$

where C_{ov} is the source/drain unit-area overlap capacitance, L_{ov} is the overlap length at source/drain, and $Q_{ov \to A,1}^{driv}$ is the Q_{ov} of T_1 on the side of node A.

III. ANALYSIS, SIMULATION, AND MEASUREMENT

A. V_T-Shift and Charge Storage Analysis

After the pixel circuit is switched from the programming phase to the driving phase, the following equations can be

978-1-4673-6145-3/13 $31.00 © 2013 IEEE

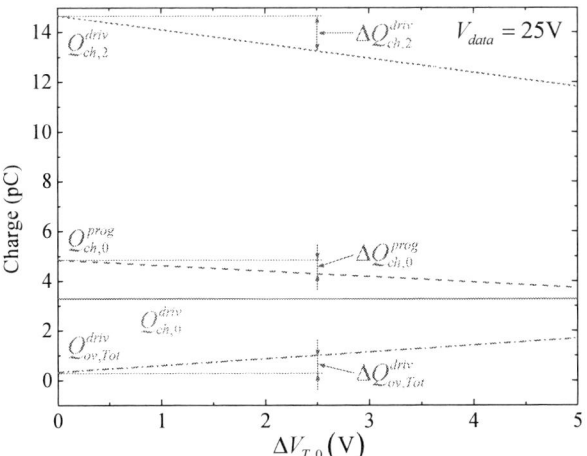

Fig. 2. Simulation results of charges vs. $\Delta V_{T,0}$ of the proposed pixel circuit.

derived based on the charge conservation on node A,

$$\frac{dQ_{ch,0}^{prog}}{dV_{T,0}} = \frac{dQ_{ch,0}^{driv}}{dV_{T,0}} + \frac{dQ_{ch,2}^{driv}}{dV_{T,0}} + \frac{dQ_{ov,Tot}^{driv}}{dV_{T,0}}, \qquad (8)$$

$$Q_{ov,Tot}^{driv} = Q_{ov,0}^{driv} - Q_{ov\to A,1}^{driv} + Q_{ov,2}^{driv} + Q_{ov,3}^{driv}. \qquad (9)$$

Note that, although other charge components also contribute to the charge conservation on node A, they do not vary with $V_{T,0}$ and thus not included in Eq. (8).

When electrical stress is applied, compensating $\Delta V_{T,0}$'s impact on $I_{DS,0}^{driv}$ requires $dV_{GS,0}^{driv}/dV_{T,0} = 1$, which is equivalent to $dQ_{ch,0}^{driv}/dV_{T,0} = 0$ (refer to Eq. (2) and Appendix). This means that if $\Delta V_{T,0}$ is fully compensated by $\Delta V_{GS,0}^{driv}$, the channel charge of T_0 in the driving phase does not change with $\Delta V_{T,0}$. Substituting $dQ_{ch,0}^{driv}/dV_{T,0} = 0$ into Eq. (8) yields

$$\frac{dQ_{ch,0}^{prog}}{dV_{T,0}} = \frac{dQ_{ch,2}^{driv}}{dV_{T,0}} + \frac{dQ_{ov,Tot}^{driv}}{dV_{T,0}}. \qquad (10)$$

Substituting Eq. (1)-(7) and Eq. (9) into Eq. (10) and then using the relevant formulas in Appendix yield

$$W_2 = \frac{2/3 C_i W_0 L_0 + 2 C_{ov,0} + C_{ov,1} + 2 C_{ov,3}}{1/2 C_i L_2 - 2 C_{ov} L_{ov}}, \qquad (11)$$

where $C_{ov,n} = C_{ov} W_n L_{ov}$ ($n = 0, 1, 2, 3$). Eq. (11) indicates the optimal W_2 for $\Delta V_{T,0}$-compensation.

B. $\Delta V_{T,0}$-Compensation Mechanism

The $\Delta V_{T,0}$-compensation mechanism is explained by analyzing the charge terms in Eq. (8). First, $Q_{ch,0}^{prog}$ reduces when $V_{T,0}$ increases (see Eq. (1)), i.e., the increase of $V_{T,0}$ results in less channel charge stored in T_0 during the programming phase. Second, to compensate $\Delta V_{T,0}$, $V_{G,0}^{driv}$ should increase as much as $V_{T,0}$. Since $Q_{ov,Tot}^{driv}$ increases with $V_{G,0}^{driv}$, more positive charge must be provided to the TFT source/drain overlap capacitors, otherwise $V_{G,0}^{driv}$ can not increase with $V_{T,0}$. Third, $Q_{ch,2}^{driv}$ decreases when $V_{T,0}$ increases, because $V_{T,2}$ increases faster than $V_{GS,2}^{driv}$ (see Eq. (3) and Appendix).

Designing W_2 as specified in Eq. (11) validates Eq. (10). When $V_{T,0}$ increases, the decrease of $Q_{ch,2}^{driv}$ is so large that it not only cancels out the decrease of $Q_{ch,0}^{prog}$ but also provides

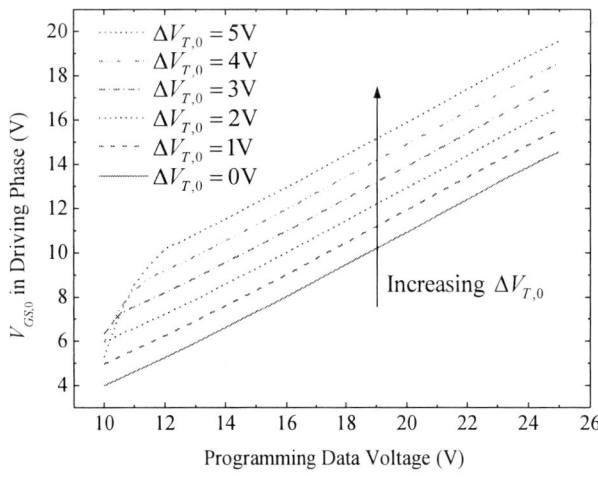

(a)

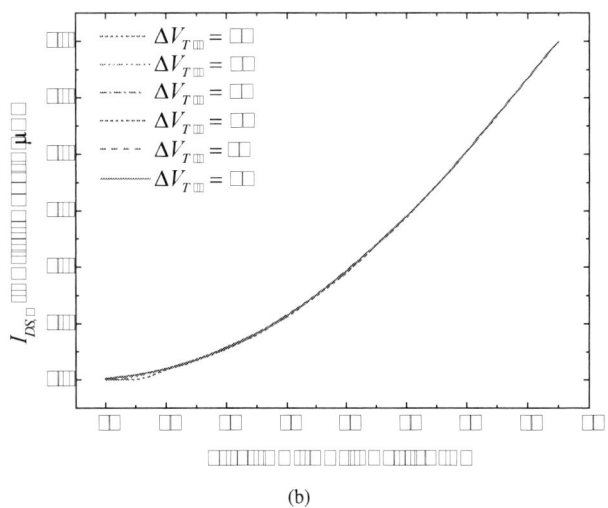

(b)

Fig. 3. Simulation results of (a) $V_{GS,0}^{driv}$ vs. V_{data} and (b) $I_{DS,0}^{driv}$ vs. V_{data} of the proposed pixel circuit.

the extra positive charge needed by the increase of $Q_{ov,Tot}^{driv}$. Therefore, $Q_{ch,0}^{driv}$ does not change with $V_{T,0}$ (see Eq. (8)), so $\Delta V_{GS,0}^{driv} = \Delta V_{T,0}$ (see Eq. (2)). Since $\Delta V_{T,0}$ is fully compensated by $\Delta V_{GS,0}^{driv}$, it does not affect $I_{DS,0}^{driv}$.

C. Simulation Results

To verify the proposed pixel circuit, simulations were conducted using TFT and OLED models [6], [7]. $V_{DD} = 30$V was assumed. The simulations were carried out before the sample fabrication, TFT parameter values were not available, so default parameter values in the TFT model file were used in simulations. To verify the $\Delta V_{T,0}$-compensation, $\Delta V_{T,0}$ was swept. As explained in Appendix, $\Delta V_{T,2}$ was set as $3/2\Delta V_{T,0}$, and the degradations of T_1, T_3 and OLED were neglected.

The simulation results of the four charge terms vs. $\Delta V_{T,0}$ in Eq. (8) are shown in Fig. 2. When $\Delta V_{T,0}$ increases, $\Delta Q_{ch,2}^{driv}$ is so large that it equals $(\Delta Q_{ch,0}^{prog} - \Delta Q_{ov,Tot}^{driv})$, so $Q_{ch,0}^{driv}$ does not change with $\Delta V_{T,0}$, as explained in section III-B.

The simulation results of $V_{GS,0}^{driv}$ and $I_{DS,0}^{driv}$ vs. V_{data} for different $\Delta V_{T,0}$ values are shown in Fig. 3. When V_{data} is relatively high, $\Delta V_{GS,0}^{driv} = \Delta V_{T,0}$, so $\Delta V_{T,0}$ does not affect $I_{DS,0}^{driv}$. When V_{data} is small while $\Delta V_{T,0}$ is large, if $V_{T,2}$

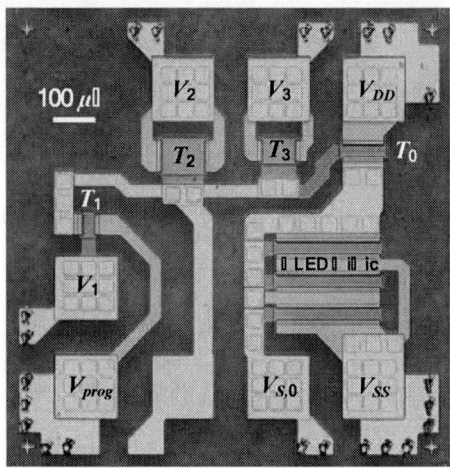

Fig. 4. Fabricated sample of the proposed pixel circuit.

is larger than $V_{GS,2}^{driv}$, T_2 can not turn on to provide $\Delta V_{T,0}$-compensation, so $\Delta V_{GS,0}^{driv} < \Delta V_{T,0}$. This issue only exists around the lowest V_{data} (correspondingly, lowest $I_{DS,0}^{driv}$) levels, so it does not affect the overall stability of $I_{DS,0}^{driv}$.

D. Measurement Results

The samples of the proposed pixel circuit were fabricated in Giga-to-Nanoelectronics (G2N) Centre at University of Waterloo. Fig. 4 is the optical microphotograph of the fabricated circuit. $\Delta V_{T,0}$-compensation was verified by aging tests. The sample was driven by the scheme shown in Fig. 1(b). Due to instrument limitations, $V_{DD} = 20\text{V}$ and $V_{data} = 15\text{V}$ were used. The low and high voltage levels were zero and 20V, respectively. Since the primary test purpose was to verify $\Delta V_{T,0}$-compensation, the OLED mimic (the diode-connected TFT) was excluded by setting V_{SS} as open-circuit. $I_{DS,0}^{driv}$ was measured from pad $V_{S,0}$, whose level was kept at virtual ground. The measured $I_{DS,0}^{driv}$ before the aging test was $1.05\mu A$.

For comparison, a stress test was carried out on another sample whose $\Delta V_{T,0}$-compensation was disabled by fixing V_2 and V_3 at 20V. Therefore, the sample acted like a simple 2-TFT pixel circuit [1]. To make it also have the initial $I_{DS,0}^{driv} = 1.05\mu A$, its V_{data} was set to 15.5V. The environment temperature of the stress tests was 30°C. The measurement results shown in Fig. 5 verify that the overall stability of the $I_{DS,0}^{driv}$ with $\Delta V_{T,0}$-compensation is much better than the one without $\Delta V_{T,0}$-compensation. Note that even with $\Delta V_{T,0}$-compensation, $I_{DS,0}^{driv}$ still has some residual instability. It could be caused by some second-order effects, including non-zero $\Delta V_{T,1}$ and $\Delta V_{T,3}$, and/or minor variations of $(\Delta V_{T,2}/\Delta V_{T,0})$ throughout the test. To minimize the residual instability, further investigations are needed.

The assumptions about the ΔV_T of TFTs used in analysis were verified by the $C-V$ measurement on the pixel sample stressed with $\Delta V_{T,0}$-compensation. As shown in Fig. 6, after the 240h-stress test, $\Delta V_{T,0} \approx 2.15\text{V}$ and $\Delta V_{T,0} \approx 3.15\text{V}$, so their ratio is close to the assumed 2/3. Besides, $\Delta V_{T,1}$ and $\Delta V_{T,3}$ are much smaller than $\Delta V_{T,0}$ and $\Delta V_{T,2}$, so it is fine to neglect $\Delta V_{T,1}$ and $\Delta V_{T,3}$ in the first-order analysis.

Fig. 7 shows the measured transfer curves of T_0 in the pixel sample stressed with $\Delta V_{T,0}$-compensation. The extracted

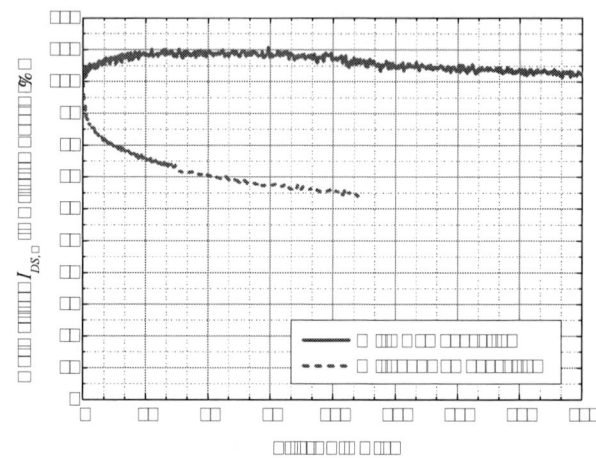

Fig. 5. Measurement results of the normalized $I_{DS,0}^{driv}$ vs. stress time for the proposed pixel circuit with and without $\Delta V_{T,0}$-compensation.

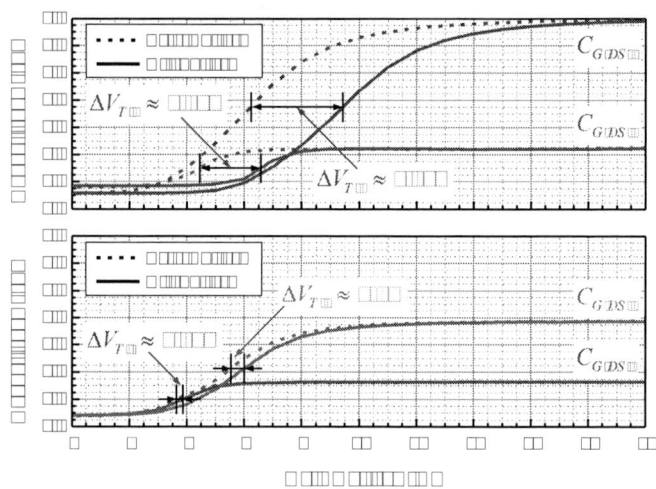

Fig. 6. $C-V$ measurement results of the TFTs in the pixel sample stressed for 240h with $\Delta V_{T,0}$-compensation.

$\Delta V_{T,0} \approx 2\text{V}$ is close to the result from $C-V$ measurement. One can see that $\Delta V_{T,0}$ results in significant $I_{DS,0}$ drops if $\Delta V_{T,0}$-compensation is not used. By contrast, Fig. 8 shows that, when $\Delta V_{T,0}$-compensation is used, except for the lowest $I_{DS,0}^{driv}$ levels, the $I_{DS,0}^{driv}$ differences caused by $\Delta V_{T,0}$ are insignificant. The under-compensation issue on the lowest $I_{DS,0}^{driv}$ levels (zero to $0.15\mu A$) is discussed in Section III-C.

As shown in Fig. 9, the programming speeds of the pixel sample with $\Delta V_{T,0}$-compensation were measured before and after the 240h-stress test, for V_{prog} going from 5V to 15V ($V_{prog} : L \rightarrow H$) and from 15V to 5V ($V_{prog} : H \rightarrow L$). In all cases, $I_{DS,0}$ in the programming phase settled down within 95% of the final value in $250\mu s$. The minimum channel length in the sample was selected as $25\mu m$ to guarantee the yield of the fabrication process. If it is reduced to $5\mu m$ (typically used in industry), the programming speed should be even faster.

IV. CONCLUSION

A voltage-programmed pixel circuit with a novel ΔV_T-compensation mechanism is proposed. It utilizes the ΔV_T-dependent charge transfer from the drive TFT to a TFT-based MIS capacitor to compensate the ΔV_T of drive TFT.

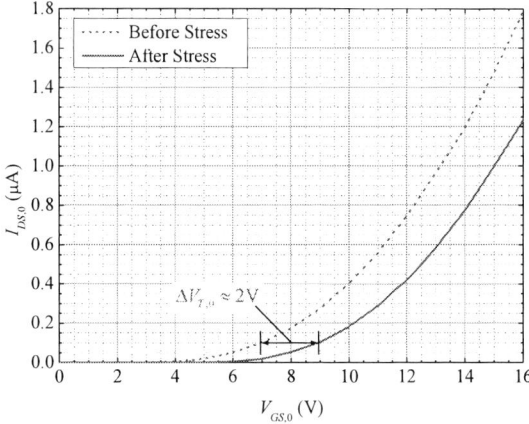

Fig. 7. Measurement results of the transfer curves of T_0 in the pixel sample stressed for 240h with $\Delta V_{T,0}$-compensation.

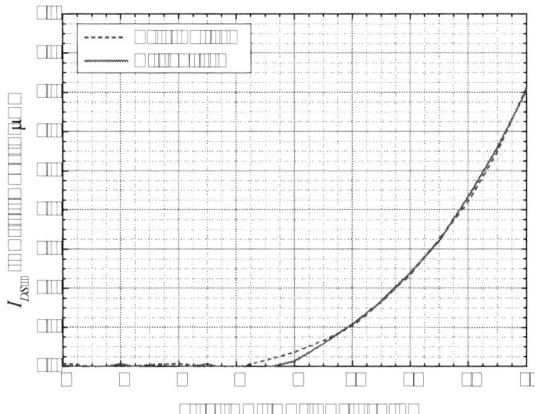

Fig. 8. Measurement results of $I_{DS,0}^{driv}$ vs. V_{data} of the pixel sample stressed for 240h with $\Delta V_{T,0}$-compensation.

Measurement results verify the effective ΔV_T-compensation and the fast programming speed of the proposed pixel circuit.

APPENDIX

Assuming the drive TFT (T_0) and OLED have typical sizes and the same stress condition, according to [8], [9], the OLED has much smaller degradation than the TFT. So, due to the steep OLED $I - V$ curve, $dV_{OLED}^{prog}/dV_{T,0} \approx 0$. Besides, assuming $\Delta V_{T,0}$ is already fully compensated, $I_{DS,0}^{driv}$ and thus V_{OLED}^{driv} do not change with $V_{T,0}$, so $dV_{OLED}^{driv}/dV_{T,0} = 0$.

Note that the programming phase is much shorter than the driving phase. For a practical refresh rate of 60Hz or higher, the effects of negative pulse gate-to-channel stress voltages on ΔV_T are much smaller than those of the positive ones [10]. Therefore, only the TFTs stressed by positive gate-to-channel voltages in the driving phase have significant ΔV_T, so $\Delta V_{T,1}$ and $\Delta V_{T,3}$ can be assumed negligible comparing to $\Delta V_{T,0}$.

The TFT degradations due to $V_{GS,0}^{prog}$ and $V_{GS,2}^{prog}$ are negligible because of the short duration of the programming phase. In the driving phase, T_2 is stressed in triode mode, but T_0 is stressed in saturation mode. V_2^L can be designed as close to V_{OLED}^{driv}, so $V_{GS,2}^{driv} \approx V_{GS,2}^{driv}$. Therefore, according to [11], $dV_{T,0}/dV_{T,2} \approx 2/3$. Substituting this formula and $V_{GS,0}^{driv} \approx V_{GS,2}^{driv}$ into $dV_{GS,0}^{driv}/dV_{T,0} = 1$ yields $dV_{GS,2}^{driv}/dV_{T,2} \approx 2/3$.

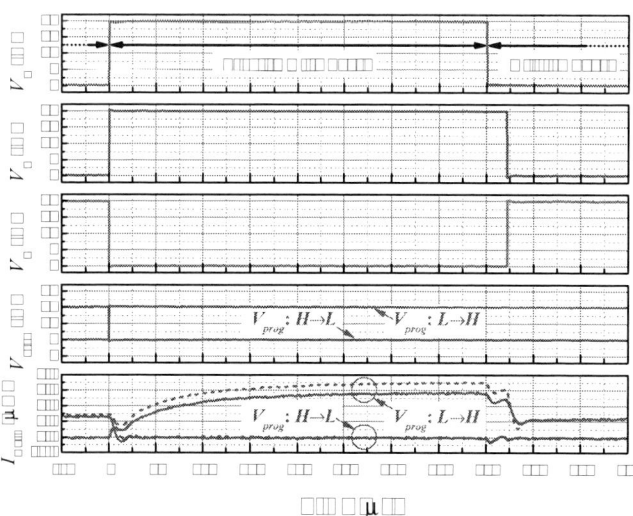

Fig. 9. Measurement results of control signals and $I_{DS,0}$ vs. time of pixel sample with $\Delta V_{T,0}$-compensation. For $I_{DS,0}$, dash curves are the data before the 240h-stress test, and solid curves are the data after the 240h-stress test.

ACKNOWLEDGMENT

The authors would like to thank Melissa Chow and Bright Chijioke Iheanacho (University of Waterloo) for discussions, and Dr. G. Reza Chaji (IGNIS Innovation Inc.) for discussions and the a-Si:H TFT model file. This work was funded by the Ontario Centers of Excellence Project No. 10402.

REFERENCES

[1] A. Nathan, G. R. Chaji, and S. J. Ashtiani, "Driving schemes for a-Si and LTPS AMOLED displays," *JOURNAL OF DISPLAY TECHNOLOGY*, vol. 1, no. 2, pp. 267–277, 2005.

[2] N. P. Papadopoulos, A. A. Hatzopoulos, and D. K. Papakostas, "An improved optical feedback pixel driver circuit," *IEEE TRANSACTIONS ON ELECTRON DEVICES*, vol. 56, no. 2, pp. 229–235, 2009.

[3] G. R. Chaji, C. Ng, A. Nathan, A. Werner, J. Birnstock, O. Schneider, and J. Blochwitz-Nimoth, "Electrical compensation of OLED luminance degradation," *IEEE ELECTRON DEVICE LETTERS*, vol. 28, no. 12, pp. 1108–1110, 2007.

[4] G. R. Chaji, S. Alexander, J. M. Dionne, Y. Azizi, C. Church, J. Hamer, J. Spindler, and A. Nathan, "Stable RGBW AMOLED display with OLED degradation compensation using electrical feedback," *2010 IEEE International Solid-State Circuits Conference - (ISSCC)*, vol. 17, no. 3, pp. 118–119, 2010.

[5] H. C. Slade, *Device and material characterization and analytic modeling of amorphous silicon thin film transistors*. PhD thesis, University of Vtrginia, 1997.

[6] P. Servati, *Amorphous silicon TFTs for mechanically flexible electronics*. PhD thesis, University of Waterloo (Canada), 2004. Ph.D.

[7] A. J. Campbell, D. D. C. Bradley, and D. G. Lidzey, "Space-charge limited conduction with traps in poly(phenylene vinylene) light emitting diodes," *Journal of Applied Physics*, vol. 82, no. 12, pp. 6326–6342, 1997.

[8] S. M. Jahinuzzaman, A. Sultana, K. Sakariya, P. Servati, and A. Nathan, "Threshold voltage instability of amorphous silicon thin-film transistors under constant current stress," *Applied Physics Letters*, vol. 87, no. 2, p. 023502, 2005.

[9] I. D. Parker, Y. Cao, and C. Y. Yang, "Lifetime and degradation effects in polymer light-emitting diodes," *Journal of Applied Physics*, vol. 85, no. 4, pp. 2441–2447, 1999.

[10] C. S. Chiang, J. Kanicki, and K. Takechi, "Electrical instability of hydrogenated amorphous silicon thin-film transistors for active-matrix liquid-crystal displays," *JAPANESE JOURNAL OF APPLIED PHYSICS PART 1-REGULAR PAPERS SHORT NOTES and REVIEW PAPERS*, vol. 37, no. 9A, pp. 4704–4710, 1998.

[11] K. S. Karim, A. Nathan, M. Hack, and W. I. Milne, "Drain-bias dependence of threshold voltage stability of amorphous silicon TFTs," *IEEE ELECTRON DEVICE LETTERS*, vol. 25, no. 4, pp. 188–190, 2004.

978-1-4673-6145-3/13 $31.00 © 2013 IEEE

Design for Manufacturing Layout Analyses Correlate Layout to Physico-Chemical Yield Loss Mechanisms

Christine P. Tan, Congshu Zhou, Yi Tian, Chang Liu, Hein-Mun Lam, Jian Zhang, and Mark Lu
GLOBALFOUNDRIES Singapore Pte. Ltd.
60 Woodlands Industrial Park D Street 2, Singapore 738406

Abstract – **We introduce a case-based learning workflow in the foundry for managing layout weakpoints and implementing layout analyses checks. In this work, we describe case studies that demonstrate how layout analyses can be used to detect layout weakpoints and correlate them to actual physico-chemical mechanisms behind defects observed on silicon.**

I. INTRODUCTION

Layout analysis is emerging as a useful toolbox in design for manufacturing (DFM) for detecting weakpoints in design layout with process marginalities. Broadly, weakpoints can be defined as locations where there is a higher probability of defect occurrence during processing, which may lead to yield loss. Although the design layouts have passed the foundry's basic design rule checks, we had observed that certain types of design patterns are more prone to defect formation.

We adopted an "active foundry" approach towards learning from and managing such weakpoints through the case studies we have encountered [1-2]. The workflow for our approach is shown in Fig. 1. Starting from a yield loss issue, we performed layout analyses to detect and correlate layout weakpoints to process marginalities and physico-chemical mechanisms behind defect formations. The learning from a new case study can be stored in a library and furthermore, the detection of such weakpoints can be hard-coded into refined layout analyses checks. In the future, any repeated occurrence of these defect-prone weakpoints and similar can be detected. The workflow is an iterative cycle of case-based feedback and layout analyses checks, enabling DFM.

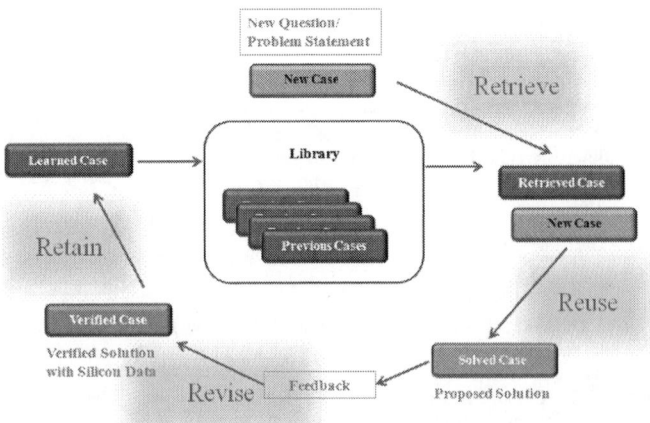

Fig. 1. Workflow for case-based learning and for managing layout weakpoints in the foundry.

In this work, we showcased two example case studies where layout analyses were instrumental in revealing the physico-chemical mechanisms behind defect formation. Incoming reports of yield loss triggered a series of studies into the failure modes. By analyzing the design layout, we were able to hypothesize and verify the root cause of i) voids formed in first copper layer metal (M1) after copper (Cu) chemical mechanical polishing (CMP), and ii) cracks formed in ultra-low-k dielectric after accumulated layers of metal stacks.

Our work demonstrates the utility of DFM layout analyses in elucidating the failure mechanisms behind yield loss, as well as the future potential for detecting layout weakpoints at an early design stage.

II. ELECTROCHEMICAL M1 VOID FORMATION

A. Background

Copper CMP is a key process step for enabling back-end-of-line (BEOL) planarization in integrated circuits (IC) manufacturing. However, the complex relationship between CMP process and design layout remains to be fully understood. In particular, both chemical and mechanical factors during CMP process, such as the diversity of CMP slurries/additives available, pH, polishing pad lifetime *etc.* can all affect CMP performance. In addition, the design layout of metal lines, *e.g.* local pattern density, can also play a role in determining the uniformity of planarization on the wafer.

In this case study, layout analyses were used to detect a particular type of void defects that systematically occurred in M1 layer after CMP. Based on our layout analyses findings, we formed a hypothesis on physico-chemical mechanism – galvanic corrosion – that could cause the M1 voids. Physical failure analysis on silicon verified that the M1 voids indeed occurred on our detected weakpoints locations. As a result, process solutions could be proposed for the M1 CMP step to mitigate void formation.

B. Layout Analyses

We performed layout analyses checks using standard EDA software to locate the M1 void weakpoints. We specified the checks to locate points that met our criteria – M1 metal island on top of a n+/N-well having an adjacent neighboring metal island on top of a p+/P-well. The weakpoints of interest are the metal island pairs with the distance separating them being <10x the smallest width of the metal island. This distance would be within the diffusion radius, as each metal island behaves as a microelectrode.

978-1-4673-6145-3/13 $31.00 © 2013 IEEE

C. Results and Discussions

The left image in Fig. 2 was highlighted as a M1 void weakpoint by our layout analyses. The right image in Fig. 2 was the aerial defect scan corresponding to the same location. We observed that the void occurred on the metal island on top of the n+/N-well.

Based on our results, we hypothesized that the M1 voids could be formed based on an electrochemical effect. This galvanic mechanism is illustrated in Fig. 3. The metal island on the p+/P-well behaved as a cathode, while the metal island on the n+/N-well behaved as an anode. During CMP and/or post-CMP cleans, the aqueous environment on top of the wafer provides a conducting medium for the movement of Cu^{2+} ions. A positive potential was formed at the p+/P-well region after the CMP process completely exposed and separated the two metal islands [3]. The potential difference between the two metal islands is typically around 0.2-0.5V [4], driving Cu oxidation and reduction reactions at the n+ and p+ metal islands respectively.

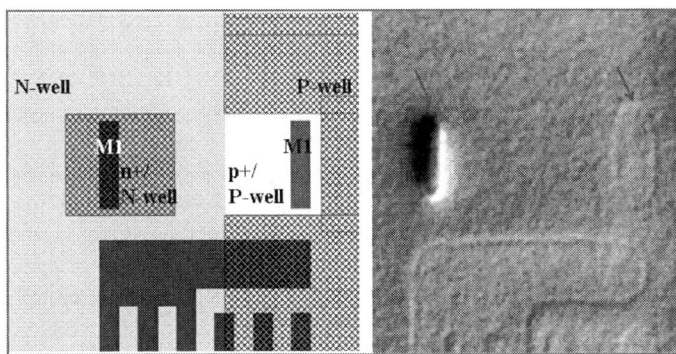

Fig. 2. Example of M1 void layout weakpoint detected by layout analyses. The left image is the design layout, and the right image is the corresponding M1 defect scan with red arrows pointing to the metal islands. Note that M1 void occurred on the metal island on top of n+/N-well.

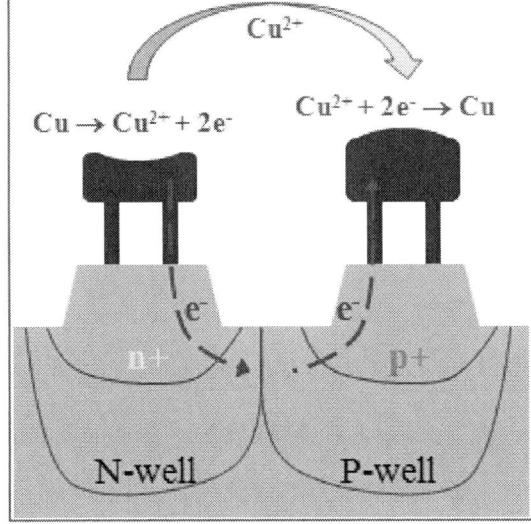

Fig. 3. Schematic showing the electrochemical mechanism for M1 void formation.

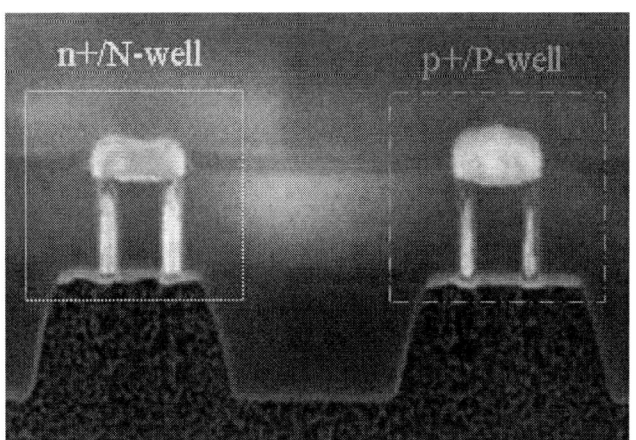

Fig. 4. SEM cross-section of layout weakpoint, confirming M1 void formation at the n+/N-well metal island and Cu deposited at the p+/P-well metal island.

Cross-sectional scanning electron micrograph (SEM) image in Fig. 4 showed that the p+ metal island was significantly thicker (>10s nm) than our baseline, confirming that Cu was deposited. Additionally, the void seen at the n+ metal island suggested that Cu was being consumed in an oxidation reaction. Our observations are consistent with previously reported mechanisms of layout pattern-specific and photo-induced galvanic corrosion of Cu described previously [3-4].

Process solutions to reduce galvanic include passivation of the exposed Cu, performing post-CMP cleans in a dark environment, adding inhibitors such as benzotriazole to the CMP slurry [5].

The extent of this M1 void defect formation is dependent on the time that the wafer is exposed to aqueous solutions, as well as the area of the metal islands. Our layout analyses can be useful for detecting the M1 void weakpoints even before wafer processing. This DFM toolbox can help to alert process owners to focus on these weakpoint locations for yield defect density scans.

III. STRESS-INDUCED ULTRA LOW-K DIELECTRIC CRACKING

A. Background

Low-k materials have been widely used as dielectrics in IC manufacturing, due to their advantages of reduced parasitic capacitance, reduced cross-talk and faster switching speeds. However, it can be challenging to incorporate these low-k and ultra-low-k materials into the process integration scheme due to their mechanical properties. For instance, low-k materials have a lower elastic modulus, lower hardness, and higher coefficient of thermal expansion (CTE) compared to silicon dioxides. As a result, during BEOL process integration, these low-k materials will encounter mismatches in these mechanical properties compared to the other materials used (e.g. Cu metal lines, cap layers), which may lead to film cracks and delamination.

In this case study, layout analyses were used to detect the weakpoints where cracks are likely to occur in ultra-low-k dielectric after processing multiple metal layers.

978-1-4673-6145-3/13 $31.00 © 2013 IEEE

B. Layout Analyses

We specified our checks to look for regions with a narrow space filled with ultra-low-k dielectric, flanked by multiple (>=3) wide metal lines stacked on top of each other. We identified the weakpoints as spaces with width < 1/5 of the adjacent metal stack width. We enhanced our layout analyses by adding a condition of a top thick metal line (M7) straddling across the space and the metal stacks, as this would represent the worst case scenario of stress mismatches in the film stack.

C. Results and Discussions

An example of a crack weakpoint we found in the design layout is shown in Fig. 5. The narrow space filled with ultra-low-k dielectric, flanked by wide metal stacks, was prone to developing cracks.

Fig. 6 illustrates a proposed model for stress-induced cracks in this type of weakpoint. The ultra-low-k dielectric and cap layers can be modeled as springs with stiffness proportional to the elastic moduli and thicknesses, and inversely proportional to the space between the metal stacks:

$$Stiffness \propto Elastic\ modulus \times Thickness \times \frac{1}{Space} \quad (1)$$

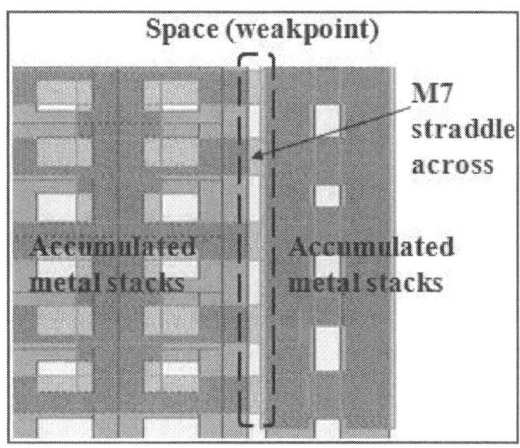

Fig. 5. Example of layout weakpoint in the ultra-low-k dielectric flanked by accumulated metal stacks on both side of the space.

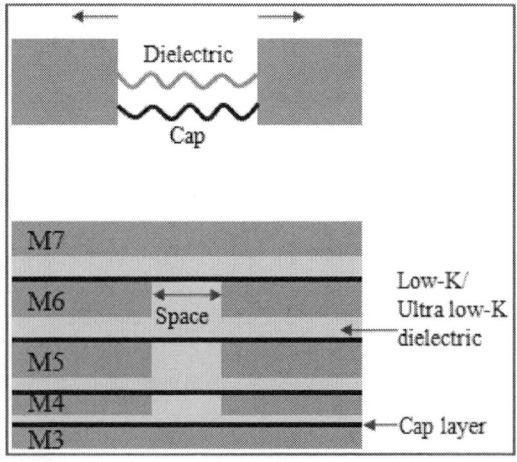

Fig. 6. Schematic showing stress-induced cracking in ultra low-k dielectric with accumulated topography.

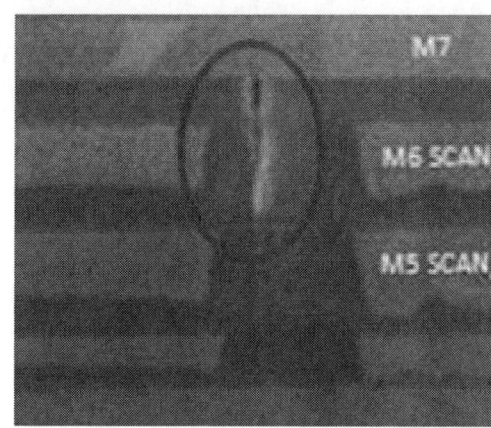

Fig. 7. SEM cross-section showing stress-induced cracking in the ultra-low-k dielectric at the weakpoint space we identified.

Fig. 7 is a SEM cross-sectional image of the weakpoint we detected, which showed a crack that propagated in the dielectric down to the cap layer on top of M5. Our observations were consistent with suggested mechanisms in literature.

Generally, the driving force of cracking (G), which determines the propagation of a crack in a film, is proportional to the stress and thickness in the film layer. Cracks were likely to form in this type of weakpoint because of the large G due to the large mismatch in stress (as a result of stiffness differences between the cap and dielectric layers). G increases when i) dielectric constant k decreases, ii) the number of metal layers increases [6-8], which were fulfilled in the layout conditions we set for this weakpoint. The built up stress then relieves itself by forming cracks in the ultra-low-k dielectric filled spaces between metal lines.

Other possible mechanisms are related to thermal processing during BEOL integration. One scenario could be crack forming in the ultra-low-k dielectric due to different extent of thermal expansion (CTE mismatched of the different materials in the film stack). Another possible scenario for the crack formation could also be the outgassing of moisture in the organosilicate-based ultra-low-k dielectric.

Layout-based solutions to reduce the stress due to the elastic moduli mismatch between the ultra-low-k dielectric and the rest of layers in the composite film stack include: i) adding non-functional metal structures (dummy fill) in the space between the wide metal stacks, and ii) widening the space between the metal stacks. Process solutions include optimizing the thickness of the dielectric, cap, and metal layers.

In the future, the learning from this case study can enable checks for this type of weakpoints in the design layout, to be integrated into the manufacturing environment. Awareness of stress-induced cracks in ultra-low-k dielectrics may help guide designers to avoid such defect-prone designs and improve yield and reliability.

IV. CONCLUSIONS

In this work, we demonstrate that layout analyses can be useful tools to help correlate design layout with physical and electrochemical failure mechanisms. Understanding of these root causes can help to debug yield loss and implementing process or layout solutions. We also introduced a case-based workflow for managing and implementing weakpoint checks in the foundry. The learning from these case studies can form key standard checks in the future to enable yield enhancement, and to bridge design to manufacturing.

ACKNOWLEDGMENT

The authors are grateful to Rina Lin and Mingli Yang for helpful discussions.

REFERENCES

[1] M. Lu, et al., "Novel customized manufacturable DFM solutions," *Proc. SPIE Photomask Technology 2012,* vol. 8522, pp. 852223, December 2012.

[2] G. Finnie, and Z. Sun, "R^5 model for case-based reasoning," *Knowledge-Based Systems,* vol. 16, pp. 59-65, 2003.

[3] H. Aoki, D. Watanabe, S. Hotta, C. Kimura, and T. Sugino, "Corrosion suppression during wet processes in FEOL and BEOL for 45nm node and beyond," *ECS Transactions,* vol. 11 (2), pp.19-30, 2007.

[4] Y. Homma, et al., "Corrosion control in copper damascene process," *Chemical Mechanical Planarization in IC Device Manufacturing,* vol. III, , R. L. Opila et al., Eds. New Jersey: The Electrochemical Society, 2000, pp. 83-93.

[5] Y. Ein-Eli, D. Starosvetsky, "Review on copper CMP and post-CMP cleaning in ULSI – an electrochemical perspective," *Electrochem. Acta,* vol. 52, pp. 1825-1838, 2007.

[6] S. Balakrishnan, R. Brain, and L. Zhao, "Integration and electrical properties," *Advanced Interconnects for ULSI Technology,* M. Baklanov et al., Eds. United Kingdom: Wiley, 2012, pp. 241-247.

[7] X. H. Liu, et al., "Low-k BEOL mechanical modeling," *Advanced Metallization Conference,* pp. 361-367, 2004.

[8] X. H. Liu, et al., "Mechanical reliability outlook of ultra-low-k dielectrics," *Advanced Metallization Conference,* 2010.

A Split-Foundry Asynchronous FPGA

Benjamin Hill, Robert Karmazin, Carlos Tadeo Ortega Otero, Jonathan Tse, and Rajit Manohar

Computer Systems Laboratory, Cornell University

Ithaca, NY, 14853, U.S.A.

{ben,rob,cto3,jon,rajit}@csl.cornell.edu

Abstract—**We present the first published measurements of a complex digital integrated circuit fabricated in both standard and split-foundry processes. Our 1.3-million-transistor asynchronous FPGA operates at over 300MHz in 130nm. We discuss the challenges inherent in split design and our automated layout tools that address them.**

I. INTRODUCTION

The semiconductor industry relies heavily on access to cost-effective integrated circuit production at state-of-the-art semiconductor foundries, which have become a global resource providing service to a wide range of customers. For these customers, design intellectual property (IP) must leave their control for fabrication. This exposes the IP to certain risks such as reverse engineering [1], hardware piracy [2], and malicious modification [3]. Foundry customers, especially those involved in national defense applications, must take steps to mitigate these risks to protect their IP, economic position, and security interests.

Split-foundry fabrication has been proposed as a technique to enable use of state-of-the-art semiconductor foundries while minimizing the risks to IP or reducing production costs [4,5]. Split manufacturing separates a design into Front End of Line (FEOL) and Back End of Line (BEOL) portions for fabrication by separate foundries. An untrusted foundry performs FEOL manufacturing, then ships wafers to a trusted foundry for BEOL fabrication.

Split manufacturing introduces additional complexity to the design process, such as FEOL/BEOL mask alignment and unknown differences in electrical characteristics of structures made by two different foundries. We had a unique opportunity to evaluate these challenges by fabricating two versions of the same design: one manufactured normally in a single foundry and the other in a split-foundry process.

In this work we describe the challenges in designing for split manufacturing (Section II) and our split-foundry capable synthesis flow (Section III). We also provide quantitative analysis of the energy and performance impacts of the split manufacturing process we used (Section VI).

The design chosen for this case study is an asynchronous field programmable gate array (Section V). FPGAs can provide an additional layer of IP protection, since the application is not introduced until after the manufacturing process [6]. Our asynchronous design methodology (Section IV) is robust to

gate and wire delays, accommodating any variation introduced by the split manufacturing process.

To the best of our knowledge, this work is the first to present measured results from a complex digital system fabricated in a split-foundry process.

II. PHYSICAL DESIGN

We fabricated our FPGA design using two different 130 nm processes: a complete FEOL/BEOL fabrication at an untrusted foundry (Foundry A), and a split-foundry process in which the FEOL and first metal level were fabricated at Foundry A followed by BEOL manufacturing in a trusted foundry (Foundry B).

Fabrication in a split-foundry process brings a number of challenges, especially due to different design rules and electrical characteristics. While the exact details will depend on the choices of foundries and technology node, we present the challenges we faced as a case study of the design considerations inherent to split-foundry fabrication.

To select the FEOL Foundry A, a designer simply chooses the process with the most appropriate semiconductor device characteristics for the application. FEOL design rules are unaffected by any design rules from Foundry B.

However, any differences in BEOL design rules and actual BEOL implementations between Foundries A and B will result in different electrical characteristics. Common metallization stacks in state-of-the-art processes provide three to six *thin* wiring layers. Some foundries offer up to three *thick* metal layers, which have lower RC parasitics per unit length of wiring. To make the most efficient use of planar wiring resources, thin wiring layers are often well-suited for local, dense interconnect whereas the thick layers are better for long distance interconnect.

Oftentimes, most switching activity is confined to the lower level metals—typically comprised of the thin metallization layers. In our FPGA case study, the thicker, higher-level metals are used mostly for power distribution nets, so the effects of the thin BEOL characteristics dominate our performance measurements. For our processes, Foundry B offers 10% to 15% worse RC characteristics per unit area for thin wiring, but approximately 5% better RC characteristics for thick wiring.

To ease implementation, we used the union of the most restrictive design rules from Foundry A and B to ensure that our design would pass DRC in both foundries. Table I shows examples of the BEOL design rule differences between Foundries A and B, as well as our composite rule set.

This work was supported by IARPA award N66001-12-C-2009

978-1-4673-6145-3/13 $31.00 © 2013 IEEE

TABLE I: Design rules normalized to Foundry A dimensions

Design Rule	Foundry B	Composite
Manufacturing grid	2.00	2.00
M1-M1 spacing	0.90	1.00
M6 min width	2.00	2.00
M7-M7 min spacing	0.80	1.00

Unlike metal wiring, via cuts have exact size and shape requirements which differ for each foundry. Our composite ruleset for vias implements the most strict rules for metal overhang, but we use a placeholder cell for via cuts. When emitting the final layout, we simply substitute the appropriate foundry's via cut geometry.

III. CELLTK: NON-STANDARD CELL LAYOUT

Our split-foundry toolflow is based on *cellTK* [7]. *cellTK* is an on-demand cell generator that transforms a transistor-level netlist into a design mask layout by clustering transistors into cells and producing the physical layout for each cell. Each of these "non-standard" cells is generated using two-step process: a transistor placement phase followed by an intra-cell routing phase. The cells produced by *cellTK* are compatible with standard cell place and route tools, which we use to assemble the final design.

Split manufacturing alone may not be enough to guarantee security. Attackers can use device proximity and standard cell pin placement information from the FEOL to infer BEOL connectivity [8]. *cellTK* does not use standard library cells, so it is less vulnerable to this approach. Further, it allows for full designer control over placement and routing of FEOL devices and metallization if additional obfuscation is required.

cellTK gracefully handles differences in the manufacturing grid by aligning geometry to the most conservative grid. An off-grid design complicates or outright prevents alignment of the FEOL/BEOL geometry when assembling the final wafer. Common hierarchical file formats such as GDS-II can exacerbate the grid problem if different grids are used throughout the hierarchy.

Our approach to our dual-foundry fabrication was to generate layout using geometry intended for Foundry A, then replace the BEOL with that of Foundry B. This requires layer translation steps as well as the instantiation of the appropriate vias as described earlier. During the conversion, we verify that all portions of the design satisfy our composite DRC ruleset and that all geometry is on the most conservative grid.

IV. ASYNCHRONOUS DESIGN METHODOLOGY

Our FPGA was designed using Martin synthesis [9], a procedure that decomposes a sequential program description into fine-grained parallel hardware processes. These processes are Quasi Delay-Insensitive (QDI) [10]: they operate correctly in the presence of arbitrary wire or gate delays[1]. As a result, QDI circuits are intrinsically robust and tolerant of variations

[1]With the exception that the relative delay on certain wire forks must be bounded

in fabrication process, voltage and temperature. For example, reducing the operating voltage simply causes processes to operate more slowly. This adaptability makes self-timed circuits a good fit for BEOL variations introduced by split manufacturing.

Instead of using a global clock to synchronize data transfers, asynchronous processes communicate by passing tokens over delay-insensitive, point-to-point channels using a handshaking protocol. Processes are naturally event-driven, waiting in a quiescent state with no switching activity until a data token arrives. This is the equivalent of perfect clock-gating in a synchronous system. Inactive processes consume only leakage current.

Finally, the local synchronization behavior of self-timed circuits enables average-case system performance. Synchronous systems define their clock period by the slowest pipeline stage (even if it is rarely exercised), throttling the entire system. In asynchronous systems, system throughput is determined only by *active* hardware units. Thus, each execution path runs at maximum local throughput, yielding average-case performance over all possible paths.

V. ASYNCHRONOUS FPGA

A. Architecture

Our FPGA is based on the one described in [11] and [12]. It has an "island-style" architecture, as shown in Figure 1. The FPGA fabric is organized as a symmetric 2D array with "islands" of logic in a "sea" of programmable routing.

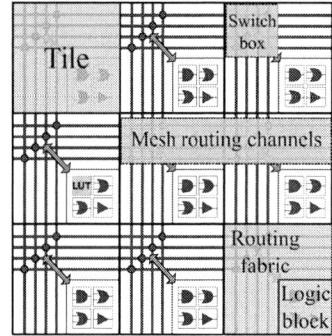

Fig. 1: "Island-style" FPGA architecture

Each tile in the array contains both a logic block and routing fabric. The FPGA is asynchronous, so all logic is implemented as QDI processes and all communication occurs via point-to-point delay-insensitive channels, as described in Section IV.

The logic block contains four lookup tables (LUTs), each of which can be programmed to implement any four-input logic function. LUT outputs can be used as inputs to other LUTs in the same tile without using any mesh routing resources. The logic block also includes units to source and sink data tokens.

Within the routing fabric, a programmable switch box statically connects inter-tile channels in a 2D mesh network. This FPGA has 32 channels/tile in each direction. Each tile also has local routing connecting the logic block to the mesh. Unlike a synchronous FPGA, the routing fabric is pipelined to

978-1-4673-6145-3/13 $31.00 © 2013 IEEE

support high throughput. The asynchronous channel protocol permits this pipelining without retiming problems.

B. Physical implementation

For this chip we fabricated a 5x5 array of tiles, shown in Figure 2a. This FPGA fabric is large enough to allow us to characterize the split-foundry process.

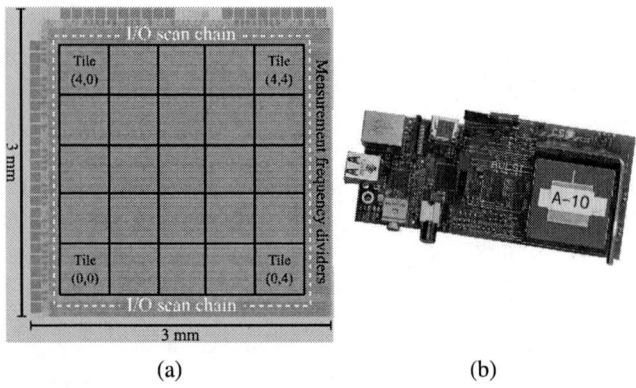

(a)   (b)

Fig. 2: FPGA layout floorplan (a) and test platform (b)

Because of pin limitations on this test chip, we use a scan chain at the array periphery to pass data to and from the FPGA. This allows us to test far more functionality than would otherwise be possible, albeit at greatly reduced throughput.

The FPGA is programmed by writing to a configuration memory, implemented as a linear scan chain. In addition to asynchronous data flow graphs, we can also map synchronous designs to the fabric using a simple conversion tool. Design place-and-route can be performed using VTR [13] (formerly VPR) or specified manually for the greatest control.

Each tile uses approximately 52k transistors. The configuration memory consumes 58% of those, routing fabric another 32%, and logic block 10%. This routing-to-logic ratio is typical of FPGAs and reflects the high cost of reconfigurability. Die area for the chip is $9\,mm^2$, with 1.33 million transistors.

C. Using Asynchronous FPGAs for Measurement

Self-timed circuits naturally run at highest possible throughput for a given environmental condition (Section IV). This behavior allows us to experimentally characterize each process technology by simply measuring the performance of a given benchmark.

Our FPGA was designed with measurement in mind. There are on-die frequency taps placed throughout the routing fabric that observe the switching activity on the mesh channels and pass it off-chip for measurement. For example, Figure 3 shows an example configuration for measuring the maximum FPGA operating frequency. Tokens are generated in one logic block, routed through a channel where they are measured, and consumed in another tile. When measuring frequency, we do not use the periphery scan chain to avoid artificially throttling the FPGA.

In addition, the programmable FPGA fabric allows for highly detailed experimentation. It is possible to choose the

number of transistors active at a given time and their location on the die. This allows us to characterize both performance and power at a fine granularity.

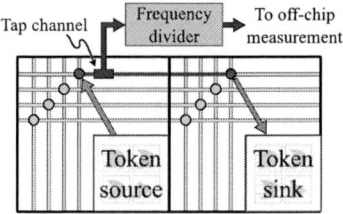

Fig. 3: Test to determine maximum FPGA operating frequency

VI. EVALUATION

To evaluate the impact of split manufacturing on functionality, power, and performance, we compared split-foundry chips with those fabricated entirely by Foundry A.

Figure 2b shows our custom-designed automated testing platform, which loads configurations into the FPGA, runs test procedures, and collects data. All current measurements were performed on ceramic-packaged die at room temperature, using a separate high-precision source-measure unit.

A. Functionality/Yield

We constructed a benchmark exercising the entire configuration memory and all LUTs within the FPGA. All die[2] from both processes were found to be completely functional. This suggests that at least for our particular FEOL/BEOL pair, split manufacturing is a viable option.

B. Static power

To measure static power, we loaded the FPGA with an empty configuration. Due to the event-driven nature of self-timed circuits, the absence of data inputs guarantees there will be no switching activity.

Static power consumption depends only the FEOL transistors[3], so we should expect identical results for both fabrication techniques. Figure 4a shows that this is indeed the case: measured leakage current is closely matched across the full range of supply voltage. The average static power consumption of Foundry A chips was $1.57\,mW$ at $1.5\,V$ and $2.09\,\mu W$ at $0.7\,V$, versus $1.58\,mW$ and $2.00\,\mu W$ for the split-foundry.

C. Dynamic power and performance

Maximum operating frequency[4] for the FPGA was measured using the configuration shown in Figure 3. We instantiated 15 of these test configurations at different locations within the fabric, in order to capture intra-die variation.

Unlike static power, maximum operating frequency depends strongly on the BEOL wiring, which has 10-15% higher parasitic capacitance for thin wiring in Foundry B. It takes

---

[2]12 chips from Foundry A alone; 13 split-foundry chips
[3]Assuming power networks are adequate to handle the very small current
[4]More precisely throughput, measured in asynchronous tokens per second. This has units of Hz and is roughly equivalent to synchronous frequency.

978-1-4673-6145-3/13 $31.00 © 2013 IEEE

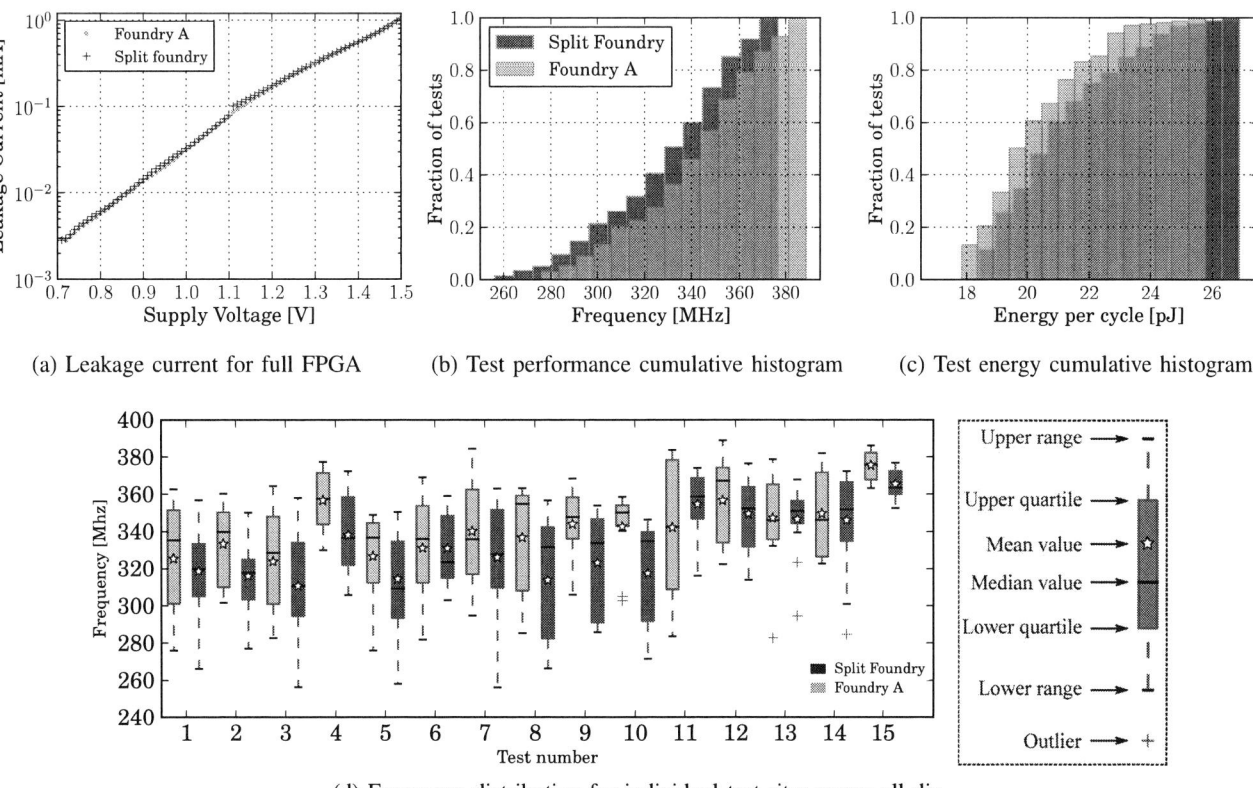

(a) Leakage current for full FPGA   (b) Test performance cumulative histogram   (c) Test energy cumulative histogram

(d) Frequency distribution for individual test sites across all die

Fig. 4: Measured static and dynamic characteristics for each manufacturing process

longer to charge and discharge this capacitance, so we would expect the split process to be somewhat slower. Figure 4d shows the frequency distribution for each test site across all chips. The split-foundry chips are generally slower overall, but the variation between die is too large to allow a definitive conclusion.

The trend is more evident in a cumulative histogram of all tests, pictured in Figure 4b. The distribution for Foundry A chips is shifted higher in frequency, with a mean of 342 MHz versus 311 MHz for the split-foundry[5].

Figure 4c shows a similar shift, with split-foundry chips consuming more energy per operation on average[6]. This too follows from what we know about the BEOL, since each signal transition in the split process must charge and discharge a larger capacitance.

VII. Summary

In this paper we presented a comparative study of an asynchronous FPGA fabricated in both a standard 130 nm process and a split manufacturing process. We also described our methodology for creating split layout, as well as the challenges posed by split-foundry design.

Measurements found all split-foundry FPGAs to be fully functional, with a mean peak throughput over 300 MHz. Compared to chips from the standard process, they showed a 10% decrease in frequency and a 5% increase in energy, both

likely attributable to higher capacitance BEOL wiring. Due to the inherent variation tolerance of asynchronous circuits, we were able to use the same netlist without modification for both processes. Our results demonstrate that split manufacturing is a viable technique.

References

[1] R. Torrance and D. James. "The state-of-the-art in semiconductor reverse engineering." *IEEE DAC*, 2011.
[2] J. A. Roy, *et al.* "EPIC: ending piracy of integrated circuits." *IEEE DATE*, 2008.
[3] R. S. Chakraborty and S. Bhunia. "HARPOON: An Obfuscation-Based SoC Design Methodology for Hardware Protection." *IEEE TCAD*, 2009.
[4] "IARPA Trusted Integrated Circuits (TIC) program announcement." https://www.fbo.gov/utils/view?id=b8be3d2c5d5babbdffc6975c370247a6.
[5] R. Jarvis and M. G. McIntyre. "Split manufacturing method for advanced semiconductor circuits." *US Patent 10/305,670*, 2007.
[6] S. Trimberger. "Trusted Design in FPGAs." *IEEE DAC*, 2007.
[7] R. Karmazin, *et al.* "cellTK: Automated Layout for Asynchronous Circuits with Nonstandard Cells." *IEEE ASYNC*, 2013.
[8] J. Rajendran, *et al.* "Is Split Manufacturing Secure?" *IEEE DATE*, 2013.
[9] A. Martin. "Compiling Communicating Processes for Delay-Insensitive VLSI Circuits." *Distributed Computing*, 1986.
[10] A. Martin. "The Limitations to Delay-Insensitivity in Asynchronous Circuits." *6th MIT Conference on Advanced Research in VLSI*, 1990.
[11] J. Teifel and R. Manohar. "An asynchronous dataflow FPGA architecture." *IEEE Computers*, 2004.
[12] R. Manohar. "Reconfigurable Asynchronous Logic." *IEEE CICC*, 2006.
[13] J. Rose, *et al.* "The VTR project: Architecture and CAD for FPGAs from Verilog to routing." *ACM FPGA*, 2012.

---

[5]Foundry A: n=165, s=28.0 MHz; split-foundry: n=195, s=29.2 MHz
[6]Foundry A: $\bar{x}$=20.3 pJ, s=1.71 pJ; split-foundry: $\bar{x}$=21.2 pJ, s=1.97 pJ

978-1-4673-6145-3/13 $31.00 © 2013 IEEE

A 40-nm 8T SRAM with Selective Source Line Control of Read Bitlines and Address Preset Structure

S. Yoshimoto[1], S. Miyano[2], M. Takamiya[3], H. Shinohara[2], H. Kawaguchi[1], and M. Yoshimoto[1]

[1]Kobe University, [2]Semiconductor Technology Academic Research Center, [3]University of Tokyo

Email: yoshipy@cs28.cs.kobe-u.ac.jp

Abstract- **This paper presents a 40-nm 8T SRAM in which bitlines are partially discharged by a selective source line control (SSLC) for low-power operation. The proposed SSLC scheme reduces a read bitline voltage swing in an unselected column with a floating source line (SL) of dedicated read ports. The SL is controlled by an additional NMOS switch that is turned on in a selected column, but the switch is kept off in the remaining unselected columns. The proposed scheme is effective for power reduction in successive address readouts through a single column. Furthermore, this paper introduces an address preset structure. The preset address enables the SRAM to be read out with no access time penalty for preferred use of the SSLC scheme. We fabricated a 16-Kb 8T SRAM test chip in a 40-nm CMOS process and observed that the proposed SSLC scheme with the address preset structure saves 38.1% of the readout power on average.**

I. INTRODUCTION

The scaling of CMOS processes has continually increased chip density and has enhanced SoC functionality. The ITRS predicts that the total memory size of SoCs will increase by a factor of ten until 2022, with the memory consuming up to 65% of operating power in a mobile processor [1]. Near a threshold voltage (V_t), an operating circuit is expected to be a good candidate to decrease the total power consumption [2], which would expand its battery charging cycle. A logic circuit can operate at a lower supply voltage near the threshold voltage so that energy optimization of the logic circuits is expected to be conducted for an extremely low-power SoC [3]. However, process scaling increases threshold voltage variation and degrades the operating margin [4]. The scaled SRAM cannot lower the operating voltage further because of the process variation. For this reason, near-threshold computing (NTC) is realized with a cluster of higher-voltage caches and lower-voltage processing units in an optimized multi-core processor [5].

Recent works have specifically examined reduction of bitline swing in the SRAM, not lowering of the operating voltage [6–8]. One report [7] describes that the transistor variation increases the dynamic energy because faster cells fully discharge bitlines by a sense-amplifier-enable timing. The dynamic energy increased by the transistor variation is estimated as 82% at a supply voltage of 0.5 V. Charge collector circuits [6] leverage a charge on unselected local bitlines to drive a global bitline with charge sharing. Bitline amplitude limiting schemes [7, 8] reduce the unnecessary bitline swing of a faster cell. The limiter does not degrade its cell current, but stops discharging on the bitline when the bitline level decreases to a threshold voltage of an NMOS. As

earlier works have explained, it is important to decrease the bitline swing to achieve low-power SRAM.

Instead of the conventional 6T SRAM, a single-ended 8T SRAM is widely used even as a single-port memory leveraging disturb-free dedicated read ports [9]. The 8T SRAM presents advantages in designing write and read circuits separately, for which a half-select-free write-back scheme [10, 11] is proposed. Another advantage of the 8T SRAM is that a "1" readout consumes no dynamic power because a read bitline maintains a precharging voltage [12].

An 8T sub-Vt SRAM [13] employs a footer line (= source line: SL) shared in the same row to achieve low-power operation. This SRAM can eliminate a leakage path through unselected rows. However, read bitlines are still discharged in unselected columns, which degrades power efficiency.

As described in this paper, a partially discharging 8T SRAM with a selective source line control (SSLC) scheme is proposed. The proposed scheme cuts off SLs of the dedicated read ports selectively according to a column address. The proposed SSLC improves energy efficiency in a successive read operation of an instruction cache or video processing. In the incremental address access, only a row address (= less significant address bits) is frequently changed, as presented in Fig. 1, where a column address is changed only slightly. Our proposed work improves the energy efficiency of successive read operations.

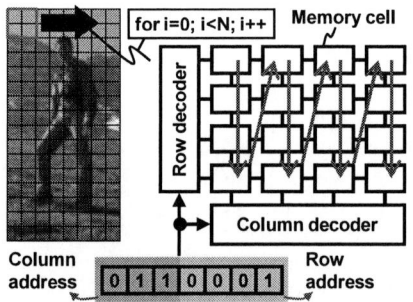

Fig. 1. Successive memory access in video processing.

II. PROPOSED 8T SRAM

A. Selective Source Line Control (SSLC) Scheme

Figure 2 illustrates an array of the 8T cells with a commonly used interleaving structure. The SLs of the dedicated read port are always grounded in the conventional structure. Although a selected local read bitline (RBL) is merely connected to a global RBL by a multiplexer, the other

978-1-4673-6145-3/13 $31.00 © 2013 IEEE

local RBLs in the unselected columns are discharged, which is not necessary for the read operation.

Figure 3 illustrates the concept of the proposed SSLC scheme. The SL is a shared virtual ground line of the dedicated read ports in a single column of the 8T SRAM array. An NMOS switch and an OR gate are inserted in every column. The switch is turned on selectively or is kept off according to a column address. In a standby mode, the SLs are grounded to prepare upcoming random access; the SL might, however, be floated if one-clock wakeup is not needed. In the write operation, the SLs are grounded because the 8T SRAM employs a disturb mitigation scheme with write back to eliminate a half-select problem [14]. For that reason, the OR gate has a write enable (WE) input. Although the SSLC circuit must be implemented in every column, the area overhead is 0.7% in our design. Figures 4(a)–4(c) respectively show schematic, FEOL, and BEOL layouts of the proposed 8T cell with an SL. In the conventional 8T cell, a ground line of the dedicated read port can be shared with an adjacent cell. However, in the proposed 8T cell, the SL must be separated. In contrast, no area overhead exists in adding the SL. In our design, the cell size is 1.01 μm^2 in a logic rule, which is slightly larger than the conventional one because the transistor length is relaxed for low-leakage operation.

Figure 5 shows operating waveforms of wordline, RBL, and SL of a selected and unselected columns, in which cells have all "0" data. In the selected columns, the bitlines are pulled down and discharged. They are then precharged for

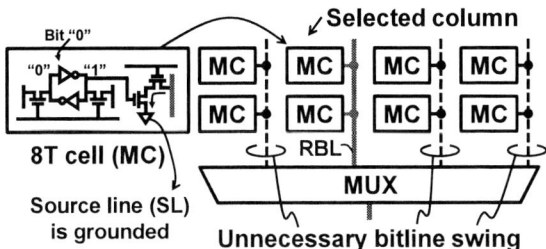

Fig. 2. Conventional 8T memory cell array. Bit "0" discharges a read bitline (RBL).

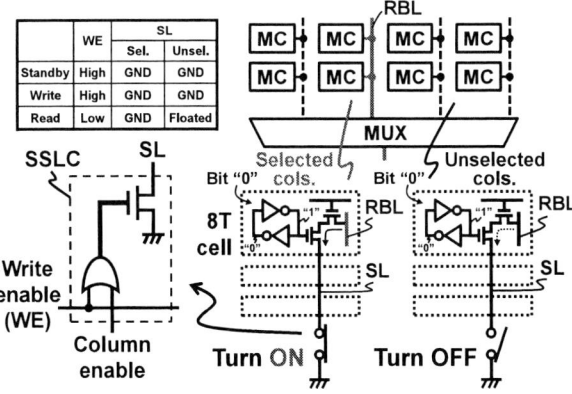

Fig. 3. Conceptual diagrams showing the proposed partially discharging 8T SRAM with the selective source line control (SSLC) scheme in read operation.

subsequent operations. In unselected columns, the bitline is not fully discharged. Its swing is suppressed because the SL is floated by the SSLC.

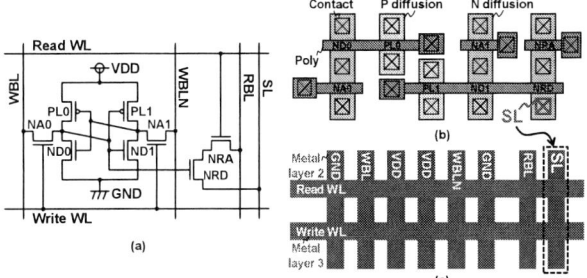

Fig. 4. (a) Schematic, (b) FEOL, and (c) BEOL layouts of the proposed 8T cell with a separated source line (SL).

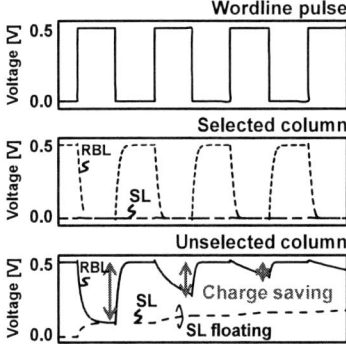

Fig. 5. Waveforms of wordline, read bitline (RBL), and source line (SL) of selected and unselected columns in consecutive "0" read operations.

B. Address Preset Structure

Figure 6 shows an important shortcoming of the SSLC scheme: the access time penalty. Before read operations, the SL must be grounded in the selected column. This SL activation demands extra access time. In this subsection, an address preset structure is presented to eliminate the access time penalty caused by SL activation. The proposed structure leverages an access address (= an address accessed at the present cycle: ADD_{acc}) and a preset address (= an address accessed at the next cycle: ADD_{pre}) as shown in Figs. 7(a) and 7(b). In particular in a video memory or a memory shared by many cores, an address accessed in the next cycle can be preset because the memory access is algorithmic or is stored in a queue. In such a case, the ADD_{pre} can be fed in a negative edge of the clock and the SL in the column accessed at the next cycle can be grounded preliminarily to prepare for the next positive edge. The address preset structure eliminates the access time penalty in the SSLC mode when the address accessed at the next cycle can be preset. The area overhead of the address preset structure is less than 1% in the SRAM macro.

In Fig. 7(a), the ADD_{acc}, which is the present address, receives an ADD_n on the first positive edge. It can then preset an ADD_{n+1} for the next cycle because it is fixed on the

978-1-4673-6145-3/13 $31.00 © 2013 IEEE   613

negative edge of the clock. The SL is always grounded before access in the successive read operation. Consequently, the SSLC with the address preset structure improves the energy efficiency with no access time penalty.

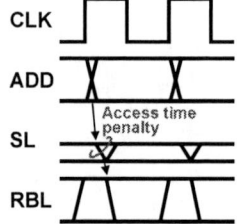

Fig. 6. Access time penalty in the SSLC scheme.

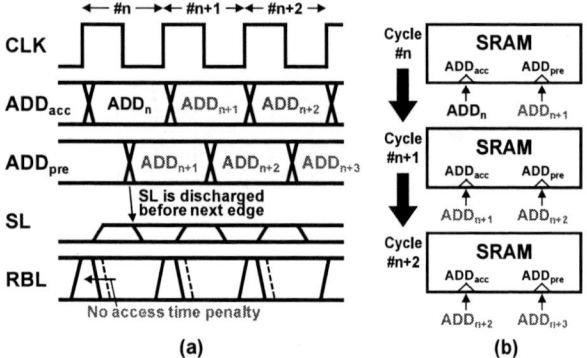

Fig. 7. (a) Waveforms and (b) timing behavior of the SSLC scheme with the address preset structure.

III. CHIP IMPLEMENTATION AND MEASUREMENT RESULTS

We implemented a 16-Kb 8T SRAM test chip using a 40-nm CMOS process as presented in Fig. 8. The macro size is 128×280 μm^2. The 8T SRAM consists of 16 bits / word × 1 K words (128 rows × 128 columns). An SRAM sub-array (128 rows × 8 columns) has a multiplexer that selects a column for an input or output datum. Figure 9 presents a schematic of the proposed 8T SRAM with the SSLC and the low-energy disturbance mitigation scheme [14]. A pair of write bitlines (WBL/WBLN) and an SL are shared by 128 cells in a column. A local RBL is shared by 16 cells and a NAND gate transports the readout datum to a global RBL driver. In the write operation, the write-back driver drives the WBL pair as to the original readout data to prevent the half-select issue. In a write cycle, all SLs are grounded for the write-back operation, as described in the previous section.

Figures 10 and 11 respectively present access patterns in the measurements and measurement results in the successive read operation. The gray and the black bars in Fig. 11 respectively show active and leakage energies per cycle. In the measurements, four data and access patterns are used.
- In the all-zero (ALL0) data with a fixed address pattern, only selected RBLs are merely discharged because an access address is fixed. The other RBLs remain floating because they are always unselected. In this case, the proposed SSLC effectively reduces the read energy by 57.2%.

- The RBL remains "1" in the all-one (ALL1) data pattern. One fixed address is accessed continuously. In this case, the SSLC does not work because all the RBLs keep the precharged voltage. Therefore, the power reduction is 0.0%.
- The checkerboard pattern using incremental row address (CKB X+) has 50% "0" data. The measurement result demonstrates the SSLC decreases the read energy by 45.0% in this case.
- In the CKB using incremental column address (CKB Y+), the column address is changed at every cycle. The SLs cannot be floated for a long time; the power reduction is less effective than ALL0 and CKB X+ patterns. The reduction is 28.5% in the pattern.

On average of the four patterns, the proposed SSLC reduces the energy consumption by 38.1% in the successive read operation. Table 1 presents the test chip characteristics.

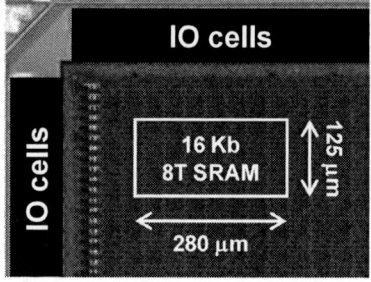

Fig. 8. 16-Kb 8T SRAM test chip.

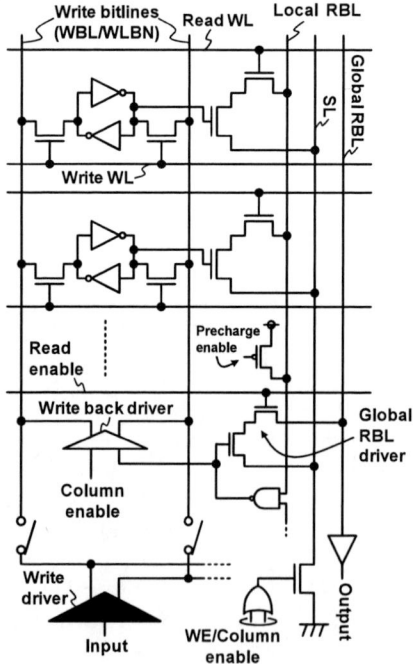

Fig. 9. Schematic of the proposed 8T SRAM with the SSLC and the disturbance mitigation scheme [14].

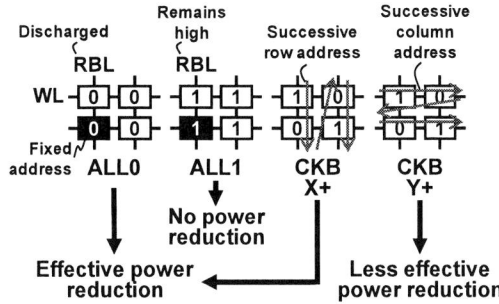

Fig. 10. Access patterns in the energy measurement. The proposed SSLC is effective in the all-zero (ALL0) and the checkerboard X address increment (CKB X+) patterns than the other two patterns.

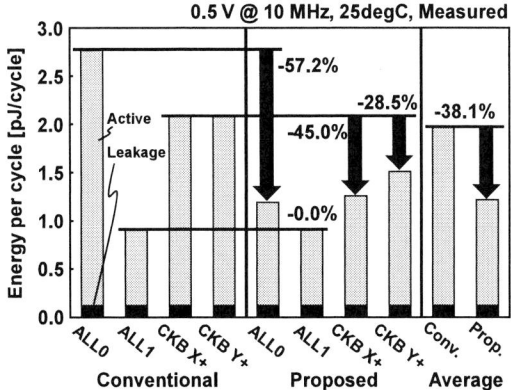

Fig. 11. Measurement results of the implemented test chip in read operation.

Table 1 Features of a test chip.

Technology	40 nm bulk CMOS
Macro size	125 μm × 280 μm
Macro configuration	16 Kb (16 bits/word, 1 K words)
Cell size	1.01 μm² (logic rule)
# of cells / BL, SL	16 (RBL), 128 (WBL), 128 (SL)
Density	457 Kb/mm²
Write active energy (CKB)	2.18 pJ @ 0.5 V, 10 MHz, RT
Read active energy (CKB)	1.14 pJ @ 0.5 V, 10 MHz, RT
Leakage energy (CKB)	0.12 pJ @ 0.5 V, 10 MHz, RT

IV. CONCLUSION

As described in this paper, we presented the selective source line control (SSLC) scheme for an 8T SRAM. The RBL swing is suppressed in an unselected column because the SSLC disconnects the source line (SL) of the dedicated read ports and therefore does not fully discharge the unselected read bitlines (RBL). In addition to the SSLC, the paper introduced the address preset structure to address the access time penalty, which best matches with the SSLC. The 16-Kb 8T SRAM test chip implemented in a 40-nm bulk CMOS technology demonstrates that the SSLC with the address preset structure reduces read energy consumption by 57.2%, 0.0%, 45.0%, and 28.5% in ALL0, ALL1, and CKB0 row address increments, and the CKB0 column address increment, respectively. On average, the proposed scheme exhibits a 38.1% energy reduction in successive addresses accessed, compared with a conventional 8T SRAM.

ACKNOWLEDGMENTS

This work was conducted as a part of the Extremely Low Power (ELP) project supported by METI and NEDO. The authors would like to thank Mr. Y. Yamamoto, Mr. Y. Okuma and Mr. K. Hirairi with STARC, Prof. T. Sakurai and Prof. T. Hiramoto with The University of Tokyo, Prof. K. Takeuchi and Dr. K. Miyaji with Chuo University.

REFERENCES

[1] ITRS Report 2009, http://www.itrs.net/
[2] Y. Pu, X. Zhang, J. Huang, A. Muramatsu, M. Nomura, K. Hirairi, H. Takata, T. Sakurabayashi, S. Miyano, M. Takamiya, and T. Sakurai, "Misleading Energy and Performance Claims in Sub/Near Threshold Digital Systems," *IEEE International Conference on Computer-Aided Design,* pp. 625-631, 2010.
[3] B. Zhai, R. G. Dreslinski, D. Blaauw, T. Mudge, and D. Sylvester, "Energy Efficient Near-threshold Chip Multi-processing," *IEEE International Symposium on Low Power Electronics and Design,* pp. 32-37, 2007.
[4] R. Heald and P. Wang, "Variability in Sub-100 nm SRAM Designs," *IEEE International Conference on Computer-Aided Design,* pp. 347-352, 2004.
[5] D. Fick, R. G. Dreslinski, B. Giridhar, G. Kim, S. Seo, M. Fojtik, S. Satpathy, Y. Lee, D. Kim, N. Liu, M. Wieckowski, G. Chen, T. Mudge, D. Sylvester, and D. Blaauw, "Centip3De: A 3930DMIPS/W Configurable Near-Threshold 3D Stacked System with 64 ARM Cortex-M3 Cores," *IEEE International Solid-State Circuits Conference,* pp. 190-191, 2012.
[6] S. Moriwaki, Y. Yamamoto, A. Kawasumi, T. Suzuki, S. Miyano, T. Sakurai, and H. Shinohara, "A 13.8pJ/Access/Mbit SRAM with Charge Collector Circuits for Effective Use of Non-Selected Bit Line Charges," *IEEE Symposium on VLSI Circuits Digest of Technical Papers,* pp. 60-61, 2012.
[7] A. Kawasumi, T. Suzuki, S. Moriwaki, and S. Miyano, "Energy Efficiency Degradation Caused by Random Variation in Low-Voltage SRAM and 26% Energy Reduction by Bitline Amplitude Limiting (BAL) Scheme," *IEEE Asian Solid-State Circuits Conference,* pp. 165-168, 2011.
[8] S. Yoshimoto, M. Terada, Y. Umeki, S. Okumura, A. Kawasumi, T. Suzuki, S. Moriwaki, S. Miyano, H. Kawaguchi, and M. Yoshimoto, "A 40-nm 256-Kb Sub-10 pJ/Access 8T SRAM with Read Bitline Amplitude Limiting (RBAL) Scheme," *IEEE International Symposium on Low Power Electronics and Design,* pp. 85-90, 2012.
[9] R. Houle, K. Batson, D. Rodko, P. Patel, W. Huott, R. Franch, Y. Chan, D. Plass, S. Wilson, and P. Wang, "6.6+ GHz Low Vmin, read and half select disturb-free 1.2 Mb SRAM," *IEEE Symposium on VLSI Circuits Digest of Technical Papers,* pp. 14-16, 2007.
[10] Y. Morita, H. Fujiwara, H. Noguchi, Y. Iguchi, K. Nii, H. Kawaguchi, and M. Yoshimoto, "An Area-Conscious Low-Voltage-Oriented 8T-SRAM Design under DVS Environment," *IEEE Symposium on VLSI Circuits Digest of Technical Papers,* pp. 256-257, 2007.
[11] J. Wu, Y. Chen, M. Chang, P. Chou, C. Chen, H. Liao, M. Chen, Y. Chu, W. Wu, and H. Yamauchi, "A Large σVTH/VDD Tolerant Zigzag 8T SRAM with Area-Efficient Decoupled Differential Sensing and Fast Write-Back Scheme," *IEEE Symposium on VLSI Circuits Digest of Technical Papers,* pp. 103-104, 2010.
[12] N. Verma and A. P. Chandrakasan, "A 65 nm 8T Sub-Vt SRAM Employing Sense-Amplifier Redundancy," *IEEE International Solid-State Circuits Conference,* pp. 328-329, 2007.
[13] H. Fujiwara, K. Nii, J. Miyakoshi, Y. Murachi, Y. Morita, H. Kawaguchi, and M. Yoshimoto, "A Two-Port SRAM for Real-Time Video Processor Saving 53% of Bitline Power with Majority Logic and Data-Bit Reordering," *IEEE International Symposium on Low Power Electronics and Design,* pp.61-66, 2006.
[14] S. Yoshimoto, M. Terada, S. Okumura, T. Suzuki, S. Miyano, H. Kawaguchi, and M. Yoshimoto, "A 40-nm 0.5-V 20.1-uW/MHz 8T SRAM with Low-Energy Disturb Mitigation Scheme," *IEEE Symposium on VLSI Circuits Digest of Technical Papers,* pp. 72-73, 2011.

978-1-4673-6145-3/13 $31.00 © 2013 IEEE

AOT-Controlled Dual-Mode AVP Buck Regulator with AEAF Mechanism

Hsin-Lun Li[*], Chia-Cheng Pao[**], Bo-Ming Chen[***], Chien-Hung Tsai[****]

Department of Electrical Engineering /
Advanced Optoelectronic Technology Center (AOTC) / Green Energy Electronics Research Center (GREERC)
National Cheng Kung University, Tainan, Taiwan, R.O.C.
E-mail：n26004896@mail.ncku.edu.tw*, n26014752@mail.ncku.edu.tw**, chtsai@ee.ncku.edu.tw****

Abstract—A novel adaptive voltage positioning (AVP) buck regulator using adaptive on-time (AOT) control targeted for applications with low-ESR output capacitors is proposed. In this work, AOT control is adapted to keep the system's switching frequency quasi-fixed or independent of the input supply voltage and the AVP mechanism is realized without the need to use conventional error amplifier compensator or extra current-sensing circuit. For ensuring the system's switching frequency not entering the range of acoustic frequency at light load, an AEAF (avoid entering acoustic frequency) circuit is also proposed. For comparison purpose, the implemented buck regulator can be set to operate under AVP or non-AVP mode. This work has been fabricated and verified with a standard 0.18μm CMOS technology. Experimental results show excellent transient recovery time of 4μs (under AVP mode), ±0.11% switching frequency variation (for the specified input voltage range), and 91% peak conversion efficiency.

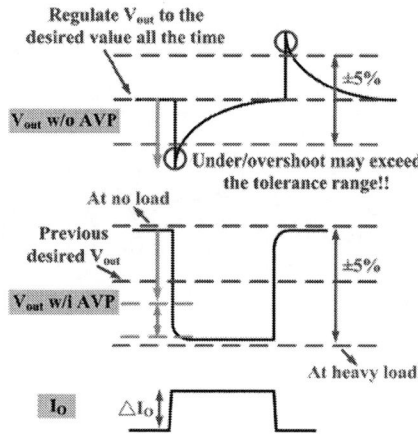

Fig. 1. The concept of AVP.

I. INTRODUCTION

As the clock rate of microprocessors is developed to be faster than GHz, a lower operation voltage is better for data processing efficiency. Conversely, the current demand of microprocessors is increasing because of the high-density semiconductor integration. This trend poses a stringent challenge on transient response, many capacitors are required in the DC-DC converter output, which increases size and cost. Adaptive voltage positioning (AVP) is a very popular technique in VRM applications to reduce the output capacitors [1]. Instead of regulating the output voltage to the desired value all the time, AVP control has a converter whose output voltage depends on load conditions. The AVP concept is shown in Fig. 1. By utilizing the maximum allowable tolerance window, the extra green region means the potential reduced cost on output capacitors. Achieving constant output impedance over the widest possible frequency range is the necessary condition to achieve ideal AVP [1]. The high current slew rate requires the power supply to possess extreme fast response time. The conventional PWM controls fail to deliver the desired performance mainly due to their relying on the error amplifier speed to react the load transition. The fast-transient ripple-based controls become a better choice to replace the conventional PWM controls. The rippled-based control does not need error amplifier and external compensation components and has natural pulse-frequency modulation (PFM) function in discontinuous conduction mode (DCM) make them suitable for portable electronic devices.

Constant on-time [2-4] and hysteretic [5-6] control both are very popular in fast-transient applications, but the more unpredictable switching frequency variation characteristic in hysteretic control makes adaptive on-time (AOT) control a better choice. In the previously reported AOT AVP solutions [2-3], error amplifier compensator or differential difference amplifier (DDA) are needed to achieved constant output impedance which cause the loss of low-cost characteristic of rippled-based control. The AOT AVP solution in [4] does not need error amplifier compensator, but lack of a constant output impedance design flow. On the other hand, the AVP implementation in [4] needs an extra current-sensing loop similar to "active-droop" reported in traditional PWM AVP [1] which is complex and power-consumed.

Considering the continuous load decrease, the system operates in the discontinuous conduction mode (DCM). As a result, the switching frequency becomes smaller than 20kHz and causes the noise interference. In order to avoid the noise interference, a new AEAF technique is proposed. The system on-time length is adjusted by monitoring the system switching frequency. Facing the different specifications, the DC-DC converter should have the ability to provide different solutions. When the current slew rate requirement of the application is not high, the proposed regulator can be configured as a Non-AVP mode.

This paper proposed a low-cost dual-mode AOT AVP without using error amplifier compensator and extra current sensing circuits with a standard constant output impedance design flow. We will review the conventional

AOT AVP schemes in Section II. System architecture, operation principle and circuit implementation will be addressed in Section III. The chip measurement results will be presented in Section IV. Finally, we will make a conclusion in Section V.

II. CONVENTIONAL ADAPTIVE ON-TIME AVP SCHEME

In order to handle the stringent challenge on transient response, the fast-transient ripple-based controls become a better choice to replace the conventional PWM controls. The rippled-based control does not need error amplifier and external compensation components and has automatic pulse-frequency modulation (PFM) function in discontinuous conduction mode (DCM) make them suitable for portable electronic devices. However, the original rippled-based control can't operate properly in employing ceramic capacitor [5]. Fortunately, the solutions of the requirement of large ESR have become mature in recent years [5]. One of the common methods is shown in Fig. 2. An RfCf network which behaves like an integrator is introduced to eliminate the large ESR requirement [5].

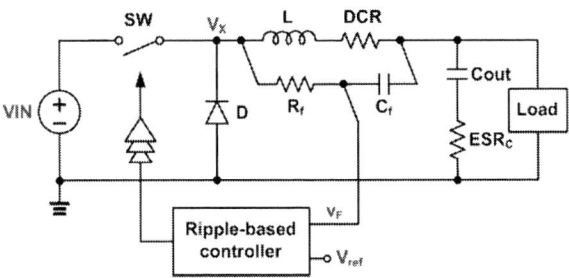

Fig. 2. A common way to eliminate the large ESR requirement

The more unpredictable switching frequency variation characteristic in hysteretic control makes adaptive on-time control (AOT) [2-4] a better choice. Fig. 3 shows the previously reported AOT AVP solutions [2-4]. The AVP implementation in Fig. 3(a) is to sense the in-phase inductor current information from an extra series resistor R_s through a high-bandwidth current sensing amplifier. This scheme needs an error amplifier compensator and external passive components to achieve constant output impedance which is similar with peak current-mode AVP. The second type AOT AVP solution is shown in Fig. 3(b). Utilize a pair of power-consumed DDAs to increase the noise immunity significantly. There is no high-gain loop in this scheme so the AVP function can be achieved by using the dc current information from R_rC_r network. The achievement of constant output impedance is to adjust the gain of the DDAs which causes the need of hard tuning. Another type of AOT AVP in Fig. 3(c) is similar with the active droop control in old PWM control [1] except for there is no error amplifier compensator. However, there is no constant output impedance design flow in this scheme and the required extra current-loop (red-line) makes the system become a complex tri-loop system.

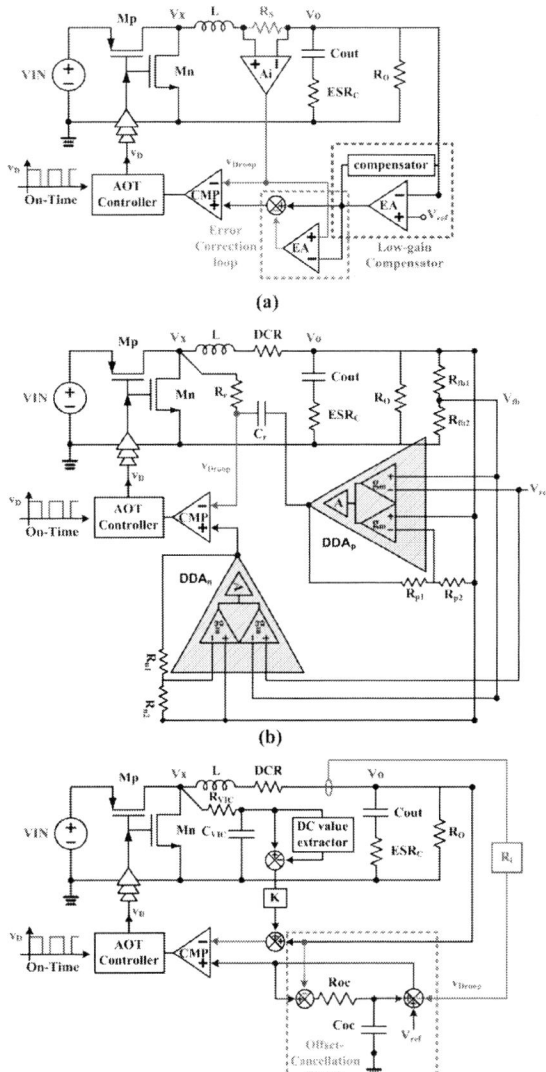

Fig. 3. Conventional adaptive on-time AVP

(a) Type I [2] (b) Type-II [3] (c) Type-III [4]

III. CONCEPTS OF PROPOSED ADAPTIVE ON-TIME AVP

A. Constant Zoc Design Without EA Compensator

Fig.4 shows the architecture of the proposed low-cost dual-mode AOT AVP. Employ the method shown in Fig. 2 to eliminate the large ESR requirement. The drawback of this method is that there is an output voltage dc offset caused by the parasitic resistor of the inductor (DCR). The dc offset which is proportional to the load current is usually eliminated with the aid of high-gain error amplifier in ordinary power system. However, the dc offset can be utilized to achieve droop function in power system with AVP function. The DC values of the signals at the inputs of the comparator are approximately equal at steady-state. The equation can be derived:

$$<Vs> = I_L \cdot DCR + Vo = Vref \cdot \frac{Ra1}{Ra1+Ra2} + Vo \cdot \frac{Ra2}{Ra1+Ra2} = <Vp> \quad (1)$$

978-1-4673-6145-3/13 $31.00 © 2013 IEEE   617

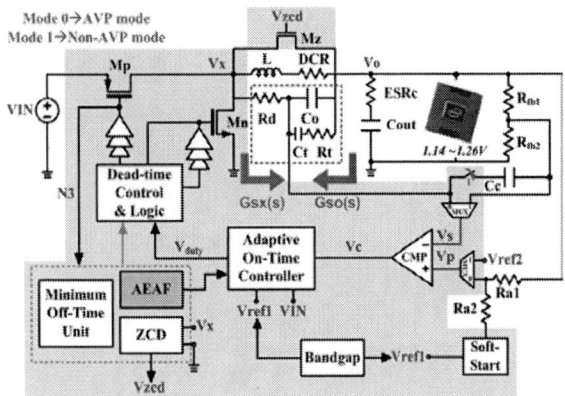

Fig. 4. Proposed low-cost dual-mode adaptive on-time AVP

Therefore,

$$Vo = Vref - I_L \cdot \frac{Ra1+Ra2}{Ra1} \cdot DCR \qquad (2)$$

According to eq. (2), the DC value of output voltage now is a function of load current. The AVP function is achieved. In order to achieve the nearly ideal transient response, the constant output impedance design flow based on [6] is derived below:

$$Zo(s) = (DCR + sL) \cdot Gox(s) \qquad (3)$$

$$Zocl(s) = \frac{Zo(s)}{1+T(s)}, T(s) = Gox(s) \cdot \frac{Gso(s)}{Gsx(s)} \qquad (4)$$

,where Zocl(s) is the closed loop output impedance, Zo(s) is the open loop output impedance, Gox(s) is the open loop gain transfer function, and T(s) is the system closed-loop gain transfer function. According to eq. (4), in order to achieve the constant Zocl(s), we have to keep the shape of Zo(s) and [1+ T(s)] the same as shown in the small signal simulation result in Fig. 5. This issue can be achieved easily by designing the loss-less RC filter (green region in Fig.4) related transfer function, Gso(s) and Gsx(s). Without using traditional error amplifier compensator, the Zocl(s) is compensated from the current sensing node V_S by extra pole-zero pair produced by Rt and Ct. The condition to achieve constant closed-loop output impedance is shown below:

$$Rd \cdot Co = (1 - \frac{DCR}{ESRc}) \cdot \frac{L}{DCR} \qquad (5)$$

$$Rd \cdot Ct = (\frac{L}{DCR} - DCR \cdot Cout) \qquad (6)$$

$$Rt \cdot Ct = ESRc \cdot Cout \qquad (7)$$

The constant Zocl(s) verification result by using SIMPLIS is shown in Fig. 6. The phase difference between output voltage and load current is kept about 180 degree over a wide frequency range. Without using error amplifier and extra current sensing circuit, the proposed AOT AVP preserves the simple low-cost nature of the ripple-based control. When the current slew rate requirement of the application is not high, the proposed regulator can be configured as a Non-AVP mode. The "Ripple Injection" technique proposed by [8] eliminates the dc offset caused by DCR. The in-phase AC inductor current information is "injected" through Cc on the output feedback signal in Fig.4.

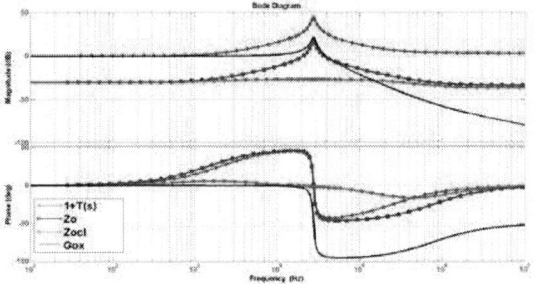

Fig. 5. The constant Zocl design concept

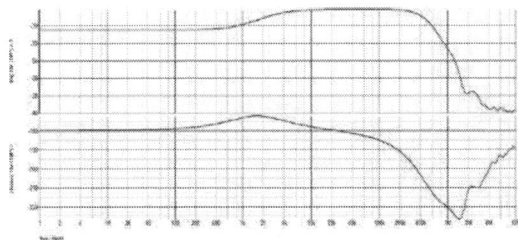

Fig. 6. The constant Zocl verified by using SIMPLIS.

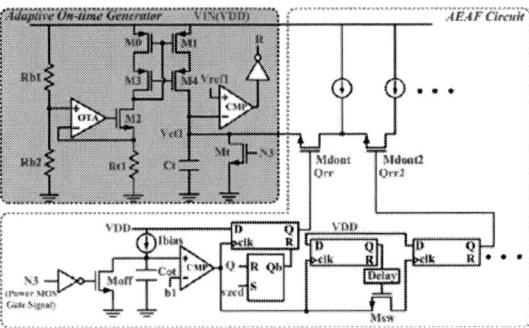

Fig. 7. AOT generator and the proposed AEAF.

B. AEAF Mechanism

In order to avoid the noise interference, a new AEAF technique is proposed. The AOT generator and the proposed Avoid Entering Audio Frequency (AEAF) region circuit are shown in Fig. 7. By feeding the input voltage information to the v-to-i converter, the system on-time length become longer when input voltage is decreased in order to reducing the switching frequency variation caused by input voltage changes. On the right-side of Fig. 7, the proposed AEAF avoids the system falling into noisy audio frequency region by compare the system switching frequency with the internal clock set around audio frequency. If the system switching frequency is below 20kHz, the comparator will send out "high" signal to force the D flip-flop to turn on the first current source to shorten the on-time length for keeping system away from audio region. If the system switching frequency is still below 20kHz after a designed delay, the second current source will turn on and the third will then turn on, too. In order to prevent AEAF mechanism from affecting the switching frequency in CCM, an easy DCM/CCM detector composed of a SR latch and the signal vzcd from zero current detector (ZCD) is required. When the system operates in CCM, the DCM/CCM

978-1-4673-6145-3/13 $31.00 © 2013 IEEE   618

detector will send a signal to reset the first D flip-flop. As a result, the current from AEAF will not flow into Ct in CCM. When the system operates in no load condition, the AEAF will stop functioning for saving the power which is not required.

IV. EXPERIMENTAL RESULTS

The proposed system was fabricated with a standard 0.18µm CMOS process. The comparison result is summarized in Table I. The chip micrograph is shown in Fig. 8. Fig. 9 shows the measurement result of the transient response. The load current was changed between 200mA and 900mA. The recovery time is about 4µs for the step up load of 700mA. At heavy load, the output voltage is set at lower value which verifies the AVP function. There is no any under/overshoot in the output voltage during load variations. Fig. 10 shows the Non-AVP mode transient response. The undershoot is about 80mV and the recovery time is about 19µs for the same step up load. Obviously, AVP function improves the transient response of the system significantly.

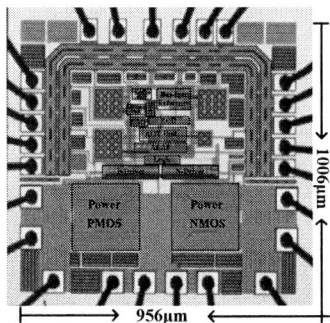

Fig. 8. Chip micrograph.

V. CONCLUSION

A high efficiency low-cost dual-mode AOT AVP without using error amplifier compensator is proposed. Measurement results show that this converter can operate under load current of 1-1200mA for supply voltage from 2.7V to 3.6V, and an output voltage of 1.2V. Excellent transient response makes the proposed regulator suitable in low-voltage and high-current applications.

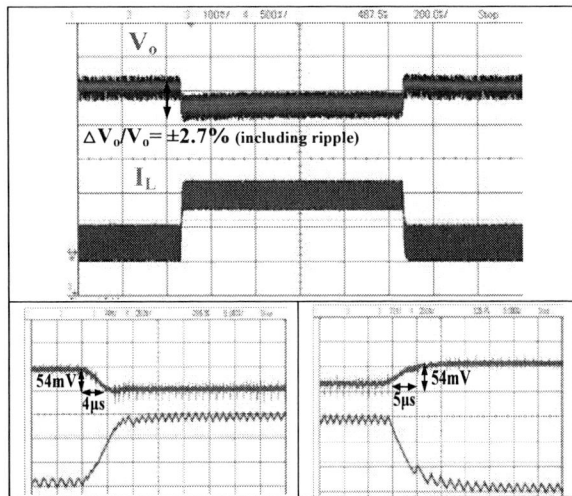

Fig. 9. Measured AVP-mode load transient response

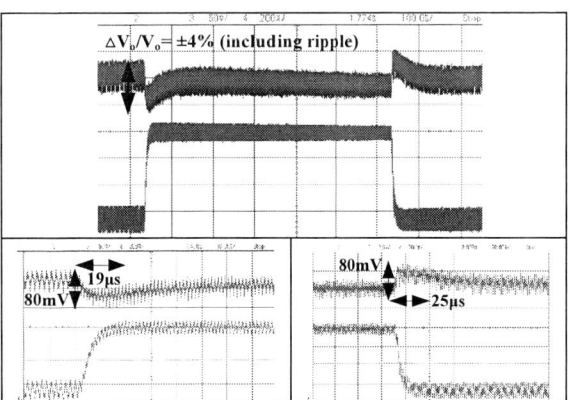

Fig. 10. Measured Non-AVP mode load transient response

TABLE I. PERFORMANCE COMPARISON

	Ref.	This work	[2]	[3]	[4]
Comparison	Control	AOT	AOT	AOT	AOT
	Solution for fixed fsw@ΔV_{IN}	YES	YES	YES	YES
	Solution for low ESR	YES	YES	YES	YES
	EA or DDA	NO	YES	YES	NO
	Extra I_L sensing circuit	NO	NO	NO	YES
	Zoc Design	YES	YES	YES	NO
Specifications	Technology	0.18µm	0.5µm	0.5µm	N/A
	V_{IN} (V)	2.7-3.6	5-20	3-6	5
	V_O (V)	1.2	1.2	0.6-2.6	1
	AVP droop R_{OC}	77mΩ	3mΩ	5mΩ	50mΩ
	Max I_{Load} (A)	1.2	40	1.1	4
	fsw (kHz)	730	600	1980	55
	Δfsw/fsw@ΔV_{IN}	±0.1%	±2.5%	±6.2%	N/A
	Transient (µs)	4	N/A	5	N/A
	Peak Efficiency	91%	83%	86.8%	N/A

ACKNOWLEDGMENT

The authors would like to thank the Chip Implementation Center for chip fabrication. This work was supported by the Green Power Electronics Project, NCKU.

REFERENCES

[1] K. Yao, Y. Meng, P. Xu, and F. C. Lee, "Optimal design of the active droop control method for the transient response," in *Proc. IEEE APEC*, 2003, pp.718-723.

[2] X. Duan, J. Park, and A. Q. Huang, "Current-Mode Variable-Frequency Control Architecture for High-Current Low-Voltage DC–DC Converters," *IEEE Trans. Power Electronics*, vol. 21, pp.1133-1137, Jul. 2006.

[3] J. Fan, X. Li, J. Park, and A. Huang, "A monolithic buck converter using differentially enhanced duty ripple control," in *Proc. IEEE CICC*, 2009, pp.527-530.

[4] Y. C. Lin, C. J. Chen, D. Chen, and B. Wang, "A Ripple-Based Constant On-Time Control With Virtual Inductor Current and Offset Cancellation for DC Power Converters," *IEEE Trans. Power Electronics*, vol. 27, pp. 4301-4310, Oct. 2012.

[5] F. Su and W.-H. Ki, "Digitally Assisted Quasi-V2 Hysteretic Buck Converter with Fixed Frequency and without Using Large-ESR Capacitor," in *IEEE International Solid-State Circuits Conference*, 2009, pp. 446-447,447a.

[6] M. Castilla, L. G. d. Vicuna, J. M. Guerrero, J. Matas, and J. Miret, "Simple Low-Cost Hysteretic Controller for Single-Phase Synchronous Buck Converters," *IEEE Trans. Power Electronics*, vol. 22, pp. 1232-1241, Jul. 2007.

[7] "D-CAPTM With All-Ceramic Output Capacitor Application," TI Application Report, 2011

Switched-Capacitor Filter based Type-III Compensation for switched-mode Buck Converters

G. Bawa[1,2] and A. Q. Huang[1]

[1]North Carolina State University, NC, USA; [2]Texas Instruments Inc., TX, USA.

Abstract— **In this paper, we present a novel switched-capacitor filter based Type-III compensation architecture for closed-loop regulation of fixed-frequency switched-mode Buck converters. Compared to the conventional all-analog filter, the proposed compensator can be fully-integrated onto the die resulting in reduced footprint and cost. In addition, the filter time constants scale linearly with the Buck converter's switching time-period, resulting in increased programmability and ease-of-use. A prototype of a voltage-mode PWM controller with programmable Buck Converter switching frequencies of 0.5 and 1 MHz, has been implemented and validated in a 0.36-μm BCD process, consumes 1.1 mA of static current from a 3.3 V supply, and occupies ~ 0.65 mm² of active area on-chip.**

I. INTRODUCTION

In contemporary battery or line powered computing applications, the source voltage is typically higher than the target load's operating voltage. Inductive switched-mode step-down (or Buck) converters (Fig. 1) can convert the power highly efficiently (> 90 %) for high load current requirements (> 1 A). The closed-loop regulation can be performed using fixed-frequency voltage-mode pulse-width modulation (PWM) controllers to ensure a power-supply with low jitter and controlled EMI, by synchronization with external clock.

Typical power controller Integrated Circuits (ICs) offer several degrees of programmability through external components or a digital communications interface [1]. External components include the compensation components that take up board area and need to be carefully selected. It is advantageous to integrate these compensation components to decrease the size and improve the IC usability. This also reduces the risk and development time for new power supplies based around the integrated circuit.

The integration of the passive components of the conventional Type-III analog filter (Fig. 2) for voltage-mode PWM control is prohibitive at the switching frequencies ($f_{SW} \le 1$ MHz) of interest due to prohibitively large component values. In addition, the RC time constants vary by ~ 40 % over process, voltage and temperature (PVT) variations.

In contrast, a Switched-Capacitor Filter (SCF) time constant is given as [2]:

$$\tau_{SCF} = \frac{1}{2\pi \times f_{SCF}} = T_S \times \frac{C_I}{C_S} = \frac{1}{f_S} \times \frac{C_I}{C_S} \qquad (1)$$

where, f_S is the sampling rate; C_S and C_I are the sampling and

This project was supported by the research funding from Texas Instruments Inc., USA.

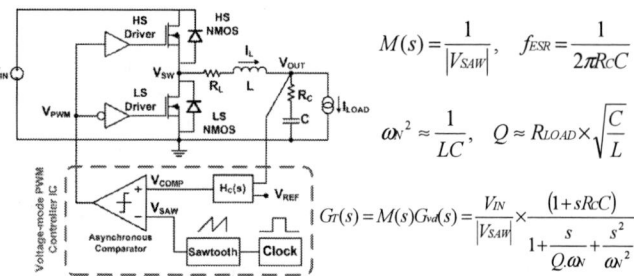

$$M(s) = \frac{1}{|V_{SAW}|}, \quad f_{ESR} = \frac{1}{2\pi R_C C}$$

$$\omega_N^2 \approx \frac{1}{LC}, \quad Q \approx R_{LOAD} \times \sqrt{\frac{C}{L}}$$

$$G_T(s) = M(s)G_{vd}(s) = \frac{V_{IN}}{|V_{SAW}|} \times \frac{(1+sR_CC)}{1 + \frac{s}{Q.\omega_N} + \frac{s^2}{\omega_N^2}}$$

Fig. 1. Circuit schematic of closed-loop switched-mode Buck converter with voltage-mode PWM control [4].

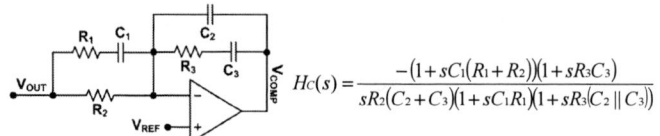

$$H_C(s) = \frac{-(1+sC_1(R_1+R_2))(1+sR_3C_3)}{sR_2(C_2+C_3)(1+sC_1R_1)(1+sR_3(C_2 \| C_3))}$$

Fig. 2. Conventional Type-III filter circuit for $H_C(s)$ in Fig. 1.

integrating capacitors, respectively. It can be seen that by controlling the clock (and hence sampling) frequency accuracy (< 1 %) and capacitor matching accuracy (< 1 %), τ_{SCF} can be made highly accurate (< 2 %), even without on-chip trimming. In addition, it can be deduced from (1) that f_{SCF} scales linearly with f_S, which can be a scaled-up version of the Buck converter's switching frequency ($f_{SW} = 1/OSR \times f_S$). Here, OSR can be understood to be the SCF's Over-Sampling Ratio.

With these concepts in mind, we proceed to the development of a fully-integrated Type-III compensation for a voltage-mode PWM controller. The salient features include frequency-scalability, improved ease-of-use, reduced board size and power consumption comparable to the conventional analog filter implementation [1]. Furthermore, the modular SCF architecture lends itself to Type-II compensation and potentially other variants of Type-III compensation [3].

II. TYPE-III SCF ARCHITECTURE

The use of conventional Type-III filter (Fig. 2) is necessitated by ceramic output capacitors, as the high ESR-zero frequency ($f_{ESR} \sim f_{SW} = 1$ MHz) provides negligible phase-lead. Thus, the filter places two real zeroes ($f_{Z1,2} \sim 20$ KHz) close to the LC double-pole frequency ($f_{LC} \sim 29$ KHz) for phase-boost and Closed-Loop Bandwidth ($f_{CL} \sim 100 - 200$ KHz) extension. Finally, it places a pole-at-origin for high DC regulation; and two poles ($f_{P1,2} \sim 550$ KHz) beyond f_{CL} and below f_{SW}, for finite gain margin (GM > 10 dB) and

978-1-4673-6145-3/13 $31.00 © 2013 IEEE

attenuation of high-frequency switching harmonics. It must be understood that while the foregoing discussion considers $f_{SW} = 1$ MHz, all the aforementioned frequencies will scale linearly with f_{LC} for stability, which in-turn scales linearly with f_{SW} in order to ensure the same V_{OUT} ripple [4].

The development of the Type-III SCF architecture begins by judicious partitioning and sequencing of the conventional filter's transfer function into three 1st order cascaded sections: an Integrator followed by two High Pass Filters (HPFs) (see Fig. 3). Now, if we treat the output voltage of the Buck (V_{OUT}

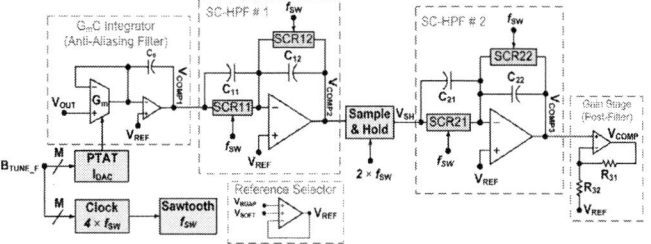

Fig. 3. Proposed Type-III SCF Architecture (SCR = Switched-Capacitor Resistor)

in Fig. 1) as the input to the SCF, we realize that it has a desirable low-frequency component at a cutoff ~ f_{LC}, and undesirable high-frequency switching harmonics at multiples of f_{SW}. If the worst-case ripple magnitude is 5 % (SNR = 26 dB), it does not meet the Signal-to-Noise Ratio (SNR > 40 dB) requirements for SCF's robustness. Thus, the SCF sampler cannot be directly exposed to this signal, and an Anti-Aliasing Filter (AAF) is required. In this prototype, we have realized the 1st stage as a G_mC integrator, and implemented both the Type-III and AAF functions simultaneously. More details on the G_mC integrator design will be discussed in section III-A.

As a next step, we need to choose the SCF OSR (and hence f_S) carefully. If the OSR is made too low, the realization of high-frequency poles is difficult. If OSR is made too high, it would lead to increased capacitor spread for a target time constant and unit sampling capacitor (C_S) size (see (1)). In addition, it would lead to increased loading and faster settling requirements for the SCF's amplifiers at the same time, making the design extremely power hungry.

In this prototype, we have employed Bi-Linear Transformation [5] for realizing the high-frequency poles for the two SC-HPFs. BLT offers the advantage of mapping the entire analog frequencies (-∞, ∞) within the Nyquist frequency (-f_S/2, f_S/2). By choosing f_S/2 = f_{SW} (OSR = 2), we make a reasonable choice for synthesizing the aforementioned Type-III filter poles $f_{P1,2}$. In addition, as can be deduced from the foregoing analysis, we have laid the foundation for highly area- and power-efficient SCF realization.

Now, a low OSR makes the AAF requirements for both the SC-HPFs challenging, since there is no implicit low-pass filtering in the filters' transfer functions [2]. For SC-HPF#1, these requirements are met by making the integrator bandwidth (f_I) ≤ f_{SW}/10, if the worst-case ripple (5 %) is considered (see Fig. 4). However, if SC-HPF#2 is directly cascaded with the previous stage (SC-HPF#1), severe aliasing distortion is observed. The only remedy is to introduce a Sample-and-Hold Amplifier (SHA) operating at f_S between

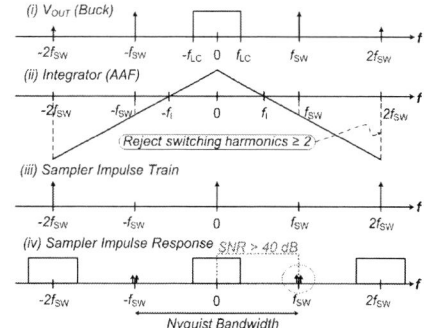

Fig. 4. Anti-aliasing analysis for input sampler of SC-HPF # 1.

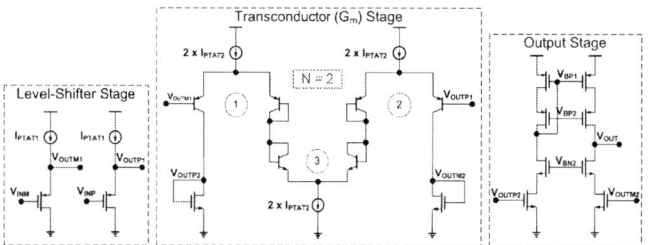

Fig. 5. OTA for implementing G_m stage in Fig. 3.

the cascaded SCF stages. The SHA breaks the direct capacitive path from V_{COMP1} to V_{COMP3}, and prevents aliasing distortion, while also resulting in an undesirable yet deterministic phase-lag of T_S/2 [2]. In our prototype, we have countered this phase-lag by decreasing the target zero frequency (f_{Z2} ~ 15 KHz). We have also used a reference selection buffer, which selects the lower of the bandgap (V_{BGAP}) and soft-start (V_{SOFT}) voltage (Fig. 3). Finally, it is clear to see from Fig. 3, that a Type-II filter can be realized by utilizing just the first two filter stages.

III. CIRCUIT IMPLEMENTATION

Unless otherwise mentioned, all the active elements have been implemented as PMOS-input based folded-cascode class-AB amplifiers to account for low Input Common-Mode Range (ICMR), high gain-bandwidth and fast settling requirements. The notable circuit techniques are discussed as follows:

A. G_mC Integrator

The integrator needs to maintain PVT independence and linear scalability of the Type-III function with f_{SW}. Thus, as shown in Fig. 3, we use the same bits (B_{TUNE_F}) for configuring the integrator f_I and f_{SW} (and hence f_S).

$$H_I(s) = -\frac{G_m}{C_0} \times \frac{1}{s}; \quad f_I = \frac{1}{2\pi} \times \frac{G_m}{C_0} \quad (2)$$

The OTA topology of Fig. 5 is used to implement the G_m stage (in Fig. 3). The 1st stage is a level shifter to allow operation at low-ICMR during soft-start. The 2nd stage is a novel all-Bipolar core, with a small-signal $G_m = I_{PTAT2}/(2N - 1)V_T$. By choosing a PTAT current source and scaling it with B_{TUNE_F} using a current-DAC, we can achieve PVT independence and linear scalability for G_m. In addition, by increasing N, we can make the G_m highly-linear. This is essential to prevent slewing distortion when large differential-input signal (≤ 500 mV ensured by design at worst case temperature of –40 °C) is presented at the input during soft-

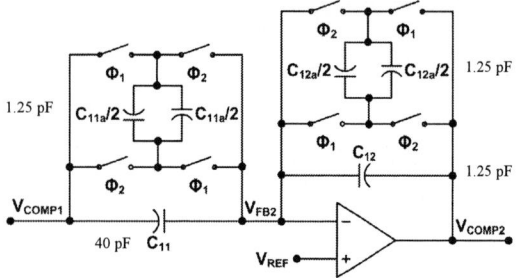

Fig. 6. Circuit schematic for SC-HPF # 1.

start and line/load transients. It was observed that prevention of slewing is a sufficient criterion to prevent any transient performance degradation; and requirement of a flat linear region as described in [6] is not necessary. This is a direct consequence of our Type-III partitioning/sequencing scheme. Finally, we can see from (2) that C_0 variations still need to be compensated via trimming.

B. SC-HPF # 1

The schematic of SC-HPF#1 is shown in Fig. 6. The pole/zero pair is implemented using the BLT switched-capacitor circuit element described in [5]. It can be seen that the capacitor samples in both clock half-cycles, hence the sampling rate is doubled. Thus, the clock frequency need only be f_{SW}(see Fig. 3). The SCF transfer function is given as:

$$H_{SCF1}(z) = -\frac{C_{11a}}{C_{12a}} \times \left[1 + \frac{C_{11}}{C_{11a}} \times \frac{(1-z^{-1})}{(1+z^{-1})}\right] \times \left[1 + \frac{C_{12}}{C_{12a}} \times \frac{(1-z^{-1})}{(1+z^{-1})}\right]^{-1} (3)$$

This SCF design is parasitic-sensitive, and the bottom-plate ($\sim$ 5 % of the main capacitance) of the switching capacitor can introduce asymmetry between the two sampling paths. To minimize the mismatch, we have divided the switching capacitor into two equal halves and flipped them to provide the same (halved) parasitics on the two switching nodes.

It was observed that the finite amplifier Gain-Bandwidth (f_{GBW}) can limit the high-frequency gain of this SC-HPF, by introducing a complex-pole at the geometric mean of "effective" f_{GBW} and f_{ZI} (unchanged).

$$f_{P1\_NEW} = \sqrt{\frac{f_{GBW} \times f_{Z1}}{A_{CL\_DC}}}; \quad \zeta \approx \frac{1}{2 f_{P1}} \sqrt{\frac{f_{GBW} \times f_{Z1}}{A_{CL\_DC}}} \quad (4)$$

Here, A_{CL_DC} is the closed-loop DC gain of the SCF. f_{P1} now only controls the damping factor (ζ) of the complex pole (f_{P1_NEW}). f_{P1_NEW} can be accurately controlled by designing

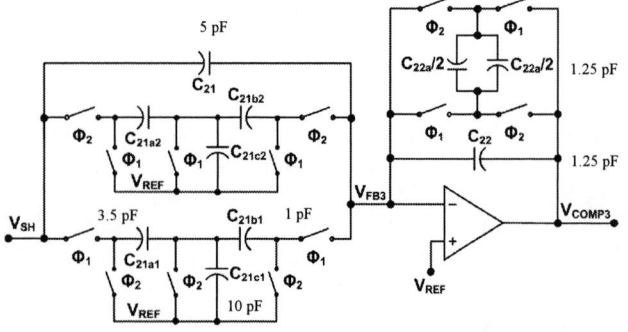

Fig. 7. Circuit schematic for SC-HPF # 2.

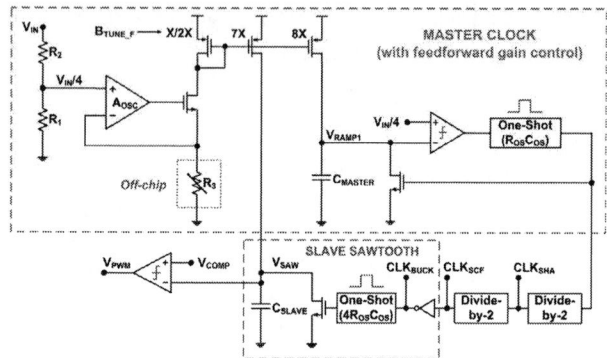

Fig. 8. Proposed clocking scheme for the Type-III SCF Architecture.

amplifiers for a target f_{GBW} (= 20 MHz), carefully chosen for the highest operating f_{SW} (= 1 MHz).

C. SC-HPF # 2

To maximize the dynamic range of the filter, the SHA is designed to have a rail-to-rail architecture, and has fast settling requirements. This can make the SHA very power hungry especially if it has to drive a large input capacitance for the succeeding SCF. To alleviate this problem, we have implemented the double-sampling version of the T-network approach in [7] to implement the low-frequency zero (f_{Z2}), as shown in Fig. 7. The effective loading of the SHA is reduced by 5X. The transfer function of novel SC-HPF # 2 is given as:

$$H_{SCF2}(z) = -\frac{C_{21}}{C_{22}} \times \left[1 + \frac{C_{21a}}{C_{21}} \cdot \frac{C_{21b}}{C_{21c}} \cdot \frac{1}{\left(\frac{C_{21b}}{C_{21c}} + \frac{C_{21a}}{C_{21c}} + 1\right)} \cdot \frac{1}{1 - z^{-1}}\right] \times \left[1 + \frac{C_{22}}{C_{22a}} \cdot \frac{1 + z^{-1}}{1 - z^{-1}}\right]^{-1}$$

$$(5)$$

D. Clocking Scheme

A central clocking scheme is employed which drives the signal-chain and sawtooth generator for PWM control. Since the SCFs employ double-sampling, the duty-cycle inaccuracy of the clock must be low (< 1 %) to minimize path-mismatch [8]. To ensure this, we have generated a master relaxation-oscillator clock ($4 \times f_{SW}$) and divided-by-2 to provide the SHA clock ($2 \times f_{SW}$). This is further divided-by-2 to drive the two SCFs with a clock rate = f_{SW} (see Figs. 3 and 8).

Now, since the SCF architecture is DC-coupled from V_{COMP1} to V_{COMP} (Fig. 3), there is only a range of V_{COMP} within the rail that can be accommodated (for a given V_{REF}), without saturating one or more amplifiers in the signal chain. It is difficult to design a sawtooth that "tracks" this range, and the situation is exacerbated during soft-start and line/load transients. To maximize the dynamic range, we have used a near rail-to-rail sawtooth with fixed feed-forward gain ($V_{IN}/|V_{SAW}| = 8/7$). The target open-loop DC gain is then set by the continuous-time gain stage (Fig. 3), which also provides post-filtering for high-frequency switching transients emanating from the SCF to generate V_{COMP}.

IV. MEASUREMENT RESULTS

The voltage-mode controller IC is implemented in TI's 0.36-μm BCD process, occupies $\sim$ 0.7 mm^2 of active area (Fig. 9) and bonded onto a 32-pin QFN package. Highly

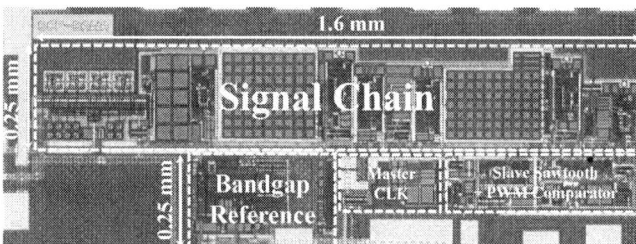

Fig. 9. Die Photo of the controller IC in 0.36-μm BCD process.

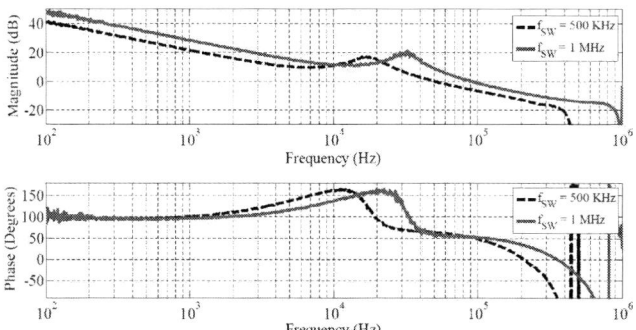

Fig. 10. Measured loop response of the complete system.

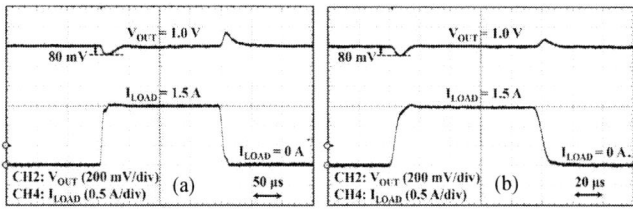

Fig. 11. Load step transient response at f_{SW} = (a) 0.5 and (b) 1.0 MHz

linear Poly2-Oxide-Poly1 (1.5 fF/μm^2) capacitors are used for implementing all the capacitors described in Figs. 3, 6 and 7. The total static current drawn from a 3.3 V supply is 1.1 mA, while static current consumption of the proposed Type-III SCF core is only 0.63 mA. A large portion of the static current is spent in the associated bias circuitry (0.24 mA), as the design was not very optimized.

For this prototype, we have designed the controller for V_{IN} = 3.3 V, V_{OUT} = 1.0 V, I_{LOAD} = 1.5 A, and 1-bit configuration for f_{SW} = 0.5, 1 MHz (B_{TUNE_F} = [0, 1]). The power-stage (drivers and MOSFETs) and LC filter were implemented off-chip on a PCB. For f_{SW} = 0.5 MHz case, L = 1.5 μH and C = 64 μF (f_{LC} ~ 16 KHz), while for f_{SW} = 1 MHz case, L = 1.0 μH and C = 30 μF (f_{LC} ~ 29 KHz).

Fig. 10 shows the measured open-loop frequency response for the two cases of f_{SW}. It can be seen that the filter transfer function scales linearly with f_{SW}, and no distortion is observed. At higher frequencies, close to f_{SW}, we can observe slight gain peaking in the magnitude response (as described in Section III-B), followed by a notch, due to the sample-and-hold effect of the fixed-frequency PWM control scheme. As can be inferred from Fig. 10, the loop is stabilized for both values of f_{SW}, with Phase-Margin (PM) = 63°, Gain Margin (GM) = 14 dB and f_{CL} = 52 KHz when f_{SW} = 500 KHz; and PM = 55°, GM = 12 dB and f_{CL} = 95 KHz when f_{SW} = 1 MHz.

TABLE I
TYPE-III SCF CONTROLLER IC SPECIFICATIONS

Parameter	Value
Fabrication Technology	0.36-μm, BCD process
Active die area (total/filter)	~ 0.65/0.4 mm^2
Switching Frequency (f_{SW})	0.5 and 1.0 MHz
$V_{DD}/V_{IN}/V_{OUT}$	3.3/3.3/1.0 V
I_{LOAD} (max.)	1.5 A
Output Filter (L/C)	1.5 μH/64 μF @ f_{SW} = 0.5 MHz
	1.0 μH/30 μF @ f_{SW} = 1.0 MHz
Analog/Filter core static current	800/630 μA
Bias + Bandgap static current	300 μA
Total static current	1.1 mA

Fig. 11 shows the measured load step transient response of the Buck converter in closed-loop configuration. A load step of 0 – 1.5 A is applied and removed at 0.3 A/μs. At f_{SW} = 500 KHz, the settling time is 50 μs, while at f_{SW} = 1 MHz, it is 20 μs. To conclude, we were able to validate the closed-loop stability of the proposed compensator via both time- and frequency-domain experiments. The specifications of the proposed IC are enlisted in Table I.

V. CONCLUSIONS

We have conceived and implemented novel switched-capacitor filter based Type-III compensation architecture for closed-loop regulation of Buck Converters. The compensation is fully-integrated with minimal trimming requirements, is frequency-scalable, and has moderate power dissipation. Thus, compared to the conventional analog filter, it can result in lower area/cost and higher programmability/ease-of-use.

As a larger perspective, we have laid the theoretical foundations for designing sampled-data analog filter based Type-III compensation of fixed-frequency switched-mode power converters employing linear PWM control. These fundamentals can have wider applicability in the future.

REFERENCES

[1] "3-A Step-Down Regulator with Integrated Switcher," Mar. 2011, Texas Instruments, datasheet of chip no. TPS53311.

[2] R. Gregorian and G. Temes, "Analog MOS integrated circuits for signal processing," Wiley, 1986.

[3] P.Y. Wu, S.Y.S. Tsui and P.K.T. Mok, "Area- and Power-Efficient Monolithic Buck Converters with Pseudo-Type III Compensation," IEEE JSSC, vol. 45, no. 8, pp. 1446-1455, Aug. 2010.

[4] R.W. Erickson and D. Maksimovic, "Fundamentals of Power Electronics," Kluwer Academic Publishers, 2nd Ed., 2000.

[5] G. Temes, H.J. Orchard and M. Jahanbegloo, "Switched-Capacitor filter design using the Bilinear z-transform", IEEE Trans. Cir. and Sys., vol. 25, no. 12, pp. 1039-1044, Dec. 1978.

[6] B. Gilbert, "The multi-tanh principle: a tutorial overview," IEEE JSSC, vol. 33, no. 1, pp. 2-17, Jan. 1998.

[7] W.M.C. Sansen and P.M.V. Peteghem, "An area-efficient approach to the design of very-large time constants in switched-capacitor integrators," IEEE JSSC, vol. 19, no. 5, pp. 772 – 780, Dec. 1984.

[8] J.J.F. Rijns and H. Wallinga, "Spectral analysis of double-sampling switched capacitor filters," IEEE JSSC, vol. 38, no. 11, pp. 1269 – 1279, Nov. 1991.

Estimation of Passive Mixer Output Bandwidth Using Switched-Capacitor Techniques

Essam S. Atalla[*†], Frank Zhang[†], Abdellatif Bellaouar[†], Poras T. Balsara[*]

[*] Department of Electrical Engineering, The University of Texas at Dallas, Richardson, TX 75080

[†] NVIDIA, Richardson, TX 75081

Email: essam.atalla@student.utdallas.edu

Abstract—**Passive mixers have become an essential component of SAW-less receivers. It is well known that the passive mixer behaves as a switched-capacitor circuit (SC) but to the authors knowledge, there is no reported analysis of the mixer impedance that truly accounts for the SC behavior. In this paper, we present for the first time a closed form of the passive mixer output impedance based on SC techniques. We prove that the fundamental lower limit of the mixer impedance is proportional to the well-known switched capacitor resistor $1/(f_{LO}C)$ and different from the previously reported mixer switch ON resistance. We also explain that the equation is useful in estimating output bandwidth of passive mixer based front-ends with general LNA load impedance. We finally show that our bandwidth estimation matches measured results of two receiver front-ends.**

Index Terms—**receivers, mixer, voltage mode, current mode, mixer impedance, mixer bandwidth, switched capacitor**

I. INTRODUCTION

Passive mixers have become an essential building block in state of the art wireless transceivers. They have replaced active mixers in most implementations due to superior linearity, power consumption and noise. In contemporary SAW-less frequency duplex (FDD) receiver implementations, the mixer directly loads the LNA and therefore mixer exhibits no reverse isolation. In other words, the LNA becomes the transconductor driving the mixer. Interaction between the input and output ports of the passive mixer have been recently studied by analyzing the mixer impedance [1]. There has been a lot of attention to the mixer input impedance and how it impacts the gain, noise figure, and linearity of the receiver front-end [1] [2]. It has been well established that the mixer input exhibits high quality filter (Q) bandpass filtering centered at the frequency of local oscillator (LO) signal that drive the mixer switches [1] [2] [3]. In [1], the authors attributed this high Q filtering to the upconversion of the low-pass baseband impedance. They theoretically proved that using steady-state analysis ignoring the impact of transient effects within each LO cycle. A different approach was given in [3] where the authors proposed a passive mixer first receiver that has no LNA and used the baseband impedance translation to provide 50 ohms input impedance across the band of interest. Both concluded that the lower limit of the mixer impedance is the switch ON resistance (r_{sw}). The impedance is maximum at the LO frequency and reduces as frequency departs from LO frequency until it reaches r_{sw}. In this paper, we prove using circuit analysis and simulations that the lower limit

of a double-balanced passive mixer impedance is a switched capacitor (SC) equivalent resistor proportional to $1/(f_{LO}C)$ and not r_{sw} as reported in [1]. Although passive mixers can be operated in current mode or voltage mode, in this paper we focus on voltage mode operation where the baseband amplifier exhibits high input impedance and therefore the mixer output impedance and load capacitor set the output mixer bandwidth and decide the blocker filtering (see Figure I). We report a closed form equation for the mixer output impedance. This equation captures the dependency on the LO switching frequency, the LO duty cycle, and the switch ON resistance r_{sw}. The paper is organized as follows. Section II discusses the analogy between the passive mixer and the SC resistor. In section III we present time-domain analysis of the passive mixer output impedance and present our novel equation. We then extend the equation to include the LNA load impedance. In section IV we present simulation results of two receiver front-end designs that verify the analysis given in section III. In section V we present measurement results and compare them to simulations and calculations. We finally conclude the paper in section VI.

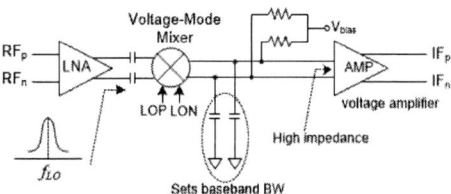

Fig. 1. A SAW-less receiver utilizing passive mixer operating in Voltage Mode

II. ANALOGY OF PASSIVE MIXER AND SWITCHED-CAPACITOR RESISTOR

Figure 2 shows a simplified receiver front-end. The fully-differential LNA is modeled by two voltage controlled current sources having transconductance G_m and Z_T is LNA load impedance which can be either inductive or resistive. The capacitors C_p model the parasitics of the RF lines going from the LNA to the mixer. The mixer is driven by the differential signals LOP and LON. The AC coupling capacitors C_c isolate the DC operating points of the LNA and mixer and filter any low frequency content (especially 2nd order non-linearity products). The mixer bias is set by a bias voltage via resistors

978-1-4673-6145-3/13 $31.00 © 2013 IEEE

(see Figure I). The parasitic plate capacitance of C_c is lumped into C_p. In order to illustrate the SC resistor analogy, we will assume that $Cc \gg C_p$ and can be considered short circuit, and that Z_T can be considered open circuit. Consider the mixer

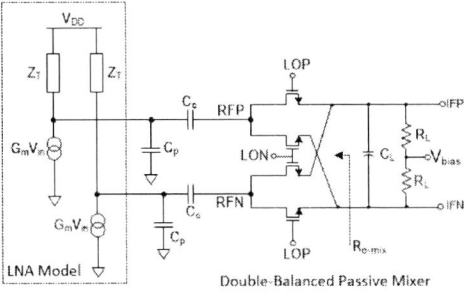

Fig. 2. A simplified model of a receiver front-end using passive mixer

half circuit shown in the left side of Figure 3. By unfolding this circuit, we get the SC resistor shown in the right side of Figure 3 which is a well known building block of SC circuits and has $R_{eq} = 1/(f_{clk}C)$, where f_{clk} is the frequency of the clock signal driving the switches, and C is the capacitance at the intermediate node. This simple R_{eq} formula assumes that the capacitor is fully charged to the final voltages (V_1 and V_2) in each clock cycle which implies that $T_{clk}/2$ (or $T_{LO}/2$) is much longer than the time constant of the circuit $\tau = r_{sw}C$. This requires sizing the switches such that r_{sw} is very small which is feasible in low frequency switched-capacitor circuits but may not be always feasible in high frequency mixer circuits due to other considerations such as the power consumption of the LO signal buffers that drive the mixer switches. In addition, any resistance that appears in series with the switches (such as LNA load as will be shown later) can be modeled as part of r_{sw} and make the approximation invalid. In the following section we will present a novel formula for estimating the mixer impedance at low frequencies and determining the mixer output bandwidth while accounting for the ON resistance of the switches.

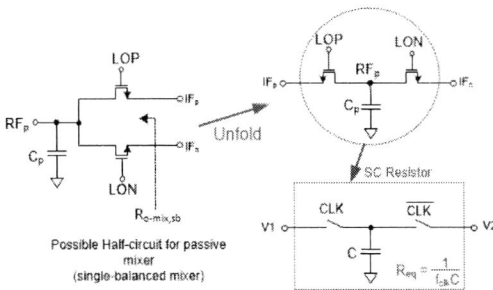

Fig. 3. An example of half circuit of double-balanced mixer and its analogy to SC resistor

III. ANALYSIS OF PASSIVE MIXER OUTPUT IMPEDANCE

In the previous section we have shown that the single-balanced mixer resembles a SC resistor (see Figure 3). In down-conversion mixers, only low frequency content is of

interest at the output port. Therefore the impedance looking into the output port (IF) of the mixer is simply the equivalent resistance of the SC resistor. We will show in section IV that this is true from DC up to tens of MHz. This implies that analyzing the circuit at DC is good enough to estimate the mixer output impedance. Figure 4 shows the circuit we used in our analysis. The switches are modeled as linear resistors with value r_{sw}. We assumed that the LO has sharp edges and has ON time t_D that corresponds to a general duty cycle D. This can help us account for duty cycle effects on the mixer impedance. We analyze the circuit in the two LO phases independently and solve for the currents and voltages. The interaction between the two phases is modeled by the initial voltage on the capacitor which represents the accumulated charge on the capacitor C_p at the end of the previous phase. Since V_1 and V_2 are DC inputs, the capacitor voltage will alternate between two values only V_P and V_N. These two values are the initial voltages on the capacitor in the LOP and LON phases respectively. By applying conservation of charge as a boundary condition we can find V_P and V_N. Using SC concepts, $R_{o-mix,sb} = (V_1 - V_2)/I_{av}$, where I_{av} is the average current flowing from V_1 to V_2 given by $I_{av} = \Delta Q/T_{LO}$, and ΔQ is the net accumulated charge stored on the capacitor C_p in one LO cycle and can be stated as $\Delta Q = C_p(V_N - V_P)$. Therefore $R_{o-mix,sb}$ can be given by the following formula:

$$R_{o-mix,sb} = \frac{1}{f_{LO}C_p} \cdot \frac{1 + e^{-\alpha t_D}}{1 - e^{-\alpha t_D}} \qquad (1)$$

where $\alpha = 1/(r_{sw}C_p)$ is the inverse of the circuit time constant. Equation (1) is a modified version of the well known SC resistor equation. The extra factor depends on the ratio of the LO ON time (t_D) and the time constant of the circuit. If ($t_D \gg r_{sw}C_p$), the equation reduces to the well known form $1/(f_{LO}C_p)$. The equation also reveals the complicated dependency of the mixer impedance on the LO frequency which has an explicit term $1/f_{LO}$ and an implicit term $\exp(-\alpha t_D) = \exp(-\alpha D T_{LO})$ where D is the duty cycle. It is quite clear that the mixer impedance is neither equal to r_{sw} nor equal to $1/(f_{LO}C)$. Therefore it is different from the results reported in [1] that suggested that the lower limit for mixer impedance is r_{sw}. It is worth mentioning that Equation (1) is more accurate than the intuitively derived equation $R = [f_{LO}C_p(1 - exp(-\alpha T_{LO}))]$ reported in [4]. Equation (1) represents the impedance of the single-balanced mixer which is only half of the mixer circuit. Since the other half is similar and appears in parallel (see Figure 2), the output impedance of the full mixer circuit (double-balanced) can be written as: $R_{o-mix} = 0.5 R_{o-mix,sb}$. The above analysis does not account for finite rise and fall times of LO signals. However, for relatively sharp LO edges, Equation (1) provides good agreement with simulations.

A. Impact of LNA load and AC Coupling Capacitor

So far in the analysis we assumed that $Cc \gg C_p$ and can be considered short circuit, and that Z_T can be considered open circuit. In other words, we ignored the effects of the AC coupling capacitor and the LNA load impedance. The LNA load appears as a composite impedance of Lp, C_T, and R_p in parallel with C_p and in series with C_c as shown in Figure 5. With all these components, Equation (1) does not hold anymore, and in order to estimate $R_{o-mix,sb}$ we need to repeat the analysis and account for all of them which leads to complicated mathematical formulation. An intuitive alternative is to use impedance transformation as shown in Figure 5 to represent the composite impedance by a series resistor R_x and capacitor C_x. This allows us to use Equation (1) by replacing r_{sw} with $r_{sw} + R_x$ and C_p with C_x. R_x and C_x are frequency dependent, and in order to compute R_{o-mix} we have to choose a single frequency to evaluate R_x and C_x accordingly. Since R_x and C_x reside at the RF port of the mixer (or at LNA output), and we are interested in output impedance close to DC (after down-conversion), it is necessary to use the input RF frequency (or LO frequency) in evaluating R_x and C_x.

B. Estimation of the mixer output bandwidth

The mixer output bandwidth is set by the impedance looking into the output port of the mixer R_{o-mix} and the total load capacitance at the output of the mixer. In addition, any resistance connected to the mixer output port should be combined with R_{o-mix} in order to compute the bandwidth. An example of such resistor is the bias resistor R_L shown in Figure 2. We can estimate the output 3dB bandwidth of the double-balanced passive mixer using the half circuit shown in Figure 5 and is formulated in Equation (2).

$$BW_{3dB} = [\pi C_L (R_{o-mix,sb} || 4R_L)]^{-1} \qquad (2)$$

IV. SIMULATION RESULTS

Figure 6 shows simulation results of the output impedance of a single-balanced mixer circuit switching at f_{LO} of 1GHz and 2GHz with 25% duty cycle. We assumed inductive load LNA and all the component values are shown in the figure. The peaking in the impedance close to the LO frequency is the well known high-Q filtering feature of the passive mixer [1] [3]. The output impedance is flat (frequency independent) over a wide range of frequencies (up to 100MHz) which supports our argument that DC inputs are sufficient to analyze and compute the mixer output impedance. Calculated impedances based on Equation (2) (highlighted in Figure) are very close to the simulated values. Figure 7 shows the simplified schematic of the front-end circuit we implemented. We had two different

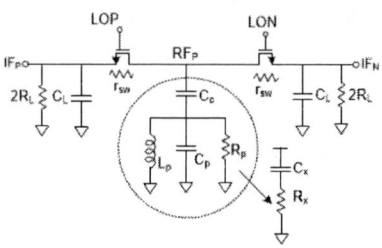

Fig. 5. Modeling the impact of the series AC coupling capacitor (C_c)

LNA implementations, one with resistive load, and the other with inductive load. Inductive degeneration was used in both cases to linearize and match the input. C_p models the parasitic capacitance of the RF lines going from the LNA to the mixer which is about $55fF$ for resistive LNA and $160fF$ for the inductive LNA. C_p also includes the plate capacitance of the AC coupling capacitance C_c and the drain-bulk capacitance of the LNA cascode devices. The resistive load LNA is designed to cover cellular band 1 which extends from 2120MHz to 2170MHz [5] and used R_p of about 150Ω. The inductive load LNA covers cellular band 7 which extends from 2620MHz to 2690MHz [5]. It uses a differential load inductor of $9.5nH$ which corresponds to $L_p = 4.75nH$ (see Figure 7). The capacitor C_T resonates with L_p and controls the center frequency of the load tank circuit. At the center of the band (2655MHz), $C_T = 216fF$. The double balanced mixer consists of 4 NMOS switches sized such that $r_{sw} \approx 10\Omega$. The LO differential signals driving the mixer switches each swing between 0 and 1.3V with 25% duty cycle. The LNA output is AC coupled to the mixer inputs via $1pF$ capacitors (C_c). The mixer output is biased at $1V$ through the two $80k\Omega$ resistors (R_L). The single-ended capacitor C_s ($3pF$) filters high frequency products at the output of the mixer. C_L is a programmable differential capacitor that sets the mixer bandwidth. Resistors r_{p1} and r_{p2} are parasitic resistors of the traces found by parasitic RC extraction as 10Ω and 33Ω. Figure 8 shows the post layout simulation results of the mixer output bandwidth for both inductive and resistive load LNAs. The trace between

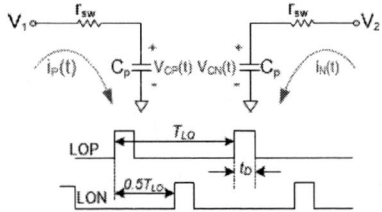

Fig. 4. Schematics and timing diagram used in the derivation of R_{o-mix}

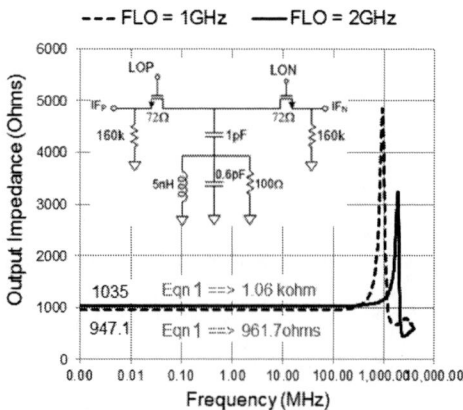

Fig. 6. Simulated output impedance of single-balanced mixer assuming inductive load LNA

978-1-4673-6145-3/13 $31.00 © 2013 IEEE

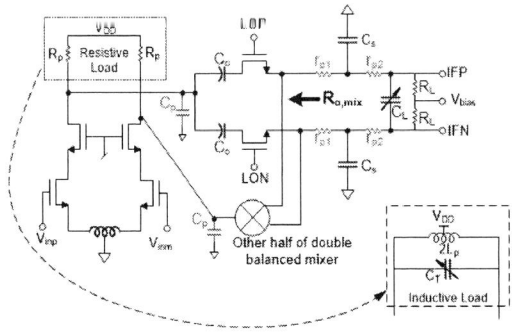

Fig. 7.   Schematic of the receiver front-end

C_s and C_L was not extracted and therefore effect of r_{p2} is not captured. The mixer bandwidth varies as we change the load capacitor (C_L) setting as shown at the bottom of Figure 8. The results of inductive load and resistive load LNAs are different due to differences in the LNA load component values and due to the different mixer LO frequencies. As C_L reduces, the bandwidth increases. The simulated data is also listed in Table I where we compare it to measurement and calculation results.

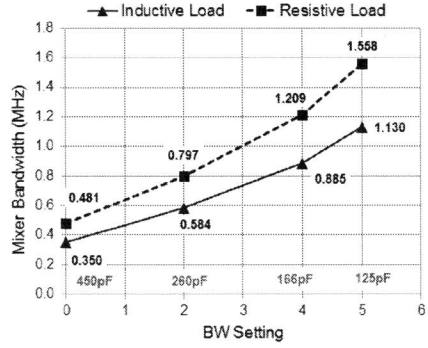

Fig. 8.   Post layout simulations of the Mixer BW for both front-ends

V. MEASUREMENT RESULTS

Figure 9 shows a micrograph of the fabricated front-ends in 65nm technology. Table I lists the measured, simulated and calculated mixer poles. The lowest mixer bandwidth (setting 0) was measured using a spectrum analyzer by sweeping the RF input tone frequency and measuring the down-converted output tone power. Wider bandwidths were measured by comparing the output power for different input frequencies compared to setting 0 and taking an average. The error between the post layout simulations and the measurements reaches about 8% for the widest bandwidth. This can be attributed to not extracting r_{p2} trace (see Figure 7). Calculated results using Equations (1) and (2) are also given in Table I under the column (Calc. w/o trace). It can be shown that the error in these calculations ranges from 6% to about 18% for different bandwidths. We attribute this error to the impact of the trace resistances r_{p1} and r_{p2}. By accounting for them in our calculations, the results are within −3% to 4% error from measurements as given in Table

I. The bandwidth formula is modified as given by Equation (3) based on open circuit time constant approximation.

$$BW_{3dB} = \Big[\frac{C_s}{2} \left((R_{o-mix} + 2r_{p1}) \,\|\, (2R_L + 2r_{p2}) \right)$$
$$+ C_L \left(2R_L \,\|\, (2r_{p1} + 2r_{p2} + R_{o-mix}) \right) \Big]^{-1} \tag{3}$$

Fig. 9.   Micrograph of the chip (Mixer capacitor is not shown)

VI. CONCLUSIONS

In this paper we presented a closed form equation for the mixer output impedance that can be used to estimate its output bandwidth. It correctly captures the SC transient effects for the first time and expresses dependency on LO frequency, LO duty cycle, and ON resistance of mixer switch. We also extended the formula to account for effect of LNA load and AC coupling capacitor on mixer output impedance and bandwidth. We finally presented simulation and measurement results and showed that our formula can estimate the mixer bandwidth within 4% accuracy compared to measured results.

REFERENCES

[1] A. Mirzaei, H. Darabi, J. C. Leete, X. Chen, K. Juan, and A. Yazdi, "Analysis and optimization of current-driven passive mixers in narrow-band direct-conversion receivers," *IEEE Journal of Solid-State Circuits*, vol. 44; 44, no. 10, pp. 2678–2688, 2009.

[2] B. W. Cook, A. Berny, A. Molnar, S. Lanzisera, and K. Pister, "Low-power 2.4-ghz transceiver with passive RX front-end and 400-mv supply," *IEEE Journal of Solid State Circuits*, vol. 41, no. 12, pp. 2757–2766, 2006.

[3] C. Andrews and A. C. Molnar, "Implications of passive mixer transparency for impedance matching and noise figure in passive mixer-first receivers," *IEEE Transactions on Circuits and Systems I: Regular Papers*, vol. PP, no. 99, pp. 1 –12, 2010.

[4] N. Kim, L. E. Larson, and V. Aparin, "A highly linear SAW-less CMOS receiver using a mixer with embedded tx filtering for cdma," *IEEE Journal of Solid-State Circuits*, vol. 44; 44, no. 8, pp. 2126–2137, 2009.

[5] "User equipment (UE) radio transmission and reception (FDD)," 3rd Generation Partnership Project (3GPP), User Equipment (UE) radio transmission and reception (FDD) TS 25.101, October 2010, release 10.

Front-end with Inductive Load LNA, f_{LO} 2655MHz				
Setting	Measured	Sim	Calc. w/o trace	Calc. w/ trace	Error(%)
0	0.339	0.35	0.3585	0.3302	-2.49
2	0.572	0.584	0.612	0.564	-1.37
4	0.886	0.885	0.964	0.8883	0.23
5	1.157	1.13	1.277	1.177	1.74
Front-end with Resistive Load LNA, f_{LO} 2120MHz				
Setting	Measured	Sim	Calc. w/o trace	Calc. w/ trace	Error(%)
0	0.461	0.481	0.5225	0.464	-2.57
2	0.754	0.797	0.892	0.792	0.26
4	1.134	1.209	1.405	1.249	4.05
5	1.437	1.558	1.862	1.655	3.74

TABLE I
COMPARISON OF MEASURED, SIMULATED, AND CALCULATED MIXER
BANDWIDTHS IN MHZ

How to Reduce Power in 3D IC Designs: A Case Study with OpenSPARC T2 Core

Moongon Jung[1], Taigon Song[1], Yang Wan[1], Young-Joon Lee[1], Debabrata Mohapatra[2], Hong Wang[2], Greg Taylor[3], Devang Jariwala[4], Vijay Pitchumani[4], Patrick Morrow[4], Clair Webb[4], Paul Fischer[4], and Sung Kyu Lim[1]

[1] School of ECE, Georgia Institute of Technology, Atlanta, GA, USA
[2] Intel Labs, Intel, Santa Clara, CA, USA   [3] Intel Labs, Intel, Hillsboro, OR, USA
[4] Technology and Manufacturing Group, Intel, Hillsboro, OR, USA
moongon@gatech.edu, paul.fischer@intel.com, limsk@ece.gatech.edu

Abstract—**Low power is considered by many as the driving force for 3D ICs, yet there have been few thorough design studies on how to reduce power in 3D ICs. In this paper, we discuss design methodologies to reduce power consumption in 3D IC designs using a commercial-grade CPU core (OpenSPARC T2 core). To demonstrate power benefits in 3D ICs, four design techniques are explored: (1) 3D floorplanning, (2) metal layer usage control for intra-block-level routing, (3) dual-Vth design, and (4) functional unit block (FUB) folding. With aforementioned methods combined, our 2-tier 3D designs provide up to 52.3% reduced footprint, 25.5% shorter wirelength, 30.2% decreased buffer cell count, and 21.2% power reduction over the 2D counterpart under the same performance.**

I. INTRODUCTION

Power reduction has been one of the most critical design considerations for IC designers. Minimizing both dynamic and leakage power is imperative to meet power budgets for portable devices (low power applications) as well as server farms (high power applications). The power efficiency also directly affects ICs packaging and cooling costs. In addition, the power of an IC has a significant impact on its reliability and manufacturing yield.

Because of the increasing challenges in achieving efficiency in power, performance, and cost beyond 32-22nm, industry began to look for alternative solutions. This has led to the active research, development, and deployment of thinned and stacked 3D ICs with TSVs. Black et al. studied the potential to achieve 15% power reduction as well as 15% performance gain of a high performance microprocessor by a 3D floorplan [1]. Kang et al. demonstrated 25% dynamic and 50% leakage power reduction in 3D DRAM [2].

In this work, we present four physical design techniques that are shown to significantly reduce power consumption in 3D ICs. Our study is based on the OpenSPARC T2 core design database [3] and a PDK that are both available to the academic community. We build GDSII-level 2D and 2-tier 3D layouts, analyze and optimize designs using the standard sign-off CAD tools. Based on this design environment, we first discuss how to rearrange functional unit blocks (FUB) into 3D to reduce power. Next, we study how the number of intra-block-level routing layer used affects routing congestion and power consumption in 2D and 3D designs differently. We also examine the impact of dual-Vth design technique on 2D and 3D power consumptions. Lastly, we demonstrate the effectiveness of functional unit block (FUB) folding, i.e., partitioning a FUB into two sub-FUBs and stacking them, in achieving power savings in the 3D design.

II. 3D FLOORPLANNING BENEFITS

In this section, we explain how we implement both 2D and 3D block-level designs in detail. Then, based on our layout simulations,

This work is supported by Intel Corporation through Semiconductor Research Corporation (ICSS Task 2293).

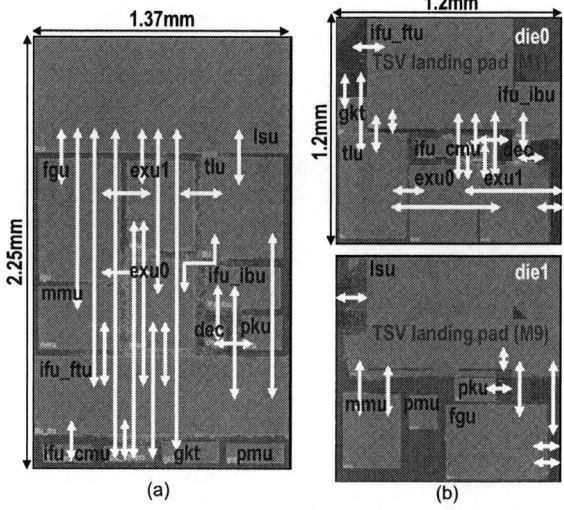

Fig. 1. 2D and 3D placement results. (a) 2D design. (b) 3D design with 2979 TSVs. Cyan dots are core-level buffers. Blue and red rectangles are TSV landing pads at M1 and M9, respectively. White arrows represent major inter-block connections.

we compare several critical design metrics such as footprint area, wirelength, and power consumption of 3D designs with the traditional 2D designs under the same performance, i.e., iso-performance comparison.

A. 2D Design

The OpenSPARC T2 core consists of 13 FUBs including two integer execution units (EXU), a floating point and graphics unit (FGU), five instruction fetch units (IFU), and a load/store unit (LSU) [3]. Each FUB is synthesized with a 28nm cell library. In our implementation, top-level logic cells, i.e., cells outside FUBs, are grouped during synthesis to form an additional block. Thus, a total of 14 FUBs are floorplanned, and special cares are taken to use both connectivity and data flow between FUBs to minimize inter-block wirelength.

With a given target timing constraint, cells and memory macros are placed in each FUB. Note that we only utilize regular-Vth (RVT) cells as a baseline. Then, we perform a static timing analysis (STA) on the placed 2D T2 core and obtain a new timing constraint for I/O pins of each FUB. With this new timing constraint, we perform FUB-level and core-level timing optimizations (buffer insertion and gate sizing) as well as power optimizations (gate sizing). We improve the design quality through iterative optimization steps such as pre-route and post-route optimizations. The 2D placement result is shown in

978-1-4673-6145-3/13 $31.00 © 2013 IEEE

TABLE I
COMPARISON BETWEEN 2D AND 3D DESIGNS WITH A TARGET CLOCK
PERIOD OF 1.5NS. NUMBERS IN PARENTHESES ARE
(INTRA BLOCK/INTER BLOCK) BREAKDOWN.

	2D	3D	diff
footprint (mm^2)	3.08	1.47	-52.3%
utilization (%)	67.8	66.8	-1.0%
# cells ($\times1000$)	504.8 (483.8/21.0)	481.0 (458.8/22.2)	-4.7%
# buffers ($\times1000$)	209.5 (188.5/21.0)	186.0 (163.8/22.2)	-11.2%
Wirelength (m)	23.3 (18.6/4.7)	20.2 (17.6/2.6)	-13.3%
Total power (mW)	**539.4 (489.3/50.1)**	**481.3 (455.5/25.8)**	**-10.8%**
Cell power (mW)	118.0 (111.9/6.1)	106.5 (100.8/5.7)	-9.7%
Net power (mW)	181.8 (150.9/30.9)	154.5 (137.3/17.2)	-15.0%
Leakage power (mW)	239.6 (226.9/12.7)	220.3 (213.3/7.0)	-8.1%

TABLE II
CELL SIZE USAGE (%) COMPARISON BETWEEN 2D AND 3D DESIGNS. X0
IS THE SMALLEST CELL SIZE.

	X0	X1	X2	X4	X8	X16	X32
2D (%)	11.3	44.4	25.9	6.7	8.2	1.8	1.7
3D (%)	13.2	51.8	21.3	6.1	6.3	0.8	0.5

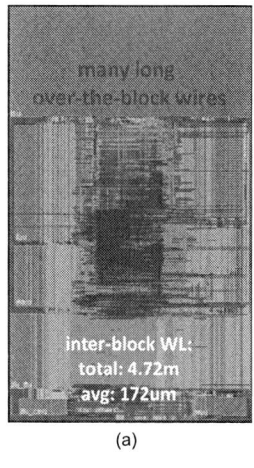 

Fig. 2. 2D and 3D inter-block routing results. Intra-block routing uses up to M5, and inter-block routing uses up to M9. (a) 2D design. (b) 3D design.

Fig. 1(a). Note that intra-block (inside FUB) and inter-block routing utilize up to M5 and M9, respectively.

B. 3D Design

The T2 core netlist is partitioned into two dies considering the area balance between dies and connectivity between FUBs. Then, the 3D floorplanner in [4] is employed with an objective of minimizing inter-block wirelength. In addition, two dies are assumed to be bonded in a face-to-back style. Note that TSV arrays are treated as additional blocks in this flow, hence all TSVs can be placed outside FUBs only. The TSV diameter, height, resistance, capacitance, and landing pad size are $3\mu m$, $25\mu m$, $50m\Omega$, $30fF$, and $3.3\times3.3\mu m^2$, respectively. The total number of TSVs is 2979 in this design. The 3D placement result is shown in Fig. 1(b).

C. 3D CAD Tools

Our RTL-to-GDSII tool chain is based on commercial tools and enhanced with our in-house tools to handle TSVs and 3D stacking. With initial design constraints, the entire 3D netlist is synthesized. The layout of each die is done separately based on 3D floorplanning result. The netlists and the extracted parasitic files are used for 3D STA, followed by the timing and power optimization with the timing constraints from the 3D timing results [5].

D. 2D vs. 3D Floorplanning

We now compare our 2D and 3D designs with a target clock period of 1.5ns (= 667MHz) as shown in Table I. Note that our designs run much slower than UltraSPARC T2, a commercial product version of OpenSPARC T2, that runs at 1.4GHz [6]. This is mainly because some custom memory blocks in T2 core such as a content-addressable memory are synthesized with cells, since a general memory compiler cannot afford this kind of memories. Unfortunately, these synthesized memories are much larger and run slower than the memory macros generated by a memory compiler.

First, interestingly, the footprint area reduction in the 3D design is more than 50%. This is largely related to the buffer count reduction in the 3D design because of shorter wirelength and hence better timing. Note that the silicon area utilization, i.e., area occupied by cells, memory macros, and TSVs (3D only), for 2D and 3D designs are 67.8% and 66.8%, respectively, which supports a fair comparison.

Second, we observe 11.2% total buffer count reduction and 13.3% total wirelength decrease in the 3D design. However, counterintuitively, inter-block level buffers (= 22.2K) in the 3D design are more than the 2D (= 21.0K) even with the much shorter inter-block wirelength. As we optimize the design iteratively in FUB level and core level, buffers can be inserted either inside or outside FUBs to optimize paths. Additionally, to drive 3D nets with a large TSV

capacitance, buffers need to be inserted. Thus, although inter-block level buffers are deployed more in the 3D design, we save a significant number of buffers in the intra-block level. In addition, we see 5.4% intra-block wirelength reduction in the 3D design mainly because of the intra-block level buffer counter reduction.

Third, most importantly, the 3D design reduces power consumption over the 2D counterpart by 10.8%. We see that cell (9.7%) and leakage (8.1%) power reduction are far more than the cell count decrease (4.7%) in the 3D design. As shown in Table II, the 3D design utilizes less larger cells than the 2D case thanks to better timing, i.e., more positive timing slack in paths. With the positive slack, we can downsize cells in the 3D design if this change still satisfies the timing constraint during power optimization stages.

This smaller cell size in the 3D design also helps reduce net power consumption. The load capacitance of a driving cell is defined as the sum of wire capacitance and input pin capacitance of the loading side, hence the net power is defined as the sum of wire and pin power. Thus, the wire power reduction is directly from shorter wirelength, and the pin power decrease is from the smaller cell size as well as the reduced cell count.

III. JUDICIOUS METAL LAYER USAGE

A. Different Routing Resource Demand of 2D and 3D

So far, each FUB is routed using five metal layers to reserve sufficient routing resources for inter-block level routing that utilizes all nine metal layers. In this setting, four high metal layers can be used for over-the-block interconnections.

However, as shown in Fig. 2, the inter-block routing demand is quite different between 2D and 3D designs. As for the 2D case, a large number of over-the-block wires are required, and this increases both total and average wirelength. Thus, more high metal layers (or global metal layers) are necessary to complete inter-block routing. On the other hand, many wires in the 3D design are connected to nearby TSVs, and this reduces over-the-block wiring demand significantly

978-1-4673-6145-3/13 $31.00 © 2013 IEEE

TABLE III
IMPACT OF METAL LAYER USAGE IN INTRA-BLOCK LEVEL ROUTING ON
POWER CONSUMPTION FOR 2D AND 3D DESIGNS. POWER IS NORMALIZED
TO THE CASE OF INTRA-BLOCK ROUTING UP TO M5.

	intra-blk M5	intra-blk M6	intra-blk M7
2D power	1.0	1.026	1.021
3D power	1.0	0.977	0.971

TABLE IV
IMPACT OF INTRA-BLOCK METAL LAYER USAGE ON INTRA-BLOCK AND
INTER-BLOCK DESIGN METRICS IN THE 3D DESIGN. TARGET CLOCK
PERIOD IS 1.5NS AND NUMBERS IN PARENTHESES ARE DIFFERENCE WITH
RESPECT TO THE CASE OF INTRA-BLOCK ROUTING UP TO M5.

		intra-blk M5	intra-blk M6	intra-blk M7
Wirelength (m)	intra block	17.6	17.3 (-1.5%)	17.1 (-3.0%)
	inter block	2.6	2.7 (+2.7%)	2.8 (+8.8%)
	total	20.2	20.0 (-1.0%)	19.9 (-1.4%)
# buffers (×1000)	intra block	163.8	149.5 (-8.7%)	145.2 (-11.4%)
	inter block	22.2	22.9 (+3.2%)	25.9 (+16.7%)
	total	186.0	172.4 (-7.3%)	171.0 (-8.1%)
Power (mW)	intra block	455.5	443.5 (-2.6%)	439.7 (-3.5%)
	inter block	25.8	27.1 (+4.8%)	27.8 (+7.5%)
	total	**481.3**	**470.6 (-2.2%)**	**467.5 (-2.9%)**

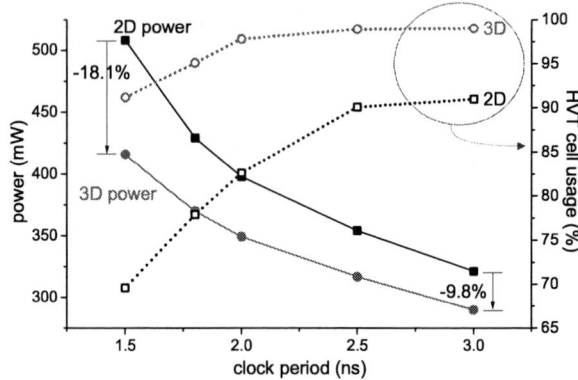

Fig. 3. Power vs. delay curves and HVT cell usage for 2D (intra-block routing up to M5) and 3D (intra-block routing up to M7) DVT designs.

to examine their impact on power consumption in 2D and 3D designs. Each HVT cell shows around 30% slower, yet 50% lower leakage and 5% smaller cell power consumption than the RVT counterpart.

A. HVT Cell Usage

To examine the 3D power benefit under different performances, we implement five designs for both 2D and 3D cases: target clock periods are 1.5ns, 1.8ns, 2ns, 2.5ns, and 3ns. In all cases, we used a dual-Vth (DVT) cell library. As shown in Fig. 3, 3D designs always use more HVT cells than 2D counterparts, and the HVT cell usage increases as the target timing decreases. Even in the fastest case (1.5ns), the HVT cell usage in the 3D design is 91.2%, while that in the 2D design is only 69.6%. Thus, better timing in 3D designs translates to higher HVT cell usage, and this further reduces leakage power.

B. Power Benefit in 3D with DVT Design

As shown in Fig. 3, with a DVT design method, 3D designs benefit more in power reduction for faster cases. This is directly related to the HVT cell usage. At 1.5ns clock period, the 3D design reduces power consumption by 18.1%. As target clock period becomes slower, 2D designs also heavily utilize HVT cells and reduce the total power consumption noticeably, which decreases the 3D power benefit. Still, the DVT design method provides higher power improvement to 3D designs than RVT only cases for all target performances.

We observe that the DVT design technique reduces power noticeably for both 2D (5.8%) and 3D (13.6%) designs compared with the RVT only design at 1.5ns clock period. However, in the 2D case, the power saving is solely from leakage power reduction (17.6%). By employing weak HVT cells, the 2D design uses 7.6% more buffers and 6% longer wirelength than the RVT counterpart, which worsens cell and net power by 0.8% and 5.6%, respectively.

On the other hand, although the 3D DVT design uses slightly more buffers (1.5%) and longer wirelength (0.5%) than the 3D RVT design, cell power decreases by 4.2% since the HVT cell power is slightly lower than the RVT cell, and net power remains similar. Most importantly, leakage power decreases by 21.4%. Thus, the 3D design benefits more from the DVT design, especially for faster cases.

V. FOLDING FUNCTIONAL UNIT BLOCKS

So far, block-level designs are implemented for both 2D and 3D designs. Thus, even in 3D designs, each FUB is located in the same die. In addition, TSVs are always outside FUBs and used only for inter-block connections. In this section, we examine the impact of FUB folding, i.e., partitioning a single FUB into two sub-FUBs

as well. Additionally, inter-block distance within a die is reduced with the reduced footprint area. As a result, the 3D design achieves a huge reduction in both total (44.7%) and average (52.3%) inter-block wirelength over the 2D design. Therefore, this 3D design may not need four high metal layers for over-the-block wiring.

B. Impact of Intra-block Metal Layer Usage on Power

Next we investigate whether we can further save power in the 3D design by allowing more metal layers for intra-block level routing. The key idea here is to reduce the amount of coupling capacitance inside FUBs by relaxing routing congestions with more metal layers and hence to reduce net power consumption. Three cases are studied in this work: intra-block routing up to M5 (baseline), M6, and M7.

The total power consumption of these three cases are shown in Table III. All power numbers are normalized to the baseline 2D and 3D. As more metal layers are available for intra-block routing (less high metal layers for over-the-block wiring in inter-block routing), the 3D design further reduces power. For example, in the case of intra-block routing up to M7, the total wirelength and wire capacitance reduce by 1.4% and 3.5%, respectively, compared with the baseline. Note that the wire capacitance reduction is much more than the wirelength decrease, which indicates less routing congestion inside FUBs. This results in 5.8% net power and 2.9% total power saving.

However, in the 2D case, the opposite trend is observed largely because of the increase in both inter-block wirelength and buffer count. Moreover, the 2D design with intra-block routing up to M7 does not even close the target timing, and thus the power number is not reliable.

The impact of intra-block metal layer usage on intra-block and inter-block design metrics of the 3D design is shown in Table IV. We see that the 3D design with more intra-block metal layers achieves power reduction by improved intra-block level wirelength and buffer count that overwhelm the degraded inter-block level metrics.

IV. DUAL-VTH BENEFITS FOR 3D ICS

Up to this point, both 2D and 3D designs utilize only regular-Vth (RVT) cells. However, industry has been using multi-Vth cells to further optimize power, especially for leakage power, while satisfying a target performance. In this section, we employ high-Vth (HVT) cells

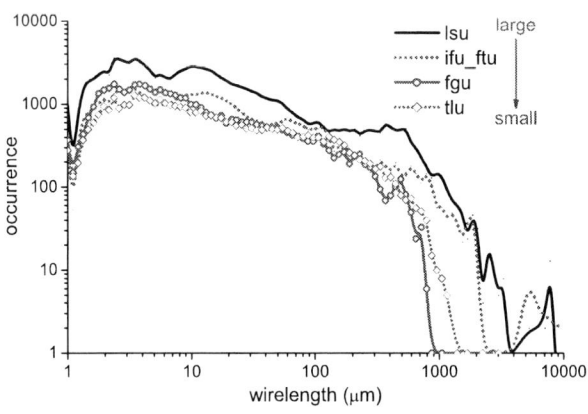

Fig. 4. Wirelength distributions of top four largest FUBs in T2 core.

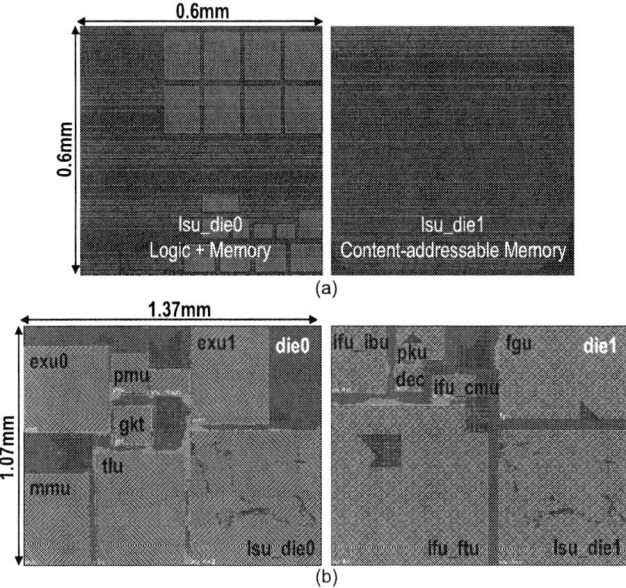

Fig. 5. Placement results of a folded FUB and a 3D block-level design with the folded FUB. (a) folded LSU block (TSV#: 596). (b) 3D design with the folded LSU (TSV#: 2411 (1815+596)).

and connect them with TSVs for intra-block connections, on power consumption.

A. Which Block to Fold?

For the FUB folding to provide power saving, certain criteria need to be met. First, the target FUB needs to contain a large number of long wires so that wirelength decrease and hence net power reduction in the folded FUB can be nonnegligible. In general, large blocks tend to contain many long wires. Wirelength distributions of top four largest FUBs in the T2 core are shown in Fig. 4. We observe that top two largest FUBs, LSU and IFU_FTU, are outstanding.

Second, the target FUB is required to consume high enough portion of the total system power. Otherwise, the power saving from the FUB folding could be negligible in the system level. In our implementations, LSU and IFU_FTU consume around 28% and 23% of the total T2 core power, respectively. Third, the net power portion of the target FUB needs to be high. If the FUB is cell and leakage power dominant, the wirelength reduction of the folded FUB may not reduce the total power noticeably. The net power portion of LSU, FGU, and TLU are about 33%, 47%, and 43%, while that

TABLE V

COMPARISON BETWEEN 2D (INTRA-BLOCK ROUTING UP TO M5), 3D WITHOUT FUB FOLDING (INTRA-BLOCK ROUTING UP TO M7), AND 3D WITH FUB FOLDING DESIGNS WITH A TARGET CLOCK PERIOD OF 1.5NS. DUAL-VTH DESIGN TECHNIQUE IS APPLIED TO ALL CASES. NUMBERS IN PARENTHESES ARE DIFFERENCE AGAINST THE 2D DESIGN.

	2D	3D w/o folding	3D w/ folding
footprint (mm^2)	3.08	1.47 (-52.3%)	1.47 (-52.3%)
utilization (%)	69.2	67.5 (-1.7%)	67.1 (-2.1%)
# cells ($\times$1000)	532.3	471.9 (-11.3%)	450.9 (-15.3%)
# buffers ($\times$1000)	225.5	173.6 (-23.0%)	157.5 (-30.2%)
# HVT cells ($\times$1000)	370.4	430.4 (+16.2%)	444.8 (+20.1%)
Wirelength (m)	24.7	20.0 (-19.0%)	18.4 (-25.5%)
Total power (mW)	**508.2**	**416.0 (-18.1%)**	**400.7 (-21.2%)**
Cell power (mW)	118.9	100.6 (-15.4%)	99.1 (-16.7%)
Net power (mW)	191.9	145.0 (-24.4%)	135.6 (-29.3%)
Leakage power (mW)	197.4	170.4 (-13.7%)	166.0 (-15.9%)

of IFU_FTU is only 17%. Therefore, in this T2 core case, LSU is the best choice for folding.

B. FUB Folding Impact on Power

The LSU block is partitioned into two dies and designed with an in-house mixed-size 3D placer as shown in Fig. 5(a). The dual-Vth design technique is also applied. This folded LSU block reduces the footprint, buffer count, and wirelength by 50.8%, 9.7%, and 7.1%, respectively, compared with the 2D LSU block. In addition, the HVT cell usage in the folded LSU is 96.8%, while that in the 2D LSU is 79.7%. More importantly, the total power of LSU is reduced by 5.4% largely due to the decreased net (9.2%) and leakage (4.9%) power.

Detailed comparisons between 2D, 3D without FUB folding (*3D w/o folding*), and 3D with FUB folding (*3D w/ folding*) designs are shown in Table V. In *3D w/ folding*, the total power reduces by 21.2% compared with the 2D design and by 3.7% compared with *3D w/o folding*.

Interestingly, the FUB folding helps both the folded block itself and the overall floorplan for power saving. The inter-block wirelength decreases significantly because of the increased flexibility of 3D floorplanning with smaller FUBs, i.e., the largest FUB is divided into two. In this design, inter-block wirelength decreases by 27.0%, which in turn reduces inter-block buffers by 29.6% compared with *3D w/o folding*. As a result, inter-block power reduces by 23.6% compared with *3D w/o folding*.

VI. CONCLUSIONS

In this paper, the power benefit of 3D ICs is demonstrated with an OpenSPARC T2 core. Four design techniques are explored to optimize power in 3D IC designs: (1) 3D floorplanning, (2) intra-block level metal layer usage control, (3) dual-Vth design, and (4) FUB folding. With aforementioned methods, the total power saving of 21.2% has been achieved against the 2D counterpart.

REFERENCES

[1] B. Black et al., "Die Stacking (3D) Microarchitecture," in *Proc. Annual Int. Symp. Microarchitecture*, 2006.

[2] U. Kang et al., "8 Gb 3-D DDR3 DRAM Using Through-Silicon-Via Technology," in *IEEE J. Solid-State Circuits*, 2010.

[3] Oracle, "OpenSPARC T2." [Online]. Available: http://www.oracle.com

[4] D. H. Kim, et al., "Block-level 3D IC Design with Through-Silicon-Via Planning," in *Proc. Asia and South Pacific Design Automation Conf.*, 2012.

[5] M. B. Healy et al., "Design and Analysis of 3D-MAPS: A Many-Core 3D Processor with Stacked Memory," in *Proc. IEEE Custom Integrated Circuits Conf.*, 2010.

[6] U. G. Nawathe et al., "An 8-Core 64-Thread 64b Power-Efficient SPARC SoC," in *IEEE Int. Solid-State Circuits Conf. Dig. Tech. Papers*, 2007.

978-1-4673-6145-3/13 $31.00 © 2013 IEEE

A General-purpose Vision Processor with 160x80 Pixel-Parallel SIMD Processor Array

Alexey Lopich   Piotr Dudek

School of Electrical and Electronic Engineering
University of Manchester
Manchester, UK
{a.lopich, p.dudek}@manchester.ac.uk

Abstract— **In this paper we present a vision processor, which incorporates a 160×80 SIMD array of pixel-processors. The processor operates with a 100MHz clock and 1.8V supply. The device provides 640 GOPS (binary) and 23 GOPS (greyscale) consuming 0.5 W. The chip occupies 50mm² and is fabricated in a standard 0.18 μm CMOS process. The I/O interface supports 200 MPixels/s (greyscale), 1.6 GPixels/s (binary) and 40 MPixels/s (address-event readout) data rate, and PE-parallel image sensing mode for embedded high-speed vision applications. Experimental results indicate that the performance of the presented chip approaches the efficiency of recently reported application-specific vision processors, while providing full programmability and thus being adjustable to a wide range of applications.**

I. INTRODUCTION

Computer vision applications demand high computational power, due to the large amount of raw image data required to be processed at real-time speeds. Low- and medium-level image processing algorithms are characterised by a high degree of data parallelism and computational locality, i.e. every value in the result image is a function of the original pixel value and its bounded neighbourhood. Conventional processor architectures have to cope with massive data-flows and frequent memory access, which result in increased power consumption and performance limitations. At the same time, massively parallel processor arrays are ideally suited to regular, local computations with inherent data parallelism. Massively parallel fine-grain computing has been explored for several decades. Particular application benefits have been realised in image pre-processing, where regularity of image data and full-scale parallelism of processing algorithms naturally map onto fine-grain processor-per-pixel architectures. Many examples of such arrays have been recently presented in literature [1-6]. Typically, the architecture of such devices is based on a cellular processor array (CPA) operating according to the SIMD paradigm. Progress in CMOS fabrication technology has led to integration of image sensing and processing on a single device, a so called "vision chip". This approach eliminates the I/O data-transfer bottleneck for high frame-rate applications as sensory data is fed directly into corresponding processing elements. The output of the vision chip can be a simplified

representation of the scene in the form of abstract descriptors (e.g. object coordinate, presence indicator, etc.), thus significantly increasing useful data-throughput. A number of application-specific [7-8], and general-purpose [1-6] vision chip architectures have been recently proposed. While application-specific devices often offer better performance to area/power ratio, we concentrate our efforts on programmable general-purpose chips that can be applied in a wide range of tasks.

In this paper we present a 160×80 digital vision chip that operates both in synchronous (local pixel operations) and asynchronous (global operations) modes. The presented work builds on our previous design of a vision chip a with pixel-parallel SIMD array [6]. The new device introduces further improvements in processor functionality and performance, combined with reduction in cell area and an increase in fill-factor. The chip facilitates high frame-rate processing, by enabling the output of global image descriptors (e.g. pixel coordinates, global summation, etc.), thus reducing the required output data rate down to several bytes per frame. Real-time feedback control systems can then be created based on sensed image data. Basic object segmentation and coordinate extraction takes 1.62 μs at 100 MHz operating frequency. The device is implemented in a standard 180 nm CMOS process and provides 46 GOPS/W and 650 MOPS/mm², approaching the efficiency of recently reported application-specific vision processors [7-8], while providing a general-purpose programmable processor architecture.

II. ARCHITECTURE

The outline of the presented chip architecture is depicted in Fig. 1. It is based on a massively parallel fine-grain cellular processor array that operates according to the SIMD paradigm, i.e. all cells receive and execute identical instructions issued by a central controller (located off chip), and operate on local data. All cells are locally interconnected with 4 nearest neighbours. The I/O interface between every PE and its four neighbouring cells consists of four 8-bit input ports for greyscale values, four 1-bit input ports for binary asynchronous propagations, one 8-bit output port for greyscale data (connects to corresponding input ports of four neighbours) and one 1-bit input and one 1-bit output port in

978-1-4673-6145-3/13 $31.00 © 2013 IEEE

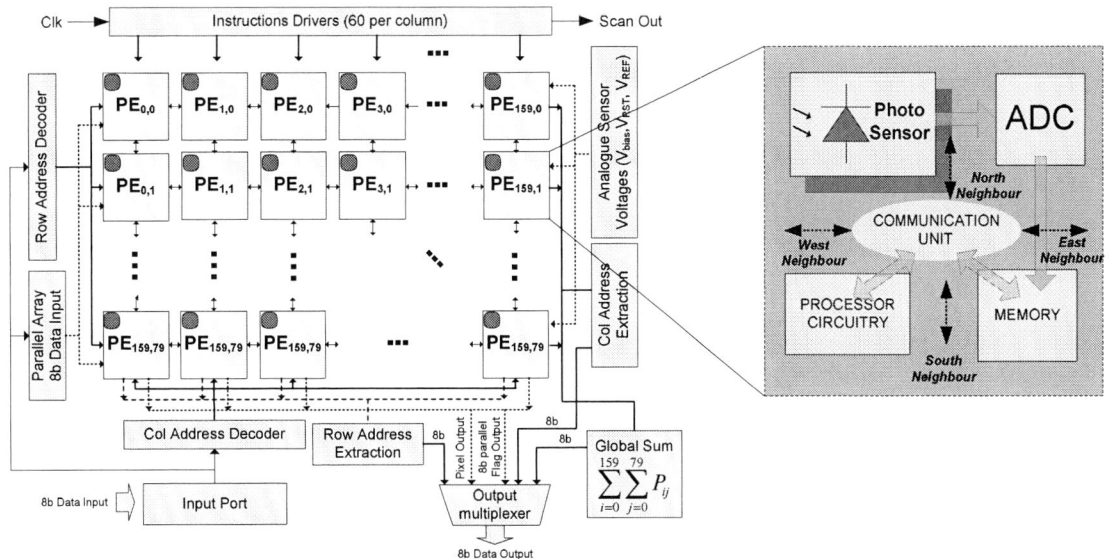

Figure 1. Block diagram of the ASPA2 architecture

the ALU for global asynchronous addition. The latter port is connected only to the east neighbour, so that global addition is enabled row-wise. Altogether, each cell has 37 digital inputs (nine from every neighbour and one sum input from the west) and 10 digital outputs. The local autonomy of each cell is provided by flags, which specify whether the cell executes an instruction or omits it. The flags can also be used to constrain the topology of the asynchronous network for continuous-time operation so that the link between neighbours is defined in each pixel individually. Instructions are latched inside the chip at a 100MHz rate and distributed by periphery drivers placed at the boundary of the array. Pixels can be addressed for I/O individually in a random-access fashion, as a block or in a flexible pattern, providing simultaneous writing to many

locations and global logic operations on readout. The chip can output various types of data, which are multiplexed onto a single 8-bit output port. This port can be configured to output single 8-bit greyscale pixel value, eight binary values from 8 cells in parallel, a global sum value, X/Y pixel coordinates (enabling Address Event Representation (AER) readout mode [9]) and finally, the result of logic OR operations of selected pixels. The I/O interface supports 200 MPixels/s (greyscale) and 1.6 GPixels/s (binary) data rate.

III. PROCESSING CELL

Each processing element (PE) operates as a simple bit-serial digital microprocessor (Fig. 2). The chip is programmable and software development is based on a custom

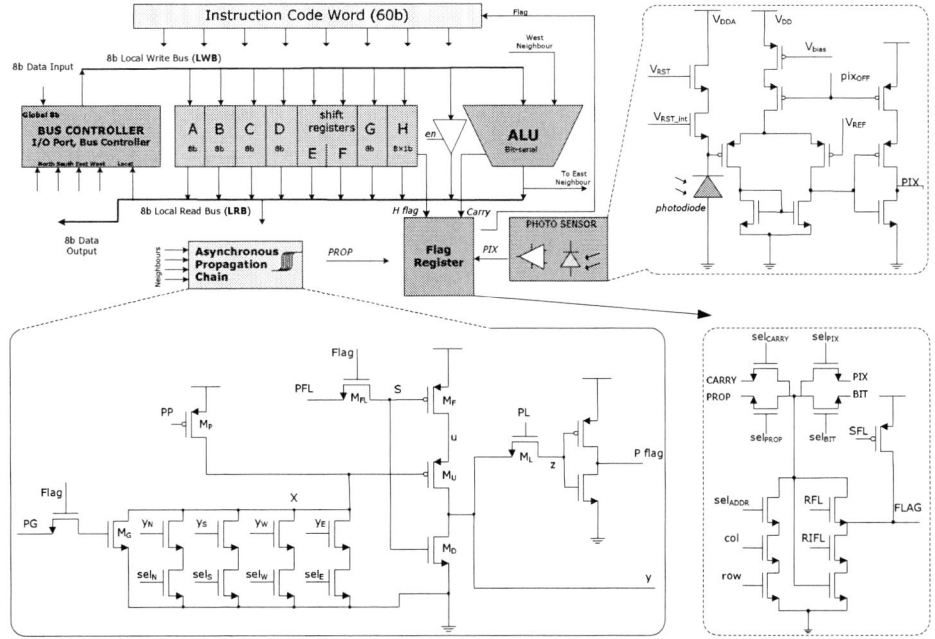

Figure 2. Schematic diagram of the processing cell with detailed schematic of photo sensor, asynchronous propagation chain and flag register.

978-1-4673-6145-3/13 $31.00 © 2013 IEEE   633

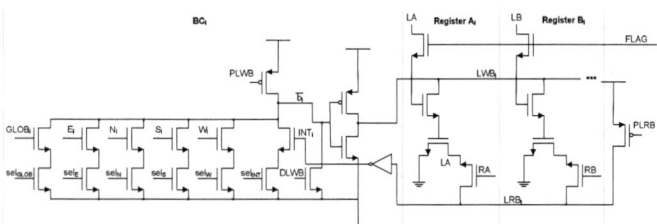

Figure 3. Bus controller and memory bit-slice.

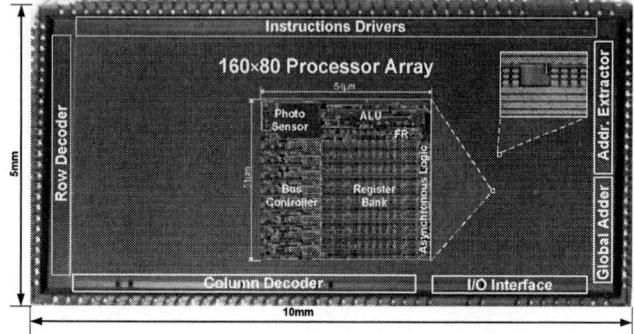

Figure 4. Chip photomicrograph, with inset showing the layout of a single PE with highlighted functional blocks and a photomicrograph of the PE.

built Assembly language. Every processor executes various instructions ranging from basic logic operations to multiplication/division and global data-flow processing. Auxiliary program flow aspects such as global variables, function calls and loops are handled by an off-chip central controller.

A. Memory and data-path

Every PE comprises eight 8b general purpose registers (GPR), two of which are shift registers. Because image pre-processing algorithms constantly deal with both greyscale and binary data, the PE was set to support fast access to binary image descriptors as well as logic and conditional operations on this data. Therefore in addition to GPRs there are also eight 1b registers, which can be used either separately for storing intermediate binary values or as a single 8b register (register H). The value of this register can either be transferred to the Local Read Bus (LRB) as an 8-bit greyscale value, or a selected bit can be passed as a single binary value (H flag) to the flag register. The binary output is the result of a logic OR operation on bits, selected in the instruction word. The total PE memory capacity is therefore 72 bits. All local memory is based on 3-transistor dynamic registers. Inputs of memory cells are connected to Local Write Bus (LWB) while outputs are connected to a precharged LRB. LRB and LWB are connected via the Bus Controller (BC), responsible for multiplexing input data (local, neighbour or global) to the LWB. The BC also performs the logic OR operation on input data. One bit-slice of the 8-bit BC and two registers is presented in Fig. 3. In the current implementation, the retention time of such dynamic registers ranges from 100 µs to 1 ms. Such a large range appears to be the result of unmatched coupling between the storage node, ground plane and instruction lines. It results in the necessity to refresh values algorithmically for temporal inter-frame processing.

Basic register-transfer operations comprise two clock cycles: precharge and transfer. During the precharge cycle the LRB and an internal node b_i (Fig. 3) in the BC are precharged so that values on the LRB and the LWB are 0xFF and 0x00 correspondingly. During the transfer cycle, the data is read from any register(s) or accumulator to the LRB, transferred through the BC to the LWB and then loaded into another register(s) or accumulator. The data appears inverted on LRB and node b_i of the BC. At the end of the transfer cycle, when the load signal (e.g. LA and LB in Fig. 3) goes to logic '0', the data is latched in the dynamic register. The communication between local neighbours is identical to local register-transfer operations and also performed within two clock cycles.

Additionally, each cell contains an 8-bit parallel three-state buffer, in the form of two pass transistors per bit, which connects LWB and LRB (see Fig. 2). It is used to perform the NOT operation on binary data. This operation is bit-controlled, so it enables inversion of only specified bits.

B. ALU

The 18-transistor bit-serial ALU performs NOT, ADD, SUB, XOR and XNOR logic operations. The design of the adder is based on compact XOR/XNOR gates coupled with dynamic pull-up/pull-down transistors and a transmission gate multiplexer, resulting in an area of 3.1×8.9 µm^2. It has a single non-accumulative V_{th} loss on the CARRY output and full voltage swing on data output, which enables an unlimited number of such arithmetic cells to be used in an asynchronous addition/subtraction chain. ALUs in neighbouring cells can be chained together to form a single multi-bit ripple-carry adder. This enables asynchronous row-wise global summation. The final column-wise addition is performed at the periphery of the array and the result of the global summation is stored in a dedicated register at the periphery of the chip, whereas row-wise sub-sums are stored in the PEs in the most right column. Global addition is extensively used to gather various image statistics, e.g. pixel count, histograms, etc.

C. Data I/O

In addition to digital I/O, the chip supports in-cell sensing in each PE, so that it can operate as a 'vision chip'. The sensor is based on a n+/p-subst. photodiode followed by a simple voltage comparator (as shown in Fig. 2). The photodetector works in integration mode. The voltage across photodiode V_{PD} is compared to the reference voltage V_{ref}. The inverted value of the comparator output is transferred to the PIX input of the flag register. Based on this flag indicator a digital value that corresponds to light intensity can be written to a corresponding GPR. For greyscale conversion the binary threshold is performed at different times. The A/D conversion is set, but not limited, to 8-bit resolution. Every cell can locally control its integration time.

D. Cell layout

The microphotograph of the chip with highlighted functional block is depicted in Fig. 4. A summary of measured characteristics of the chip is provided in Table I. Compared to the previously reported chip [6], the area of the PE cell is decreased by 76% while the overall functionality of the cell

978-1-4673-6145-3/13 $31.00 © 2013 IEEE

TABLE I. ASPA2 VISION CHIP CHARACTERISTICS

Parameter	Value
Fabrication process	0.18μm, 1P6M CMOS
Power supply	1.8V
Array size	160×80
PE Size	54×51μm^2
Performance	640 GOPS (binary) 23 GOPS (greyscale) 2.9 GOPS (mult/div)
Clock frequency	100MHz
Power efficiency	46 GOPS/W
Area efficiency	650 MOPS/mm^2
PE Memory	72b

has been improved. With 5 routing layers placed above active circuitry, less than 1% of the silicon space is occupied solely by routing, therefore further increase in the number of routing layers will not result in area optimization. As compared with [6], The area of the photodiode increased by 40% and the fill-factor has improved from ~2% to 5.6%. The "complexity" of the cell (number of transistors) has also increased by 27%.

IV. PERFORMANCE

Because the chip has a software-controlled architecture it is capable of executing a wide range of image processing algorithms. The support of logic and arithmetic operations in every cell enables straightforward implementation of basic convolutions and filters. Examples of pre-processing algorithms implemented on the fabricated device are illustrated in Fig. 5 (a)-(c).

A significant number of useful morphological image processing algorithms deal with images represented in binary format (binary dilation, erosion, skeletonization). The chip can execute many of these tasks in an efficient way, as some of the logic operations are performed on data buses during data transfers, i.e. at no additional cost. In addition to local operation, the support for global binary trigger-wave propagations enables fast geodesic reconstructions and hole filling, which can be a part of more complex algorithms. An example of using these operations for object segmentation and coordinate extraction is presented in Fig. 5 (d)-(i).

The performance of the chip on a set of several common image processing tasks is presented in Table II.

TABLE II. ASPA2 PROCESSING PERFORMANCE

Operation	Frame processing time @ 100 MHz
Sobel Edge Detection	5.2 μs
3×3 Erosion(Dilation)	60 ns
3×3 Median Filter	9.3 μs
Frame min/max value	1.1 μs
Binary object reconstruction	180 ns
Closed contour detection	200 ns
Object bounding box extraction (4 coordinates)	1.42 μs
Output rate	greyscale: 200 MPixel/s binary: 1.6 GPixel/s [x,y] pair: 40 MPixel/s

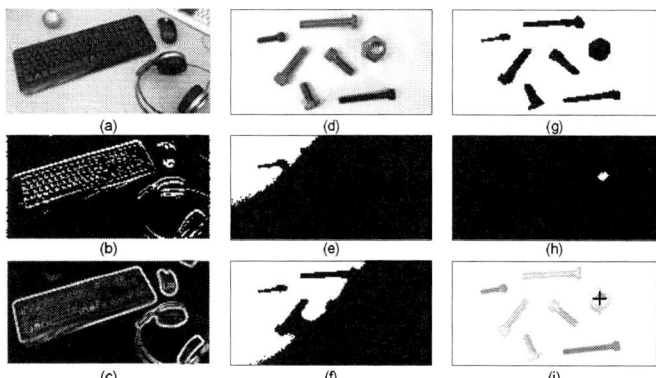

Figure 5. Image processing in ASPA2: a – original; b – high pass filter with threshold; c – sobel edge detection; d – original; e-g – asynchronous propagation triggered in the top left corner; h – closed curve detection; i – coordinate extraction of the closed area (nut).

V. CONCLUSIONS

In this paper we presented a general-purpose vision chip, based on fine grain cellular processor-per-pixel array that operates in the focal-plane. The 160×80 chip is fabricated in 0.18 μm standard CMOS technology and measures 10×5 mm^2. The overall architecture is scalable to larger dimensions. The chip operates at 100MHz with V_{DD} 1.8V. At a 100MHz clock, each cell provides 50 MOPS (binary), 1.8 MOPS (greyscale) and 0.23 MOPS (unsigned products and quotients). The chip is suitable for low- and medium-level image processing and its main application areas include high-speed industrial inspection, robotics and control systems.

ACKNOWLEDGMENT

This work has been supported by the EPSRC under UK grant no EP/D029759/1.

REFERENCES

[1] W. Zhang, Q. Fu, and N. Wu, "A Programmable Vision Chip Based on Multiple Levels of Parallel Processors", IEEE Journal of Solid-State Circuits, vol. 46, pp. 2132-2147, 2011

[2] A. A. Abbo, R. P. Kleihorst, et,. al., "Xetal-II: A 107 GOPS, 600 mW Massively Parallel Processor for Video Scene Analysis," IEEE Journal of Solid-State Circuits, vol. 43, pp. 192-201, 2008

[3] D. Ginhac, J. Dubois, M. Paindavoine, and B. Heyrman, "A high speed programmable focal-plane SIMD vision chip", Analog Integrated Circuits and Signal Processing, vol 65 (3), pp. 389-398, 2010.

[4] C.-C. Cheng, C.-H. Lin, C.-T. Li, et.al.: "iVisual: An Intelligent Visual Sensor SoC with 2790fps CMOS Image Sensor and 205GOPS/W Vision Processor" in ISSCC Dig.Tech.Papers, pp. 306-615, 2008

[5] S.J.Carey, A.Lopich, D.R.W.Barr, P.Dudek, "A 100,000 fps Vision Sensor with Embedded 535 GOPS/W 256x256 SIMD Processor Array", VLSI Circuits Symposium, Kyoto, 12-14 June 2013

[6] A. Lopich and P. Dudek, "A SIMD Cellular Processor Array Vision Chip With Asynchronous Processing Capabilities," IEEE Transactions on Circuits and Systems I, vol 58 (10), pp. 2420-2431, 2011.

[7] Jinwook Oh, Junyoung Park, Gyeonghoon Kim, Seungjin Lee, et.al, "A 57mW Embedded Mixed-Mode Neuro-Fuzzy Accelerator for Intelligent Multi-core Processor", ISSCC Dig.Tech.Papers, pp. 130-131, 2011

[8] Jae-Sung Yoon, et.al. " A Graphics and Vision Unified Processor with 0.89μW/fps Pose Estimation Engine for Augmented Reality", ISSCC Dig.Tech.Papers, pp. 336-337, 2010

[9] T. Serrano-Gotarredona, and B.Linares-Barranco, "A 128x128 1.5% Contrast Sensitivity 0.9% FPN 3 μs Latency 4 mW Asynchronous Frame-Free Dynamic Vision Sensor Using Transimpedance Preamplifiers", IEEE J. Solid-State Circuits, vol. 48, pp. 827-838, 2013

A Programmable Analog Frequency-Locked Loop for VCO Characterization and Test with 8 ppm Resolution

Sadok Aouini, Jean-François Bousquet, Naim Ben-Hamida,
Lukas Jakober, John Wolczanski, Christopher Kurowski
Ciena Corporation
3500 Carling Avenue, Lab 10, Ottawa, CANADA, K2H 8E9
{saouini, jbousque, nbenhami, ljakober, jwolczan, ckurowsk}@ciena.com

Abstract— This article presents a digitally controlled analog frequency-locked loop used for VCO characterization and test. The proposed scheme allows a frequency tuning better than 8 parts per million (ppm). The AFLL is implemented in 32nm CMOS technology and standard CMOS library cells are used for all the digital blocks. The AFLL comprises a 17-bit frequency counter running at 5GHz, a 1st order sigma-delta modulator used for dithering the correction signal, a charge-pump and capacitance used as integrator and a VCO. The frequency counter generates a count difference between the VCO clock and a reference clock. This difference is then pulse-density modulated and applied to a charge-pump feeding a capacitor that acts as an integrator. The generated output voltage is applied to the VCO tuning port and adjusts its oscillating frequency accordingly. An offset value added to the frequency difference allows the VCO to settle to a proportional frequency offset. Using this architecture, the VCO frequency can accurately be tuned digitally without having to change the frequency of a reference clock or sweeping its tuning voltage. Hence, the proposed AFLL can serve as a design-for-test (DFT) solution allowing characterization and testing of the VCO in an all-digital environment such as for digital automated test equipment (ATE).

I. INTRODUCTION

VCO's are the heart of any clock generation scheme. In transmitters, they are used within a phase-locked loop for synthesizing the sampling clocks used to modulate the data to be sent. In receivers, they are the main block in any clock recovery scheme where the received clock signal is extracted from the data. When the VCO is locked in a PLL or CDR loop, its transfer characteristic (frequency gain vs. input voltage) directly impacts the bandwidth of the system and the total generated jitter. Thus, in order to optimize the jitter, it is paramount to ensure the VCO is operating at its optimal performance. Integrated phase noise measurements have been realized on-chip [1]. However, such a DFT structure consumes a large area [1] and does not have a digital interface. In [2], a loopback DFT scheme for testing VCO-based transceivers has been proposed. However, this scheme only provides go/no-go classification. Other on-chip VCO DFT schemes with a digital interface have been proposed in the literature [3,4]. For example, a digital technique for synthesizing accurate frequencies was proposed in [4] as a test stimulus generator. However, a long bit-stream with the desired frequency is required and its frequency resolution is directly proportional to the length of the bit-stream. In [5], a digital on-chip instrument capable of measuring the jitter transfer function of a PLL in closed-loop is described. In this paper, a programmable analog frequency-locked-loop (AFLL) capable of accurately synthesizing frequencies with a resolution better than 8ppm is demonstrated. The proposed AFLL has a complete digital interface to the outside world and can thus be easily controlled. In fact, the proposed architecture can be used in any testing environment, from characterization to production testing using a digital ATE

The paper is divided as follows: in Section II, a top-level description of the AFLL is provided together with a system level model in Matlab/Simulink. In Section III, the circuit implementation of the various blocks is described along with post-layout simulations. In Section IV, experimental data validating the operation of the AFLL and demonstrating its use as a digital-friendly VCO testing and characterization instrument is presented. Finally, conclusions are outlined in Section V.

II. SYSTEM-LEVEL DESCRIPTION

In order to observe the VCO performance under different tuning conditions, a high-resolution analog frequency-locked loop is used. For example, accurately determining the gain (frequency vs. tuning voltage) requires very fine tuning of the VCO frequency. Similarly, accurate frequency tuning is required in assessing the internal jitter performance and how the quality factor of the VCO varies. The AFLL loop presented here shows an architecture realizing frequency placement with an accuracy of 7.6ppm, which corresponds to an equivalent frequency resolution of 17 bits.

A. Analog FLL

As depicted in Figure 1, the analog FLL implementation consists of a 17-bit counter running from a 5GHz clock (divided from 20GHz VCO), a pulse-density modulator DAC with adjustable gain, and an integrator (which can be implemented as a charge pump feeding a capacitor). Assuming a reference clock frequency f_{REF} and a VCO clock frequency f_{VCO}, the frequency count difference Δf_{cnt} is:

$$\Delta f_{cnt} = T_S(f_{VCO} - f_{REF}) \qquad (1)$$

Therefore the resolution of the frequency counter is $\Delta f_{min}=1 / T_S$ and the maximum range it can lock to is $\Delta f_{max}=MAX\_COUNT / T_S$. Given that the counter runs at

978-1-4673-6145-3/13 $31.00 © 2013 IEEE

5GHz, the refresh rate of the of the 17 bit counter is thus equal to:

$$F_S = \frac{5GHz}{2^{17}} \cong 38.1 kHz \tag{2}$$

Hence, the equivalent frequency resolution is:

$$\frac{10^6}{2^{17}} = 7.6 ppm \tag{3}$$

The objective is to pre-load the counter with an offset which would make the AFLL lock to the corresponding frequency value.

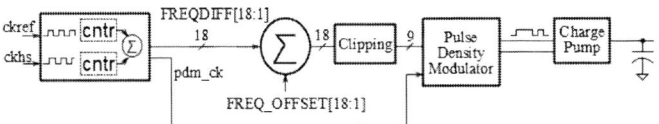

Figure 1: AFLL time-domain model. Note that charge-pump/cap output is connected to VCO input and that ckhs is divided clock from the VCO.

A Matlab/Simulink transient model of the analog FLL has been implemented and used to investigate the AFLL loop. The model is used to ensure it settles within the specified resolution even in the presence of a 200GHz/s frequency drift. Note that the role of the sigma-delta modulator is to spread the frequency counter difference (pulse-density encoding) such that the integrator generates a ramp proportional to the input code. As depicted in Figure 2, for a counter offset of zero, the frequency variation is less than 6ppm once the AFLL loop has settled.

To verify the stability and bandwidth of the loop, a linear model of the A-FLL is developed in Matlab. The loop filter is simply a capacitor with $C_{filt} = 30\ nF$, the charge pump current is $I_{cp} = 0.15\ mA$, and the VCO gain is assumed to be $K_{VCO} = -250\ MHz/V$. Also, because the PDM modulator maximum differential pulse density is ±25%, its gain K_{PDM} is 0.25. The frequency counter update rate is T_s. Using the standard control theory terminology, the forward path gain is:

$$A(s) = \frac{T_S K_{PDM}}{MAX\_CNT} I_{CP} K_{VCO} \frac{1}{C_{filt} s} \tag{4}$$

The feedback path is simply $\beta = 1$. For this system, the closed loop transfer function is shown in Figure 3. Since the transfer function is that of a first order system, the A-FLL remains always stable and the 3dB bandwidth is 10.2 kHz.

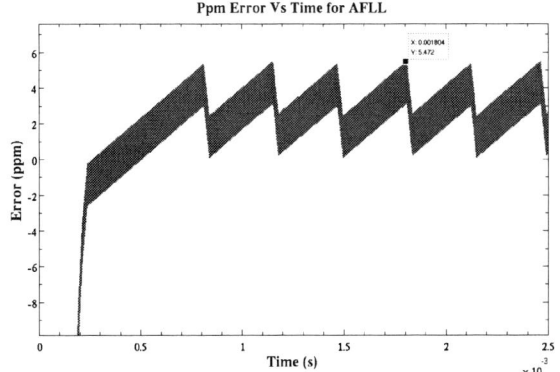

Figure 2: Frequency output vs time in ppm.

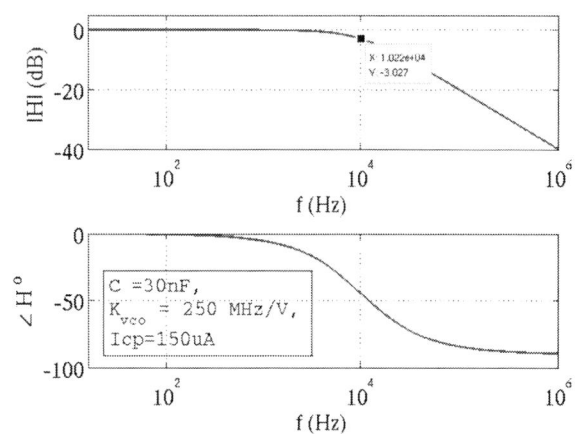

Figure 3: Linear model of the A-FLL

III. Circuit Implementation and Post-Layout Simulation

A. Frequency Counter

To provide the frequency difference, the output of a counter driven by a reference clock *ckref* is subtracted from the output of a second counter driven by the high-speed clock *ckhs*, as depicted in Figure 4. The counters are reset simultaneously and are frozen after a period T_s. For a 13 bit counter, the maximum frequency difference would be +/-312.5 MHz, which corresponds to 1%. However, the VCO can drift to greater values than this, thus the counter length must be extended. For this purpose the nominal value for the high speed clock is fixed at 5 GHz. The output of each divider stage serves in the count representation of the VCO output. By properly timing the reset and reading of the counts on the two different clock domains (*ckref* and *ckhs*), the resolution of the frequency difference is accurate to the resolution of the counter running on the *ckhs* domain. This results in an additional 4 bit resolution, for a total number of bits equal to 17.

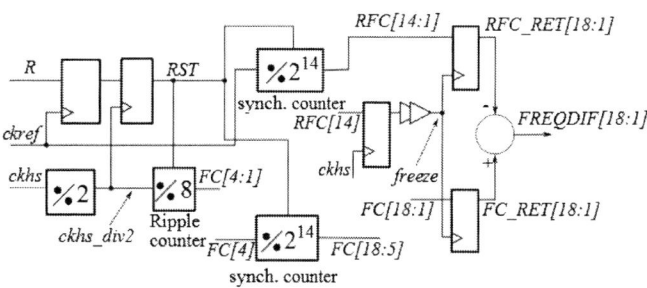

Figure 4: Counter control block diagram.

A simple toggle flip-flop serves to divide the high speed VCO clock from 5 GHz to 2.5 GHz. Further, a ripple counter divides the 2.5 GHz clock by eight. The ripple counter is chosen to maximize the speed of operation. Its output is re-timed using matched gate delays to align all counter outputs. The 2.5 GHz input to the ripple counter is the VCO clock frequency count LSB, *FC[1]*. The ripple counter outputs *Q[2:0]* are *FC[4:2]*. The last stage of the ripple counter provides a nominal clock frequency equal to 312.5 MHz. This clock drives a 14-stage synchronous counter. The output of the

978-1-4673-6145-3/13 $31.00 © 2013 IEEE

synchronous counter *Q[13:0]* is the frequency count MSBs, specifically *FC[18:5]*.

The output of the reference clock is connected to a 14-stage synchronous counter. The output values *OUT[13:0]* of the synchronous counter are the MSBs of the reference frequency count *RFC[18:5]*. *RFC[4:1]* are set to 0.

The timing diagram for the counters is shown in Figure 5. To adequately use the additional resolution of the 4 LSBs, the counters must be reset simultaneously. A RST pulse is generated from a user reset request (*R*) or once the previous count has been registered. The *RST* pulse is registered on the reference clock *ckref*. It is also re-timed on *ckhs_div2* which operates at a nominal frequency of 2.5 GHz. This re-time is necessary to define the start of the VCO output counter. When the MSB of the reference counter rises, the value of both counters is registered. To ensure that the VCO counter output is sampled in the middle of the eye, the *freeze* event is firstly re-timed on the 5 GHz clock. It is further delayed relative to this event by the delay that the VCO counter has been subject to. Because of the *RST* re-time, depending on the phase between *ckref* and *ckhs_div2*, there are +/-1 counts of uncertainty. This error will be averaged out over a long period.

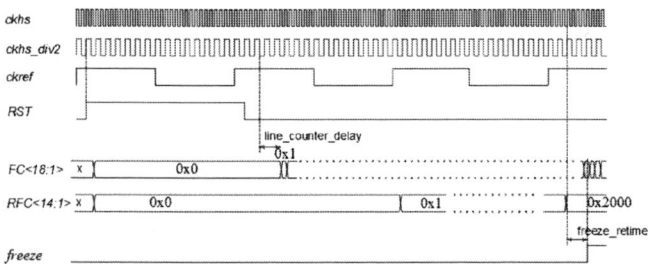

Figure 5: Frequency counter timing diagram.

The eye diagram obtained from the simulation is shown in Figure 6 for the worst case corner, i.e. when the setup margin is the smallest. The data is re-timed on the rising edge of the freeze signal. Note that the frequency counter along with its control circuitry all fits in a 95um by 95um square area.

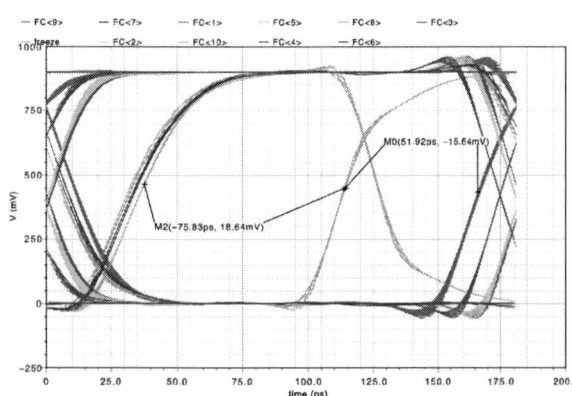

Figure 6: Eye diagram for RCMin, T = 0° C , Vdd = 0.9 V, SS.

B. Pulse-Density Modulator

The pulse density modulator (PDM) circuit is shown in Figure 7. It is implemented using standard CMOS logic. The input data is coded using 2's complement representation. The PDM provides two differential pulse modulated signals to the charge pump: UPP/UPN and DNP and DNN. The pulse density for DNP/DNN is constant at 50% of V_{DD}. For the UPP/UPN signal, the pulse density is proportional to the value of *CPTUNE*, where the minimum pulse density is 25% of V_{DD} and the maximum is 75% of V_{DD}.

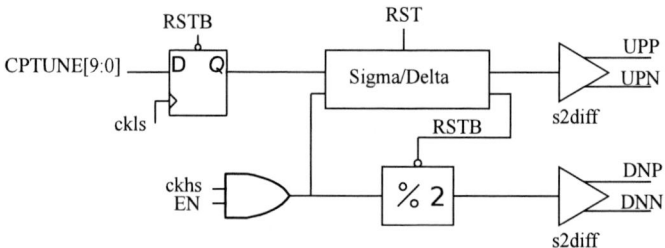

Figure 7: Pulse-density modulator implementation and controls.

A first order sigma-delta architecture [6] is implemented using CMOS logic cells. The MSB at the output of the integrator is the final digital pulse pattern. This binary signal is fedback to the input of the sigma-delta using a simple DAC. To implement a 25% to 75% pulse-density for input values between -2^9, and $+2^9$, the DAC conversion table is shown in Table 1. As can be seen, only the MSB of the DAC output changes, thus minimizing hardware complexity.

Table 1: DAC conversion table.

Input	Output (dec.)	Output (hex.)
1	-2^{10}	0xC00
0	2^{10}	0x400

The sigma-delta is implemented using a 12-bit adder and a 12-bit subtractor; this avoids overflow. Note that an inverter is required at the output of the sigma-delta modulator to flip the logic of the PDM. This is necessary because the polarity of the ADC is inverted. The pulse density modulator output for $CPTUNE = 2^9$ is shown in Figure 8.

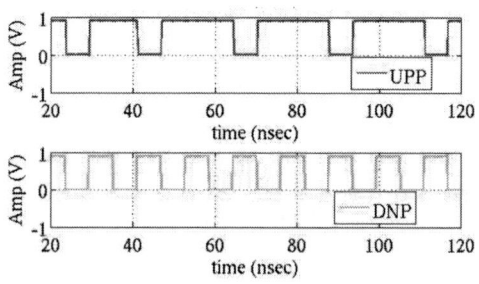

Figure 8: PDM output pulses for CPTUNE = 512. The up pulse duty cycle is 75%.

C. Charge Pump

As mentioned earlier, the integration function in the AFLL is implemented using a charge-pump charging/discharging a capacitor. The charge-pump is a traditional differential charge-pump, such as in [7] that generates up/dn single ended currents of 150uA (300uA differential).

978-1-4673-6145-3/13 $31.00 © 2013 IEEE

D. VCO Core

An example of a VCO measured by this method is a process-model test circuit comprising an LC differential CMOS cross-coupled VCO core [7]. The voltage-controlled oscillator (VCO) is a self-contained block that is intended to operate in the frequency range of 18.7 GHz to 20.7 GHz. It is designed to provide the fundamental clock for the PLL and its characteristics are digitally verified using the AFLL. A block diagram of the VCO is depicted in Figure 9.

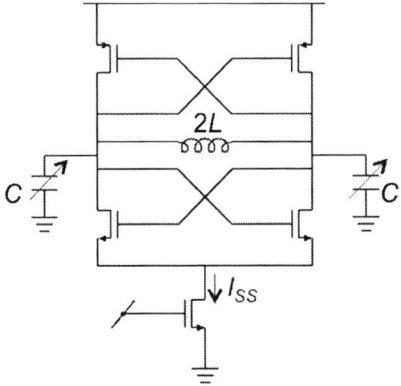

Figure 9: Implemented LC tank VCO with varactor-based tuning.

IV. EXPERIMENTAL RESULTS

In this section, experimental results validating the operation of the AFLL are presented. Also, it is demonstrated that the proposed AFLL is used to characterize the VCO. To observe the VCO spectrum, a divided version of its output (divided by 64) is connected to an output bump. When the VCO frequency is locked in the AFLL loop, its frequency can be adjusted anywhere within its linear range of operation. In the case shown in Figure 10, the reference clock runs at 307.7MHz and the synthesized frequency offset is adjusted such that the VCO locks to 64×307.715 MHz, i.e., the VCO oscillates at 19.6938 GHz.

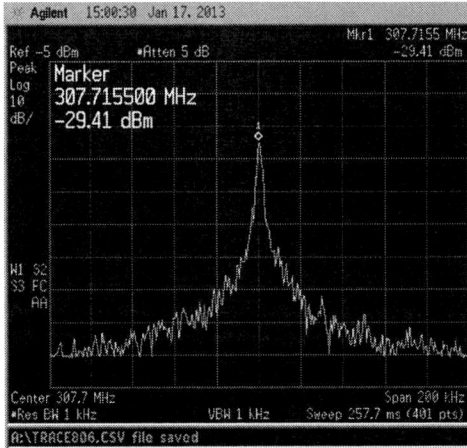

Figure 10: VCO frequency offset from the lock frequency (19.6938GHz) vs the differential charge pump tuning voltage.

In order to investigate the VCO transfer characteristic, its locking frequency has been swept and the corresponding DC differential tuning voltage (voltage at the VCO tuning port, or at the charge pump output) is measured using an on-board digitizer. Note that on a digital ATE, the tester multi-meter is used to measure the differential tuning voltage. The transfer characteristic curve for the VCO under test is provided in Figure 11. As can be observed, the VCO tuning curve has the expected "S" curve. The experimentally measured VCO gain is equal to 193MHz/V; however, the gain from the post-layout simulation is equal to ~250MHz/V for a typical corner at 85°C. Such a difference in VCO tuning gain is attributed to the inaccuracy of the passive components for this range of frequencies. This should be taken into account in the design process.

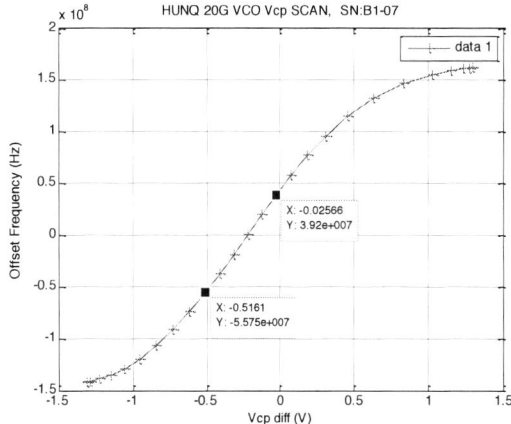

Figure 11: Tuning sensitivity in the linear range = 1.9348e8 Hz/V.

V. CONCLUSION

In this work, a programmable analog frequency-locked loop capable of synthesizing frequencies with a resolution of 8ppm has been presented. The digital interface of the proposed architecture makes it an attractive DFT solution for characterizing VCOs. Indeed, it has been shown that the entire VCO transfer characteristic curve can be measured without having to sweep its tuning port or clock reference. In addition, this scheme is compatible with an all-digital ATE; hence, facilitating its production test.

REFERENCES

[1] S. Godet, E. Tournier, O. Llopis, A. Cathelin, "An Integrated Phase Noise Measurement Bench for On-Chip Characterization of Resonators and VCOs," IEEE Joint Conference on Frequency Control and Eurepean Frequency & Time Forum, May 2011, pp. 1-5.

[2] G. Srinivasan, F. Taenzler, A. Chatterjee, "Loopback DFT for Low-Cost Test of Single-VCO-Based Wireless Transceivers," IEEE Design and Test of Computers Magazine, vol. 25, no. 2, March 2008, pp. 150-159.

[3] M. Ouda, E. Hegazi, H. F. Ragai, "Digital On-Chip Phase Noise Measurement" IEEE Design and Test Workshop, November 2009, pp.1-5.

[4] T.Y. Tsai, S. Aouini, G.W. Roberts, "High-Speed On-Chip Signal Generation for Debug and Diagnosis," Springer Journal of Electronic Testing, vol.28, no. 5, October 2012, pp. 625-640.

[5] B. R. Veillette, G. W. Roberts, "On-Chip Measurement of the Jitter Transfer Function of Charge-Pump Phase-Locked Loops," IEEE Journal of Solid-State Circuits, vol. 33, no. 3, March 1998, pp. 483-491.

[6] R. Schreier, and G. C. Temes, *Understanding Delta-Sigma Data Converters,* IEEE Press, 2005.

[7] J. Rogers, C. Plett, F. Dai, *Integrated circuit design for high-speed frequency synthesis,* Artech House, 2006.

978-1-4673-6145-3/13 $31.00 © 2013 IEEE

Detection of Early-Life Failures
in High-K Metal-Gate Transistors and Ultra Low-K Inter-Metal Dielectrics

Young Moon Kim[1], Jun Seomun[2], Hyung-Ock Kim[2], Kyung-Tae Do[2], Jung Yun Choi[2],
Kee Sup Kim[2], Matthias Sauer[3], Bernd Becker[3], Subhasish Mitra[1,4]

[1] Dept. of Electrical Engineering and [4] Dept. of Computer Science, Stanford University, Stanford, CA 94305 USA
[2] Design Technology Team, System LSI Division, Samsung Electronics, Kiheung 466-711 Korea
[3] Dept. of Computer Science, University of Freiburg, Freiburg 79110 Germany

Abstract – **Using 28nm test chips, we derive signatures for early-life failures (ELF) in both high-K/metal-gate transistors and ultra low-K inter-metal dielectrics. We also demonstrate that the derived ELF signatures can be successfully detected using a clock control technique. Our results can be utilized to overcome scaled-CMOS reliability challenges in several ways: 1. Low-cost ELF detection during on-line operation of robust systems without requiring expensive redundancy-based error detection techniques; 2. Effective ELF screening during production test while reducing stress time and/or stress levels associated with stress tests such as burn-in.**

I. INTRODUCTION

Early-life failures (*ELF*, also referred to as *infant mortality failures*) are caused by defective integrated circuits (ICs) that pass manufacturing tests but fail early in the field, much earlier than expected product lifetime. ELF causes include defects in transistor gate dielectrics and inter-metal dielectrics (IMD) [Carulli 06, Malandruccolo 11]. Traditionally, burn-in is used to screen ELF candidates during manufacturing test. Burn-in accounts for a significant portion of test cost [Sumikawa 12, Vassighi 08]. Moreover, burn-in is getting increasingly difficult in advanced technologies [Borkar 05, Bulter 06, Zakaria 06]. Burn-in challenges include power dissipation, cost, and serious concern about potentially reduced effectiveness of burn-in in the future. Other ELF screening techniques such as various flavors of high voltage stress testing, Iddq testing and its variants, Very Low Voltage (VLV) and minVdd testing, and outlier analysis techniques also face significant challenges: coping with circuit leakage, process variations, and the related problem of overkill. Hence, low-cost alternatives for effective detection of ELF are necessary. One such approach is to design robust systems that can detect ELF during system operation (in the field) using error detection techniques. Upon error detection, appropriate error recovery [Bowman 11, Meaney 05] and self-repair actions [Powell 09, Li 13] are invoked.

Traditional error detection techniques, e.g., those based on redundancy, logic parity [Mitra 00] and redundant execution [Reinhardt 00] are expensive in terms of power/performance/area. On-chip sensors [Karl 08, Keane 09, Tschanz 09], inserted at specific locations (selected during design time) inside an IC, can reduce power/performance/area costs of traditional error detection. However, ELF may not be detected using such on-chip sensors since defects can occur anywhere inside an IC.

To overcome these challenges, we derive device- and digital logic circuit-level ELF signatures, and demonstrate that the derived signatures can be effectively utilized to provide early indications of ELF before functional failures appear. Such an ELF detection technique can be successfully implemented using a special clock control activated during periodic on-line self-test and diagnostics [Inoue 08, Li 08, 10]. Unlike expensive redundancy, this approach introduces minimal power/performance impact since it is activated only during periodic on-line self-test and diagnostics in robust systems. At the same time, it can detect ELF anywhere inside an IC using special test patterns for ELF detection (detailed later in this paper) with high test coverage. Since we provide early indications of ELF before functional failures appear, we can also enable efficient system self-repair [Li 13] with minimal reliance on expensive (and complex) rollback recovery. Our ELF signatures can also be used during manufacturing test to reduce stress time and/or stress levels.

In this paper, we make the following contributions:

1. Using 28nm test chips, we derive device- and digital logic circuit-level ELF signatures for defects in gate dielectric of HK/MG transistors. We demonstrate that a HK/MG transistor with gate dielectric ELF defect in a digital logic circuit causes changes in delays over time before functional failures appear. These changes in delays are distinct from those induced by aging mechanisms such as Bias Temperature Instability.

2. We characterize the distribution of delay changes caused by 28nm HK/MG dielectric ELF in digital logic circuits.

3. We derive device- and digital logic circuit-level ELF signatures for defects in 28nm ultra low-K (ULK) IMD. For the first time, we demonstrate that the digital logic circuit-level behavior of ULK IMD ELF is similar to that of HK/MG dielectric ELF: changes in delays over time before functional failures appear.

4. We demonstrate the effectiveness of a clock control technique in detecting ELF-induced delay changes in digital circuits. This technique does not rely on expensive error detection (discussed before) or flip-flop modifications (e.g., [Bowman 11]). In fact, delay fault-detection flip-flops are not suitable for detection of our ELF signatures (when ELF defects occur on short paths having large timing slacks).

5. We demonstrate the practicality of special test pattern generation to detect ELF-induced delay changes for the industrial OpenSPARC T2 design [Oracle 13].

These results extend far beyond [Kim 10b] which focused only on 90nm gate-oxide (SiO_xN_y) ELF. Understanding HK/MG ELF is important for advanced technologies since HK/MG plays a key role in achieving high-performance and low-power CMOS circuits. IMD ELF was not analyzed in [Kim 10b]. IMD reliability is of great concern with aggressive scaling of interconnect dimensions and introduction of new low-K materials [Chen 09]. In [Kim 10b], delay changes due to 90nm gate-oxide ELF were measured on a single die, and the distribution of delay changes was not characterized.

II. TEST CHIP DETAILS

A. HK/MG Dielectric ELF Test Structure

Fig. 1 shows our HK/MG dielectric ELF test structure consisting of a 33-stage inverter chain, and summarizes the configurations of four different modes. It can be configured as a ring oscillator (*RO*) for off-chip frequency measurement. During stress modes, a single NMOS (T1) or PMOS (T2) (nominal supply 1V) is stressed using high stress voltage of 2.5V to emulate an ELF transistor. Other transistors are protected by the thick-oxide transistors (nominal supply 1.8V). During a stress phase, either T1 or T2 is stressed (using T1 or T2 stress configuration) for a certain amount of time. After each stress phase, the gate leakage of the stressed transistor is measured in quiescent state. Then, the oscillation frequency of *RO* is measured off-chip, and the rising and falling delays are separately measured on-chip using a clock control technique (detailed in Sec. III.B).

978-1-4673-6145-3/13 $31.00 © 2013 IEEE

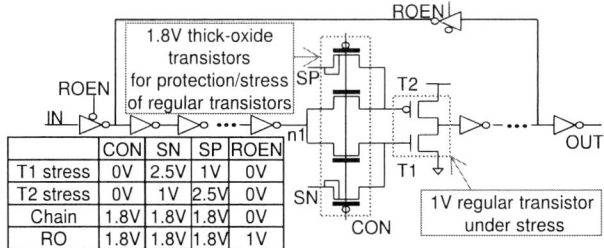

	CON	SN	SP	ROEN
T1 stress	0V	2.5V	1V	0V
T2 stress	0V	1V	2.5V	0V
Chain	1.8V	1.8V	1.8V	0V
RO	1.8V	1.8V	1.8V	1V

Fig 1. HK/MG dielectric ELF test structure and configurations.

B. ULK IMD ELF Test Structure

Fig. 2 shows our ULK IMD ELF test structure consisting of a 17-stage inverter chain, and summarizes the configurations of three different modes. A comb-serpent-comb IMD structure used in [Chen 09, Jeong 12] is placed between two adjacent inverters and is stressed to emulate IMD ELF defects. The comb-serpent-comb IMD structure provides a large area of IMD under stress so that IMD defects can be easily generated. Metal-2 layer is arbitrarily chosen for this IMD structure since IMD defects can occur on any layer. High voltage of more than 20V is applied for stress. In order to protect other transistors from such high stress voltage, T3 and T4 in Fig. 2 are utilized. During a stress phase, the IMD structure is stressed (using the IMD stress configuration in Fig. 2) for a certain amount of time. During stress, current compliance is set by an external parameter analyzer. The sizes of T3 and T4 are carefully chosen to ensure that the individual on-current is greater than the current compliance level. As a result, the voltage values at various nodes in the inverter chain are less than 1V during stress in order to avoid any damage to the transistors. After each stress phase, the leakage of the IMD structure is measured in quiescent state. Then, the oscillation frequency of *RO* is measured off-chip (detailed in Sec. III.B).

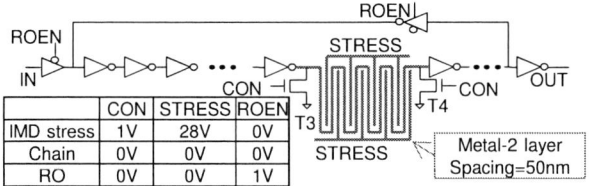

	CON	STRESS	ROEN
IMD stress	1V	28V	0V
Chain	0V	0V	0V
RO	0V	0V	1V

Fig 2. ULK IMD ELF test structure and configurations.

C. Clock Control to Detect ELF-induced Delay Changes

Fig. 3 shows our on-chip clock control circuit to detect ELF-induced delay changes. Using a system clock *SC* and its delayed version *DSC*, it creates scan-based launch and capture pulses whose timing is controlled by the delay between *SC* and *DSC* (Fig. 3b, similar to [Kim 10b]). In our experiments, the delay between *SC* and *DSC* is controlled externally by a pulse generator (5ps resolution). In actual systems, on-chip high-resolution phase shifters [O'Mahony 10] or configurable delay lines [Kim 10b] can be used.

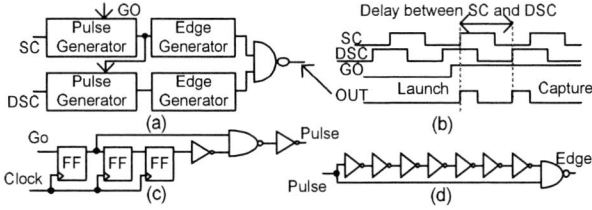

Fig 3. (a) Clock control circuit. (b) Timing diagrams. (c) Pulse generator in (a). (d) Edge generator in (a). [Kim 10b]

Fig. 4 shows the overall test chip architecture. The system operates in three modes. In *normal mode*, *SC* runs through the clock tree. In *test mode* (on-line self-test and diagnostics), test patterns are scanned in, launch and capture pulses are applied, and captured values are scanned out (repeated for varying launch-to-capture delays). In *stress mode*, the circuit is in quiescent state, and ELF defects are emulated using voltage stress. The stress mode is for our ELF experiment purposes to accelerate degradation and it may not be required in real systems.

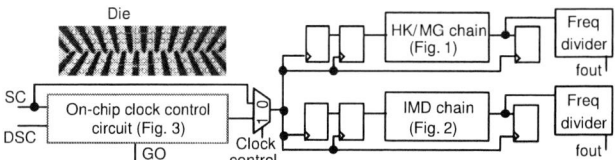

Fig 4. Overall test chip architecture. Clock control circuit (Fig. 3) integrated with test structures in Figs. 1 and 2.

III. EXPERIMENTAL RESULTS

A. Device-level Results

HK/MG dielectric ELF: Fig. 5a (5b) shows measured on-current (I_{on}) and gate leakage (I_g) of an arbitrarily chosen stand-alone NMOS (PMOS) over stress time. Constant current compliance was set during stress to limit the severity of gate-dielectric defects. The stress was applied in the same way as the experimental flow in Fig. 7a for circuit-level results in Sec. III.B. Fig. 5 demonstrates device-level HK/MG dielectric ELF signature for both NMOS and PMOS: **changes in I_{on} and I_g over time before functional failure**. Increased I_g values indicate gate-dielectric defects, i.e., changes in I_{on} are not solely caused by BTI. Simulations using calibrated BTI models confirm this point as well. Our DC stress ensures minimal HCI degradation. Table 1 shows the distribution of I_{on} degradation and I_g increase over stress time for multiple transistors.

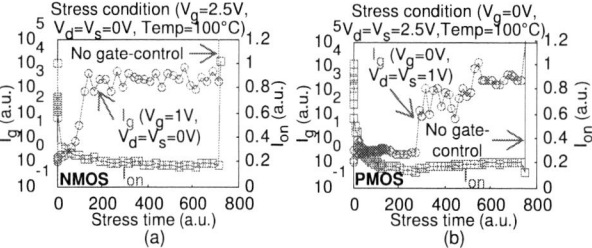

Fig 5. HK/MG dielectric ELF device-level results. (a) I_g and I_{on} over stress time (NMOS). (b) I_g and I_{on} over stress time (PMOS).

Table 1. I_{on} degradation and I_g increase for 7 NMOS and 7 PMOS. 100% stress time: point at which no gate-control is observed.

	At 5% of stress time			At 10% of stress time			At 100% of stress time		
	I_{on} degradation		I_g incr.	I_{on} degradation		I_g incr.	I_{on} degradation		I_g incr.
	avg.	std. dev.	avg.	avg.	std. dev.	avg.	avg.	std. dev.	avg.
NMOS	47.2%	10.9%	1.06X	70.4%	15.4%	433X	72.6%	14.5%	3,541X
PMOS	29.3%	19.4%	1.74X	56.9%	12.0%	155X	66.6%	13.2%	4,686X

ULK IMD ELF: Fig. 6a shows leakage current during 28V DC stress while *CON* of Fig. 2 is set to 1V. Since the thickness of IMD is much thicker than that of gate-dielectric, it requires much higher voltage to emulate defects. During stress, current compliance is set to control the severity of breakdown as in [Chen 09]. Unlike in [Chen 09], current compliance was progressively increased over stress time to observe the progressive IMD breakdown efficiently. In Fig. 6a, current compliance was initially set at 0.6a.u., and then it was increased to 0.7a.u., and then to 1a.u.. The stress was applied in the same way as the experimental flow in Fig. 8a for circuit-level results in Sec. III.B. After each stress mode, *STRESS* was swept for leakage measurement from 0.4V to 5V in steps of 0.5V. Fig. 6b shows IMD leakage current after each stress mode, and demonstrates the device-level ULK IMD ELF signature: **changes in IMD leakage over time (during normal operation at 1V) before hard breakdown**. In this work, hard breakdown is defined as the state where the IMD leakage reaches the current compliance level at the *STRESS* voltage of 0.4V. This device-level ULK IMD ELF signature is different from sudden IMD breakdown observed in older technologies (before ULK IMD) [Chen 09]. If the severity of degradation in ULK IMD is not enough to incur functional failures, circuit may continue to function correctly with additional delay not exceeding the margin of the circuit. We report the circuit-level data in the following Sec. III.B.

978-1-4673-6145-3/13 $31.00 © 2013 IEEE 641

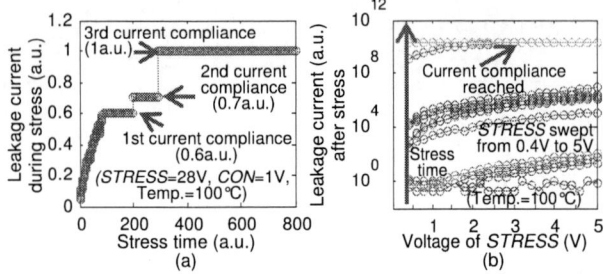

Fig 6. ULK IMD ELF device-level results. (a) IMD leakage current through *STRESS* (Fig. 2) during stress. (b) IMD leakage current through *STRESS* (Fig. 2) after each stress mode.

B. Circuit-level Results

HK/MG dielectric ELF: Fig. 7a shows the experimental flow. After each stress mode, the oscillation period is measured off-chip (RO mode), and gate leakage of T1 is measured in quiescent state. Then, rising and falling delays of the inverter chain are separately measured using the on-chip clock control circuit. On-chip delay measurement data is reported later in Sec. III.C. Fig. 7b shows measured values of RO *period shift(t)* from an arbitrary HK/MG dielectric ELF test structure where an NMOS transistor (T1 in Fig. 1) is stressed until functional failure (Fig. 7b). Here, *period shift* is defined as:

RO *period shift(t)* = (RO period measured at time *t*) – (fresh RO period measured at time *0*).

The results clearly demonstrate the circuit-level HK/MG dielectric ELF signature: **changes in delay of the chain over time before functional failure**. The distribution of ELF-induced changes in RO periods (Table 2) confirm consistent circuit-level HK/MG dielectric ELF signature, with changes in RO periods exceeding 20ps, for all the seven measured test structures. Fresh RO measurements over time confirm that the RO period shifts are indeed caused by ELF.

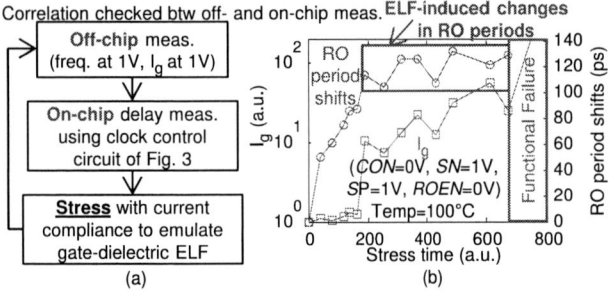

Fig 7. HK/MG dielectric ELF circuit-level results. (a) Experiment flow. (b) Off-chip measurement results after each stress mode (T1 in Fig. 1 stressed): I_g of T1 through SN (Fig. 1) and RO period shifts over stress time.

Table 2. Distribution of ELF-induced changes in RO periods over time for 7 HK/MG dielectric ELF test structures.

ELF-induced change in RO period (ps)	< 0	0 to 20	20 to 40	40 to 60	≥ 60	Total
Chain 1	2	4	1	0	0	7
Chain 2	4	5	7	3	1	20
Chain 3	0	4	4	1	0	9
Chain 4	0	3	2	5	2	12
Chain 5	0	4	4	0	1	9
Chain 6	1	3	1	6	9	20
Chain 7	2	8	4	2	1	17

ULK IMD ELF: Fig. 8 shows circuit-level results for the ULK IMD ELF test structure (same structure as in Fig. 6). Fig. 8b shows the measured RO *period shift(t)*. The data demonstrates the circuit-level ULK IMD ELF signature: **changes in delay of the chain over time before functional failure** (similar to HK/MG dielectric ELF).

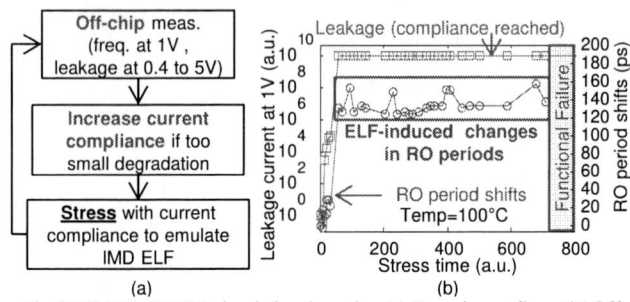

Fig. 8. ULK IMD ELF circuit-level results. (a) Experiment flow. (b) Off-chip measurement results: IMD leakage current through STRESS (Fig. 2) and RO period shifts over stress time.

C. Robust System-level Results: Low-cost ELF Detection

Fig. 9 demonstrates that delay changes due to HK/MG dielectric ELF can be successfully detected (before functional failures) using our clock control technique. *Shift in passing delay* is defined as:

Shift in passing delay(t) = (smallest delay between *SC* and *DSC* for "Pass" at time *t*) – (smallest delay between *SC* and *DSC* for "Pass" at time *0*).

Fig 9a shows that the rising transition delay at n1 of Fig. 1 slows down over stress time compared with that at time *0* (represented by the positive values of "shift in passing delay"). This slowed-down transition is attributed to weaker pull-down of NMOS (T1) due to emulated HK/MG dielectric ELF under stress. In the same way, the speed-up of the falling transition delay at n1 in Fig. 9b can be explained (negative values of "shift in passing delay"). The amount of shifts in slowed-down transition (Fig. 9a) is observed to be greater than that in speed-up transition (Fig. 9b), which is consistent with [Kim 10a]. Fig. 9c shows the difference between on-chip and off-chip measurements. *Difference* is defined as:

Difference(t) = [Shift in passing delay(t) in Fig. 9a + Shift in passing delay (t) in Fig. 9b] – [RO period shift(t) in Fig. 7b].

The values differ by less than 20ps showing high correlation between on-chip and off-chip measurements. This further confirms that the "shift in passing delay" values in Figs. 9a and 9b are real and not due to measurement artifacts. Repeated on-chip measurements of fresh chains also confirm that the reported delay changes are indeed ELF-induced.

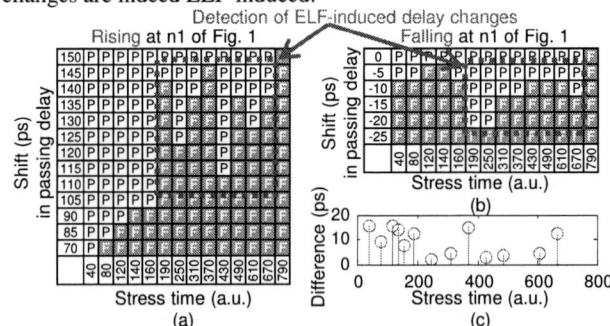

Fig 9. On-chip detection of HK/MG dielectric ELF-induced delay changes (same structure as Fig. 7b) using clock control technique (Fig. 3). Shmoo plots over time. (a) Rising transition at n1. (b) Falling at n1. (c) Difference between on-chip and off-chip measurements.

D. Impact of Voltage and Temperature on ELF-induced Delay Changes

Since circuit delays can vary over time due to changes in supply voltage and temperature, it is important to understand how these factors influence ELF-induced changes in delays. Similar to the off-chip frequency measurements in Fig. 7a, we measure RO frequencies at two VDD values (0.95V and 1V) and two temperature values (65°C and 100°C) **after each stress phase.** The temperature value is controlled using a thermal chuck in the probe station with an accuracy of 0.5°C.

978-1-4673-6145-3/13 $31.00 © 2013 IEEE

Fig. 10a shows the measured values of RO *period shift(t)* at 0.95V and 1V for an arbitrary HK/MG dielectric ELF test structure where an NMOS transistor (T1 in Fig. 1) is stressed until functional failure. At a lower VDD value, ELF-induced delay changes are magnified, which is consistent with the simulation results in [Kim 10a]. Fig. 10b shows normalized RO frequencies (for the same structure as in Fig. 10a) for the two different VDD values and two different temperature values. The plotted frequencies correspond to two different stress times: time *0* – no stress and time *665* a.u.. The 28nm technology used in this paper has a positive temperature coefficient unlike older technologies, which explains why we obtain higher frequencies at 100°C than at 65°C. Fig. 10b shows that HK/MG dielectric ELF-induced delay changes are more sensitive to environmental variations. For example, when VDD changes from 1V to 0.95V at 65°C, frequency slows down by 10.2% at time *0* (fresh) vs. 16.4% at time *665* a.u.. When the temperature changes from 100°C to 65°C at 0.95V, frequency slows down by 3.2% at time *0* (fresh) vs. 6.7% at time *665* a.u..

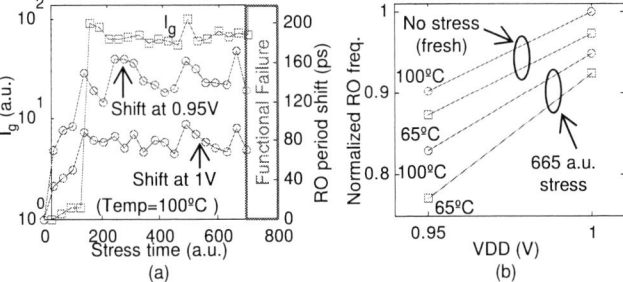

Fig. 10. (a) I_g of T1 through SN (Fig. 1) and RO period shifts over stress time at two VDD values (0.95V and 1V) at 100°C. (b) RO frequency at two VDD values (0.95V and 1V) and two temperature values (65°C and 100°C).

E. Test Pattern Generation to Detect ELF-induced Delay Changes

Since ELF defects are localized, detection of ELF-induced delay changes does not require expensive path delay tests. However, conventional transition fault test patterns [Waicukauski 87] do not guarantee the detection of ELF-induced delay changes. Fig. 11 shows such an example. If the ELF-induced delay change (Δ) due to the defective gate (*g1*) is less than *D* (the difference in arrival times at *n1* and *n2* in Fig. 11), the transition at "*out*" is not affected by the defective gate. Robust path delay test patterns [Cheng 93] face similar issues. (Note that, the pattern shown in Fig. 11 is also a valid robust path delay test pattern). Hence, special attention must be paid to various Boolean conditions to be satisfied in conjunction with clock control (for varying launch-capture delays). A detailed explanation of such Boolean conditions is beyond the scope of this paper. Table 3 demonstrates that it is practical to use commercial tools [Cadence 13] special ways to generate such test patterns that detect ELF-induced delay changes for the industrial OpenSPARC T2 design [Oracle 13]. In Table 3, a gate is assumed to have two different ELF-induced faults: one affecting rising transition at the output of the gate and another affecting falling.

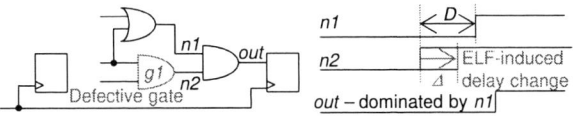

Fig. 11. A test escape example of a conventional transition test pattern and robust path delay test pattern for detecting ELF-induced delay changes.

Table 3. Test pattern generation results to detect HK/MG dielectric ELF-induced changes in delay for OpenSPARC T2 [Oracle 13].

Block name	Gate count	Test pattern count	Test coverage
SPC	300,450	19,333	94.0%
NCU	49,778	1,265	97.0%
SIU	34,890	1,963	92.6%
PIU	149,583	8,198	94.1%

Test coverage = [(Number of detected faults) / (Total number of faults)] x 100
Total number of faults = 2 x (Gate count)
Detected fault = Fault for which additional delay of 1ps or more is detected

IV. CONCLUSION

Our 28nm ELF signatures enable low-cost techniques to overcome reliability challenges in robust SoCs. We demonstrate that the derived ELF signatures can be successfully detected using a clock control technique, activated during periodic on-line self-test and diagnostics, without requiring expensive concurrent error detection. These results can be successfully utilized to overcome scaled-CMOS reliability through effective ELF screening during production test or on-line during system operation with built-in self-healing. Future directions include: 1. Exploration of optimized design methodologies for seamless integration of our low-cost ELF detection techniques for robust system design; 2. Analysis of cost trade-offs associated with various ways of using our ELF signatures to reduce stress time and/or stress levels during manufacturing test.

ACKNOWLEDGMENTS
We thank FCRP C2S2, Intel, NSF, and Samsung for support.

REFERENCES

[Borkar 05] Borkar, S., "Designing reliable systems from unreliable components: The challenges of transistor variability and degradation," *IEEE Micro*, vol. 25, issue 6, pp. 10-16, Nov.-Dec., 2005.

[Bowman 11] Bowman, J., *et al.*, "A 45 nm Resilient Microprocessor Core for Dynamic Variation Tolerance," *IEEE Journal of Solid-State Circuits*, vol. 46, no. 1, pp. 194-208, 2011.

[Butler 06] Butler, K.M., et al., "Successful Development and Implementation of Statistical Outlier Techniques on 90nm and 65nm Process Driver Devices," *Proc. Intl. Reliability Physics Symp.*, pp. 552-559, 2006.

[Cadence 13] Cadence Encounter True-Time APTG , 2013

[Carulli 06] Carulli, J.M., and T.J. Anderson, "The impact of multiple failure modes on estimating product field reliability," *IEEE Design & Test of Computers*, vol. 23, issue 2, pp. 118-126, Mar., 2006.

[Chen 09] Chen, F., "Critical Ultra Low-k TDDB Reliability Issues For Advanced CMOS Technologies," *Proc. Intl. Reliability Physics Symp.*, pp. 464-475, 2009.

[Cheng 93] Cheng, K.-T., and H.-C. Cheng, "Delay testing for non-robust untestable circuits," *Proc. Intl. Test Conf.*, pp. 954-961, 1993.

[Inoue 08] Inoue, H., *et al.*, "VAST: Virtualization-Assisted Concurrent Autonomous Self-Test," *Proc. Intl. Test Conf.*, pp. 1-10, 2008.

[Jeong 12] Jeong , T.-Y., *et al.*, "Effective line length of test structure and its effect of area scaling on TDDB characterization in advanced Cu/ULK process," *Proc. Intl. Reliability Physics Symp.*, pp. BD.3.1-BD.3.4, 2012.

[Karl 08] Karl, E., *et al.*, "Compact In-Situ Sensors for Monitoring Negative-Bias-Temperature-Instability Effect and Oxide Degradation," *Proc. Intl. Solid-State Circuits Conf.*, pp 410-411, 2008.

[Keane 09] Keane, J., *et al.*, "An all-in-one silicon Odometer for separately monitoring HCI, BTI, and TDDB," *Proc. VLSI Circuits Symp.*, pp. 108-109, 2009.

[Kim 10a] Kim, Y.M., *et al.*, "Gate-oxide early-life failure identification using delay shifts", *Proc. VLSI Test Symp.*, pp. 69-74, 2010.

[Kim 10b] Kim, Y.M., *et al.*, "Low-cost gate-oxide early-life failure detection in robust systems," *Proc. VLSI Circuits Symp.*, pp. 125-126, 2010.

[Li 08] Li, Y., *et al.*, "CASP: Concurrent Autonomous Chip Self-Test Using Stored Test Patterns," *Proc. Design Automation and Test in Europe*, pp. 885-890, 2008.

[Li 10] Li, Y., *et al.*, "Concurrent autonomous self-test for uncore components in system-on-chips", *Proc. VLSI Test Symp.*, pp. 232-237, 2010.

[Li 13] Li, Y., *et al.*, "Self-Repair of Uncore Components in Robust System-on-Chips: An OpenSPARC T2 Case Study," *Proc. Intl. Test Conf.*, 2013.

[Malandruccolo 11] Malandruccolo et al., " Design and Experimental Characterization of a New Built-In Defect-Based Testing Technique to Achieve Zero Defects in the Automotive Environment," *IEEE Trans. Device and Materials Reliability*, vol. 11, issue 2, pp. 349-357, 2011.

[Meaney 05] Meaney, P., *et al.*, "IBM z990 Soft Error Detection and Recovery," *IEEE Trans. Device and Materials Reliability*, vol. 5, issue 3, pp. 419-427, 2005.

[Mitra 00] Mitra, S., and E.J. McCluskey, "Which Concurrent Error Detection Schemes to Choose?" *Proc. Intl. Test Conf.*, pp. 985-994, 2000.

[O'Mahony 10] O'Mahony , F., *et al.*, "A programmable phase rotator based on time-modulated injection-locking," *Proc. Symp.VLSI Circuits*, pp. 45-46, 2010.

[Oracle 13] http://www.opensparc.net, 2013.

[Powell 09] Powell, M.D., *et al.*, "Architectural Core Salvaging in a Multi-Core Processor for Hard-Error Tolerance," *Proc. Intl. Symp. Computer Architecture*, pp. 93-104, 2009.

[Reinhardt 00] Reinhardt, S.K., *et al.*, "Transient fault detection via simultaneous multithreading," *Proc. Intl. Symp. Computer Architecture*, pp. 25-36, 2000.

[Sumikawa 12] Sumikawa, N., *et al.*, "An experiment of burn-in time reduction based on parametric test analysis ," *Proc. Intl. Test Conf.*, pp. 1-10, 2012.

[Tschanz 09] Tschanz, J., *et al.*, "Tunable replica circuits and adaptive voltage-frequency techniques for dynamic voltage, temperature, and aging variation tolerance," *Proc. VLSI Circuits Symp.*, pp 112-113, 2009.

[Vassighi 08] Vassighi, A., *et al.*, "Characterizing infant mortality in high volume manufacturing," *Proc. Intl. Reliability Physics Symp.*, pp. 717-718, 2008.

[Waicukauski 87] Waicukauski, J.A., *et al*, "Transition Fault Simulation," *IEEE Design & Test of Computers*, Vol. 4, Issue 2, pp. 32-38, 1987.

[Zakaria 06] Zakaria, M.F., et al., "Reducing burn-in time through high-voltage stress test and Weibull statistical analysis," *IEEE Design & Test of computers*, vol. 23, ssue 2, pp. 88-98, 2006.

978-1-4673-6145-3/13 $31.00 © 2013 IEEE

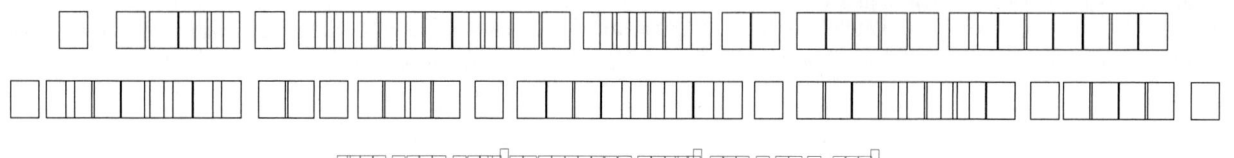

1. School of Electrical and Computer Engineering, Georgia Institute of Technology, Atlanta, GA 30332 USA

2. Toshiba Corporation, Kawasaki, 212-8520 Japan

Abstract— This paper presents an ultra-compact transformer-based quadrature generation scheme, which converts a differential input signal to fully differential quadrature outputs with low passive loss, broad bandwidth, and robustness against process variations. A new layout strategy is proposed to implement this 6-port transformer-based network within only one inductor-footprint for significant area saving. A 5 GHz quadrature generation design is implemented in a standard 65 nm CMOS process with a core area of only 260 μm by 260 μm, achieving size reduction of over 1,600 times compared to a 5GHz λ/4 branch-line coupler. This implementation achieves 0.82 dB signal loss at 5 GHz and maximum 3.8° phase error and ±0.5dB amplitude mismatch within a bandwidth of 13% (4.75 GHz to 5.41 GHz). Measurement results over 9 independent samples show a standard phase deviation of 1.9° verifying the robustness of the design.

$$Z_2 = (j\omega L) || (\frac{1}{j\omega C_1}) || (50\ ohm)$$

$$Z_3 = (50\ ohm) || (\frac{1}{j\omega C_2})$$

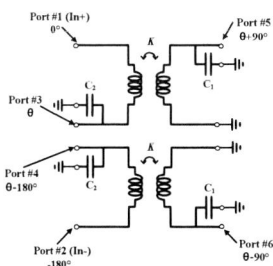

$$\frac{Z_2}{Z_3} = \frac{25 + j25}{25 - j25}$$
$$Z_2 + Z_3 = 50 \; ohm$$

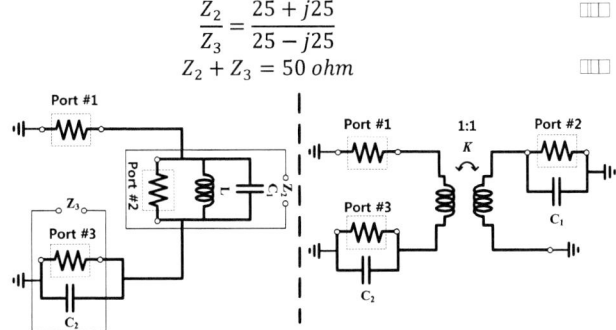

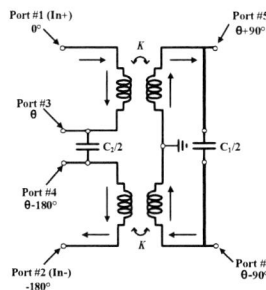

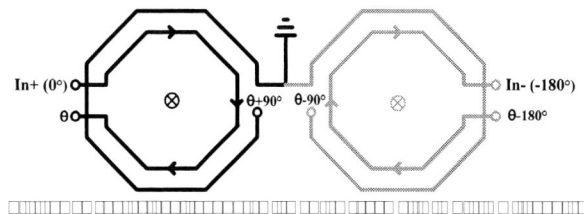

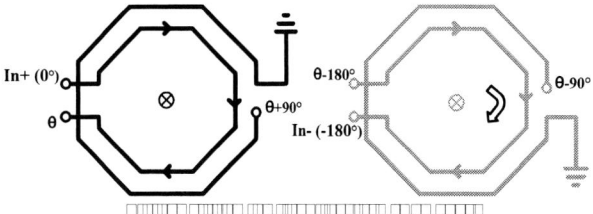

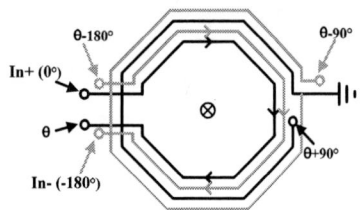

$$Trace\_Length_{in^+\_to\_in-phase^+} = N + 0.5 \ (turns) \quad (5)$$
$$Trace\_Length_{in^-\_to\_in-phase^-} = N + 0.5 \ (turns) \quad (6)$$
$$Trace\_Length_{quadrature^+\_to\_quadrature^-} = 2N + 1 \ (turns) \quad (7)$$

N ▯▯▯▯▯▯▯▯▯▯▯▯▯▯▯▯▯▯▯▯▯▯▯▯▯▯▯ $N=1$. ▯▯

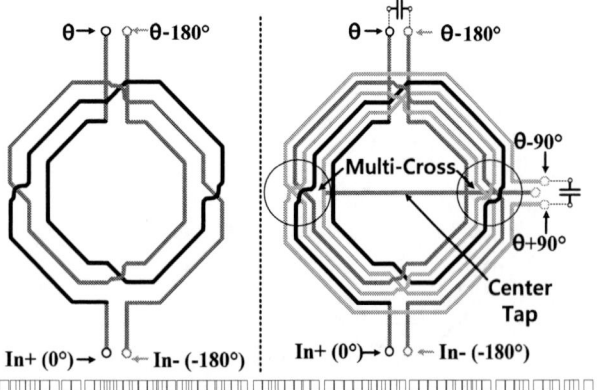

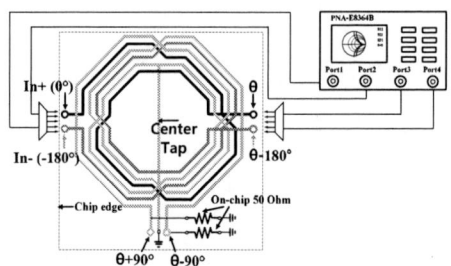

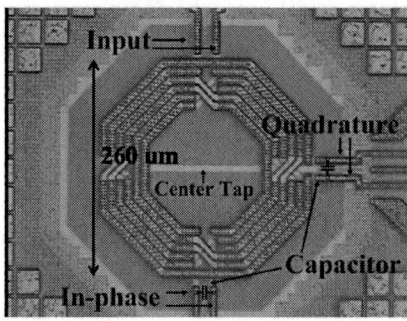

With a core size of 260 μm by 260 μm, our design achieves a drastic size reduction of over 1,600 times compared to a λ/4 branch-line coupler at 5 GHz.

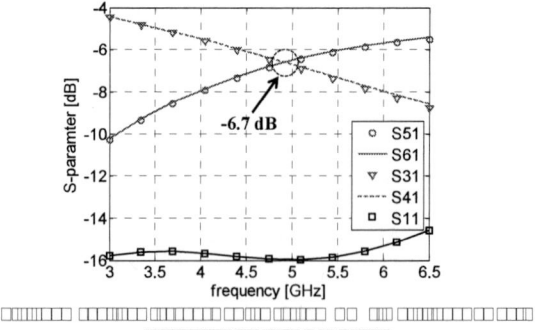

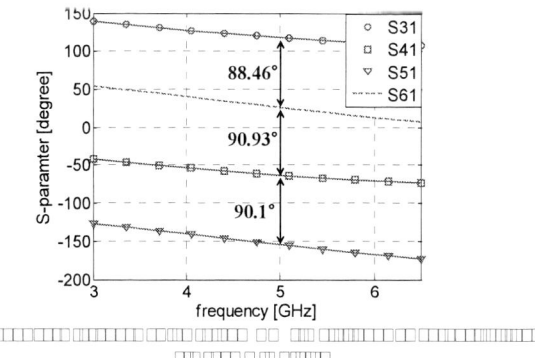

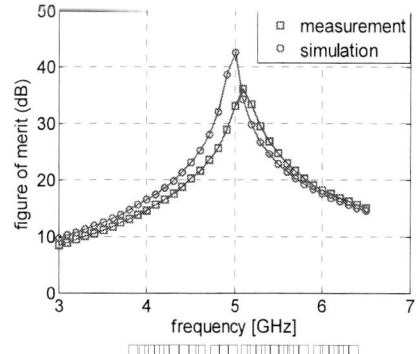

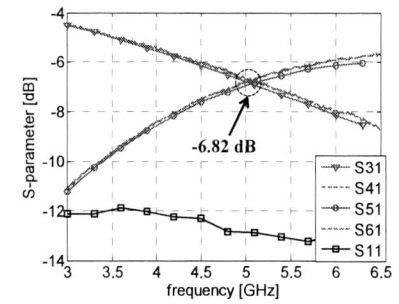

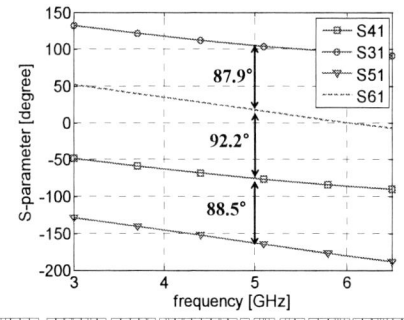

$$\gamma = 20 \log_{10}\left(\frac{S_{21}(\omega) - jS_{31}(\omega)}{S_{21}(\omega) + jS_{31}(\omega)}\right).$$

A -173 dBc/Hz @ 1 MHz offset Colpitts Oscillator using AlN Contour-Mode MEMS Resonator

Jabeom Koo[*], Augusto Tazzoli[†], Jeronimo Segovai-Fernandez[†], Gianluca Piazza[†], and Brian Otis[*]

[*]Department of Electrical Engineering, University of Washington, Seattle, WA 98195-2500

[†]Department of Electrical and Computer Engineering, Carnegie Mellon University, Pittsburgh, PA 15213-3890

Abstract - **A differential Colpitts oscillator using AlN MEMS CMR designed in 0.13 um CMOS is presented in this work. The oscillator operates at 1.16 GHz, with a total power consumption of 4.2 mW at 1 V supply. It achieves a phase noise of -143.6 dBc/Hz, -173.3 dBc/Hz at 100 kHz and 1 MHz offset frequency respectively with a figure of merit (FOM) of 228.3 dB. Current-based temperature compensation was employed to reduce oscillator drift across temperature.**

I. INTRODUCTION

Low power and low noise RF frequency generators are essential for applications such as high performance ADCs, high speed serial data links, and low power radios [1]. Micromachined Aluminum Nitride (AlN) Contour-Mode-Resonators (CMRs) show an enormous opportunity in high performance RF frequency generation. These resonators have a Q-factor over 100 times that of on-chip inductors. Their integration on wafer or package can effectively replace large off-chip crystals and SAWs.

Due to their small-form factor, high frequency of operation, and capability to be co-integrated with CMOS circuits, MEMS resonators are a candidate for the implementation of compact and multi-frequency banks of high-quality-factor mechanical elements that can be used for the synthesis of next-generation reconfigurable local oscillators for radio frequency (RF) transceivers [2-3]. In this work, we focus on AlN Contour-Mode Resonators (CMRs). The high transduction efficiency of the piezoelectric film translates to low values of achievable motional resistance (tens of ohms).

Modern telecommunication systems require oscillators that are stable over a wide range of parameters, including vibration and temperature [4]. In high precision commercial oscillators, either temperature compensated crystals (TCXO) [5] or oven stabilized devices (OCXO) [6] are used. Other work demonstrates an embedded ovenization technique to decrease temperature dependence of the resonance frequency down to less than a few ppm (< 2ppm) [12]. However, most of the existing ovenization techniques are aimed for compensating the temperature dependence of the resonator only. To achieve less than 1 ppm frequency variation, any contributions from CMOS and passive devices in the circuit should be taken into account as well.

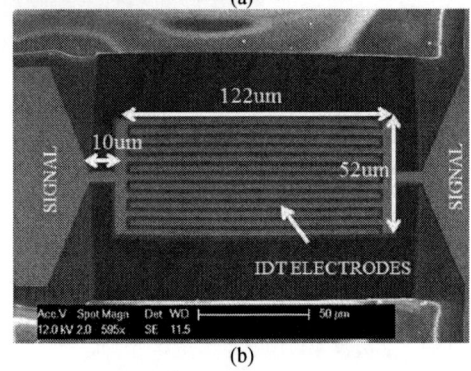

ALUMINIUM
Mechanical
E=70 GPa
ρ=2.7 g/cm^3
Electrical
ρ=51.1 nΩm (thin film)

Fig. 1 1GHz Resonator : (a) AlN contour mode resonator with aluminium top metal electrodes and (b) SEM pictures of it.

There are two key contributions of this work. First, we demonstrate an extremely low oscillator phase noise floor resulting from a co-design of the MEMS resonator with a CMOS oscillator. Secondly, we introduce a bias technique to reduce the oscillator temperature coefficient when using an ovenized resonator. The resonator and sustaining oscillator are presented in section II. Section III explains in more detail how the current source compensates the frequency drift caused by the components in the oscillator. The test result follows in section IV.

II. DIFFERENTIAL COLPITTS OSCILLATOR & AlN MEMS CMR

The resonator in this work is formed by a 4 μm thick piezoelectric plate. It is covered by the IDT electrodes layer, which is made of 100nm Al thin film deposited straight on the silicon substrate. The dimensions of the resonator plate are 52x122 um^2. For the fabrication, high resistivity Si wafer is

978-1-4673-6145-3/13 $31.00 © 2013 IEEE

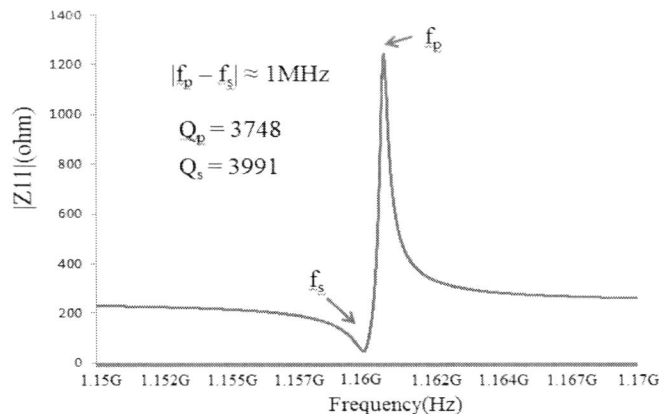

Fig. 2 Impedance magnitude of the resonator when loaded by oscillator capacitance.

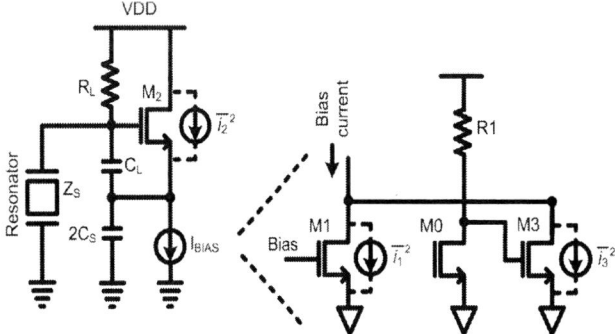

Fig.4 Single ended model of the differential Colpitts oscillator

$$w_{osc} \cong \sqrt{\frac{gmL_{MEMS} + R_L(C_s + C_L)}{L_{MEMS}C_SR_LC_L}}\,(gmr_o \gg 1) \quad gmR_B > \frac{C_L}{2C_S} \quad (1)$$

In order to decrease the phase noise of the oscillator, the noise contribution of each active device needs to be carefully analyzed. The differential oscillator is converted to a single ended, as shown in Fig.4, thus leading to equation (2) [8].

$$\frac{\overline{v_n^2}}{\Delta f} = \frac{4kT\delta g_{m1}}{g_{m2}^2} + \frac{4kT\delta g_{m3}}{g_{m2}^2} + \frac{4kT\delta}{g_{m2}} \quad (2)$$

Based on this equation, the transconductance of M2 needs to be increased to minimize the noise.

III. CURRENT SOURCE FOR TEMPERATURE COMPENSATION

The dominant source of the resonance frequency drift over temperature variation is the resonator (-22ppm/C). In the field of crystal oscillators, temperature compensated crystal oscillators (TCXO) [5] and oven stabilized devices (OCXO) [6] are often employed. In contrast, this work is targeted at a microscale heater contained in the resonator itself. This heater reduces the temperature drift from the resonator to 2ppm [12]. However, to achieve less than one ppm frequency variation, the CMOS circuit related contribution must be considered as well since the heater does not stabilize the temperature of the oscillator. Fig.6 shows the frequency variation, when the temperature coefficient of current source is zero and ideal resonator (no temperature dependent) are used, which reveals the inherent frequency drift due to temperature resulting from the CMOS circuitry.

To reduce the drift due to CMOS, we must design a current source that compensates for the temperature drift of the oscillator. The addition-based current source is a good candidate to stabilize oscillator frequency variation due to the CMOS temperature coefficient [10]. Though its aim is for compensating bias current variation caused by process variation, we can modify the structure to allow temperature compensation. The proposed technique allows us to tune the temperature coefficient of the current reference to cancel residual frequency drift.

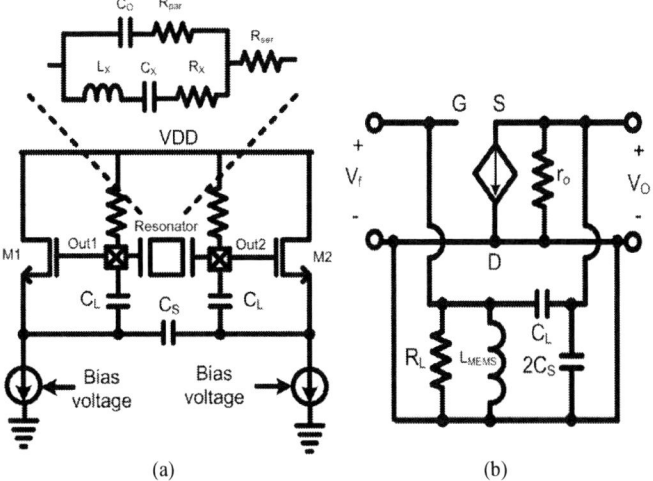

Fig.3 (a) Differential Colpitts Oscillator and (b) its small signal model

employed. Fig.1(a) depicts the electrode of the resonator and its properties. An SEM of the entire structure is shown in Fig.1(b).

Fig. 2 shows the magnitude of impedance of the resonator when loaded with the oscillator capacitance. In this figure k_t^2 is proportional to the difference between the parallel (f_p) and series (f_s) resonance frequency and has a value of 0.99%. The quality factor (Q) of the unloaded parallel resonance is around 3,700.

In an unloaded condition, when R_p is 48 kΩ, in parallel resonance mode, the Colpitts oscillator senses the voltage from the tank of the resonator. Due to the series and parallel parasitic capacitance, the frequency difference between f_s and f_p is decreased from 5 MHz to 1 MHz in the case of loaded condition. The effects to the phase noise performance due to the decrease in k_t^2 will be explained further in section IV.

To utilize this AlN CMR, the differential Colpitts oscillator was designed as shown in Fig.3. The equivalent small signal model allows calculation of the conditions for oscillation, leading to the equation (1). To sustain oscillation, C_S is chosen to have a large value and hence the capacitance ratio between C_L and C_S becomes smaller than $g_m \cdot R_B$.

978-1-4673-6145-3/13 $31.00 © 2013 IEEE

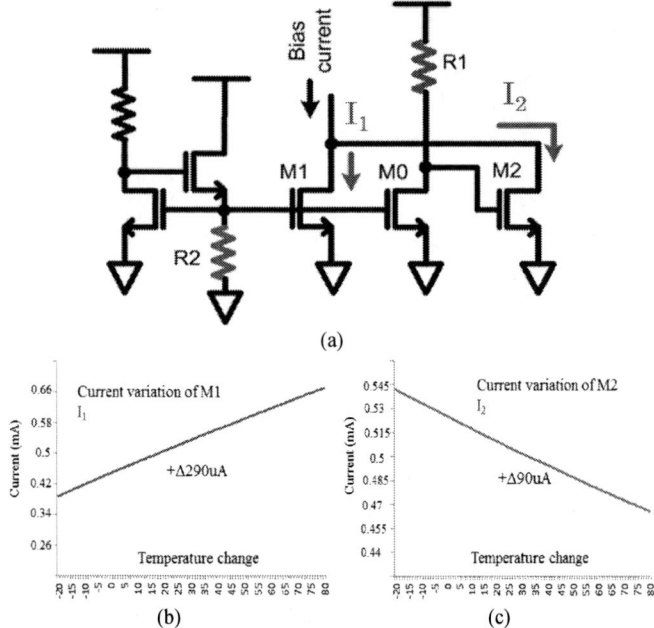

(a)

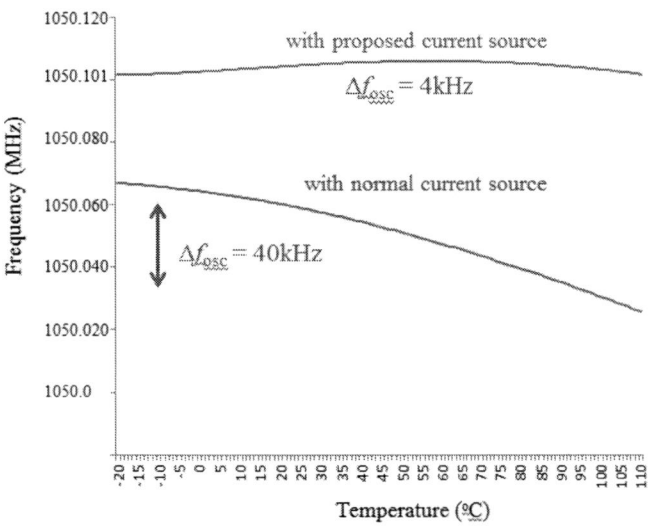

(b)   (c)

Fig.5 (a) Proposed current source and the current variation of M1(b), M2(c) under the temperature change (Simulation).

Fig.6 Simulation result of resonance frequency variation with and without proposed current source.

Fig.5(a) is a block diagram of the proposed current source to compensate the frequency variation caused by oscillator. The two poly bias resistors (R1 and R2) are identical in value and layout. The resistance variation dominates the temperature coefficient of the current reference. The resistors have a positive temperature coefficient, leading to a decrease in current in M2 as temperature increases. Conversely, the current through M1 increases, allowing cancellation and tuning of the current temperature coefficient. Fig.5(b) and (c) shows these behaviors. The simulated frequency drift of the oscillator, assuming the resonator temperature is completely stabilized and current source has zero temperature coefficient is 40 kHz (simulated). With this proposed current source having total bias

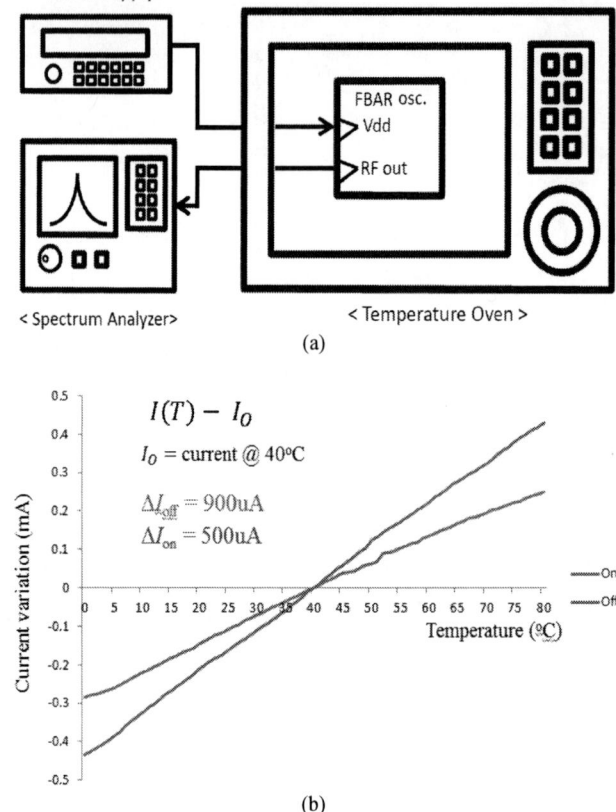

(a)

(b)

Fig. 7 (a) Test bench for the proposed current source and (b) test result

current with a positive temperature coefficient, the oscillation frequency variation is decreased to less than 4 kHz (simulated). To verify this effectiveness, the test bench is set up as shown in Fig.7(a). As the temperature is swept from 0 °C to 80 °C, the total frequency change is -1777 ppm which is equal to -22ppm/°C when the temperature compensating function is off. But, when the modified current source is on, the total frequency variation is decreased by 25 kHz over the whole temperature range, which means the frequency variation caused by oscillator decreases. Of course, the temperature variation due to the resonator remains. However, as mentioned above, using an ovenized resonator should allow us to get expected temperature coefficients around 1 ppm.. Fig.7(b) shows the test result of a current variation with and without this current source, demonstrating the ability to tune the current temperature coefficient.

IV. TEST RESULT

To test the phase noise of this high Q (low noise) oscillator, the noise floor inherent in the test equipment should be considered. Fig. 8 is the measured phase noise result. The dots in the plot represent the noise floor of the test equipment (Agilent Technology E5052B). At around 1 MHz offset frequency, there is a dip in the phase noise response. This feature is consistent with the difference in f_s and f_p (roughly 1 MHz). We believe this dip in phase noise is due to a reduction in tank impedance at f_s. Fig. 2(b) shows that the

978-1-4673-6145-3/13 $31.00 © 2013 IEEE

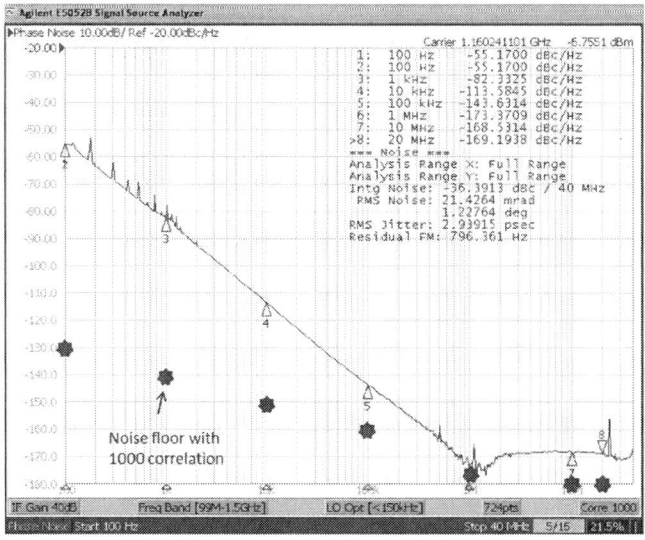

Fig.9 Die photo of the oscillator and resonator.

Fig.8 Phase noise measurement with 1000 correlations.

	f_{osc} (GHz)	Power (mW)	Tech.	Phase noise @10kHz (dBc/Hz)	Phase noise @100kHz (dBc/Hz)	Phase noise @1MHz (dBc/Hz)	FOM
This work	1.16	4.2	0.13um CMOS, MEMS	-113.5	-143.6	-173.3	**228.3**
[11]	1.55	11.3	0.13um CMOS, FBAR	-	-	-144.3	197
[12]	0.585	10	GaAs p-HEMT	-120	-145	-155	200.3
[13]	2	0.025	0.18um CMOS, FBAR	-	-121	-140	222
[14]	0.6	5.6	0.13um CMOS, FBAR	-126	-140	-150	198
[15]	1.5	1	0.35um CMOS, FBAR	-112	-133	-147	210

Table 1 Performance comparison with the previous work.

impedance of the resonator is at a local minimum at f_s. Table 1 shows the performance comparison with the previous works based on the phase noise at 1 MHz offset frequency and power consumption. The FOM is described in equation below. $L(\Delta f)$ represents the absolute value of the phase noise, and f_0, Δf are center frequency and offset frequency respectively.

$$\text{FOM} = L(\Delta f) + 20 \log \left(\frac{f_o}{\Delta f}\right) - 10 \log P(mW)$$

Fig. 9 shows the die photo of the realization. The AlN MEMS CMR on the left side is wirebonded directly to the differential Colpitts oscillator.

V. CONCLUSION

This work exploits the high Q of AlN MEMS CMRs to achieve good phase noise performance. The phase noise at 1 MHz offset frequency is -173.3 dBc/Hz. The measured jitter (integrated from 12 kHz to 40 MHz) is 2.9 ps rms. The center frequency of this oscillator is 1.16 GHz, and it is fabricated using 130nm CMOS technology. The next step will be employing an ovenized resonatorin order to decrease the temperature stability of the overall circuit in the entire temperature rangeto sub-ppm levels.

REFERENCE

[1] D. Ruffieux, J. Chabloz, C. Muller, F.-X. Pengg, P. Tortori, and A. Vouilloz, "A 2.4GHz MEMS-Based Transceiver," in Solid-State Circuits Conference, 2008. ISSCC 2008. Digest of Technical Papers. IEEE International, pp. 522 – 523, Feb. 2008.

[2] C. Y. – C. Nguyen, "MEMS technology for timing and frequency control," *IEEE Trans. Ultrasonic Ferroelectric Frequency Control*,vol. 54, n. 2, pp. 251 - 270, Feb. 2007.

[3] G. Piazza, P. J. Stephanou, A. P. Pisano, "Piezoelectric aluminum nitride vibrating contour-mode MEMS resonators," *Journal of Microelectromechanical Systems*, vol. 15, n. 6, pp. 1406 - 1418, Dec. 2006.

[4] J. R. Vig, "Military applications of high accuracy frequency standards and clocks", Ultrasonics, Ferroelectrics and Frequency Control, IEEE Transactions on, vol. 40, n. 5, pp. 522-527, 1993.

[5] Datasheet from SiTime Corporation web-site: http://www.sitime.com/products/datasheets/sit5000/SiT5000-datasheet.pdf

[6] Datasheets from Vectron International web-site: http://www.vectron.com/products/ocxo/ocxo_index.htm

[7] J. Segovia-Fernandez, N.-K. Kuo, G. Piazza, "Impact of Metal Electrodes on the Figure of Merit ($k_t^2 \cdot Q$) and Spurious Modes of Contour Mode AlN Resonators," to be published in Proceeding of International Ultrasonics Symposium, 2012.

[8] Ali Hajimiri and Thomas Lee, "The design of Low Noise Oscillators," Kluwer Academic Publishers, 1999.

[9] Thomas H. Lee and Ali Hajimiri, "Oscillator Phase Noise: A Tutorial" *IEEE Journal of Solid-State Circuits,* Vol.35, No.3, Mar. 2000

[10] Xuan Zhang, Alyssa B. Apsel, "A Low Variation GHz Ring Oscillator with Addtion-based current Source," *IEEE ESSCIRC*, pp. 216-219, Sept. 2009

[11] M. Nagaraju, K. Sankaragomathi, S. Gilbert, Brian Otis and Richard Ruby, "A Low Noise 1.5GHz VCO with a 3.75% tuning range using Coupled FBAR's," *IEEE* International Ultrasonics Symposium (IUS), October 2012

[12] A. Tazzoli, M. Rinaldi, G. Piazza, "Ultra High Frequency Temperature Compensated Oscillators Based on Ovenized AlN Contour-Mode MEMS Resonators", IEEE Frequency Control Symposium(FCS), May 2012.

[13] Andrew Nelson, Julie Hu, Jyrki Kaitila, Richard Ruby, Brian Otis, "A 22uW, 2.0GHz FBAR Oscillator," IEEE Radio Frequency Integrated Circuits Symposium (RFIC), June 2011

[14] Julie Hu, Lori Callaghan, Richard Ruby, Brian Otis, A 50ppm 600MHz Frequency Reference Utilizing the Series Resonance of an FBAR," IEEE Radio Frequency Integrated Circuits (RFIC), June 2010

[15] S. Rai, Ying Su, W. Pang, R. Ruby, B. Otis, "A Digitally Compensated 1.5GHz CMOS/FBAR Frequency Reference," IEEE Transactions on Ultrasonics, Ferroelectrics, and Frequency Control, March 2010

978-1-4673-6145-3/13 $31.00 © 2013 IEEE

An Ultra-Broadband Compact Mm-Wave Butler Matrix in CMOS for Array-Based MIMO Systems

Jong Seok Park, Taiyun Chi and Hua Wang

School of Electrical and Computer Engineering, Georgia Institute of Technology

Atlanta, GA 30332 USA

Abstract — **This paper presents the design scheme of an ultra-broadband compact mm-wave Butler Matrix. The design employs new transformer-based swapped-port couplers and lumped LC-based π-network phase shifters to achieve an extremely compact chip size and ultra-broad bandwidth. This scheme is implemented as a 4×4 Butler Matrix in a standard 65nm CMOS technology with a core area of 0.335×0.215 mm² and a broad bandwidth of 9.8 GHz with the center frequency of 63 GHz. This Butler Matrix design also achieves small insertion loss of 2.77 dB and high-quality matching among the ports with less than 0.6 dB amplitude mismatch and less than 5° phase imbalance at 63 GHz. Based on the measured S-parameters, the four concurrent electrical array patterns of the Butler Matrix achieve its array peak-to-null ratio of better than 17dB between 57 GHz and 67 GHz.**

Index Terms — **Butler Matrix, transformer-based swapped-port quadrature coupler, beam-forming, CMOS integrated circuits, MIMO, phased-array.**

I. INTRODUCTION

Recently, array-based MIMO systems are receiving increasing attentions particularly in the emerging mm-wave wireless systems due to their advantageous properties, such as multi-beam forming, beam steering, spatial filtering, and array signal-to-noise enhancement. Butler Matrix is a key building block in many array-based MIMO systems, since it is capable of forming multiple beams concurrently and enabling high system throughput [1] [2].

A 4×4 Butler Matrix, as a widely used configuration, is typically composed of four quadrature couplers and two 45° phase shifters, which together form four array beams spaced equally in the beam angle. As the most critical component in a Butler Matrix, the quadrature coupler is conventionally realized using parallel line couplers or branch line couplers, which generally demand λ/4 sizing and result in large chip area and excessive passive loss [3]. On the other hand, quadrature couplers with lumped elements or bifilar transformers are demonstrated in CMOS Butler Matrix, which unfortunately also require extensive areas [4] [5].

In this paper, we propose a new design scheme of Butler Matrix using transformer-based swapped-port quadrature couplers to achieve extremely small footprint. Moreover, our Butler Matrix also achieves low loss and excellent amplitude/phase balance over an ultra-broad operation bandwidth. This further enables broadband beam-forming and enhanced system functionality and diversity, once the Butler Matrix is employed in an array-based MIMO system.

This paper is organized as follows. Section II introduces the design of the transformer-based swapped-port quadrature

coupler and its implementation in our proposed Butler Matrix scheme. Comprehensive measurement and simulation results are presented in Section III to characterize a Butler Matrix design example at 63GHz. These include the amplitude/phase mismatch, insertion loss, operation bandwidth, and beam patterns. A comparison with recently reported on-chip Butler Matrix designs is also summarized.

II. DESIGN OF THE PROPOSED BUTLER MATRIX SCHEME

Figure 1 shows the schematic of a 4×4 Butler Matrix. The port numbers and the correspondingly formed beams are highlighted.

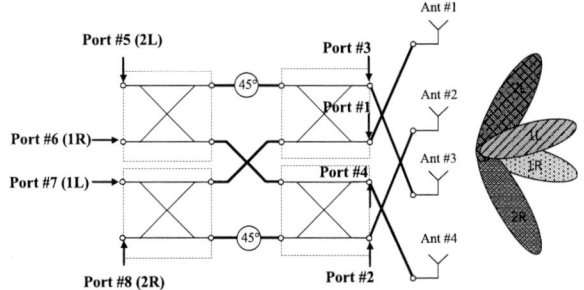

Fig.1. Schematic of a 4×4 Butler Matrix. The beams and their corresponding output ports are highlighted.

To achieve significant size reduction, we choose to employ swapped-port quadrature hybrid couplers, instead of conventional quadrature coupler, as the base-line coupler design. It has been shown that swapped-port quadrature couplers directly achieve 50% size reduction [6]. To further decrease the size, we coil the swapped-port coupler to form an on-chip transformer. As a result, the diameter of our 63 GHz quadrature coupler is 85 μm, whose size is more than 65 times smaller than a λ/4 branch line coupler at 63 GHz. Figure 2 shows the 3D EM simulation model of our transformer-based swapped-port coupler. The simulated amplitude and phase responses are shown in Figure 3. Different from conventional quadrature couplers, the swapped-port coupler has its coupled port leading the through port by 90°, which accounts for the double-cross at the Butler Matrix input (Fig.1).

Since ideal equal power dividing results in a 3 dB loss, our simulation therefore shows an extra loss of 0.8 dB with a phase error of 2.4°. Shunt capacitors are placed with the transformer to achieve further size reduction together with quadrature imbalance compensation [7]. In addition, our transformer-based coupler layout also ensures perfect symmetry and easy routing access to all the four ports.

978-1-4673-6145-3/13 $31.00 © 2013 IEEE

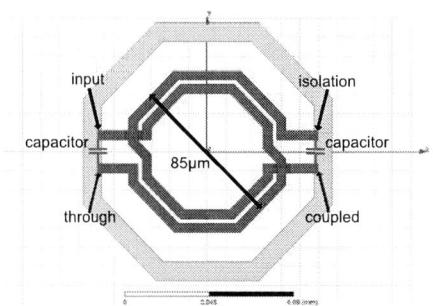

Fig.2. 3D EM model of the transformer-based swapped-port coupler.

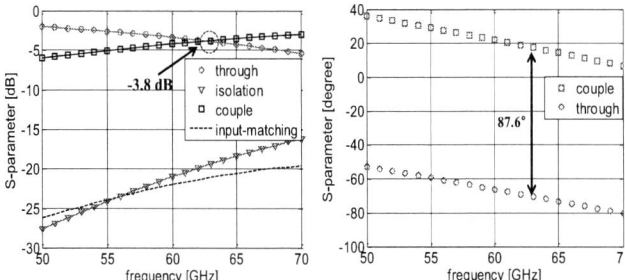

Fig.3. Simulated magnitude and phase responses of the designed transformer-based swapped-port quadrature coupler.

For the 45° phase shifter, we employ a lumped π-network with two shunt capacitors and one series inductor. The desired phase shift β can be achieved with the equations below. The quantities Z_{C1}, Z_{C2} and Z_L represent the impedances of the capacitors and the inductor, while Z_1 and Z_2 stand for the two port impedances.

$$Z_{C1} = jZ_1Z_2 \sin\beta / (Z_2 \cos\beta - \sqrt{Z_1Z_2}) \qquad (1)$$

$$Z_{C2} = jZ_1Z_2 \sin\beta / (Z_1 \cos\beta - \sqrt{Z_1Z_2}) \qquad (2)$$

$$Z_L = j\sqrt{Z_1Z_2} \sin\beta \qquad (3)$$

Assuming that the π-network's input and output port impedances are both 50Ω ($Z_1=Z_2=R_0=50Ω$), the design equations can be further simplified as

$$Z_{C1} = Z_{C2} = jR_0 \sin\beta / (\cos\beta - 1) \qquad (4)$$

$$Z_L = jR_0 \sin\beta. \qquad (5)$$

We choose to utilize a 2-turn coiled inductor with two shunt capacitors by considering the design trade-off among inductor quality factor, self-resonant frequency, and layout compactness. The full 3D EM simulations show that the 45° phase shifter achieves its insertion loss of 0.41 dB and phase error of 1° at 63 GHz (Fig. 4).

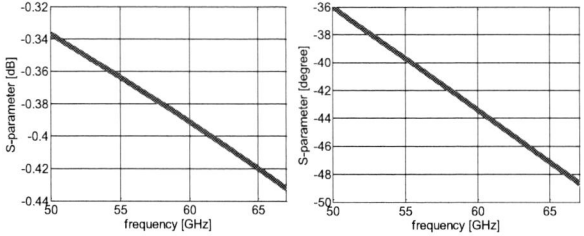

Fig.4. Simulated magnitude and phase responses of the 45°phase shifter.

Figure 5 shows the lumped element model of our proposed Butler Matrix in a 4×4 configuration. The port1, port2, port3, and port4 can be connected to four uniformly spaced antennas to form an array-based MIMO system, while port5, port6, port7, and port8 are the output ports for beams incident in different directions (Fig.1). In order to capture all the parasitic effects, a full 3D EM model of the complete 4×4 Butler Matrix is established and characterized using HFSS®.

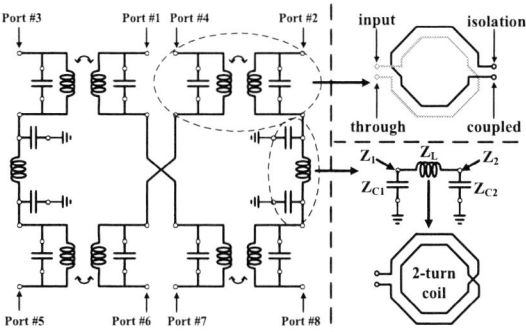

Fig.5. Lumped element model of the proposed butler matrix scheme.

III. MEASUREMENT AND SIMULATION RESULTS

As a proof-of-concept demonstration, a 4×4 Butler Matrix with the proposed scheme is implemented in a standard 65 nm CMOS process. The center frequency is 63 GHz.

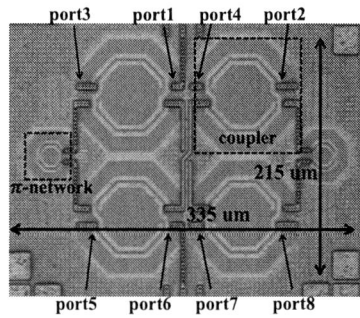

Fig.6. Chip microphotograph of the designed 4×4 butler matrix.

Figure 6 shows the chip microphotograph of the complete 4×4 Butler Matrix with its core chip area of only 335 μm by 215 μm (excluding the pads). *This is actually 5.4 times smaller than a single λ/4 branch line coupler at 60GHz.* To enable the testing, we implement a complete set of test structures, each of which allows probing of two ports with all the other ports terminated on-chip. The measurement is performed using an Agilent 67 GHz 2-port network analyzer (PNA-E8361C).

The measured input-to-output amplitude responses are shown in Fig.7 and Fig. 8 together with the simulation results. Since the input signal is equally divided into 4 output ports, the input-output S-parameter magnitude for an ideal and lossless network is –6 dB. Figure 7 shows the additional passive loss of the entire Butler Matrix network is 2.77 dB and the amplitude mismatch is 0.6 dB among the output ports when port1 is the input port. Based on the simulations, the bandwidth with ±1.5 dB amplitude mismatch is 15.5%

978-1-4673-6145-3/13 $31.00 © 2013 IEEE   653

(from 59 GHz to 68.8 GHz) with the center frequency of 63 GHz. The measurements verify this performance up to 67 GHz due to the limitation of the network analyzer. Figure 8 shows the corresponding amplitude responses when port2 is the input port.

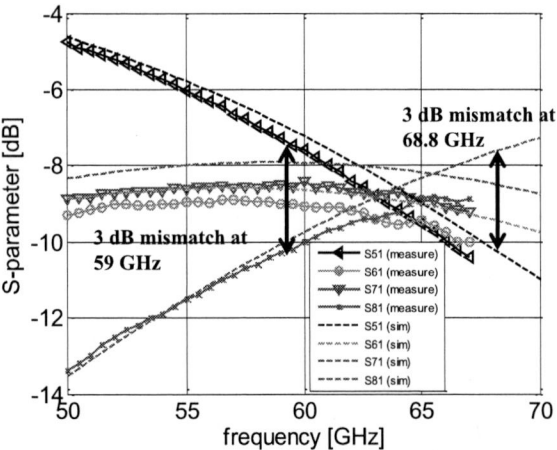

Fig.7. Measured and simulated input-output amplitude responses when port1 is the input port.

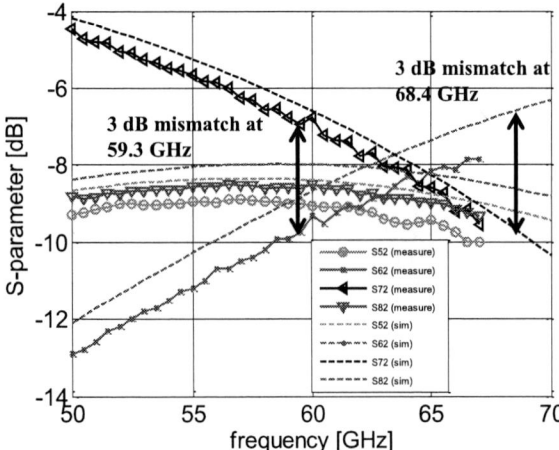

Fig.8. Measured and simulated input-output amplitude responses when port2 is the input port.

For any input port, the ideal phase differences among the four output ports are -45°, 135°, -135° and 45°. Figure 9 shows the absolute values of the measured phase differences between output ports, indicating the maximum phase error of 5° at the 63 GHz operating frequency.

In order to evaluate the beam forming capability of the Butler Matrix, the electrical beam patterns are synthesized and characterized based on the measured S-parameters, assuming that the port1 to port 4 of the Butler Matrix are connected to four antennas with λ/2 spacing.

If the incident beam presents 45° phase difference between adjacent antennas, the output signals at the four output ports, i.e. port5 to port8, are shown in Fig. 10. Computed based on the measured S-parameters, the additional loss on the desired output signal is 2.7 dB, while

the signal rejections at the undesired output ports are better than 33dB at 63GHz center frequency.

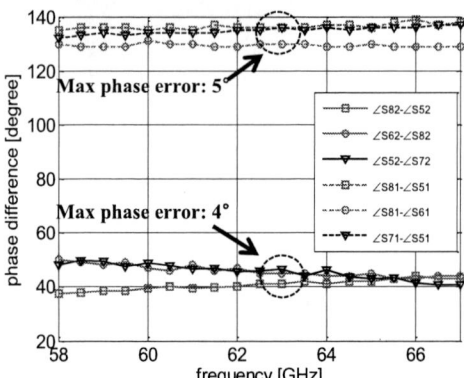

Fig.9. Measured input-output phase responses.

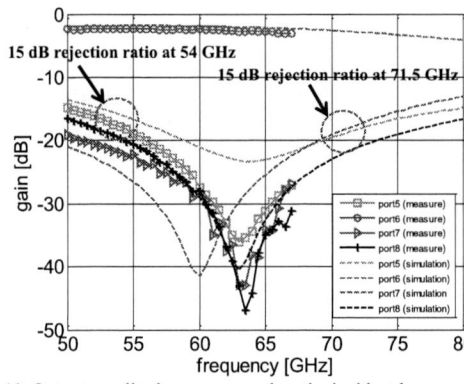

Fig.10. Output amplitude responses when the incident beam presents 45° phase difference between adjacent antennas.

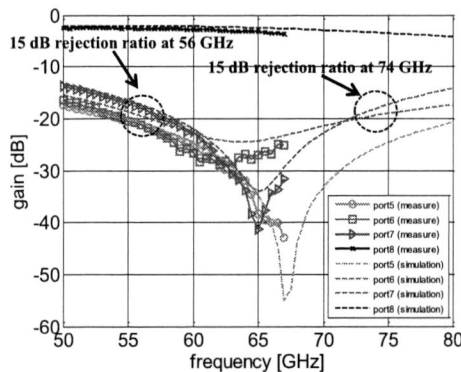

Fig.11. Output amplitude responses when the incident beam presents 135° phase difference between adjacent antennas.

Figure 11 shows the output results when the incoming beam presents a 135° phase difference between adjacent antennas. Computed based on the measured S-parameters, the loss at the desired signal output is 3 dB and the rejections at the undesired outputs are better than 26 dB at 63GHz.

Note that in Fig. 10 and 11, both the signal transmissions at the desired ports and the signal rejections at the undesired ports present ultra-wide bandwidth. By maintaining 15 dB rejection ratio, the operation bandwidth is as large as 27.8% (from 54 GHz to 71.5 GHz) with the center frequency of 63

GHz. This indicates that the designed Butler Matrix is capable of concurrently forming four beams with high-quality over a very large bandwidth. This enables frequency multiplicity in addition to the spatial multiplicity for array-based MIMO systems.

Based on the measured S-parameters, the concurrent four beam patterns of the designed Butler Matrix are synthesized with respect to phase difference between adjacent antennas. The beam patterns at 63 GHz (center frequency), 57 GHz, and 67 GHz (highest measured frequency) are shown in Figure 12, 13, and 14, respectively. At 63 GHz, the array peak-to-null ratio (PNR) is better than 27dB. The worst-case array PNR is 17dB happening at 57GHz. These results verify that our Butler Matrix is capable of forming four concurrent beams over a broad bandwidth.

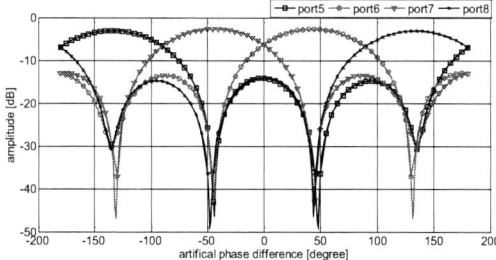

Fig.12. Synthesized electrical beam patterns based on the measured S-parameters at 63 GHz. The four beams are formed concurrently by the 4×4 Butler Matrix.

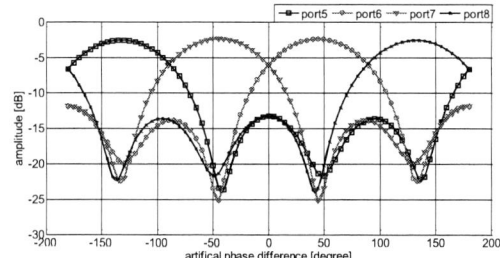

Fig.13. Synthesized electrical beam patterns based on the measured S-parameters at 57 GHz.

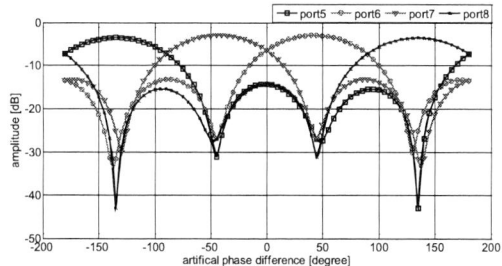

Fig.14. Synthesized electrical beam patterns based on the measured S-parameters at 67 GHz.

The performance comparison with recently reported 4×4 Butler Matrix designs is summarized in Table I.

IV. CONCLUSION

This paper presents an ultra-broadband and compact mm-wave Butler matrix design scheme in CMOS. As a design example, a 63 GHz 4×4 Butler Matrix is implemented in a standard 65nm bulk CMOS technology. A new transformer-based swapped-port coupler is proposed in the design which enables significant size reduction, leading to the core chip area of only 0.335×0.215 mm². This design also achieves an ultra-broad bandwidth with high-quality beam forming. From 59 GHz to 68.8 GHz, the amplitude mismatch across channels is below 3 dB, and phase error is less than 8.2° with an insertion loss of 2.77 dB. Based on the measured S-parameters, the synthesized electrical array pattern achieves a peak-to-null ratio of 27 dB at 63 GHz and a worst-case peak-to-null ratio of 17 dB from 57 GHz to 67 GHz.

Table I Recently Reported On-Chip 4×4 Butler Matrix Designs

	[3]	[5]	This work
Center Frequency	60 GHz	24 GHz	63 GHz
Coupler Topology	λ/4 T-line	Bifilar Transformer	Swapped-Port Transformer
Process	130 nm CMOS	180 nm CMOS	65 nm Bulk CMOS
On-Chip Switches	Yes	No	No
Chip Size	1.43×0.73 mm²	0.9×0.46 mm²	0.335×0.215 mm²
Amplitude Mismatch	1 dB	0.4 dB	0.65 dB
Phase Error at Center Frequency	±5°	6°	5°
Insertion Loss	3 dB	2.2 dB	2.77 dB
Bandwidth	10% (57-63 GHz)	8.2% (23-25 GHz)	15.6% (59-68.8 GHz)

V. ACKNOWLEDGMENT

The authors would like to thank Toshiba Corporation for foundry service and Dr. Shouhei Kousai from Toshiba and members of GT GEMS lab for their helpful technical discussions. The authors would also like to thank Dr. John Papapolymerou and GT MiRCTECH Lab for measurement support.

VI. REFERENCE

[1] J. Butler, and R. Howe, "Beamforming matrix simplifies design of electronically scanned antennas," *Electronic Design*, no. 9, pp. 170–173, Apr. 1961.

[2] C.E. Patterson, W.T. Khan, G.E. Ponchak, G.S. May, and J. Papapolymerou, "A 60-GHz Active Receiving Switched-Beam Antenna Array With Integrated Butler Matrix and GaAs Amplifiers," *IEEE Trans. Microw. Theory Tech.*, vol. 60, no. 11, pp. 3599–3607, Nov. 2012.

[3] K. Park, W. Choi, Y. Kim, K. Kim, and Y. Kwon, "A V-band switched beam-forming network using absorptive SP4T switch integrated with 4×4 Butler matrix in 0.13-μm CMOS," *IEEE MTT-S Int. Microw. Symp. Dig.*, May 2010, pp. 73–76.

[4] B. Cetinoneri, Y. A. Atesal, and G. M. Rebeiz, "An 8×8 Butler matrix in 0.13-μm CMOS for 5–6-GHz multibeam applications," *IEEE Trans. Microw. Theory Tech.*, vol. 59, no. 2, pp. 295–301, Feb. 2011.

[5] T. Chin, S. Chang, C. Chang, and J. Wu, "A 24 GHz CMOS Butler Matrix MMIC for Multi-Beam Smart Antenna Systems," *IEEE Radio Freq. Integr. Circuits Symp. Dig.*, Jun. 2008, pp. 633–636.

[6] Y.S. Jeong and T.W. Kim, "Design and Analysis of Swapped Port Coupler and Its Application in a Miniaturized Butler Matrix," *IEEE Trans. Microw. Theory Tech.*, vol. 58, no. 4, pp. 764–770, Apr. 2010.

[7] J. Park and H. Wang, "A Passive Quadrature Generation Scheme for Integrated RF Systems", *IEEE MTT-S IWS*, Apr. 2013.

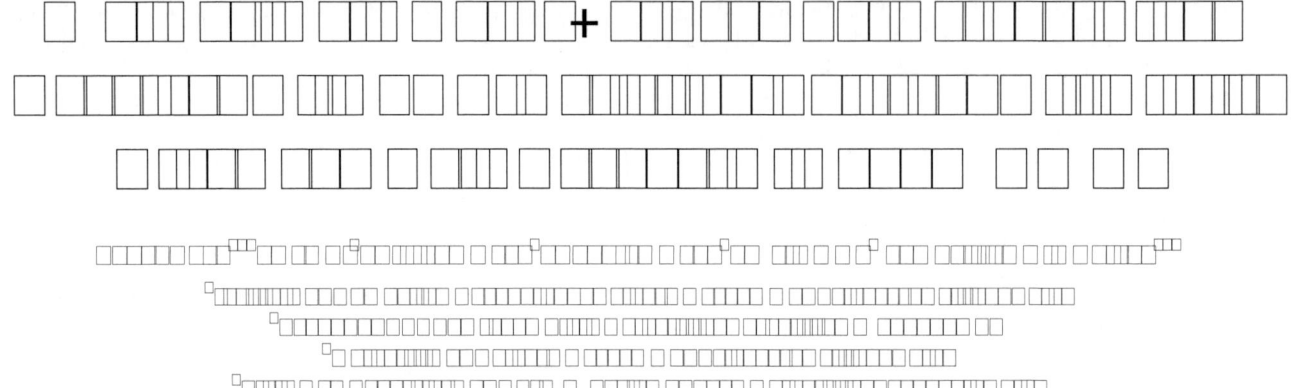

Abstract -This paper presents an energy/area-efficient forwarded-clock receiver fabricated in a 28nm CMOS process. The receiver consists of 8 data lanes plus one forwarded-clock lane, and adopts a novel all-digital clock and data recovery (CDR) using a delay-locked loop (DLL). The all-digital DLL with calibration can generate accurate multiphase clocks for both duty-cycle correction and the data recovery in the presence of process variations. The all-digital DLL-based CDR can enter into open-loop mode after lock-in to reduce power and eliminate the clock dithering phenomenon. Furthermore, the CDR can re-lock in the closed-loop mode using a proposed update algorithm to track the temperature and voltage variations without disturbing the data recovery. Measurement results show that the receiver can operate at a data rate of 6.4 Gb/s with a BER<10^{-12}, consuming 7.5 mW per lane under a 0.85 V power supply. The core receiver occupies an area of 0.02 mm^2 per lane.

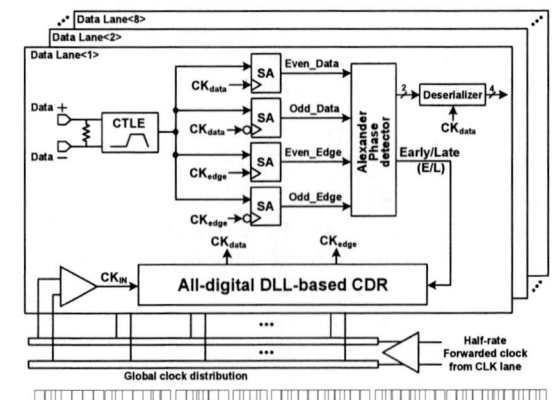

978-1-4673-6145-3/13 $31.00 © 2013 IEEE

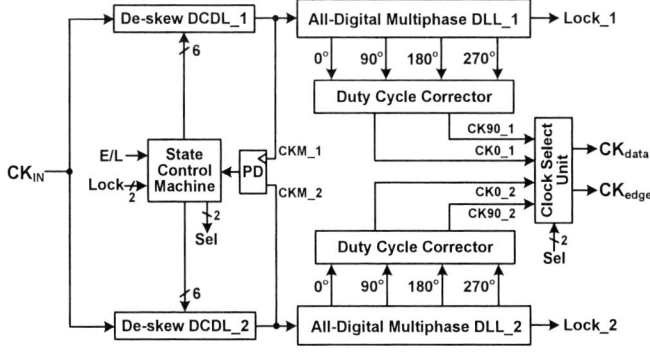

A. All-digital DLL-based CDR and the Update Strategy

State 0#:

State 1#:

State 2#:

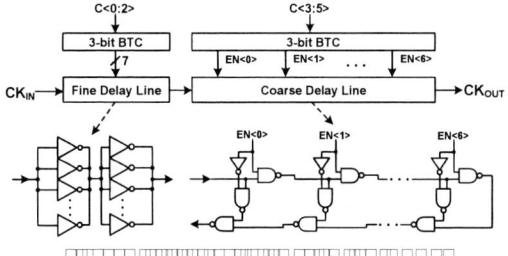

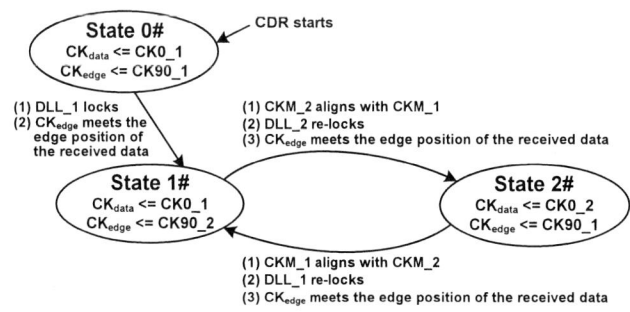

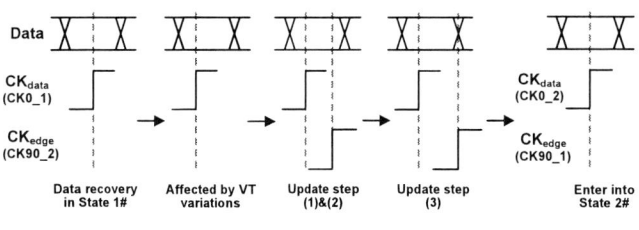

B. All-Digital Multiphase DLL and the Locking Flow

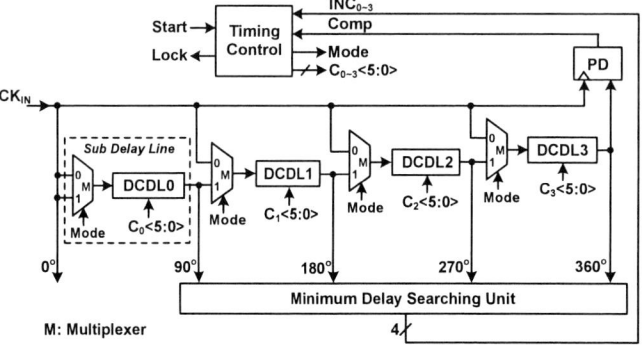

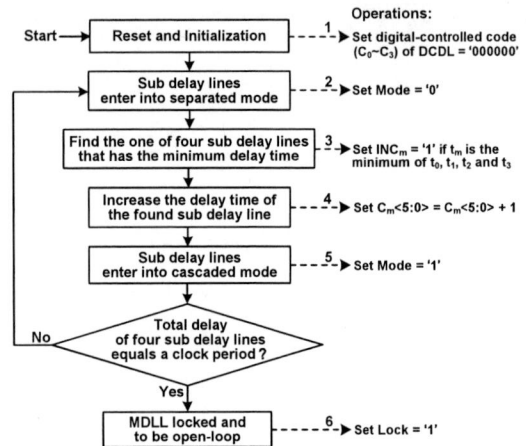

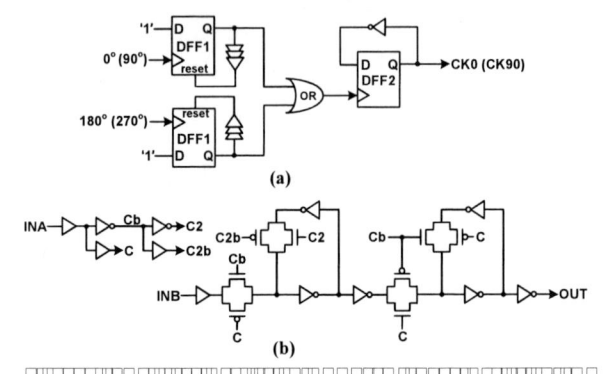

C. Phase Detector and Duty-Cycle Corrector

$$T_{DFF1,clock->Q} + T_{OR} + T_{DFF2,clock->Q}$$

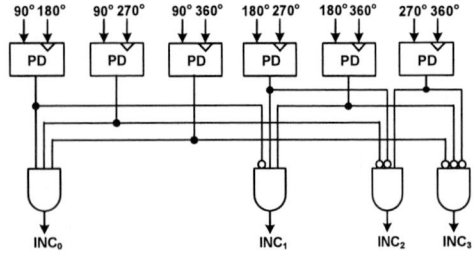

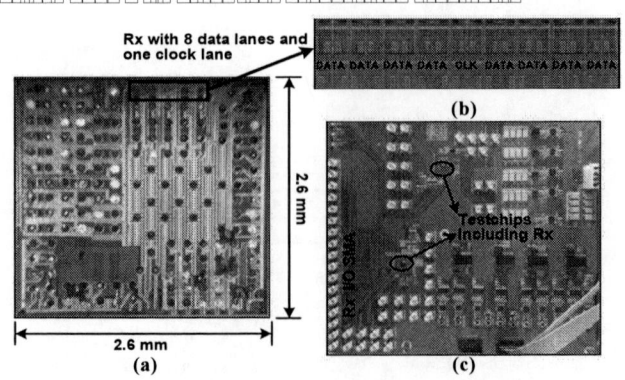

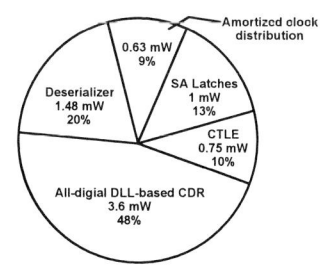

Amortized clock distribution
0.63 mW 9%
SA Latches 1 mW 13%
Deserializer 1.48 mW 20%
CTLE 0.75 mW 10%
All-digial DLL-based CDR 3.6 mW 48%

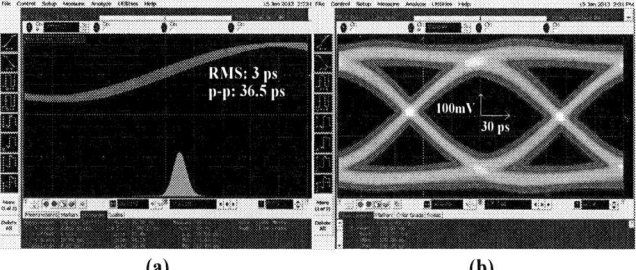

RMS: 3 ps
p-p: 36.5 ps

100mV
30 ps

(a)   (b)

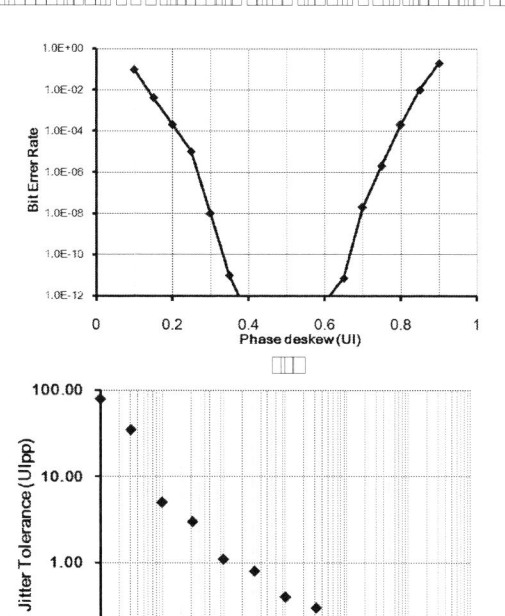

Reference				
Technology				
Voltage				
Data rate (Gb/s)				
CDR architecture				
Clock rate				
DCC				
Area*				
Power efficiency				

978-1-4673-6145-3/13 $31.00 © 2013 IEEE   659

A True 4-Cycle Lock Reference-Less All-Digital Burst-Mode CDR Utilizing Coarse-Fine Phase Generator with Embedded TDC

Tetsuya Iizuka[†], Satoshi Miura[‡], Yohei Ishizone[‡], Yoshimichi Murakami[‡] and Kunihiro Asada[†]

[†]Department of Electrical Engineering and Information Systems, University of Tokyo
7-3-1 Hongo, Bunkyo-ku, Tokyo 113-8656, Japan
[‡]THine Electronics, Inc.
9-1 Kanda-mitoshiro-cho, Chiyoda-ku, Tokyo 101-0053, Japan
Email: iizuka@vdec.u-tokyo.ac.jp

Abstract— This paper presents a reference-less all-digital burst-mode CDR using a coarse-fine phase generator with embedded TDC. It achieves true 4-cycle lock without any reference clocks and warm-ups, and eliminates dynamic power consumption in a stand-by state. Fabricated in 65nm CMOS, this CDR operates from 1.40 to 2.06Gb/s and consumes 9.6mW at 2.06Gb/s while occupying 80×80μm^2 area.

I. INTRODUCTION

Due to the growing demands for I/O bandwidth not only for enterprise applications but also for mobile consumer electronics, power-efficient I/O architecture is gaining its significance. In addition to a dynamic power efficiency in an active state, effective usage of a low-power stand-by state is essential particularly for mobile or sensor network applications which demand a long battery life. In these areas, a data link has to recover its communication immediately after activated not to waste power for state transitions, while consuming extremely low power in a stand-by state.

As illustrated in Fig. 1, several types of burst-mode CDR (BMCDR) has been proposed and widely used to realize a fast-lock to an input data stream with locking time of several UIs. The conventional BMCDR architecture shown in Fig. 1 (a) uses a replica PLL and this PLL locks to pre-determined frequency using the external reference clock[1], [2]. Then, the control voltage (V_{ctrl}) is distributed to the main gated-VCO (GVCO) to replicate the clock frequency and its phase is aligned by injecting the data transition edge through the gating circuit. This architecture is suitable for a multi-channel architecture since the replica PLL and its V_{ctrl} can be shared by several channels. However, the external reference clock is essential and the replica PLL consumes an additional area and power even when it is waiting for valid data streams. Moreover, a mismatch between main and replica GVCO often causes the frequency offset and degrades the tolerance to Consecutive Identical Digits (CID). To eliminate this mismatch issue, a BMCDR architecture shown in Fig. 1 (b) has been proposed. This architecture removes the replica GVCO and tunes the main GVCO frequency directly using the frequency locked loop (FLL)[3]. In this BMCDR architecture, however,

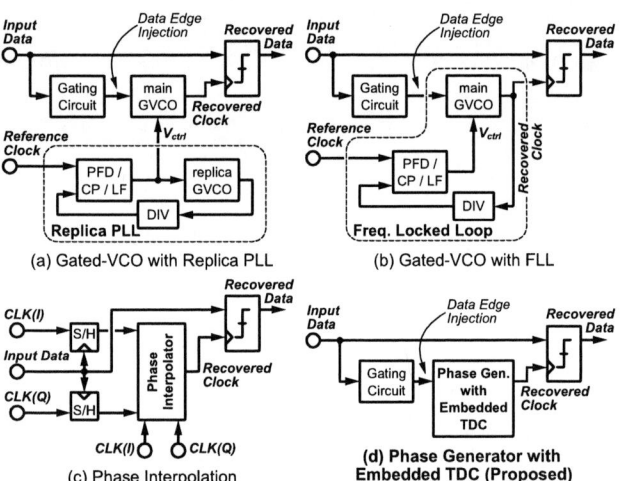

Fig. 1. Typical BMCDR architectures and the proposed BMCDR architecture.

the external reference clock is still essential, and the FLL consumes dynamic power even when waiting for valid data streams. A novel BMCDR architecture shown in Fig. 1 (c) uses a phase interpolator and produces a recovered clock by tuning the I/Q interpolation weighting factor depending on the sampled value[4]. Though this BMCDR realizes a locking time of less than 1UI with simple and small implementation, two quadrature I/Q clocks are essential and I/Q generation PLL also consumes dynamic power even when waiting for valid data streams.

As shown in Fig. 1 (a), (b) and (c), almost all of the conventional BMCDRs require an external reference clock and warm-up operations to achieve lock of local PLL or DLL, and they consume dynamic power even when they are waiting for valid data streams. In this paper, we propose a reference-less all-digital BMCDR circuit which realizes true 4-cycle lock without any warm-up operations after power-on, and eliminates dynamic power consumption in a stand-by state. As shown in Fig. 1 (d), the proposed architecture utilizes

978-1-4673-6145-3/13 $31.00 © 2013 IEEE

a phase generator with embedded time-to-digital converter (TDC) to detect a frequency and to recover a correct clock phase. A simple architecture and its all-digital implementation reduces the occupation area and relaxes the CDR design burden. Its coarse-fine hierarchical architecture demonstrates a continuous-rate operation from 1.40 to 2.06Gb/s in a 65nm CMOS with $80{\times}80\mu m^2$ area.

II. Proposed CDR Architecture and Implementation

Figure 2 illustrates a block diagram of the proposed half-rate CDR architecture and its timing diagram. Although a single-ended block diagram is shown, a differential implementation is adopted for an actual circuit. The proposed CDR circuit is composed of TDC-Embedded Phase Generator(TE-PG), Polarity Detector(PD)[1] and Edge Detector(ED)[1]. During a stand-by state without any input data transitions, the *Feedback Clock* signal is suspended by TE-PG and there is no active block in the CDR core. Thus this CDR consumes only a static leakage current during a stand-by state. Once a transition is fed into the *Data In* port, ED detects the data transition and generates the *Edge Detect* pulse to inject the transition from the *Data In* into TE-PG after PD aligns the data transition polarity by the *Inv* signal[5]. To measure a valid 1UI period using TE-PG, this CDR requires at least 2bits "10" preamble signal at start-up. After consecutive 0-to-1 and 1-to-0 transitions are injected, a delay-line-based simple TDC embedded in TE-PG detects the time interval and generates the *Phase Select* signal. According to the value of the *Phase Select*, Phase Selector(PS) selects the proper timing of the *Feedback Clock* from multi-phase outputs of Shared Delay Line(SDL). The *Recovered Clock* for sampling the *Delayed Data* is selected through another PS by dividing the value of the *Phase Select* by 2. When there is no data transition in the data input, the inverted *Feedback Clock* is fed into SDL and SDL forms a closed loop oscillator through PS and Edge MUX. Since this loop delay must be tuned to be equal to 1UI, the fixed delays of PS and Edge MUX should be compensated from the 1UI measurement result by the TDC. This delay compensation and the divide-by-2 operation for the *Recovered Clock* selector are simply implemented by bit shifting and OR operations respectively, because the *Phase Select* is one-hot-coded. TDC Enable block enables a TDC clock and updates a *Phase Select* code only when there is a valid 1UI time interval in the data stream, which corresponds to consecutive 0-to-1 and 1-to-0 transitions in the *Edge* signal, to prevent the TDC from acquiring the time interval with the *Feedback Clock* transitions which include the time quantization error of the TDC itself.

The TDC quantization error results in the period difference between the closed loop oscillation and the actual 1UI, and this difference causes phase error. Since this phase error is accumulated during Consecutive Identical Digits(CIDs), the TDC time resolution should be as fine as possible to improve the CID tolerance. However, if we use a fine-pitch delay line with single-stage manner as shown in Fig. 2, a large number of stages are required to realize a wide lock range and they increase layout area. To realize a fine time

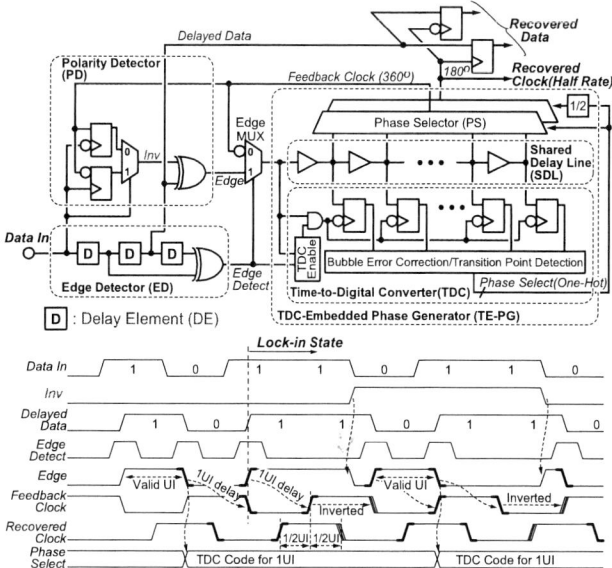

Fig. 2. Block and timing diagrams of the proposed CDR architecture.

resolution and a small area implementation simultaneously, the CDR architecture with coarse and fine TDC-embedded phase generators shown in Fig. 3 is proposed. The same ED and PD circuits as those in Fig. 2 are used. The TDCs and PSs in the coarse and fine phase generators use the same circuits except for their number of stages. A start-up timing diagram of the CDR is also shown in Fig. 3. When 2bits "10" is fed into the *Data In* port as a preamble, Coarse Phase Generator(C-PG) generates coarsely tuned clocks. Then Fine Phase Generator(F-PG) compares the 360° phase output from C-PG with a 1-to-0 data transition using an additional consecutive 2bits "10" and generates the finely tuned *Feedback Clock*. Thus, the proposed coarse-fine CDR requires 4bits "1010" preamble to achieve lock.

The circuit implementations of the coarse and fine delay lines are shown in Fig. 4. A differential delay element(DDE) is designed using 4 inverters and the coarse delay line is implemented using 18-stage DDEs. The fine delay line uses delay interpolation with resistors to realize a fine delay step and a 12-step delay line with 6-stage DDEs is implemented. In this paper, all the circuit blocks except for these resistors are implemented using logic cells in a standard-cell library to enhance process portability. As shown in Fig. 4, a coarse delay step of about 36ps and a fine delay step of about 6ps on an average are verified by simulation.

III. Experimental Results

The proposed CDR is fabricated in a 65nm standard CMOS technology and occupies $80{\times}80\mu m^2$ area as shown in Fig. 5. The CDR core consumes 6.8mA from a 1.4V supply with 2.06Gb/s data communication and consumes only static leakage current about 6.1μA during a stand-by state due to its reference-less and warm-up-less properties.

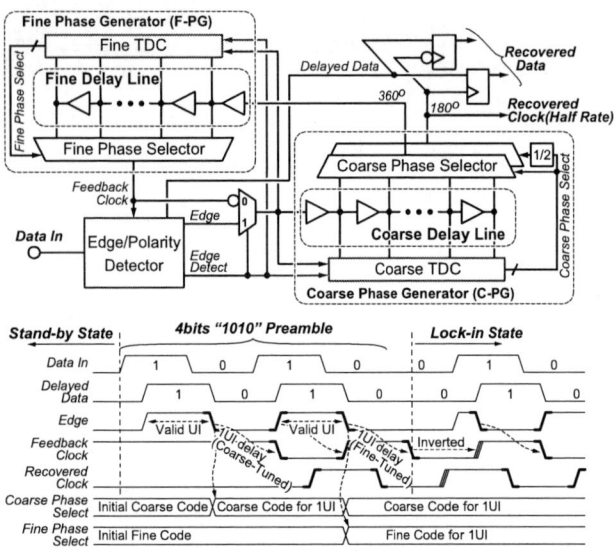

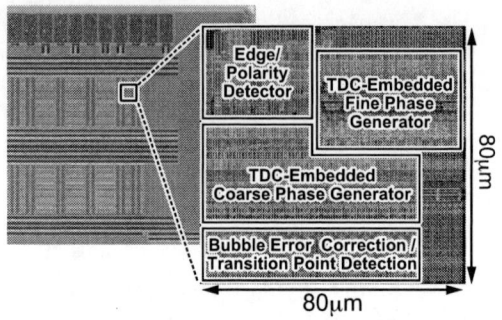

Fig. 5. A chip micrograph and a layout of the CDR.

Fig. 3. A block diagram of the proposed true 4-cycle lock CDR with coarse-fine phase generator and a lock-in sequence with 4bits "1010" preamble.

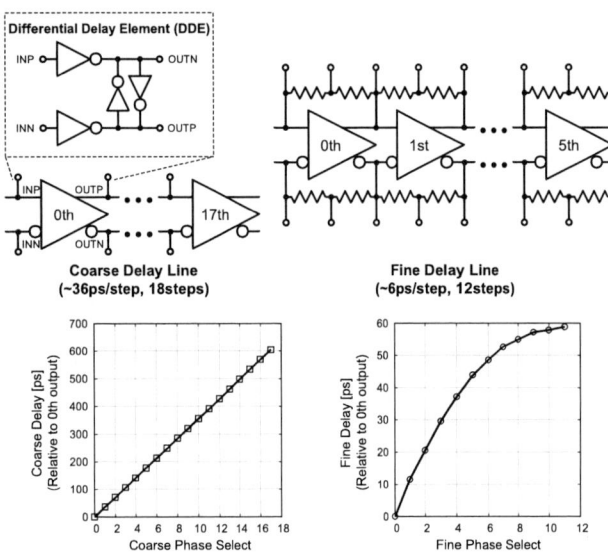

Fig. 4. Coarse and fine delay line schematic diagrams and their delay step simulation results.

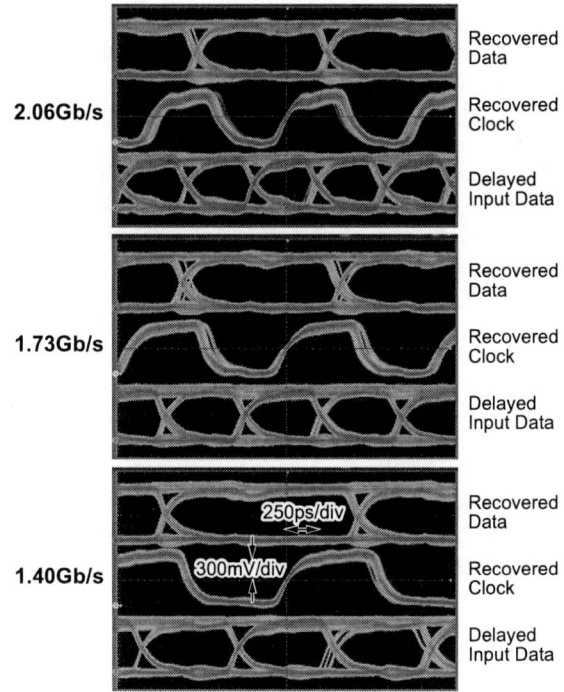

Fig. 6. Measured waveforms of several different data rates.

Figure 6 demonstrates the continuous-rate clock recovery operation from 1.40 to 2.06Gb/s data rate and Fig. 7 shows measured waveforms of a lock-in sequence and a CID input. As explained in Fig. 3, the proposed CDR recovers a finely tuned sampling clock after 4bits preamble "1010". The CID tolerance measurement result shows a 13bits CID tolerance at 2.06Gb/s and verifies that an input data transition correctly realigns the recovered clock phase. Considering the worst-case time resolution ~12ps of the fine delay line, a theoretical CID tolerance should be about 20bits at 2.06Gb/s. This 7bits degradation in the measured CID tolerance is caused by an oscillation frequency offset from the data rate due to

a mismatch between the fine phase selector delay and the compensated delay by the fine TDC. This error can be tuned by design optimization and the CID tolerance is further improved by using a finer time resolution considering the trade-off with occupation area and power consumption.

A jitter tolerance measurement result and waveforms of the recovered clock and data at 2.06Gb/s are shown in Fig. 8. Although the jitter tolerance measurement results in 100k-8MHz region are limited by the test equipment, the tolerance of the CDR at 1MHz is estimated about 8UIpp. The rms jitter of the recovered clock is measured as 12.8ps using 2^7-1 PRBS data input. Since the signal source and the oscilloscope contribute 2.0ps jitter, the rms jitter contribution of the proposed CDR is estimated as 12.6ps.

Table I gives a performance comparison with the previously-

TABLE I

PERFORMANCE COMPARISON WITH PRIOR-BMCDRs.

Reference	[2]	[3]	[4]	[5]	**This Work**
CDR Scheme	GVCO w/ Replica PLL	GVCO w/ FLL	Phase Interpolator	GVCO w/ FLL	**TDC-embedded Phase Generator**
Technology[nm]	40	250 (SiGe)	65	55	**65**
Supply Voltage[V]	1.1	3.3/1.8	1.2	3.3/1.2	**1.4**
Data Rate[Gb/s]	1.296 - 5.184	10.3125	1 - 6	2.5	**1.40 - 2.06**
Power[mW]	12.4	856 (w/ I/O)	22	13.68	**9.5**
Area[mm^2]	0.12×0.14	3.0×3.0 (whole chip)	0.25×0.07 (w/ peripheral)	0.180×0.225	**0.08×0.08**
RMS-Jitter[ps]	0.82 @5.184Gb/s	2.4	N/A	28.89 (p-p)	**12.6 @2.06Gb/s**
Tolerable CIDs	N/A	160bits	N/A	253bits	**13bits**
Locking Time	<20UIs	1UI	<1UI	1UI	**4UIs**
Reference Clock	Yes	Yes	Yes	No	**No**
Warm-Up Prior to Valid Data	Yes (PLL lock)	Yes (FLL lock)	Yes (I/Q generation)	Yes (<2µs for DAC code update)	**No**

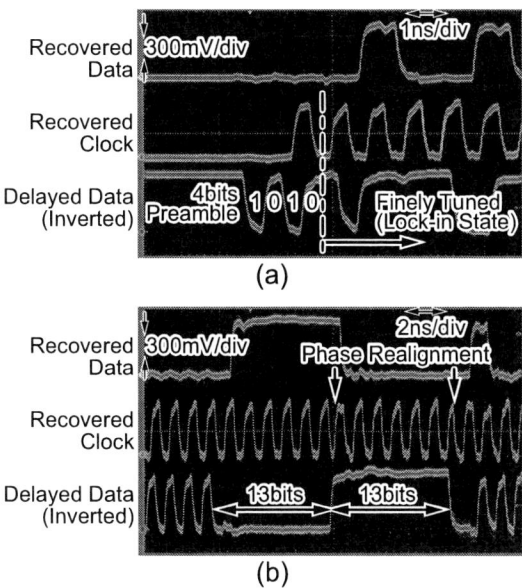

Fig. 7. (a) a 4-cycle lock-in sequence and (b) a CID tolerance verification at 2.06Gb/s.

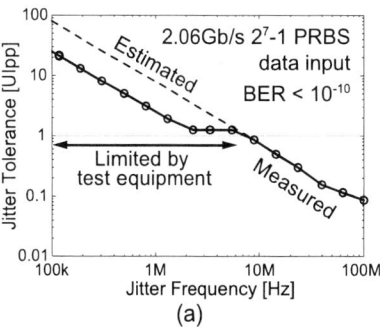

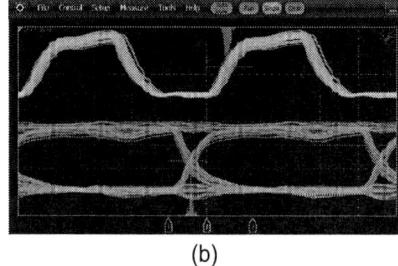

Fig. 8. (a) Jitter tolerance measurement result with BER < 10^{-10} and (b) measured waveforms of recovered clock and data using 2.06Gb/s 2^7-1 PRBS data input.

published BMCDRs. Among them, the proposed CDR achieves true 4-cycle lock without any external reference clocks and warm-ups while occupying the smallest area with all-digital implementation.

IV. CONCLUSIONS

This paper proposed a reference-less all-digital burst-mode CDR using a coarse-fine phase generator with embedded TDC. The proposed BMCDR was fabricated in 65nm CMOS and its measurement results demonstrated that it achieves a true 4-cycle lock operation without any reference clocks and warm-up operations and also eliminates dynamic power consumption in a stand-by state. This CDR operates from 1.40 to 2.06Gb/s and consumes 9.6mW at 2.06Gb/s while occupying 80×80µm^2 area.

ACKNOWLEDGMENTS

The VLSI chip in this study has been fabricated in the chip fabrication program of VLSI Design and Education Center(VDEC), the University of Tokyo in collaboration with STARC, e-Shuttle, Inc., and Fujitsu Ltd.

REFERENCES

[1] S. Kaeriyama and M. Mizuno, "A 10Gb/s/ch 50mW 120x130µm^2 Clock and Data Recovery Circuit," *IEEE ISSCC Dig. Tech. Papers*, pp. 70–71, 2003.

[2] K. Maruko et al., "A 1.296-to-5.184Gb/s Transceiver with 2.4mW/(Gb/s) Burst-mode CDR using Dual-Edge Injection-Locked Oscillator," *IEEE ISSCC Dig. Tech. Papers*, pp. 364–365, 2010.

[3] J. Terada et al., "A 10.3125Gb/s Burst-Mode CDR Circuit using a ΔΣ DAC," *IEEE ISSCC Dig. Tech. Papers*, pp. 226–227, 2008.

[4] B. Abiri et al., "A 1-to-6Gb/s Phase-Interpolator-Based Burst-Mode CDR in 65nm CMOS," *IEEE ISSCC Dig. Tech. Papers*, pp. 154–155, 2011.

[5] C.-F. Liang et al., "A Reference-Free, Digital Background Calibration Technique for Gated-Oscillator-Based CDR/PLL," *IEEE Symp. VLSI Circuits Dig. Tech. Papers*, pp. 14-15, 2009.

Thermal Noise Modeling of Nano-scale MOSFETs for Mixed-signal and RF Applications

Chih-Hung Chen[*, **], David Chen[*], Ryan Lee[*], Peiming Lei[***], and Daniel Wan[***]

[*]Advanced Technology Development Division, United Microelectronics Corporation, Hsinchu, Taiwan, R.O.C.
[**]Department of Electrical and Computer Engineering, McMaster University, Hamilton, ON, L8S 4K1, Canada
[***]United Microelectronics Corporation Group, CA, U.S.A.

Abstract—**This paper presents the thermal noise in nano-scale MOSFETs – from measurement, characterization, modeling, and potential technology enhancement for future low power, mixed-signal, and radio-frequency (RF) applications. Experimental data from five CMOS technology nodes, namely 180 nm, 130 nm, 90 nm, 65 nm, and 40 nm nodes are presented and discussed.**

Keywords—*channel thermal noise; nano-scale MOSFETs; thermal noise characterization; thermal noise modeling; low-noise technology; mixed-signal circuit; RF circuit*

I. INTRODUCTION

Thermal noise is the undesired random fluctuation from the electronic devices in circuits and added onto a signal. In analog circuits, it sets the fundamental lowest limit at which the desired signal can be detected. Thermal noise becomes an increasingly important issue, especially when the power supply voltage reduces in modern technology for low power, mobile wireless applications [1].

For circuits working at high frequencies, the thermal noise in a transistor is the dominant noise source, determines the noise performance of a circuit like low noise amplifiers - the first stage in a receiver, and therefore governs the overall noise performance of a receiver. On the other hand, for analog circuits like operational amplifiers (op-amp), although they work in the frequency range where the $1/f$ noise dominates, designers can apply certain techniques such as autozeroing, correlated double sampling, and chopper stabilization to effectively remove the impact of the $1/f$ noise in the analog circuits [2]. However, the thermal noise is still unresolved and therefore becomes the main issue limiting the performance of these analog circuits for low power applications.

In this paper, we start with a brief review of high-frequency noise measurement in Section II for readers to appreciate the higher-than-expected measurement uncertainties in the thermal noise characterization, when compared to their experiences in I-V, C-V or even scattering (*s*-) parameter measurements. Keeping these high uncertainties in mind, we demonstrate in Section III, for the first time, the methodology to capture the enhanced channel thermal noise using these measured noise data to ensure its bias and geometry scalability. This is particularly important when we want to extend our technique to develop the corner noise models with limited numbers of noise

data. Finally, in Section IV, we use the noise sheet resistance R_{nsh} described in [3] with experimental demonstration down to 40 nm technology node to discuss the potential solutions for the innovation of future low power, low noise technologies.

II. THERMAL NOISE CHARACTERIZATION

A high-frequency noise measurement system, in general, includes a noise source, a vector network analyzer (VNA), a noise figure analyzer (NFA), a microwave tuner, and a low noise amplifier (LNA) [4],[5]. The microwave tuner provides different source impedances for the noise receiver and the device-under-test (DUT) during the system calibration and the device measurement. The LNA reduces the noise figure of the receiver and therefore increases the measurement accuracy, especially when Friis' equation is used to de-embed the noise contribution from the noise receiver [6]. Other than using the Friis' equation, the noise contribution from the measurement system can also be de-embedded if we know the receiver gain G_o and the four noise parameters of the receiver at the noise reference plane. After IRE defined the representation of noise and the standards on methods of measuring noise in linear two ports in 1960 [7],[8], two approaches were developed to obtain the noise factors and the noise parameters, namely the minimum noise factor (or minimum noise figure in dB) NF_{min}, the equivalent noise resistance R_n, and the optimized source reflection coefficient Γ_{Sopt}. The first approach obtains four (or more) noise factors at different source admittances using the Y-factor method [9]. The four noise parameters are then obtained by solving the linearized noise equations with different algorithms to take care of various experimental errors and the uncertainties in noise factors and source admittances [10]-[21]. In the second approach, noise power is expressed as a function of source admittances with the noise parameters included as their coefficients. The noise parameters are extracted using the measured noise power at different source admittances, and the noise factors are then calculated based on the noise parameters obtained from the power equation [22],[23]. This approach forms the basis of the so-called "cold-only" method, in which only the cold noise power is measured in the device measurement [24],[25].

Based on the noise theory [22],[23], the receiver gain G_o can be experimentally extracted by one "hot" noise power P_h (measured in the hot state) and one "cold" noise power P_c

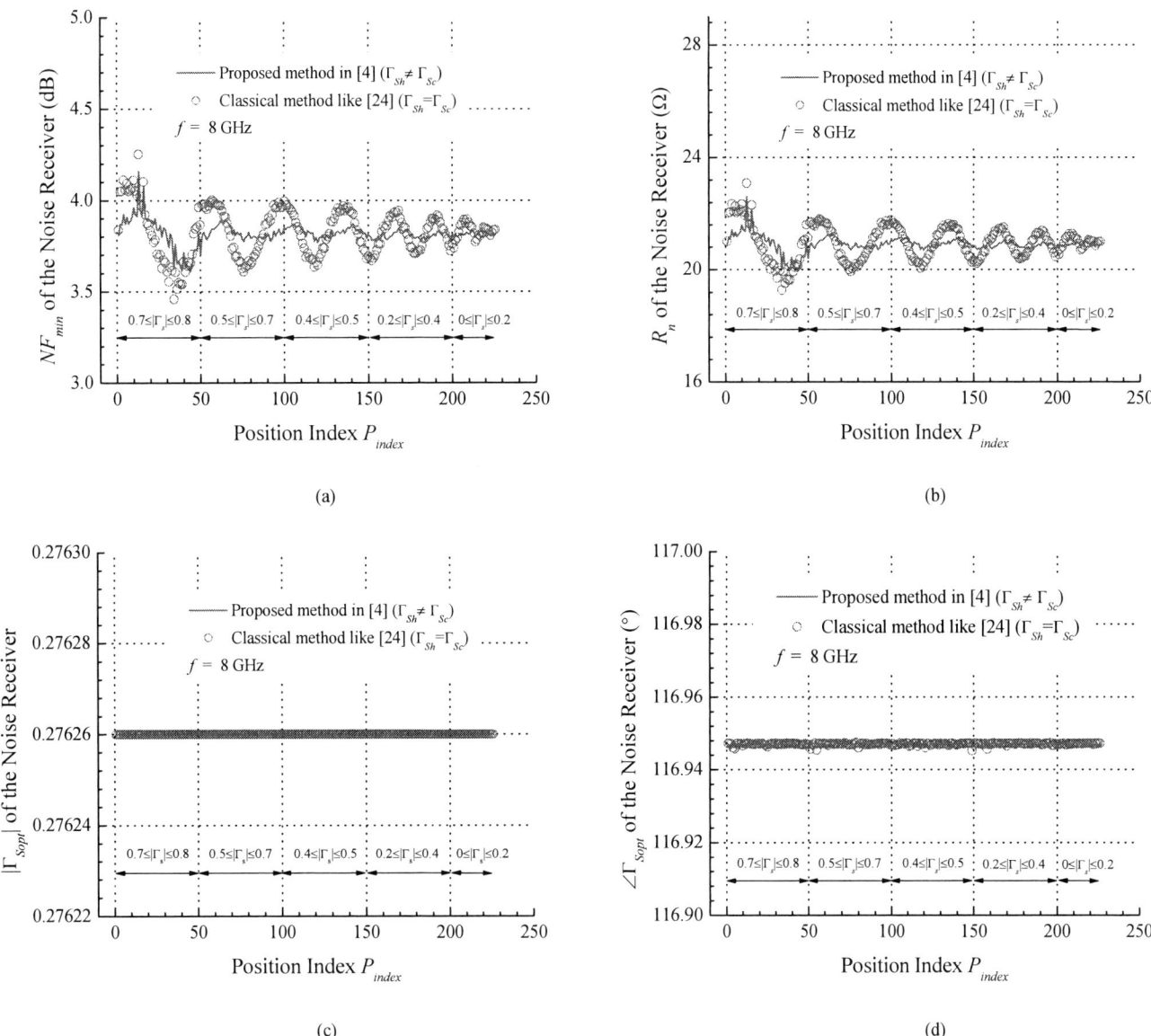

Fig. 1. Extracted four noise parameters of the noise receiver as a function of source reflection coefficients Γ_S using the proposed method in [4] (read lines) and the classical methods (blue circles) like [24].

(measured in the cold state) at "any" available source reflection coefficient Γ_S. However, due to the uncertainty in the linearity of the noise receiver, the Γ_S used does have a big impact on the extracted G_o [4], NF_{min}, and R_n as shown in Fig. 1, especially in the high Γ_S region. Although the proposed method in [4] improved the variations in the extracted NF_{min} and R_n (red lines in Fig. 1) by taking care of the Γ_S difference between the hot (Γ_{Sh}) and the cold (Γ_{Sc}) states, the variations still remain. From theory or experiment, there is no clear evidence for us to determine one G_o extracted at one Γ_S being more reliable than the other obtained at different Γ_S. Therefore, the variations in the extracted noise parameters with G_o obtained at different Γ_S become the major uncertainties for us, and these uncertainties become higher when the operation frequency f increases. After de-embedding the noise contributions from the measurement

system [4] and the parasitics of probe pads and metal interconnections [26], respectively, we extracted the power spectral density (PSD) of the channel thermal noise S_{id} [27], and found these uncertainties eventually appeared in the experimental S_{id} as shown in Fig. 2. Fig. 2 shows the extracted S_{id} of the channel thermal noise vs. frequency characteristics for an n-type Metal Oxide Semiconductor Field Effect Transistors (MOSFET) with channel length $L = 90$ nm and channel width $W = 128 \times 1$ µm biased at $V_{GS} = 1.0$ V and $V_{DS} = 0.8$ V with $I_{DS} = 36.8$ mA. Since the channel thermal noise is frequency independent (or white), we can then calculate the mean value using the extracted S_{id} at all frequencies to obtain the nominal value (4.64×10^{-21} A^2/Hz, green dashed line), its upper bound (5.41×10^{-21} A^2/Hz, red dashed line), and its lower bound (4.16×10^{-21} A^2/Hz, blue dashed line). The nominal mean

978-1-4673-6145-3/13 $31.00 © 2013 IEEE

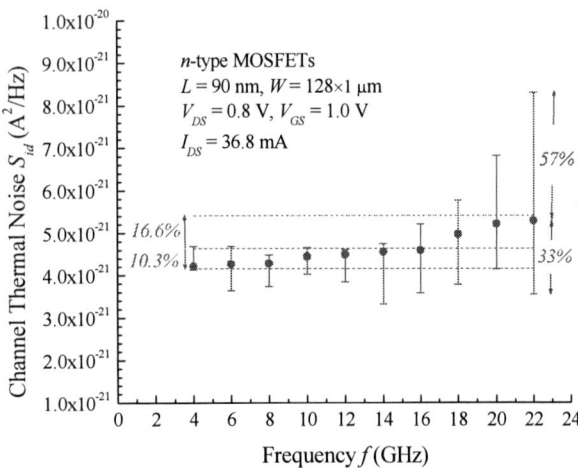

Fig. 2. Extracted chennal thermal noise S_{id} vs. frequency characteristics for an n-type MOSFET with channel length L = 90 nm and channel width W = 128×1 μm biased at V_{DS} = 0.8V and V_{GS} = 1.0V with I_{DS} = 36.8 mA [5].

value is calculated using all nominal values (symbols), which are obtained using the Γ_S closest to 0 when extracting G_o, while the upper and lower bounds are obtained using other Γ_S. We observed that the extracted S_{id} varies between -10% and +17% from its nominal value at this particular bias condition. Depending on how the noise data is recorded, the uncertainties could vary between -33% and +57% as seen at 22 GHz. This could be another reason to explain the significant spread in the published noise factors summarized in [28]. Keeping this high level of uncertainty in the experimental data in mind, in the next section, we present our methodology to develop the channel thermal noise model and its corner models, scalable in both bias and geometry, for mixed-signal and RF applications.

III. THERMAL NOISE MODELING AND IMPLEMENTATION

Because of the high system-level integration capability, the enhanced high-frequency performance up to hundreds of GHz [1], and the competitive fabrication cost in CMOS technology, noise characterization and modeling of nano-scale MOSFETs attract lots of attentions in recent years [29]-[36], particularly after Jindal reported the higher-than-expected channel thermal noise in 1985 [37]. In this section, we present the physics-based thermal noise model, the implementation of the scalable noise model to account for the enhanced channel thermal noise in circuit simulator, and the development of the corner models. Although we use BSIM4 compact model [38] and Spectre simulator [39] in this paper to demonstrate our methodology, these modeling strategies are general and can also be applied to any other compact model (e.g., PSP [40], HiSIM [41], or EKV [42]) and circuit simulator (e.g., HSPICE [43] or Eldo [44]).

Channel thermal noise S_{id} is the major noise source of interest in future nano-scale MOSFETs. Hot carrier effect [30] and the channel-length modulation (CLM) effect [31] are the main physical mechanisms responsible for the excess channel thermal noise before the impact ionization-induced shot noise occurs at the high drain bias [34]. On the other hand, to our best knowledge, degradation in carrier's mobility, caused by

the lateral electrical field in the gradual channel region, is the only mechanism reported in the literature which reduces the channel thermal noise [32]. To investigate the impacts of the hot electron, the CLM, and the lateral field effects on S_{id} in Section IV, we used the physics-based analytical expressions presented in [3] to predict the power spectral density (PSD) of the channel thermal noise S_{id}, which is given by

$$
\begin{aligned}
S_{id} &= \frac{\overline{i_d^2}}{\Delta f} \\
&= 4kT \frac{\mu_{eff}^2 W^2 C_{ox}^2}{L^2 I_D}[(V_{GS}-V_{TH})^2 V_{DS} - \alpha(V_{GS}-V_{TH})V_{DS}^2 + \frac{\alpha^2}{3}V_{DS}^3] \\
&\quad - 4kT \frac{\mu_{eff} W C_{ox}}{L^2 E_C}[(V_{GS}-V_{TH})V_{DS} - \frac{\alpha}{2}V_{DS}^2] + \delta \frac{4kT I_D}{L^2 E_C^2}V_{DS}
\end{aligned}
\tag{1}
$$

for $V_{DS} \leq V_{DSAT}$ or

$$
\begin{aligned}
S_{id} &= \frac{\overline{i_d^2}}{\Delta f} \\
&= 4kT \frac{\mu_{eff}^2 W^2 C_{ox}^2}{L'^2 I_D}[(V_{GS}-V_{TH})^2 V_{DSAT} - \alpha(V_{GS}-V_{TH})V_{DSAT}^2 + \frac{\alpha^2}{3}V_{DSAT}^3] \\
&\quad - 4kT \frac{\mu_{eff} W C_{ox}}{L'^2 E_C}[(V_{GS}-V_{TH})V_{DSAT} - \frac{\alpha}{2}V_{DSAT}^2] + \delta \frac{4kT I_D}{L'^2 E_C^2}V_{DSAT}
\end{aligned}
\tag{2}
$$

for $V_{DS} > V_{DSAT}$, where V_{DSAT} is the voltage potential at the "pinch-off" point (or at the position in the channel where carriers travel at their saturation velocity v_{SAT}) [29],[45]. Here, T is the lattice temperature, and δ is a fitting parameter to capture the hot electron effect [31]. The distance L' in (2) is the channel length for the gradual channel region, i.e., $L' = L - \Delta L$, where ΔL is given by [46]

$$
\Delta L = L_{itl} \cdot \ln\left[\frac{(V_{DS}-V_{DSAT})/L_{itl} + E_D}{E_C}\right], \tag{3}
$$

$$
E_D = E_C\sqrt{1 + \frac{[(V_{DS}-V_{DSAT})/L_{itl}]^2}{E_C^2}}, \tag{4}
$$

and

$$
L_{itl} = \frac{1}{\lambda}\sqrt{\frac{x_j \varepsilon_{si}}{C_{ox}}}. \tag{5}
$$

Here x_j is the junction depth, ε_{si} is the permitivity of silicon, and λ is the fitting parameter to account for the CLM effect on the dc drain current. In (1) and (2), the first term models the enhanced channel thermal noise by the CLM effect, the second term accounts for the channel noise reduction due to the lateral field effect, and the third term captures the enhanced channel noise due to the hot electron effect.

In general, all physics-based models (e.g., [3] and [36]) for the channel thermal noise introduce two new fitting parameters - one like δ in (2) to account for the hot carrier effect and the other like λ in (5) to take care of the CLM effect. However, these two fitting parameters increase the complexity of the parameter extraction routine when applying the noise model to cover all regions of operational conditions and geometries of interest. Therefore, it would be preferred if we can remove one fitting parameter with reasonable justification in physics for the

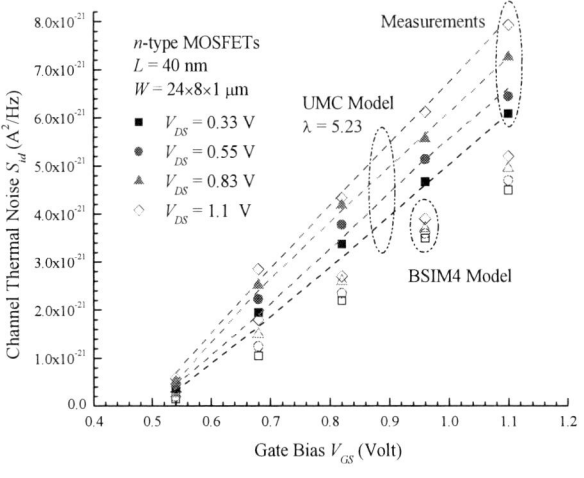

(a)

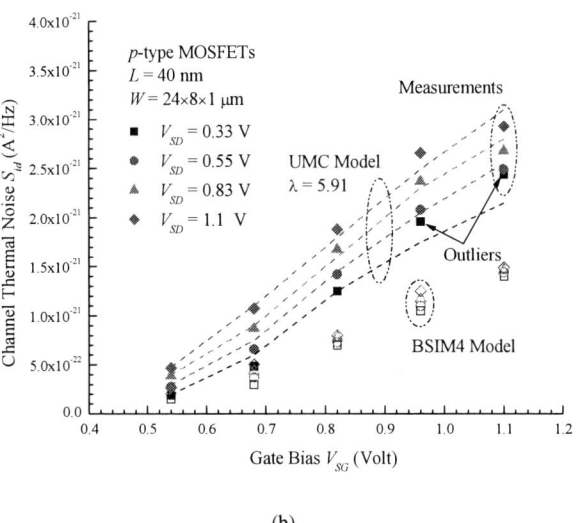

(b)

Fig. 3. Measured (solid symbols) and simulated channel thermal noise S_{id} using UMC model (dashed lines) and BSIM4 model (empty symbols) as a function of gate bias V_{GS} for the (a) *n*-type and (b) *p*-type MOSFETs with channel length L = 40 nm and width W = 24×8×1 μm biased at $|V_{DS}|$ = 0.33V, $|V_{DS}|$ = 0.55V, $|V_{DS}|$ = 0.83V, and $|V_{DS}|$ = 1.1V, respectively.

To include the extra channel thermal noise into the circuit simulator, the sub-circuit approach [48],[49] with Verilog-A codes are used to capture the bias and geometry dependence. Before implementing the enhanced channel thermal noise into our models, all of the dc, ac, and RF related model parameters in BSIM4 are already extracted using different sets of dc, CV, and RF test structures. To be physically consistent with the dc, ac, and RF models already developed, the parameter values used in (1) to (5) are calculated using the model parameters in BSIM4. Because of the high level of uncertainties in the extracted S_{id} as described in Section II, the parameter λ is extracted by minimizing the total errors between the simulated and the measured S_{id} at all bias conditions. Fig. 3 shows the measured (solid symbols) and simulated channel thermal noise S_{id} using UMC model (dashed lines) and BSIM4 model (empty symbols) as a function of gate bias V_{GS} for the (a) *n*-type (λ = 5.23) and (b) *p*-type (λ = 5.91) MOSFETs with channel length L = 40 nm and width W = 24×8×1 μm biased at $|V_{DS}|$ = 0.33V, $|V_{DS}|$ = 0.55V, $|V_{DS}|$ = 0.83V, and $|V_{DS}|$ = 1.1V, respectively. This can prevent the impact from some outliers (e.g., the two data points in Fig. 3(b) at V_{SD} = 0.33V) in the λ extraction.

After we implement the enhanced channel thermal noise in the TT models, we also want to include it in our corner noise models, which are important for circuit designers to estimate the impact of process variations on their circuit performance. The challenge to provide the corner models is the availability of enough noise data. Compared to other experiments like I-V or C-V measurements, high-frequency noise measurement is very time consuming due to the necessity of changing at least four (eight to twelve in practice) different source impedances at one biasing condition as mentioned in Section II. Before we can measure enough devices to collect statistically meaningful data, we used the corner model parameters in BSIM4 and the same λ value used in the TT models to predict and develop the corner noise models. At this point, we assume that the fitting parameter λ is a process independent parameter, unless we observe the evidence in future experiment. Fig. 4 shows the measured (symbols from the golden devices) and simulated channel thermal noise S_{id} for FF, FNSP, TT, SNFP, and SS corner models, respectively as a function of gate bias V_{GS} for the (a) *n*-type and (b) *p*-type MOSFETs with channel length L = 40 nm and width W = 24×8×1 μm biased at $|V_{DS}|$ = 1.1V.

IV. FUTURE LOW-NOISE TECHNOLOGY

With the capabilities of predicting the physical behavior of the channel thermal noise and providing designers the TT and corner noise models, how these physics-based noise models can guide semiconductor foundries to develop advanced low noise processes for mixed-signal and RF IC designs becomes our next question. When the feature size of the devices scales down, channel thermal noise S_{id} increases in nature due to the higher local conductance [31]. On the contrary, a stronger signal can also be achieved by these advanced devices due to their higher transconductance g_m. Therefore, when evaluating different technologies for their noise performance, the figure-of-merit used should include the information of both S_{id} and

removal of this particular parameter. It had been reported and experimentally verified in a 130 nm technology node that the noise reduction due to the mobility degradation effect (caused by the lateral electrical field) can be completely cancelled out by the noise increment due to the hot carrier effect [36]. In addition, the authors in [36] concluded that the channel thermal noise model with the CLM effect only is sufficient to predict the noise performance of deep sub-micron MOSFETs for all of the operating conditions. Another group, on the other hand, ran TCAD simulations for a 40 nm technology node [35], and showed that the thermal noise model including only the CLM effect predicts the closet noise parameter γ [47] to experimental data compared to the models with other effects included. Therefore, we only incorporate the CLM effect using λ in our thermal noise modeling to account for the enhanced channel thermal noise for devices down to 40 nm technology nodes.

978-1-4673-6145-3/13 $31.00 © 2013 IEEE

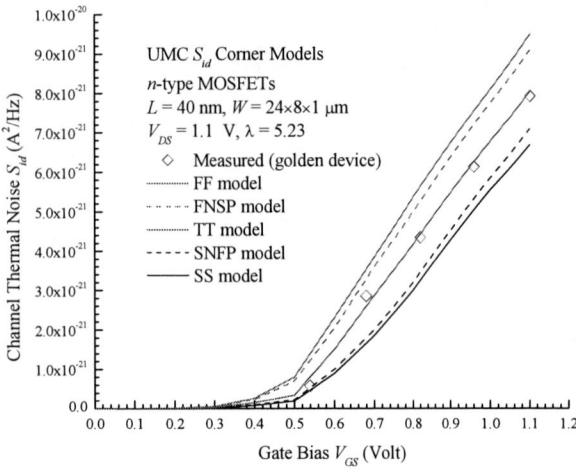

(a)

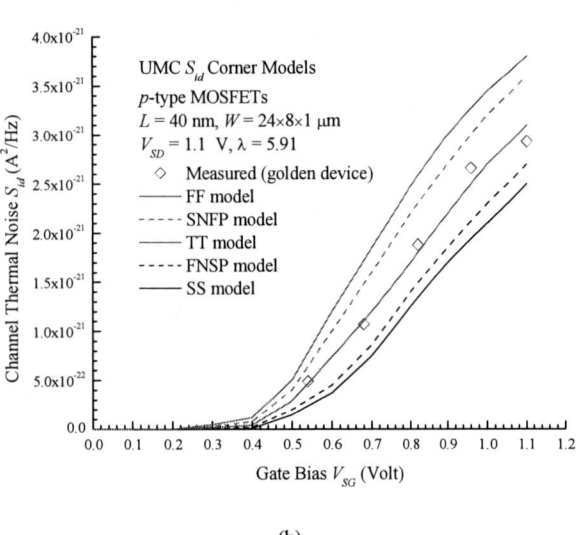

(b)

Fig. 4. Measured (symbols for the golden devices) and simulated channel thermal noise S_{id} using FF, FNSP, TT, SNFP, and SS models, respectively as a function of gate bias V_{GS} for the (a) n-type and (b) p-type MOSFETs with channel length L = 40 nm and width W = 24×8×1 µm biased at $|V_{DS}|$ = 1.1V.

g_m to demonstrate the full picture of technology evolution. In this paper, we used the noise sheet resistance R_{nsh} of an intrinsic device, which is defined as [3]

$$R_{nsh} = \frac{S_{id}}{4kT_o g_m^2} \cdot \left(\frac{W}{L}\right) \qquad (6)$$

where k is Boltzmann's constant, T_o is the standard temperature 290 K, S_{id} is the power spectral density of the channel thermal noise (A^2/Hz), g_m is the transconductance (S), and W/L is the aspect ratio of the DUT. The beauty of this parameter is that the term $S_{id}/4kT_o g_m^2$ (in Ω) monitors both the channel thermal noise S_{id} and the transconductance g_m in one parameter, and physically it equals to the portion of the equivalent noise resistance R_n due to S_{id} [33]. As discussed in [3], normalizing this parameter by (L/W) gives a meaningful comparison for devices with different geometries fabricated in different

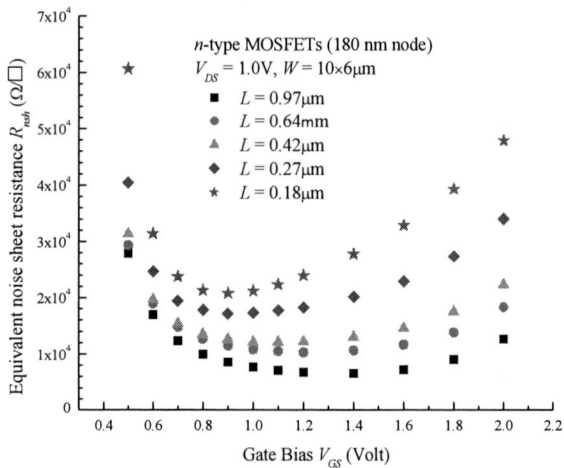

Fig. 5. Experimental data for R_{nsh} as a function of gate bias V_{GS} for the n-type MOSFETs fabricated in 180 nm node [27] with channel width W=10×6µm and channel lengths L = 0.97 µm (■), 0.64 µm (●), 0.42 µm (▲), 0.27 µm (◆), and 0.18 µm (★), respectively.

technologies. After normalizing to their geometries, R_{nsh} has a unit of $\Omega/\square$, and can be viewed as a process-related parameter because it only depends on the physical parameters such as the effective mobility, the critical electrical field along channel, the threshold voltage, and the gate-oxide capacitance. This parameter can also provide new insights for reliability study like [50]-[52] for future process innovation.

Before investigating the impacts of the hot electron effect, the channel-length modulation effect, and the lateral field effect, we would like to see the behavior of R_{nsh} predicted by the long-channel noise theory. In strong inversion, the channel thermal noise S_{id} of long-channel transistors can be modeled by [53],[54]

$$S_{id} = \frac{8kTg_m}{3} = \frac{8kT}{3} \cdot \frac{W}{L} \mu_{eff} C_{ox}(V_{GS} - V_{TH}) \qquad (7)$$

if we choose α = 1. Therefore, R_{nsh} can be predicted by

$$R_{nsh} = \frac{2T}{3T_o \mu_{eff} C_{ox}(V_{GS} - V_{TH})}, \qquad (8)$$

which suggests that R_{nsh} is not a function of channel length. To verify this prediction, we first use previously published data in [27] and show the measured R_{nsh} as a function of gate bias V_{GS} for n-type MOSFETs fabricated in a 180 nm technology node with channel width W = 10×6 µm and channel lengths L = 0.97 µm (■), 0.64 µm (●), 0.42 µm (▲), 0.27 µm (◆), and 0.18 µm (★), respectively in Fig. 5. When sweeping V_{GS} bias, (8) could explain the decreased R_{nsh} in the low V_{GS} region due to the increasing V_{GS} bias, and the increased R_{nsh} in the high V_{GS} region due to the stronger mobility degradation effect in μ_{eff} from V_{GS} [54]. However, (8) fails to predict the strong channel length dependence, which becomes even more pronounced when the channel length smaller than 42 µm.

Another interesting prediction shown in (8), especially for semiconductor foundries is that for low noise applications,

978-1-4673-6145-3/13 $31.00 © 2013 IEEE

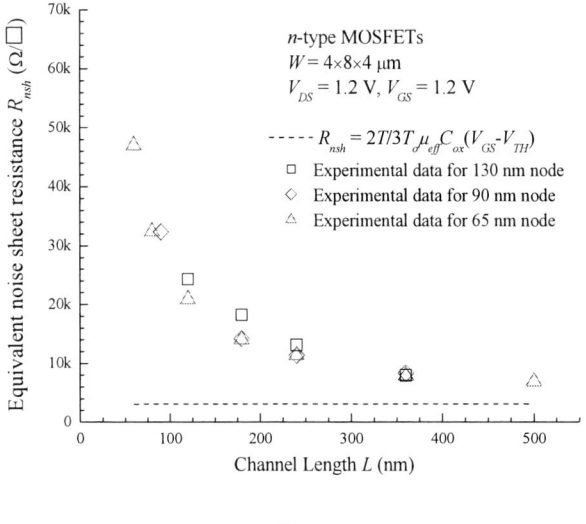

(a)

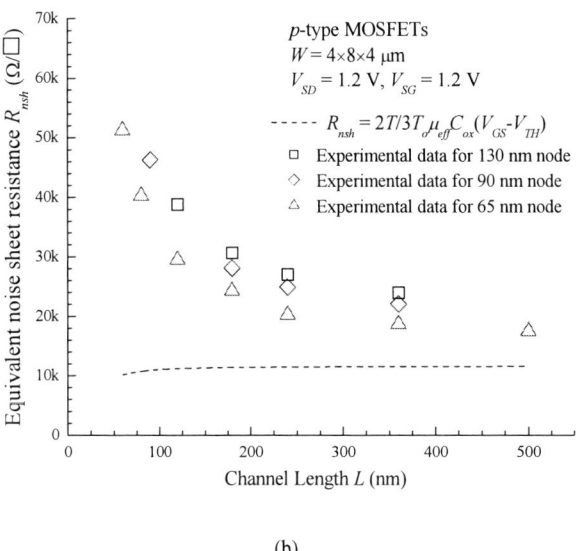

(b)

Fig. 6. Measured (symbols) and calculated R_{nsh} using (8) (dashed lines) as a function of channel length L for the (a) n-type and (b) p-type MOSFETs fabricated in 130 nm ($\square$), 90 nm ($\diamond$), and 65 nm ($\triangle$) CMOS technology nodes and biased at $|V_{DS}| = 1.2$V and $|V_{GS}| = 1.2$V [3].

R_{nsh} can be improved with materials/devices having larger μ_{eff} or thinner oxide thickness (i.e., larger C_{ox}). To verify these predictions in the technology innovation, we conducted the noise measurements for the devices fabricated in UMC's 130 nm, 90 nm, and 65 nm technology nodes, respectively. From 130 nm to 90 nm node, the technology innovation used is the reduction of the oxide thickness. Besides, starting from the 65 nm node, strain engineering was applied to enhance the effective mobility. For each device-under-test, multi-finger structures were designed to reduce the parasitic gate resistance, and the total channel width W equals $M \times N_f \times W_f$ μm, where M is the number of multi-finger structures connected in parallel, N_f is the number of transistor fingers in a multi-finger

structure, and W_f is the finger width of each transistor. After the intrinsic noise parameters were obtained, we used the procedures in [27] to extract the experimental data for S_{id} and g_m and calculate their R_{nsh}. Fig. 6 shows the experimental results for the R_{nsh} as a function of channel length L for the (a) n-type and (b) p-type MOSFETs in 130 nm ($\square$), 90 nm ($\diamond$), and 65 nm ($\triangle$) CMOS technology nodes biased at $|V_{DS}| = 1.2$V and $|V_{GS}| = 1.2$V. We noticed that from the 130 nm to the 90 nm node, R_{nsh} reduces for both n- and p-type MOSFETs. However, R_{nsh} in the 65 nm node becomes comparable to that in the 90 nm node for n-type MOSFETs. For p-type MOSFETs, R_{nsh} keeps improved in the 65 nm node.

It is crucial to understand the physics behind the increasing trend of R_{nsh} in the short-channel transistors, particularly for processing engineers to improve their technologies and for circuit designers to select the proper technology for their low noise applications. For this purpose, we first studied the impact of μ_{eff} on R_{nsh} using (2) to calculate the channel thermal noise S_{id} and the dc model in [3] to calculate the transconductance g_m numerically (i.e., $g_m = \partial I_D / \partial V_{GS}$). Instead of using E_C in (2), we changed this variable using the carrier's drift velocity v_{drift} given by [45]

$$v_{drift} = \begin{cases} \dfrac{\mu_{eff}}{1 + E/E_C} E & \text{for } E \leq E_C \\ v_{SAT} & \text{for } E > E_C \end{cases} \qquad (9)$$

and replaced E_C by $2v_{SAT}/\mu_{eff}$ [55]. Here we use v_{SAT} as an independent variable instead of E_C because v_{SAT} is directly related to the material property, and is hard to change by engineering techniques. For example, when using the strained Si in devices, the effective mobility μ_{eff} can be enhanced by about 2 times, but the saturation velocity remains the same [56]. In this case, E_C is actually reduced by a half. Fig. 7 shows the calculated R_{nsh} as a function of channel length L under different μ_{eff} and v_{SAT} for (a) n-type and (b) p-type MOSFETs in 65 nm technology node biased at $|V_{DS}| = 1.2$V and $|V_{GS}| = 1.2$V, and the results for the strained Si are plotted in the lines with solid squares (-■-). We noticed that the strained Si improves the R_{nsh} for long-channel devices due to the enhanced effective mobility. However, for short-channel devices, the reduction in the critical field E_C results in much higher R_{nsh}. Therefore, improving the effective mobility without increasing the saturation velocity at the same time actually degrades the noise performance of nano-scale MOSFETs. It needs not to mention that the devices made by materials with higher mobility but lower saturation velocity (such as strained Ge [57],[58]) have even worse noise performance. Another common technique to boost the (silicon) device performance uses different channel planes and orientation (e.g., (110) for p-channel transistors [59]). We also noticed in Fig. 7 that we need to improve the saturation velocity by about 10 times for the doubled effective mobility (lines with solid triangles, -▲-). In reality, the doubled hole mobility using stressors in the [110] direction only enhances the saturation velocity by about 2 times [60], and this is not sufficient to simultaneously improve its noise performance (see the lines with solid diamonds, -◆-, in Fig. 7). Materials with higher saturation velocity such as

978-1-4673-6145-3/13 $31.00 © 2013 IEEE

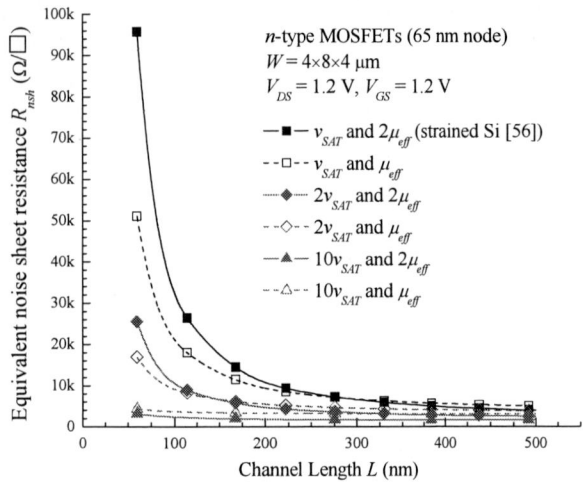

(a)

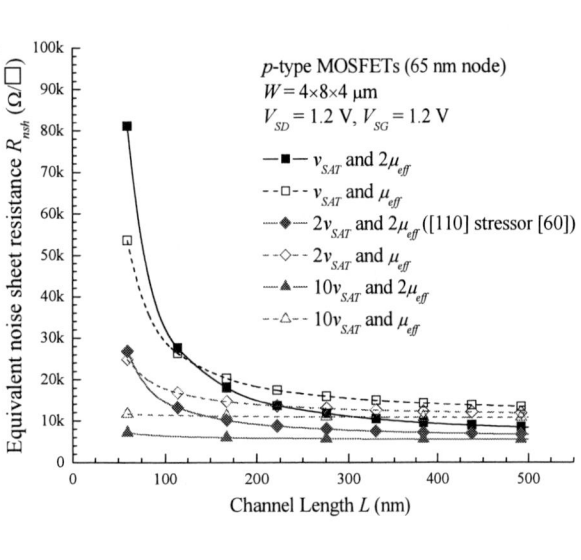

(b)

Fig. 7. Calculated R_{nsh} as a function of channel length L under different μ_{eff} and v_{SAT} for (a) n-type and (b) p-type MOSFETs biased at $|V_{DS}| = 1.2$V and $|V_{GS}| = 1.2$V including the hot electron effect in (2) (i.e., $\delta = 1$) [3].

InSb or In$_{0.7}$Ga$_{0.3}$As in [61] might be the potential solution for future low noise technologies. However, if the enhancement in its effective mobility is more than that in its saturation velocity, at the same channel length, the III-V transistors could be even noisier than Si-based transistors. Therefore, based on (8), we conclude that increasing C_{ox} by reducing the oxide thickness or using high-k materials becomes the only possible solution for future low noise technologies.

Finally, we want to study the impact of power supply reduction in V_{dd} on R_{nsh} in future nano-scale MOSFETs. Fig. 8 shows the measured R_{nsh} as a function of channel length L for the n-type MOSFETs fabricated in 65 nm ($\square$) and 40 nm ($\star$) CMOS technology nodes, respectively. The data are obtained at $V_{DS} = V_{GS} = V_{dd} = 1.2$V for 65 nm devices, and at $V_{DS} = V_{GS} = V_{dd} = 1.1$V for 40 nm devices. We observed that with lower power supply voltage V_{dd}, R_{nsh} degrades in the 40 nm node

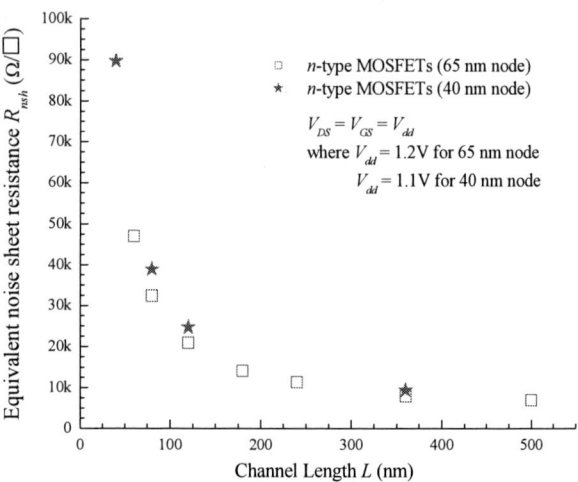

Fig. 8. Measured R_{nsh} as a function of channel length L for the n-type MOSFETs fabricated in 65 nm ($\square$) and 40 nm ($\star$) CMOS technology nodes, respectively. The data are obtained at $V_{DS} = V_{GS} = V_{dd} = 1.2$V for 65 nm devices, and at $V_{DS} = V_{GS} = V_{dd} = 1.1$V for 40 nm devices.

compared to that in the 65 nm node at the same channel length. This implies that the reduction in V_{dd} results in higher degradation in g_m than that in S_{id}, which makes future low noise design even more challenging when using future nano-scale MOSFETs for low power applications.

ACKNOWLEDGMENT

Authors are grateful to National Sciences and Engineering Research Council of Canada (NSERC) for its support in part.

REFERENCES

[1] International Technology Roadmap for Semiconductors (ITRS), 2011. [Online]. Available: http://www.itrs.net/reports.html.

[2] C. Enz and G. C. Temes, "Circuit techniques for reducing the effects of op-amp imperfections: autozeroing, correlated double sampling, and chopper stabilization," *IEEE Proc.*, vol. 84, pp. 1584–1614, Nov. 1996.

[3] C. H. Chen, R. Lee, G. Tan, D. C. Chen, P. Lei, and C. S. Yeh, "Equivalent sheet resistance of intrinsic noise in sub-100nm MOSFETs," *IEEE Trans. Electron Devices*, vol. 59, no. 8, pp. 2215–2220, 2012.

[4] C. H. Chen, Y. L. Wang, M. Bakr, and Z. Zeng, "Novel noise parameter determination for on-wafer microwave noise measurements," *IEEE Trans. Instrum. Meas.*, vol. 57, issue 11, pp. 2462–2471, Nov. 2008.

[5] C. H. Chen, P. Lei, and C. W. Liang, "Impact of impedance-dependent receiver gain on the extracted channel thermal noise in sub-100nm MOSFETs," *Mathematical and Computer Modeling*, in press, 2012.

[6] H. T. Friis, "Noise figures of radio receivers," *Proc. IRE*, vol. 32, no. 7, pp. 419–422, Jul. 1944.

[7] IRE Subcommittee on Noise, 59 IRE 20. S1, IRE standards on methods of measuring noise in linear two ports, 1959, *Proc. IRE*, vol. 48, no. 1, pp. 60–68, Jan. 1960.

[8] IRE Subcommittee 7.9 on Noise, Representation of noise in linear two ports, *Proc. IRE*, vol. 48, no. 1, pp. 69–74, Jan. 1960.

[9] Fundamentals of RF and microwave noise figure measurements. Agilent application note 57-1.

[10] R. Q. Lane, "The determination of device noise parameters," *Proc. IEEE*, vol. 57, no. 8, pp. 1461–1462, 1969.

[11] M. S. Gupta, "Determination of the noise parameters of a linear 2-port," *Electron. Lett.*, vol. 6, no. 17, pp. 543–544, 1970.

[12] G. Caruso and M. Sannino, "Computer-aided determination of microwave two-port noise parameters," *IEEE Trans. Microw. Theory*

978-1-4673-6145-3/13 $31.00 © 2013 IEEE

Tech., vol. MTT-26, no. 9, pp. 639–642, Sept. 1978.

[13] M. Sannino, "On the determination of device noise and gain parameters," *Proc. IEEE*, vol. 67, no. 9, pp. 1364–1366, Sept. 1979.

[14] G. I. Vasilescu, G. Alquie, and M. Krim, "Exact computation of twoport noise parameters," *Electron. Lett.*, vol. 25, no. 4, pp. 292–293, 1988.

[15] J. M. O'Callaghan and J. P. Mondal, "A vector approach for noise parameter fitting and selection of source admittances," *IEEE Trans. Microw. Theory Tech.*, vol. 39, no. 8, pp. 1376–1382, Aug. 1991.

[16] J. W. Archer and R. A. Batchelor, "Fully automated on-wafer noise characterization of GaAs MESFET's and HEMT's," *IEEE Trans. Microw. Theory Tech.*, vol. 40, no. 2, pp. 209–216, Feb. 1992.

[17] A. Boudiaf and M. Laporte, "An accurate and repeatable technique for noise parameter measurements," *IEEE Trans. Instrum. Meas.*, vol. 42, no. 2, pp. 532–537, Apr. 1993.

[18] W. Wiatr and D. K. Walker, "Systematic errors of noise parameter determination caused by imperfect source impedance measurement," *IEEE Trans. Instrum. Meas.*, vol. 54, no. 2, pp. 696–700, Apr. 2005.

[19] N. J. Kuhn, "Curing a subtle but significant cause of noise figure error," *Microw. J.*, vol. 27, no. 6, pp. 85–98, Jun. 1984.

[20] A. C. Davidson, B. W. Leake, and E. Strid, "Accuracy improvements in microwave noise parameter measurements," *IEEE Trans. Microw. Theory Tech.*, vol. 37, no. 12, pp. 1973–1978, Dec. 1989.

[21] L. F. Tiemeijer, R. Havens, R. Kort, and A. J. Scholten, "Improved Y-factor method for wide-band on-wafer noise parameter measurements," *IEEE Trans. Microw. Theory Tech.*, vol. 53, no. 9, pp. 2917–2925, 2005.

[22] V. Adamian and A. Uhlir, "A novel procedure for receiver noise characterization," *IEEE Trans. Instrum. Meas.*, vol. IM-22, no. 2, pp. 181–182, Jun. 1973.

[23] M. N. Tutt, Low and high frequency noise properties of heterojunction transistors, Ph.D. dissertation, Dept. Elect. Eng., Comput. Sci., Univ. Michigan, Ann Harbor, MI, 1994.

[24] R. Meierer and C. Tsironis, "An on-wafer noise parameter measurement technique with automatic receiver calibration," *Microw. J.*, vol. 38, no. 3, pp. 22–37, Mar. 1995.

[25] M. Kantanen, M. Lahdes, T. Vähä-Heikkilä, and J. Tuovinen, "A wideband on-wafer noise parameter measurement system at 50–75 GHz," *IEEE Trans. Microw. Theory Tech.*, vol. 51, pp.1489–1495, 2003.

[26] C. H. Chen and M. J. Deen, "A general noise and s-parameter de-embedding procedure for on-wafer high-frequency noise measurements of MOSFETs," *IEEE Trans. Microw. Theory Techn.*, vol. 49, no. 5, pp. 1004–1005, May 2001.

[27] C. H. Chen, M. J. Deen, Y. Cheng, and M. Matloubian, "Extraction of the induced gate noise, channel thermal noise and their correlation in sub-micron MOSFETs from RF noise measurements," *IEEE Trans. Electron Devices*, vol. 48, no. 12, pp. 2884–2892, Dec. 2001.

[28] S. Dronavalli and R. P. Jindal, "CMOS device noise considerations for terabit lightwave systems," *IEEE Trans. Electron Devices*, vol. 53, no. 4, pp. 623–630, Apr. 2006.

[29] D. P. Triantis, A. N. Birbas, and D. Kondis, "Thermal noise modeling for short–channel MOSFETs," *IEEE Trans. Electron Devices*, vol. 43, no. 11, pp. 1950–1955, Nov. 1996.

[30] P. Klein, "An analytical thermal noise model of deep submicron MOSFETs," *IEEE Electron Device Lett.*, vol. 20, pp. 399–401, 1999.

[31] C. H. Chen and M. J. Deen, "Channel noise modeling of deep sub-micron MOSFETs," *IEEE Trans. Electron Device*, vol. 49, no. 8, pp. 1484–1487, Aug. 2002.

[32] A. S. Roy and C. Enz, "Compact modeling of thermal noise in the MOS transistor," *IEEE Trans. Electron Device*, vol. 52, pp. 611–614, 2005.

[33] M. J. Deen, C.-H. Chen, S. Asgaran, G. Ali Rezvani, J. Tao, and Y. Kiyota, "High-frequency noise of modern MOSFETs: compact modeling and measurement issues," *IEEE Trans. Electron Devices*, vol. 53, no.9, pp. 2062–2081, Sept. 2006.

[34] A. O. Adan, M. Koyanagi, and M. Fukumi, "Physical model of noise mechanisms in SOI and bulk-silicon MOSFETs for RF applications," *IEEE Trans. Electron Device*, vol. 55, no. 3, pp. 872–880, Mar. 2008.

[35] V. M. Mahajan, P. R. Patalay, R. P. Jindal, H. Shichijo, S. Martin, F.-C. Hou, C. Machala, and D. E. Trombley, "A physical understanding of RF noise in bulk nMOSFETs with channel lengths in the nanometer regime," *IEEE Trans. Electron Device*, vol. 59, no. 1, pp. 197–205, 2012.

[36] S. N. Ong, K. S. Yeo, K. W. J. Chew, L. H. K. Chan, X. S. Loo, C. C. Boon, and M. A. Do, "Impact of velocity saturation and hot carrier effects on channel thermal noise model of deep sub-micron MOSFETs,"

Solid-State Electronics, vol. 72, pp. 8–11, Jun. 2012.

[37] R. P. Jindal, "High frequency noise in fine line NMOS field effect transistors," in *IEDM Tech Dig.*, 1985, pp. 68–71.

[38] BSIM4 Compact Model. [Online]. Available: http://www-device.eecs. berkeley.edu/bsim/?page=BSIM4

[39] Virtuoso Spectre Circuit Simulator. [Online]. Available: http://www.cadence.com/products/rf/spectre_circuit/pages/default.aspx

[40] PSP Compact Model. [Online]. Available: http://www.nxp.com/wcm_documents/models/mos-models/model-psp/psp103p2_summary.pdf

[41] HiSIM Compact Model. [Online]. Available: http://www.hisim.hiroshima-u.ac.jp/

[42] EKV Compact Model. [Online]. Available: http://ekv.epfl.ch/

[43] HSPICE Circuit Simulator. [Online]. Available: http://www.synopsys.com/tools/Verification/AMSVerification/CircuitSimulation/HSPICE/Pages/default.aspx

[44] Eldo Circuit Simulator. [Online]. Available: http://www.mentor.com/products/ic_nanometer_design/analog-mixed-signal-verification/eldo/

[45] C. G. Sodini, P. K. Ko, and J. L. Moll, "The effect of high fields on MOS device and circuit performance," *IEEE Trans. Electron Devices*, vol. ED-31, no. 10, pp. 1386–1393, Oct. 1984.

[46] P. K. Ko, R. S. Muller, and C. Hu, "A unified model for the hot–electron currents in MOSFETs," in *IEDM Tech. Dig.*, 1981, pp. 600–603.

[47] A. van der Ziel, *Noise: Sources, Characterization, Measurement.* Englewood Cliffs, NJ: Prentice-Hall, 1970.

[48] C. H. Chen, F. Li, and Y. Cheng, "MOSFET drain and induced-gate noise modeling and experimental verification for RF IC design," in *IEEE International Conference on Microelectronics Test Structures*, Awaji, Japan, 2004, pp. 51-56.

[49] Y. Cheng, M. J. Deen, and C. H. Chen, "MOSFET modeling for RFIC design," *IEEE Trans. Electron Devices*, vol. 52, pp. 1286–1303, 2005.

[50] W. S. Kwan, C. H. Chen, and M. J. Deen, "Hot-Carrier Effects on RF Noise Characteristics of LDD NMOSFET," *Journal of Vacuum Science and Technology A*, vol. 18(2), pp. 765–769, March/April 2000.

[51] S. Naseh, M. J. Deen, and C. H. Chen, "Effects of hot-carrier stress on the performance of CMOS low noise amplifiers," *IEEE Trans. Device and Materials Reliability*, vol. 5, no. 3, pp. 501–508, Sept. 2005.

[52] S. Naseh, M. J. Deen, and C. H. Chen, "Hot-carrier reliability of submicron NMOSFETs and integrated NMOS low noise amplifiers," *Microelectronics Reliability*, vol. 46, pp. 201–212, 2006.

[53] H. E. Halladay and A. van der Ziel, "On the high frequency excess noise and equivalent circuit representation of the MOS-FET with *n*-type channel," *Solid-State Electronics*, vol. 12, pp. 161–176, 1969.

[54] Y. Tsividis and C. McAndrew, *Operation and Modeling of the MOS Transistor*, 3rd ed., New York, Oxford University Press, 2011.

[55] M. V. Dunga, X. Xi, J. He, W. Liu, K. M. Cao, X. Jin, J. J. Ou, M. Chan, A. M. Niknejad, and C. Hu, *BSIM4.6.0 MOSFET Model – User's Manual*, Department of Electrical Engineering and Computer Sciences, University of California, Berkeley, CA, 2006.

[56] T. Yamada, Z. Jing-Rong, H. Miyata, and D. K. Ferry, "In-plane transport properties of Si/Si Ge structure and its FET performance by computer simulation," *IEEE Trans. Electron Devices*, vol. 41, no. 9, pp. 1513–1522, Sept. 1994.

[57] T. Krishnamohan, Z. Krivokapic, K. Uchida, Y. Nishi, and K. C. Saraswat, "High-mobility ultrathin strained Ge MOSFETs on bulk and SOI with low band-to-band tunneling leakage: experiments," *IEEE Trans. Electron Devices*, vol. 53, no. 5, pp. 990–999, May 2006.

[58] S. M. Sze, *Physics of Semiconductor Devices*. John Wiley & Sons, 2nd edition, 1985, p. 46.

[59] L. Chang, M. Ieong, and Min Yang, "CMOS circuit performance enhancement by surface orientation optimization," *IEEE Trans. Electron Devices*, vol. 51, no. 10, pp. 1621–1627, Oct. 2004.

[60] S. Mayuzumi, S. Yamakawa, D. Kosemura, M. Takei, K. Nagata, H. Akamatsu, K. Aamari, Y. Tateshita, H. Wakabayashi, M. Tsukamoto, T. Ohno, M. Saitoh, A. Ogura, and N. Nagashima, "Comparative study between Si (110) and (100) substrates on mobility and velocity enhancements for short-channel highly-strained PFETs," in *Proc. 2009 Symposium on VLSI Technology*, pp. 14–15.

[61] G. Dewey, M. K. Hudait, K. Lee, R. Pillarisetty, W. Rachmady, M. Radosavljevic, T. Rakshit, and R. Chau, "Carrier transport in high-mobility III–V quantum-well transistors and performance impact for high-speed low-power logic applications," *IEEE Electron Device Lett.*, vol. 29, no. 10, pp. 1094–1097, Oct. 2008.

978-1-4673-6145-3/13 $31.00 © 2013 IEEE

A Model-Agnostic Technique for Simulating Per-Element Distortion Contributions

Nagendra Krishnapura and Rakshitdatta K. S.

Department of Electrical Engineering, Indian Institute of Technology Madras, Chennai 600036, India.

Abstract—**The nonlinearity of an element can be altered while retaining the original operating point and first-order terms by appropriately combining two instances of the nonlinear element with complementary scaling factors for incremental voltages above the operating points. Per-element distortion contributions in a circuit can then be determined by altering the nonlinear terms by known factors and simulating the output distortion in each case. This technique can be used in a standard circuit simulator with the appropriate nonlinear device models but requires no knowledge of the device model details on the part of the circuit designer. The technique is demonstrated by applying it to a common source amplifier with a nonlinear load and a two stage fully differential opamp.**

I. MOTIVATION

Distortion due to nonlinearity and noise due to inherent randomness are the most important disturbances in signal processing circuits. Circuits must be designed such that these are kept below certain specified levels. It would be convenient to determine individual contributions of these disturbances from different blocks or components to the overall output so that the circuit can be suitably optimized. Determining noise contributions from individual components is routinely done in standard circuit simulators. For distortion though, no such facility is available. The total output distortion can however be determined easily by running a transient or periodic steady state analysis with nonlinear device models.

To determine the distortion or noise contribution of each device, one needs to know the equivalent nonlinear distortion or noise source in each component, and the transfer function from that source to the output. The difficulty in resolving individual distortion contributions is that, unlike in the case of noise, it is not straightforward to calculate the equivalent distortion source at the device level. If the nonlinear device were described by a Taylor or Volterra series in the port variables, the nonlinear source would consist of the higher order terms in the series. In practice, the device models are a lot more complicated and not described in closed form.

In this paper, we present a technique that bypasses both these steps of explicitly computing the distortion source of the device or the transfer functions to the output. It is shown that by running multiple simulations of the total output distortion of a circuit with slightly changed nonlinear characteristics in the relevant element, the contribution of the element to the output distortion can be determined. *Obtaining a device with changed*

E-mail: nagendra@iitm.ac.in; This work was supported in part by Texas Instruments, Bangalore, India.

nonlinear characteristics is based on elementary circuit theory and requires no knowledge of the device model.

In the next section, we review previously available techniques for determining individual distortion contributions. In Section III, we show how to synthesize a new nonlinear element which has the same operating point and linear characteristics, but different nonlinear characteristics. In Section IV, we show how to use this element with scalable nonlinear terms to obtain individual distortion contributions. In Section V the proposed technique is verified by applying it to several examples. Section VI concludes the paper.

II. EXISTING METHODS FOR DETERMINING INDIVIDUAL CONTRIBUTIONS TO DISTORTION

In case of a cascade of open loop stages, one can simulate the distortion of the stages individually to determine their contributions. In cases where a stage is significantly loaded by the following one, it is harder to isolate the contributions. For closed loop systems, which are frequently used for low distortion applications, this method is not applicable.

The probing method described in [1] successively evaluates higher order nonlinear contributions by injecting additional nonlinear sources to the linear equivalent circuit. This is further developed or simplified for analog integrated circuits in, e.g., [2], [3], [4], [5] for analysis and to gain insight into distortion behavior of circuits. All of these are based on Taylor or Volterra series descriptions of the circuit components, which, as pointed out earlier, are not readily available, and have to be extracted by the designer. The specific question of systematically determining per-element distortion contributions is addressed in [6]. This describes an algorithm that can be used in a simulator to determine per-element contributions and cannot be used by a circuit designer running a conventional SPICE-like simulator.

[7] circumvents the extraction of nonlinear device models by using an appropriate multi-sine excitation with which one can determine the equivalent additional distortion source of each element. These sources are used in conjunction with small-signal transfer functions determined by ac/noise analysis to identify the distortion contribution from that particular element. This is essentially a missing tone test, and it may not be easy to relate this to conventional single-tone harmonic distortion and two-tone intermodulation distortion tests. It also involves choosing an appropriate multi-tone input signal which entails additional labor.

978-1-4673-6145-3/13 $31.00 © 2013 IEEE

III. Obtaining a Device with Scaled Nonlinearity

The proposed technique is based on substituting the nonlinear element in the circuit by another nonlinear element whose operating point and first order behavior are the same but whose nonlinearity is different. For simplicity, the principle is

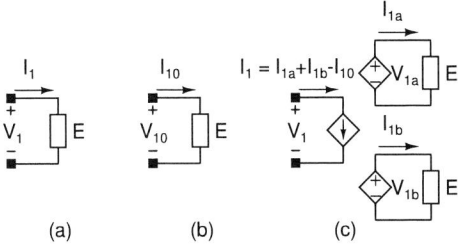

Fig. 1. (a) Nonlinear element E, (b) Operating point, (c) Nonlinear one port element constructed from two instances of E driven by V_{1a} and V_{1b}.

first illustrated with a memoryless one port element. Fig. 1(a) shows a nonlinear element E with a current-voltage relationship $I_1 = f(V_1)$. Fig. 1(b) shows the same element E at a certain operating point (V_{10}, I_{10}). Defining incremental voltage v_1 and current i_1 respectively as $v_1 = V_1 - V_{10}$ and $i_1 = I_1 - I_{10}$ and expanding the nonlinear relationship in a Taylor series around the operating point, we get

$$I_1 = f(V_{10}) + \left.\frac{df}{dV_1}\right|_{V_{10}} v_1 + \frac{1}{2!}\left.\frac{d^2 f}{dV_1^2}\right|_{V_{10}} v_1^2 + \frac{1}{3!}\left.\frac{d^3 f}{dV_1^3}\right|_{V_{10}} v_1^3 + \cdots \tag{1}$$

The first term is the operating point, the second term is the linear part, and successive terms are nonlinearities.

Now consider the one port in Fig. 1(c) which is constructed from two instances of elements E driven with voltages V_{1a} and V_{1b}. These voltages are related to V_1, the voltage across the one port, as follows:

$$V_{1a} = V_{10} + a_1(V_1 - V_{10}), V_{1b} = V_{10} + (1 - a_1)(V_1 - V_{10}) \tag{2}$$

where a_1 is a scaling factor. In other words, the two copies of E experience differently scaled versions of the incremental voltage $v_1 = (V_1 - V_{10})$ applied to the one port. The current I_1 in the new one port element is defined as $I_1 = I_{1a} + I_{1b} - I_{10}$. Using Taylor series expansions for $I_{1a} = f(V_{1a})$ and $I_{1b} = f(V_{1b})$ we get

$$
\begin{aligned}
I_1 = {} & f(V_{10}) + \left.\frac{df}{dV_1}\right|_{V_{10}} v_1 + \left(a_1^2 + (1-a_1)^2\right)\frac{1}{2!}\left.\frac{d^2 f}{dV_1^2}\right|_{V_{10}} v_1^2 \\
& + \left(a_1^3 + (1-a_1)^3\right)\frac{1}{3!}\left.\frac{d^3 f}{dV_1^3}\right|_{V_{10}} v_1^3 + \cdots
\end{aligned}
\tag{3}
$$

It is clear from equations (1) and (3) that the nonlinear one port in Fig. 1(c) has the same operating point and linear terms as the one port in Fig. 1(a), but scaled nonlinear terms. The Nth order term in the series is scaled by $a_1^N + (1 - a_1)^N$.

This reasoning can be easily extended to two or more ports. Fig. 2(a) shows a two port E with voltage V_1, V_2 and current I_1, I_2. Fig. 2(b) shows the operating point condition. Fig. 2(c) shows a new two port network constructed from two

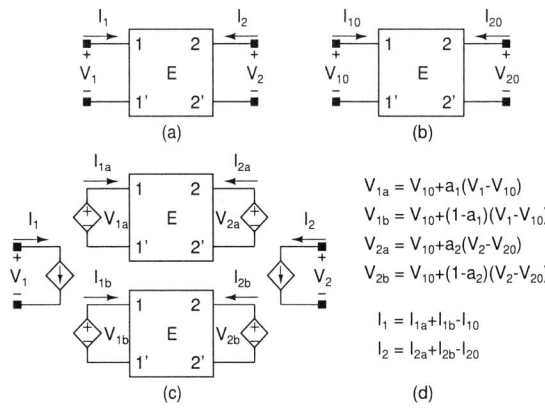

Fig. 2. (a) Nonlinear two port E, (b) Operating point, (c) Nonlinear two port constructed from two instances of E driven by $V_{1a,2a}$ and $V_{1b,2b}$.

instances of E which receive scaled versions of the incremental voltages above the operating point. The voltages applied to the two ports and the port currents are given by the relationships in Fig. 2(d). Using similar reasoning as with the one port, it is clear that I_1 and I_2 of the composite two port network in Fig. 2(c) consist of the same operating point and first order terms as in the original two port in Fig. 2(a), but have scaled higher order terms. Table I lists the scaling factors for second and third order terms in I_1 and I_2. The pattern for higher order terms is obvious.

TABLE I

Scaling Factors for Nonlinear Terms of the Two Port

Second order	Scaling factor	Third order	Scaling factor
v_1^2	$a_1^2 + (1-a_1)^2$	v_1^3	$a_1^3 + (1-a_1)^3$
$v_1 v_2$	$a_1 a_2 + (1-a_1)(1-a_2)$	$v_1^2 v_2$	$a_1^2 a_2 + (1-a_1)^2(1-a_2)$
v_2^2	$a_2^2 + (1-a_2)^2$	$v_1 v_2^2$	$a_1 a_2^2 + (1-a_1)(1-a_2)^2$
		v_2^3	$a_2^3 + (1-a_2)^3$

In an M port network, one would need M scaling factors $a_1, \ldots a_M$ to scale the incremental port voltages $v_1, \ldots v_M$. The Nth order nonlinear term in the scaled network will be of the form $\prod_{k=1}^{M} v_k^{l_k}$ where $0 \le l_k \le N$. The scaling factor for this term would be $\prod_{k=1}^{M} a_k^{l_k} + \prod_{k=1}^{M} (1 - a_k)^{l_k}$. Though Taylor series expressions are used above for simplicity, the method is equally applicable to nonlinearity with memory. An Nth order term of the Volterra series of an M port network in which v_k appears l_k times $(0 \le l_k \le N)$ will be scaled by the same factor $\prod_{k=1}^{M} a_k^{l_k} + \prod_{k=1}^{M} (1 - a_k)^{l_k}$.

IV. Determining an Element's Contribution

Fig. 3(a) shows a circuit with a nonlinear two-port element E whose contribution to distortion has to be determined. For clarity of discussion, we consider a single sinusoidal input voltage V_s, a voltage V_{out} as the output, and a two port E which has nonlinear terms only up to the second order. But the technique is general and works in the same way for multi-tone

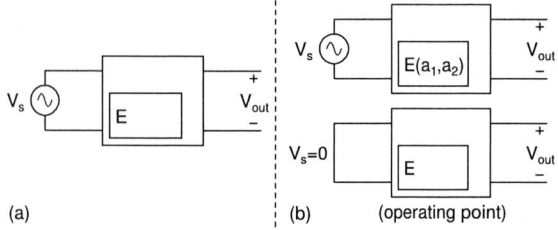

Fig. 3. (a) Original circuit with element E, (b) Circuit with E replaced by its scaled version and a copy of the original circuit at the operating point.

inputs, currents instead of voltages, and multi-port nonlinear elements with higher orders of nonlinear terms.

First, the distortion H_{out}^0 of the circuit in Fig. 3(a) is simulated. H_{out}^0 stands for any distortion component of interest in the output V_{out}—harmonic distortion of a given order, or the total harmonic distortion, or in case of multi-tone excitation, the relevant intermodulation component(s). H_{out}^0 could be the frequency domain representation (Fourier transform magnitude and phase) of the distortion components or the time domain distortion waveform. Here we will assume the former.

Then, the schematic in Fig. 3(b) is generated from the original schematic. It consists of the circuit with the nonlinear element E replaced by its scaled version with scaling factors a_1, a_2. The scaled element E requires the operating point information. Therefore, a copy of the original circuit in quiescent condition is included in the schematic from which the operating point information is extracted[1]. The distortion $H_{out}^{a1,a2}$ is simulated in this scaled circuit.

Let the distortion contributed to the output in the original circuit (Fig. 3(a)) by the v_1^2, $v_1 v_2$, and v_2^2 terms of the element E be denoted by $H_{v_1^2}$, $H_{v_1 v_2}$, and $H_{v_2^2}$ respectively. Let H_{rest} denote the distortion contributed by the rest of the circuit. Then, the total distortion H_{out}^0 in the original circuit is given by

$$H_{out}^0 = H_{v_1^2} + H_{v_2^2} + H_{v_1 v_2} + H_{rest} \tag{4}$$

When the nonlinear element E is scaled by a_1, a_2, the distortion contributed by the v_1^2, $v_1 v_2$, and v_2^2 terms will be scaled as shown in Table I. The distortion contributed by the rest of the circuit is not changed (This assumption holds when the elements are weakly nonlinear and the distortion is small). The output distortion $H_{out}^{a1,a2}$ is therefore:

$$H_{out}^{a1,a2} = \left(a_1^2 + (1-a_1)^2\right) H_{v_1^2} + \left(a_2^2 + (1-a_2)^2\right) H_{v_2^2} + (a_1 a_2 + (1-a_1)(1-a_2)) H_{v_1 v_2} + H_{rest} \tag{5}$$

Simulating the circuit in Fig. 3(b) for three different combinations of a_1, a_2, yields us four equations—(4) and three cases of

---

[1]For convenience, the technique is illustrated with a duplicated circuit for the operating point. But this duplication is not essential. As an alternative, the operating point could be simulated first and appropriate information could be fed to the scaled network E. Alternatively, one could, in transient simulation, initially deactivate the input source and set the scaling factors to unity. This yields the operating point information of the original circuit. After a certain delay, these values could be sampled and held and fed to the scaled network, the input signal activated, and the scaling factors set to the desired values.

(5)—from which the four unknowns $H_{v_1^2}, H_{v_2^2}, H_{v_1 v_2}$, and H_{rest} can be determined.

If one is interested only in the total contribution from second order terms $H_{v_1^2, v_2^2, v_1 v_2} = H_{v_1^2} + H_{v_2^2} + H_{v_1 v_2}$ of E and not in individual contributions from each second order term, one can set $a_2 = a_1$. This results in

$$H_{out}^{a1,a1} = \left(a_1^2 + (1-a_1)^2\right) H_{v_1^2, v_2^2, v_1 v_2} + H_{rest} \tag{6}$$

In this case, the circuit in Fig. 3(b) needs to be simulated only for one value of a_1 to can determine $H_{v_1^2, v_2^2, v_1 v_2}$ and H_{rest}.

In the above, we assumed that the element E has nonlinear terms only up to the second order. If the element E has nonlinear terms up to N^{th} order, N circuit simulations are required to determine the total contribution from each of the $N-1$ nonlinear terms. If it is further required to resolve the distortion into separate terms in each order, $N(N+3)/2 - 1$ simulations with distinct combinations of a_1, a_2 are required. The number of significant nonlinear terms has to be initially determined by trial and error. Apriori knowledge that some terms are insignificant (e.g. even order terms in a fully differential two port) can reduce the number of simulations.

V. EXAMPLES

A. Common source amplifier with a resistive load

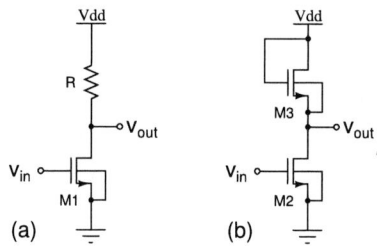

Fig. 4. Common source amplifier with (a) resistive load (b) diode connected load.

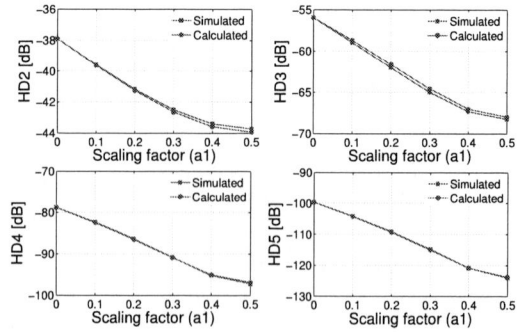

Fig. 5. Output distortion components in Fig. 4(a) with scaling.

Fig. 4(a) shows a common source amplifier with a resistive load. This circuit has a single nonlinear component, the transistor, and is used to verify that components of different order scale as described in the previous section. For input amplitudes such that there is negligible compression, a nonlinear

978-1-4673-6145-3/13 $31.00 © 2013 IEEE

term of a given order contributes only to a harmonic of the same order. Fig. 5 shows the output distortion components for a 20 mV peak input signal for different values of a_1. Good agreement is seen between the expected scaling factor $a_1^N + (1 - a_1)^N$ and that obtained from simulation.

B. Common source amplifier with a MOS transistor load

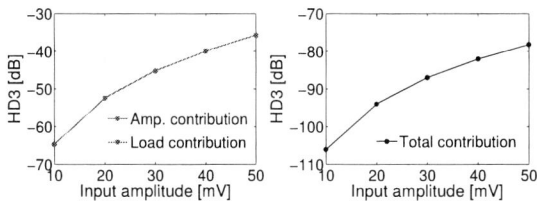

Fig. 6. Distortion contributions in Fig. 4(b).

Fig. 4(b) shows a common source amplifier with an nMOS diode connected load. Since the load is a replica of the amplifier, nonlinearities should cancel. The amplifier device is scaled by $a_1 = 0.01$ and the distortion contributions from the amplifier and the load are calculated using the method in Section IV. Fig. 6 shows the total (third harmonic) distortion and the contributions from each device. It can be seen that the contributions from the two components are almost equal in magnitude. Simulation results show that these contributions have opposite phase, which leads to cancellation. This can also be seen from the fact that the total output distortion is $\sim$40 dB below the contribution from either device. Extracting the contributions using a different value of the scaling factor $a_1 = 0.02$ yields the same results.

C. Opamp in closed loop

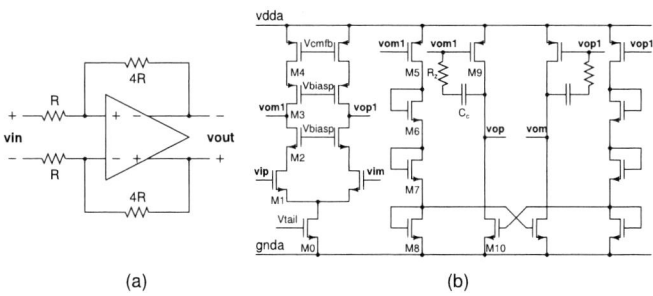

Fig. 7. (a) 4$\times$ inverting amplifier (b) Circuit diagram of the opamp in [8].

Fig. 7(a) shows an inverting amplifier of gain 4 with a bandwidth of the $\sim$3 MHz. Fig. 7(b) shows the opamp used in the amplifier [8]. Distortion of the second stage is extracted by scaling M_{5-10} by the same factor $a_1 = 0.01$. Similarly, contributions from different sections of the circuit are extracted by scaling all the transistors in the corresponding section. Fig. 8 shows the original distortion (third harmonic), contributions from different stages, and the corresponding sum versus input frequency for a 10 mV peak input signal. It can be seen that the second stage and the common mode feedback circuitry for the first stage are the major contributors to the output distortion.

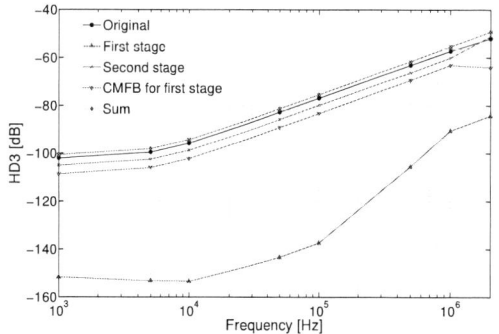

Fig. 8. Distortion versus frequency for the amplifier in Fig. 7.

VI. CONCLUSIONS

A method has been proposed by which a circuit designer can conveniently determine distortion contributions from different elements and from different terms of each element without going into device model details. It does not require the extraction of Taylor or Volterra series models for individual circuit elements. The technique is demonstrated by applying it to a common source amplifier with linear and nonlinear loads and a closed loop amplifier.

Since the modified element is based on instances of the original element, and not an abstracted model, the proposed technique also allows one to determine distortion contributions with process or temperature variations. This is in contrast to [2], [3], [4], [5] where the Taylor/Volterra model at each corner has to be determined. The technique can be applied to blocks (e.g. stages of an opamp, or the entire opamp) as well as to individual transistors to produce a hierarchical listing of distortion contributions. Because of its ease of use, the proposed technique can serve as a convenient tool for optimizing distortion or investigating the robustness of distortion cancellation schemes.

REFERENCES

[1] J. J. Bussgang et al., "Analysis of nonlinear systems with multiple inputs," *Proc. IEEE*, vol. 62, no. 8, pp. 1088-1119, Aug. 1974.

[2] P. Wambacq and W. Sansen, *Distortion Analysis of Analog Integrated Circuits*, Kluwer, 1998.

[3] P. Wambacq et al., "Symbolic simulation of harmonic distortion in analog integrated circuits with weak nonlinearities," *Proc. 1990 ISCAS*, May 1990, pp. 536-539.

[4] B. Hernes and W. Sansen, "Distortion in single-,two- and three-stage amplifiers," *IEEE Transactions on Circuits and Systems-I: Regular Papers*, vol. 52, no. 5, May 2005.

[5] S. O. Cannizzaro et al., "Distortion analysis of miller-compensated three-stage amplifiers," *IEEE Transactions on Circuits and Systems-I: Fundam. Theory Appl.*, vol. 53, no. 5, May 2006.

[6] P. Li and L. T. Pileggi, "Efficient per-nonlinearity distortion analysis for analog and RF circuits," *IEEE Transactions on Computer-Aided Design of Integrated Circuits and Systems*, vol. 22, no. 10, pp. 1297-1309, Oct. 2003.

[7] A. Cooman et al., "Determining the dominant nonlinear contributions in a multistage op-amp in a feedback configuration", *Proc. 2012 SMACD*, Sep. 2012, pp. 205-208.

[8] S. Pavan et al., "A power optimized continuous-time delta-sigma modulator for audio applications," *IEEE Journal of Solid State Circuits*, vol. 43, no. 2, pp. 351-360, Feb. 2008.

Corner Models: Inaccurate at Best, and it Only Gets Worst ...

Colin C. McAndrew, Ik-Sung Lim, Brandt Braswell, and Doug Garrity
Freescale Semiconductor, Tempe, AZ 85284

Abstract— Corner (best- and worst-case) models have been a mainstay of integrated circuit design for decades. Obviously they can be effective, especially for digital CMOS design. However, there are significant inaccuracies that arise when digital CMOS corner models are used for analog circuits, or any types of circuits or measures of circuit performance they were not targeted for. This paper details what corner models can and cannot do, and shows their inadequacies for analog CMOS circuits.

I. INTRODUCTION

Semiconductor manufacturing processes are inherently stochastic; the electrical performance of devices depends on global variation (between die, wafers, and lots), local variation (i.e. mismatch, uncorrelated atomistic variation of each device instance), and layout dependent effects (which are not considered here). Historically, the first of these was recognized and understood earliest and was mitigated in design by analysis using "corner" or "worst-case" models. The relative contribution of local variation for minimum geometry devices has been greater than that of global variation since about the 130 nm technology node, and although this should invalidate the use of corner models, which assume perfect correlation between devices (i.e. zero local variation), they are still widely used— the local variability averages out over the large number of transistors in digital CMOS circuits.

Several approaches are commonly used to generate corner models: by introducing fixed, often $\pm 3\sigma$, variations in model parameters; by adjusting process parameters so models give specified values of device electrical performances [1]; or by computing model parameters to give extreme values of one or more circuit performances [2].

Whatever methodology is used to generate the corner models, there is a widespread lack of understanding of what they can, but more importantly cannot, do. This is especially true when digital corner models are used to simulate analog circuits. The quantitative inadequacies of digital corner models for analog were presented in [3], but were not formally published, and for the average duty cycle of a VCO in [4]. In this paper we: build on the analysis approach of [3], providing more extensive circuit level results; detail theoretically the underlying cause of the inadequacies; and provide examples that show the limitations of corner models for resistors and for MOS transistor gate capacitance.

II. CORNER MODELS FOR DIGITAL CMOS CIRCUITS

Digital CMOS circuits are topologically similar, in general use minimum channel length devices, and historically were designed to run at as high a clock frequency as possible. The slew rate for the switching transitions is

$$\frac{\partial V}{\partial t} = \frac{I}{C} \tag{1}$$

so corner models that give extreme values for this, i.e. that maximize I and minimize C, and vice versa, can do a reasonable job of bracketing the manufacturing variation of digital circuit speed.

Serendipitously, the other key performance measures that are important for digital circuits, leakage and power, are highly correlated with speed: fast chips have high leakage and power, slow chips have low leakage and power. Digital corner models that target speed variation therefore also do a reasonable job for leakage and power.

III. FUNDAMENTAL ISSUE WITH CORNER MODELS

The most important limitation of corner models is that they are defined by *fixed* variations in the parameters p_i. The variation in an arbitrary measure of circuit performance e_m is

$$\delta e_m = \sum_i \frac{\partial e_m}{\partial p_i} \delta p_i . \tag{2}$$

The sensitivities $\partial e_m / \partial p_i$ vary with circuit topology, device geometry, and device bias. Using fixed variations in δp_i independent of the sensitivities gives models that guarantee *nothing* about what the σ-level variation in e_m will be. Even if the linearization in (2) is not sufficiently accurate and higher order terms must be included that conclusion does not change.

Fundamentally, for each performance measure for each circuit the sensitivities $\partial e_m / \partial p_i$ are different, therefore corner models should be generated on a circuit performance-by-circuit performance basis, and can change as device geometries and biases change even for one circuit topology. Techniques for doing this are known [1], [2], [5], [6] but are not commonly used; fixed corner models are still the norm in the industry, and they have the basic limitation just described.

What are the practical consequences of this limitation? For digital CMOS circuits the short answer is: not much. As discussed above digital CMOS circuits have similar topologies and use minimum device channel lengths, and have a small number of important measures of circuit performance that are highly correlated. Not surprisingly, the corner models that target these can be quite reasonable, with the caveat that these days care must be taken to appropriately account for both global and local variation. For analog circuits the answer is: it depends. We now show why.

978-1-4673-6145-3/13 $31.00 © 2013 IEEE

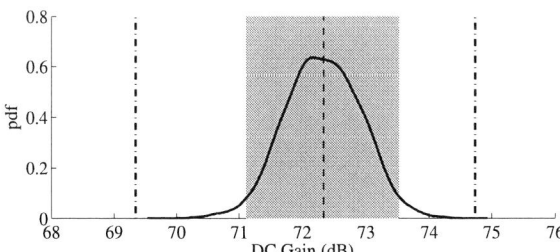

Fig. 1. DC gain. Solid curve is MC simulation result; shaded region is digital CMOS corners; dash lines are capacitance-only corners; dash-dot lines are MOS transistor plus capacitance plus resistance corners (the same line styles are also used for Figs. 2 through 8).

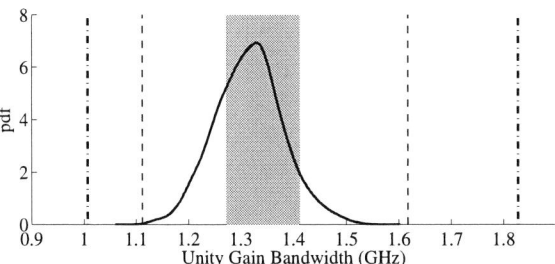

Fig. 2. Unity-gain bandwidth.

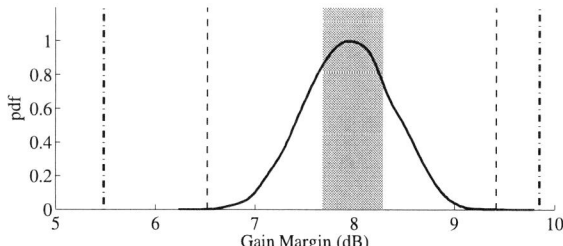

Fig. 3. Gain margin.

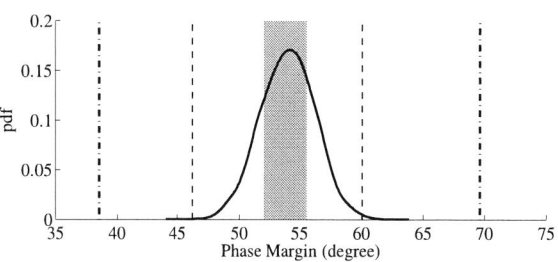

Fig. 4. Phase margin.

IV. OPERATIONAL AMPLIFIER ANALYSIS

To evaluate the accuracy of digital corner models for analog circuits we have analyzed a 2-stage CMOS differential operational amplifier in a 90 nm CMOS technology. The circuit comprises 48 MOS transistors, with RC Miller compensation and resistively sensed common mode feedback. The models included 3 separate corners each (min/typ/max) for resistors and capacitors, and 9 MOS transistor corners, giving a total of 81 possible corner combinations. We ran all of these corner combinations, and also a 1000 sample Monte Carlo (MC) simulation for global-only variation. MC simulation was also done including local variation but the results were similar (the devices in the circuit are all fairly large) and so are not shown. Layout dependent effects were not taken into account, but do not affect conclusions drawn from comparison of corner and MC simulations. Figs. 1 through 8 show the results for a variety of important circuit performance measures. The probability density functions are kernel density estimates [7] from the MC samples, and lower and upper values are shown for the range predicted by: all 81 MOS transistor plus resistor plus capacitor corner combinations; the capacitance-only corners; and the MOS transistor-only (i.e. digital CMOS) corners.

The DC gain (Fig. 1) is sensitive to the transconductance g_{m} and output resistance r_{o} of the MOS transistors, and so the MC distribution is reasonably well bracketed by the MOS-only corner simulations. For most other figures of merit, settling time being the exception, the MOS-only corners, which target digital circuit speed, do a woefully inadequate job of predicting the variability. For many figures of merit (unity-gain bandwidth UGBW, gain margin, phase margin, and output IP$_3$) the variability is reasonably well predicted by just using capacitance-only corners; for all but the last of these this

could expected because they are primarily determined by the value of the compensation capacitors. Conversely, the DC gain and common-mode rejection ratio have no sensitivity to the capacitance corners, also as expected. In all cases the extreme values from all corner combinations significantly overestimate the variability cf. the MC simulated distribution, but by a variable amount; there is not a fixed "over-design" margin.

The results in [3] did not include slew rate, settling time, common-mode rejection ratio, or IP$_3$, but did include DC gain, gain margin, phase margin, and UGBW (Figs. 1 through 4). For the first three of these our results are consistent with those of [3], but for UGBW our results indicate that the digital corners significantly underestimate the variation cf. MC, whereas in [3] they significantly overestimate it.

The conclusion is clear: as claimed in the previous section, corner models defined by fixed variations δp_i in the process parameters, independent of the sensitivities $\partial e_m / \partial p_i$, guarantee nothing about the level of variation they predict for figures of circuit merit e_m that they were not targeted for: Digital CMOS corners are not accurate for most performances for analog circuits.

V. PROBLEMS WITH CORNER MODELS FOR RESISTORS

Intuitively, the situation would not seem to be as dire for resistors; they are simple, nearly linear devices, so it should be easy to generate corner models that exactly bracket $\pm 3\sigma$ variations in resistance, right?

A simple model for the resistance R of a resistor is

$$R = \rho_{\mathrm{s}} \frac{L_{\mathrm{m}} + \Delta L}{W_{\mathrm{m}} + \Delta W} \qquad (3)$$

where ρ_{s} is sheet resistance, L_{m} and W_{m} are the mask (or drawn) length and width, respectively, and ΔL and ΔW are the difference between the mask and effective electrical length and width, respectively. In manufacturing there is statistical variability in ρ_{s}, ΔL and ΔW. Linearizing (3) and applying

978-1-4673-6145-3/13 $31.00 © 2013 IEEE

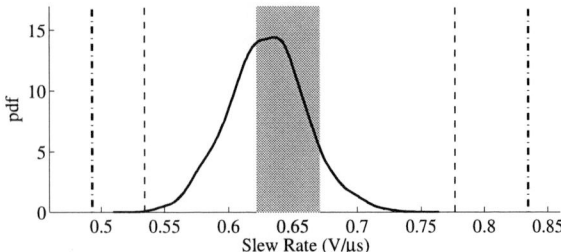

Fig. 5. Slew-rate.

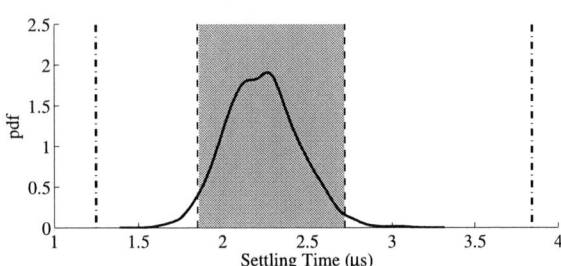

Fig. 6. Settling time.

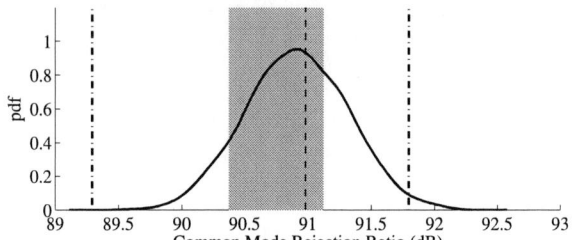

Fig. 7. Common-mode rejection ratio.

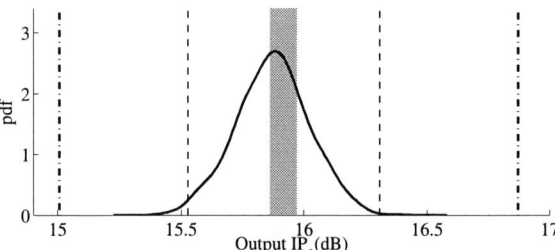

Fig. 8. Output 3rd order intercept point.

propagation of variance gives

$$\sigma^2_{\delta R/R} = \sigma^2_{\delta \rho_s/\rho_s} + \frac{\sigma^2_{\delta \Delta L}}{(L+\Delta L)^2} + \frac{\sigma^2_{\delta \Delta W}}{(W+\Delta W)^2} \quad (4)$$

so the variability in R is geometry dependent. This is what makes it impossible to generate exact corner models for resistors: It can be done for up to 3 geometries, but cannot be done for all geometries simultaneously.

Fig. 9 shows corner model and MC simulation results for a resistor in a 0.18 μm RF BiCMOS technology [8]. The corner models for the device were generated by setting each of ρ_s, ΔL, and ΔW to $\pm 3\sigma$ values. Clearly, this will give the correct $\pm 3\sigma$ bounds for R for wide and long resistors, whose resistance depends essentially only on ρ_s, but will overestimate the variability for wide or short resistors (by a factor of 35% for the narrowest resistor in Fig. 9).

A better approach to generating resistor corner models is to, from (4), compute $\sigma_{\delta R/R}$ for each of three different geometries (wide and long, narrow and long, and wide and short, for example) and use the backward propagation of variance procedure of [1] to compute $\delta \rho_s/\rho_s$, $\delta \Delta L$, and $\delta \Delta W$ so that the $\pm 3\sigma$ resistance values of the three geometry devices are fitted. This was the procedure that was used to generate the corner models for simulation of the results of Fig. 10, which are from a 0.25 μm power BiCMOS technology [9]; clearly the corner models accurately capture the overall resistance variation of both short and wide resistors, but there are still discrepancies (up to 12%) cf. accurate MC simulation results for intermediate widths.

VI. PROBLEMS WITH CORNER MODELS FOR C_{GG}

Some important circuit performances depend on capacitance; what do corner models say about the variability of MOS transistor gate capacitance C_{GG}? Consider operation in strong inversion nonsaturation, with $V_{\mathrm{DS}}=0$, so the channel is fully inverted along the length of the transistor and C_{GG} is

approximately its maximum value of LWC'_{ox}, where L and W are the effective electrical length and width of the transistor and $C'_{\mathrm{ox}} = \epsilon_{\mathrm{ox}}/t_{\mathrm{ox}}$ is the oxide capacitance per unit area, where ϵ_{ox} is the oxide permittivity and t_{ox} its thickness. For long channel transistors variations in C_{GG} are primarily from variations in t_{ox}; the worst-case corner has a thicker oxide (giving higher threshold voltage V_{T0} and lower g_{m}) and lower C_{GG} than the best-case corner. For short channel transistors, variations in C_{GG} are primarily from variations in effective channel length; the worst-case corner has a longer L (giving lower g_{m}) and higher C_{GG} than the best-case corner.

This implies that at some intermediate channel length the worst-case and best-case corners for C_{GG} could cross. Fig. 11 shows corner and MC simulation results for C_{GG} for a 55 nm CMOS technology. As predicted, for L_{m} slightly greater than 0.4 μm the corner models have absolutely no variability (the corner models for this transistor were specified to hit $\pm 3\sigma$ targets in t_{ox}, V_{T0}, g_{m}, and saturated drain current I_{Dsat}). On seeing this result the design team requested that the modeling team adjust the best- and worst-case corner models so that C_{GG} variation was properly bracketed for all channel lengths. But this is just not possible—adjusting the variations in δp_i to match the MC simulation limits for C_{GG} completely blew out the modeled windows for the other device performances to unrealistically large values.

VII. VARIANCES ADD, STANDARD DEVIATIONS DO NOT

Sometimes to gauge the overall variability of a circuit mismatch simulations are done on top of best- and worst-case (bcs and wcs, respectively) corner models. However, this is really adding global and local standard deviations together, which is incorrect. In fact variances add, and

$$\sigma_{\mathrm{total}} = \sqrt{\sigma^2_{\mathrm{global}} + \sigma^2_{\mathrm{local}}} < \sigma_{\mathrm{global}} + \sigma_{\mathrm{local}} \quad (5)$$

therefore running mismatch simulations on top of corner models will overestimate the variability, see Fig. 12.

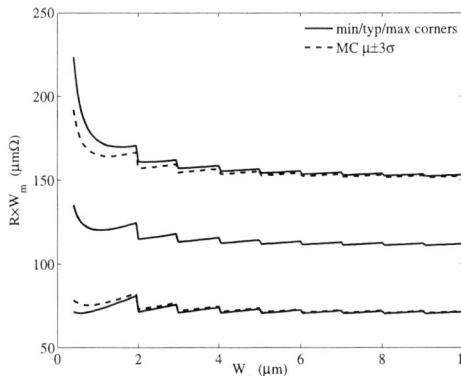

Fig. 9. Normalized resistance vs. width for corner models defined by $\pm 3\sigma$ perturbations in ρ_s, ΔL, and ΔW (the "cusping" is from the number of contacts incrementing discretely as the width increases; $L_m = 10.0$ μm).

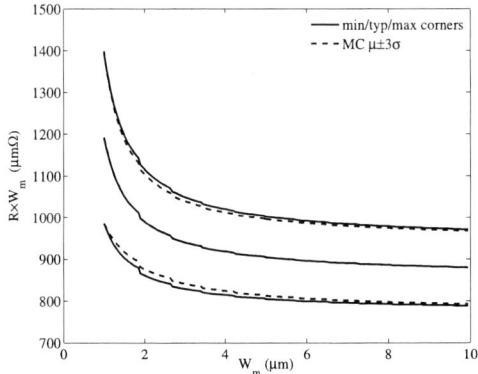

Fig. 10. Normalized resistance vs. width for corner models defined by $\pm 3\sigma$ perturbations in R (cusping is from the number of contacts incrementing discretely as the width increases; $L_m = 10.0$ μm.

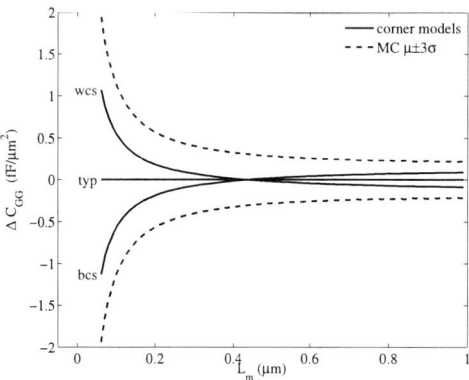

Fig. 11. Normalized capacitance vs. channel length; $\Delta C_{GG} = (C_{GG} - C_{GG,typ})/(L_m W_m)$, wcs and bcs denote worst- and best-case, respectively.

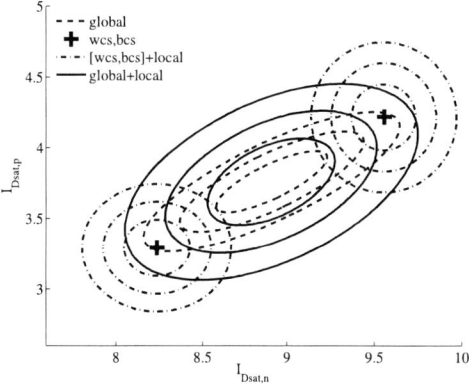

Fig. 12. Mismatch simulation on top of corner models overestimates variability; ellipses are 1-, 2-, and 3-σ yield regions computed from MC simulations.

A similar situation exists when a circuit performance depends on more than one type of device, for example resistors and capacitors as well as MOS transistors. Using combinations of bcs and wcs corners for all devices will over estimate the variability both because in effect this adds standard deviations, and not variances, and because even if variances were added then $\pm 3\sigma$ variations in n different types of devices leads to $\pm 3\sqrt{n}\sigma$ overall variation.

VIII. CONCLUSIONS

Corner models have been a mainstay of IC design for decades, so obviously are useful and can be effective. However, they *cannot* accurately bracket $\pm 3\sigma$ variation in every performance measure for every circuit; "appropriate" corner models are not just circuit dependent, they also vary with the device sizes and biases used within a single circuit, and can be different for different measures of circuit performance for the same circuit. It is often too time consuming to run MC simulations to characterize circuit variability, so corner models are and will remain fundamental to IC design. However, designers need to understand the limitations of digital CMOS corner models for measures of performance of analog and RF circuits; if necessary circuit performance specific corner models should be generated using the techniques presented in [1], [2], [5], [6]. Designer's should also have realistic expectations of what modeling teams are able to provide and not request physically unrealizable corner models; they should ask for additional, not wider, corners.

REFERENCES

[1] C. C. McAndrew, "Statistical modeling using backward propagation of variance," in *Compact Modeling: Principles, Techniques and Applications*, G. Gildenblat (Ed), Springer, pp. 491-520, 2010.

[2] F. Y. Chang, "Generation of 3-sigma circuit models and its application to statistical worst-case analysis of integrated circuit designs," *Rec. Asilomar Conf. Circuits, Systems and Computers*, pp. 29-34, 1977.

[3] J. Krick, "Statistical transistor SPICE modeling in advanced CMOS technologies," presentation at *Compact Modeling of Variability Workshop, IEEE/ACM ICCAD*, 2008.

[4] P. Yao, R. Trihy, J. Ge, K. Breen, and T. McConaghy, "Understanding and designing for variation in GLOBALFOUNDRIES 28-nm technology," *DAC*, 2013.

[5] S. R. Nassif, "Statistical worst-case analysis for integrated circuits," in *Statistical Approach to VSLI*, S. W. Director and W. Maly (Eds), North-Holland, pp. 233-253, 1994.

[6] A. N. Lokanathan and J. B. Brockman, "Efficient worst case analysis of integrated circuits," *Proc. IEEE CICC*, pp. 237-240, 1995.

[7] M. P. Wand and M. C. Jones, *Kernel Smoothing*, Chapman and Hall, 1995.

[8] J. P. John, F. K. Chai, D. Morgan, T. Keller, J. Kirchgessner, R. Reuter, H. Rueda, J. A. Teplik, J. White, S. Wipf, and D. Zupac, "Optimization of a SiGe:C HBT in a BiCMOS technology for low power wireless applications," *Proc. IEEE BCTM*, pp. 193-196, 2002.

[9] V. Parthasarathy, R. Zhu, V. Khemka, T. Roggenbauer, A. Bose, P. Hui, P. Rodriguez, J. Nivision, D. Collins, Z. Wu, I. Puchades, and M. Butner, "A 0.25μm CMOS based 70V smart power technology with deep trench for high-voltage isolation," *IEDM Tech. Digest*, pp. 459-462, 2002.

Energy Centric Model of SRAM Write Operation for Improved Energy and Error Rates

Swaroop Ghosh

Computer Science and Engineering, University of South Florida, Tampa, FL-33647, sghosh@cse.usf.edu

Abstract—We propose an energy centric model of SRAM operation. The model provides useful insights about energy and write error rates. We introduce the concept of intrinsic energy margin induced errors. The proposed model is employed for evaluating various write assist mechanisms and their potential in reducing the intrinsic memory error rates. We also demonstrate that this model can be used for optimizing energy of memory arrays.

I. INTRODUCTION AND MOTIVATION

Cache size is growing over the technology generations in order to boost the performance and maintain a constant miss rate in the era of multi-core design. In todays' microprocessor (both server and desktop) the amount of on-chip cache (typically 6T SRAM) is in the order of 50MB [1]. Naturally cache power is becoming an important component of total power consumption. Lowering minimum operating voltage (also known as "active Vmin") is an effective technique to reduce the power consumption. Under this operating model, read and write assist techniques are employed to lower the Vmin beyond the limits set by process variations. Few examples of read and write assist mechanisms [2-8] are wordline underdrive, bitcell voltage modulation (for read assist), negative bitline, WL boosting and supply voltage collapse (for write assist). The assist techniques suppress the read/write errors under process variation to extend Vmin scaling. Although effective, these techniques don't consider the impact of energy on errors. This may result into cases where energy spent on assist mechanism is more than the energy benefit obtained from Vmin reduction.

In this paper, we provide a novel energy centric perspective of the memory operation. Specifically, we propose an analytical framework to model intrinsic energy of memory during write operation in presence of noise sources (e.g., process variations). The model is validated with detailed Hspice simulation in 22nm PTM [9]. We introduce the concept of *intrinsic energy margin induced errors*. We demonstrate that energy centric perspective can provide us crucial insights for maximizing the energy efficiency of memories and reducing error rates. The proposed model answers several non-trivial optimization questions such as (a) how energy and error rates are related? (b) what is the best assist mechanism to make right tradeoff between energy and error rates? (c) how to improve the error rate under a given energy budget or how to improve the energy for a given error rate? *To the best of our knowledge this is the first treatment of SRAM optimization that incorporates energy perspective of the bitcell.*

The paper is organized as follows. The modeling methodology is explained in Section 2. The intrinsic energy model is presented in Section 3. Simulation results along with comparison of existing write assist techniques are also described in this section. Energy optimization methodology under a target write error rate is presented in Section 4. The conclusions are drawn in Section 5.

2. MODELING METHODOLOGY

In this section, first we present the intrinsic energy model of SRAM for write operation. Next we discuss the mathematical background for modeling of intrinsic energy and write speed.

2.1 Intrinsic Energy Model:

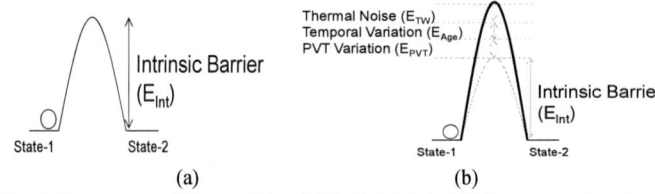

(a)   (b)

Fig. 1 Energy centric model of SRAM (a) intrinsic energy barrier in absence of noise and, (b) intrinsic energy barrier in presence of noise (e.g., PVT, aging and thermal noise). Intrinsic energy barrier increases due to noise making the bitcell hard to write.

In the intrinsic energy model of SRAM, the memory or storage element contains two states separated by an intrinsic energy barrier (E_{int}) as shown in Fig. 1(a). The barrier appears as a result of internal node capacitance and feedback mechanism (that is inherent in SRAM bitcell). Note that E_{int} is the barrier in absence of external noise. Write operation is performed by supplying sufficient energy to allow the bit to overcome its intrinsic barrier and jump to the other state. Read is performed by supplying sufficient energy to the bitcell in order to develop differential at the senseamp input that can be resolved correctly. The presence of noise sources (e.g., process variations, aging, thermal noise, voltage and temperature variations) impacts the memory operations due to fluctuations in energy barrier between memory states. Fig. 1(b) shows the worst case effective intrinsic energy barrier in presence of noise. Due to elevated energy barrier, some bits would require more energy to complete successful write operation. In other words, insufficient amount of write energy would result into write failures. In section 3, we propose analytical model for intrinsic energy barrier in presence of variations. In this analysis, we have only considered noise due to process variations.

2.2 Key Mathematical Background:

Let us consider a function $y=f(x_1,...,x_n)$ where $x_1,...,x_n$ are independent Gaussian random variables with mean $\mu_1,..., \mu_n$ and standard deviation (STD) $\sigma_1,...,\sigma_n$. The mean (μ_y) and the STD (σ_y) of the random variable y can be estimated by using multivariable Taylor-series expansion [12] as follows:

$$\mu_y = f(\mu_y, ..., \mu_y) + \sum_{i=1}^{n}\left[\frac{\partial^2 f(x_1,...,x_n)}{\partial (x_i)^2}\bigg|_{\mu_i}\frac{\sigma_i^2}{2}\right] \tag{1}$$

$$\sigma_y^2 = \sum_{i=1}^{n}\left[\left(\frac{\partial f(x_1,...,x_n)}{\partial (x_i)}\bigg|_{\mu_i}\right)^2\sigma_i^2\right] \tag{2}$$

The probability of $(y>Y_0)$ for a Gaussian probability distribution function (PDF) $[N_y(y:\mu_y,\sigma_y)]$ is given by

$$P(y > Y_0) = \int_{y=Y_0}^{\infty} N_y(y: \mu_y, \sigma_y)dy = 1 - \int_{y=-\infty}^{Y_0} N_y(y)dy \tag{3}$$

The mean and STD of N Gaussian random variables with mean μ_y and STD σ_y, is given by

$$\mu = N\mu_y; \sigma = \sigma_y\sqrt{N} \tag{4}$$

3. WRITE OPERATION MODEL

In this section, we employ the mathematical background presented above for modeling of intrinsic energy, write time and intrinsic energy induced error rate under process variations.

978-1-4673-6145-3/13 $31.00 © 2013 IEEE

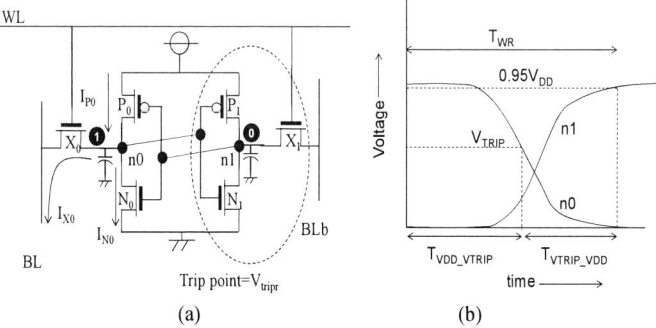

(a)   (b)

Fig. 2 (a) Schematic of SRAM during write operation (b) timing diagram showing two step write process

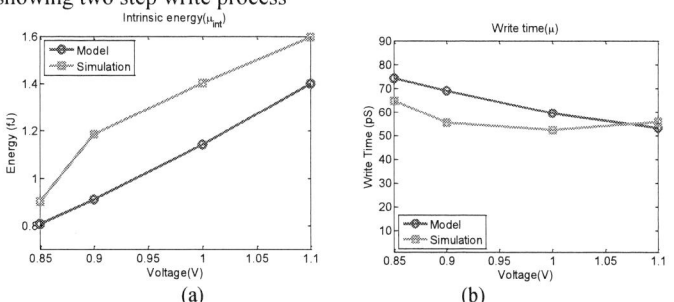

(a)   (b)

Fig. 3 Comparison of model and simulation for (a) Mean intrinsic energy and (b) mean write time.

3.1 Alpha Power Law Current Model:

We employ alpha power law model [10] to find the drain current through individual transistors of the SRAM cell. According to this law, the drain current I_D is given by

$$I_D = \begin{cases} 0 & (V_{GS} \leq V_{TH}: cutoff\ region) \\ (I'_{D0}/V'_{D0}) & (V_{DS} < V'_{D0}: triode\ region) \\ I'_{D0} & (V_{DS} \geq V'_{D0}: pentode\ region) \end{cases} \quad (5)$$

where

$$I_D = I'_{D0} \left(\frac{V_{GS}-V_{TH}}{V_{DD}-V_{TH}}\right)^{\alpha} \ and \ V'_{D0} = V_{D0} \left(\frac{V_{GS}-V_{TH}}{V_{DD}-V_{TH}}\right)^{\alpha/2} \quad (6)$$

The values of α, V_{D0}, I_{D0}, V_{TH} for pmos and nmos are extracted from 22nm PTM [9] devices of size 1u/0.03u.

3.2 Intrinsic Energy Barrier Model during Write:

Intrinsic energy (during write) is the amount of energy that is needed to charge one side of the cell all the way from 0 to V_{DD} and energy supplied when the other side is discharged from V_{DD} to 0. Alternatively, it can also be thought of as the energy that is supplied to discharge one side of the cell to the trip point of the inverter (V_{tripr}) that is driving the other side (Fig. 2(a)). Beyond this point, the feedback mechanism kicks-in and cell changes state. From Fig. 2(a) it

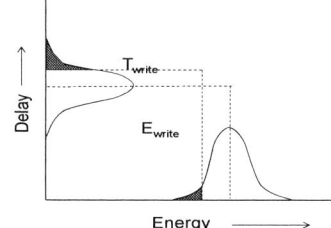

Fig. 4 Distribution of energy and delay. Errors during write operation can be due to insufficient write time or insufficient write energy (error regions are shaded in red).

can be observed that there are two contention mechanisms associated in this operation (i) contention from P0 and (ii) contention from N1. Finally the internal node needs to be charged beyond V_{tripr} to complete the switching. Note that the intrinsic energy is infinite if node n0 fails to reach the trip point under DC condition. The write time can be divided into two steps (a) T_{tripr} which is time required by node n0 to reach the trip point and (b) T_{tripr_VDD} which is the time required by node n1 to charge from trip point to approximately V_{DD}. Fig. 2(b) illustrates these two components. If node capacitance is C_n then T_{tripr} is given by

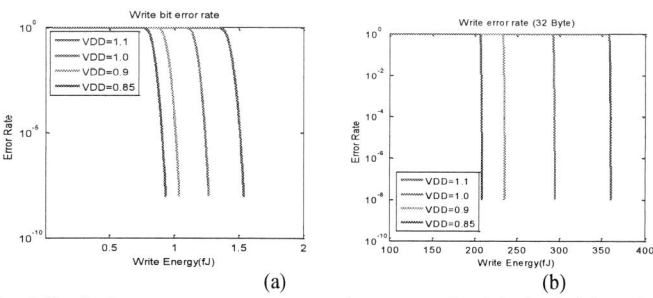

(a)   (b)

Fig. 5 Intrinsic energy error rate vs write energy for (a) single bit and (b) 32Byte.

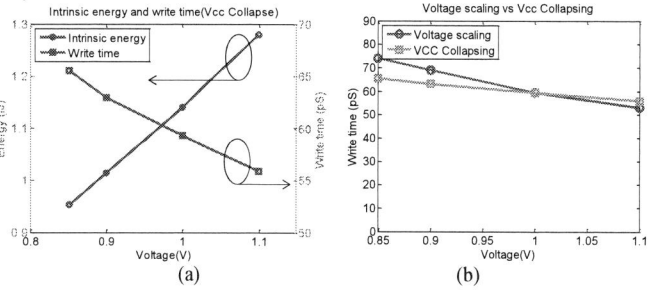

(a)   (b)

Fig. 6 (a) Intrinsic energy and write time with Vcc collapse. Energy barrier reduces but write time increases with Vcccol, (b) comparison of write time for V_{DD} scaling vs Vcc collapse. Vcccol performs better due to less timing impact compared to V_{DD} scaling.

$$T_{tripr} = \int_{V_{DD}}^{V_{tripr}} \frac{C_n dV}{I_{X0} - I_{P0} - I_{N0}} \quad (7)$$

In the above expression V_{tripr} is obtained by solving following equation

$$I_{P1}(V_G = V_D = V_{tripr}, V_S = V_{DD}) + I_{X1}(V_G = V_D = V_{DD}, V_S = V_{tripr}) = I_{N0}(V_G = V_D = V_{tripr}, V_S = V_{DD}) \quad (8)$$

The time required for P1-N1 to charge from V_{tripr} to V_{DD} is given by

$$T_{tripr\_VDD} = \int_{V_{trip1}}^{V_{DD}} \frac{C_n dV}{I_{P1} + I_{X1} - I_{N1}} \quad (9)$$

The energies supplied at node n0 and n1 are given by

$$E_{cnt\_n0} = V_{DD} \int_0^{T_{tripr}} I_{P0} dt + V_{DD} \int_{T_{tripr}}^{T_{tripr\_VDD}} I_{P0} dt$$

$$E_{cnt\_n1} = V_{DD} \int_0^{T_{tripr}} (I_{P1} + I_{X1}) dt + V_{DD} \int_{T_{tripr}}^{T_{tripr\_VDD}} (I_{P1} + I_{X1}) dt \quad (10)$$

The total energy (E_{int}, Fig. 1(a)) and write time (T_{WR}, Fig. 2(b)) is given by

$$E_{int} = E_{cnt\_n0} + E_{cnt\_n1} \ if \ V(n0) > V_{tripr}$$

$$= \infty \ otherwise \quad (11)$$

$$T_{WR} = T_{tripr} + T_{tripr\_VDD} \ if \ V(n0) > V_{tripr}$$

$$= \infty \ otherwise \quad (12)$$

The value of V(n0) in (11) and (12) is determined by solving KCL at node n0 and n1 in self consistent manner. In this work, we only consider the case when V(n0)>V_{tripr} under DC condition which is a reasonable assumption otherwise there will be hard functional failures. The success of write operation depends on providing sufficient amount of energy to overcome E_{int} within the allocated write time T_{WR}.

The framework for computing write intrinsic energy and write time is implemented in MATLAB and the results are compared against Hspice simulation. Fig. 3(a)-(b) compares the mean of write intrinsic energy and write time obtained from model and simulation. The error is found to be approximately 10%. We believe that the source of error is the mismatch in node n0/n1 capacitance that can be tuned further to improve the model accuracy.

978-1-4673-6145-3/13 $31.00 © 2013 IEEE   681

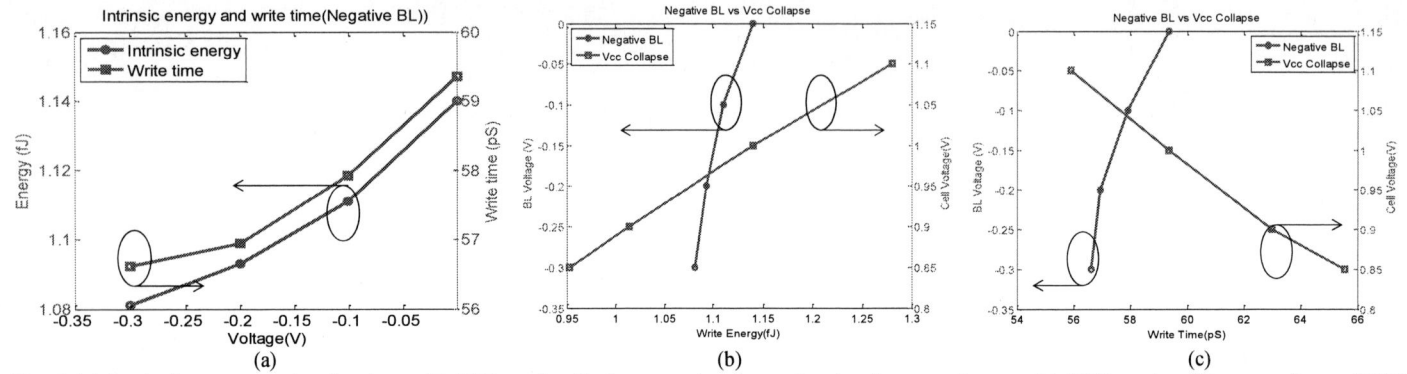

Fig. 7 (a) Intrinsic energy and write time with NBL assist. Both energy barrier and write time gets better with NBL assist, (b) comparison of NBL an Vcccol in terms of intrinsic energy barrier, (c) comparison of NBL and Vcccol in terms of write time.

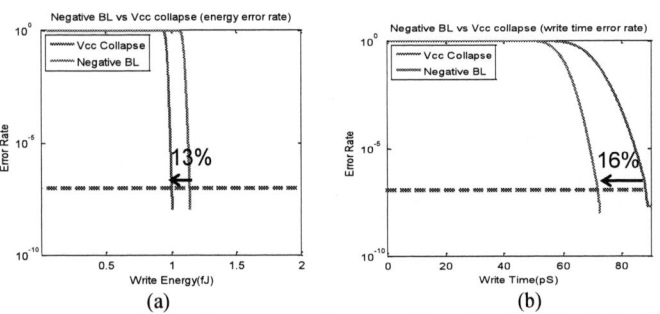

Fig. 8 Comparison of error rates between Vcccol and NBL (a) intrinsic energy bit error rate and (b) write time error rate. Vcccol is better at reducing intrinsic energy errors while NBL is good at reducing timing errors.

It is interesting to observe from Fig. 3(a) that the intrinsic energy barrier reduces with global supply voltage-- suggesting that the cell becomes easy to write at lower voltages. This is a simple yet intuitive result that provides insight on write assist mechanisms (supply voltage scaling in this case) from energy perspective. The write time increases with reduction in V_{DD} indicating that although the energy barrier has reduced, overcoming that barrier takes additional time due to less flow of electrons.

3.3 E_{int} and Error Rates under Process Variation:

The mean and STD of write intrinsic energy is estimated by using (1) and (2) in presence of process variation. The effect of process variation is modeled as V_{TH} variation. Variation in channel length, width, oxide thickness is lumped in the V_{TH} variation model. The mean and STD of V_{TH} variation for each of the transistors in 6T SRAM is assumed to be (0, 30mV). The mean and STD is computed by sweeping V_{TH} of one transistor at a time and using (1) and (2). There are two sources of error during write operation (a) insufficient energy to overcome the intrinsic energy barrier. We refer to this type of error as "intrinsic energy error" and, (b) insufficient time to complete the write operation. We refer to this error as "write time error". These two error types are illustrated in Fig. 4 where T_{write} is maximum time allocated by operating frequency to complete the write and E_{write} is minimum energy needed to overcome the intrinsic energy barrier. In this paper, these two quantities have been assumed to be independent of each other for the sake of simplicity. However, it is possible to determine correlation between them for a more accurate modeling of write error rates. The intrinsic energy and write time error rates are computed by using (3). Fig. 5(a) shows the intrinsic energy error rate wrt write energy for a single bit at different supply voltages. It can be observed that write energy error rate can be reduced by providing more energy to the bitcell. The mean and STD of array energy error rate is computed by using (4) and the results for 32 Byte memory write access is illustrated in Fig. 5(b).

3.4 Evaluation of Write Assist Mechanisms:

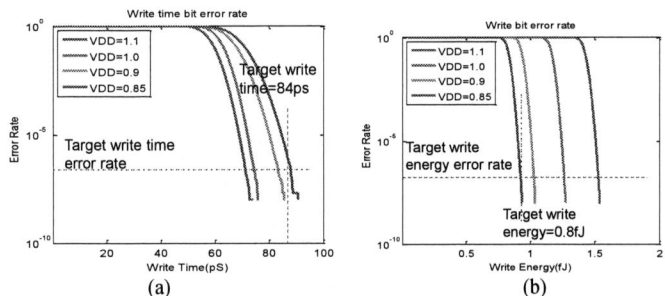

Fig. 9 Determination of (a) target write time and (b) target write energy for fixed error rate. At error rate=10^{-7} the target is found to be (84ps, 0.8fJ).

Table-1 Parameters used for write energy optimization

Param	Value	Param	Value
Array dim	512Rx1024C	C(BL)	75fF
Col mux	4:1	Vcccol swing	0.25V
C(Vcccol)	150fF	Bitcell dim(P:N:AX)	0.1u:0.2u:0.2u
WR Activity Factor	1%-100%	Leakage condition	110C

In order to gain further insight on the efficacy of assist mechanisms in reducing the intrinsic write barrier, we evaluated two popular write assist techniques namely column based supply voltage collapse (Vcccol) [11] and negative bitline (NBL) [4-5].

Fig. 6 shows the results for Vcc collapse assist technique. It can be observed from Fig 6(a) that intrinsic energy decreases at lower cell voltages with increase in write time. The magnitude of intrinsic energy reduction between V_{DD} scaling and Vcccol is identical however, the write time is better with Vcccol (Fig. 6(b)) making it a better candidate for write assist in frequency constrained systems.

Fig. 7(a) shows the evaluation of NBL assist. It is interesting to observe that this technique is not so effective in lowering the energy barrier however, it improves the write speed with higher amount of assist. This is in contrast with Vcccol that worsens the write speed with higher assist. Therefore, the speed dominated write failures get corrected after application of NBL assist both due to lower energy barrier and improved speed. Fig. 7(b) compares effectiveness of NBL and Vcccol in lowering intrinsic energy barrier. It can be noted that Vcccol can outperform NBL in terms of intrinsic energy error rate. However, it is poor in fixing the speed dominated errors where NBL can be useful (Fig. 7(c)). Fig. 8 compares Vcccol and NBL in terms of write energy and write time error rates. For the sake of simplicity, graphs are plotted at maximum assist condition for both techniques (i.e., Vcc=0.85V for Vcccol and V_{BL}=-0.3V for NBL). As noted before, NBL is effective in suppressing the timing errors whereas Vcccol is effective in suppressing intrinsic energy induced errors. At a fixed error rate (10^{-7}), NBL can reduce write time by ~13%. Under the same conditions, Vcccol can reduce intrinsic energy barrier by ~16%.

978-1-4673-6145-3/13 $31.00 © 2013 IEEE

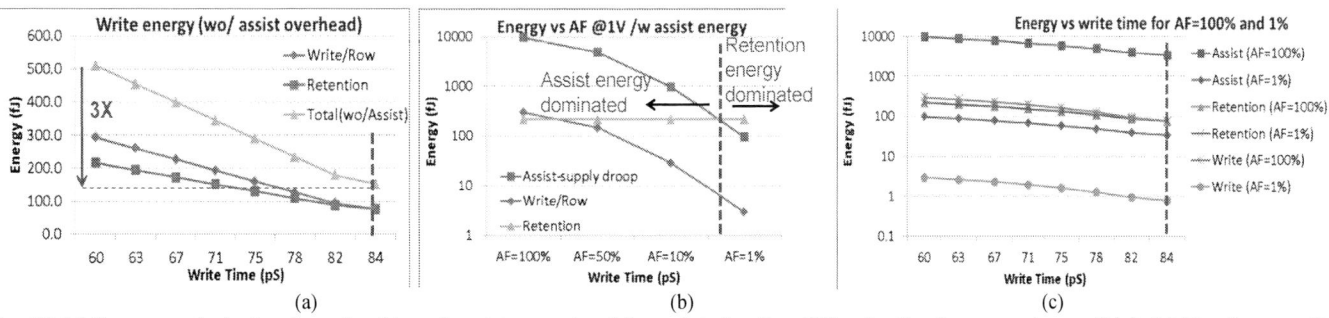

Fig. 10 (a) Energy optimization by using Vcccol assist assuming idle assist circuitry (3X reduction in energy is possible) (b) Dominance of assist energy consumption at high activity factors. Leakage dominates the low activity factor region. (c) Components of total energy consumption and their dependencies on activity factor.

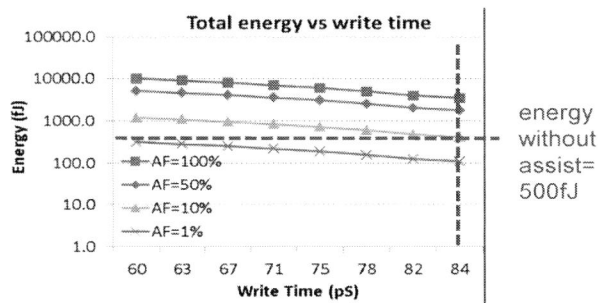

Fig. 11 Total energy with write time under different activity factors. In order to get positive rate of return from assist, the activity factor should be lower than 10%.

4. WRITE ENERGY OPTIMIZATION

In the previous section, we provided intrinsic energy model and compared two assist mechanisms in terms of write error reduction. In this section, we apply the proposed model for write energy optimization of memory array. In order to study the write energy optimization, we consider two scenarios: (a) assuming energy overhead from assist mechanism to be negligible (which is the ideal case) and (b) with appropriate energy estimates of assist circuitry. In the following paragraphs, we describe both of the above conditions. Furthermore, we consider Vcc collapse as an example assist mechanism for optimizations.

4.1 Experimental Setup:

For energy estimation, we have assumed the memory array parameters described in Table-1. The target write time and write energy is chosen by fixing the bit error rate (Fig. 9(a)-(b)). Although the resulting target values (84ps, 0.8fJ) are pessimistic, it still provides vital insights on energy optimization. Note that this methodology is different than conventional methodology that is purely based on write time error rate and may result into a sub-optimal solution. For fair estimation of array level energy tradeoff, we have considered write energy as well as retention energy. Read energy is not considered because the optimization for write is assumed to be independent of real read operation (ignoring dummy read). Column based Vcc collapse [5,11] is assumed for write assist. Activity factor of 1% to 100% for write operation is considered for fair evaluation of write assist energy overhead.

4.2 Without Energy Overhead of Assist:

In absence of energy overhead from assist circuitry, the energy barrier of the bitcell could be reduced freely by scaling the supply voltage and collapsing the bitcell voltage by an extra 250mV (Table-1). Fig. 10(a) plots the write energy, retention and total energy consumption with respect to write time. It can be observed that ~3X reduction in total energy is possible for the target delay of 84ps.

4.3 With Assist Energy Overhead:

At 100% activity factor (i.e., write access every cycle), the energy overhead from assist circuitry becomes dominant component of total energy consumption. It can be noted from Fig. 10(b) that using assist activity factors, both assist and write energy reduces significantly but leakage energy of idle bitcells starts dominating. Fig. 10(c) illustrates the energy optimization with respect to activity factor and Vcc collapse. It can be observed that assist provides energy savings only when write activity factor is very low (which can be a fair assumption in real operation). Fig. 11 depicts the total energy with respect to activity factor and assist. The energy consumption of SRAM without any assist mechanism is ~500fJ (from Fig. 10(a)). Therefore, in order to make assist technique attractive, the activity factor needs to be restricted below 10%. At 1% activity factor, the energy consumption at optimal point is 110fJ which is ~5X better than nominal design (i.e., without assist).

5. CONCLUSIONS AND FUTURE WORK

We proposed an intrinsic energy model of SRAM write operation. The model provides useful insight towards understanding assist mechanisms in terms of their ability to improve intrinsic error rates and energy consumption. We investigated two commonly used assist techniques (Vcc collapse and negative bitline) and compared them in terms of intrinsic energy induced error and write timing errors. We also studied array level energy optimization with the aid of proposed energy centric model and drew solid conclusions for energy optimizations. The future work will involve developing intrinsic energy models for read and retention modes of operation.

ACKNOWLEDGEMENTS

The author would like to acknowledge Dr. Arijit Raychowdhury for initial discussions and useful inputs.

REFERENCES

[1] F. Hamzaoglu et al. "Bit cell optimizations and circuit techniques for nanoscale SRAM design." Design & Test of Computers, 2011.

[2] H. Nho, et al. "A 32nm High-k-metal gate SRAM with adaptive dynamic stability enhancement for low-voltage operation." ISSCC, 2010.

[3] R. Mann, et al. "Impact of circuit assist methods on margin and performance in 6T SRAM." Solid-State Electronics, 2010.

[4] H. Pilo et al, "An SRAM design in 65nm and 45nm technology nodes featuring read and write-assist circuits to expand operating voltage," VLSI Circuits, 2006.

[5] E. Karl et al., "A 4.6GHz 162Mb SRAM design in 22nm tri-gate CMOS tech with integrated active VMIN-enhancing assist circuitry," ISSCC 2012.

[6] Y. Wang, "A 4.0 GHz 291Mb voltage-scalable SRAM design in 32nm high-k metal-gate CMOS tech with integrated power management"JSSC, 2010.

[7] K. Nii et al, "A 45-nm bulk CMOS embedded SRAM with improved immunity against process and temperature variations," JSSC, 2008.

[8] S. Ghosh et al, "Parameter variation tolerance and error resiliency: New design paradigm for the nanoscale era." Proceedings of the IEEE, 2010.

[9] Predictive technology model, ASU, http://www.asu.edu/~ptm.

[10] T. Sakurai et al, "Alpha-power law MOSFET model and its applications to CMOS inverter delay and other formulas," JSSC, 1990.

[11] Y. Wang, et al. "Dynamic behavior of SRAM data retention and a novel transient voltage collapse technique for 0.6 V 32nm LP SRAM." IEDM, 2011.

[12] A. Papoulis. "Probability, Random Variables and Stochastic Processes", McGrawHill, 2002.

SRAM Read Current Variability and its Dependence on Transistor Statistics

Sriramkumar Venugopalan[1], Vivek Joshi[2], Luis Zamudio[2], Matthias Goldbach[3],
Gert Burbach[3], Ralf VanBentum[3], Sriram Balasubramanian[2]

[1]Dept of EECS, University of California, Berkeley, Email[1]: sriram@eecs.berkeley.edu
[2]GLOBALFOUNDRIES, Sunnyvale, CA, USA, Email[2]: Sriram.Balasubramanian@globalfoundries.com
[3]GLOBALFOUNDRIES, Dresden, Germany

Abstract— **Our study breaks down the dependence of SRAM read current (I_{read}) variability (σI_{read}) into constituting pass-gate (PG) and pull down (PD) NMOS transistor variability. We report a bottoms-up model for σI_{read} including feedback in stacked transistors and discuss its implications on SRAM performance.**

I. INTRODUCTION

I_{read} and its statistics is a key performance metric that limits voltage scaling in SRAMs [1]. Traditional methods have depended on running millions of Monte-Carlo (MC) simulations but don't address the fundamental relation between transistor and I_{read} variability.

In this work we develop a unified σI_{read} model from its fundamental components – namely, the variability of the PG and PD devices, Fig. 1. We define the concept of effective stack variability and clarify the role of negative feedback present in the SRAM read stack that results in σI_{read} being lower than the variability of an equivalent single PG transistor, Fig. 2. We validate this model using both MC simulations and 32/28nm technology (tech) SRAM hardware bit-cell data. With continued scaling, 6σ-I_{read} validation needed for large memories are a significant challenge in early tech development. Using this σI_{read} model we verify the empirical Voltage Acceleration Method (VAM) [2] and derive the value of the VAM voltage shift (σV_{AM}) based on PG and PD variability.

II. SRAM READ CURRENT VARIABILITY MODEL

I_{read} is the current that flows from a pre-charged bit-line to ground through stacked PG (in saturation) and PD (in linear region) NMOS devices that are 'ON', Fig. 1. During an I_{read} measurement, the read stack internal node voltage rises above ground to V_{read} with PG in saturation and PD in linear regions of operation, Fig. 1. σI_{read} is then attributed to variability of the PG and PD devices respectively. The equations for I_{read}, σI_{read} with its individual component breakdown are summarized in Table 1. A dominant source of PG variability is from its threshold voltage, $\sigma V_{TH,sat,PG}$ [3]. It is important to note that the overall contribution of $\sigma V_{TH,sat,PG}$ to σI_{read} is reduced owing to the negative feedback of V_{read} through gate overdrive ($\Delta V_{TH,sat,PG}$ and ΔV_{read} are negatively correlated,

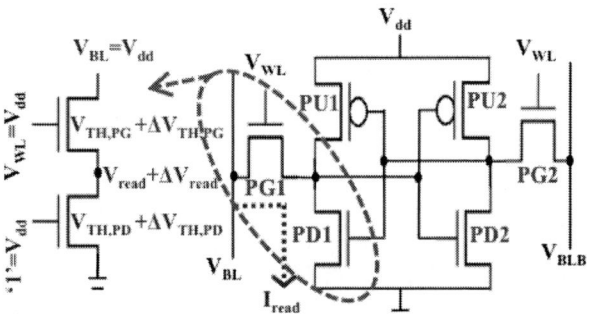

Figure 1. 6T-SRAM cell showing PG-PD read stack voltage conditions for I_{read} measurement.

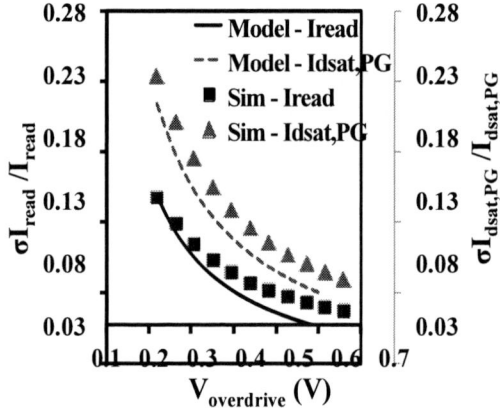

Figure 2. Negative feedback of V_{read} on the read stack variability gives rise to a lower σI_{read} than the constituting PG device variability at the same gate overdrive condition.

Fig. 3(a)). On the other hand, PD threshold voltage variability, $\sigma V_{TH,lin,PD}$ impact on σI_{read} is affected in two distinct ways – a small positive feedback through the (V_{dd}-$V_{TH,lin}$-$V_{read}/2$) term and larger negative feedback through the drift field term (V_{read}) ($\Delta V_{TH,lin,PD}$ and ΔV_{read} are positively correlated, Fig, 3(b)). Overall the negative feedback component of V_{read} on both PG and PD variability effectively reduces σI_{read} as compared to the inherent variability of both the PG and PD devices, Fig. 2. We note from the derived

978-1-4673-6145-3/13 $31.00 © 2013 IEEE

$$I_{read} = K_{PG} \cdot (V_{DD} - V_{TH,sat,PG} - V_{read})^{\alpha_{sat}} = K_{PD} \cdot \left(V_{DD} - V_{TH,lin,PD} - 0.5V_{read}\right)^{\alpha_{lin}} \cdot V_{read}$$

$$\left(\frac{\Delta I_{read}}{I_{read}}\right)_{PG} = -\frac{\alpha_{sat} \cdot \Delta V_{TH,sat,PG}}{(V_{DD} - V_{TH,sat,PG} - V_{read})} - \frac{\alpha_{sat} \cdot \Delta V_{read}}{(V_{DD} - V_{TH,sat,PG} - V_{read})}$$

$$\left(\frac{\Delta I_{read}}{I_{read}}\right)_{PD} = -\frac{\alpha \cdot \Delta_{\;\;,}}{(\;\;-\;\;_{,\;}-0.5\;\;)} - \frac{0.5 \cdot \alpha \cdot \Delta}{(\;\;-\;\;_{,\;}-0.5\;\;)} + \frac{\Delta V_{read}}{V_{read}}$$

$$\left(\frac{\Delta I_{read}}{I_{read}}\right)_{Tot} \approx -\frac{\alpha_{sat} \cdot \Delta V_{TH,sat,PG}}{(V_{DD} - V_{TH,sat,PG} - V_{read})} \cdot \frac{\beta}{1+\beta} - \frac{\alpha \cdot \Delta_{\;\;,}}{(\;\;-\;\;_{,\;}-0.5\;\;)} \cdot \frac{1}{1+\beta}$$

$$= A \cdot \Delta V_{TH,sat,PG} + B \cdot \Delta V_{TH,lin,PD}$$

$$\frac{\sigma I_{read}}{I_{read}} = \sqrt{(A \cdot \sigma V_{TH,sat,PG})^2 + (B \cdot \sigma V_{TH,lin,PD})^2} = \frac{\alpha_{sat} \cdot \sigma V_{TH,eff}}{(V_{DD} - V_{TH,sat,PG} - V_{read})} \qquad (1)$$

$$\frac{\sigma I_{dsat,PG}}{I_{dsat,PG}} = \frac{\alpha_{sat} \cdot \sigma V_{TH,sat,PG}}{(V_{DD} - V_{TH,sat,PG} - V_{read})} = \frac{\alpha_{sat} \cdot \sigma V_{TH,sat,PG}}{V_{PG,overdrive}} \qquad (2)$$

Table 1. Equations describing SRAM read current variability as a function of PG and PD device variability. Single transistor variability representation of σI_{read} through an effective stack variability, $\sigma V_{TH,eff}$ is derived as a function of $\sigma V_{TH,sat,PG}$ and $\sigma V_{TH,lin,PD}$.

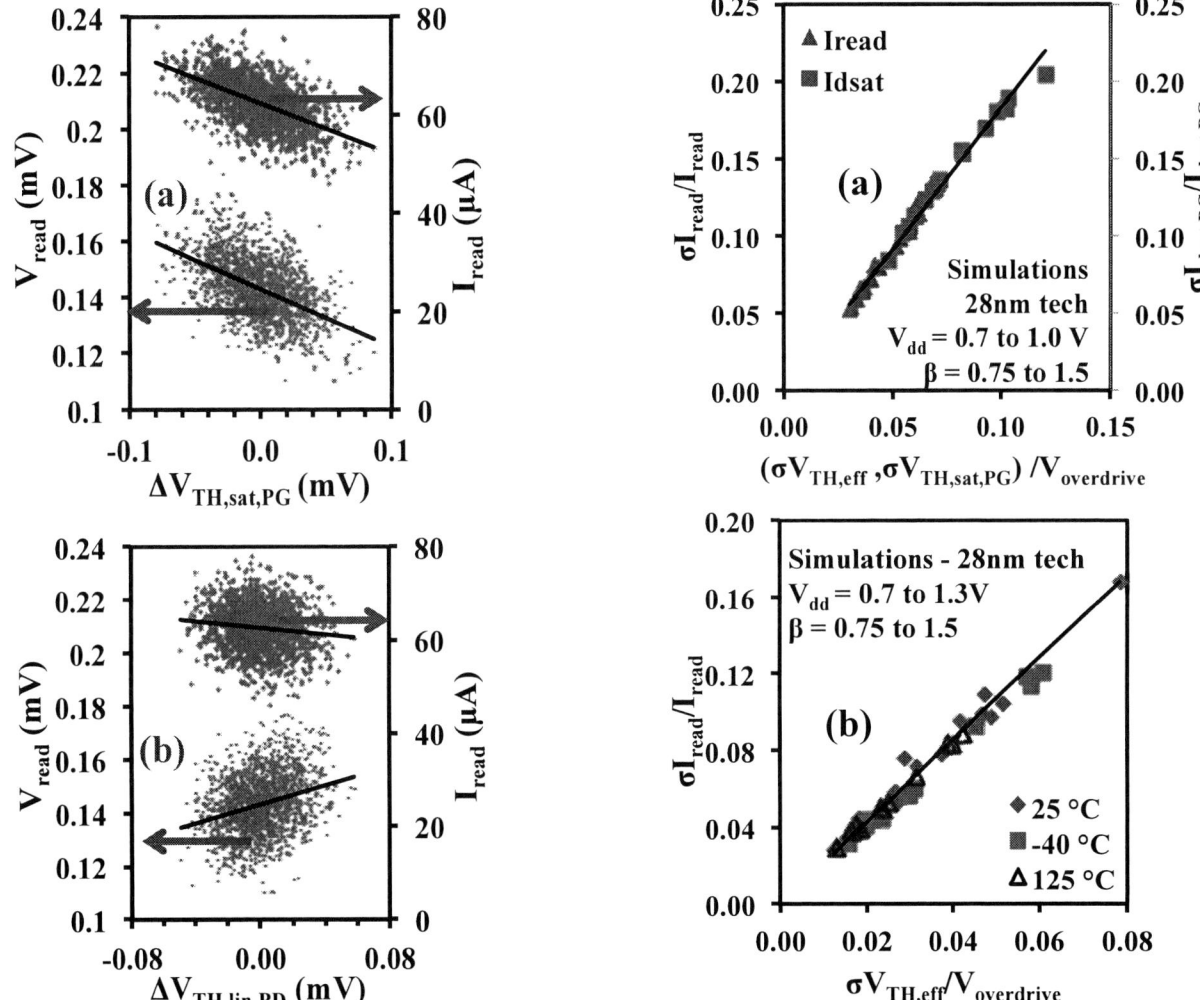

Fig 3. SRAM bit-cell hardware data from a 32nm tech show (a) negative correlation of V_{read} vs. $V_{TH,sat,PG}$ (b) positive correlation of V_{read} vs. $V_{TH,lin,PD}$. Both result in negative feedback of V_{read} on σI_{read}.

Fig 4. MC simulation results of a 28nm tech validate the effective stack variability model showing σI_{read} as an extrapolation of $\sigma I_{dsat,PG}$ across (a) V_{dd}, SRAM design points (β) and (b) temperatures.

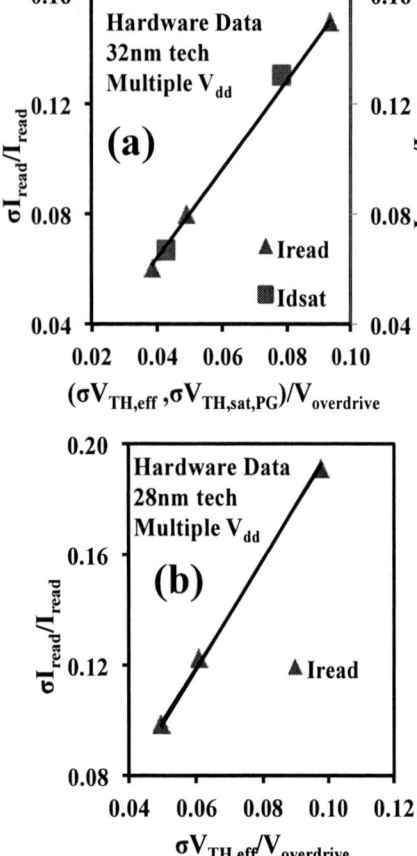

Figure 5. SRAM bit-cell level measurements of (a) $\sigma I_{dsat,PG}$ and σI_{read} for a 32nm tech, and (b) σI_{read} for a 28nm tech across multiple V_{dd} lie on the same unified straight line as predicted in Eq.(1,2).

model $\sigma I_{read}/I_{read}$ in Eq.(1) is purely a function of the gate-overdrive, $\sigma V_{TH,sat,PG}$, $\sigma V_{TH,lin,PD}$ and PD/PG size ratio (β).

III. EFFECTIVE READ STACK I_{READ} VARIABILITY

A convenient way to represent σI_{read} is to represent it as the variability of an equivalent single transistor (with $V_{source}=V_{read}$) with an effective stack variability, $\sigma V_{TH,eff,}$ which is dependent on both $\sigma V_{TH,sat,PG}$ and $\sigma V_{TH,lin,PD}$, Eq.(1). Fig. 4 shows validation of this model by unifying normalized $\sigma I_{dsat,PG}$ and σI_{read} against $\sigma V_{TH,sat,PG}$ and $\sigma V_{TH,eff}$ normalized to PG device gate-overdrive. Results from MC simulations of a 28nm tech SRAM for a wide range of V_{dd}, temperatures and design explorations (varied β) all fall on a straight line with unified slope, Eq.(1,2). Effective stack variability plots for 32nm and 28nm measured Si hardware data across cell sizes and V_{dd} reveal that $\sigma I_{dsat,PG}$ and σI_{read} lie on the same straight line and the unified slope is relatively independent of technology as they are functions of $\sigma V_{TH,sat,PG}$, $\sigma V_{TH,lin,PD}$ and gate-overdrive, Fig. 5.

IV. IMAPCT ON CHOICE OF CELL-RATIO (β)

While the choice of 6T-SRAM β is often co-optimized to meet read-disturb/write/performance margins, 8T read stacks are often tuned for performance (CV_{dd}/I_{read}). In order to validate the σI_{read} model, we explore one possible optimization to minimize σI_{read}, by varying the sizes of PG

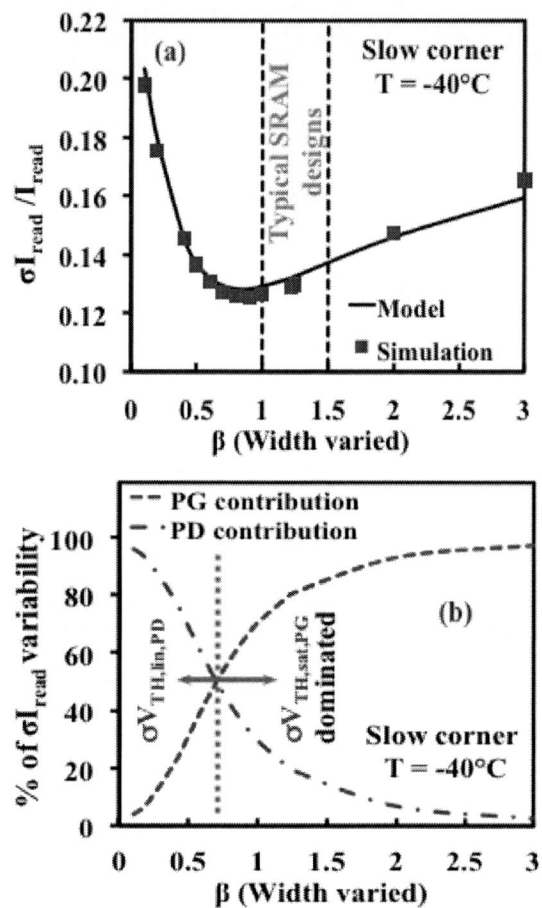

Figure 6. At constant I_{read}, 28nm tech (a) Both simulations and the model, Eq.(1) indicate $\beta \approx 1$ designs are preferred for lower σI_{read}. (b) Relative contributions of PG and PD device variability towards σI_{read} shows that PG contribution dominates even for $\beta=1$.

$$I_{read} = f(V_{dd} - V_{THeff,read})$$

$$\frac{\Delta I_{read}}{I_{read}} = A \cdot \Delta V_{TH,sat,PG} + B \cdot \Delta V_{TH,lin,PD}$$

$$\frac{\Delta I_{read}}{I_{read}} = \frac{\partial I_{read}}{\partial V_{dd}} \cdot \frac{\Delta V_{AM}}{I_{read}} = (A+B) \cdot \Delta V_{AM}$$

$$\sigma V_{AM} = \frac{\sqrt{\left(A \cdot \sigma V_{TH,sat,PG}\right)^2 + \left(B \cdot \sigma V_{TH,lin,PD}\right)^2}}{A+B} \quad (3)$$

$$\sigma V_{AM} = \sigma V_{TH,eff} \cdot \left(\frac{A}{A+B}\right) \cdot \left(1 + \frac{1}{\beta}\right) \quad (4)$$

Table 2. Equations describing the sensitivities of I_{read} and σV_{AM} required for estimating 6σ-I_{read}.

and PD devices (β varied by varying width) while maintaining a constant I_{read}. The simulation results show (and the model agrees) that $\beta <1$ is optimal for minimal $\sigma I_{read}/I_{read}$, Fig. 6(a). The variability model provides insights into the relative sensitivities of PG vs. PD (prefactors *A* vs. *B*) devices towards σI_{read}. The PG limits σI_{read} (for $\beta>0.8$) with the PD starting to limit σI_{read} only when the PD device gets really small (for $\beta<0.8$), Fig. 6(b). At $\beta=1$, PG still dominates due to its larger sensitivity (*A>B*) and so for minimum

978-1-4673-6145-3/13 $31.00 © 2013 IEEE

$\sigma I_{read}/I_{read}$, optimum $\beta < 1$, Fig. 6(a). Also from a similar exercise, 2% improvement in σI_{read} for every 10% increase in

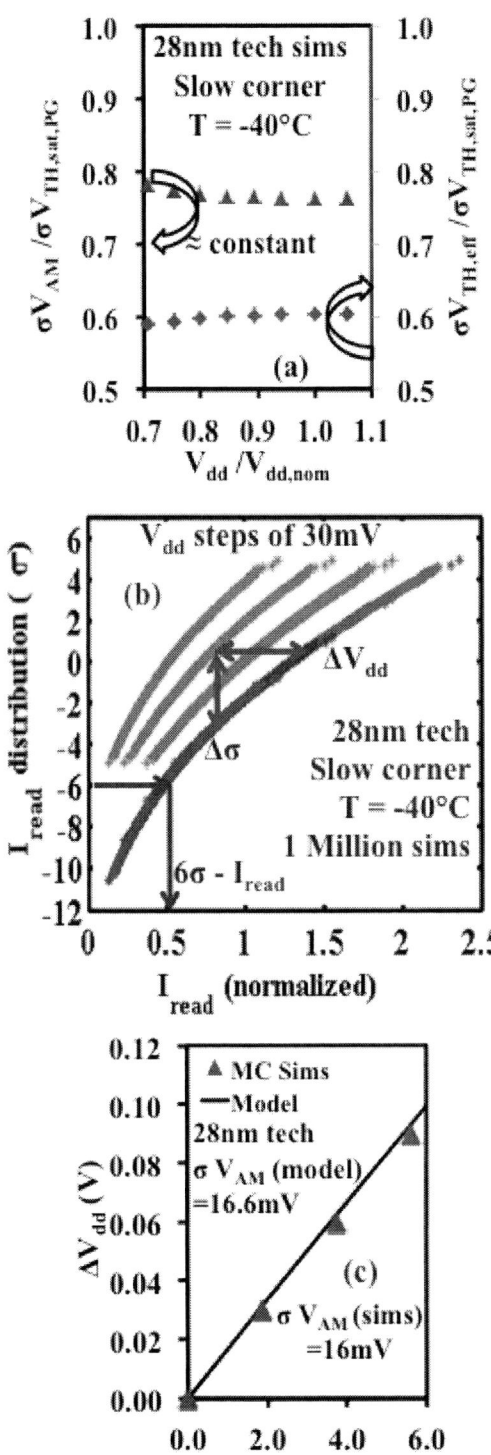

length of the PD device was found, again favoring lower β values (figure not shown).

V. VAM METHOD FOR 6σ READ CURRENT ESTIMATION

The VAM method proposed in [2] uses I_{read} measurements at lower V_{dd} to predict 6σ-I_{read}. By equating the I_{read} sensitivity to V_{dd} and V_{TH} shifts in the σI_{read} model, we estimate the V_{dd} reduction ($=V_{dd}$-σV_{AM}) to be applied to the read stack that would produce a σI_{read} offset in the stack current, Table 2. We observe that σV_{AM} (like $\sigma V_{TH,eff}$) is only a function of $\sigma V_{TH,sat,PG}$, $\sigma V_{TH,lin,PD}$, β and gate overdrive and is almost constant with V_{dd}, (for a given β, **A/B** is $\approx$ constant) allowing for extraction of 6σ-I_{read} by applying V_{dd}-$6*\sigma V_{AM}$ to the read stack, Fig. 7(a). We recreate the VAM plot with simulation results for a 28nm tech by overlaying the I_{read} distribution plots at different V_{dd} by an I_{read} sigma shift $\Delta\sigma$ (= $\Delta V_{dd}/\sigma V_{AM}$) in Fig. 7(b), thereby permitting the estimation of 6σ-I_{read} value. Given that σV_{AM} is relatively constant, $\Delta\sigma$ is also relatively constant up to 6σ-I_{read} projections, which agrees with the observations in [2]. Both the MC simulation extracted and the model estimated $\Delta\sigma$ are in excellent agreement in Fig. 7(c), which validates the formulation that σV_{AM} is indeed related to PD/PG variability.

VI. CONCLUSIONS

We describe the dependence of SRAM I_{read} variability on fundamental transistor variability parameters through a simple Si-validated model that includes the negative feedback of V_{read} on σI_{read}. We illustrate the concept of effective stack variability that unifies σI_{read} and single transistor variability into a single plot. Study indicates a preference for lower than typical β values for SRAM to achieve lower σI_{read}. This has a profound impact on design choices for 8T-SRAM cells where writability is not a constraint. We evaluate the VAM method for accelerated I_{read} distribution tail analysis and explain the theoretical basis of the VAM voltage shift that can be used to extract 6σ-I_{read} value.

REFERENCES

[1] E. Grossar et al, "Read Stability and Write-Ability Analysis of SRAM Cells for Nanometer Technologies" *IEEE Journal of Solid-State Circuits*, Vol. 41, No. 11, Pg. 2577, 2006.

[2] J. Wang et al, "Non-Gaussian distribution of SRAM read current and design impact to low power memory using Voltage Acceleration Method", Symposium on *VLSI Technology Digest*, Pg. 220, 2011.

[3] A. Asenov, "Simulation of Statistical Variability in Nano MOSFETs", Symposium on VLSI Technology Digest, Pg. 12, 2007

Figure 7. (a) Estimated σV_{AM} in Eq.(4) and $\sigma V_{TH,eff}$ in Eq.(1) are $\approx$ constant across V_{dd} of interest (b) 6σ-I_{read} estimated from smaller number of MC simulations at lower V_{dd} (VAM) using Eq.(3). (c) Extracted σV_{AM} from Fig. 7(b) and calculated σV_{AM} using Eq.(3) agree with each other.

Mismatch Characterization of Small Metal Fringe Capacitors

Vaibhav Tripathi and Boris Murmann
Department of Electrical Engineering, Stanford University
420 Via Palou Mall, Stanford, CA
vaibhavt@stanford.edu, murmann@stanford.edu

Abstract— **Even though small metal fringe capacitors are important for the realization of low-energy A/D converters, present literature is lacking experimental data on their mismatch characteristics. This paper describes a test structure and measurements results pertaining to the characterization of single-layer, lateral-field, 0.45-fF and 1.2-fF unit metal capacitors in a 32-nm SOI CMOS process. The measurement-inferred average standard deviations for these capacitances are 1.2% and 0.8%, respectively, confirming area scaling according to Pelgrom's matching law.**

I. INTRODUCTION

The ability to control the lateral dimensions in a modern CMOS process with high precision has provided an attractive means to realize very small metal fringe capacitors. Capacitors as small as 0.5 fF are now being used routinely [1], and even values as small as 0.05 fF [2] have been utilized in successive approximation register (SAR) ADCs. Unfortunately, little to no information on the matching properties of such capacitors is available in the present literature. In this paper, we describe a SAR ADC-like test structure to infer capacitance mismatch for 0.45-fF and 1.2-fF lateral field capacitors, built on the Metal 7 layer of a 32-nm SOI CMOS process.

The remainder of this paper is structured as follows. Section II reviews prior-art methods for capacitor mismatch measurement along with their limitations. Section III describes the proposed technique along with its circuit realization and Section IV summarizes our measurement results.

II. PRIOR ART MISMATCH MEASUREMENT TECHNIQUES

The presence of pad capacitances and the limited resolution of bench top measurement equipment make the direct probe-based measurement of small capacitors impractical. Most of the previously published works address this issue by translating the capacitance mismatch information into another domain where it can be more easily measured. Prior art techniques can be broadly divided into two categories: (a) voltage-based measurements and (b) frequency-based measurements.

A. Voltage-Based Measurement

Fig. 1 shows an example of a voltage-based technique. A voltage ramp is applied at node A when B is grounded (and vice-versa), so that the mismatch information is contained in the difference of the output slopes for the two cases (S_1 and S_2) [3, 4]. This technique is plagued by the analog nature of the measurement, and the measured data will typically be impaired by a variety of noise sources and noise coupling mechanisms.

B. Frequency-Based Measurement

Reference [5] identified the shortcomings of voltage-based measurements [3, 4] and proposed a mismatch characterization method based on frequency. The key idea is to translate the relative mismatch between two capacitors to a relative change in frequency of oscillation. Fig. 2 shows the circuit that was used to characterize mismatch of capacitance values down to 8 fF. While this approach is substantially more robust and also provides a "quasi-digital" readout, it requires undesired de-embedding of circuit parameters that lead to measurement uncertainty. First, reliable simulation data is needed to account for the loading effects of large resistances that are added to suppress switch mismatch. Second, these simulation data must be combined with measurements from an unloaded replica of the structure to scale the observed values.

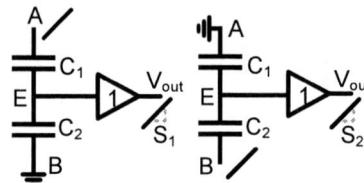

Fig. 1. Voltage-based mismatch measurement.

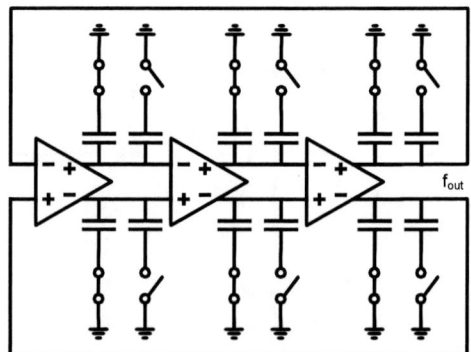

Fig. 2. Ring oscillator for frequency-based mismatch measurement.

III. PROPOSED TECHNIQUE

The test structure described in this paper was designed to provide a fully digital readout of mismatch data without requiring extensive (manual) measurement calibration. This is accomplished by using a SAR ADC-like test structure that

978-1-4673-6145-3/13 $31.00 © 2013 IEEE

translates capacitance mismatch information into a digital bit stream. Furthermore, we utilize a switching scheme that allows us to distinguish random mismatch from variations that are common to neighboring unit elements.

A. Circuit Description and Capacitance Modeling

Fig. 3 shows a simplified schematic of the test structure. The actual implementation is differential, but ϕ_1 is exercised only in the positive-half circuit (as explained further below). The structure resembles a SAR ADC, except that the converter takes no input and the capacitor array is unary (64 same-size capacitors). Except for V_{REF} and $V_{REF,dac}$, all I/Os of this circuit are digital. Fig. 4 shows the layout of the 64 unit elements. 32 dummy elements (not shown in Fig. 4) are added left and right of the linear capacitive array to reduce potential systematic errors due to metal density fluctuations. The dimensions of a single unit element are listed in Table I.

We model each of the unit capacitors as a Gaussian random variable with mean C_u and standard deviation σ_u. Due to very small dimensions and proximity of the unit elements, we also consider correlation through a nearest neighbor model. Every unit element is assumed to have finite correlation with its nearest neighbors and is independent of the rest. As a result, the assumed covariance matrix (Σ) is tri-diagonal and given by

$$\Sigma = \begin{bmatrix} \sigma_u^2 & \sigma_{ab}^2 & 0 & 0 & \cdots \\ \sigma_{ab}^2 & \sigma_u^2 & \sigma_{ab}^2 & 0 & \cdots \\ 0 & \sigma_{ab}^2 & \sigma_u^2 & \sigma_{ab}^2 & \ddots \\ 0 & 0 & \sigma_{ab}^2 & \sigma_u^2 & \ddots \\ \vdots & \vdots & \ddots & \ddots & \ddots \end{bmatrix} \quad (1)$$

TABLE I.   UNIT CAPACITOR DIMENSIONS

Dimensions	0.45 fF	1.2 fF
G (μm)	0.65	0.65
H (μm)	3.6	9.5
S (μm)	0.13	0.13
W (μm)	0.13	0.13
P (μm)	0.5	0.5

B. Capacitance Switching Scheme

A mismatch-proportional error voltage is generated at the comparator input by switching the unit capacitances in the positive-half of the circuit (via ϕ_1) as shown in Fig. 5. First, the top plate (node T) is connected to V_{REF} while the bottom plates are connected to V_{REF} or GND alternately, in groups of two. Then, the top plate switch is turned off and the bottom plate connections are reversed. This generates an error voltage at the input of the comparator

$$V_\Delta = V_{REF} \frac{\sum_{i=1,2,5,6\ldots} C_i - \sum_{j=3,4,7,8,\ldots} C_j}{\sum_{k=1}^N C_k} = V_{REF} \frac{X}{Y} \quad (2)$$

For N unit elements with the above-assumed covariance matrix, the mean and variance of V_Δ (see Appendix) are given by

$$E(V_\Delta) = -\frac{V_{REF}}{[E(Y)]^2} Cov(X,Y) \quad (3)$$

$$Var(V_\Delta) = \frac{V_{REF}^2}{N}\left[\left(\frac{\sigma_u}{C_u}\right)^2 + \frac{2}{N}\left(\frac{\sigma_{ab}}{C_u}\right)^2\right] \quad (4)$$

From (3), we can see that the mean of the error voltage will be negative in presence of finite neighbor correlations and this is confirmed by our measurement results in Section IV. More importantly, it is clear from (4) that for large N, the described switching scheme suppresses the effect of finite nearest neighbor correlation and allows us to infer the coefficient of variation (relative unit element mismatch) $c_v = \sigma_u/C_u$ from the variance of V_Δ. With sufficient digital resolution in the subsequent SAR conversion, V_Δ can be replaced by its digitized value ($D\{V_\Delta\}$) and (4) can be rearranged as

$$c_v = \frac{\sigma_u}{C_u} \approx \sqrt{N}\frac{V_\Delta}{V_{REF}} Stdev(D\{V_\Delta\}) \quad (5)$$

In the circuit of Fig. 3, V_Δ is digitized using a SAR A/D conversion, which is described next.

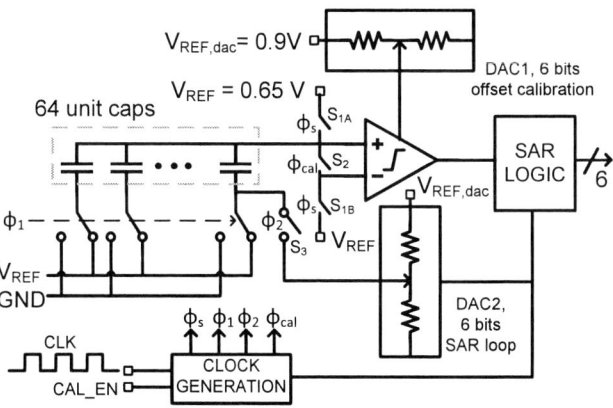

Fig. 3. Proposed test structure for capacitance mismatch measurement.

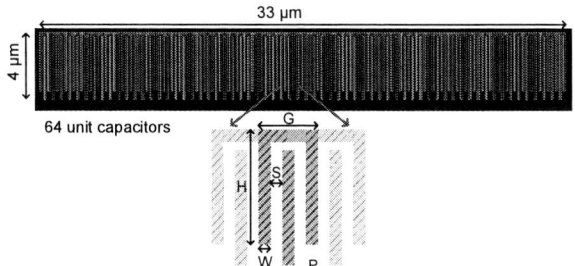

Fig. 4. Capacitor layout.

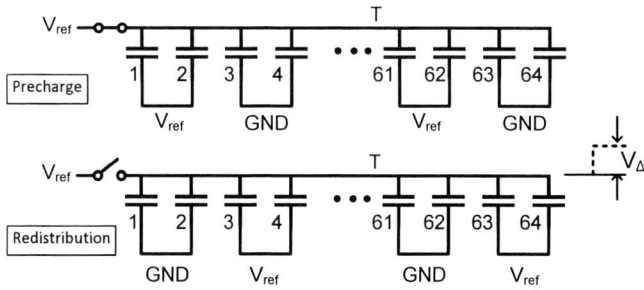

Fig. 5. Switching scheme.

C. Error Voltage Measurement

Fig. 6 shows the steps involved in calculating the coefficient of variation. The error measurement begins with a coarse offset calibration cycle (ϕ_{cal}), where the comparator input is shorted and its offset is minimized via DAC1. Next, the residual offset is measured through SAR conversions via DAC2. S_3 is turned on and DAC2 is coupled to the comparator via one unit capacitor, which gives an attenuation and LSB size reduction of approximately 64.

Finally, the unit capacitors are switched via ϕ_1 and generate the error voltage (V_Δ), which is then quantized via the same SAR loop. The earlier digitized residual offset is subtracted from this measurement. Unlike the capacitance switching via ϕ_1 (which is done only in one half circuit), the SAR A/D conversion is fully differential for improved robustness. With 6 bits of resolution in DAC2, the LSB size of the measurement corresponds to that of a 12-bit ADC. The measurements are repeated and averaged to reduce the kT/C noise to within an LSB at the 12-bit level.

Even though the proposed procedure combines several measurements and includes a calibration step, it avoids the key shortcomings of [6], where the characterization routine relies on simulation data and measurement results from a separate reference structure.

D. Accuracy Limits

It is important to consider error sources that may affect the accuracy of the mismatch characterization. One imperfection to consider in our circuit is the mismatch in the coupling capacitance that injects the DAC2 signal. Analysis shows that this error merely scales the measurement by the mismatch factor; i.e., the percent uncertainty in the measurement is equal to the mismatch percentage, which is on the order of one percent and thus negligible.

In addition, the proposed test structure relies upon the accuracy of DAC2 for the SAR A/D conversion and the minimum mismatch that can be measured is limited by the integral nonlinearity (INL) of this DAC. Detailed analysis shows that the INL of DAC2 must be less than its LSB (at the 6-bit level). The expected INL from process data lies within one fifth of this requirement.

IV. MEASUREMENT RESULTS

The designed chips were fabricated in IBM's 32nm CMOS SOI process and contain 72 of the test structures shown in Fig. 3. Half of the test structures are built with capacitances of 0.45 fF, while the other half uses 1.2 fF. The dimensions of one structure are 90 μm x 55 μm (0.005 mm^2). The die photo is shown in Fig. 7 along with the chip layout.

A 7-bit decoder (on chip) is used to enable the test structures one at a time and a PC controls the measurement and data collection. kT/C noise is suppressed by averaging 131,072 measurements for each test structure. Fig. 8 shows mismatch histograms for all test structures on a single die. From the histogram, it can be seen that the measured mean is negative. This is confirmed by equation (3), which predicts a negative mean for (2) in presence of finite neighbor correlation among the unit elements. Fig. 9 shows our measurement results across 8 dies. The calculated sample coefficient of variation is 1.2% for 0.45 fF, and 0.8% for 1.2 fF unit capacitors, respectively. The sample variance is $1/\sqrt{N} = 0.167$ or 16.7% (N = 36), which is in line with the variations seen from chip-to chip in Fig. 10.

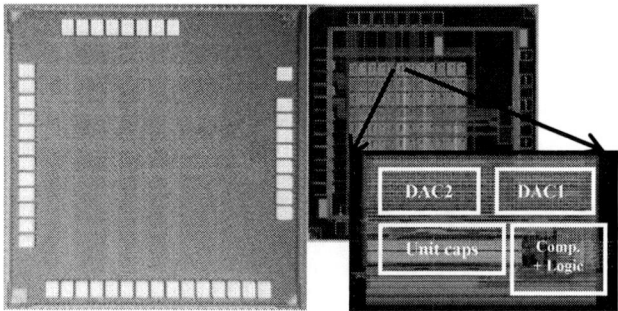

Fig. 7. Die photo and layout.

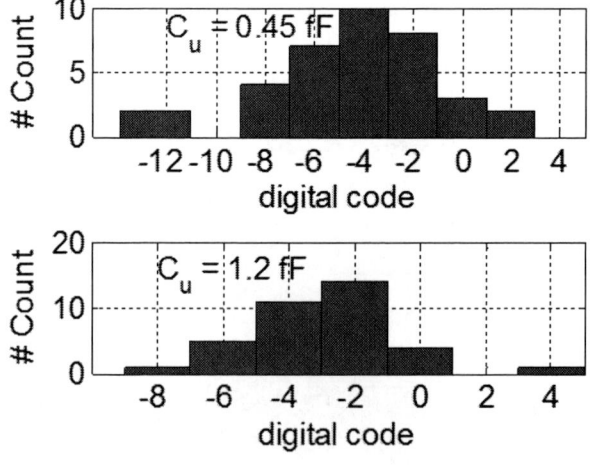

Fig. 8. Measured mismatch histogram on a single die.

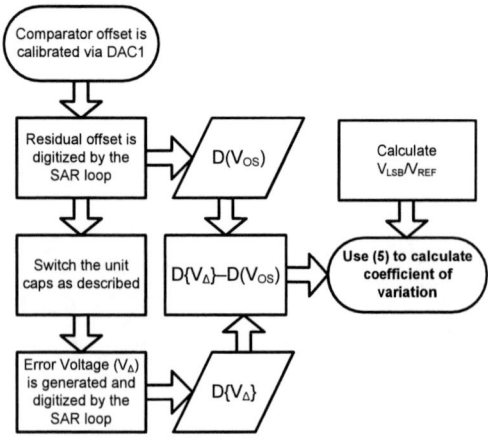

Fig. 6. Characterization procedure.

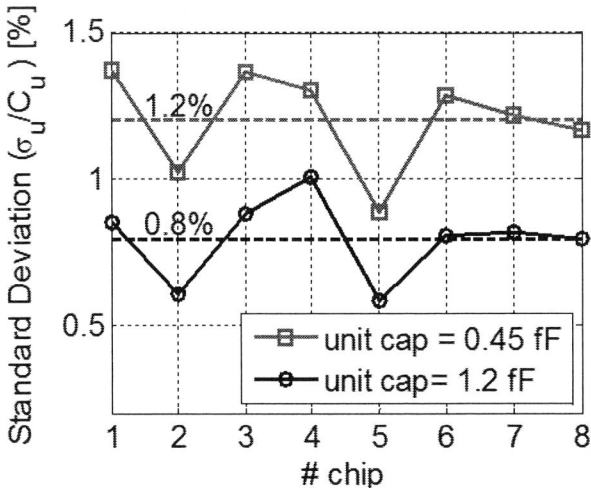

Fig. 9. Mismatch measurement across 8 dies.

V. CONCLUSION

In summary, this paper presented a test structure for the characterization of small fringe capacitances in an advanced CMOS process. The salient features of the proposed test structure include: (1) no analog inputs (except for DC reference voltages) and (2) a purely digital output bit-stream that contains the mismatch information, thus eliminating sources of error like instrumentation inaccuracy and noise. For each chip, 72 test structures were measured (36 each for 0.45 fF and 1.2 fF unit capacitance) and the obtained results confirm good matching despite small geometries. Moreover, the matching improvement was found to be in accordance with Pelgrom's area scaling law ($\sigma^2 \sim 1/\text{Area}$).

ACKNOWLEDGMENT

This work was funded by the Semiconductor Research Corporation under agreement 1836.078. Chip fabrication was provided by IBM and supported by the DARPA LEAP program. We thank Silicon Frontline for providing access to layout parasitic extraction tools.

APPENDIX

The variance of the error voltage V_Δ, given in (2), can be found using the statistics of the sums in the numerator (X) and denominator (Y)

$$Var(V_\Delta) \approx V_{REF}^2 \left(\frac{\mu_X}{\mu_Y}\right)^2 \left[\left(\frac{\sigma_X}{\mu_X}\right)^2 + \left(\frac{\sigma_Y}{\mu_Y}\right)^2 - \frac{2C_{OV}(X,Y)}{\mu_X \mu_Y}\right]$$

where

$$\mu_X = 0 \quad and \quad \mu_Y = NC_u.$$

Thus,

$$Var(V_\Delta) \approx \frac{V_{REF}^2}{N^2 C_u^2} \sigma_X^2.$$

For the remaining unknown, it can be show that

$$\sigma_X^2 = N\sigma_u^2 + 2\sigma_{ab}^2$$

With this substitution, we obtain

$$Var(V_\Delta) = \frac{V_{REF}^2}{N}\left[\left(\frac{\sigma_u}{C_u}\right)^2 + \frac{2}{N}\left(\frac{\sigma_{ab}}{C_u}\right)^2\right].$$

REFERENCES

[1] P. Harpe, C. Zhou, X. Wang, G. Dolmans, and H. de Groot, "A 12fJ/conversion-step 8bit 10MS/s asynchronous SAR ADC for low energy radios," Proc. ESSCIRC, pp. 214-217, Sept. 2010.

[2] D. Stepanovic and B. Nikolic, "A 2.8GS/s 44.6mW Time-Interleaved ADC Achieving 50.9dB SNDR and 3dB Effective Resolution Bandwidth of 1.5GHz in 65nm CMOS," Symp. on VLSI Circuits, pp. 84-85, June 2012.

[3] H. P. Tuinhout, H. Elzinga, J. T. Brugman, and F. Postma, "Accurate Capacitor Matching Measurements using Floating Gate Test Structures," Proc. International Conference on Microelectronic Test Structures (ICMTS), pp. 133-137, Mar. 1995.

[4] C. Kortekaas, "On-Chip Quasi-static Floating-gate Capacitance Measurement Method," Proc. International Conference on Microelectronic Test Structures (ICMTS), pp. 109-113, Mar. 1990.

[5] A. Verma and B. Razavi, "Frequency-Based Measurement of Mismatches Between Small Capacitors," Proc. IEEE CICC, pp. 481-484, Sept. 2006.

A 1GHz Hardware Loop-Accelerator with Razor-based Dynamic Adaptation for Energy-Efficient Operation

Shidhartha Das, Ganesh Dasika, Karthik Shivashankar and David Bull

ARM Ltd., Cambridge, U.K.

Abstract— We describe the implementation and silicon measurement results from a Razor-based hardware loop-accelerator (RZLA), implementing the Sobel edge-detection algorithm. We demonstrate robust operation with a large Dynamic Voltage Scaling (DVS) range achieved using 50% of the clock-period for timing-speculation. At 1GHz operating frequency, Razor DVS enables 34% energy-efficiency improvement on a per-device basis and 33% overall on the entire batch of devices.

I. INTRODUCTION

Dynamic adaptation using Razor-based detection and correction of timing errors has demonstrated substantial improvements in performance and energy-efficiency in microprocessors [1-5]. Razor relies on combination of *in situ* error-detecting circuits and micro-architectural recovery mechanisms to reclaim unused voltage guardbands. Error-detection occurs using so-called Razor flip-flops (RFF) that explicitly check for late-arriving transitions at critical-path endpoints. Error correction is performed by the system either through correct-data substitution [1] or instruction replay from a check-pointed state [2-5].

In this work, we apply Razor to application-specific hardware loop-accelerators (LA) that represent an entirely different category of compute engines. Dedicated LAs find increasing application in energy-constrained system-on-chips (SoCs) where they trade-off flexibility for energy-optimal execution of compute-intensive kernels in wireless and signal-processing algorithms. In contrast with microprocessors, LAs are a class of co-processors that accelerate a particular function and as such do not need to maintain an internal architectural state. Instead, queues are used in a dataflow-like

manner to transfer transient data between functional units. This makes the LAs extremely amenable for implementing Razor recovery, as simply extending existing queues provides the necessary storage for the speculative state.

We present silicon measurement results from the first application of Razor to a hardware accelerator. The accelerator (RZLA) is optimized for the dominant loop of the Sobel edge-detection algorithm (Sec. 2). We describe the design of a pulsed-latch based RFF deployed in conjunction with level-sensitive latches, automatically inserted in the combinational logic, to address the minimum-delay constraint. This algorithm enables the use of 50% of the clock-period for timing-speculation. Fabricated in 65nm CMOS, the RZLA reclaims voltage margins to demonstrate 34% energy-efficiency improvements on a per-device basis and 33% overall, for the entire batch of devices at 1GHz operation.

II. MICRO-ARCHITECTURAL DESIGN OF THE RZLA

Fig. 1(a) shows the baseline micro-architecture of the RZLA. The RZLA is a hardware realization of a modulo scheduled loop [7-8]. Modulo-scheduling is a software pipelining technique that achieves high levels of parallelism by overlapping successive iterations of a loop. The RZLA exploits this parallelism through the use of multiple functional units (FUs), each dedicated to a specific operation in the loop. Unlike microprocessors, the RZLA does not require explicit support for mechanisms such as exception handling. Therefore, state is primarily maintained primarily in Shift Register Files (SRF) to be consumed when required and then immediately discarded. A Central Register File (CRF) contains constants and configuration information and the local

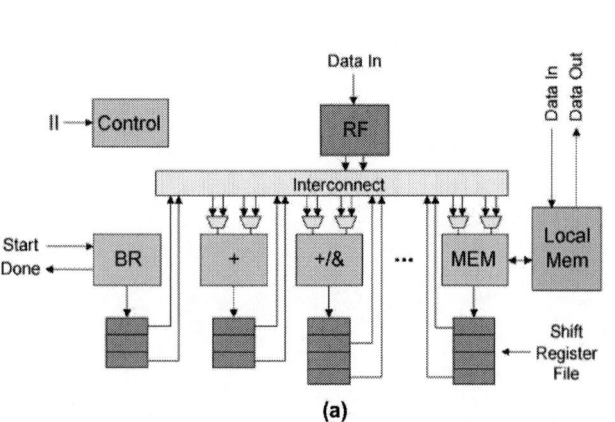

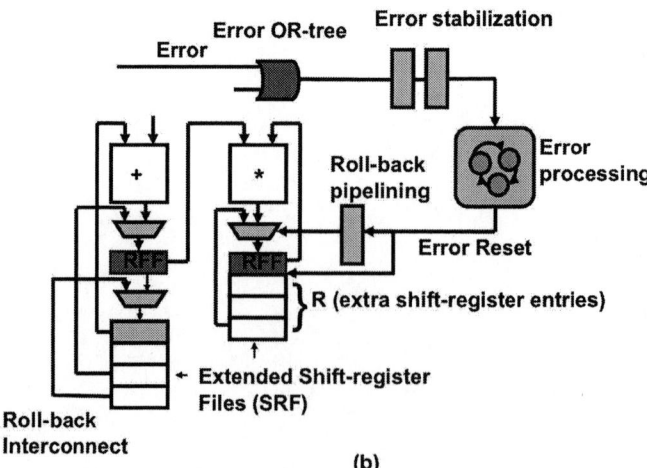

Fig. 1: RZLA Micro-architecture design a) shows the baseline micro-architecture and b) shows the modifications necessary for Razor-based timing-error detection and correction

978-1-4673-6145-3/13 $31.00 © 2013 IEEE

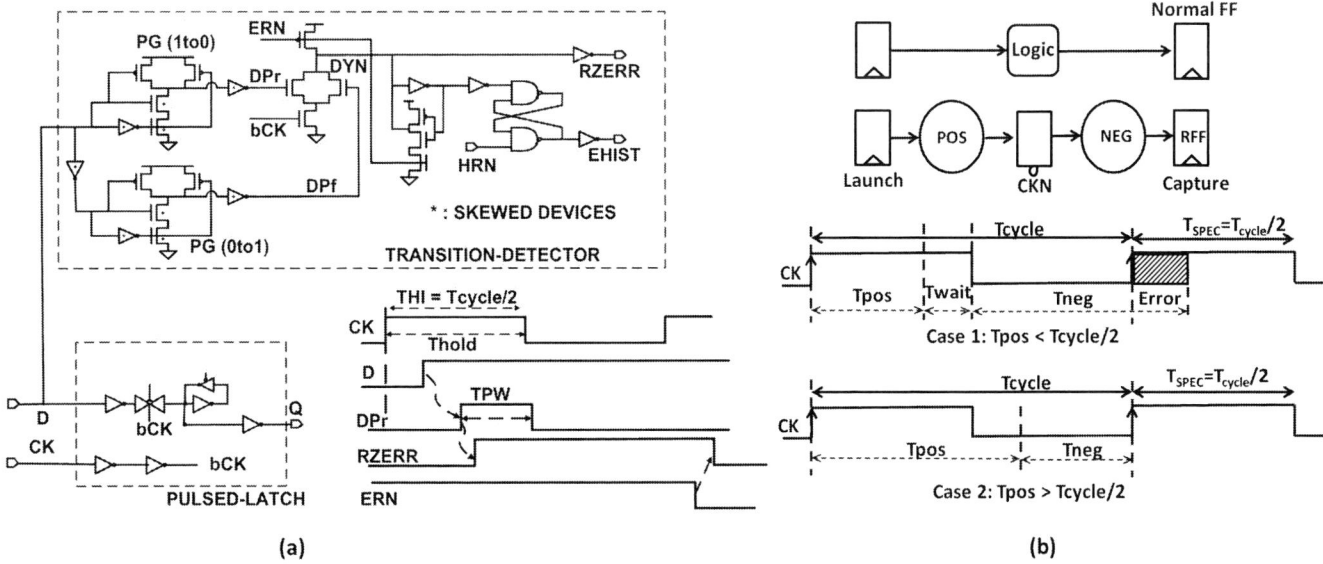

Fig. 2: a) Shows the circuit schematic of the RFF and associated timing diagrams. b) Shows the latch-insertion algorithm to satisfy the minimum-delay constraint, enabling the use of 50% of the cycle time for timing-speculation

memory provides the primary input and output data storage.

Fig. 1(b) shows the structural modifications to the baseline micro-architecture required for Razor support. The top entries of each SRF employ RFFs that flag timing violations at the outputs of the functional units (FUs). Individual RFF error-flags are OR-ed together to generate a composite error signal for the entire design. The error OR-tree output is double-latched to mitigate against potential metastability risk. When a timing-error is detected, the error controller asserts the ERROR_RESET signal that resets the error-states of all RFFs in the design. Recovery occurs by overwriting the top-entries of the SRFs with known correct values from a previous computation. Consequently, SRFs are extended by additional entries so that correct pipeline state is still in flight and available for recovery. The number of extra SRF entries, R, is dependent upon the error-response latency (interval between timing-error detection and assertion of ERROR_RESET) and is at least 2 cycles to account for double-latching the output of the error OR-tree. The error-controller tracks the error-rate and adjusts the voltage and frequency accordingly.

III. RFF TRANSISTOR-LEVEL IMPLEMENTATION

The RFF (Fig. 2(a)) augments a positive-phase transparent pulsed-latch with a transition-detector (TD) that flags late-arriving transitions on the input, D. Rising-edge triggered flip-flop operation is enforced by flagging any transition on D, in the clock high-phase, as a timing error. The transition-detector incorporates explicit generators to create wide pulses, DPr and DPf, out of rising and falling transitions on D. The overlap of the data-pulse with the clock high-phase discharges the dynamic node, DYN, flagging the error signal. The error-reset signal, ERN, precharges DYN in preparation for the next error event. The RFF includes a RS-latch that acts as an error history (EHIST) diagnostic bit, set whenever the error signal is flagged. Reading out the EHIST bits allows precise identification of individual RFFs that triggered during a test.

Unlike in [3], the RZLA design considerably relaxes the sizing requirements of the TD by using a pulsed-latch, instead of a flip-flop, which eliminates any setup constraints at the rising edge. The pulsed-latch designs in [2,4,6] incur the additional power overhead of the suppression pulse-generator in [2] and the shadow flip-flop in [4]. The design in [6] enables time-borrowing, although at the expense of double-sided constraints on the clock high-phase. Narrow pulse-width causes spurious error-detection while wide pulse-width prevents a significant portion of the high-phase from being monitored.

A. Latch-insertion based hold-fixing algorithm

Disambiguating between early- and late- arriving transitions imposes a minimum-delay constraint in all Razor systems [1]. This is especially exacerbated in the pulsed-latch architecture wherein the minimum-delay constraint spans the entire high-phase of the clock. In previous work [2,4,6], this is addressed by distributing an asymmetric duty-cycle clock that is invariant with frequency. The truncated high-phase trades-off a smaller speculation window for reduced power penalty due to delay buffers on the short-paths. However, at advanced process nodes, significant minimum-delay margining is required on the short-paths of the design which adversely impacts the energy-efficiency improvements due to Razor.

The RZLA addresses the minimum-delay constraint by the explicit instantiation of level-sensitive latches (henceforth referred to as "hold-fixing latches"). Combinational logic within each pipeline stage is divided into two blocks with approximately equal critical-path delay (Fig. 2(b)). Negative-phase transparent latches are inserted between the two logic blocks thus created. The opaque latches prevent short-path computations from updating inputs to the RFFs until the negative clock-phase. The latches are only required in the fan-in logic of the RFFs, unlike in [5] where latches are inserted in the entire design which significantly increases the clocking load. The key advantage of this technique is that it enables the use of conventional 50% duty cycle clocking, thereby greatly simplifying clock-generation and propagation. A large window for timing-speculation (Tspec=Tcycle/2) mitigates against late-transitions catastrophically escaping detection.

978-1-4673-6145-3/13 $31.00 © 2013 IEEE

Technology	UMC 65SP
Die Area	690um x 340um
Total RFF (DFF)	223 (926)
Total Latches	709
VDD Range	0.8V – 1.3V
Signoff Fmax	667MHz SS/0.9V/125C

Power Overheads over Baseline @ 1V/85C/TT	
Clock buffers due to hold-fixing latches	1.68%
Hold-fixing Latches	1.47%
Extra SRF Entries to support recovery	3.31% (2 cycles latency) 11.7% (8 cycles latency)
Power overhead of RFF compared to conventional DFF	
State static (RFF/DFF)	7.03fJ/6.98fJ (0.7%)
State toggles (RFF/DFF)	42.15fJ/24.26fJ (74%)
Total RFF overhead	**1.7%**
Total Power Overhead	**8.16% (2 extra SRF entries)**

Fig. 3: Die photograph and implementation details

Further, by eliminating buffer-insertion, this technique is resilient to process variability in short-paths.

Insertion of latches increases the critical-path delay and creates phase-paths that can potentially impact the error-rate (Fig. 2(b)). Negative-phase logic cannot begin computation until after the falling clock-edge, even when the positive-phase logic (Tpos) completes execution early within in the high-phase (Case 1 in Fig. 2(b)). This introduces an additional "wait time" (Twait) in the path that causes previously non-critical paths to fail timing. However, in practice the contribution to the error-rate due to latches is minimal because the error-rate is almost entirely dominated by multiple RFFs failing as a consequence of the balanced nature of the pipeline. Measured results in Fig. 6 illustrate this in further detail.

Hardware accelerators are typically characterized by high switching activities when enabled. This leads to significantly higher contribution of the combination logic to the total power of the design, compared to microprocessors. The combinational logic power amortizes the increased clocking power attributed to the hold-fixing latches. For example, in the RZLA, combinational logic consumes 56% of the total power. In this design, 223 RFFs required 709 level-sensitive latches. The latches and the additional clock-buffers contribute 1.68% and 1.47% overhead, respectively, to the baseline power.

IV. CHIP IMPLEMENTATION DETAILS

The die photograph and the implementation details of the RZLA are shown in Fig. 3. The signoff frequency is 667MHz at the slow corner (SS/0.9V/125C). The RZLA has 926 pipeline registers (not including the flip-flop based local memory and the central register file), of which 223 registers are RFFs. The latch-insertion algorithm inserts 709 level-sensitive latches. The pulsed-latch architecture of the RFF eliminates the master-latch, leading to low overheads when

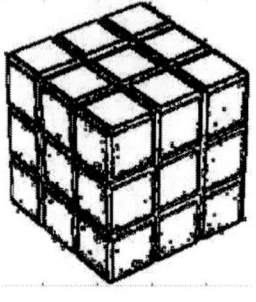

a) Original 8-bit grayscale image **b) Edge-Detected RZLA Output**

Fig. 4: Sobel Edge Detection on the RZLA.

RFF state does not toggle at the rising edge of clock (0.7%). This takes advantage of the higher activity-rate of the clock signal compared to the data, in order to amortize the extra switching in the transition-detector when state toggles. The power overhead of the Razor recovery logic is dominated by additional (SRF) entries. A minimum 2-cycle latency contributes 3% to the baseline power that increases to 11.8% for 8 entries of the SRF. The total power overhead of Razor circuitry is measured to be 8.16% (for 2 extra SRF entries).

V. MEASUREMENT RESULTS

Fig. 4 shows the RZLA output for an 8-bit grayscale input, used as the test workload. The original (256x256 pixels) image is split into multiple frames of size 62x3 pixels each, which can be storage within the RZLA internal frame-buffer. The execution output from each frame is combined together to create the final edge-detected image.

Fig.5 shows the response of the Razor adaptive voltage controller, executing an individual frame from the test-image at 1GHz operating frequency, on 3 dies from the typical (TT), fast (FF) and slow (SS) silicon lots. The controller monitors the on-die error-register and tunes the supply voltage until the Point of First Failure (PoFF) of individual die. For the SS die, the PoFF at 1GHz occurs at 1.17V compared to 1.1V for TT. For 100% parametric yield, operation without Razor requires sufficient safety margins guaranteeing the slowest die to operate correctly under worst-case variation conditions. Hence, we add 100mV (~10% VDD) as margin for VDD variations and a bare minimum 30mV (~3%) as margin for measurement uncertainty. This results in 1.3V being the

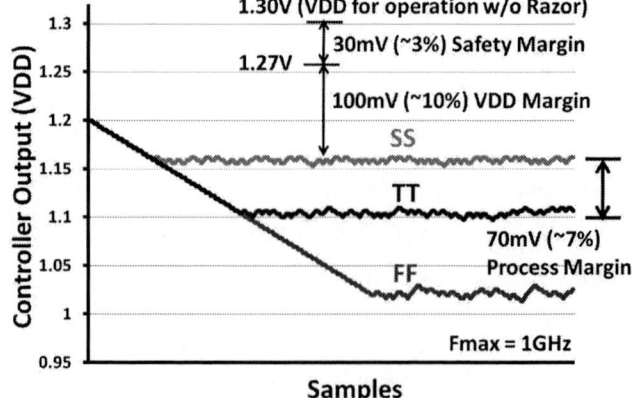

Fig. 5: Razor DVS Controller Output

978-1-4673-6145-3/13 $31.00 © 2013 IEEE

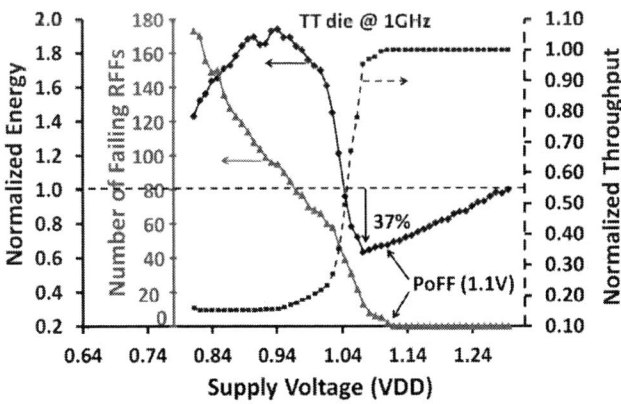

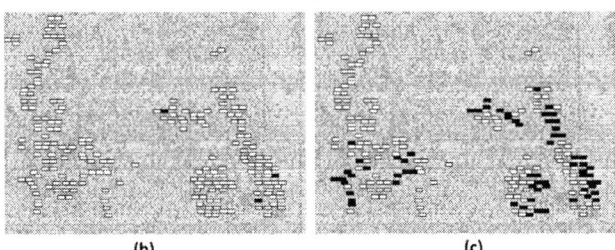

Fig 6: a) Normalized Energy/Throughput as function of VDD for TT die at 1GHz frequency b) and c) show the map of failing RFFs (represented by black rectangles) at 1.1V(b) and 1V (c) at1GHz operation, respectively

worst-case safe voltage, for correct operation without Razor. The efficiency gains using Razor are potentially greater due to margins for temperature variations and ageing effects that have not been budgeted for in this work.

Fig. 6 shows the impact of timing errors on normalized (with respect to execution at 1.3V) energy and throughput, plotted as a function of the supply voltage. Energy consumption reduces quadratically with VDD, until the PoFF. Further VDD reduction incurs exponentially increasing recovery penalty due to multiple RFFs detecting timing errors, leading to the overall energy consumption to increase. Razor-based DVS enables 37% energy savings at the optimal point and 34% energy saving at the PoFF. Correct operation with successful recovery is achieved until 0.8V representing a 300mV error-detection range. Correctness at a particular voltage point is determined by comparing the execution output against that at 1.3V. Fig. 6 (b) shows the map of failing RFFs generated from EHIST information of individual RFFs. At the PoFF (1.1V), there are only 3 RFFs that fail timing compared to 66 at 1.0V.

Fig. 7 shows the total energy savings obtained using Razor. The first set of bars show the case where Razor is turned off and the dies operate at the worst-case safe voltage (1.3V). The additional power due to margins for process, voltage and safety are labeled separately. The margin for process variation is computed by the extra power consumption due to operating at the zero-margin point or the PoFF of the slowest die, "SS" (1.17V) versus operating at their native PoFF. Similarly, power due to voltage margin is computed by the extra overhead of operating at 1.27V (Fig.5) versus operating at 1.17V. The TT die consumes 137mW when

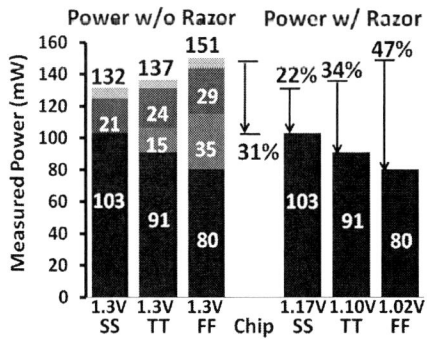

Fig. 7. Energy savings with Razor

operating at 1.3V out of which 15.5mW is due to process variation margins, 24.7 is due to supply voltage margins and 5.6mW is due to safety margins. Razor enables correct operation at the native PoFF for individual die, eliminating excess margins. Margin elimination enables 34% saving for the TT die (22% for the SS and 47% for FF). Fig. 8 compares the statistical distribution of the power consumption with Razor enabled versus the worst-case operation at 1.3V. Razor enables 33.3% savings on the worst-case power-consumption for the entire batch of devices. It is known that canary-circuits based adaptive techniques and "binning" of devices can reclaim some of the process margins. However, unlike these techniques, Razor can address margins due to rapidly-changing voltage transients, local process variations and test and measurement uncertainties.

VI. CONCLUSION

We presented the design and silicon measurement results of a Razor-enabled hardware loop-accelerator (RZLA). We exploit unique micro-architectural feature in loop-accelerators to make different trade-offs in the implementation of Razor, compared to microprocessors. Gains due to Razor illustrate the efficacy of error-tolerant designs to address PVT variations at advanced process nodes.

REFERENCES

[1] S. Das et al., JSSC 2006

[2] S. Das et al., JSSC 2009

[3] D. Bull et al., JSSC 2011

[4] K. Bowman et al., JSSC 2011

[5] M. Fojtik et al., JSSC 2012

[6] P. Whatmough et al., ISSCC 2013

[7] G. Dasika et al., DAC 2008

[8] Rau et al., http://www.hpl.hp.com/techreports/92/HPL-92-4

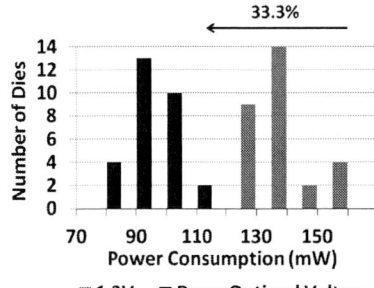

Fig. 8: Distribution of power consumption on 30 dies

Energy-Efficient Recognition and Mining Processor using Scalable Effort Design

Vinay K. Chippa, Hrishikesh Jayakumar, Debabrata Mohapatra, Kaushik Roy, Anand Raghunathan

School of Electrical and Computer Engineering, Purdue University

{vchipp,hjayakum,dmohapat,kaushik,raghunathan}@purdue.edu

Abstract—A domain-specific processor for energy-efficient execution of Recognition and Data Mining (RM) workloads is presented. The processor consists of a 2-D array of processing elements and a streaming memory hierarchy and interconnect network that are customized to efficiently execute dominant computational kernels (matrix-vector multiplication, vector dot product, L1 norm, and L2 norm) from a wide range of RM algorithms. To achieve further energy efficiency, the RM processor utilizes scalable effort design, a technique that exploits the inherent resilience of algorithms to inexactness in their constituent computations. The scalable effort RM processor adopts a cross-layer approach by combining scaling mechanisms at the algorithm, architecture, and circuit levels, to create a desirable trade off between energy consumption and output quality. Measurements from the implemented chip in 65nm CMOS indicate processing efficiencies of 569 GOPS/W - 4.68 TOPS/W. The use of scalable effort design achieves energy savings of 1.2-2.3X with no loss in output quality, and 2X-20X with modest reduction in quality.

I. INTRODUCTION

The need to understand, interpret, and make inferences from various forms of digital data is increasingly driving the usage of computing platforms. In data centers, the explosion in the creation and consumption of digital data has led to a need for technologies that efficiently organize, analyze and search through large datasets of text, images, and videos. On the other hand, mobile and embedded computing systems increasingly need to understand events and inputs from the physical world or human users. An emerging class of applications, called Recognition and Mining, addresses these problems using mathematical techniques and algorithms derived from machine learning and data mining [1]. Some examples of RM applications include semantic text analysis, document and web search, handwriting and speech recognition, object detection/recognition/tracking from images and video, unstructured data analytics, business intelligence, and recommendation engines.

RM applications are highly demanding computing workloads, and are expected to only get more demanding due to the growth in datasets and the use of more complex algorithms. Fortunately, they contain abundant parallelism, which can be exploited on general-purpose multicore and many-core processors. However, achieving high performance under stringent power/energy constraints requires hardware that better exploits the unique characteristics of RM applications. Previous hardware implementations of specific RM algorithms [2], [3] and applications [4], [5] have demonstrated great promise. However with the increasing diversity and ubiquity of RM applications, programmable domain-specific processors that can execute a range of RM algorithms (like GPUs for graphics or network processors for network protocol processing) have drawn increasing interest [6]. Such an RM processor may be integrated into System-on-chips (SoCs) for mobile or embedded platforms, or may be deployed as an accelerator card in servers (Fig. 1).

We describe an energy-efficient RM processor that consists of a 2D array of processing elements, and a streaming memory hierarchy and interconnect network. The architecture of the RM processor is optimized for the computation kernels present in RM algorithms and their corresponding memory access and communication patterns. These computation kernels, including matrix-vector multiplication, vector

---

This work was supported in part by the National Science Foundation under grant no. 1018621

dot-product, distance computation (L1/L2 norm), *etc.*, expose fine-grained parallelism that is best exploited by many small processing elements, and require communication patterns such as reduction and nearest-neighbor communication, which are not efficiently supported by current many-core accelerators such as GPUs.

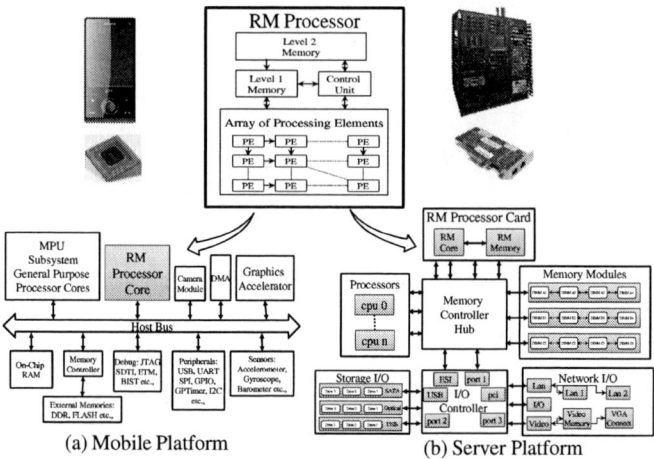

Fig. 1: RM processor deployed as (a) On-Chip accelerator in mobile platforms and (b) External daughter card in server platforms

Key to the energy efficiency of the RM processor is a design technique called *scalable effort hardware design*. This technique exploits a unique characteristic of RM algorithms, *viz.* their ability to produce acceptable outputs even in the presence of significant inexactness or approximations in their constituent computations [7]. This inherent resilience is due to several factors, including redundancy in the input data, the use of algorithms that are robust noisy inputs, and computation patterns that attenuate or cancel out errors in computations. Moreover, for most RM applications, the eventual objective is to produce output that meets a certain quality (*e.g.* a given classification accuracy) rather than compute an exact "golden" numerical output. The objective of scalable effort hardware design is therefore to obtain the best possible energy-quality tradeoff, *i.e.*, the lowest energy for a given quality or conversely the highest quality for a given energy. This is achieved by (i) implementing hardware with inbuilt *scaling mechanisms* or knobs that may be used to tradeoff energy efficiency for output quality, and (ii) synergistically utilizing these scaling mechanisms to obtain the most desirable energy-quality tradeoff.

Inherent application resilience has been exploited in hardware implementations of digital signal processing algorithms through the use of techniques such as voltage overscaling [8], [9]. However, the more complex nature of RM algorithms enables a richer space of scaling mechanisms at the algorithm, architecture, and circuit levels, and implies that cross-layer optimization across scaling mechanisms is necessary to realize the best energy-quality tradeoffs.

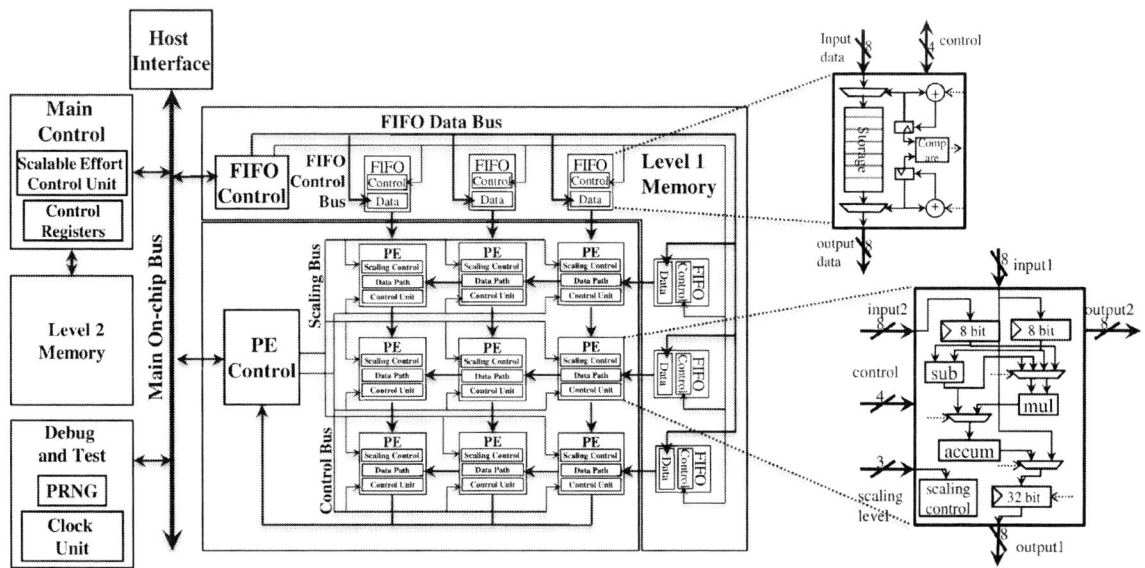

Fig. 2: Block diagram of the RM processor and its components

II. SCALABLE EFFORT RM PROCESSOR

In this section, we first describe the architecture of the RM processor and its constituent components. We then describe how scalable effort design techniques were applied to the RM processor to improve its energy efficiency, and discuss the scaling mechanisms that were chosen at the algorithm, architecture, and circuit levels in order to provide a desirable energy-quality tradeoff.

Recognition and Mining applications, like many compute-intensive workloads, are dominated by computation kernels that account for most of their execution time. The dominant kernels for seven widely-used RM applications, and their contributions to the respective application run times, are reported in Table I. The RM processor is designed to execute these kernels (matrix-vector multiplication, vector dot-product, and distance computation using L1 and L2 norm), which may be offloaded from applications executing on a host processor.

TABLE I: Dominant kernels in RM applications and the fraction of execution time spent in them

Application (Algorithm)	Dominant Kernel	Contribution to Run-Time
Document Search (Semantic Search Index)	Matrix Vector Multiplication	86%
Image Search (Feature Extraction)	Dot Product Computation	71%
Hand Written Digit Recognition (Support Vector Machines)	Dot Product Computation	94%
Image Segmentation (K-means Clustering)	Distance Computation (L2)	66%
Eye Detection (Generalized Learning Vector Quantization)	Distance Computation (L2)	89%
Optical Character Recognition (K-Nearest Neighbours)	Distance Computation (L1)	92%
Online Data Clustering (Stream Cluster)	Distance Computation (L2)	68%

A. RM Processor: Design and Operation

The architecture of the implemented RM processor is shown in Figure 2. It contains 9 processing elements (PEs) arranged in a 3x3 systolic array, with each PE containing a datapath and control logic to perform the scalar operations that constitute the computation kernels. A 2-level memory hierarchy is used, consisting of 6 streaming FIFOs (each FIFO has 15 entries of 8 bits) as the first level memory and a second level memory that is connected to a wide on-chip bus. The FIFOs are connected to PEs at the top and right borders of

the systolic array. Streaming nearest-neighbor connections between PEs are used to transfer data through the PE array. The proposed architecture can be scaled to an $m{\times}n$ array of PEs with $m+n$ FIFOs, with proportional increase in the second level memory bandwidth, to suit varying power/performance targets.

A PE array controller broadcasts control inputs to all the PEs to: (i) initialize the PEs and specify the kernel computation, (ii) orchestrate the streaming data transfer across PEs. Similarly, control inputs to the FIFOs, supplied by the FIFO controller (i) orchestrate data transfer between the second level memory and the FIFOs, and between FIFOs and PEs and (ii) keep track of the full/empty status of the FIFOs.

The RM processor communicates with the host processor through a host interface. The host processor transfers the data into the second level memory and downloads the program to be executed into the main controller. This program consists of a series of instructions that configure the PE controller and the FIFO controller to realize the desired kernel computation. Once the program is downloaded, the host initiates the kernel computation by writing to a special memory-mapped control register in the main controller. The completion of the kernel computation is detected by the host polling a memory mapped status register in the main controller, after which the output of the kernel computation is transferred back from the second level memory to the host.

B. Energy Efficiency Through Scalable Effort Design

Scalable effort design is a promising approach to achieving energy efficiency in hardware implementations of inherently resilient applications. A conceptual illustration of scalable effort design is

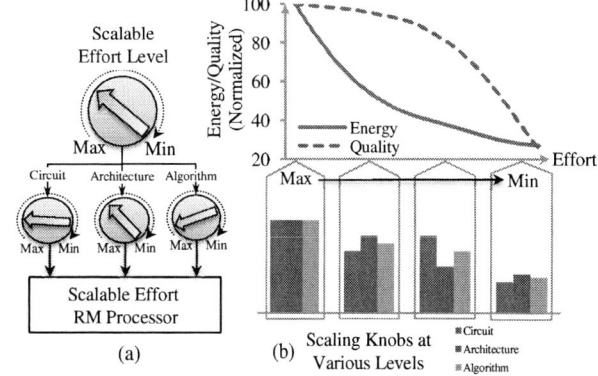

Fig. 3: Scalable Effort Design: (a) Run-time energy-quality tradeoff through scaling mechanisms at different levels of design abstraction (b) Energy consumption vs. output quality at different effort levels

978-1-4673-6145-3/13 $31.00 © 2013 IEEE   697

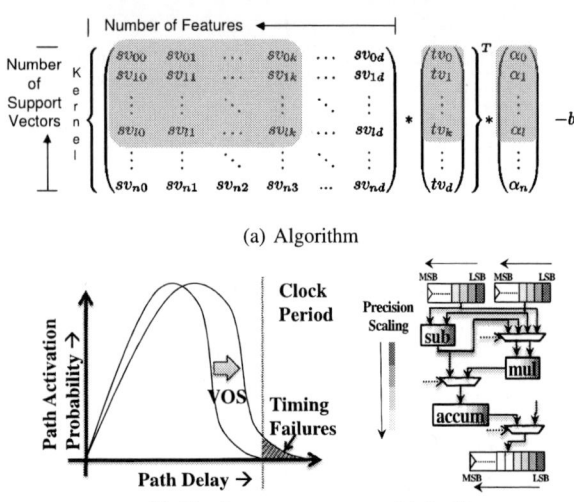

(a) Algorithm

(b) Circuit (c) Architecture

Fig. 4: Scaling mechanisms at different levels of design abstraction

provided in Figure 3. Scalable effort hardware implementations are augmented with scaling mechanisms that can be regulated at run-time to achieve significantly improved energy efficiency with little impact on the application output quality. Scaling mechanisms can be implemented at the algorithm, architecture, and circuit levels (Figure 3(a)). While these individual scaling mechanisms themselves present energy-quality trade-offs (each scaling mechanism is associated with a level that determines how aggressively it is applied), the overarching objective of scalable effort design is to synergistically co-optimize these scaling mechanisms in order to achieve the best overall energy-quality tradeoff (Figure 3(b)).

We next describe the algorithm, architecture, and circuit level scaling mechanisms that are implemented in the RM processor.

a) Algorithm Level: At the algorithm level, we identify parameters that control the number of operations performed in each kernel, and vary them to tradeoff computation complexity with output quality. These parameters typically involve limiting the number of iterations that a loop is allowed to execute (early termination), or the number of dimensions of a vector that are used in a vector operation (computation skipping). For example, Fig. 4a shows that, in the case of Support Vector Machine (SVM) classification, we could reduce the number of support vectors, or the number of features, which correspond to the number of rows or columns utilized in the matrix-vector multiplication. Rather than indiscriminately eliminating computations, we prioritize computations based on their impact on the output quality to obtain a better tradeoff. For SVM classification we sort support vectors based on their significance (quantified by the Lagrangian coefficient values) and skip the less significant support vectors.

b) Architecture Level: At the architecture level (Figure 4c), we scale effort by reducing the precision with which the variables of the algorithm are represented and processed by the PEs. We employ clock gating to obtain energy savings in the bit-slices of the datapath

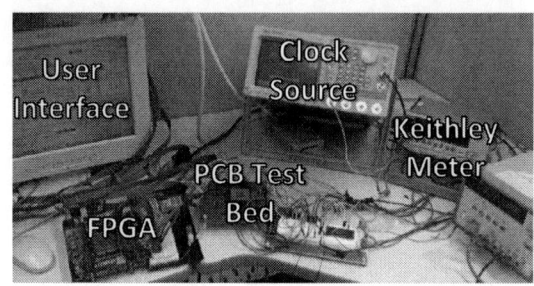

Fig. 6: Validation setup

that correspond to the unused bits. Other techniques such as power gating and approximate datapath circuits may also be used [7].

c) Circuit Level: At the circuit level, we scale the electrical effort expended by the hardware, by overscaling the supply voltage to the PEs and FIFOs without varying the clock frequency. This leads to timing violations in near-critical paths of the circuit as shown in Figure 4b. To ensure gradual degradation in output quality, we designed the PEs such that the timing errors are limited only to the data path by ensuring adequate timing margins in the control logic. Further benefits are obtained by employing processing elements that are designed to behave gracefully under voltage overscaling [10].

A centralized scalable effort control unit (in the main controller) supplies scaling controls to the PEs and FIFOs through their respective controllers. The scaling mechanisms can be controlled at run-time by writing into special registers present in the scalable effort control unit.

III. MEASUREMENT RESULTS

In this section, we first describe the implementation and measurement setup used to evaluate the scalable effort RM processor. Next, the results of various experiments that demonstrate the energy efficiency of the implemented RM processor are presented.

A. Implementation and Experimental Setup

The RM processor chip is implemented in TSMC's 65nm technology node. The details of the chip implementation are shown in Figure 5. The RM processor occupies 1.3sq.mm. on a 6mm x 6mm chip shared by 4 other designs. It achieves power efficiencies ranging from 569 GOPS/W - 4.68 TOPS/W depending on the settings of the scaling mechanisms (the energy-quality tradeoff resulting from the scaling mechanisms is discussed in the next sub-section). In order to validate the RM processor chip, we emulated the scenario where it is integrated into a SoC. An Altera NIOS2 core operating at 100MHz on an Altera DE3 FPGA board was used to emulate the host processor, and it communicated with the test board for the RM Processor through GPIO pins. Due to the frequency limitations of the GPIO pins, data was transferred in and out of the RM processor at a slower clock frequency (1 MHz), while the computation in the RM Processor was performed at a higher clock frequency (250 MHz). A suite of four RM applications (described in Table II) was first ported to the NIOS2 processor, and the applications were then modified to offload their

Process Technology	TSMC 65nm rVt
Supply Voltage	0.6-1.0 Volts core 3.3Volts I/O
Operating Frequency	250MHz
Peak Power Consumption	0.96mW to 7.9mW
Area	1.3 mm²
Gate Count	10,000

(a)

(b) (c)

Fig. 5: (a)Die micrograph (b) Layout and (c) Chip statistics

TABLE II: Applications used for evaluation of the scalable effort RM Processor Chip

Application (Dataset)	Algorithm	Dimensionality	Classes (Clusters)
Digit Recognition	Support Vector Machines	784	10
Checkerboard Classification	Support Vector Machines	3	2
Image Segmentation	K-Means Clustering	3	4
Eye Detection	Generalized Learning Vector Quantization (GLVQ)	512	2

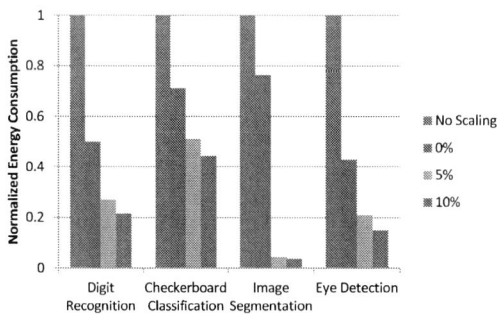

Fig. 7: Energy savings for various applications on the scalable effort RM Processor for varying output quality

computation kernels to the RM processor. These applications were repeatedly executed after setting the scaling mechanisms of the RM processor to all possible levels, and the energy consumption in the RM processor and quality of the application results were measured. The output quality was measured in terms of % of correctly clustered points for image segmentation and % of correctly classified points for the other applications.

B. Experimental Results

In this section, we first present results that demonstrate the benefits of scalable effort design in the RM processor. Figure 7 shows the normalized energy consumption of different applications at varying levels of loss in output quality. Energy consumption is normalized to the case in which no scaling mechanisms are utilized. Energy savings ranging from 1.3X-2.3X are obtained with no noticeable impact on the output quality. The energy savings increase to 2X-20X if a modest (5%) loss in output quality is allowed.

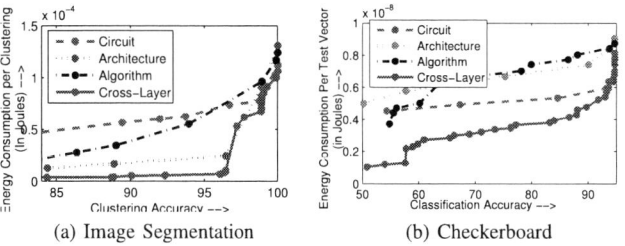

(a) Image Segmentation   (b) Checkerboard

Fig. 8: Cross-Layer Optimization for different applications

Figures 8a and 8b illustrate the benefits of cross-layer optimization of the scaling mechanisms for checkerboard classification and image segmentation. Significantly improved energy *vs.* quality trade offs are obtained by utilizing the optimal combinations of all the scaling mechanisms as compared to the application of individual scaling mechanisms.

In order to understand the effects of cross-layer optimization on application output quality, we visually depict the impact of different scaling mechanisms on the checkerboard classification output in Fig. 9. The objective is to classify points on a checkerboard into white or black squares based on their (x,y) co-ordinates. The ideal application output is shown in Fig. 9(a). Even with no scaling applied (Fig. 9(b)), the application output differs from the ideal case due to the inherent limitations of the underlying algorithm. Figs. 9(c)-(e) show that different scaling mechanisms affect the application output in profoundly different ways. Algorithm level scaling results

(a) Ideal   (b) No Scaling (c) Algorithm (d) Architecture   (e) Circuit

Fig. 9: Output quality of `Checkerboard` classification application for different scaling mechanisms

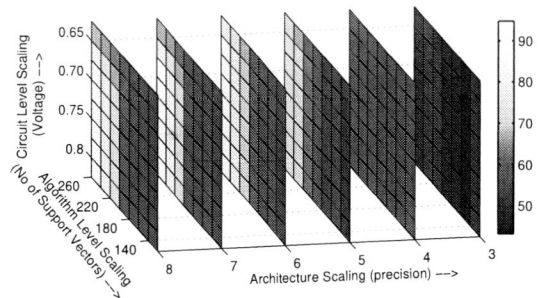

Fig. 10: Interaction between different scaling mechanisms

in a gradual deformation in the classification boundary (Fig. 9(c)). Architecture level scaling results in a quantization of the classification boundary (Fig. 9(d)). Finally, circuit level scaling results in random errors across the input space (Fig. 9(e)). Due to this diversity in their effects, different scaling mechanisms can be applied together to obtain additional benefits in energy.

To gain further insights into the effects of the scaling mechanisms, we exhaustively apply all possible levels of different scaling mechanisms to the checkerboard classification application and present the impact on output quality in Fig. 10. The three axes in Fig. 10 represent the three scaling mechanisms, and the cells in the 2D slices are color-coded to represent the output quality. The figure suggests that only a weak dependence exists between different scaling mechanisms. Once scaling is applied along one of the axes to reach a desired output quality, further scaling along a different axis is possible, leading to additional energy benefits without further degrading output quality .

In summary, our results demonstrate the efficiency of the proposed RM processor and the utility of scalable effort hardware design in improving the energy efficiency of RM applications.

IV. Conclusions

A Recognition and Mining (RM) Processor that exploits the inherent application resilience of RM applications through scalable effort design has been implemented in 65nm technology. The RM processor utilizes scaling mechanisms that trade-off output quality *vs.* energy consumption at the algorithm, architecture, and circuit levels. Measurement results for 4 representative RM applications demonstrate 1.3X-20X energy improvements though scalable effort design and establish the benefits of cross-layer optimization of the scaling mechanisms.

References

[1] Yen-Kuang Chen *et al.* Convergence of recognition, mining, and synthesis workloads and its implications. *Proceedings of the IEEE*, 96(5):790–807, 2008.

[2] J. Misra and I. Saha. Artificial neural networks in hardware: A survey of two decades of progress. *Neurocomputing*, 74(13):239–255, 2010.

[3] Magdy A. Bayoumi. *Learning on Silicon: Adaptive VLSI Neural Systems.* Kluwer Academic Publishers, Norwell, MA, USA, 1999.

[4] S. Lee *et al.* A 345mw heterogeneous many-core processor with an intelligent inference engine for robust object recognition. In *Proc. ISSCC*, pages 332–333, 2010.

[5] J. Park *et al.* Online reinforcement learning noc for portable hd object recognition processor. In *Proc. CICC*, 2012.

[6] A. Majumdar *et al.* A massively parallel, energy efficient programmable accelerator for learning and classification. *ACM Trans. Archit. Code Optim.*, 9(1), March 2012.

[7] V.K. Chippa, D. Mohapatra, A.Raghunathan, K. Roy, and S.T. Chakradhar. Scalable effort hardware design: Exploiting algorithmic resilience for energy efficiency. In *Proc. DAC*, 2010.

[8] R. Hegde and N.R. Shanbhag. A low-power digital filter IC via soft DSP. In *Proc. CICC*, pages 309–312, 2001.

[9] Amit Sinha, Alice Wang, and Anantha Chandrakasan. Energy scalable system design. *IEEE Trans. VLSI Syst.*, 10(2):135–145, 2002.

[10] Debabrata Mohapatra *et al.* Design of Voltage Scalable Metafunctions for Multimedia, Recoginition and Mining Applications. In *Proc. DATE*, pages 950–955, 2011.

978-1-4673-6145-3/13 $31.00 © 2013 IEEE

An Energy-Efficient Coarse-Grained Dynamically Reconfigurable Fabric for Multiple-Standard Video Decoding Applications

Leibo Liu[1], Chenchen Deng[1], Dong Wang[1]*, Min Zhu[1], Shouyi Yin[1], Peng Cao[2], Shaojun Wei[1]

[1]Institute of Microelectronics, Tsinghua University, Beijing, P. R. China, [2]Nation ASIC System Engineering Research Center, Southeast University, Nanjing, P. R. China.*Email: wangdong1981@tsinghua.edu.cn

Abstract-In this paper, we introduce a coarse-grained dynamically reconfigurable fabric, named Reconfigurable Processing Unit (RPU), which is implemented on a 5.4x3.1 mm^2 silicon with TSMC 65 nm LP1P8M technology. This fabric consists of 16x16 multi-functional Processing Elements (PEs) interconnected by an area-efficient Line-Switched Mesh Connect (LSMC) routing. A Hierarchical Configuration Context (HCC) organization scheme is proposed to reduce the scale of the context memory and enhance configuration efficiency. Two reconfigurable processors are then designed and fabricated to verify the proposed techniques. One processor (called REMUS_HPP) integrates two RPUs, targeting the high performance applications. REMUS_HPP could decode 1920x1080@30fps H.264 streams with 280mW under 200MHz, achieving a performance gain of 1.81x and a 14.3x energy efficiency improvement over XPP-III. The other processor (called REMUS_LPP) integrates only one RPU, targeting the low power applications. REMUS_LPP could decode 720x480@35fps H.264 streams with 24.81mW under 75MHz, achieving a 76% power reduction and a 3.96x energy efficiency improvement compared with ADRES. More importantly, RPU is not only limited to video decoding applications. It can also be used to process some other computation-intensive applications and the corresponding analysis is given in this paper as well.

I. INTRODUCTION

Reconfigurable computing fabrics fall in between the instruction driven processor, such as GPPs (General Purpose Processor), DSPs, etc. and hardwired logic, such as ASICs (Application Specific Integrated Circuit). When dealing with computation-intensive applications, coarse-grained reconfigurable computing fabrics have substantial advantages in performance and power over instruction driven processors, while still possessing a very high function flexibility compared with the hardwired logic.

A lot of coarse-grained reconfigurable fabrics have been presented to explore the hardware/software architectures during the last decade [1][2][3]. A common focus of these works is how to utilize the computing capability of the fabrics to improve the system performance. And these works have proved that the computation-intensive tasks could be effectively boosted if the multiple levels of parallelism of these tasks are fully exploited. However, there is still quite a distance for the reconfigurable techniques from being widely used, because some key problems are still unresolved, including compiling issues, programming paradigms, etc. Also, one of the important metrics of reconfigurable fabrics, the energy efficiency (i.e. the number of operations per unit energy consumption), has not yet been deeply considered or extensively discussed.

In this paper, we would like to propose an energy-efficient architecture of a coarse-grained Reconfigurable Processing Unit (RPU). Measured results show that the proposed architecture has great improvements in energy efficiency compared with the state-of-art designs.

II. PROPOSED VLSI ARCHITECTURE

The proposed RPU architecture is shown in Fig. 1, which can be divided into two parts: data processing path and configuration path. The data processing path includes Processing Element Array (PEA), Input Control logic, Output Control logic and Inner Buffering memory. The configuration data path refers to Context Storing & Controlling Logic.

A. Data Processing Path

The data processing path is responsible for fetching, processing, storing and exporting of data streams. The core part is a 16x16 coarse-grained PEA and it is organized into four groups referred as the Reconfigurable Cell Array (RCA). As shown in Fig.2, each RCA contains 8x8 reconfigurable Processing Elements (PEs). These four RCAs can work independently thus providing a higher level of parallelism and can be turned off individually to save power. Each PE contains a 16-bit Arithmetic Logic Unit (ALU), integrating up to 26 different operators: logical operators, such as NOT, AND, OR, XOR, etc., and arithmetic operators, such as Adder, Multiplier, Shifter, Comparator, Saturation, Absolute, etc. Considering the area overhead, only 1 out of the 8 lines in each RCA contains Multipliers (i.e. 8 multipliers in total reside in one RCA), which makes PEA a heterogeneous array.

The PEs within each RCA are organized in a line-to-line manner, i.e. each PE could be connected to any PEs in the adjacent upper and lower lines through the inter-layer mesh interconnections as shown in Fig. 2 (the last connects to the first). For I/O data, the RCA is connected to a pair of FIFOs through an I/O bus. Fig.3 (a) shows a traditional full mesh connection scheme where PEs within the RCA are equally connected to the I/O FIFO's data ports (D1, D2, ... , D32). In order to save the interconnection area, a Line-Switched Mesh Connect (LSMC) structure is proposed as shown in Fig. 3 (b). In this structure, only one line of PEs and a corresponding line of FIFO data ports are selected and connected to the I/O bus at one configuration. After that, a full mesh network is utilized to transmit data

978-1-4673-6145-3/13 $31.00 © 2013 IEEE

between the I/O FIFO and PEs in the I/O bus. For a 16x16 PEA, the area ratio of interconnection to PEs is therefore reduced from 80% to 20%, which substantially reduces the power consumed by the interconnection as well. We have also verified that there is little downside to performance as long as the input/output data streams are well pipelined.

B. Configuration Path

In the configuration path, Hierarchical Configuration Context (HCC) representation scheme is proposed to reduce the context storage and transmission overhead. As shown in Fig. 4, the context is logically organized in three levels: core context, group context and complete context. Core context (Level1) contains 5-bit ALU OPCODEs, inter-lay interconnection configuration bits and data loading/storing commands. For H.264 decoding case, the total scale of the core contexts of the 16×16 PEA would exceed 1Kbits even without counting the contexts for the interconnections; the indexes of core contexts are grouped into a higher level, called group context (Level2), which represents a serial of data-dependent RCA level calculations. This idea is based on the fact that for certain applications, such as an iterative algorithm, the computation requires similar operations and the processed data tend to be re-used. Therefore, RCA is likely to be reconfigured using the same structures. In such scenario, repetitive storage of core contexts will waste the memory resource since most of them are the same. Consequently, using indexes of core contexts can greatly reduce such redundancy. The highest level of context representation is referred as the complete context (Level3), and it is an extension of the group context with synchronization command to control the flow of the implemented algorithm. The complete contexts are dynamically generated by the external host processor.

The architecture of the configuration path is illustrated in Fig.5. Two dedicated context memories – Global Group Context Memory (GGCM) and Global Core Context Memory (GCCM) are designed to store the group and core contexts. These memories can be initialized at system startup or reprogrammed at run time through configuration interface. During the execution, the complete contexts are fed from external host processor and transmitted to the RPU context parser, which translates the complete contexts and reads the corresponding group contexts from GGCM. The data loading or storing commands are executed by configuring the external load/store controllers and the indexes of the core contexts are then sent to RCA context parser to perform the defined operations.

By utilizing HCC, 76% of the overall contexts are reduced compared with the non-hierarchical one (for H.264 decoding case, reduced from 452KB to 104KB). As illustrated in Fig. 6, the whole execution time (Configuration + Data access & Calculation) for running the computation-intensive sub-tasks of H.264 decoding on RPU are substantially decreased compared with those running on XPP40 (the core of XPP III) [4]. It is worth noting that the configuration time for these sub-tasks only accounts for 4.3%~13.2% of the whole execution time, which is much lower than that of XPP40. In other words, the computing potential of the fabric has been utilized extensively. This is one of the major reasons for the proposed architecture to achieve better performance and power results, as given below.

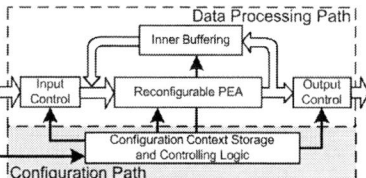

Fig. 1. RPU architecture.

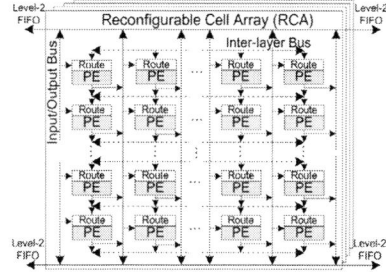

Fig. 2. Architecture of PEA and RCA.

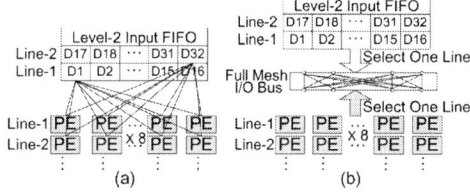

Fig. 3. (a) Traditional full mesh connection and (b) Proposed LSMC.

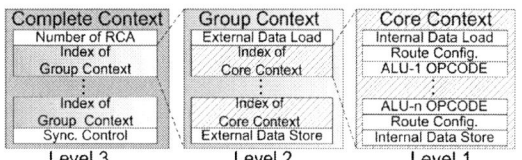

Fig. 4. Hierarchical representation of context.

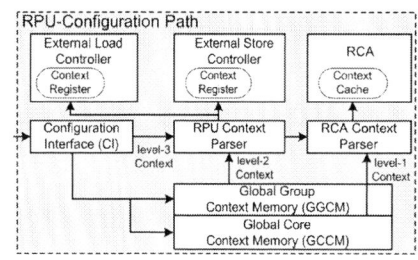

Fig. 5. Architecture of configuration path.

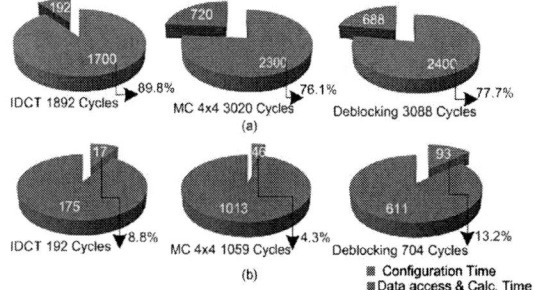

Fig. 6. Configuration time vs. Data access & Calculation time for (a) XPP40 and (b) RPU.

978-1-4673-6145-3/13 $31.00 © 2013 IEEE   701

III. IMPLEMENTATION AND MEASUREMENTS

We first implement RPU as a single chip, called CHAMELEON (die photo shown in Fig. 7). This prototype chip (seriously pad-limited) is used for massive algorithm testing and profiling. In order to verify the efficiency of RPU comprehensively, two reconfigurable processors, called REMUS_HPP (integrating 2 RPUs) and REMUS_LPP (integrating 1 RPU), are designed and integrated into two SoCs, targeting high performance applications (called RHINOCEROS) and low power applications (called REINDEER) respectively. The relationships of the cores, processors and SoCs are depicted in Fig.8 and their physical numbers are listed in Table I.

The block diagram and die photo of REMUS_HPP are shown in Fig. 9. There are four main functional parts in REMUS_HPP: RPU micro-controller (RMC) is used dedicatedly to control the configuration and data I/O of RPU; Master micro-controller (MMC) initializes RMC and other peripherals; Video pre-processor (VPP) is used to perform bit-level subtasks, while all the computation-intensive tasks (e.g. inverse discrete cosine transformation, deblocking, reconstruction, intra-prediction, motion compensation, etc.) are carried out on RPUs. We use four 128Mb DDR2 memories to serve as one single off-chip memory, each providing an 8-bit wide data bus and 4 banks. A dedicated off-chip memory controller, called EMI, is designed to connect with a 32-bit system AHB (Advanced High performance Bus) on one side, and to connect RPUs with a 64-bit user-defined data interface on the other side. To improve the communication efficiency with the off-chip DDR2 memory, we map different lines of the decoded frame data into different banks and access multiple banks (interleaved multi-bank accessing) to hide the long precharging time. In addition, a block buffer inside EMI is built to further reduce the data accessing latency [5].

REMUS_HPP can support the whole multi-standard video decoding tasks independently, and the other modules in RHINOCEROS are only responsible for the routine tasks such as operating system running, peripheral controlling and video display, etc. Therefore, we only provide the comparison data for REMUS_HPP in the following discussion. Measured results show that REMUS_HPP can decode 1920x1080 H.264 (High Profile), AVS (Jizhun Profile), MPEG-2 (Main Profile) at 30fps, 39fps and 41fps under 200 MHz, respectively. As listed in TableII, REMUS_HPP achieves a 1.81x faster decoding speed and a 14.3x energy efficiency improvement over XPP-III[6]. When normalized to the same process (full-scaling approach), REMUS_HPP outperforms XPP-III 6.98x in energy efficiency.

The major difference between REMUS_LPP and REMUS_HPP is that REMUS_LPP has only one RPU, as shown in Fig.10. Measured results show that REMUS_LPP can decode 720x480 H.264(Baseline), AVS(Jizhun Profile), MPEG-2 (Main Profile) at 35fps, 65fps and 67fps under 75 MHz respectively. As listed in Table III, REMUS_LPP achieves a 3.67x faster decoding speed, 76% power reduction and a 3.96x energy efficiency improvement compared with ADRES [2]. When normalized to the same process, REMUS_LPP still outperforms ADRES 1.58x in energy efficiency.

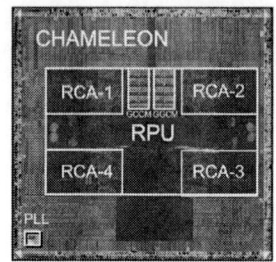

Fig. 7. Die photo of CHAMELEON.

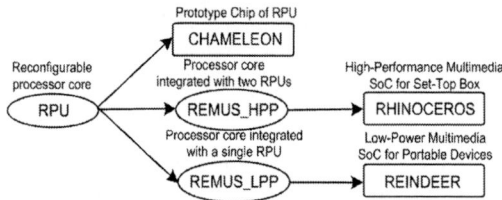

Fig. 8. Relation map of the implementations.

TABLE I
PHYSICAL INFORMATION OF CHAMELEON, RINOCEROS AND REINDEER

Chip Name	CHAMELEON	RHINOCEROS	REINDEER
Core Name	RPU	REMUS_HPP	REMUS_LPP
Chip Area(mm²)	6.2×6.2	9.28×9.28	6.9×6.5
Core Area (mm²)	16.7	48.9	21.6
PE Array Size	16×16	(16×16)×2	(16×16)×1
Technology	TSMC 65 nm LP1P8M		

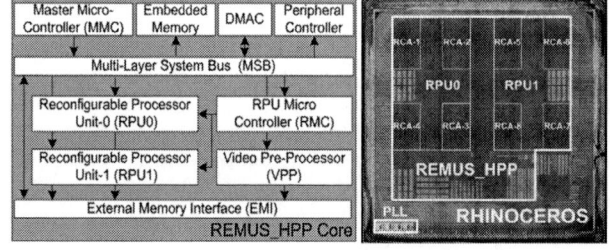

Fig. 9. Block diagram and Die photo of REMUS_HPP.

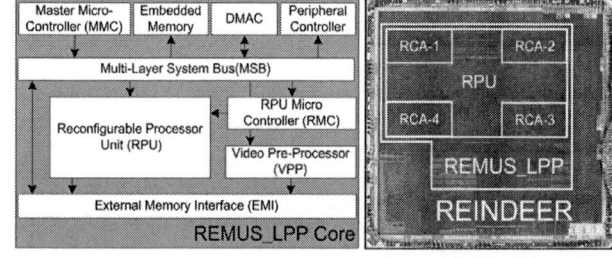

Fig. 10. Block diagram and Die photo of REMUS_LPP.

Compared with the state-of-the-art many-core processor [7] and DSP [8] (from which the results of H.264 decoding could be obtained), as expected, the proposed architecture shows considerable advantages in many aspects as shown in Table II and III. This is partly because the computing efficiency of the instruction driven processor is intrinsically limited by the scheme to sequentially fetch-and-execute instructions. Also, it is worth mentioning again that the above improvements mainly come from RPUs, since all the computation-intensive tasks are carried out on them.

IV. CONCLUSIONS AND EXTENDED DISCUSSIONS ON FLEXIBILITY

TALBE II

COMPARISONS WHEN PERFORMING H.264 DECODING

	REMUS_HPP (Reconfigurable Processor)	XPP-III [6] (Reconfigurable Processor.)	[7] (Many-core Processor)
Technology(nm)	65	90	40
Area(mm²)	48.9	75	210
Frequency(MHz)	200	450	333
Resolution [a]	1920x1080	1920x1080	1920x1080
Performance(fps)	30	24	30
Power(mW)	280	3420	500
Normalized Performance[b] (MBs/s/MHz)	1224	435	735
Energy Efficiency [c] (MBs/s/mW)	874	57	490

[a]Totally 8160 Macro Blocks (MBs) per 1920x1080 frame; [b]Number of MBs per second per MHz; [c]Number of MBs per second per mW.

TABLE III

COMPARISONS WHEN PERFORMING H.264 DECODING

	REMUS_LPP (Reconfigurable Processor)	ADRES [2] (Reconfigurable Processor.)	[8] (DSP)
Technology(nm)	65	90	130
Area(mm²)	21.6	3.6	529
Frequency(MHz)	75	300	600
Resolution	720x480	720x480	720x576
Performance(fps)	35	30	25
Power(mW)	24.5	105.5	1900
Normalized Performance(MBs/s/MHz)	630	135	67.5
Energy Efficiency (MBs/s/mW)	1904	384	21.3

LSMC and HCC are exploited to enhance the energy efficiency of the proposed reconfigurable fabric. Measured results demonstrate that great improvements have been achieved in terms of performance, power and energy efficiency when performing video decoding applications.

However, the objective of this paper is to explore the effectiveness of the reconfigurable computing technologies in various computation-intensive applications rather than only focusing on video decoding. Such computation-intensive applications include computer vision, communication baseband processing, encryption / decryption, etc. The reasons we select video decoding algorithms as benchmarks in this paper are that, a) firstly, these algorithms are famous for their heavy computing-load; b) secondly, these algorithms are most widely studied and the comparison data from the reference designs (especially from the reported reconfigurable fabrics) are easy to obtain. In fact, from the design space exploration phase, the function flexibility issue has been taken into account. In hardware side, we have integrated up to 26 operators into each PE. All of these operators are commonly used ones and are not custom-designed or optimized for video decoding algorithms. Although LSMC simplifies the structure of PEA interconnection, this interconnection is still flexible enough to transmit data among different PEs efficiently. Also, according to our statistics, most applications would benefit from HCC scheme because their configuration contexts could be organized into hierarchical patterns as well. All of these efforts make the proposed fabric

satisfy the function requirements of versatile applications. In software side, we have already proposed the template-based compilation flow and several mapping optimization techniques [9] [10] [11]. With the aid of certain manual operations (such as creating domain-specific template libraries, inserting directives to the source code, etc.), the applications programmed in high-level languages could be efficiently implemented onto the fabric. Obviously, these software works have extended the application scope of the proposed hardware.

We have tried the AdaBoost (face detection algorithm) on REMUS_HPP, achieving 17fps@640x480 with more than 95% detection rate [12]. SIFT (feature detection algorithm) has been implemented on REMUS_HPP as well, achieving 33% and 50% speed-up over the multi-core platform and FPGA [13]. Also, we have demonstrated the baseband processing algorithm of GPS (Global Positioning System) and AES/DES/SMS4/A5 (symmetric cryptographic algorithms) on REMUS_LPP platform. All of the above experiments have proved that proposed reconfigurable fabric can not only process video decoding applications efficiently, but also some other heavy computing-load domain-specific applications.

REFERENCES

[1] D. Rossi, F. Campi, S. Spolzino, S. Pucillo and R. Guerrieri, "A heterogeneous digital signal processor for dynamically reconfigurable computing," *IEEE J. Solid-State Circuits*, vol.45 no. 8, pp.1615-1626, Aug. 2010.

[2] B. Mei, B. D. Sutter, T. V. Aa, M. Wouters, A. Kanstein, and S. Dupont, "Implementation of a coarse-grained reconfigurable media processor for AVC decoder," *J. Signal Process Syst.*, vol.51, no.3, pp.225-243, June 2008.

[3] S. Gao, T. Kihara, S. Shimizu, Y. Arakawa, N. Yamanaka, and A. Watanabe, "A novel traffic engineering method using on-chip diorama network on dynamically reconfigurable processor DAPDNA-2," *2009 International Conference on High Performance Switching and Routing (HPSR 2009)*, pp.1-6, 2009.

[4] White Paper of Video Decoding on XPP-III, PACT, 2006.

[5] X. Liu, C. Mei, P. Cao, M. Zhu, and L. Shi, "Date flow optimization of dynamically coarse grain reconfigurable architecture for multimedia applications," *IEICE Trans. Inf. Syst.*,vol.E95-D, no.2, pp. 374-382, 2012.

[6] M. K. A. Ganesan, S. Singh, F. May, and J. Becker, "H.264 decoder at HD resolution on a coarse grain dynamically reconfigurable architecture," *2007 International Conference on Field Programmable Logic and Applications, FPL 2007*, pp.467-471, 2007.

[7] H.Xu, J. Tanabe, H. Usui, S. Hosoda, T. Sano, K. Yamamoto, et al.,"A low power many-core SoC with two 32-core clusters connected by tree based NoC for multimedia applications," *IEEE Symp. VLSI Circuits*, pp. 150-151, 2012.

[8] F. Pescador, C. Sanz, M. J. Garrido, E. Juarez, and D. Samper, "A DSP based H.264 decoder for a multi-format IP set-top box," *IEEE Trans. Consum. Electron.*, vol.54, no.1, pp.145-153, Feb. 2008.

[9] D. Liu, S. Yin, L.Liu, and S. Wei, "Polyhedral model based mapping optimization of loop nests for CGRA," *Design Automation Conf.2013*, in press.

[10] D. Liu; S. Yin; C. Yin; L. Liu; and S. Wei, "Mapping optimization of affine loop nests for reconfigurable computing architecture," *IEICE Trans. Inf. Syst.*, vol.E95-D, no.12,pp. 2898-2907,Dec. 2012.

[11] C. Yin, S. Yin, L. Liu, and S. Wei, "Compiler framework for reconfigurable computing architecture," *IEICE Trans. on Electro.* vol.E92-C, no.10, pp.1284-1290, Oct. 2009.

[12] J. Xiao, J. Zhang, M. Zhu, J. Yang, and L. Shi, "Fast AdaBoost-based face detection system on a dynamically coarse grain reconfigurable architecture," *IEICE Trans. Inf. Syst.*, vol.E95-D, no.2, pp.392-402, 2012.

[13] P. Ouyang, S. Yin, H. Gao, L. Liu, and S. Wei. "Parallelization of computing-intensive tasks of SIFT algorithm on a reconfigurable architecture system," *IEICE Trans. Inf. Syst.*, in press.

978-1-4673-6145-3/13 $31.00 © 2013 IEEE

SURFEX: A 57fps 1080P resolution 220mW Silicon Implementation for Simplified Speeded-Up Robust Feature with 65nm Process

Leibo Liu, Weilong Zhang, Chenchen Deng*, Shouyi Yin, Shanshan Cai and Shaojun Wei

Institute of Microelectronics, National Laboratory for Information Science and Technology
Tsinghua University, Beijing, P.R.China. Email: chenchendeng@tsinghua.edu.cn

Abstract—Speeded Up Robust Feature(SURF) is widely used in computer vision applications. In many recent applications like mobile devices and vision sensor network, it is extremely difficult to meet both the performance and power consumption requirements of SURF implementations, especially for CPU, GPU, DSP or FPGA based solutions. In this paper, the SURF algorithm is simplified and optimized for hardware implementation. To increase the throughput, procedures like orientation assignment and descriptor extraction are re-organized while maintaining enough accuracy; the memory accesses have also been improved to increase the bandwidth and reduce repeated data accesses; the workload of each stage in the pipeline is analyzed and balanced to reduce the pipeline bubble. Furthermore, a method called Word Length Reduction (WLR) is adopted to compress the integral image, which reduces the on-chip memory by 40%. In addition to that, the corresponding power consumptions are reduced significantly. The Simplified SURF is implemented onto a $3.4 \times 4.0 \ mm^2$ chip called SURFEX using TSMC 65nm process. The chip is able to process 57 frames of 1080p(1920×1080) video per second with a 200MHz working frequency while dissipating 220mW. This throughput is 6 times of the ones reported in the latest literatures and the power consumption is less than half of the most outstanding implementations.

I. INTRODUCTION

In computer vision applications, it is very challenging to find the correspondence between two scenes with different illumination conditions, viewing points and noises. However, SURF [1], inspired by SIFT [2] can manage this type of tasks very well and has shown great competitiveness in 3D reconstruction, image registration, camera calibration, object recognition [3], [4], and image retrieval [5].

There had been several SIFT and SURF implementations using traditional CPU, GPU, DSP, or FPGA based solutions. However, due to the intrinsic drawbacks of the hardware platform, it is still difficult to meet the performance and power requirements [6]–[10]. In [6], the author implemented SIFT on a Altera Stratix II FPGA and achieved a performance of QVGA (320×240) at 30fps. In [9], ROI (Region of Interest) is calculated first to narrow down the processing areas, and the visual engine is able to extract SIFT feature in a 60fps VGA (640×480) video while dissipating 496mW. In recent years an increasing number of computer vision applications like mobile robot, vehicle-mounted device or autonomous flight require both a high resolution and a high frame rate which tends to consume more computation energy. On the other hand, the energy budget is becoming increasingly limited.

Moreover, for video processing applications, data storage requirement is usually large and memory bandwidth limits the final throughput, which are also the main difficulties for high-performance low-power silicon implementation of such algorithms.

In order to solve these problems mentioned above, the SURF algorithm is customized for hardware implementation and a high performance, low power SURF ASIC is implemented in this paper. The contributions of this work are as follows: To increase the throughput, firstly, orientation assignment and descriptor calculation are arranged in a more efficient way to minimize the times of memory accesses. The quantization will introduce accuracy deterioration but the performance benefits are far more significant. Secondly, the integral image memory is divided into four parts according to the address parity and therefore the memory bandwidth quadruples. Thirdly, the operations are pipelined both in and among the proposed hardware modules and the workload of each stage is balanced to reduce the number of idle cycles. Moreover, a storage reduction method called WLR [11] is adopted to decrease the chip area as well as the power consumption.

II. SURF ALGORITHM CUSTOMIZATION AND HARDWARE ARCHITECTURE

The SURF algorithm mainly consists of two parts, i.e. IPoints(Interest Points) detection and IPoints description. Fig. 1 shows the architecture of SURFEX which includes four computing modules: Integral Image Generator(IIG), IPoint Detector(IPDT), IPoint Orientation Assigner(IPOA), and IPoint Feature Vector Extractor(IPVX).

A. Integral Image Generation

The IIG module receives the image source input from the image sensor and outputs the integral image pixel by pixel serially. The advantages of using integral image is that it is convenient to calculate the sum of the intensities over any upright rectangle area using integral image. Besides, the calculation time is independent of the rectangle size. Take a 1080P integral image as an example, the total cycles to generate it is $2.07 \times 10^6 (1920 \times 1080)$. However, the integral image requires a large binary word length to store the accumulated sums. Therefore, special measures are taken to reduce the length of integral image word without affecting its performance. The

978-1-4673-6145-3/13 $31.00 © 2013 IEEE

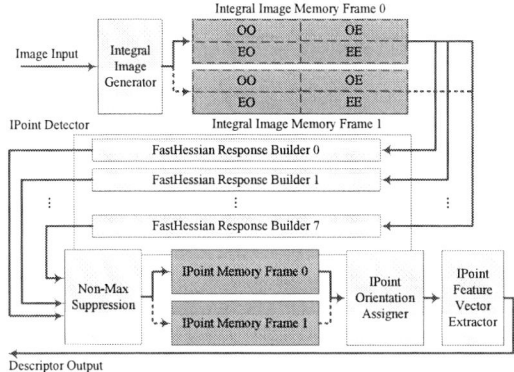

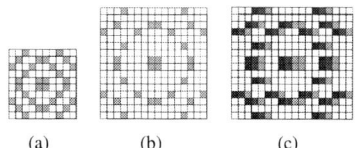

Fig. 2. (a)(b) Memory access footprint of scale 9 and 15 when building hessian response, (c) Three sequent memory accesses of scale 15. Different colors represent memory accesses of different IPoint candidates.

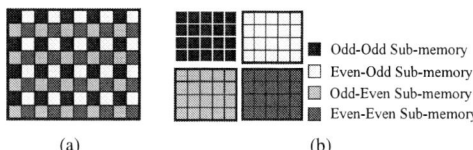

Fig. 3. Division of integral image according to the parity.

Fig. 1. The hardware architecture of SURFEX.

demonstration in [11] has shown that the integral image binary word length L_{ii} satisfies (Equ. 1), where L_i is the input image pixel depth.

$$\left(2^{L_{ii}} - 1\right) \geqslant \left(2^{L_i} - 1\right) WH \qquad (1)$$

In most algorithms the maximum size of the rectangle is a priori known. The size of the first three octaves in SURF is 100×66 which determines the minimum word length of the rectangle sum is $20.69(\log_2{(100 \times 66)} + 8)$ bits. Since the final result of the rectangle pixel sum is the difference between the four vertexes, several most significant bits can simply be discarded. This truncation will inevitably bring overflow into intermediate procedures but the final results are still correct. Furthermore, since the rounding error is uncorrelated, several least significant bits can be discarded as well. In this way, the final rounding error of the rectangle sum is small and therefore in our approach the last bit is discarded. With these two methods the integral image word length is reduced from 29 bits (without WLR) to $20(\lceil 20.69 - 1 \rceil)$ bits which saves us 40% memory resources without any side-effects.

B. Hessian-matrix Based IPoint Detection

The IPDT module locates IPoints by detecting local maximum gradient and SURF algorithm approximates Laplace of Gaussian L_{xx}, L_{yy}, and L_{xy} with box filters D_{xx}, D_{yy}, and D_{xy}. The determinant of the hessian-matrix or hessian response is the criterion of whether the candidate point should selected as an IPoint. Eight accesses of the integral image are needed to calculate D_{xx} and D_{yy} respectively, while D_{xy} takes 16 memory accesses. Thus the IPDT takes 32 memory accesses altogether for an IPoint candidate at a specific location and scale (Fig. 2). A true dual-port block memory can finish any two read or write operations in one cycle.

In OpenSURF [12], the IPoint candidate intervals are 2, 4 and 8 pixels for the first, the second and the third octave respectively. In a 1080P image the total number of IPoint candidates is $6.804 \times 10^5 (1920 \times 1080 \times (1/4 + 1/16 + 1/64))$. Each candidate could be at 4 scales in one octave. All the

scale factors are odd which makes the length and the width of the rectangle odd as well. The coordinates of the four vertexes are (odd,odd),(odd,even),(even,even) and (even,odd). With this pattern, the integral image memory could be divided into four parts and each of them corresponds to a specific parity as illustrated in Fig. 3. Therefore the pixels at four vertexes can be accessed at the same time. This division quadruples the bandwidth of the integral image memory at the cost of a slight increase in area. The total number of clock cycles IPDT takes is $1.09 \times 10^7 (6.804 \times 10^5 \times 4 \times 32 \times 1/8)$ cycles. To achieve a higher throughput, a parallel mechanism is explored. Although hessian response building presents an irregular memory access pattern, neighboring candidates share the same pattern. In Fig. 2, it can be seen that sequent memory accesses are along a straight line allowing sequent integral image entries to be accessed simultaneously and fed to several FastHessian response builders in the meanwhile (Fig. 1).

In our approach, eight paralleled hessian response builders are implemented instead of only one. Thirty-two integral image entries will be fetched at the same time. Since the sequent memory accesses suffer from a misalignment issue, only half of the 32 entries will be used. As the candidate interval increase from 2 for octave 1 to 8 for octave 3, the number of the entries that can be fed to the response builders decreases from 16 to 4. To save the power consumption, those unused response builders will be powered off. The total number of clock cycles for the paralleled IPDT is $1.81 \times 10^6 (640 \times 480 \times (1/4 \times 1/8 + 1/16 \times 1/4 + 1/64 \times 1/2) \times 4 \times 32 \times 1/8)$ cycles. This means that with our method, the throughput of IPDT has been increased by almost an order of magnitude. Although for each individual module the ratio of acceleration to hardware increment is below 1, the overall system benefits since the workload is balanced and the pipeline works more efficiently with much less idle cycles.

C. Orientation Assignment

In order to be invariant to rotations, every IPoint is assigned to a reproducible reference orientation. Haar wavelet responses

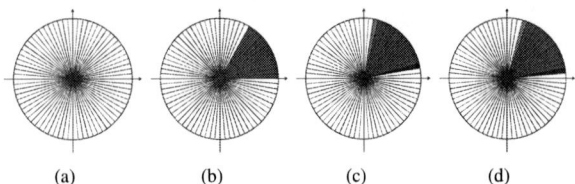

Fig. 4. When the sector window slides, the new sum of Haar response can be calculated by adding the new 5 degrees and removing the last 5 degrees.

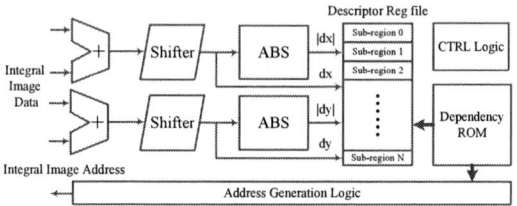

Fig. 5. The architecture of IPVX module.

at specific locations in the circle window are calculated. In OpenSURF the slide step of the sector window is 0.15 radians (8.59 degrees). To ease the quantization of the angle value, we set the slide step to 5 degrees, and in Fig. 4 the whole 360 degrees are quantized to 72 discrete values. The $\pi/3$ sector window covers 12 of the 5 degrees. Thus, every time the sector window rotates by 5 degrees, the last 5 degrees is subtracted from the Gaussian weighted sum and the new 5 degrees is added to it. All the in-between values are not needed to be recalculated [13]. Further, in our approach, the sector window's radius is set to 3σ to discard the low-weighted outer responses, where σ is the scale factor. In this way, the number of sample locations decreases from 109 to 49. The absence of the low-weighted samples and the quantization of the turning angle will decrease the accuracy by about 5 percent in the rotation transformed scenes. However, the performance is nearly twice of the original module. Each Haar response relies on 8 integral image entries which takes 8 memory accesses. The total number of clock cycles elapses in orientation assignment is $49(49 \times 8/8)$ cycles/IPoint.

D. SURF Descriptor Extraction

To calculate the descriptor of the IPoint , a 20σ-size square is centered on the IPoint and oriented along the orientation of the IPoint, where σ is the scale factor of the IPoint as mentioned above. The 20σ-size square is then divided into 4×4 5σ-size sub-regions, each containing 5×5 sample locations. In OpenSURF, every sub-region samples Haar wavelet responses at 9×9 rather than 5×5 locations (more distinctive). By looking into the memory access addresses we find that the sub-regions share a large amount of data on the integral image. The data dependency is a prior known and therefor could be stored in the dependency ROM in advance (Fig. 5). Every time one haar wavelet response is calculated, the IPoint descriptor elements belonging to several adjacent sub-regions will be updated. By avoiding this repetition, memory access times per IPoint decreases from 1296 to 576 without any disadvantages. The feature descriptor extractor calculates 24×24 Haar responses along X and Y directions and the total number of clock cycles for the descriptor generation is $576(24 \times 24 \times 8/8)$ cycles/IPoint.

For a 1080P resolution, in typical cases 5000 IPoints per frame is a reasonable estimation. 5000 IPoints will cost IPoint Descriptor (IPDS) including IPOA and IPVX $3.13 \times 10^6(5000 \times (576+49))$ cycles. Cycles taken to process a 1080P

image by each module is listed in Table. I. Although there is only one set of computation modules, a pair of memories are used for two sequent frames. Thus SURFEX is able to process $57(200\text{MHz} \times 2/7.01 \times 10^6)$ frames per second. Only in some extreme cases when there are too much texture in one frame, the number of the IPoints will be tremendous and the throughput will decrease.

III. EXPERIMENT RESULTS AND COMPARISON OF SURFEX

SURFEX die parameters are listed in Table. II and the chip photo is shown in Fig. 6. To prove the repeatability and the distinctiveness of SURFEX, it is tested on the commonly-used INRIA Graffiti images [14]. The recall v.s. 1-precision [15] graphs are shown in Fig. 7. The two surfaces above are original SURF recall rate surface and SURFEX recall rate surface while the surface below is the difference between these two. The comparisons in Fig. 7 show that SURFEX is capable to accomplish all the tasks and its performance is almost as good as SURF itself with less than 10% deterioration of accuracy in most cases. Only for the tough cases such as significantly out-of-plane rotated Graffiti and heavily scaled and rotated Boat, the recall rate decreases by less than 20%. Overall this evaluation confirms the accuracy of SURFEX and in addition to that the resulting acceleration is significant.

Comparisons with the state-of-art architectures are listed is Table III. As a fair comparison, the throughput is normalized to pixels per second at the same clock frequency. Although the architectures for comparison accomplish different tasks, Table III clearly shows that SURFEX outperforms most of the architectures reported in the latest literatures. The throughput is approximately 6 times of best know results and the power consumption is less than half of them.

TABLE I

CYCLES TO PROCESS A 1080P IMAGE

Moudle	IIG	IPDT	IPDS	Total
Cycles	2.07×10^6	1.81×10^6	3.13×10^6	7.01×10^6

TABLE II

SURFEX DIE PARAMETERS

Process	TSMC65nm1p10m	Freq	200MHz
Size	$4.0 \times 3.3 mm^2$	Gates	0.6M Gates, 0.4MB MEM
Voltage	3.3V I/O, 1V Core	Power	220mW@200MHz

TABLE III
COMPARISON BETWEEN RELATED WORKS

	Algorithm	Task	Resolution	FPS	Architecture	Power	Clock freq	Pixels/s@1MHz
TCSVT08 [6]	SIFT	IPDT IPDS	320×240	30	Altera Stratix II NIOS II	-	100MHz	23.04k
FPT09 [10]	SIFT	IPDT	640×480	30	Xilinx Virtex-5	-	100MHz	92.16k
TePRA09 [7]	SURF	IPDT IPDS	1024×768	10	Xilinx Spartan-6 PowerPC-440	<10W	100MHz(FPGA) 200MHz(CPU)	78.64k
JSSC10 [9]	SIFT	ROI IPDT IPDS	640×480	60	SoC	496mW	200MHz	92.16k
FCCM10 [8]	SURF	IPDT	640×480	56	Xilinx Virtex 5	<20W	200MHz	86.02k
SURFEX	SURF	IPDT IPDS	1920×1080	57	ASIC	220mW	200MHz	591k

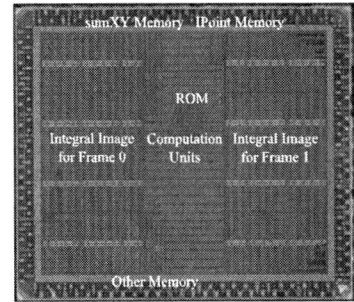

Fig. 6. Photo of SURFEX Chip.

IV. CONCLUSION

This paper presents an efficient silicon implementation of the speeded-up robust feature extraction for high resolution and high frame rate video. The whole system achieves a high throughput with limited hardware resource and low power consumption making it attractive for many computer vision applications. In addition to that, the memory access optimization and reorganization can also be utilized in other vision algorithm implementations.

REFERENCES

[1] H. Bay, T. Tuytelaars, and L. Van Gool, "Surf: Speeded up robust features," *Comput. Vision. 9th Eur. Conf. on*, pp. 404–417, 2006.

[2] D. Lowe, "Object recognition from local scale-invariant features," in *Comput. Vision. Proc. Seventh IEEE Int. Conf. on*, vol. 2, 1999, pp. 1150–1157.

[3] B. Besbes, A. Apatean, A. Rogozan, and A. Bensrhair, "Combining surf-based local and global features for road obstacle recognition in far infrared images," in *Intell. Transp. Syst. 13th Int. IEEE Conf. on*, 2010, pp. 1869–1874.

[4] D. Jang and M. Turk, "Car-rec: A real time car recognition system," in *Appl. Comp. Vision. IEEE Workshop on*, 2011, pp. 599–605.

[5] A. Wang, Z. Wang, D. Lv, and Z. Fang, "Research on a novel non-rigid registration for medical image based on surf and apso," in *Image. Signal. Process. 3rd Int. Congr. on*, vol. 6, 2010, pp. 2628–2633.

[6] V. Bonato, E. Marques, and G. Constantinides, "A parallel hardware architecture for scale and rotation invariant feature detection," *Circuits. Syst. Video. Technol. IEEE Trans. on*, vol. 18, no. 12, pp. 1703–1712, 2008.

[7] J. Svab, T. Krajnik, J. Faigl, and L. Preucil, "Fpga based speeded up robust features," in *Technol. Pract. Rob. Appl. IEEE Int. Conf. on*, 2009, pp. 35–41.

[8] D. Bouris, A. Nikitakis, and I. Papaefstathiou, "Fast and efficient fpga-based feature detection employing the surf algorithm," in *Field-Programmable Custom. Comput. Mach.18th IEEE Annual Int. Symp. on*, 2010, pp. 3–10.

[9] J. Kim, M. Kim, S. Lee, J. Oh, K. Kim, and H. Yoo, "A 201.4 gops 496 mw real-time multi-object recognition processor with bio-inspired neural perception engine," *IEEE J. Solid-State Circuits*, vol. 45, no. 1, pp. 32–45, 2010.

[10] L. Yao, H. Feng, Y. Zhu, Z. Jiang, D. Zhao, and W. Feng, "An architecture of optimised sift feature detection for an fpga implementation of an image matcher," in *Field-Programmable Technol. Int. Conf. on*, 2009, pp. 30–37.

[11] H. Belt, "Word length reduction for the integral image," in *Image. Process. 15th IEEE Int Conf. on*, 2008, pp. 805–808.

[12] C. Evans, "Notes on the opensurf library," *University of Bristol, Tech. Rep. CSTR-09-001, January*, 2009.

[13] B. Han, Y. Wang, and X. Jia, "Fast calculating feature point's main orientation in surf algorithm," in *Comput. Mechatro. Control. Electro. Eng. Int. Conf. on*, vol. 6, 2010, pp. 165–168.

[14] INRIA, "Affine Covariant Features," http://www.robots.ox.ac.uk/ vg-g/research/affine/, 2007, [Online; accessed July, 2007].

[15] Y. Ke and R. Sukthankar, "Pca-sift: A more distinctive representation for local image descriptors," in *Comp. Vision. Pattern. Recognit. Proc. IEEE Comput. Soc. Conf. on*, vol. 2, 2004, pp. II–506.

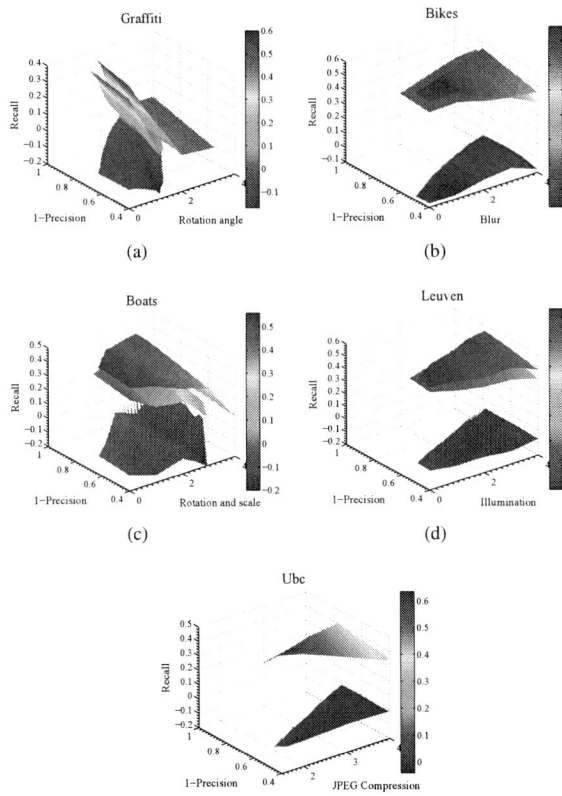

Fig. 7. The recall v.s. 1-precision graphs. (a) Graffiti (out-of-plane rotation), (b) Bikes (image defocus), (c) Boats (scale and in-plane rotation), (d) Leuven (light change), (e) Ubc (JPEG compression).

Supply-Noise Resilient Adaptive Clocking for Battery-Powered Aerial Microrobotic System-on-Chip in 40nm CMOS

Xuan Zhang, Tao Tong, David Brooks, Gu-Yeon Wei
Harvard University, Cambridge, MA 02138
Email: xuanzhang@eecs.harvard.edu

Abstract—A battery-powered aerial microrobotic System-on-Chip (SoC) has stringent weight and power budgets, which requires fully-integrated solutions for both clock generation and voltage regulation. Supply-noise resilience is important yet challenging for such SoC systems due to a non-constant battery discharge profile and load current variability. This paper proposes an adaptive-frequency clocking scheme that can tolerate supply noise and improve performance when implemented with an integrated voltage regulator (IVR). Measurements from a 'brain' SoC, implemented in 40nm CMOS, demonstrate $2\times$ performance improvement with adaptive-frequency clocking over conventional fixed-frequency clocking. Combining adaptive-frequency clocking with open-loop IVR extends error-free operation to a wider battery voltage range (2.8 to 3.8V) with higher average performance.

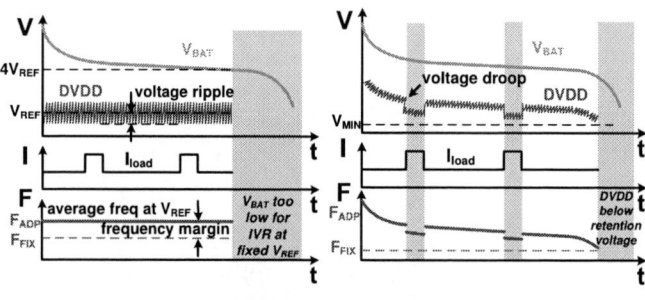

(a) closed-loop regulation (b) open-loop operation

Fig. 1: Illustrations of two SC-IVR modes of operation versus typical battery dicharge profile.

I. INTRODUCTION

System-on-chip (SoC) for aerial microrobots faces stringent weight and size constraints; at the same time, it requires considerable performance and power to handle a variety of tasks such as image sensing, navigation, and flight control. Conventional digital systems typically operate at a fixed frequency with respect to an external clock source and off of a constant voltage supplied by an external voltage regulator. However, in order to demonstrate an autonomous flying robotic insect with limited lift force generated by its wings [6], a 'brain' SoC designed for the Harvard RoboBee must be powered directly off of a battery with no external components. In order to meet this stringent constraint, the SoC system employs an integrated voltage regulator (IVR) and an internal voltage-tracking clock generator for its digital logic and memories.

One of the main objectives of this work is to maximize the RoboBee's flight time with respect to the total energy available in its battery and the associated battery discharge profile. In this vein, this paper explores the relative merits of different operational modes offered by the IVR and on-chip clock generator. The IVR is a 4:1 switched-capacitor (SC) converter that cascades two 2:1 SC stages individually tuned to maximize conversion efficiency [7]. This SC-IVR converts the 3.6-4V battery voltage down to 0.9V and below for the digital SoC load. It can operate in either open- or closed-loop regulation modes. Fig. 1 illustrates these two possible modes of operation with respect to a discharge profile, typical of lithium-based batteries. In closed-loop operation (regulated voltage), the SC-IVR works to provide a constant supply voltage that is resilient to input battery (V_{BAT}) and output load (I_{LOAD}) conditions. However, for a target output voltage level (V_{REF}), the SC-IVR's operating range is limited to $V_{BAT} > 4V_{REF}$. In contrast, open-loop operation (unregulated voltage) exhibits an entirely different set of attributes; the SC-IVR's output voltage is roughly 1/4th the input battery voltage, but varies

with both the discharge profile and load fluctuations. While open-loop SC-IVR mode allows the system to operate over a wider range, down to the minimum voltage limit of the digital load, performance and efficiency of energy utilization depends on the clocking strategy used. Hence, we also explore two clocking schemes: *fixed-frequency* and *adaptive-frequency* clocking. Out of the four total combinations, this paper compares the following three: (1) regulated voltage, fixed frequency; (2) regulated voltage, adaptive frequency; and (3) unregulated voltage, adaptive frequency. With closed-loop voltage regulation, fixed-frequency clocking (F_{FIX}) requires extra timing margins to account for non-negligible voltage ripple, which is an artifact of the SC-IVR's feedback loop. Alternatively, an adaptive-frequency clocking (F_{ADP}) scheme offers higher average frequency . Adaptive-frequency clocking also works well for open-loop SC-IVR mode, because it maximizes performance with respect to battery and load conditions.

In order to explore the relative tradeoffs and merits associated with the different operational modes described above, the remainder of the paper first describes the SC-IVR and clock generator designed into the prototype 'brain' SoC implemented in TSMC's 40nm process. Then, Section III presents experimental results that verify the performance advantages of using adaptive clocking with both regulated and unregulated voltages generated by the SC-IVR.

II. SYSTEM ARCHITECTURE

The brain SoC designed for the RoboBee, shown in Fig. 2, contains a fully-integrated two-stage 4:1 switched-capacitor voltage regulator (SC-IVR), a 32-bit ARM Cortex-M0 general-purpose processor, two identical 64KB memories, and a programmable digitally-controlled oscillator (DCO) that generates the voltage-tracking adaptive-frequency clock. The test chip also includes numerous blocks for test and debug purposes: a built-in self test (BIST) block allows thorough testing of the two memory blocks; a scan chain configures the digital blocks;

978-1-4673-6145-3/13 $31.00 © 2013 IEEE

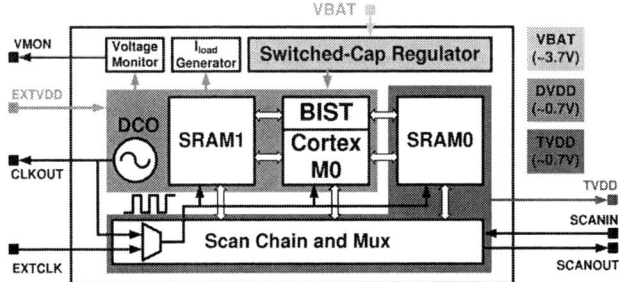

Fig. 2: Block diagram of the fully-integrated SoC.

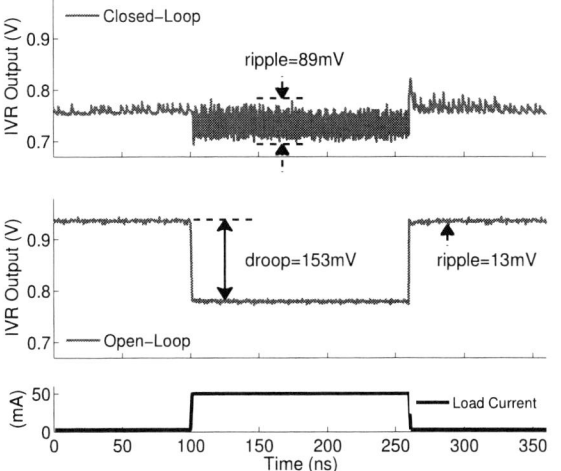

Fig. 3: Transient response of SC-IVR output voltage to load current steps.

a voltage monitor block probes internal supply voltages; and a current-load generator enables different IVR testing scenarios. Lastly, the test chip allows for external power and clock sources to investigate different operating modes.

A. Switched-capacitor integrated voltage regulator

The SC-IVR converts the battery voltage ($V_{BAT} \approx 3.7V$) down to the digital supply (DVDD $\approx 0.7V$). It consists of a cascade of two 2:1 switched-capacitor converters that are respectively optimized for high input voltage tolerance and fast load response, and each converter stage employs a 16-phase topology to reduce voltage ripple. A low-boundary feeback control loop can regulate DVDD to a desired voltage level. A thorough discussions of the SC-IVR and its implementation details can be found in [7].

While this SC-IVR achieves high conversion efficiency, 70% at its optimal operating point, it is subject to the inherent limitations of any switched-capacitor based DC-DC converter; efficiency varies with respect to input and output voltages and the load. Extensive measurement results in [7] show that open-loop operation consistently offers higher conversion efficiency. Fig. 3 plots transient behavior of the IVR with respect to load current steps between 3mA and 50mA for both open- and closed-loop operation. Closed-loop operation quickly responds to avoid the steep voltage droop otherwise seen in the open-loop case; however, it exhibits larger steady-state voltage ripple, especially for higher load currents, due to the control loop topology and feedback delay.

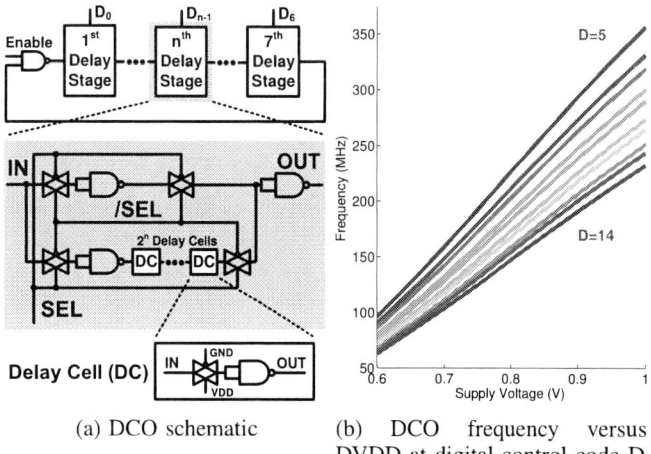

(a) DCO schematic (b) DCO frequency versus DVDD at digital control code D

Fig. 4: Digital-controlled oscillator schematic and measured characteristics.

B. Digitally-controlled oscillator

Our proposed adaptive-frequency clocking scheme needs a clock generator whose frequency tracks closely with changes in supply voltage. This allows the operating frequency of the digital load circuitry to appropriately scale with voltage fluctuations, providing intrinsic resilience to supply noise. There are numerous examples of critical-path-tracking circuits for local timing generation [3], [2]. Instead, we use a programmable digitally-controlled oscillator (DCO) to generate the system clock. The DCO contains a ring of programmable delay cells comprising transmission and NAND gates that approximate a typical fanout-of-4 inverter delay. Such designs have been found to deliver decent tracking accuracy, acting as a proxy for critical path delay in complex digital logic [1], [4], [8].

As shown in Fig. 4(a), a digital code, $D = D_7...D_1D_0$, sets the DCO frequency by selecting the number of delay cells in the oscillator loop. While our implementation uses 7 bits of control code, measurement results show that the lower 4 bits are sufficient for the normal operating range of the digital load. Fig. 4(b) plots DCO frequency versus supply voltage (DVDD) across a range of the digital control codes. Across the measured voltage range (0.6 to 1V), frequency scales roughly linearly with voltage, but slightly flattens out for voltages below 750mV. The uneven frequency spacing with respect to D results from the delay cell's asymmetric design.

C. Cortex-M0 and memory

The Cortex-M0 microprocessor and one of the 64KB memories (SRAM1) share the voltage domain (DVDD) with the DCO. A built-in self-test (BIST) module performs at-speed test of the SRAMs. To make sure all of the memory cells are thoroughly tested, the BIST performs a modified MARCH-C routine that writes and reads different data patterns to and from every address in the SRAMs. The BIST module raises a pass/fail flag at the conclusion of the MARCH-C routine, and the address and data of the last failure are recorded for post-test analysis.

SRAM0 shares the same physical design as SRAM1, but operates off of a separate voltage domain (TVDD) along with other test peripherals such as a voltage monitor circuit that captures fast nanosecond-scale supply transients. The SoC is

978-1-4673-6145-3/13 $31.00 © 2013 IEEE

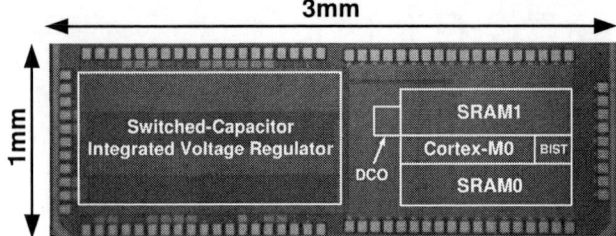

Fig. 5: Die photo of the fully-integrated system-on-chip

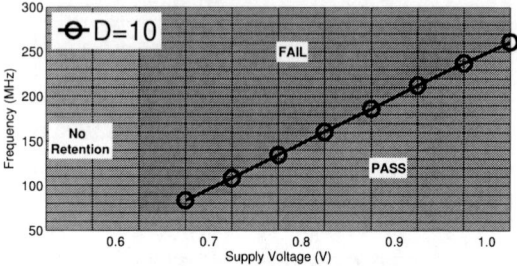

(a) External clock at fixed frequencies

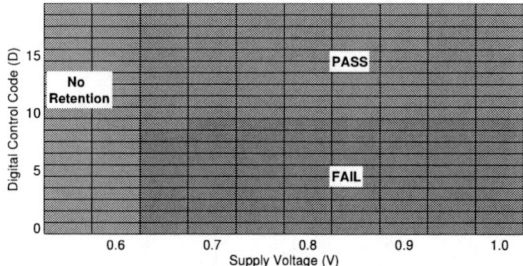

(b) Internal adaptive clock generated by the DCO at different control code (D)

Fig. 6: Shmoo plots for two different clocking schemes

capable of operating directly off of the battery without any external supply or clock reference. TVDD, EXTVDD, and EXTCLK were added for testing purposes only.

III. EXPERIMENTAL RESULTS

To demonstrate improved resilience and performance of the proposed adaptive clocking across a wide range of supply voltage, measurement results were obtained from a prototype SoC chip (Fig. 5) fabricated in TSMC's 40nm CMOS technology. We use the maximum error-free operating frequency of the memory performing built-in self-test as a proxy metric, because it is often the on-chip SRAM sharing the same voltage domain with the digital logic that limits the system performance at lower supply levels. Also, the retention voltage of the SRAM cells typically determines the minimum operating voltage of the system [5].

This section presents the following set of experimental results: First, we characterize the voltage versus frequency relationship of the SRAMs using external sources in order to determine the efficacy of using the DCO for adaptive-frequency clocking. Then, we compare the fixed- and variable-frequency clocking schemes with a regulated voltage generated by the SC-IVR in closed-loop operation. Lastly, we present the advantages of combining adaptive clocking with a variable voltage provided by operating the SC-IVR in open loop.

A. Frequency vs. voltage characterization

The on-chip SRAMs were characterized at static supply voltage levels provided externally via EXTVDD, in order to determine the SRAM's voltage to frequency relationship under quiet supply conditions. Using an external clock (EXTCLK) at different fixed frequencies, we obtained the Shmoo plot in Fig. 6(a). It shows (1) the minimum retention voltage of SRAM cell is between 0.6V and 0.65V; (2) the maximum SRAM frequency scales roughly linear with supply voltage and ranges from 68MHz at 0.65V to 256MHz at 1.0V; and (3) the maximum SRAM frequency closely correlates with DCO frequency plot for control code D=10. This correlation suggests the FO4-delay-based DCO tracks the critical path delay in the SRAM across a wide supply range and should enable error-free memory BIST for control word D above 10. We experimentally verified this by turning on the internal DCO to provide the system clock instead of using EXCLK and sweeping the same voltage range via EXTVDD. The resulting Shmoo plot in Fig. 6(b) further demonstrates the DCO's ability to track SRAM delay at different static supply voltage levels.

B. Fixed vs. adaptive clocking with regulated voltage

Having verified the DCO, we now compare fixed- and adaptive-frequency clocking schemes for a system that op-

erates off of a regulated voltage with the SC-IVR operating in closed loop. We also emulate noisy operating conditions using the on-chip I_{LOAD} generator that switches between 0 and 15mA at 1MHz. Measurements made via the on-die voltage monitoring circuit showed approximately $\pm 70mV$ worst-case ripple about a mean voltage of 0.714V.

For the conventional fixed-freqeuency clocking scheme, the maximum operating frequency ought to depend on the worst-case voltage droop, measured to be 0.647V. Using the measured relationship in Fig. 6(a), the maximum frequency cannot exceed 68MHz. To measure the actual maximum error-free frequency, we performed 100 independent BIST runs using the external clock, EXTCLK, set to a fixed frequency and recorded the failure rate. Fig. 7(a) summarizes the measured failure rates across different externally driven operating frequencies. These results show that the maximum error-free frequency is below 55MHz for the fixed-frequency clocking scheme, which is even lower than the anticipated 68MHz, perhaps attributable to the additional noise injection.

Using the same IVR configuration and test conditions, Fig. 7(b) plots the failure rates versus different digital control codes of the DCO. At D=10, there were intermittent failures, attributable to the additional noise not present in the prior experimental results of Fig. 6(b). The adaptive-frequency clocking scheme delivers consistent and reliable operation at D=11. Based on the average DCO frequency measured during the tests and also plotted in Fig. 7(b), D=11 corresponds to an average frequency of 111MHz, which is 2× the fixed-frequency clocking scenario.

We can attribute this large frequency difference between the two clocking scenarios to a couple of factors. Fixed-frequency clocking requires sufficiently large guardbands to guarantee operation under the worst-possible voltage droop condition. In contrast, adaptive-frequency clocking allows both the clock period and load circuit delays to fluctuate together as long as both vary with voltage in a similar manner. Hence,

978-1-4673-6145-3/13 $31.00 © 2013 IEEE

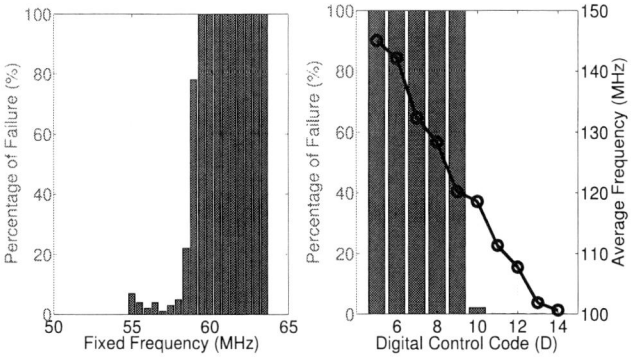

(a) Fixed-frequency externally (b) Adaptive-frequency from DCO

Fig. 7: Comparison of memory BIST failure rates for fixed- vs. adaptive-frequency clocks, but under the same IVR closed-loop regulation.

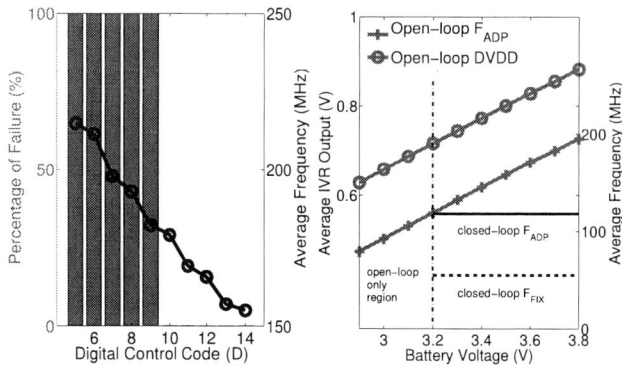

(a) Failure rate and average fre- (b) Measured average clock frequency vs. DCO settings   quency and output voltage

Fig. 8: Performance under unregulated voltage from open-loop SC-IVR operation.

the guardband must only cover voltage-tracking deviations between the DCO and load circuit delay paths across the operating voltage range of interest, and can be built into the DCO. Another factor that penalizes the performance of the fixed-frequency clocking comes from the additional noise on the external clock signal for crossing the TVDD to DVDD boundary.

C. Adaptive-frequency clocking with unregulated voltage

We now turn our attention to how adaptive-frequency clocking performs with an unregulated voltage generated by the SC-IVR operating in open loop. We used the same test setup with $V_{BAT} = 3.7V$ and noise injection via the on-chip I_{LOAD} generator. The failure rates and the average frequencies are captured in Fig. 8. Compared to the measured results in Fig. 7(b), average frequencies are much higher, because DVDD settles to higher values ($\approx 0.8V$) when the SC-IVR operates in open loop. Despite the high susceptibility to fluctuations on DVDD to load current steps as seen in Fig. 3, Fig. 8(a) shows zero errors occured even for D=10. The higher DVDD voltage provides more cushion to avoid intermittent retention failure.

In order to illustrate the extended operating range offered by running the SC-IVR in open loop, Fig. 8(b) plots the average DCO frequency and average DVDD voltage for error-free operation versus battery voltage. These measurements were again made with 0 to 15mA current load steps. As expected, the open-loop SC-IVR's average output voltage scales proportional to the battery voltage. Moreover, the system can operate error-free even for battery voltages below 3V, which approaches the 2.5-2.7V lower discharge limit of Li-ion batteries. In comparison, assuming a target SC-IVR regulated voltage of 0.7V, the system would only operate down to a battery voltage of 3.2V and at a lower frequency across the battery discharge profile even with the adaptive-frequency clocking scheme. A fixed-frequency clocking scheme would lead to even lower performance.

In addition to validating the resilience and the performance advantages of adaptive-frequency clocking, our experimental results also reveal the synergistic properies between the clocking scheme and the IVR design in a battery-powered SoC system. The supply-noise resilience provided by an adaptive clock alleviates design constraints imposed by voltage ripple and voltage droop. Therefore, the IVR can trade off its transient response for better efficiency or smaller area when co-designed with adaptive-frequency clocking.

IV. CONCLUSION

An adaptive-frequency clocking scheme offers several advantages when combined with an IVR in an SoC, fabricated in 40nm CMOS, for battery-powered aerial microrobotic applications. For regulated voltage operation via closed-loop SC-IVR, adaptive-frequency clocking enables $2\times$ performance improvement, compared to conventional fixed-frequency clocking. Combining adaptive-frequency clocking with an unregulated voltage via open-loop IVR extends the operating range across a wider portion of the battery discharge profile. Finally, the noise resilience demonstrated by the adaptive-frequency clocking scheme calls for co-design and co-optimization of clock generation and voltage regulation in weight-and-power constraint integrated systems.

ACKNOWLEDGMENT

The authors thank the TSMC university shuttle program for chip fabrication. This work was supported in part by the National Science Foundation (NSF) Expeditions in Computing Award #: CCF-0926148.

REFERENCES

[1] T. D. Burd et al. A dynamic voltage scaled microprocessor system. *JSSC*, 35(11), 2000.

[2] I. J. Chang et al. Exploring asynchronous design techniques for process-tolerant and energy-efficient subthreshold operation. *JSSC*, 45(2), 2010.

[3] M. E. Dean. *STRiP: a self-timed RISC processor*. PhD thesis, Stanford University, 1992.

[4] Y. Ikenaga et al. A 27% active-power-reduced 40-nm CMOS multimedia SoC with adaptive voltage scaling using distributed universal delay lines. *JSSC*, 47(4), 2012.

[5] S. Jain et al. A 280mV-to-1.2V wide-operating-range IA-32 processor in 32nm CMOS. In *ISSCC*, 2012.

[6] M. Karpelson et al. Energetics of flapping-wing robotic insects: Towards autonomous hovering flight. In *IROS*, 2010.

[7] T. Tong et al. A fully integrated battery-connected switched-capacitor 4:1 voltage regulator with 70% peak efficiency using bottom-plate charge recycling. Submitted to CICC, 2013.

[8] G.-Y. Wei and M. Horowitz. A fully digital, energy-efficient, adaptive power-supply regulator. *JSSC*, 34(4), 1999.

Distributed clock generator for synchronous SoC using ADPLL network

E. Zianbetov[1], D. Galayko[1], F. Anceau[1], M. Javidan[1], C. Shan[1], O. Billoint[3],
A. Korniienko[2], E. Colinet[3], G. Scorletti[2], J. M. Akré[1], J. Juillard[4]

[1]UPMC, LIP6 lab, Paris, France [2]Ampère lab, Lyon, France [3]CEA-LETI, Grenoble, France [4]Supélec, Gif-Sur-Yvette, France
eldar.zianbetov@lip6.fr

Abstract—**This paper presents a novel architecture of on-chip clock generation employing a network of oscillators synchronized by the distributed all-digital PLLs (ADPLLs). The implemented prototype has 16 clocking domains operating synchronously in a frequency range of 1.1-2.4 GHz. The synchronization error between the neighboring clock domains is less than 60 ps. The fully digital architecture of the generation offers flexibility and efficient synchronization control suitable for use in synchronous SoCs.**

I. INTRODUCTION

Clock generation and distribution are one of the main challenges in the design of modern large scale SoC [1]. The increase of relative dimensions of digital SoCs, together with power limitations, make the techniques of centralized clocking prohibitive. While long transmission lines are needed for chip-wise clock distribution, the associated delays must be perfectly mastered. This is very difficult to obtain with acceptable power consumption costs. It is the main reason for the popularity of globally asynchronous locally synchronous (GALS) SoC architecture, which allows to use many small size clocking domains with asynchronous communication between them.

However, GALS presents a number of fundamental drawbacks related with reliability and verification issues, as well as reduced communication speed. The solution proposed in this paper guarantees a synchronization between the clocks of the neighboring local zones, and in this way, makes possible the synchronous communication between the neighboring zones and even between zones situated at some distance on the chip. This solution, called "distributed clock generator", uses a network of local clock oscillators distributed over the chip area similarly with the GALS architecture . However, the local clock generators are mutually coupled with their immediate neighbors in the phase domain so that all local clocks have the same phase and frequency. Here, long clock distribution lines employed by conventional architectures are replaced by local short network links which connect small local clock trees. The network of local clock generators is sufficiently dense, so that: (i) geometric distance between each couple of neighboring oscillators is small enough and delays associated with the network oscillator links are negligible, (ii) distribution of the clock signal inside of each clocking domain is done by conventional techniques and (iii) synchronous communication between neighboring zones is possible as far as the corresponding local oscillators are synchronous in phase.

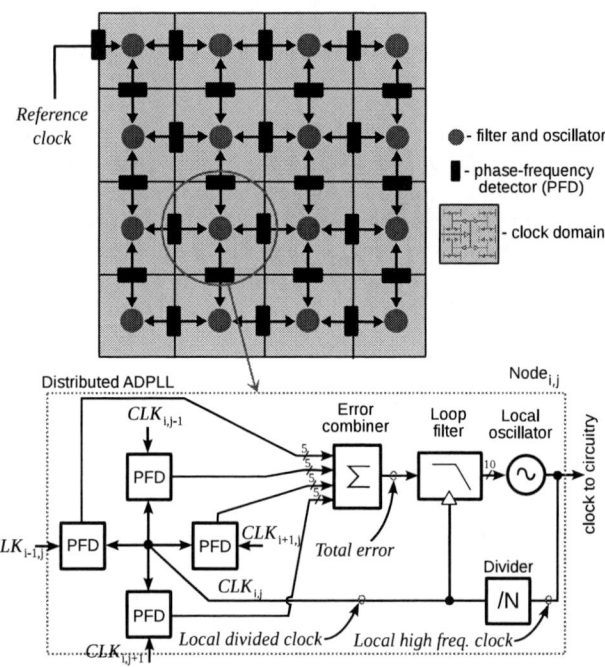

Fig. 1. Architecture of the ADPLL network and of a single node.

In the past, a few analog implementations of such clocking systems were presented [2], [5]. However, since the clocking circuits are tightly integrated in the core digital system on chip, the analog nature of these solutions made them weakly suitable for practical use.

This paper presents a first fully digital implementation of an array of 16 oscillators coupled through a network of all-digital phase locked loops (ADPLLs) intended for distributed clock generation. The proof-of-concept chip generating 1.1-2.4 GHz clock is implemented in 65 nm CMOS technology. The theoretical basis for this system was provided by the studies [2], [3], [4], while the only priorly existing silicon implementation of this concept employed a network of analog PLLs [5].

II. DISTRIBUTED CLOCKING ARCHITECTURE

A. System description

The structure of the clocking network is presented in Fig. 1. The local clocks are generated by digitally controlled oscil-

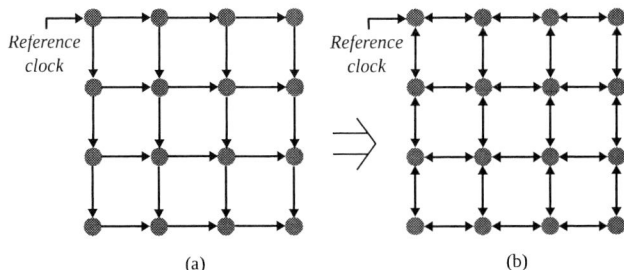

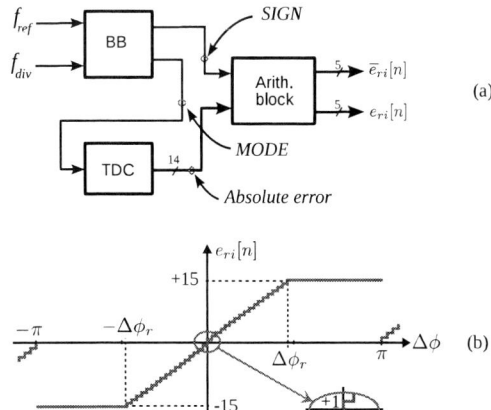

Fig. 2. Mode-lock elimination technique: (a) power up in unidirectional configuration, (b) bidirectional configuration.

lators (DCOs). 24 digital Phase-Frequency Detectors (PFD) measure the timing error between each couple of neighboring DCOs. The network is coupled with the external reference clock through a PFD placed in upper left corner of the network. The digital error signals from PFDs are processed by the digital proportional-integral (PI) loop filters. Each filter processes a weighted sum of errors from up to 4 PFDs (depends on topological position of the node) and generates a digital control code for the DCO. The control objective of the filter is to maintain the sum of the errors close to zero. Such a network, if properly designed, synchronizes at the phase of the reference clock.

The paramount question about the stability of such a complex dynamical system was addressed in theoretical studies through control theory tools [3], [4]. A formal proof of stability, together with an algorithm of choice of the block parameter were proposed.

B. Desirable mode selection

In difference with single PLLs, in PLL networks there are exist multiple synchronized modes in which all oscillators have the same frequency and a fixed (zero or not) phase error. Only the mode with a zero phase error is required for the clocking application. However, the actual synchronization mode depends on initial conditions on which in most cases the system has no control. Several methods have been proposed in the past for the selection of the desirable mode [2].

Our method exploits the ability of the digital PLL network to modify its topology (connectivity) on the fly [8]. The idea is to artificially define the initial conditions of the network inside of the attraction basin of the desired synchronization mode. The start-up procedure consists of two steps:

Step 1. The clocking network is powered up and programmed into a unidirectional configuration. This is achieved by disabling or enabling the feedback links between the nodes. For example, each node receives the information about errors from upper and left neighbors (Fig. 2(a)). This mode excludes the cycles of propagation of information, hence eliminates the possibility of undesired locking. However, in such an operation mode the suppression of perturbations is weak [8], and this mode is not suitable for reliable clocking.

Step 2. Once the network is synchronized with small timing errors, it is re-programmed into a bidirectional configuration

(Fig. 2(b)). In this mode the reverse links are activated, and the network operates in a fully synchronous mode with distributed feedback (coupling) maintaining the synchronization in the desired mode with near-zero phase errors.

III. NODE ARCHITECTURE

A. Phase-frequency detector overview

The PFD is an analog-to-digital converter quantifying the synchronization error into a digital 5-bit signed number. According to its transfer function, showed in Fig. 3(b), its range is limited by the boundaries $\pm\Delta\phi_r$, which are derived from the constraints of precision and hardware complexity. The detail block diagram of the PFD is shown in Fig. 3(a). The PFD consists of a bang-bang phase detector (BB) measuring the sign of the phase error [7] and a time-to-digital converter (TDC) for the quantification of the absolute time error between two clocks. The arithmetic block combines the signals from these blocks and produces two binary signed signals (straight and inverted) thereafter used by the local and neighboring nodes.

The TDC is based on a tapped delay line followed by the sampling register. Its resolution is 32 ps.

B. Digitally controlled oscillators

The implemented DCOs are the ring CMOS oscillators employing width-modulated technique for the digital frequency tuning [6], [7]. Their structure is based on a 7-stage ring oscillator (Fig. 4) with parallelly connected tuning inverters (Fig. 4, CTI0-CTI6 and FTI0-FTI2) to each stage of the oscillator. The main inverters (Fig. 4, MI0-MI6) are always active and define the lowest oscillation frequency. The tuning inverters are distributed over all 7 stages of the oscillator and divided in two arrays: 256 coarse tuning (CTI) and 3 fine tuning (FTI) inverters. They provide respectively 6 MHz and 1.5 MHz frequency tuning steps with a total of 256×4=1024 steps. The cells are controlled by three thermometer codes obtained from the binary-to-thermometer decoders. The monotonicity of the code-frequency characteristic is guaranteed by an appropriate

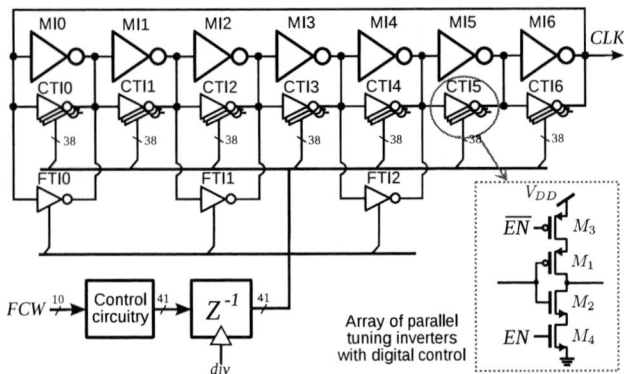

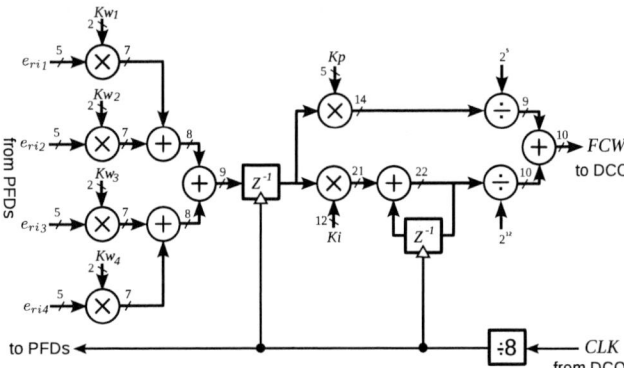

Fig. 4. Digitally controlled oscillator based on a 7 stage CMOS ring oscillator.

Fig. 5. The error combiner and proportional-integral filter.

choice of the control algorithm. The oscillator designed in 65 nm technology has a frequency tuning range of 1-2.5 GHz.

C. Error processing

The error processing in node is performed in two steps by an error combining block and a loop filter (Fig. 5).

The first block receives up to four 5-bit 2-complement coded errors. They are passed through four variable gain blocks and then summed using a four-input adder. The weighting coefficients of the variable gain $Kw_1 - Kw_4$ are programmable. Each gain can take independently a value from the set $\{0,1,2,4\}$ and implemented as a binary shift, so introducing a very small delay. Programming these coefficients, we can control the connectivity between the nodes of the network. Then the four-input adder operates with 7-bit operands and produces a 9-bit sum. The output of the adder is buffered with a register. We mention that each node is an auto-sampled system: the filter is sampled with the generated local clock divided by 8 and PFDs compare the clocks at this rate.

The PI filter processes the 9-bit sum of the errors. It has coefficients K_p and K_i that can be programmed by respectively 5 and 12-bit words. The programmability of the filter is essentially intended for the testing purposes and for the theory validation.

IV. TEST CIRCUIT DESIGN

A prototype of the distributed clock generator with 16 nodes has been designed and manufactured in 65 nm CMOS technology. It has an area of ≈ 2 mm^2 where the clock network itself occupies 0.8×0.9 mm^2 (Fig. 6). Besides the clocking network, the on-chip digital circuitry includes design-for-test block, the PFD and bang-bang detector for their characterization. The microphotograph of the fabricated silicon prototype is presented in Fig. 6.

V. MEASUREMENT RESULTS

The goals of the experiments were a characterization of the phase synchronization between the DCOs and an investigation of the sensitivity of the network to different perturbations.

The initial frequencies of the 16 DCOs of the fabricated network are distributed within a 47 MHz range, which gives good conditions for the fast start-up of the network. This range is explained mainly by the sensitivity of the oscillators to the supply voltage, which is measured to be ≈ 900 MHz/V. However, even with this mismatch, the frequency adjusting range of the network is 1100-2380 MHz, is guaranteed for ± 10 % supply voltage variation.

Fig. 7 presents the captured waveforms of divided by 16 local clocks when the network is synchronized. The observed timing errors between neighboring clocks were in the range of 30-60 ps for 1.6 GHz local frequency. This corresponds to 2 steps of the PFD resolution and is in a good agreement with the theory and simulation. The obtained phase error is less than 10% of the clock period. This result can be improved by increasing the PFD resolution. As predicted by the theory, the error is a zero-centered random process, i.e. the static skew is zero.

Fig. 8 shows the transient process in one of the nodes. The perturbation has been introduced in a network (@ $t = 8$ μs) in order to study the robustness of the network. After this

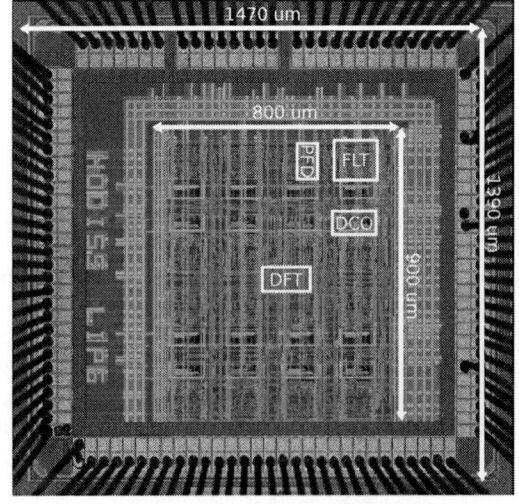

Fig. 6. Die microphotograph.

978-1-4673-6145-3/13 $31.00 © 2013 IEEE

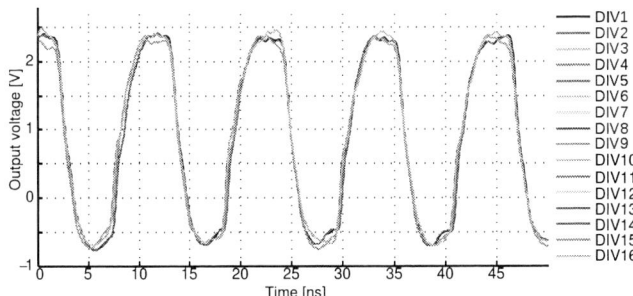

Fig. 7. Captured synchronous divided clocks.

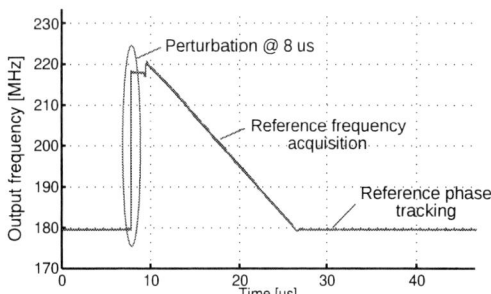

Fig. 8. Transitional process in Node 11.

TABLE I
NETWORK TEST CHIP MEASUREMENTS SUMMARY AND COMPARISON

Parameter	[5]	This work
Number of nodes	16	16
Frequency range, MHz	1100-1300	1100-2380
Timing error, ps	30	< 60*
Power consumption, mW	390@1.2 GHz	186.2@1.6 GHz
Technology, nm	350	65
Clocking core area, mm²	-	≈0.72
Chip area, mm²	≈9	≈2.04
Circuit family	analog	digital

* between neighbor nodes

perturbation, the clocking network is resynchronized after 17 μs. The frequency acquisition speed can be increased by employing special techniques more efficient than a simple PI filtering.

In order to study the synchronization in the undesired modes we have repeated the cycle of global perturbation (all nodes affected) of the network 500 times. In all test cases we have not observed the mode-locking. In fact, this result is in contradiction with modeling and the theory: the possibility of mode-locking is one of particular properties of the PLL networks which is always mentioned in theoretical studies [2] and reproduced in prototype [9]. Therefore, in order to verify the proposed technique of desirable mode selection we have repeated the experiment with reduced (2×2) network configuration, where theoretically mode-locking must occur with high probability. In such a configuration, for 500 cycles of global perturbation we have observed the mode-locking 4 times and the proposed method showed to be efficient.

In order to check the robustness of the network operation in presence of variation of the block parameters, several experiments were done. In particular, the network was tested under 10% variation of the filter coefficients: no degradation in the quality of the oscillator synchronization was observed.

The power consumption of the clocking network has been measured for 1.6 GHz oscillation frequency under 1.2 V supply voltage. The PFDs and PI filters consume 32 mW (≈2 mW per node). The DCO consumption is 9.8 mA/node (≈6.15 mW/GHz). We note that the power optimization of the DCO was not an objective of this prototype and better results can be obtained by a more involved design.

Table I shows the summary of measured results and comparison with existing implementation of the distributed clock generator.

VI. CONCLUSION

A distributed clock generator for synchronous SoC based on the network of phase-coupled oscillators has been demonstrated. The synchronization of the oscillators is achieved by the ADPLL network. The problem of undesirable synchronization modes is solved by a dynamic reconfiguration of the network interconnection topology at the start-up stage. The advantage of the proposed system is the compatibility with the digital environment, its flexibility of reconfiguration and the possibility of advanced control over the clock generation. The fabricated prototype has proved the reliability of the proposed clock generation methodology. It has 16 nodes and operates in a frequency range of 1.1-2.4 GHz. The measured timing accuracy between neighboring clocking domains of the circuit is less than 60 ps. This result can be improved by more involved design of the phase-frequency detector.

ACKNOWLEDGMENTS

This work was supported by French National Agency of Research in a framework of HERODOTOS research project.

REFERENCES

[1] N. Kurd, P. Mosalikanti, M. Neidengard, J. Douglas, and R. Kumar. *"Next generation Intel Core micro-architecture (Nehalem) clocking."* IEEE JSSC vol. 44, no. 4 (2009): 1121-1129.

[2] G. Pratt and J. Nguyen. *"Distributed synchronous clocking."* Parallel and Distributed Systems, IEEE Trans. on 6, no. 3 (1995): 314-328.

[3] A. Korniienko et al., *"Control Law Synthesis for Distributed Multi-Agent Systems: Applications to Active Clock Distribution Network"*, Automatic and Control Conference, San Francisco, CA, 2010

[4] J. M. Akre, J. Juillard, D. Galayko and E. Colinet. *"Synchronization analysis of networks of self-sampled all-digital phase-locked loops."* Circuits and Systems I, IEEE Trans. on 59, no. 4 (2012): 708-720.

[5] V. Gutnik and A. P. Chandrakasan. *"Active GHz clock network using distributed PLLs."* IEEE JSCC, vol. 35, no. 11 (2000): 1553-1560.

[6] E. Zianbetov et al., *"A Digitally Controlled Oscillator in a 65-nm CMOS process for SoC clock generation."*, IEEE Int. Symp. on Circuits and Systems, pp. 2845-2848, 2011.

[7] J. A. Thierno et al., *A Wide Power Supply Range, Wide Tuning Range, All Static CMOS All Digital PLL in 65 nm SOI*, IEEE JSSCC, vol. 43, no. 1, January 2008.

[8] M. Javidan et al., *"All-digital PLL array provides reliable distributed clock for SOCs."* IEEE ISCAS conf., pp. 2589-2592, 2011.

[9] C. Shan et al., *"FPGA implementation of reconfigurable ADPLL network for distributed clock generation."* IEEE Int. Conf. on Field-Programmable Technology, pp. 1-4, 2011.

A 920MHz Quad-core Cryptography Processor Accelerating Parallel Task Processing of Public-key Algorithms

Shuai Wang, Jun Han*, Yang Li, Yifan Bo, Xiaoyang Zeng

State State-Key Lab of ASIC and System, Fudan University, Shanghai, China, 201203
*junhan@fudan.edu.cn

Abstract—**The wireless access point (AP) devices of the next generation demand low-latency, high-security, and flexible authentication and authorization for heterogeneous clients. This makes it necessary to implement the high-complexity public-key ciphers efficiently on programmable processors. Therefore, this paper presents a quad-core processor that accelerates public-key computations by enabling high-speed parallel task processing. A test chip (1.925x3 um²) of the proposed processor is implemented in 65nm process. This chip is superior to the powerful platforms like Nvidia GPU and AMD Opteron workstation with regard to the performance: 0.13ms for the decryption of RSA-1024 and 0.095ms for the scalar multiplication of ECC-256 in F_p.**

I. INTRODUCTION

With the rapid evolution of internet and wireless technologies, the demand for secure communication is growing. Therefore, in the next generation wireless applications like WiMAX, femicell and ubiquitous computing, the devices of access point or small base station need to provide high-security and low-latency authentication and authorization services for users [1]. Public key ciphers like RSA (Ron Rivest, Adi Shamir, Len Adleman), ECC (Elliptic Curves Cryptography) can separate the capacities for encryption and decryption so that they can fulfill the security requirements of the above wireless applications [2] [3].

However, it is really a big challenge to implement such public-key systems efficiently due to the high complexity of modular computations. For example, the low-cost embedded processor like MIPS32 20Kc core needs about 20ms to run 1024-bit RSA decryption [4]. This latency will greatly affect the quality of delay-sensitive applications such as streaming audio/video and IP telephony when the user walks from one AP to another AP. The powerful GPU [5], on the other hand, can get a high-throughput in a fixed latency due to owning many processing cores, but it doesn't enhance the latency of a single 1024-bit RSA decryption.

Along with the strong momentum of achieving high performance on the resource-limited wireless AP devices, this paper presents a quad-core processor for implementing public-key ciphers. With the dedicated long-integer modular arithmetic units and task-level parallel processing technology, the computation time of public-key ciphers can be significantly reduced on the proposed processor. Moreover,

aiming at parallel processing of public-key ciphers, a dedicated data switch unit and long-integer FIFOs are implemented to reduce the data broadcast and communication overhead on chip, which have not been presented in the previous related works. A test chip of this processor is fabricated in 65nm CMOS process. It can run at 920MHz and will cost 332mW when processing ECC-256 at 1.2V at room temperature.

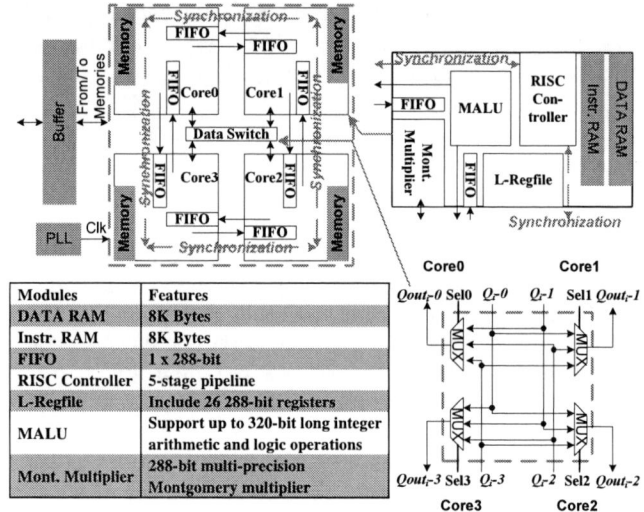

Modules	Features
DATA RAM	8K Bytes
Instr. RAM	8K Bytes
FIFO	1 x 288-bit
RISC Controller	5-stage pipeline
L-Regfile	Include 26 288-bit registers
MALU	Support up to 320-bit long integer arithmetic and logic operations
Mont. Multiplier	288-bit multi-precision Montgomery multiplier

Figure 1. Architecture of Proposed Quad-core Processor

II. OVERVIEW OF PROPOSED PROCESSOR

Fig. 1 shows the block diagram and basic information of the proposed processor involving 4 homogeneous cores. As a high-performance and domain specific multi-core processor, it has following three features.

A. Flexibility and Programmability

The 5-stage RISC pipeline in each core is shown in Fig. 2. It implements an instruction set including 30 basic ALU, memory and branch instructions, 1 synchronization instruction used to sync up the two adjacent cores, 16 MALU instructions such as 3-to-1 320-bit Add, 288-bit sub, and modular sub, and 5 dedicated instructions for Montgomery multiplier which can perform multi-precision Montgomery multiplication [6]. So the proposed processor can process public-key ciphers with

---

This paper is supported by National Natural Science Foundation of China with the grant No.61176023 and Project of State Key Laboratory of ASIC & System with grant No.11MS005.

varied word length such as RSA-1024, RSA-2048, ECC-256, and ECC-521.

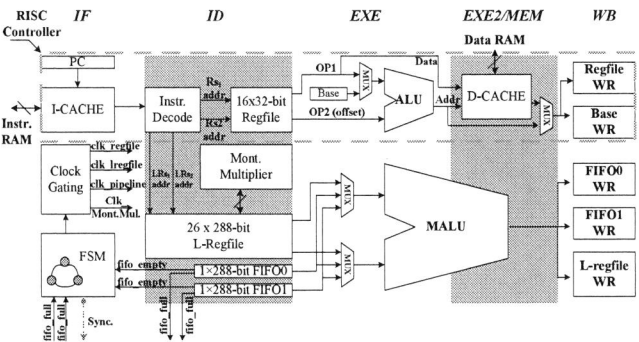

Figure 2. RSIC Pipeline of Each core

B. High-performance Montgomery Multiplier

In general, implementing modular multiplication will result in large latency and high area cost. To solve this problem, each core of the proposed processor integrates a 288-bit improved quotient pipelined Montgomery multiplier. By pipelining the Carry Save Adder (CSA) and the CSA tree, this proposed Montgomery multiplier can achieve both low latency and relative low area cost.

C. Low Inter-Core Communication Overhead

Due to the data dependency, partitioning a single public-key application into tasks on different cores will introduce lots of inter-core message exchanges. To reduce the inter-core communication overhead, the proposed processor integrates 4 couples of 1x288-bit FIFOs and one 24-bit data switch unit. As shown in Fig. 1, each couple of the FIFOs is used for discrete long-integer data exchange between two adjacent cores. On the other hand, the 24-bit data switch is dedicated to on-the-fly broadcast the quotients (Q_i) generated in the Montgomery Multiplier of one core to other cores. Fig. 1 also shows the architecture of this data switch unit. The select signals (*Sel0~Sel3*) can be dynamic configured by the cores (Core0~Core3) so that each core can get Q_i from other cores according to the partitions when processing modular multiplications more than 288 bits.

III. MONTGOMERY MULTIPLIER

A. Algorithm

The proposed multi-precision Montgomery multiplier embedded in each core is the hardware implementation of an improved Montgomery algorithm based on quotient pipelined technology proposed in [7], as shown in algorithm 1. Each long integer is separated into several *r*-bit words, in which *r* represents bit number of the radix.

This proposed algorithm can both simplify the quotient determination by introducing *P2* proposed in [7] while keep the result less than $2^{r+2} \times P$ by adding the final Montgomery reduction operation in line 6. If we choose *e* satisfying *P* less than $2^{e \times r-3}$, it is not difficult to prove that the value of result *S* will be kept less than $2^{r+2} \times P$, the same range as the inputs *A* and *B*.

Alg. 1 Improved Quotient Pipelined Montgomery Algorithm(Radix 2^r)

Input:
$P = (P_{e-1}, P_{e-2}, ..., P_0)_R$; $P''=-P^{-1} \bmod 2^{2r}$; $P2 = (P'' \times P + 1) \text{ div } 2^{2r}$
$P'=-P^{-1} \bmod 2^r$; $P1 = (P' \times P + 1) \text{ div } 2^r$, $\mathbf{P<2^{exr-3}}$, $\mathbf{R=2^r}$
$A = (A_e, A_{e-1}, ..., A_1, A_0)_R$; $B = \{B_e, B_{e-1}, ..., B_0\}_R$; $\mathbf{A,B < 2^{r+2} \times P}$

Output: S $= (AB) \times 2^{-(e+1) \times r} \bmod P$, n is the number of bits of N, $\mathbf{S < 2^{r+2} \times P}$

1 : S := 0; Q_0 := 0;
2 : **for** i:=0 to e+1 **do**
3 : Q_{i+1} := S mod 2^r;   // **Quotient determination**
4 :   S := (S >> r)+ $A_i \times B + Q_i \times P2$; //**Montgomery reduction**
5 : **end**
6 : $S := S + P1 \times Q_{e+2}$; // **Final Montgomery reduction**
7 : **return** S;

B. Hardware implementation

In algorithm 1, Montgomery reduction of line 4 is commonly implemented using a CSA tree. However, a straightforward implementation of CSA tree will inevitably make area cost large and result in a long critical path. The Montgomery multiplier implemented in the paper is developed as radix to be 2^{24} and width to be 288 considering the performance, area cost and word length of RSA and ECC. As shown in Fig. 3, this paper takes three technologies to reduce the area cost and critical path delay: (1) By taking the pre-computation of *P2 + B* [8], the number of the partial products is reduced to 24 after the data selection in the 4-to-1 Multiplexor. (2) Using two pipeline stages, the critical path is halved. (3) We replace 320-bit-wide CSA with the 160-bit-wide one, named PCSA shown in Fig. 4, which reduces the area cost by half. The operation of a conventional CSA (a) can be completed by the PCSA (b) in two clock cycles, manipulating low 160 bits and high 160 bits of operand serially. Therefore, the PCSA tree will process the Montgomery reduction in two clock cycles on average by pipelining the iterations as shown in Fig. 3. Notice that, computing low part and high part of operands using pipeline do no harm to performance because of the data dependency of algorithm 1.

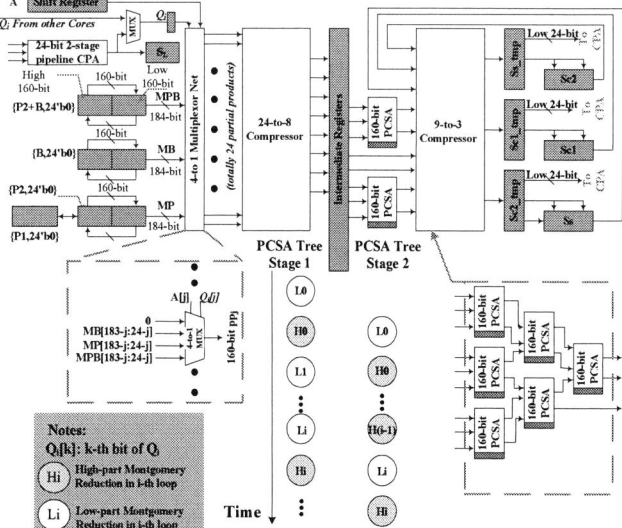

Figure 3. Architecture of Proposed Montgomery Multiplier

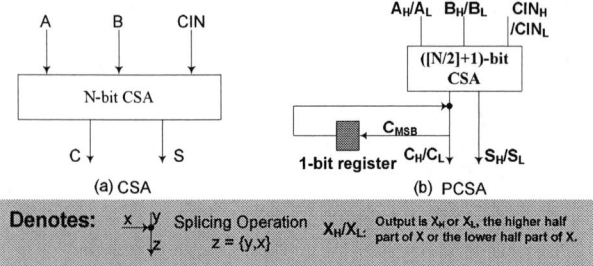

Figure 4. Proposed Pipelined-CSA Architecture.

IV. ACCELERATING PARALLEL TASK PROCESSING

In the proposed multi-core processor, the parallel processing of public-key ciphers can be largely enhanced by hardware-aided data communication and rescheduling the task-level pipeline.

A. Hardware-aided data communication

Hardware-aided data communication presented in this paper can largely reduce the data transfer latency between the parallel tasks running on different cores. Taking 1024-bit Modular multiplication (MM) which can be divided into 16 288-bit sub-MMs as an example, Fig .5 shows its parallel execution flow on four cores. Due to the data dependency, Core0 needs to transfer the on-the-fly generated Q_i to Core1~Core3. The data switch can real-time broadcast each Q_i by twelve 24-bit words, and thus the sub-MM computations in Core1~Core3 will spend little latency on waiting Q_i. On the other hand, the transfer of intermediate results like 288-bit $S^j_{i,i}$ can be accelerated by the FIFO pair between the two cores. By the hardware acceleration, the execution of 1024-bit MM can achieve a speedup rate close to 4. In fact, the proposed processer can speed up MM-521/MM-512 1.70x (184 clock cycles/108 clock cycles) on two cores, and MM-1024 3.51x (684/195) on four cores, comparing to the single-core results.

B. Rescheduling the task-level pipeline

The programmability of proposed multi-core processor makes it easy to exploit the parallelism of public-key algorithms. By partitioning these algorithms into small tasks and rescheduling them optimally, we can make full use of the parallel computing resource in this processor. Taking consecutive ECC-256 point doubling in $\mathbf{F_p}$ under *Jacobian* coordinates as an example, Fig. 6 shows its task-level pipeline on 3 cores. For clarity, it only demonstrates the time-consuming MM tasks. Due to data dependency, the straightforward implementation (a) can't make full use of the computing-resources of the 3 cores. Its critical path appears in Core1 which has 4 MMs. In fact, this critical path can be shortened by rescheduling the task pipeline: shift the first MM (Z_1^2) of Core1 from the current point doubling task to the idle states of the previous point doubling task. Fig. 6(b) shows the optimized data flow after this rescheduling. It reduces 1 MM on the critical path of point doubling, and thus enhances the performance of parallel processing. In fact, mapping the scalar multiplication on the proposed processor, the consecutive point doubling and point addition tasks can be both accelerated by this rescheduling method. As a results, the

proposed processor can finish ECC-256 point doubling in 169 clock cycles (speed up 2.6x), and needs 232 clock cycles (speed up 2.3x) for ECC-256 point addition.

It is noticed that our approach is similar to [9] which presents SIMD schemes to accelerate ECC computation. However, [9] focus on the software platforms, in which modular squaring is less computationally expensive than modular multiplication. This paper pays more attention to reduce the inter-core data transfers, since modular squaring has the same performance as modular multiplication on the proposed processor.

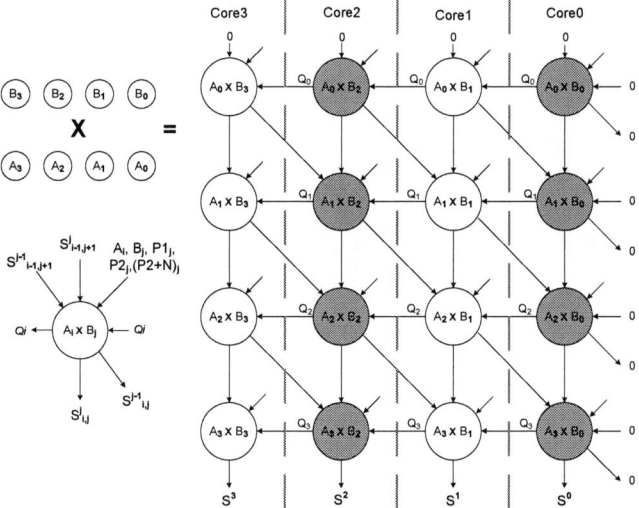

Figure 5. Parallel Processing of MM-1024

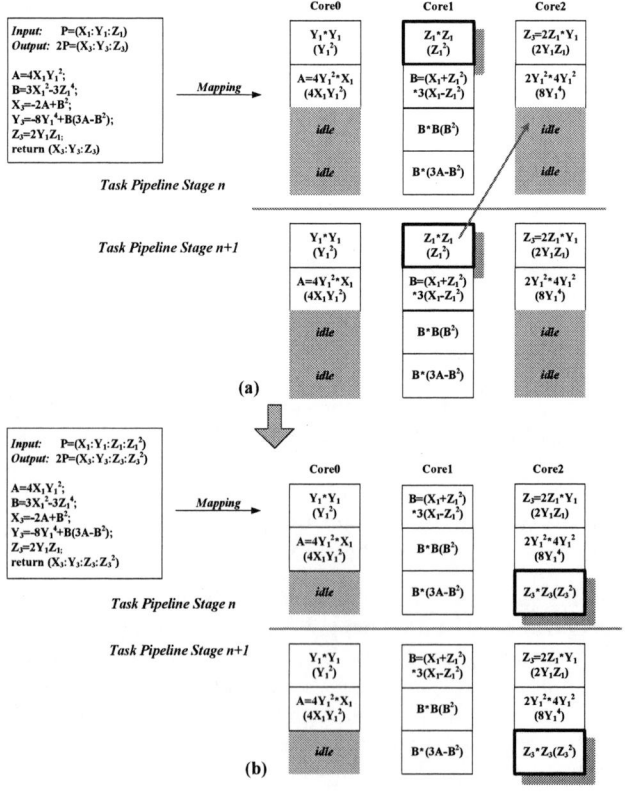

Figure 6. Rescheduling Task Pipeline of ECC-256 Point Doubling

978-1-4673-6145-3/13 $31.00 © 2013 IEEE

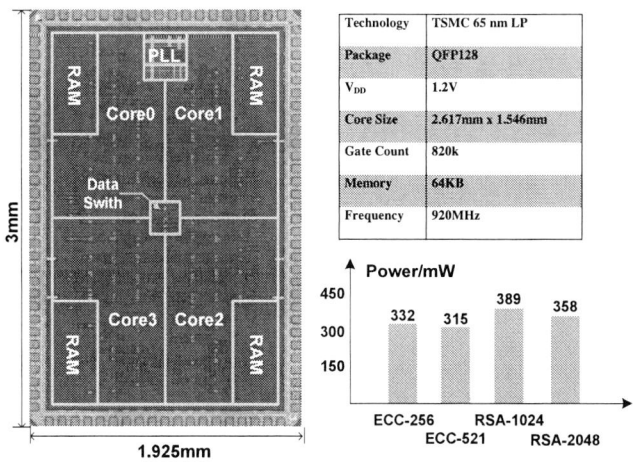

Technology	TSMC 65 nm LP
Package	QFP128
V_{DD}	1.2V
Core Size	2.617mm x 1.546mm
Gate Count	820k
Memory	64KB
Frequency	920MHz

Figure 7. Chip Die Photo and Features

V. EXPERIMENTAL RESULTS

Fig. 7 presents the chip die photo and its features. It can run at 920MHz at 1.2V. At this frequency, this chip can achieve low latencies in calculating public-key algorithms, as shown in Table 1. For example, it takes 0.095ms for the scalar multiplication of ECC-256 in F_p under *Jacobian* coordinates, less than the software implementation on AMD Opteron workstation [10]. It consumes 0.13ms for the decryption of RSA-1024, better than the result of Nvidia 8800GTX GPU [5]. It also outperforms the best previous work in calculating the squaring of F_{p12} [12]. The ASIC implementation in [11] cost less area than us by employing dedicated 128-bit data-path, but it only supports RSA-1024, lacking flexibility like ours. The 2.617mm x 1.546mm core size and less than 400mW power consumption show its costs are low, regarding the performance achieved by it.

VI. CONCLUSION

This paper presents a high-performance quad-core processor for public-key ciphers. Exploiting the parallelism of these algorithms, the proposed processor can achieve low latency even less than the powerful workstation. Only

2.617mm x 1.546mm core size and less than 400mW power consumption make it suitable for the AP devices of the next generation.

REFERENCES

[1] Daojing He, etc., "Secure and Efficient Handover Authentication Based on Bilinear Pairing Functions", IEEE Transactions on Wireless Communications, Vol. 11, Publication Year: 2012 , Page(s): 48 - 53

[2] Diffie, W., "The first ten years of public-key cryptography" Proceedings of the IEEE, Vol. 76, Publication Year: 1988 , Page(s): 560 - 577

[3] Jacobs, S., "WiMAX Subscriber and Mobile Station Authentication Challenges", Communications Magazine, IEEE, Vol.49, Publication Year: 2011 , Page(s): 166 - 172

[4] MIPS Technologies, "64-Bit Architecture Speeds RSA By 4x", June 2002, http://www.mips.com/media/files/white-papers/64bitarchitecturespeedsRSA4x.pdf.

[5] Robert Szerwinski and Tim Güneysu, "Exploiting the Power of GPUs for Asymmetric Cryptography", CHES 2008, Lecture Notes in Computer Science, Vol.5154, pp. 79–99.

[6] P. Montgomery. 1985. "Modular multiplication without trial division", Mathematics of Computation, Vol.44, Pages: 519-521.

[7] H. Orup. 1995. *Simplifying quotient determination in high-radix modular multiplication.* Proc. 12th IEEE Symp, Computer Arithmetic, pp. 193-199.

[8] M.S. Kang and F.J. Kurdahi, "A novel systolic VLSI architecture for fast RSA modular multiplication", in ASIC, IEEE Asia-Pacific Conference, pp.81-84, Aug. 2002.

[9] Patrick Longa and Ali Miri, "Fast and Flexible Elliptic Curve Point Arithmetic over Prime Fields", IEEE Transactions on Computers, Vol.57, Publication Year: 2008, Page(s): 289 – 302.

[10] Patrick Longa and Catherine Gebotys, "Efficient Techniques for High-Speed Elliptic Curve Cryptography", CHES 2010, Lecture Notes in Computer Science, Vol.6225, pp. 80–94.

[11] Miyamoto, A. Homma, N. Aoki, T. Satoh, A., "Systematic Design of RSA Processors Based on High-Radix Montgomery Multipliers", IEEE Transactions on VLSI 2011, Vol.19, Page(s): 1136 – 1146.

[12] Ray C. C. Cheung, Sylvain Duquesne, Junfeng Fan, etc., "FPGA Implementation of Pairings Using Residue Number System and Lazy Reduction", CHES 2011, Lecture Notes in Computer Science, Vol.6917, pp. 421–441.

TABLE I. COMPARISON WITH RELATED WORKS

Ref.	Technology	Freq.	Area (k-gates)	Latency				
				RSA-1024	RSA-2048	ECC-256 Scalar multiplication	ECC-521 Scalar multiplication	Squaring (F_{p12})
This work	65nm CMOS	920MHz	820 [1]	0.13ms	0.88ms	0.095ms	0.58ms	1.1us
[5]	Nvidia8800 GTX GPU	1.35GHz	N/A	1.2ms	9.6ms	0.7ms [2]	N/A	N/A
[10]	AMD Opteron 252	2.6GHz	N/A	N/A	N/A	0.105ms [4]	N/A	N/A
[11]	90nm CMOS	422MHz	154 [3]	0.24ms	Not Supported	Not Supported	Not Supported	Not Supported
[12]	Xilinx FPGA (Virtex-6)	250MHz	7032slices 32DSPs	Not Supported	Not Supported	Not Supported	Not Supported	1.2us

Note: 1: Excluding the area of memories;
2: It is the latency of scalar multiplication of ECC-224.
3: It is the data-path area of the Radix-2^{128} design.
4: We choose the author's result under *Jacobian* coordinates for a fair comparison

978-1-4673-6145-3/13 $31.00 © 2013 IEEE

Parallel Gain Enhancement Technique for Switched-Capacitor Circuits

Hariprasath Venkatram, Benjamin Hershberg, Taehwan Oh, Manideep Gande,
Kazuki Sobue[1], Koichi Hamashita[1], Un-Ku Moon

Oregon State University, Corvallis, OR, U.S.A; Asahi Kasei Microdevices, Atsugi, Japan
venkatha@eecs.oregonstate.edu

Abstract—This paper presents a unified classification model for gain enhancement techniques used in the design of high performance amplifiers. A parallel gain enhancement technique is proposed for switched capacitor circuits which combine the best features of the existing gain enhancement techniques found in continuous-time and discrete-time amplifiers. This technique utilizes two dependent closed loop amplifiers to enhance the open loop DC gain of the main amplifier. This replicated parallel gain enhancement (RPGE) technique enables a very high DC gain amplifier with an improved harmonic distortion performance. A proof of concept pipeline ADC in a 0.18 um CMOS process using RPGE technique achieves 75 dB SNDR, 91 dB SFDR, -87 dB THD at 20 MS/s. The measured 13 bit DNL and INL is +0.75/-0.36 and +0.88/-0.92 LSB respectively. The ADC operates from a supply voltage of 1.3 V, consumes 5.9 mW, occupies 3.06 mm^2 and achieves a figure of merit of 65 fJ/CS.

I. INTRODUCTION

Gain enhancement, distortion reduction and cancellation techniques are widely used in building high performance amplifiers. A unified gain enhancement classification chart is shown in Fig. 1. The gain enhancement techniques can be classified into *continuous*-time and *discrete*-time gain enhancement techniques. The continuous-time gain enhancement technique can be broadly classified into *parallel* [1] and *replica* amplifier [2, 3] based techniques. The discrete-time gain enhancement techniques can be classified as *sequential* and *parallel* gain enhancement techniques based on their operation and implementation. Using this classification method, we identify the advantages and disadvantages of the existing gain enhancement techniques and propose a parallel gain enhancement technique, for switched-capacitor circuits, combining the best features of the existing gain enhancement techniques.

In discrete-time switched-capacitor circuits, the gain enhancement technique is classified as a sequential technique when the measurement of gain error and its correction are performed in two different clock phases. This sequential discrete-time gain enhancement technique can be further classified into *input* and *output* based techniques based on the location of the gain error measurement (input or output) and the correction of the measured gain error (input or output). Under this classification, correlated double sampling (CDS) technique [4] will fall under sequential input based gain enhancement technique. The correlated level shifting (CLS)

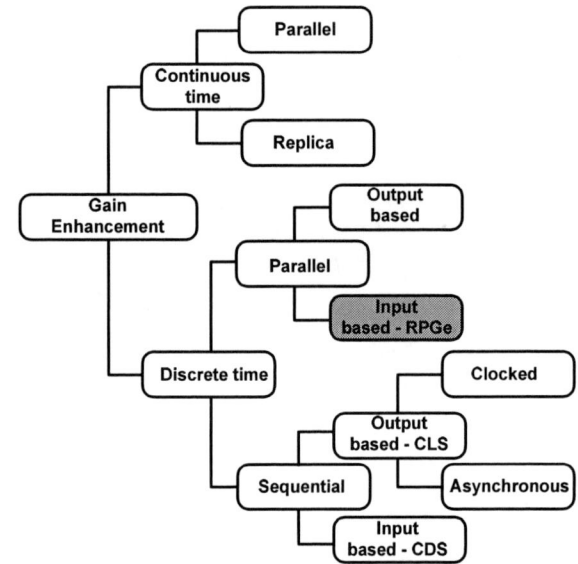

Fig. 1. Gain enhancement technique in amplifiers

and Split-CLS techniques can be classified as sequential output based gain enhancement technique [5]. The *asynchronous* sequential output based CLS technique was proposed in [6] and the parallel output based Class A+ amplifier technique was proposed in [7].

In this paper, the proposed RPGE technique is a discrete-time parallel input based gain enhancement technique for switched-capacitor circuits. Section II describes the RPGE amplifier in detail. Section III describes the high resolution pipelined ADC enabled by RPGE amplifier and Section IV concludes the paper with measurement results.

II. PARALLEL GAIN ENHANCEMENT TECHNIQUE

The replica amplifier based gain enhancement technique [1] and the switched capacitor based CLS gain enhancement technique [5] are discussed briefly to identify their strengths and weakness.

A. Replica Amplifier and CLS Amplifier

The main amplifier and the replica amplifier from [1] are shown in Fig. 2a. The replica amplifier is a scaled version of the main amplifier. The coupling transconductance supplies the majority of the output current while the main amplifier supplies only the mismatch current between the coupling transconductance current and the load current.

978-1-4673-6145-3/13 $31.00 © 2013 IEEE

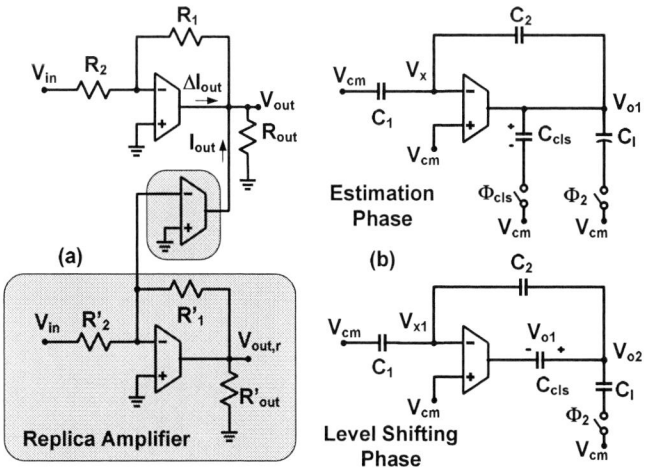

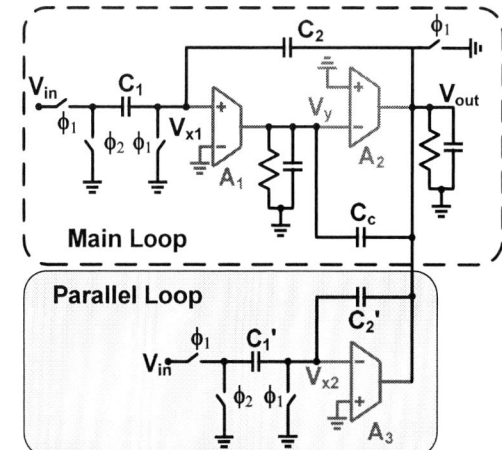

Fig. 2. (a) Replica amplifier (b) CLS estimation & level shift phase

Fig. 3. Replicated Parallel Gain Enhanced Amplifier

As described above, in addition to requiring an additional amplifier, the replica amplifier technique in [1] suffers from mismatch errors between the main amplifier and replica amplifier output impedance.

The sequential output based switched capacitor technique [5], CLS, is shown in Fig. 2b. The estimation phase and the level shifting phase are shown in Fig. 2b. The equivalent loop gain at the end of the level shifting phase is approximately the product of the loop gain in the estimation phase and the level shift phase. The opamp output is closer to virtual ground at the end of level shift phase. However, this technique requires one additional phase to complete the accurate charge transfer. The C_{cls} capacitor is an additional load capacitor at the output and its size in comparison to the load capacitor is an important design trade-off.

B. Replicated Parallel Gain Enhanced Amplifier

The parallel nature of operation, present in the replica amplifier, transferred to a switched capacitor amplifier would enable the use of an existing two-phase non-overlapping clock, with reduced bandwidth and relaxed settling time requirement for the opamp. Also, the replicated parallel load at the input can be made very small. This important feature allows us to decouple DC-gain and output swing requirements to build an amplifier optimized for gain without the burden of swing requirement. The replicated parallel gain enhanced (RPGE) amplifier, shown in Fig. 3, combines the above mentioned benefits of reduced bandwidth, settling time requirement, conventional two phase operation, smaller replicated parallel path capacitors, decoupled DC gain and swing requirements to produce an optimized input based parallel gain enhancement amplifier. Fig. 3 shows the main loop amplifier as well as the parallel loop amplifier section in the shaded region. The gain of the main loop is set by C_1/C_2 and that of the parallel loop is set by C_1'/C_2'. When the gain of the parallel path, C_1'/C_2', is made higher than C_1/C_2, the first stage output (V_y) is forced to move closer to "zero" as the parallel loop virtual node (V_{x2}) swing increases. This trend is illustrated for the internal node voltages, V_y and V_{x2}, in Fig.4a.

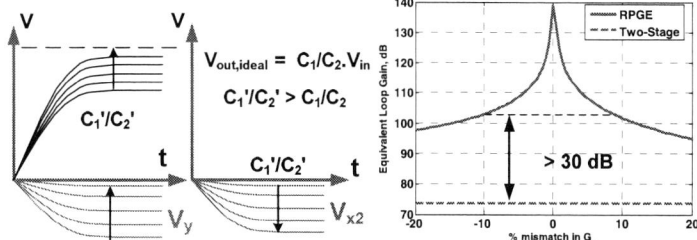

Fig. 4. (a) RPGE Trend & (b) Effective loop gain enhancement with mismatch

The improved loop gain is obtained from the gain correction/enhancement of second stage by reducing swing at node V_y. The equivalent open loop gain of RPGE amplifier can be made very high in theory (infinite) by selecting the appropriate parallel path gain and is limited only by the capacitor ratio precision (C_1'/C_2') in a given process. In reality, with non-linear transconductance and limited precision for capacitor ratio, the gain enhancement is at least the product of main loop and the parallel loop ($A_1 \cdot A_2 \cdot A_3$). The equivalent loop gain enhancement is shown in Fig. 4b. Even with +/- 10% mismatch in the parallel path gain, set by the capacitor ratio C_1'/C_2', the loop gain enhancement is at least 30 dB. This is well within the matching accuracy of the on-chip capacitor.

III. PIPELINE ADC USING RPGE AMPLIFIER

A. RPGE amplifier design for a pipeline ADC

The RPGE amplifier is an ideal candidate for the high gain, wide bandwidth and high swing amplifier required in a high resolution pipeline ADC. Fig. 5 shows a fully differential switched capacitor RPGE amplifier and a transistor level implementation using a miller compensated two-stage amplifier with RPGE amplifier embedded in the second stage. This implementation in an attractive one, among others, as it requires only switches and capacitors to implement the RPGE loop shown in the shaded region. It is instructive to note that the use of this parallel path decouples the swing at the output of the first stage of the amplifier (V_y) and it is less than 5 mV for the entire output voltage range of the amplifier. This allows us to design the first stage for DC-gain and bandwidth

978-1-4673-6145-3/13 $31.00 © 2013 IEEE   721

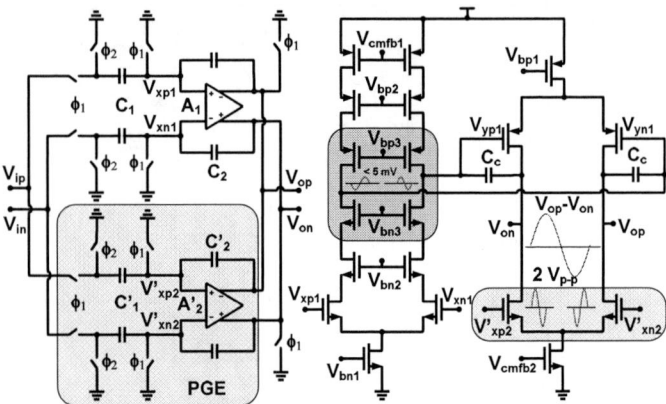

Fig.5. Switched capacitor implementation of RPGE amplifier and schematic

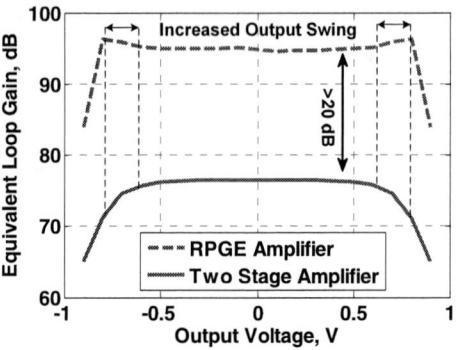

Fig. 6. Transistor level RPGE amplifier gain enhancement

without considering swing requirement. Without this benefit, it would not be possible (very difficult) to double cascode the first stage for the given supply voltage. The RPGE enabled additional cascode stage is highlighted in Fig. 5.

Fig. 6 shows the effective loop gain enhancement obtained with the transistor level design in Fig. 5. The amplifier in Fig. 5 was designed for a 2.5 bit MDAC stage in a pipeline ADC with the main path closed loop gain set to $G_M = 4$ and the parallel path closed loop gain set to $G_p = 5$. Apart from the gain enhancement obtained from the parallel path, the RPGE amplifier can provide a higher output voltage swing as compared to a conventional two-stage amplifier. This is due to the push-pull nature of RPGE output stage as shown in Fig. 6. This RPGE output stage enables a class-AB like operation of second stage, which allows 50 % reduction in the output stage power consumption and area as compared to a conventional two-stage amplifier. Table I summarizes the comparison of performance parameters of a two-stage amplifier, a feed-forward push-pull two-stage amplifier, a two stage amplifier with gain boosted (GB) first stage and the RPGE amplifier. The RPGE amplifier provides the largest DC gain, swing and low distortion without additional power consumption.

B. 13 bit Pipeline ADC using RPGE amplifier

To verify the above mentioned advantages, a 13-bit pipeline ADC was built in a 0.18 μm CMOS process. The pipelined ADC consists of six 2.5 bit MDAC stages followed by a 9-level flash ADC as shown in Fig. 7. The strong-arm dynamic comparator was used in the flash ADC. The main path (1st stage) sampling capacitor was 2 pF, $G_M = 4$ and the parallel path sampling capacitor was 500 fF (limited by the smallest unit capacitor in MDAC), $G_p = 5$. All six of the MDAC stages were gain enhanced by RPGE technique as shown in Fig. 7.

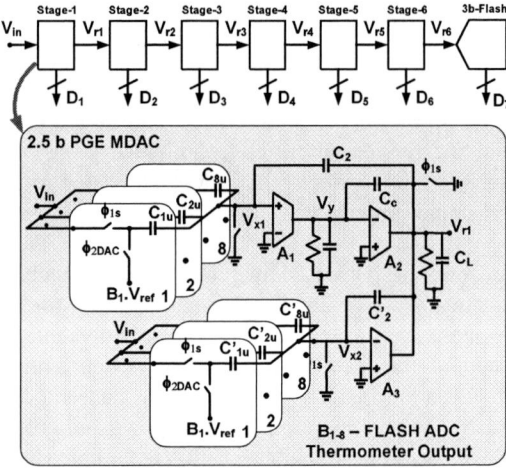

Fig. 7. Pipeline ADC using RPGE amplifier

Capacitor scaling was performed for the first four stages of the RPGE MDAC. The RPGE amplifier was also scaled down for the first four stages to maintain the same unity gain bandwidth.

IV. MEASUREMENT RESULTS

The prototype pipeline ADC occupies 3.6 mm x 0.85 mm active area and operates at 1.3 V supply voltage at 20 MHz clock frequency. Fig. 8 shows the measured spectrum for a 1 MHz input with 20 MHz clock. As shown in Fig. 9, the measured 13-bit DNL and INL were +0.75/-0.36 and +0.88/-0.92 LSB respectively.

Table I Comparison of performance parameters

Amplifier	DC Gain	Swing	Power	Distortion
Two-Stage	$A_1 \cdot A_2$	Moderate	High	Moderate
Feedforward Two-Stage	$A_1 \cdot (A_2 + A_3)$	High	Moderate	Moderate
GB Two-Stage	$A_1 \cdot A_2 \cdot A_{gb}$	Moderate	High	Low
RPGE	$\geq A_1 \cdot A_2 \cdot A_3$	High	Moderate	Low

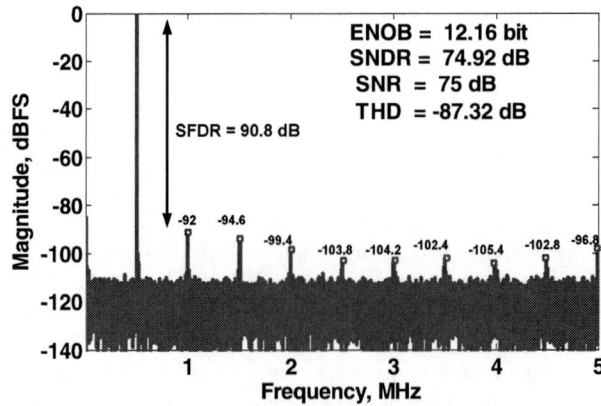

Fig. 8. Measured Spectrum at 1MHz input

978-1-4673-6145-3/13 $31.00 © 2013 IEEE

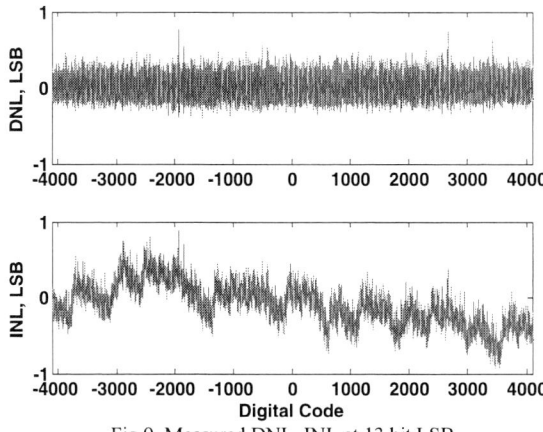

Fig.9. Measured DNL, INL at 13 bit LSB

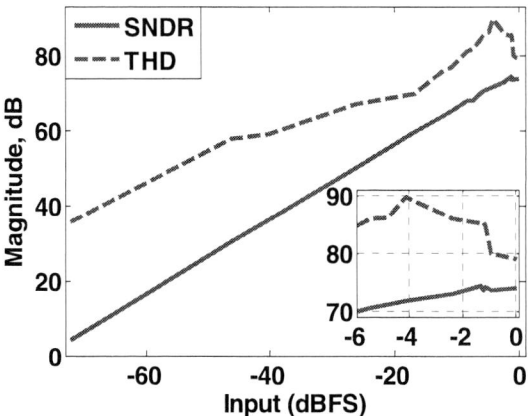

Fig.10. Input vs. SNDR and THD

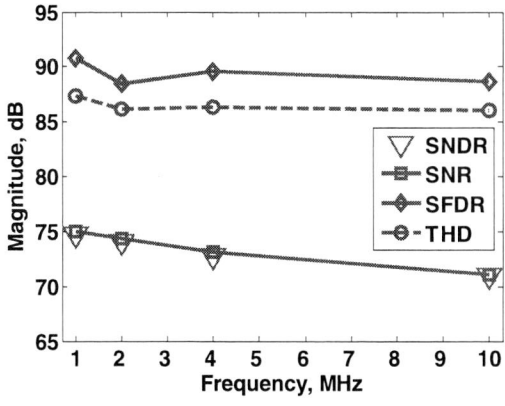

Fig. 11. Input Frequency vs. SN(D)R, SFDR and THD

Parameter	Value
Technology	1P4M 0.18 μm
Supply Voltage	1.3 V
Sampling Rate	20 MS/s
Input Voltage Range	2.4 V_{p-p}
SNDR/SFDR	74.9 / 90.8 dB
Power	5.9 mW
Active Area	3.6 x 0.85 mm^2
FoM	65 fJ/C-S
Resolution	13b

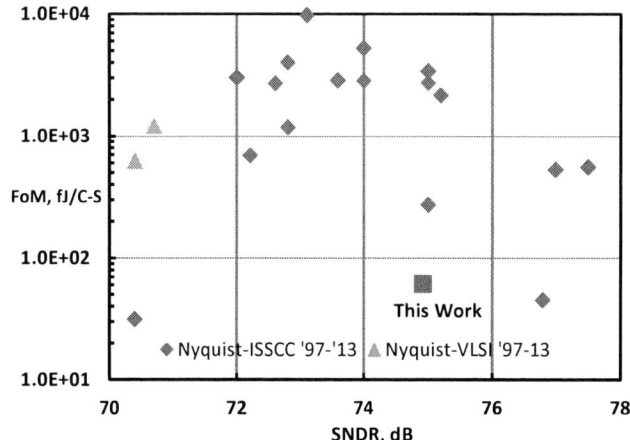

Fig.12. Performance Summary and Die Micrograph

Fig. 13. Comparison to Nyquist ADC ('97-'13) > 70 dB SNDR, >5 MHz BW

The measured SNDR, SFDR and THD were 74.92 dB, 90.8 dB and -87dB respectively. The reported measurements results are without any form of calibration. The measured SNDR, SFDR and THD at 10 MHz input were 71.4 dB, 88 dB and -86 dB. The dynamic performance is shown in Fig. 10 and Fig. 11. The analog portion of power consumption was 4 mW out of the total power consumption at 5.9 mW. Fig. 12 and Fig. 13 show the die-micrograph, performance summary and comparison table respectively.

ACKNOWLEDGMENT

The authors would like to thank AKM for supporting this work and providing fabrication.

REFERENCES

[1] P.C. Yu and H. S. Lee, "A high-swing 2-V CMOS operational amplifier with replica-amp gain enhancement," *IEEE Journal of Solid-State Circuits*, vol.28, no.12, pp.1265-1272, Dec. 1993.

[2] S. Pavan and P. Sankar, "Power Reduction in Continuous-Time Delta-Sigma Modulators Using the Assisted Opamp Technique," *IEEE Journal of Solid-State Circuits*, vol.45, no.7, pp.1365-1379, Jul. 2010.

[3] J. Chen et al., "A 62mW Stereo Class-G Headphone Driver with 108dB Dynamic Range and 600μA/Channel Quiescent Current," *IEEE Int. Solid-State Circuits Conf*, pp. 182-182, Feb. 2013.

[4] C. C Enz, G. C. Temes, "Circuit techniques for reducing the effects of op-amp imperfections: autozeroing, correlated double sampling, and chopper stabilization," *Proceedings of the IEEE* , vol.84, no.11, pp.1584,1614, Nov 1996.

[5] B. R. Gregoire and U. Moon, " An Over-60dB True Rail-to-Rail Performance Using Correlated Level Shifting and an Opamp with 30dB Loop Gain," *IEEE Int. Solid-State Circuits Conf*, pp.540-541, Feb. 2008.

[6] H. Venkatram, B. Hershberg, U. Moon, "Asynchronous CLS for Zero Crossing based circuits", *IEEE Int. Conf. Elec. Circuits Syst.*, Dec. 2010.

[7] H. Venkatram, T. Oh, J. Guerber, and U. Moon, "Class-A+ amplifier with controlled positive feedback for discrete-time signal processing circuits", *IEEE Int. Symp. Circuits Syst.*, May 2012.

978-1-4673-6145-3/13 $31.00 © 2013 IEEE

Sampling Circuits That Break the kT/C Thermal Noise Limit

Ron Kapusta, Haiyang Zhu, Colin Lyden
Analog Devices, Inc.
804 Woburn Street
Wilmington, MA 01887

Abstract – **Several circuit-level techniques are described which are used to reduce thermal noise and break the so-called kT/C limit. kT/C noise describes the total thermal noise power added to a signal when a sample is taken on a capacitor. In the first proposed technique, the sampled thermal noise is reduced by altering the relationship between the sampling bandwidth and the dominant noise source, providing a powerful, new degree of freedom in circuit design. In the second proposed technique, thermal noise sampled on an input capacitor is actively cancelled using an amplifier, so that the noise at the amplifier output can be controlled independently of input capacitor size. Measurements from two test chips are presented which demonstrate sampled thermal noise power reduction of up to 70% when compared to conventional kT/C-limited sampling.**

I. INTRODUCTION

Switched capacitor circuits are the implementations of choice for many modern mixed signal circuits, especially in CMOS technology. Inherent in any switched capacitor circuit are sampling operations; when a switch opens, freezing the charge on a capacitor, a sample is taken. The charge can then be redistributed in order to implement a variety of circuit blocks: buffers, gain blocks, filters, etc.

At the circuit design level, one of the common issues with sampling is the addition of thermal noise to a signal each time a sample is taken. While capacitors are noiseless circuit elements, the resistors or transistors used to transfer the charge onto the capacitors contribute thermal noise. In the most basic case, when a sample is taken with a single capacitor and transistor acting as a switch, the total noise power sampled is found by integrating the noise power spectral density of the transistor over frequency [1]. Equation (1) shows how this noise power v_n^2 can be approximated and then reduced to the familiar kT/C limit.

$$ v_n^2 \approx \left(4kTR_{ON}\right) * \left(\frac{1}{R_{ON}C} * \frac{\pi}{2} * \frac{1}{2\pi}\right) = \frac{kT}{C} \qquad (1) $$

In this simple case, the value of the transistor on-resistance (R_{ON}) appears in both the numerator (noise power spectral density) and denominator (noise bandwidth). Therefore, when the noise power is integrated over frequency, the on-resistance terms cancel, and only the sampling capacitor C remains. This same "cancelling" relationship also holds for more complicated sampling structures, such as when multiple switches are included, as well as when amplifiers in feedback are used to provide a virtual ground for sampling.

The cancellation can be easily seen when the noise power spectral density sampled on a capacitor is plotted versus frequency, as in Fig. 1. A lower transistor on-resistance decreases the thermal noise density, but it the noise bandwidth is increased by the same ratio. There is no obvious way to decouple the inverse proportional relationship between the noise power density and the noise bandwidth.

There are significant design implications of the kT/C limit, specifically that in order to achieve lower noise in a sampled system, larger capacitors must be used. Unfortunately, when increasing capacitor size in order to lower noise, other performance parameters suffer. The impacts can include larger die area, higher power in the sampling stage, and higher power in the amplifier that drives these increased sampling capacitors.

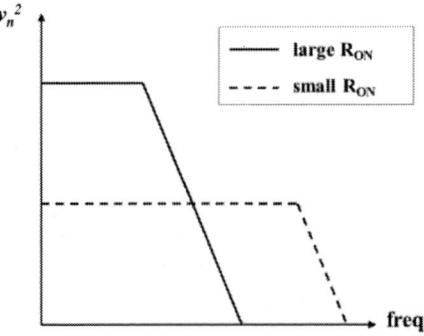

Fig. 1. Noise power spectral density for different resistances

978-1-4673-6145-3/13 $31.00 © 2013 IEEE

The remainder of this paper will be structured as follows. In Section II, a technique proposed in [2] will be reviewed and analyzed in more detail than previously described. Section III will propose another noise reduction technique that uses an amplifier to actively cancel noise sampled on an input capacitor. Finally, Section IV will contain measurement data to demonstrate the validity of the proposed techniques.

II. NOISE REDUCTION USING FEEDBACK

In order to reduce sampled thermal noise without increasing the sampling capacitance, the relationship between the dominant noise source and the impedance which limits the noise bandwidth must be broken. This can be accomplished with the feedback circuit configuration shown in Fig. 2. This configuration is very similar to a conventional auto-zero configuration, except for the addition of an explicit large feedback resistance R_{FB}.

Conceptually, the effect of R_{FB} can be explained by considering the path that current takes when flowing out of the transconductance amplifier. At low frequencies, the impedance of sampling capacitor C_S is very large, and therefore most of the current output from G_M will flow through R_L, which is no different than a conventional auto-zero. However, at high-frequencies, the impedance of C_S is small, and current output from G_M will split between to the feedback and the load paths. The effect of this current split is to reduce the high-frequency effective feedback transconductance $G_{M,EFF}$. As shown in (2), $G_{M,EFF}$ depends on the ratio between R_{FB} and R_L.

A new degree of freedom is apparent in (2), namely that $G_{M,EFF}$ can be adjusted independently of G_M by adjusting the ratio between R_L and R_{FB}. This is precisely the decoupling of dominant noise source (amplifier G_M) and bandwidth limiting impedance (R_{FB}) required in order to achieve integrated noise power lower than kT/C.

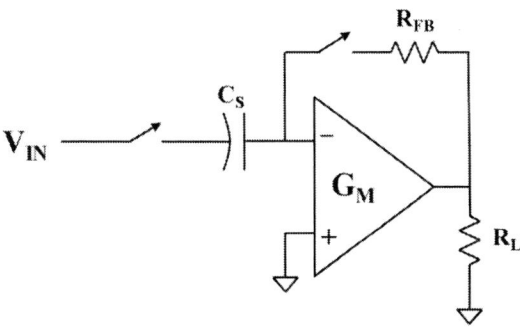

Fig. 2. Feedback configuration for reducing sampled noise

$$G_{M,EFF} = G_M \frac{R_L}{R_L + R_{FB}} \tag{2}$$

Of course, it is also important to consider the additional noise from R_{FB}, which is typically a rather large resistance and therefore might be expected to have a large noise contribution. The transfer function from thermal noise voltage associated with R_{FB} to the input capacitor C_S can be derived, and is easily simplified using the assumption that $R_{FB} \gg R_L$, which is desired to reduce $G_{M,EFF}$.

$$\frac{V_{CS}}{V_{N,RFB}} = \frac{1}{G_M R_L + 1} * \frac{1}{\frac{C_S}{G_{M,EFF}} s + 1} \tag{3}$$

Several insights can be gained from (3). First, the DC transfer function from the feedback resistor noise to the input capacitor is inversely related to the DC gain of the amplifier circuit, $G_M R_L$. This seems intuitive, as R_{FB} is at the output of the amplifier; hence, negative feedback works to reduce such noise sources when referred to the input. A second observation from (3) is that the transfer function bandwidth is exactly the same as the closed loop bandwidth of the amplifier, and is proportional to $G_{M,EFF}$. At first glance, this may appear confusing; at frequencies above the bandwidth of the amplifier, the negative feedback loop has no gain and therefore cannot reduce the R_{FB} noise contributions. However, because is R_{FB} large, the pole it forms with C_S occurs at low frequency, below the bandwidth of the amplifier. Therefore, R_{FB} effectively filters its own noise at high frequencies, and amplifier gain is not required to reduce its noise contribution at high frequencies.

When all of the noise contributions are combined, again making the assumption that $R_{FB} \gg R_L$, the total sampled thermal noise is shown in (4), and the sampling bandwidth is shown in (5). Note that (4) includes some simplifications; for example, it does not include a device noise factor, which would be 4/3 for a differential amplifier with an ideally modeled MOS input pair.

$$v_n^2 = \frac{kT}{C_S} * \left(\frac{R_L}{R_L + R_{FB}} + \frac{1}{G_M R_L} + \frac{1}{G_M R_{FB}} \right) \tag{4}$$

$$BW = \frac{G_M}{C_S} * \frac{R_L}{R_L + R_{FB}} \tag{5}$$

Equation 4 shows what conditions are needed to achieve sampled thermal noise less than kT/C. The first term represents the amplifier noise, and to reduce it, large

978-1-4673-6145-3/13 $31.00 © 2013 IEEE

R_{FB} is desired. The second term is due to feedback resistor noise and is kept small if the amplifier DC gain is large. Finally, the third term is due to load resistor noise and hasn't been discussed, but the analysis is similar to feedback resistor noise. The effect of this noise source is minimized with a large R_{FB}.

Increasing R_{FB} is not without a price, however. As can be seen in (5), increasing R_{FB} decreases the sampling bandwidth. If it is desirable to maintain a wide sampling bandwidth while decreasing noise, then G_M must be increased R_{FB} as is increased, resulting in increased power. It should be noted that increasing power to reduce noise is not specific to this technique. With the conventional sampling method, the sampling capacitor would be increased in order to lower noise, resulting in increased power in the driving circuit in order to keep the bandwidth constant.

Finally, it should also be reiterated that any real amplifier will have some additional noise factor, γ, often due to noise contributions from devices other than the input pair. This factor has not been included in the previous analysis but can simply be modeled as a scale factor in front of the amplifier noise terms.

There are some practical limitations to this noise reduction technique. First, using a very large resistance for R_{FB} can have similar area penalties as using a large sampling capacitor. Second, parasitic capacitance at the output of the amplifier creates a second pole in the system. The frequency of this pole decreases as R_{FB} and R_L are increased, and will eventually lead to loop instability.

This technique has been used in [2] as the pre-amplifier stage of a sample-and-hold circuit, as shown in Fig. 3. The technique is well suited for use in wideband pre-amplifiers. The pre-amplifier gain is typically relatively low, 16dB in [2], and does not require particularly large load resistors, ~330Ω. The feedback resistor is 1kΩ. Parasitics at the pre-amplifier output do not have a significant impact on stability, even with a closed-loop bandwidth greater than 300MHz. Given these component values, (4) predicts that the sampled thermal noise power should be ½ of the kT/C limit.

III. ACTIVE CANCELLATION OF SAMPLED NOISE

While the noise reduction technique described in the previous section focused on reducing the thermal noise sampled, the next technique will instead use active circuits to cancel the thermal noise after it has already been sampled. An implementation of this technique is shown in Fig 4.

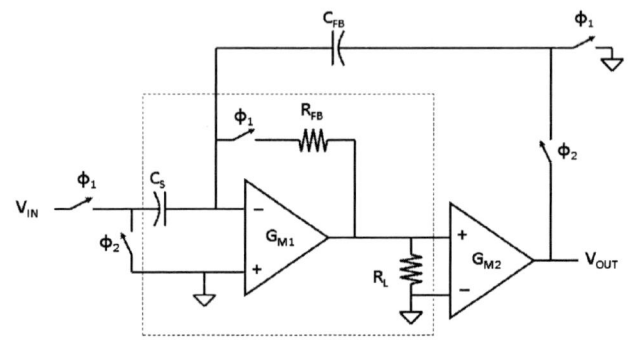

Fig. 3. Feedback noise reduction in sample-and-hold amplifier.

The circuit blocks shown in Fig. 4(a) comprise a sample-and-hold built from a two-stage amplifier with capacitive level-shifting between the two stages. As in Fig. 3, the first amplifier $A1$ is typically a low-gain preamplifier stage, though the technique would also work with a higher gain first stage.

The noise reduction is achieved through appropriate design of the switch control waveforms, as shown in Fig. 4(b). During the input sampling phase, both signals $\varphi1$ and $\varphi2$ are active (high). In this phase, the input voltage V_{IN} is stored on input capacitor C_S, the offset of amplifier $A1$ is stored on auto-zero capacitor C_2, and feedback capacitor C_{FB} is cleared. The next phase occurs when $\varphi1$ falls, freezing the charge on the input and feedback capacitors. Thermal noise charge is also sampled, and the noise power equal to $kT(C_S+C_{FB})$.

Up until this point, the operation is the same as a conventional sample-and-hold, including the thermal noise sampled. However, after $\varphi1$ falls, the thermal noise charge sampled will cause the summing node voltage to change. This change in voltage is amplified through $A1$ and stored on capacitor C_2, as signal $\varphi2$ is still active. When $\varphi2$ falls, the sampled voltage on capacitor C_2 captures both the offset of amplifier $A1$ and an amplified version of the thermal noise that was sampled on the summing node. Effectively, both offset in $A1$ and the sampled thermal noise at the summing node will be auto-zeroed out via the same mechanism during phase $\varphi3$.

During the final phase, $\varphi3$, the circuit is configured in hold mode and the input charge is transferred from C_S to C_{FB}. However, the noise charge sampled on C_S is not transferred to C_{FB}, and therefore none of this noise shows up at the amplifier output. To understand why the thermal noise isn't transferred, it is easiest to begin with the assumption that $A2$ has infinite gain. Therefore, during $\varphi3$, the feedback loop will settle with no signal (GND) at the input to $A2$, regardless of the output voltage. Further, if the

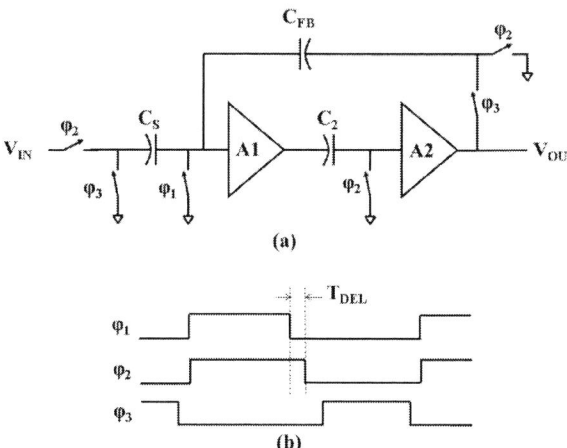

Fig. 4. Active noise cancellation of sampled noise. (a) Configured as a sample-and-hold amplifier. (b) Switch control timing diagram.

right-hand side of C_2 is at GND, then the left-hand side of C_2 and the summing node must be at the same potential as they were when $\varphi2$ sampled. Therefore, the sampled thermal noise charge stays on the summing node and is not transferred to C_{FB}. The same analysis would be used to show that the offset of $A1$ also does not appear at the amplifier output.

There are several practical limits to this noise cancellation technique. First, thermal noise is sampled on C_2, and this noise is not cancelled. The sampled thermal noise power follows the expected kT/C_2 behavior. Increasing the gain of $A1$ lessens the impact of this noise when referred back to the input. A second method to decrease the noise contribution is to increase the size of capacitor C_2, which may be acceptable as it is not driven by the input.

A second practical limitation is the time required to sample the thermal noise. In between the falling edge of $\varphi1$ and $\varphi2$, the thermal noise sampled on the summing node must propagate through $A1$ and be accurately settled on C_2. The time constant associated with the settling is determined by the output resistance of the first amplifier, R_{OUT1}, and capacitor C_2. For complete noise cancellation, the time allowed for settling, T_{DEL}, must be much longer than the settling time constant. Through simulation, it can be shown that settling for two time constants is enough to achieve the majority of the noise benefit, as shown in (6).

$$T_{DEL} > 2R_{OUT1}C_2 \qquad (6)$$

This noise cancellation technique is similar to the correlated double sampling technique proposed in [3] and used in image sensors for decades. In these applications, correlated double sampling is used to cancel the thermal

noise sampled when resetting a floating diffusion to a fixed potential. In contrast, the technique proposed here can be used to sample an arbitrary input voltage and is also able to amplify that sampled input during a second phase.

IV. MEASURED RESULTS

A. Feedback Noise Reduction

In order to confirm the effectiveness of the noise reduction technique described in Section II, the circuit shown in Fig. 5 is used. The sample-and-hold block is the same as shown in Fig.3, and it is followed by a 14-bit ADC. The clocks are applied to each block in order to measure only the input sampling noise through the use of oversampling. A single input sample is taken, as shown by clock φ_{SAMP}. This input is then held for multiple clock cycles, and the ADC repeatedly converts it. These ADC outputs are averaged together, which reduces the errors introduced by amplifier hold noise and ADC noise and results in an accurate estimate of the single input sample. Multiple input sample estimates can then be analyzed to find the variation, which is a measure of the sampling noise. For these measurements, each input sampled is converted several thousand times, reducing the effective measurement noise to less than 5μV-rms.

A test chip that includes the circuit shown in Fig. 5 has been fabricated, and is a re-configurable version of the analog front-end presented in [2]. In addition to the noise-reducing configuration, the auto-zero can be disabled and the input can be sampled in conventional fashion with a sampling switch at the summing node (not shown in Fig. 3). In this conventional sampling configuration, the sampled thermal noise power should obey the familiar kT/C limit.

Test chip #1 is implemented in a 0.18μm 1P5M CMOS technology. The ADC clock rate is 40MSPS. The size of the sampling capacitor, C_S, is 2.4pF. The other capacitance at the summing node, due to C_{FB} and some other parasitic capacitance, totals 1.5pF. Based on the calculation shown in (7), the kT/C-limited sampled thermal noise should be 75μV-rms. In (7), as compared to (1), a factor of 2 is included because the circuit implementation is differential, and the "1.5/2.4" factor is included to account for the noise sampled on the capacitance at the summing node other than C_S. As with (4), equation (7) does not include a device noise factor.

$$v_n^2 = \frac{kT}{2.4\,pF} * 2 * \left(1 + \frac{1.5\,pF}{2.4\,pF}\right) \qquad (7)$$

978-1-4673-6145-3/13 $31.00 © 2013 IEEE

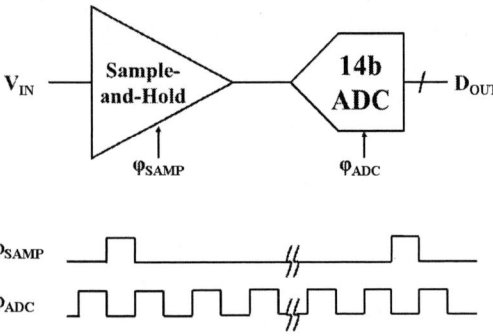

Fig. 5. Noise measurement circuit and timing diagram.

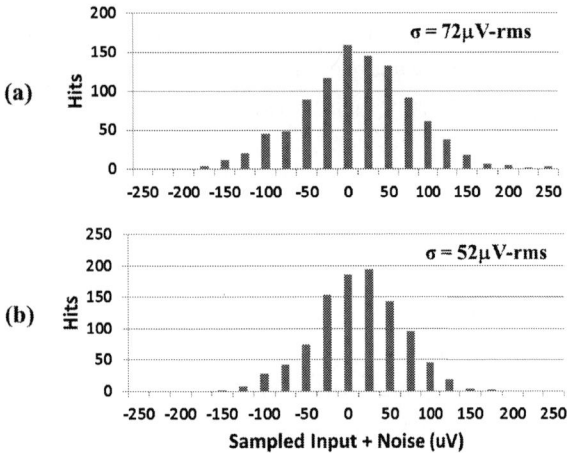

Fig. 6. Measured histograms of input samples on test chip #1, with mean value removed. (a) Test circuit configured as conventional sample. (b) Test circuit in noise-reducing feedback configuration.

Fig. 6 shows data measured from test chip #1 in two configurations. In the top plot, the circuit samples in a conventional fashion, and the measured standard deviation is 72µV-rms. As compared to the 75µV-rms noise predicted by calculation, the measured data matches quite well. There are plenty of sources for difference between measured and calculated data, not the least of which is the uncertainty in the size of the fabricated sampling capacitor.

To collect the data in the bottom plot of Fig 6, the same circuit is configured in noise-reducing feedback mode. The measured standard deviation drops to 52µV-rms. The difference between the two sets of data represents a 48% reduction in noise power. It should be noted that because all measurements were taken on the same chip, variations in circuit elements should not impact the difference in measured values.

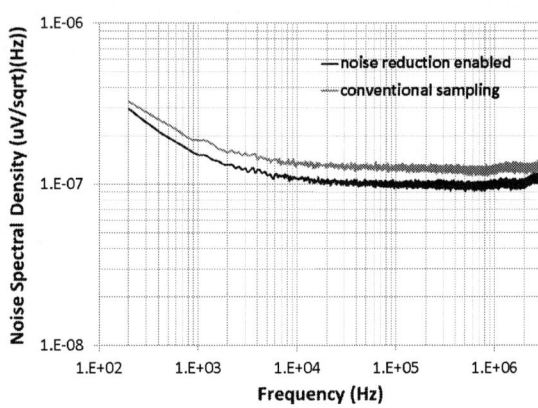

Fig. 7. Spectrum of measured data for 6MSPS sample rate, with and without noise reduction feedback enabled.

It is useful to compare the measured noise-reduced data to what would be predicted by calculation. Given that the pre-amp gain is 16dB and the feedback resistor is 3x larger than the load resistor, (4) predicts that the sampled thermal noise should be 51µV-rms. Note that this calculation also accounts for the noise increase due to extra capacitance at the summing node, just as in (7).

A spectrum of the measured output data is shown in Fig. 7. Note that the sample rate has been decreased to 6MSPS for this data, so that the low frequency portion of the spectrum is clearly visible. For this measurement, the sampling clock and ADC clock are identical. As such, the sampling noise is no longer oversampled, and it is impossible to separate the noise contributions from input sampling, amplifier hold, and A-to-D conversion. What this plot does allow is a clear view of the thermal noise floor, and how it changes when the noise-reducing feedback mode is enabled. As measured, the thermal noise floor drops by about 15%. While this seems to be a less significant decrease than the data shown Fig. 6, the addition of amplifier hold noise and A-to-D conversion noise limits the measured decrease. When those noise contributions are accounted for, the results are consistent. Similarly, it might be noted that the low frequency noise does not seem to be improved as might be expected in the noise-reducing auto-zero configuration. This is due to the 2nd stage of the sample-and-hold amplifier, whose 1/f noise dominates the low frequency spectrum, and it is not affected when the 1st stage is configured to auto-zero.

B. Active Noise Cancellation

In order to verify the second noise reduction technique, the circuit shown in Fig. 4 was implemented. Similar to test chip #1, test chip #2 includes a sample-and-hold circuit followed by an ADC. Again, the individual circuit blocks can be configured in a way that allows the noise contributed in individual phases to be measured, so that the noise sampled at the input can be extracted.

Test chip #2 is implemented in a 65nm 1P7M digital CMOS process. The ADC clock rate is 20MSPS. The size of the sampling capacitor, C_S, is 2.3pF. The other capacitance at the summing node, due to C_{FB} and some other parasitic capacitance, totals 2.4pF. Based on the calculation shown in (7), the kT/C limited thermal noise should be 87μV-rms. Note that this calculation does not account for the noise contribution from the auto-zero capacitor C_2.

Fig. 8 shows data measured from test chip #2 as the noise-cancelling settling time (T_{DEL}) is swept from 0 to 1.65ns. The two curves shown correspond to configurations in which the size of C_2 is varied. With T_{DEL} = 0ns, there is effectively no noise cancellation, as amplifier $A1$ does not have any time to respond to the sampled thermal noise. For this setting, the measured noise values of 89μV-rms and 93μV-rms match closely with expectations, especially considering that the impact of thermal noise sampled on C_2 was not included in the calculations.

Fig. 8 also shows that total sample phase noise decreases as T_{DEL} is increased, confirming that the action of amplifier $A1$ and capacitor C_2 is working to cancel the thermal noise sampled at the summing node. The measured noise decreases with increased T_{DEL}, and the relationship is a function of the time constant at the amplifier output. In test chip #2, this time constant is nominally 680ps for the "large C_2" curve shown. The estimate predicted by (6) seems to hold true, roughly 2 time constants of settling is sufficient to get the large majority of noise cancellation benefit.

Finally, the minimum achievable sample phase noise is lower with a larger C_2, as predicted in Section II, since the kT/C_2 noise contribution becomes significant when the summing node thermal noise is adequately cancelled. With the larger C_2 and adequate T_{DEL}, the total sample phase noise can be reduced from 89μV-rms to 55 μV-rms, which is a 62% reduction in thermal noise power. Furthermore, the difference between the two curves in Fig. 8 can be used to estimate the noise contribution from only the $\varphi1$ sample at the summing node. Based on this estimate, the noise from the $\varphi1$ sample was reduced from 84μV-rms to 45μV-rms and, therefore, 71% of $\varphi1$ sampled thermal noise was successfully cancelled.

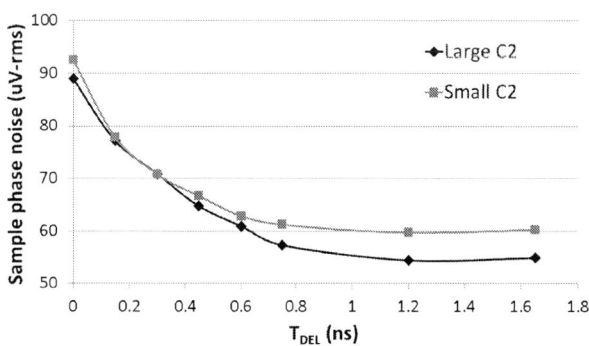

Fig. 8. Measured sample phase noise on test chip #2 vs. settling time, with varying size of auto-zero capacitor C_2.

V. CONCLUSION

While it has been commonly accepted as a fundamental limit of thermal noise when sampling on a capacitor, kT/C is, in fact, not a limit at all. This paper presented two circuit-level sampling techniques that allow the size of the input capacitor to be determined almost independently of the noise requirement. The first method broke the relationship between the sampling bandwidth and the dominant noise source. The second technique used active circuits and a second capacitor not driven by the input to cancel the noise sampled on the input capacitor. Test chip measurements were presented to demonstrate that the sampled thermal noise can be directly reduced by nearly 50% and also cancelled by more than 70% without change to the input capacitor.

ACKNOWLEDGMENTS

The authors would like to thank Peter Hurrell and Derek Hummerston for their discussions and insights into how to design around noise "limits".

REFERENCES

[1] D. Johns and K. Martin, *Analog Integrated Circuit Design*, John Wiley & Sons, New York, 1997, pp. 202-204.

[2] R. Kapusta, *et al.*, "A 4-Channel 20-to300 Mpixel/s Analog Front-End with Sampled Thermal Noise Below kT/C for Digital SLR Cameras," *ISSCC Dig. Tech. Papers*, pp. 42-43, Feb. 2009.

[3] M. White, D. Lampe, F. Blaha, I. Mack, "Characterization of surface channel CCD image arrays at low light levels," *IEEE J. Solid-State Circuits*, vol. 9, no. 11, pp. 1-12, Feb. 1974.

Blind Background Calibration of Harmonic Distortion Based on Selective Sampling

Manideep Gande, Ho-Young Lee, Hariprasath Venkatram, Jon Guerber and Un-Ku Moon
School of EECS, Oregon State University, Corvallis, OR, USA

Abstract—This paper proposes a blind calibration algorithm for suppressing harmonic distortion in analog to digital converters (ADCs). The proposed algorithm does not need any external calibration signal and is first of its kind. The proposed algorithm relies on the properties of downsampling and orthogonality of sinusoidal signals to estimate the harmonic distortion coefficients. The algorithm can be operated in both foreground and background modes to remove even and odd harmonics simultaneously. The algorithm is demonstrated on a first-order ring oscillator based $\Delta\Sigma$ ADC, whose performance is harmonic distortion limited. Built in 0.13μm, the algorithm improves the SNDR of the ADC by 39dB while improving SFDR by 45 dB.

I. INTRODUCTION

Modern day CMOS processes are characterized by scaling in terms of geometries and supply voltages. Scaling happens with digital design in perspective, thereby complicating the design of high-resolution, high-performance ADCs. Shrinking geometries tend to reduce the intrinsic gain of MOS transistors, making it difficult to realize high DC gain, wide bandwidth amplifiers. This complicates the design of high performance ADCs. To counter this problem, ADCs which are more digital in nature are being favored in smaller geometry processes. Successive approximation ADCs (SAR) [1] and digitally calibrated ADCs fit the bill perfectly [2]–[6]. In this paper, we focus on ADCs which make use of digital processing to make up for analog imperfections.

Power, linearity, bandwidth, area and process form a complex trade-off in ADC design. For example, amplifier non-linearity and capacitor mismatches in pipeline ADCs [3], capacitor mismatches in SAR ADCs [1], amplifier non-linearity, quantizer non-linearity and DAC non-linearity in $\Delta\Sigma$ ADCs lead to harmonic distortion in ADCs. These issues can be resolved in analog or digital domain. Analog domain solutions for the above issues include using larger capacitors for better matching or high loop gain, wide bandwidth and low distortion amplifiers or using feedback loops, which unfortunately are power intensive and are also becoming increasingly difficult to design in modern day CMOS process.

On the other hand, reduced gate delays and ease of portability across processes make digital calibration for analog imperfections very attractive in modern day CMOS processes. However, there are only a handful of digital calibration techniques which can remove harmonic distortion in ADCs. [1]–[6] provide a few alternatives for removing harmonic distortion in ADCs but have their own limitations, as shown in Table I. This paper proposes a, first of its kind, blind calibration algorithm, which can operate in both foreground and background

TABLE I
COMPARISON WITH EXISTING ARCHITECTURES

Reference	Notes
[1]	Calibration performed in software. Uses two ADC paths with dithered test signal. Calibration demonstrated only for odd harmonics.
[2]	Calibration performed on chip. Uses multiple pseudo-random external sequence.
[3]	Similar to structure in [2]. Implemented on a pipeline ADC.
[4]	Corrects non-linearity of only one stage using a binary random number generator. Assumes ideal back-end ADC.
[5]	Needs an external input. Normal operation of ADC is interrupted and missing sample is interpolated.
[6]	Requires a 13-bit accurate signal for calibration. Needs to be clocked at higher frequency as compared to ADC rate.
This Work	Requires no external calibration signal. Can operate in background and removes multiple harmonics simultaneously.

modes, to remove harmonic distortion in ADCs. The proposed algorithm does not use any external calibration signal and only uses the properties of downsampling and orthogonality of sinusoidal waves to calibrate for harmonic distortion in ADCs, hence the blind nature of it.

Sections II explains the background for the algorithm and illustrates the working of the algorithm in a single harmonic case in a foreground mode. Section III illustrates how the algorithm can be extended to operate in background mode. In section IV, the algorithm is extended to calibrate for multiple harmonics simultaneously. Section V briefly describes the implemented VCO based first order $\Delta\Sigma$ ADC, with section VI describing the final architecture. Measurement results are presented in section VII, with the paper being concluded in section VIII.

II. PROPOSED ALGORITHM

Figure 1 shows the single tone frequency response of an ideal ADC and non-ideal ADC which has third harmonic distortion only. As shown, when a sinusoid (of frequency f_{in}) is given to an ideal ADC, the frequency spectrum of its digital output (D_1) consists of a single tone at f_{in}. When

978-1-4673-6145-3/13 $31.00 © 2013 IEEE

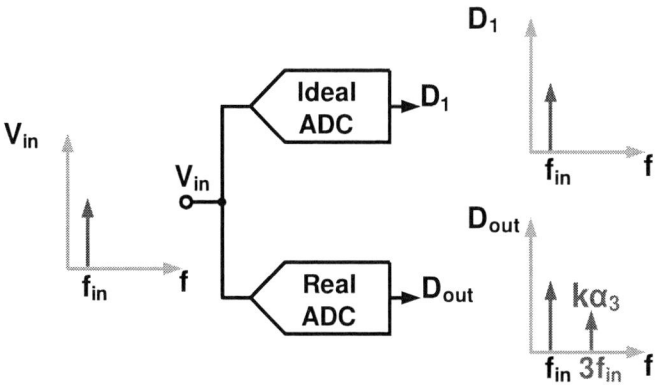

Fig. 1. Single tone frequency response of ideal and non-ideal ADC.

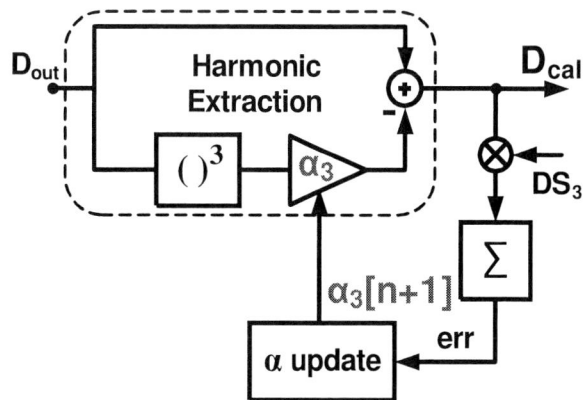

Fig. 3. LMS engine.

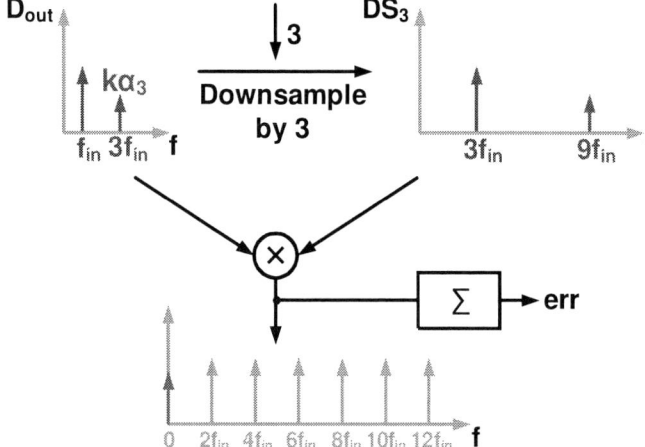

Fig. 2. 3rd harmonic extraction.

the same input is given to a non-ideal ADC which suffers from third harmonic distortion, the output (D_{out}) contains a tone at f_{in} and additional harmonic distortion component at $3f_{in}$. The assumption that the non-ideal ADC suffers only from third harmonic distortion is for simplicity reasons only. This algorithm is not limited to single harmonic tones.

For the remainder of this section, the coefficient of distortion is assumed to be α_3. Therefore, the magnitude of harmonic distortion tone at $3f_{in}$ is proportional to α_3 ($k\alpha_3$). The non-ideal ADC is modeled as follows:

$$D_{out} = D(V_{in}) + \alpha_3 \times D(V_{in}^3) \qquad (1)$$

Where D_{out} is the output of the non-ideal ADC, and $D(V_{in})$ is the ideal digital representation of V_{in}. In the above setup, the source of distortion is not important i.e. the distortion could be due to any component in the entire system.

The proposed algorithm relies on two important signal properties (1) Downsampling and (2) Orthogonality of sinusoidal waves.

A. Downsampling

Downsampling is the process of reducing the sampling rate of a signal. In time domain downsampling leads to dropping of

samples, whereas in frequency domain downsampling leads to input spectrum being spread out. Figure 2 shows the frequency spectrum of output of ADC (D_{out}) and the spectrum obtained after downsampling by 3 (DS_3). The original spectrum of D_{out} has tones at f_{in} and $3f_{in}$, while the downsampled spectrum has tones at $3f_{in}$ and $9f_{in}$. Note that $3f_{in}$ tone in D_{out} and $9f_{in}$ tone in DS_3 are due to the harmonic distortion in the ADC.

B. Orthogonality

A key property of sinusoidal waves is that they are orthogonal at different frequencies. In other words, the sum of product of two sinusoids is zero if the two frequencies are different and is non-zero if the frequencies are same. i.e.

$$\sum_{n=1}^{N} sin(f_1 n) \times sin(f_2 n) = \begin{cases} 0 & \text{if } f_1 \neq f_2, \\ \neq 0 & \text{if } f_1 = f_2 \end{cases} \qquad (2)$$

An alternative way of looking at the above property is that the product of two sinusoids of same frequencies has a component at DC, while product of two sinusoids of different frequencies does not have any component at DC. Therefore, obtaining a non-zero average for a product of two sequences implies that the both the signals have a common frequency term in it.

C. Proposed Algorithm

Combing the properties of downsampling and orthogonality of sinusoid waves, the coefficient of third harmonic distortion (α_3) is estimated. This is done by taking the running average of the product of the two digital bit streams D_{out} and DS_3. As shown in Fig. 2 when there is harmonic distortion present in the ADC, the running sum of product of D_{out} and DS_3 has a component at DC, whose magnitude is proportional to α_3. Using this information, the third harmonic distortion term can be extracted using an LMS based engine. The updated error term of the LMS engine is given as

$$err = \sum \left(D_{cal}[n] \times DS_3[n] \right) \qquad (3)$$

978-1-4673-6145-3/13 $31.00 © 2013 IEEE   731

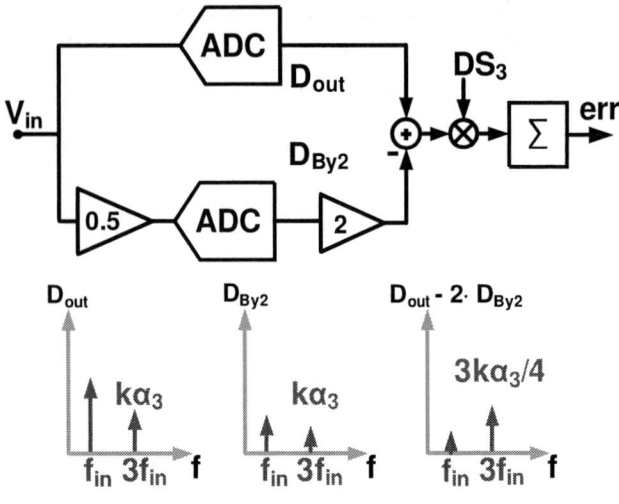

Fig. 4. Background implementation.

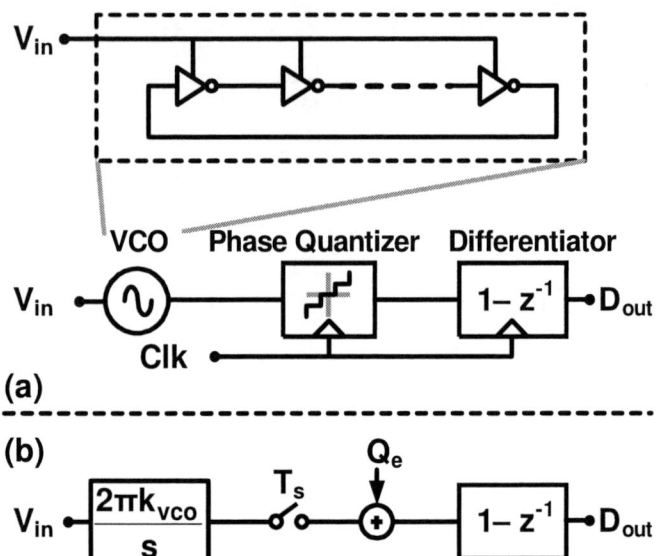

Fig. 5. (a)VCO based first order $\Delta\Sigma$ ADC. (b) Equivalent model.

Therefore the error term is zero if and only if the calibrated output D_{cal} and the downsampled output DS_3 do not have any common sinusoid in their digital streams. The simplified update equations for the LMS engine are given in equation 4 which is also illustrated in Fig. 3.

$$D_{cal}[n+1] = D_{out}[n] - \alpha_3[n] \times D_{out}^3[n], \quad (4a)$$
$$\alpha_3[n+1] = \alpha_3[n] + \mu \times err \quad (4b)$$

III. BACKGROUND CALIBRATION

The algorithm discussed so far needs a clean sinusoid input and can be operated in a foreground fashion. The above algorithm can be made to operate in a background mode by removing the signal component from the output digital stream and dealing only with the harmonics. As shown in Fig. 4, this is done by splitting the original ADC into two parts and applying input V_{in} to one ADC and an approximately scaled version of V_{in} (e.g. $V_{in}/2$) to the other. By doing so, the linear component of the input is scaled by a linear factor, whereas the harmonic distortion part is scaled by a different factor. Therefore, the digital stream $D_{out} - 2D_{By2}$ does not have any input component present in it, but only has the harmonic distortion terms present in it. Therefore, by taking the average of the product of the above obtained stream and the downsampled stream (DS_3), the third harmonic distortion coefficient can be estimated. The finite accuracy of the scaling by 2, in this example, is also merged into the LMS engine.

IV. CALIBRATION FOR MULTIPLE HARMONICS

The above algorithm can be extended to calibrate for multiple harmonics. This is done by using multiple digital output streams, which are created by downsampling the output stream by the harmonic to be estimated, and then performing the above operation in unison. For example, if the ADC suffers from both k_1 and k_2 harmonic distortion, two new digital streams (1) Digital stream downsampled by k_1 (D_{k1}) and (2) digital stream downsampled by k_2 (D_{k2}) are created. The

product of the sequences D and D_{k1} gives a term proportional to k_1, whereas the product of the terms D and D_{k2} gives a term proportional to k_2. Note that the estimation of both the coefficients happens simultaneously, i.e. the modified update equation is

$$D_{cal}[n+1] = D_{out}[n] - \alpha_{k1} \times D_{out}^{k1}[n] - \alpha_{k2} \times D_{out}^{k2}[n] \quad (5)$$

The final calibrated sequence is given by

$$D_{cal} = D_{out} - \alpha_{k1} \times D_{out}^{k1} - \alpha_{k2} \times D_{out}^{k2} \quad (6)$$

V. VCO BASED $\Delta\Sigma$ ADC

Voltage controlled oscillators (VCOs) are highly non-linear circuit elements which convert voltage information into phase. Any ADC built using VCO suffers from harmonic distortion, thereby making it an ideal candidate to test our proposed algorithm. Figure 5, shows a stand alone VCO which is reconfigured to operate as a first order $\Delta\Sigma$ ADC [2].

The VCO is a 15-cell ring oscillator built using standard cell inverter blocks. The supply of the inverter cells is directly controlled by the input of the system, thereby making it a supply controlled VCO. Also, most of the cells used in the design are standard cell blocks, thereby making the design highly scalable with modern day CMOS processes.

VI. FINAL SYSTEM ARCHITECTURE

Figure 6 shows the final architecture of the complete ADC. The ADC consists of two VCOs with input to the first VCO being V_{in} and the input to the second VCO being $V_{in}/2$. The digital outputs from the ADCs are then sent to the calibration engine, which are used to estimate the harmonic distortion coefficients, in this case 2nd and 3rd harmonic coefficients. Once the coefficients are estimated, the coefficients are stored in memory. The ADC now returns to normal operation, and the output of the ADC is corrected for harmonic distortion using

978-1-4673-6145-3/13 $31.00 © 2013 IEEE

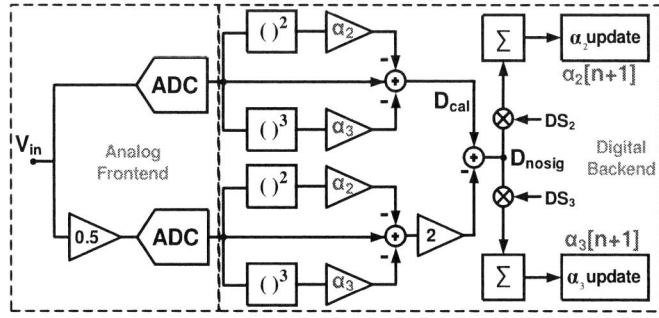

Fig. 6. Final architecture.

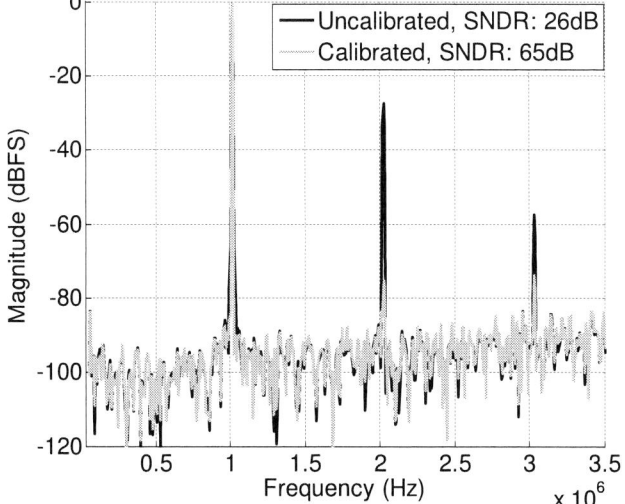

Fig. 7. Measured spectrum before and after calibration.

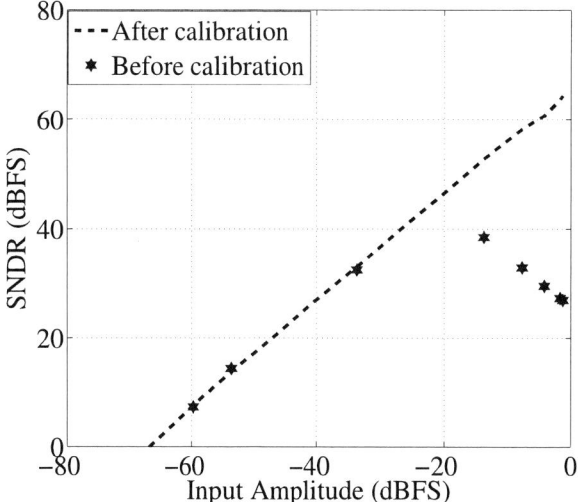

Fig. 8. SNDR vs. input amplitude.

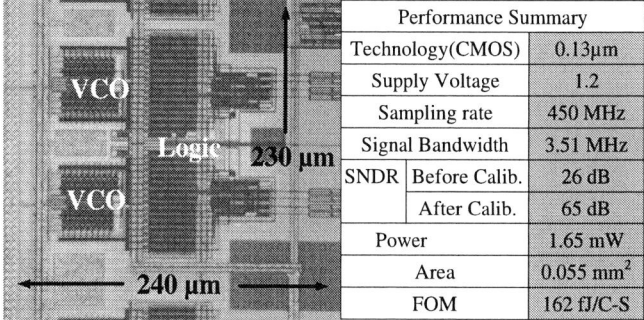

Performance Summary		
Technology(CMOS)		0.13μm
Supply Voltage		1.2
Sampling rate		450 MHz
Signal Bandwidth		3.51 MHz
SNDR	Before Calib.	26 dB
	After Calib.	65 dB
Power		1.65 mW
Area		0.055 mm²
FOM		162 fJ/C-S

Fig. 9. Chip micrograph and performance summary.

eq. (6). The digital engine is operated using 15-bit precision words and is implemented off-chip in software.

VII. MEASURED RESULTS

The performance of VCO based first-order $\Delta\Sigma$ ADC is harmonic distortion limited, thereby making it an ideal candidate for the proposed algorithm. The prototype ADC was built in $0.13\mu m$ CMOS and occupies an area of $0.055mm^2$. The ADC operates at sampling frequency of 450MHz with an OSR of 64, resulting in a signal bandwidth of 3.51MHz. A rounded square wave signal (i.e. an arbitrary signal) is given as the input and the second and third harmonic distortion coefficients are estimated. The non-linearity correction is then performed using the above obtained coefficients.

Figure 7 shows the output spectrum for a 1MHz sinusoid input before and after calibration. Before calibration, the SNDR of the ADC is harmonic distortion limited and equal to 26dB SNDR and after calibration, the SNDR is 65dB. The power consumption of this ADC was 1.65mW resulting in an FOM of 161 fJ/C-S. The limited FOM is due to a design oversight and can be improved easily. Figure 8 shows the SNDR vs. input amplitude for the ADC. The die photo and the performance summary of the ADC are shown in Fig. 9.

VIII. CONCLUSION

This paper proposes a blind calibration algorithm which can be used to reduce harmonic distortion in ADCs. The proposed algorithm can be applied to any ADC architecture in general, and can be used to remove both even and odd harmonics. The successful operation of the algorithm was demonstrated for a VCO based first order $\Delta\Sigma$ ADC.

REFERENCES

[1] W. Liu, P. Huang, and Y. Chiu, "A 12b 22.5/45MS/s 3.0mW 0.059mm2 CMOS SAR ADC achieving over 90dB SFDR," in *IEEE Int. Solid-State Circuits Conf. (ISSCC) Dig. Tech. Papers*, 2010, pp. 380–381.

[2] G. Taylor and I. Galton, "A Mostly-Digital Variable-Rate Continuous-Time Delta-Sigma Modulator ADC," *IEEE J. Solid-State Circuits*, vol. 45, no. 12, pp. 2634–2646, 2010.

[3] A. Panigada and I. Galton, "A 130 mW 100 MS/s Pipelined ADC With 69 dB SNDR Enabled by Digital Harmonic Distortion Correction," *IEEE J. Solid-State Circuits*, vol. 44, no. 12, pp. 3314–3328, 2009.

[4] B. Murmann and B. Boser, "A 12-bit 75-MS/s pipelined ADC using open-loop residue amplification," *IEEE J. Solid-State Circuits*, vol. 38, no. 12, pp. 2040–2050, 2003.

[5] B. Sahoo and B. Razavi, "A 12-Bit 200-MHz CMOS ADC," *IEEE J. Solid-State Circuits*, vol. 44, no. 9, pp. 2366–2380, 2009.

[6] C. Grace, P. Hurst, and S. Lewis, "A 12-bit 80-MSample/s pipelined ADC with bootstrapped digital calibration," *IEEE J. Solid-State Circuits*, vol. 40, no. 5, pp. 1038–1046, 2005.

978-1-4673-6145-3/13 $31.00 © 2013 IEEE

CMOS Millimeter Wave Phase Shifter Based on Tunable Transmission Lines

Wayne H. Woods[1], Alberto Valdes-Garcia[2], Hanyi Ding[1], and Jay Rascoe[1]
[1]IBM Semiconductor Research and Development, Essex Junction, VT
[2]IBM T. J. Watson Research Center, Yorktown Heights, NY

Abstract: **This paper presents a tunable transmission line (t-line) structure, featuring independent control of line inductance and capacitance. The t-line provides variable delay while maintaining relatively constant characteristic impedance using direct digital control through FET switches. As an application of this original structure, a 60 GHz RF-phase shifter for phased-array applications is implemented in a 32 nm SOI process attaining state-of-the-art performance. Measured data from two phase shifter variants at 60 GHz showed phase changes of 175° and 185°, S21 losses of 3.5-7.1 dB and 6.1-7.6 dB, RMS phase errors of 2° and 3.2°, and areas of 0.073 mm² and 0.099 mm² respectively.**

I. INTRODUCTION

With cut-off frequencies (f_T, f_{MAX}) in excess of 200GHz, deep sub-micron CMOS technology has enabled an unprecedented level of integration at millimeter-wave (MMW) frequencies. In particular, recently published examples show that 45 nm and 32 nm SOI CMOS technologies can advance the performance of MMW components [1] and even support the integration of phased arrays at frequencies approaching the sub-MMW regime [2]. Phase shifters are critical components of MMW phased arrays for communication and radar applications. State-of-the art MMW phase shifters come in several varieties [3-9] exhibiting trade-offs between phase tuning range and accuracy, loss, and size. Recently published silicon-based phase shifter designs operating in the 60 GHz band include: Reflective-type phase-shifters (RTPS) [3, 4, 8, 9], T-type switch-type phase-shifters (STPS) [5, 6], a differential varactor-loaded transmission line phase shifter (VLTL) [7], and an RTPS used with an active vector modulator [8]. Since it is possible to achieve a discrete 180° phase shift in integrated RX and TX frontends by swapping differential signals [8], most of the reported passive phase shifters have focused on attaining a phase shift range of 180°.

This paper presents an original phase shifter design implemented with fourteen contiguous sections of a new type of tunable t-line which switches inductance and capacitance independently. This type of phase shifter can replace existing t-line paths between array elements in phased-arrays. Tunable t-lines using CMOS switches have been investigated in the past [10], but only with capacitance control and limited tuning range. In this work, the tunable t-line sections were designed such that they can change delay significantly while minimizing changes in characteristic impedance, Z_0. These phase shifter sections use FET switches to control inductance, capacitance, and delay. Advanced CMOS technologies such as the IBM 32 nm SOI technology employed in these designs offer FET switches with exceptional analog performance and low power digital control. This new type of phase shifter is self-contained with all the FET switches located within the design footprint.

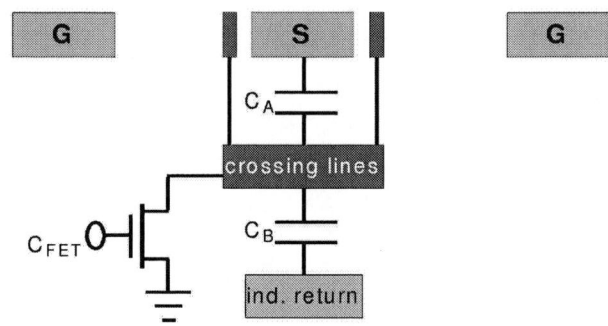

Fig. 1. Simplified cross-section of the proposed tunable t-line.

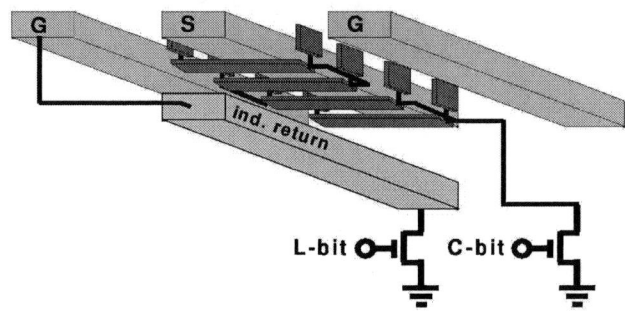

Fig. 2. Conceptual 3-D description of a tunable t-line section with independent digital control of line capacitance and inductance.

Using these FET switches, the line capacitance, C, and line inductance, L, of each tunable t-line section can be independently controlled allowing the delay to be changed:

$$Delay \approx \sqrt{LC} \qquad (1)$$

Moreover, the high and low states for line capacitance, C, and the high and low states for line inductance, L, are designed such that:

$$\frac{L_{high}}{L_{low}} = \frac{C_{high}}{C_{low}} \qquad (2)$$

As a result, ideally:

$$Z_0 = \sqrt{\frac{L_{low}}{C_{low}}} = \sqrt{\frac{L_{high}}{C_{high}}} \qquad (3)$$

This is a unique feature of these tunable t-lines and allows Z_0 variation to be minimized while significantly changing signal delay through a tunable t-line section. This allows the reflection loss to be minimized to the phase shifter and reduces reflections between individual tunable t-line sections internal to the phase shifter.

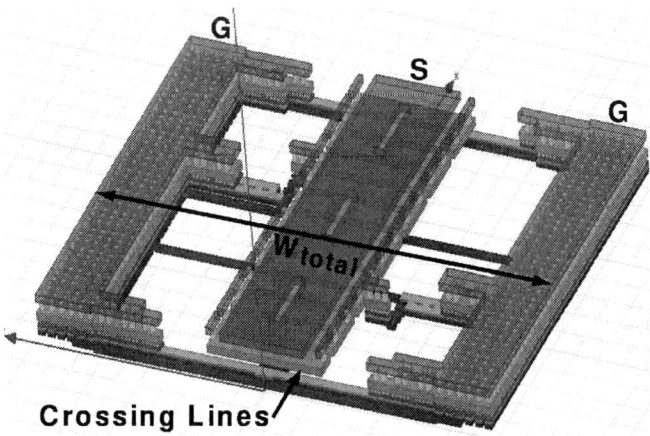

Fig. 3. Top-view of a single section of a phase shifter with W_{total} = 0.098 mm showing ground return, signal, and crossing lines.

Fig. 4. Photograph of a single 0.072 mm long section of a phase shifter with W_{total} = 0.098 mm.

II. DESCRIPTION OF THE PROPOSED STRUCTURE

A. Principle of Operation

Fig. 1 shows a simplified cross-section of a phase shifter tunable t-line section where there are three coplanar conductors labeled: G, S, G. The ground conductors, G, are always directly connected to ground. To the left, right, and below the signal line are conditional capacitance crossing lines. The crossing lines have capacitance, C_A, to the signal line and they are connected to a FET switch that allows them to be conditionally connected to ground. The conditional capacitance crossing lines do not significantly affect the signal inductance as they are largely orthogonal to the signal line as shown in the 3D drawing of a single section of the phase shifter in Fig. 2. Below the crossing lines is the conditional inductance return line labeled "ind. return". Between the crossing lines and "ind. return" is a capacitance, C_B. One end of "ind. return" is grounded as shown in Fig. 2. As a result, C_B is the capacitance between the crossing lines and ground. The other end of the conditional inductance control line in Fig. 2 is connected to an inductance-control FET with control potential, L-bit.

Fig. 2 shows clearly the independent FET controls for the line capacitance and the line inductance. When L-bit = 1 (0.9 V), return current flows in "ind. return" and signal inductance is in the low state. Likewise, when C-bit = 1 (0.9 V), the signal capacitance is equal to C_A which is the high state, and when C-bit = 0 (0 V), signal capacitance equals C_{EFF} = $C_A \times (C_B + C_{FET})/(C_A + C_B + C_{FET})$, where C_{FET} is the off-state capacitance. A single section operates as a phase shifter section when C-bit $\neq$ L-bit.

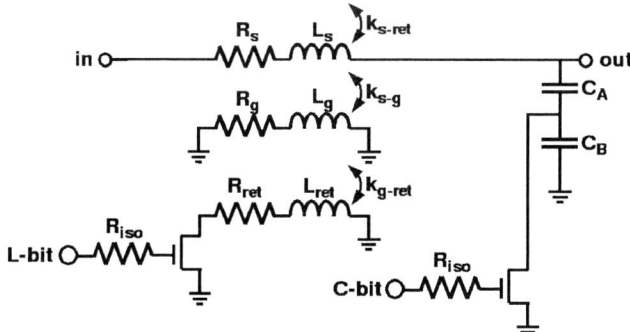

Fig. 5. Circuit schematic of a single section of the phase shifter.

Fig. 3 shows a CAD view of a single section of a phase shifter with W_{total} = 0.098 mm, where the serpentine conditional capacitance crossing lines can be seen crossing under the signal line. Fig. 4 shows a photograph of a single section of a phase shifter with W_{total} = 0.098 mm. Note that the phase shifter section shown in Fig. 4 corresponds to the phase shifter section CAD view shown in Fig. 3.

B. Circuit Modeling

A circuit schematic model for a single section of the phase shifter is shown in Fig. 5. In the circuit model, C_A and C_B correspond to the same capacitances shown in Fig. 1. There are effectively three inductance paths in each section: the signal, the ground returns, and the conditional inductance return path. All of the self and mutual inductances of these paths are solved for using a field solver. The resistance values are determined by line models from the 32 nm SOI technology, and the FETs are also modeled using the models for the 32 nm SOI technology.

C. Phase Shifter Implementation

Using the circuit schematic model above, two phase shifters, phase shifter 1, with W_{total} = 0.072 mm, and phase shifter 2, with W_{total} = 0.098 mm were designed and simulated to achieve 184° and 188° of phase change respectively. Both phase shifters are 1.008 mm long and consist of fourteen identical sections that are 0.072 mm long, and both are designed to be operated in 3-bit mode where the least-significant-bit (LSB) controls two sections, and the most-significant-bit (MSB) controls eight sections. Fig. 6 shows a schematic representation of how the fourteen tunable t-line sections are used to construct the 3-bit phase shifters that were implemented. As mentioned, when operating as a phase shifter C-bit $\neq$ L-bit; so even though there are six control bits shown in Fig.6, there are actually only three independent control bits. Note that the MSB controls eight sections, while the LSB controls two sections. Note further, that to construct 4-bit versions of these phase shifters requires only one additional section: fifteen total sections where the LSB controls a single section and the MSB controls eight sections.

Fig. 7 shows a die photograph of phase shifter 1 fabricated in the IBM 32 nm SOI CMOS process. The G-S-G RF probe pads are on the left and right sides of the phase shifter in Fig. 7. The other six pads are used for digital control of the states of the phase shifter.

978-1-4673-6145-3/13 $31.00 © 2013 IEEE

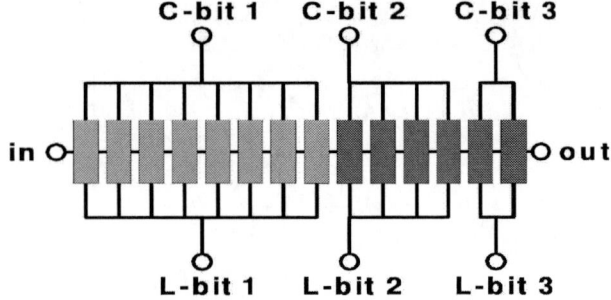

Fig. 6. Schematic representation of a 3-bit phase shifter formed from fourteen identical tunable t-line sections.

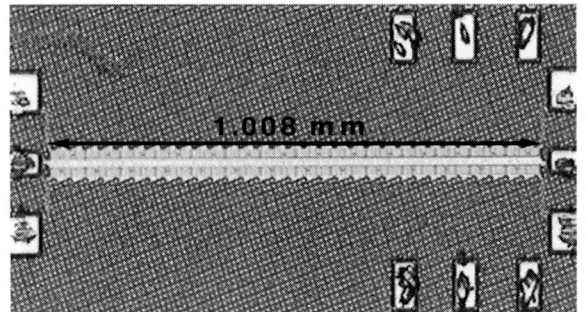

Fig. 7. Die photograph of a complete 32 nm SOI CMOS 60 GHz phase shifter consisting of fourteen identical tunable t-line sections.

The only structural difference in the two phase shifter designs (aside from geometry size differences such as total widths) was in the construction of the serpentine capacitance crossing lines. Specifically, phase shifter 1 had crossing lines that were more compact and did not extend as far away from the signal line as the design in phase shifter 2.

III. Measurement Results

Measurements were made of the two phase shifters from 10 MHz to 110 GHz with analysis primarily focused in the range of 55-65 GHz and on the target operating frequency of 60 GHz. Pad inductance and capacitance were de-embedded from the measurements with fabricated 'OPEN' and 'SHORT' structures. In Fig. 7, there are six control pads above and below phase shifter 1. The three on top control the capacitance of the various sections while the three below control the inductance of the various sections as described above and depicted in Fig. 6. The results of these measurements are shown in Figs. 8-9, in which each line represents one of the eight states of the 3-bit phase shifters, except Figs 8(d) and 9(d) which show the minimum, maximum, and RMS phase errors.

Phase shifter 1 achieved 175° phase change at 60 GHz with an insertion loss range of 3.5-7.1 dB, and reflection loss, better than -10 dB with an RMS gain error of 1.2 dB and minimum, maximum, and RMS phase errors of 0.3°, 3.8°, and 2° respectively.

Phase shifter 2 achieved 185° phase change at 60 GHz with insertion loss range of 6.1-7.6 dB and $|S11| < -10$ dB with RMS gain error of 0.5 dB and minimum, maximum, and RMS phase errors of 0.8°, 6.5°, and 3.2° respectively.

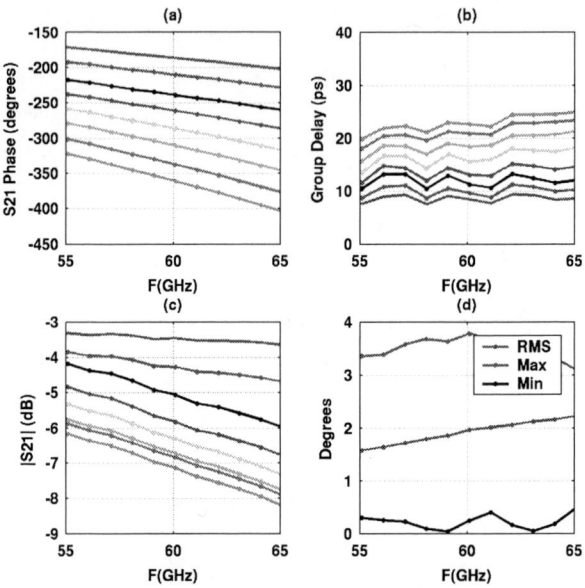

Fig. 8. Measured W_{total}=0.072 mm phase shifter performance. (a) phase shift versus frequency, (b) group delay versus frequency, (c) S_{21} versus frequency, (d) phase error versus frequency.

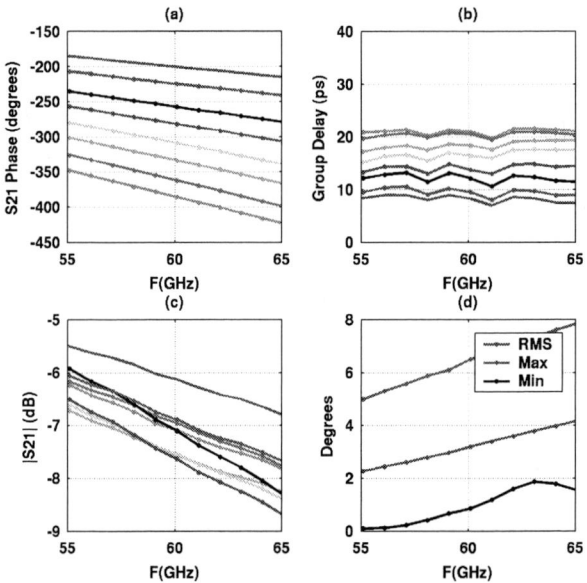

Fig. 9. Measured W_{total}=0.098 mm phase shifter performance. (a) phase shift versus frequency, (b) group delay versus frequency, (c) S_{21} versus frequency, (d) phase error versus frequency.

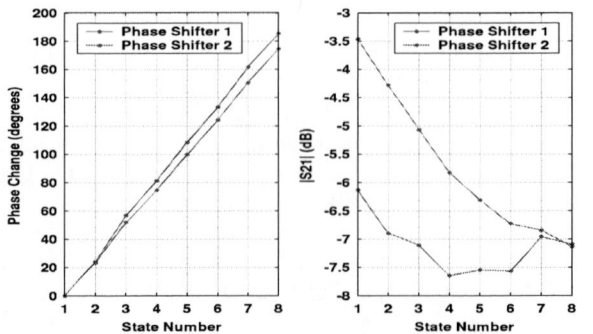

Fig. 10. Measured phase shifter performance at 60 GHz. (a) phase change versus state number, and (b) S_{21} versus state number.

978-1-4673-6145-3/13 $31.00 © 2013 IEEE

TABLE I
Performance summary and comparison to other reported 60 GHz phase shifters in silicon.

Ref.	Architecture	Technology	Loss (dB)	Phase Shift (degrees)	Area (mm^2)	Group Delay (ps)	RMS Gain Error (dB)	ΔPhase /Loss (degrees/dB)
[3]	RTPS	65nm CMOS	5-8.3	180	0.031	N/A	N/A	21.7
			3.3-5.7	147	0.048	N/A	N/A	25.8
[4]	RTPS[1]	130nm SiGe	4-6.2	156	0.33	N/A	1.1	25.2
[5]	STPS[2]	65nm CMOS	11.6-17.6	360	0.34	+/- 6	1.8	20.5
[6]	STPS[2]	65nm CMOS	11.5-15.6	360	0.28	+/- 8.5	1.3	23
[7]	VLTL	65nm CMOS	6.3-12.5	158	0.2	N/A	N/A	12.6
[8]	RTPS	130nm SiGe	4.2-7.5	180	0.18	N/A	N/A	24
[9]	RTPS	90nm CMOS	4.5-8	90	0.08	N/A	N/A	11.3
This Work	**Tunable t-line**	**32 nm SOI CMOS**	**3.5-7.1**	**175**	**0.073**	**+/-7.1**	**1.2**	**24.6**
			6.1-7.6	**185**	**0.099**	**+/-6.4**	**0.5**	**24.3**

[1]Differential design [2]57-64 GHz

The uniform sectional nature of the design of these phase shifters is helpful in producing good linearity of phase change versus state number at 60 GHz as can be seen in Fig. 10(a) for both the phase shifters. The differences in the capacitance crossing line structures of these two phase shifters was observed to cause differences in the loss variation between the two design variants as can be seen in the two example measurements of loss versus state number at 60 GHz for phase shifter 1 and phase shifter 2 shown in Fig. 10(b).

Table I summarizes measurement results from phase shifters 1 and 2, and also shows results of other reported passive phase shifter designs at 60 GHz. Of the reported phase shifters at 60 GHz, phase shifter 1 showed the lowest reported minimum, maximum, and RMS phase errors: 0.3°, 3.8°, and 2° respectively. Also, among reported passive phase shifters that achieve 180° phase change at 60 GHz, phase shifter 2 showed the highest reported phase change per loss, 24.3°/dB, the lowest RMS gain error, 0.5 dB, and the second-smallest size, 0.099 mm^2.

IV. CONCLUSION

A tunable transmission line structure capable of varying signal delay while minimizing Z_0 variation using FET switches has been introduced. Based on this structure, two CMOS 60 GHz phase shifters have been demonstrated. To the best of the authors' knowledge, these are the first MMW phase shifters implemented with tunable t-line sections that can change signal delay by separately controlling L and C. These phase shifters are self-contained, have direct digital control and require no additional circuitry such as DAC circuits which are needed in RTPS designs. Moreover, the simple t-line-like design of these phase shifters could make them ideal replacements of existing transmission line paths between antenna elements in phased array systems. The measured performance of these phase shifters shows that they are suitable for integrated phased-array radar and communication systems.

V. ACKNOWLEDGEMENT

The authors thank S. Reynolds, C. Putnam, and D. Friedman for management support. This work was partially supported by DARPA under AFRL contract # FA8650-09-C-7924 The views, opinions, and/or findings contained in this article/presentation are those of the author/presenter and should not be interpreted as representing the official views or policies, either expressed or implied, of the Defense Advanced Research Projects Agency or the Department of Defense.

VI. REFERENCES

[1] B. Cetinoneri, et al., "W-Band Amplifiers With 6-dB Noise Figure and Milliwatt-Level 170–200-GHz Doublers in 45-nm CMOS", *IEEE T-MTT*, Vol. 60, pp. 692-701, March 2012.

[2] K. Sengupta, and A. Hajimiri, "A 0.28THz 4×4 Power-Generation and Beam-Steering Array", *IEEE International Solid-State Circuits Conference*, pp. 256-257, February 2012.

[3] M. Tabesh, A. Arbabian, and A. Niknejad, "60GHz Low-Loss Compact Phase Shifters Using A Transformer-Based Hybrid in 65nm CMOS", *IEEE CICC*, pp. 1-4, September 2011.

[4] H. Krishnaswamy, A. Valdes-Garcia, and J.-W. Lai, "A silicon-based all-passive, 60 GHz, phased-array beamformer featuring a differential, reflection-type phase shifter", *Proc. IEEE Int. Phased Array Syst. Technol. Symp., pp. 225-232, October 2010.*

[5] W.-T. Li, et al., "60 GHz 5-bit Phase Shifter with Integrated VGA Phase-Error Compensation", *IEEE T-MTT*, Vol. 61, pp. 1224-1235, March 2013.

[6] Y.-C. Chiang, W.-T. Li, J.-H. Tsai, and T.-W. Huang, "A 60 GHz digitally-controlled 4-bit phase shifter with 6-ps group delay deviation", *IEEE IMS*, pp.1-3, June 2012.

[7] Y. Yu, et. al. "A 60 GHz digitally-controlled phase shifter in CMOS", *IEEE ESSCIRC*, pp. 250-253, September 2008.

[8] M.-D. Tsai and A. Natarajan, "60 GHz pasive and active RF-path phase shifters in silicon", *IEEE RFIC, pp. 223-226, June 2009.*

[9] B. Biglarbegian, et al. "Millimeter-Wave Reflective-Type Phase Shifter in CMOS Technology", *IEEE MCWL*, vol. 19, no. 9, pp. 560-562, September 2009.

[10] T. LaRocca, et al., "Millimeter-Wave CMOS Digital Controlled Artificial Dielectric Transmission Lines for Reconfigurable ICs", *IEEE RFIC*, pp. 181-184, June 2008.

978-1-4673-6145-3/13 $31.00 © 2013 IEEE

Charge Steering: A Low-Power Design Paradigm

Behzad Razavi

Electrical Engineering Department

University of California, Los Angeles

Abstract

Discrete-time charge-steering circuits consume less power than their continuous-time current-steering counterparts even at high speeds. This advantage can be exploited in the design of semi-analog circuits such as latches, demultiplexers, and CDR circuits as well as mixed-mode systems such as ADCs. Employing charge steering in 65-nm CMOS technology, a 25-Gb/s CDR/deserializer consumes 5 mW and a 10-bit 800-MHz pipelined ADC draws 19 mW.

I. INTRODUCTION

The thrust for low-power circuit design continues unabated, presenting especially tough challenges as high speeds are sought. The power-speed trade-off associated with any circuit function becomes nonlinear as the frequency of operation exceeds a certain limit, motivating efforts toward developing new techniques.

This paper presents the concept of "charge steering" as a candidate for low-power, high-speed design. Applicable to both digital and analog circuits, the concept offers a factor of 2 to 4 power saving for a given set of design constraints, and it has been demonstrated in the context of a 25-Gb/s clock and data recovery (CDR)/deserializer circuit [1] and a 10-bit 800-MHz analog-to-digital converter (ADC) [2].

Section II introduces the basic idea and Section III deals with charge-steering logic. Section IV applies the concept to CDR and deserializer design, describing how charge steering issues can be resolved at the architecture level. Section V presents charge-steering op amps and their use in pipelined ADCs.

II. BASIC IDEA

A continuous-time current-steering circuit can be transformed to a discrete-time charge-steering topology as depicted in Fig. 1: the tail current source is replaced with a charge source, and the load resistors with capacitors. Discrete-time operation requires two switches in the tail path and two at the output nodes. Shown here for a simple differential pair, the transformation can be applied to other circuit topologies as well.

Figure 2 illustrates the operation of the charge-steering stage. The circuit begins in the reset mode, with C_T discharged to ground and the output nodes precharged to V_{DD}. When CK goes high, the circuit enters the amplification mode,

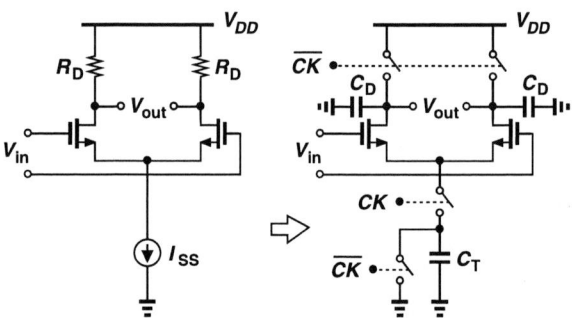

Fig. 1. Transformation from current steering to charge steering.

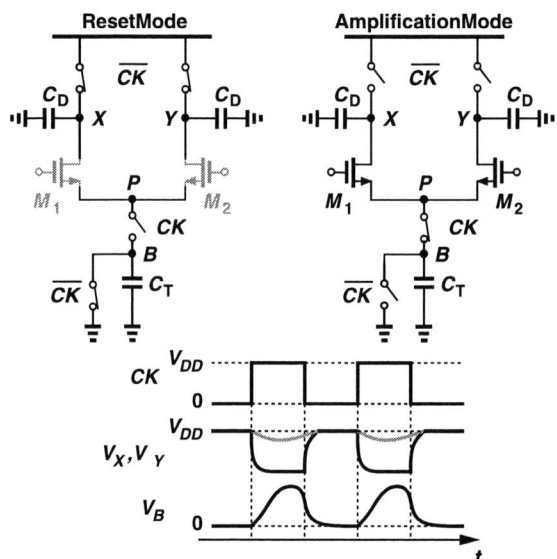

Fig. 2. Operation of charge-steering stage.

C_T pulls current from M_1 and M_2, and nodes X and Y are released. The two transistors continue to draw differential and common-mode (CM) currents from the load capacitors until C_T charges to approximately one threshold below the higher input level. As explained below, with proper choice of device parameters, the charge-steering stage can provide voltage gain.

The discrete-time nature of this topology offers three advantages over its continuous-time counterparts. First, the circuit can serve as a latch with moderate, controlled output swings, potentially running faster than rail-to-rail (CMOS) logic. Second, the stage steers current for a fraction of the clock period, thereby consuming less power than continuous-time topologies such as current-mode logic (CML) circuits. Third, since the average power consumption of the stage scales with fre-

978-1-4673-6145-3/13 $31.00 © 2013 IEEE

quency, the same design can be reused at different clock rates with no modification. For example, the ADC described in Section V has been tested with clock frequencies ranging from 100 MHz to 800 MHz, with its power consumption varying from 2.4 mW to 19 mW. This attribute is particularly attractive for applications that require a wide range of clock frequencies and typically dictate extensive programmability in the design.

Charge steering faces two issues. First, in contrast to conventional CML circuits, the switches in Fig. 2 must be driven by rail-to-rail clock swings. Nonetheless, the power savings afforded by this technique typically outweigh the clock power consumed in a CML environment. Second, as with any precharged circuit, charge-steering topologies produce a return-to-zero (RZ) output. This issue manifests itself in data communication systems, e.g., CDR circuits, but not in inherently discrete-time applications such as ADCs.

The charge-steering stage of Fig. 2 forms the foundation for the digital and analog circuits described in this paper. We will employ this building block to develop phase detectors (PDs), CDRs, and demultiplexers (DMUXes) for wireline design as well as op amps for ADC design.

It is important to distinguish this topology from other differential dynamic logic styles. In the differential precharged stage of Fig. 3(a), the absence of a tail capacitor allows the

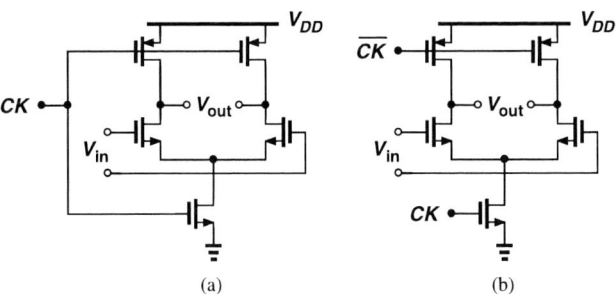

Fig. 3. Examples of differential dynamic circuits: (a) precharged, (b) dynamic amplifier.

outputs to collapse to zero, leading to a slow response. Also, in the logic style of Fig. 3(b) [3], the circuit draws current from V_{DD} for half a clock cycle and, more importantly, suffers from a lower speed because its tail current must flow from both the PMOS loads and the load capacitances.[1]

III. CHARGE-STEERING LOGIC

A. Gain and Power Consumption

For use in high-speed digital design, the charge-steering circuit of Fig. 2 must operate with moderate input and output voltage swings while providing some voltage gain so as to restore the logical levels. It can be shown that, for a small differential input, V_{in}, the voltage gain is relatively independent of the input CM level, V_{CM}, and is given by [1]

$$A_v \approx \frac{2C_T}{C_D}. \tag{1}$$

[1]But one advantage of this style is that it produces an NRZ output.

For moderate to large inputs, on the other hand, the differential output voltage depends on V_{CM} and is equal to

$$V_{out} = \frac{C_T}{C_D} \frac{(V_{CM} - V_{TH})^2 + \frac{3V_{in}^2}{4}}{V_{CM} - V_{TH} + \frac{V_{in}}{2}}, \tag{2}$$

where V_{TH} denotes the threshold voltage of M_1 and M_2 [1]. Derived using simple, rough approximations, these expressions are plotted in Fig. 4 against simulation results, exhibit-

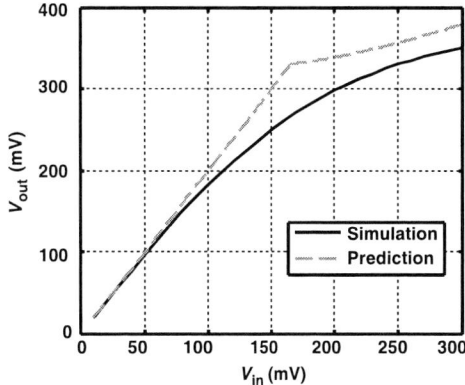

Fig. 4. Simulated and calculated characteristics of charge-steering stage.

ing modest accuracy. In this example, the circuit provides a small-signal gain of about 2 and an output swing of about 350 mV.

In order to appreciate the power swings, we perform a comparison with a continuous-time CML stage. Suppose the two topologies in Fig. 1 operate at the same rate, r_b, and deliver equal voltage swings to equal load capacitances, C_D. Note that C_D is not an explicit capacitor and simply models the parasitics at each node and the input capacitance of the following stage. The CML circuit must provide a bandwidth of about $0.7r_b$, i.e., $(2\pi R_D C_D)^{-1} = 0.7r_b$, while producing a single-ended output swing of $V = I_{SS}R_D$. In the charge-steering configuration, only one load capacitor charges to V_{DD} and discharges to $V_{DD} - V$ in each bit period, yielding an average supply current equal to

$$I_{DD} = C_D r_b V. \tag{3}$$

Eliminating r_b and R_D from the foregoing expressions, we obtain

$$\frac{I_{DD}}{I_{SS}} = \frac{1}{1.4\pi}, \tag{4}$$

predicting a factor of 4.4 reduction in power.

B. Regenerative Latch

The concept of charge steering can be applied to regenerative latches as well. Let us contemplate a discrete-time version of a standard CML latch [Fig. 5(a)], assuming that M_1-M_2 and M_3-M_4 are enabled by complementary clocks. Unfortunately, this circuit requires three phases for precharge, amplification, and regeneration. We instead consider only the cross-coupled

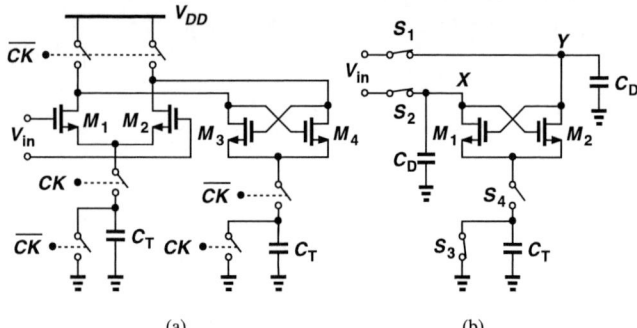

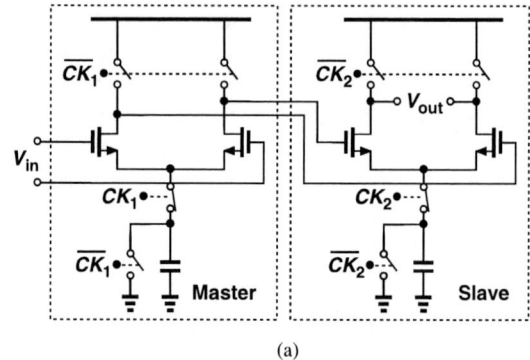

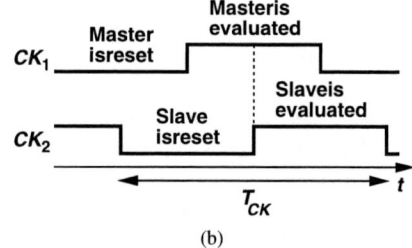

Fig. 5. (a) Charge-steering implementation of a CML latch, (b) single regenerative latch.

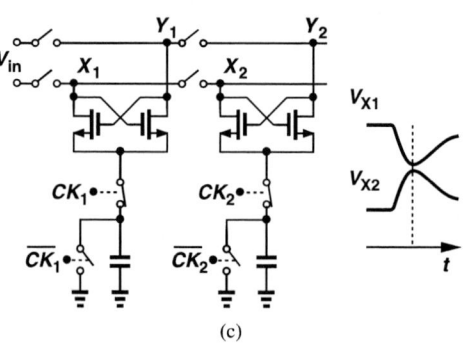

pair and convert its precharge mode to a *sampling* mode [Fig. 5(b)]. The circuit now tracks the input for half a clock cycle and regenerates for the other half, thus producing a non-return-to-zero (NRZ) output.

The NRZ latch of Fig. 5(b) can provide voltage gain. If the transistors operate in weak to moderate inversion, the gain can be approximated as [1]

$$\frac{V_{XY\infty}}{V_{XY0}} = \exp\left(\frac{C_T}{C_D}\frac{V_{CM} - V_{GS}}{2\zeta V_T}\right), \quad (5)$$

where the left-hand side represents the ratio of the final and initial voltages, V_{GS} is assumed relatively constant, and ζ denotes the subthreshold nonideality factor and is given by $1 + C_d/C_{ox}$, where C_d is the depletion capacitance under the channel. Figure 6 plots Eq. (5) with $V_{GS} = 450$ mV along with simulation results, indicating good agreement.

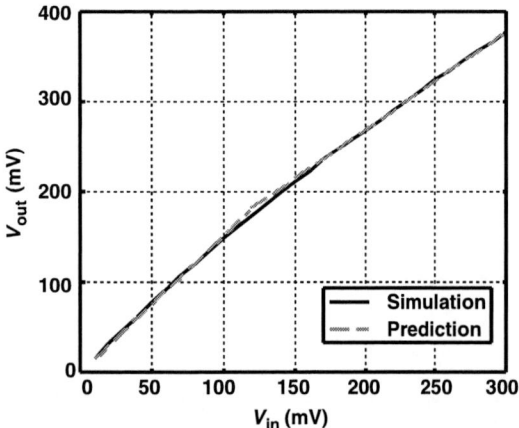

Fig. 6. Simulated and calculated characteristics of NRZ latch.

Fig. 7. (a) Cascade of two RZ latches, (b) use of quadrature clocks, (c) cascade of two NRZ latches.

C. Cascading Issues

In order to implement flipflops and more complex functions, charge-steering latches must be cascaded. We consider the topologies in Figs. 2 and 5(b) for this purpose, noting that one permutation, Fig. 5(a), has not proved practical. Figure 7(a) shows another permutation consisting of two cascaded RZ latches driven by CK_1 and CK_2. This arrangement faces an issue if the two clocks are simply complementary: when the slave begins to sense, the master enters the precharge mode

and its differential output begins to collapse. The slave may therefore generate a small output swing in some corners of the process. To avoid this race condition, one can employ quadrature phases for CK_1 and CK_2 so that the master output is held constant when the slave enters the evaluation mode [Fig. 7(b)] - but at the cost of power and complexity in clock generation.

Figure 7(c) shows another cascading permutation using two NRZ latches. In this case, the circuit contains no internal path from V_{DD} and hence provides no charge amplification. That is, only the charge deposited by V_{in} must produce the voltage swings as the signal propagates down the chain, leading to substantial corruption of random data. As illustrated in Fig. 7(c), if the states on X_2 and Y_2 are opposite of those on X_1 and Y_1, when S_3 and S_4 turn on, the master and slave capacitances experience severe charge sharing, heavily attenuating the data swings. Even if the master devices are scaled up by a factor of 5, this memory effect still produces significant intersymbol interference (ISI).

The last permutation of the RZ and NRZ latches is shown in Fig. 8. Here, the master consists of the passive sampling

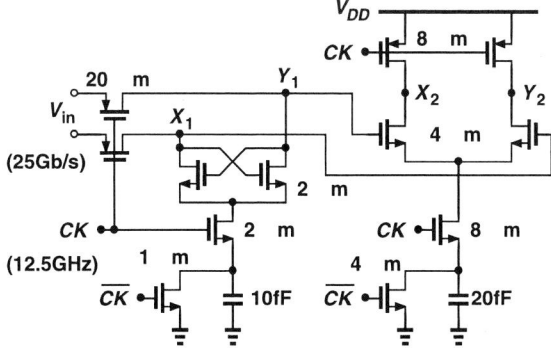

Fig. 8. Cascade of NRZ and RZ latches.

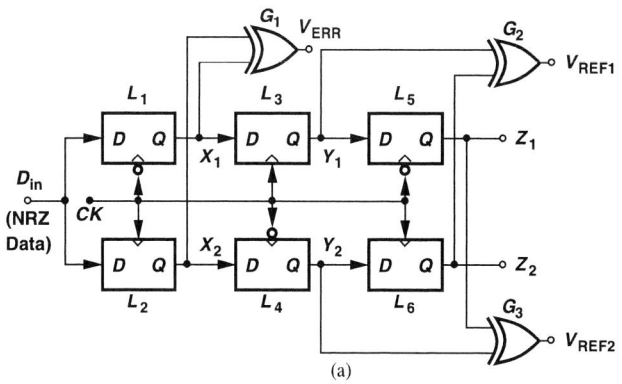

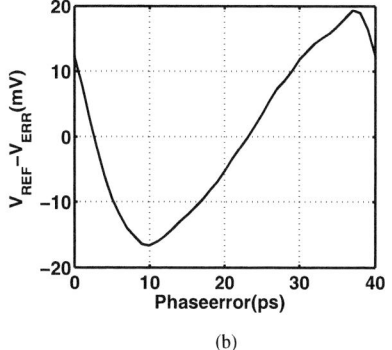

Fig. 9. PD using charge-steering latches, (b) simulated input-output characteristic.

network, S_1 and S_2, while the slave is formed by M_1-M_2 and M_3-M_4. The charge amplification provided by the latter pair avoids corruption of the state stored by the former. Clocked simultaneously, the two stages amplify the sampled signal at X_1 and Y_1, generating a single-ended swing of about 400 mV at X_2 and Y_2 with a power consumption of 158 μW at a data rate of 25 Gb/x and a clock rate of 12.5 GHz. However, the output is still in RZ form

IV. CDR AND DMUX DESIGN

The concepts developed above can be applied to the design of high-speed CDRs and (de)multiplexers. In this section, we present some examples for 25-Gb/s operation, assuming a half-rate architecture but without quadrature clock phases.

A. Half-Rate Phase Detector

A half-rate PD that does not require quadrature clocks can be realized as four latches and two exclusive-OR (XOR) gates [4]. Unfortunately, this topology does not lend itself to charge steering circuits with RZ outputs [1]. Fortunately, it is possible to modify the PD so as to operate with RZ data [Fig. 9(a)]. Here, latches L_5 and L_6 are added to insert half a cycle delay, and XOR gates G_2 and G_3 are respectively used to compare Y_1 with a delayed version of Y_2 and Y_2 with a delayed version of Y_1. The average difference between V_{ERR} and $V_{REF} = V_{REF1} + V_{REF2}$ represents the phase difference between the random data and the clock [Fig. 9(b)].

In the circuit of Fig. 9(a), the latch pairs L_1-L_3 and L_2-L_4 are constructed as shown in Fig. 8. Latches L_5 and L_6 employ the RZ topology of Fig. 2. The retimed half-rate data is available at Y_1 and Y_2 but still in RZ form.

B. Demultiplexer

In order to further demultiplex the data, we cascade RZ latches but ensure that when one is being reset, the next does not begin to sense. This is possible because the half-rate (12.5-GHz) clock can be divided by 2 so as to produce quadrature phases at 6.25 GHz. Figure 10 depicts such an arrangement and its timing diagram, where $CK_{1/2,I}$ and $CK_{1/2,Q}$ represent the quadrature phases of the 6.25-GHz clock. We note that at $t = t_1$, L_3 (one of the latches within the PD) enters the evaluation mode and, after one divider delay, so does L_7. Next,

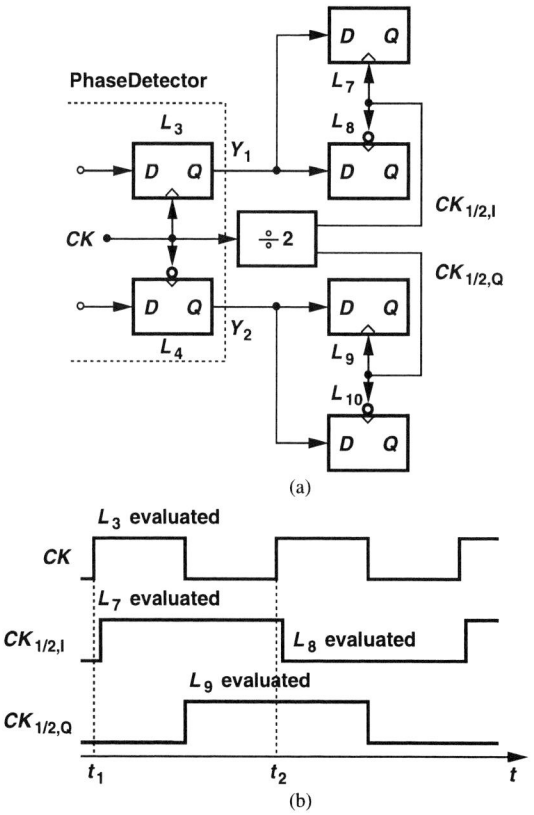

Fig. 10. DMUX realization.

at $t = t_2$, L_3 enters the evaluation mode again and so does L_8.

978-1-4673-6145-3/13 $31.00 © 2013 IEEE

The bottom DMUX path consisting of L_9 and L_{10} operates in a similar manner but with $CK_{1/2,Q}$. The four latches consume 183 μW at 6.25 GHz. According to simulations, for a given power consumption, charge-steering latches running at 6.25 GHz still outperform rail-to-rail logic in terms of the output ISI.

C. RZ-NRZ Conversion

The low-swing RZ waveforms produced by charge-steering circuits can be converted to NRZ data by means of a rail-to-rail RS latch. However, the amplification of the RZ waveform to achieve rail-to-rail swings demands substantial power. A more efficient approach incorporates a clocked dynamic comparator for amplification [1]. Exemplified by the StrongArm topology, such a comparator too produces RZ outputs when it is reset and hence resembles the dynamic circuit in Fig. 3(a). If cascaded with a charge-steering latch, the comparator thus suffers from the race condition described for the flipflop in Fig. 7(a). Since the demultiplexed data is now available at the quarter rate, we may reconsider the scenario in Fig. 7(b) and utilize the quadrature phases of the quarter-rate (6.25-GHz) clock.

Figure 11 depicts the DMUX/RZ-NRZ conversion chain.

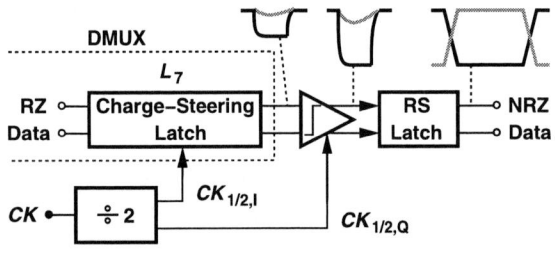

Fig. 11. RZ-to-NRZ conversion.

As mentioned earlier, L_7 is driven by $CK_{1/2,I}$. We choose $CK_{1/2,Q}$ to clock the comparator, applying the result to a CMOS RS latch. The comparator and the RS latch draw 130 μW at 6.25 GHz, far less than typical CML-CMOS converters do.

D. Clock Generation and Distribution

As pointed out in Section II, charge-steering circuits dictate rail-to-rail clock swings, a condition afforded by LC oscillators. However, the important question is whether the VCO should drive the latches, the frequency divider(s), and the wiring capacitance directly or through a buffer. The total load capacitance presented by this network is about 270 fF in our work, demanding a power of $2 \times fCV_{DD}^2 \approx 8$ mW if two inverters follow the differential VCO outputs. It is therefore beneficial to omit the buffers and absorb this capacitance in the VCO tank even at the cost of a lower tank inductance and hence a greater bias current. In this case, the higher VCO power dissipation also translates to a lower phase noise.

The critical point here is that a given power budget is more efficiently utilized in a VCO than in buffers, suggesting that buffers are generally redundant [1]! One exception is a case where the loss associated with the interconnects significantly lowers the VCO tank Q.

E. Experimental Results

A 5-mW 25-Gb/s CDR/deserializer using an LC VCO has been fabricated in 65-nm CMOS technology. The die photograph is shown in Fig. 12(a), the recovered clock phase noise in Fig. 12(b), and the jitter tolerance for two different supply voltages in Fig. 12(c) [1]. Exhibiting an rms jitter of 1.52 ps with a $2^{15} - 1$ PRBS, the prototype consumes about a factor of 20 less power than the prior art [5, 6] and demonstrates the advantages of charge steering.

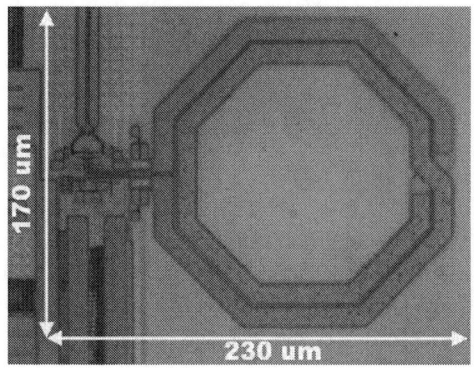

(a)

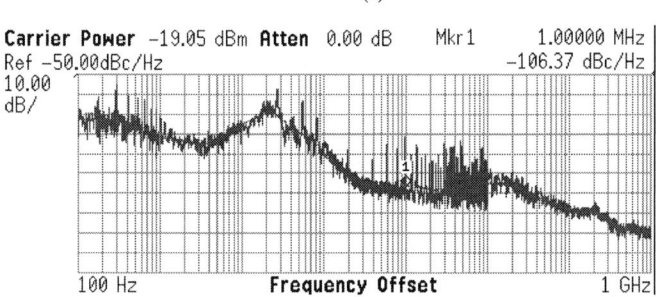

(b)

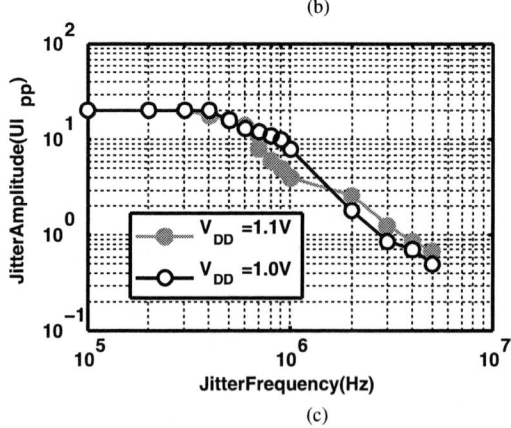

(c)

Fig. 12. CDR/deserializer's (a) die photo, (b) recovered clock phase noise, and (c) jitter tolerance.

V. CHARGE-STEERING ADCs

A. Charge-Steering Op Amps

A practical one-stage charge-steering amplifier can achieve a gain of 2 to 4, pointing to a two-stage configuration if a gain commensurate with pipelined ADC design is desired. Figure

13(a) shows a basic topology, where the tail capacitors are

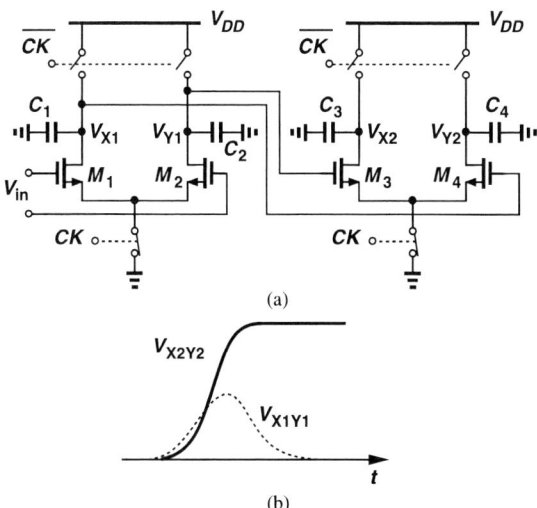

(a)

(b)

Fig. 13. (a) Charge-steering op amp, and (b) its waveforms.

removed to allow a large amount of charge to flow.[2] Here, the two stages simultaneously amplify until the outputs of the first stage collapse and the second stage turns off. During this transient, V_{X1Y1} rises to a peak and falls back to zero whereas V_{X2Y2} monotonically increases to an amplified copy of the input [Fig. 13(b)]. This two-stage design can provide an open-loop gain of about 10. The values of C_1-C_4 are dictated by kT/C noise requirements.

The charge-steering op amp exhibits a unique behavior in a closed-loop configuration [2] such as the multiplying digital-to-analog converter (MDAC) of Fig. 14(a). Owing to the large load capacitors and the absence of load resistors, each stage behaves as an integrator, incurring loss only due to the output resistance of the transistors. Simplified as shown in Fig. 14(b), the two-integrator feedback loop thus produces an underdamped output that is frozen when the second stage turns off at $t = t_1$ [Fig. 14(c)]. As a result of this overshoot, the closed-loop gain can be *greater* than C_{in}/C_F.

It is possible to design the first stage such that its CM level falls to one threshold above ground by the time V_{out} reaches its peak value at t_p. Freezing the output in its zero-slope regime, such a choice minimizes PVT-induced variations in the final value of V_{out}.

The gain, noise, speed, power dissipation, and linearity of the above op amp can be compared with those of continuous-time topologies. In a pipelined ADC environment, the MDAC stage depicted in Fig. 14(a) exercises all of these properties. We design the op amp of Fig. 13(a) as well as the two configurations shown in Fig. 15 for $C_{in} = 480$ fF, $C_F = 240$ fF, $C_L = 250$ fF, $V_{DD} = 1$ V, a differential output swing of 0.6 V_{pp}, an open-loop gain of 10, a power dissipation of 2.5 mW, and a clock rate of 1 GHz. The simulated settling time, distortion, and input-referred noise of the MDAC are

[2]Although not identified as a charge-steering op amp, a similar topology without explicit load capacitors has been used in [7] for 6-bit resolution.

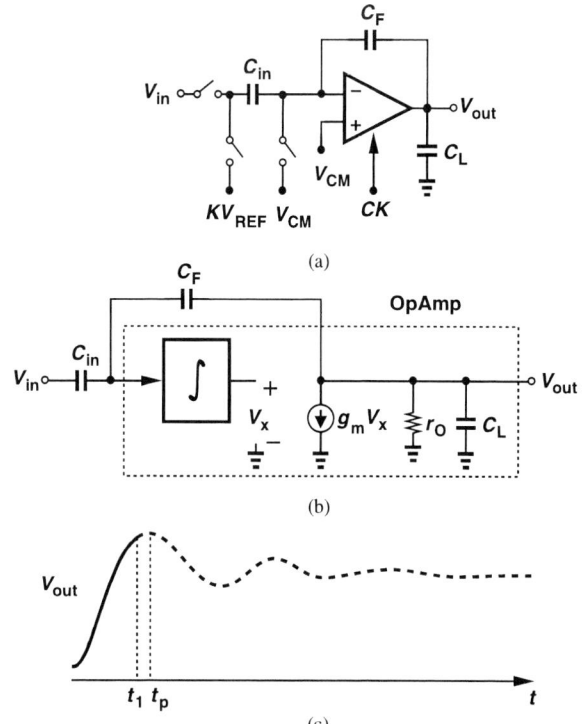

(a)

(b)

(c)

Fig. 14. (a) MDAC environment for charge-steering op amp, (b) simplified closed-loop model, (b) self-timed underdamped behavior.

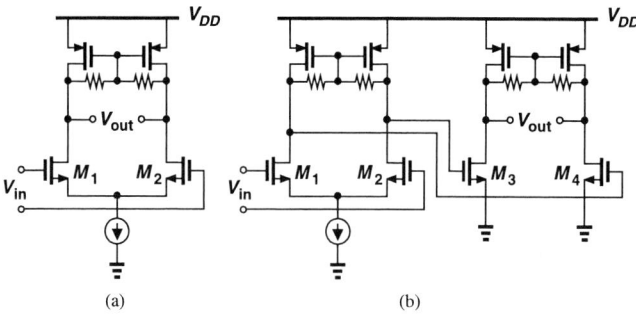

(a)   (b)

Fig. 15. One-stage and two-stage op amps studied for quantitative comparisons.

listed in Table 1, revealing the "best of all worlds" for the

	One–Stage OpAmp	Two–Stage OpAmp	Charge–St. OpAmp
SettlingTime	560ps	370ps	80ps
SDR	48dB	53dB	54dB
Input–Ref. Noise	67nV^2	138nV^2	65nV^2

Table 1. Simulated performance of MDAC using each op amp topology.

charge-steering topology.

The results depicted in Table 1 imply that, due to their low open-loop gain, charge-steering op amps do not provide adequate linearity for a 10-bit ADC, dictating nonlinearity (and gain error) calibration. For example, an accurate on-chip ladder can be utilized to calibrate the ADC in the digital domain by means of an LMS machine [8]. However, the characteristics of these op amps are somewhat sensitive to the input and output CM levels. In particular, if the output CM level, $V_{out,CM}$, shifts from its optimum value by more than 100 mV, the open-loop linearity degrades, making calibration difficult.

It is difficult to apply CM feedback to charge-steering op amps in the analog domain because the output CM level reaches its final value *after* the stages have turned off. Alternatively, the optimum value of $V_{out,CM}$ can be viewed as that which maximizes the MDAC linearity. That is, during calibration, the LMS machine can adjust $V_{out,CM}$ along with the digital coefficients so as to minimize the nonlinearity [2]. Illustrated in Fig. 16, this approach tunes $V_{out,CM}$ in discrete

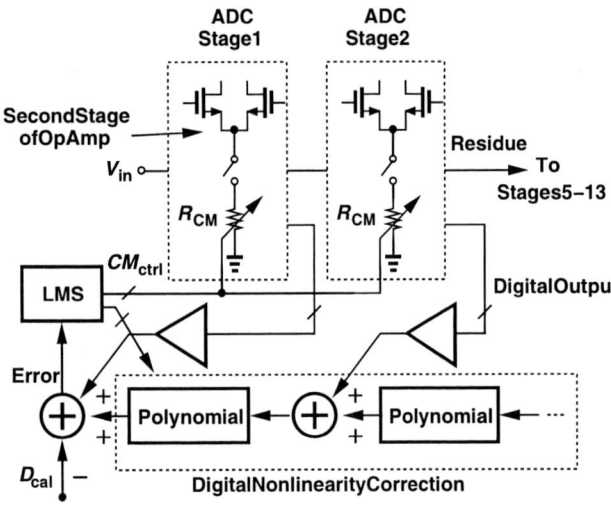

Fig. 16. Common-mode control by LMS machine.

steps by controlling the tail resistance, R_{CM}, in the second stage of each op amp.

B. Three-Stage Op Amps

The "self-timed" nature of charge-steering op amps suggests that more than two stages can be cascaded so as to increase the open-loop gain. Consider the arrangement shown in Fig. 17(a), where the outputs of the first and second stages collapse to zero at about the same time, allowing the third stage to maintain an amplified output. According to simulations, this circuit can achieve a gain of 20.

To study the circuit's behavior in a closed-loop configuration, we approximate the transfer function of each lossy integrator by $A_0/(1 + s/\omega_0)$, where $A_0 = g_m r_O$ and $\omega_0 = g_m/C_D$. [The load capacitors, C'_Ds, are not shown in Fig. 17(a) for simplicity.] With unity-gain negative feedback, the

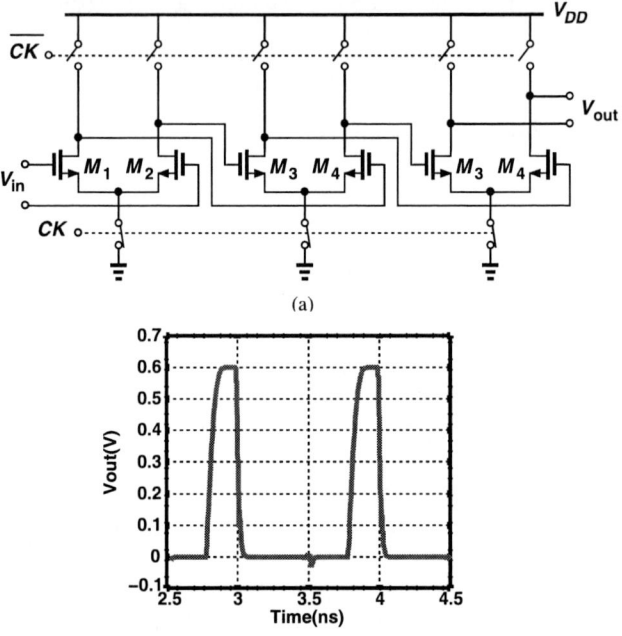

Fig. 17. (a) Three-stage charge-steering op amp, and (b) its response in an MDAC environment.

closed-loop transfer function is given by

$$H(s) = \frac{\dfrac{A_0^3}{(1 + \dfrac{s}{\omega_0})^3}}{1 + \dfrac{A_0^3}{(1 + \dfrac{s}{\omega_0})^3}}, \qquad (6)$$

and the poles are computed from

$$(1 + \frac{s}{\omega_0})^3 + A_0^3 = 0. \qquad (7)$$

The loop contains one real pole located at $-(A_0 + 1)\omega_0$ and two complex poles at $(A_0/2 - 1)\omega_0 \pm j(\sqrt{3}/2)A_0\omega_0$, which for $A_0 > 2$, fall in the right half plane and yield a growing sinusoid - just as in a three-stage ring oscillator. However, the last two stages turn off after a brief period of time, stopping the growth and producing an amplified output. Figure 17(b) plots the simulated step response of the three-stage op amp in the MDAC environment of Fig. 14(a), revealing a settling time of about 70 ps.

The foregoing study indicates that the design of charge-steering op amps markedly departs from the conventional wisdom. The closed-loop circuit is allowed to be unstable so long as the stages turn off before or at the (first) peak value of the output. It is conceivable that a larger number of stages can also be used to further increase the gain.

C. Experimental Results

A 10-bit 800-MHz pipelined ADC using two-stage charge-steering op amps has been designed and fabricated in 65-nm CMOS technology [2]. Figure 18(a) shows the die photograph

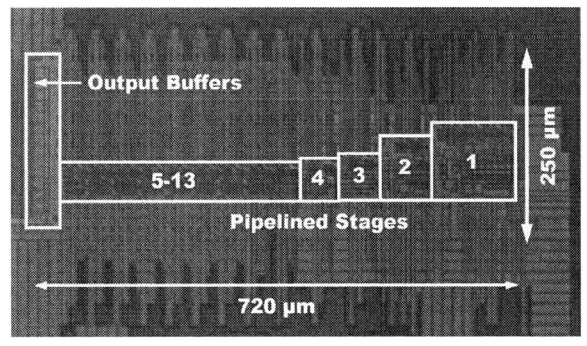

(a)

(b)

Fig. 18. (a) ADC die photograph, (b) measured DNL and INL before and after calibration.

and Fig. 18(b) the DNL and INL before and after calibration. Figure 19 plots the SNDR as a function of the input frequency,

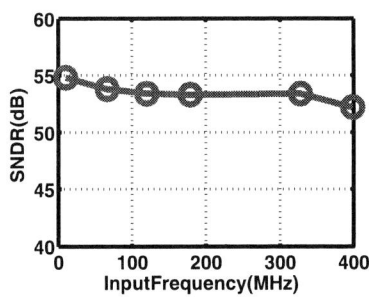

Fig. 19. Measured SNDR of the ADC at a sampling rate of 800 MHz.

exhibiting a value of 52.2 dB at Nyquist rate. The ADC draws 19 mW from a 1-V supply and provides an FOM of 53 fJ per conversion step.

VI. CONCLUSION

Charge steering holds promise for high-speed analog and mixed-signal circuits with low power consumption. The discrete-time nature of this design technique enables digital latching as well as muti-stage, nominally unstable op amps to perform in complex circuits. Issues associated with this design paradigm have been discussed and solutions have been proposed. Providing a fourfold power advantage over CML circuits, charge steering has been exploited in a 25-Gb/s CDR/deserializer dissipating 5 mW and a 10-bit 800-MHz ADC consuming 19 mW.

Acknowledgments

This work was supported by the DARPA HEALICS program, Texas Instruments, and Realtek Semiconductor. The author is grateful to the TSMC University Shuttle Program for chip fabrication.

REFERENCES

[1] J.W. Jung and B. Razavi, "A 25-Gb/s 5-mW CMOS CDR/Deserializer," *IEEE J. Solid-State Circuits,* vol. 48, pp. 684-697, Mar. 2013.

[2] S.-H. Chiang, H. Sun, and B. Razavi, "A 10-Bit 800-MHz 19-mW CMOS ADC," to be presented at *Symposium on VLSI Circuits,* Kyoto, June 2013.

[3] A. Ghilioni et al, "A 4.8mW Inductorless CMOS Frequency Divider-by-4 with more than 60% Fractional Bandwidth up to 70 GHz," *Proc. CICC,* September 2012.

[4] J. Savoj and B. Razavi, "A 10-Gb/s CMOS clock and data recovery circuit with a half-rate linear phase detector," *IEEE J. Solid-State Circuits,* vol. 36, pp. 761-768, May 2001.

[5] C. Kromer et al., "A 25-Gb/s CDR in 90-nm CMOS for high-density interconnects," *IEEE J. Solid-State Circuits,* vol. 41, pp. 2921-2929, Dec. 2006.

[6] K. Yu and J. Lee, "A 2x25-Gb/s receiver with 2:5 DMUX for 100-Gb/s Ethernet," *IEEE J. Solid-State Circuits,* vol. 45, pp. 2421-2432, Nov. 2010.

[7] B. Verbruggen et al, "A 2.6mW 6b 2.2GS/s 4-times interleaved fully dynamic pipelined ADC in 40nm digital CMOS," *ISSCC Dig. Tech. papers,* pp. 296-297, Feb. 2010.

[8] A. Verma and B. Razavi, "A 10-Bit 500-MS/s 55-mW CMOS ADC," *IEEE J. of Solid-State Circuits,* vol. 44, pp. 3039-3050, Nov. 2009.

Design for Nanoscale Patterning

CICC 2013

Puneet Gupta
http://nanocad.ee.ucla.edu

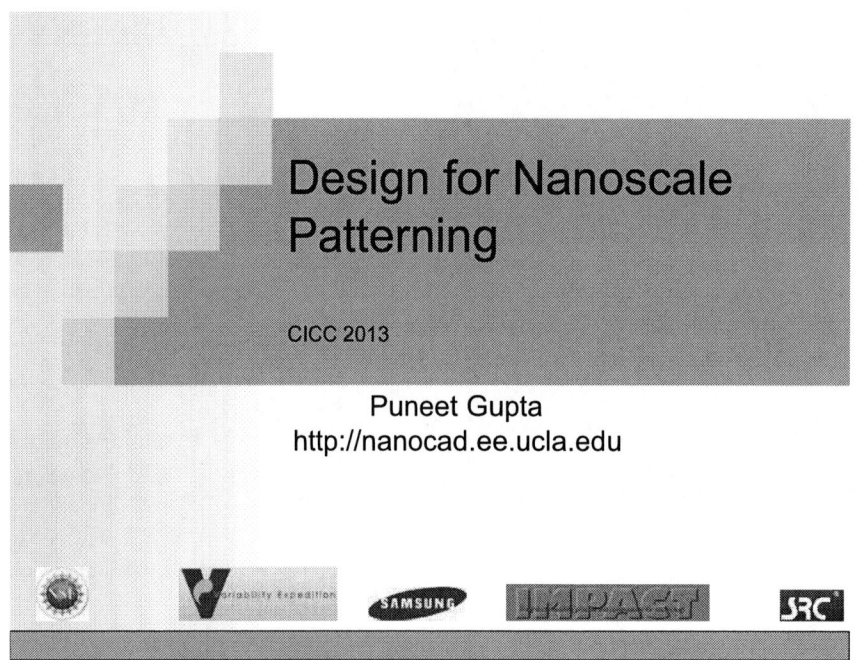

Design - Technology Co-Pathfinding

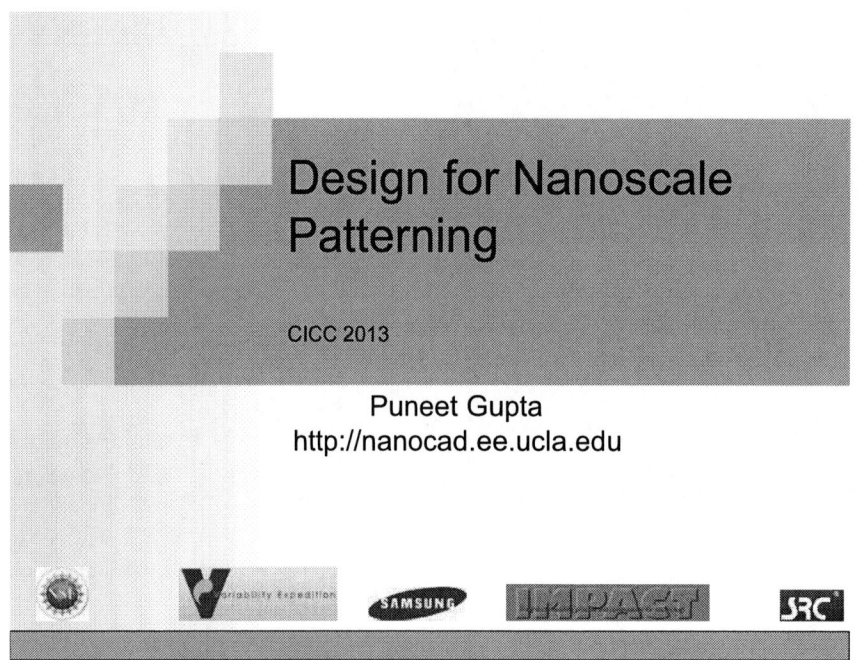

- *No systematic way to co-evaluate technology and design* → misdirected R&D of technology
  - Current state of art: either development in silos or manual, long loops

NanoCAD Lab, UCLA

Facets of Design-Patterning Interactions

[DATE'13]

- *Design-driven technology development*
  - *Exploration of manufacturable design rules/restrictions*
  - *Device/process changes to enable shrink, power/performance improvements*

- *Design-enablement of candidate technology*
  - *Changes in design tools/methodologies for manufacturable layout with new patterning*

- Design-aware manufacturing process control
  - Electrcial evaluation of lithographic control. E.g., electrical driven OPC, design-aware mask inspection, etc

---

Design Rules: The Traditional Interface

- Set of complex 1D or 2D layout restrictions
  - Modern DRs are rarely pure λ rules
  - Few 1000 DRs in modern processes
- Sources of DRs
  - Resolution
    - Width and spacing of lines on one layer
  - Overlay
    - make sure interacting layers overlap (or don't)
      - Contact enclosure
      - Poly overhang of diffusion
  - CMP
    - Min/Max metal density
  - Antennas, reliability .....
  - Example: Line End Extension Rule
    - Overlay margin
    - Line-end pullback coming from litho and etch processes

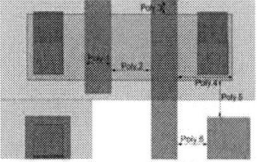

- How are DRs generated
  - Lots of process simulation and silicon experiments
  - Learning and scaling down from previous technology generations
  - Every new technology node/process adds new kind of DRs coming from changes in the process

Inherent Yield Assumption in DR Methodology

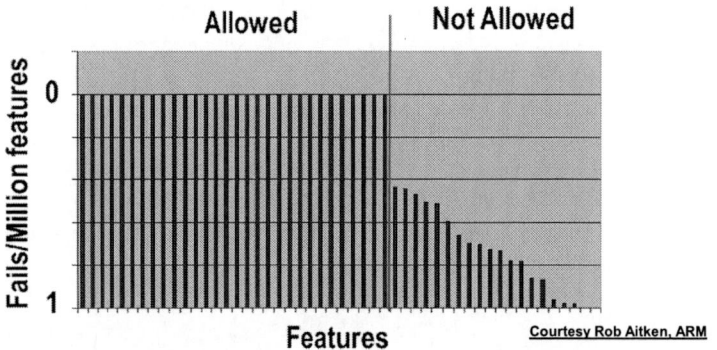

Courtesy Rob Aitken, ARM

- A sharp falloff in yield → a nice cutoff which can be used as the DR value

The Reality → Recommended Rule Mess

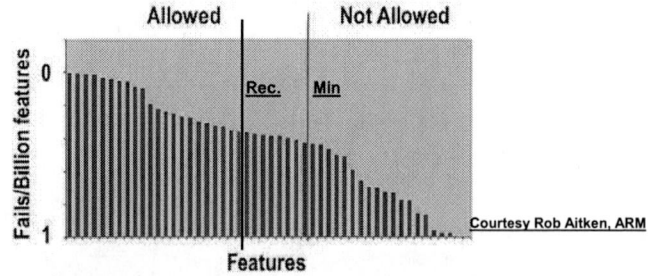

Courtesy Rob Aitken, ARM

- No nice cutoff available → have two cutoffs: minimum and recommended
  - Unfortunately the yield benefit of going to recommended is rarely specified though the area loss is obvious to the designer → recommended rules are used pretty much on a space-available basis

Optimizing DRs: DRE Pathfinding Framework

[ICCAD'09, ICCAD'12, SPIE'12, TCAD'12, SPIE'13, JM3'13]

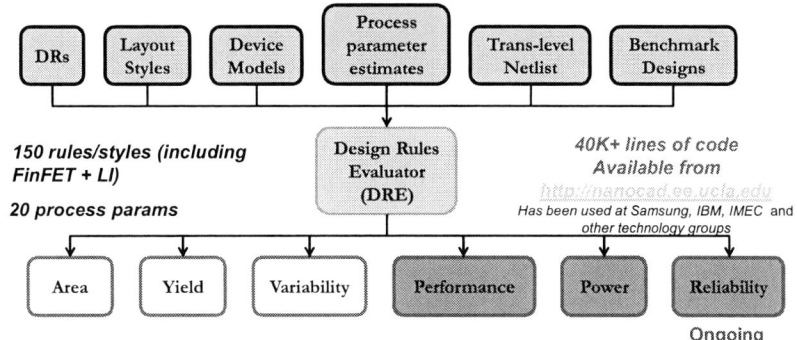

- DRE enables systematic "What-if" studies in minutes not months
  - *Better technology choices*
  - *Better layout practices*
  - *Reduced time and cost of technology development*

DRE in Technology Selection

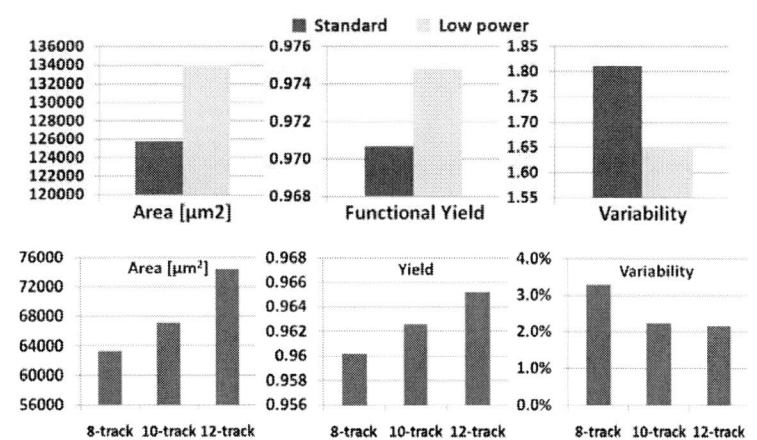

- Compare processes from different foundries
- Evaluate layout styles, cell architectures, recommended rules

978-1-4673-6145-3/13 $31.00 © 2013 IEEE

DRE in Optimizing Layout Styles

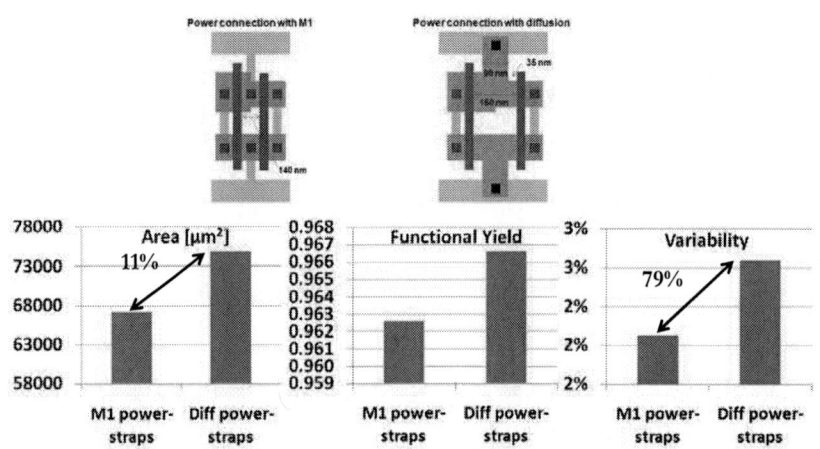

- Many such optimizations are neither "obvious" nor general
  - Results depend on layout, library, process

CURRENT LITHOGRAPHY

NanoCAD Lab, UCLA

Lithographic Process

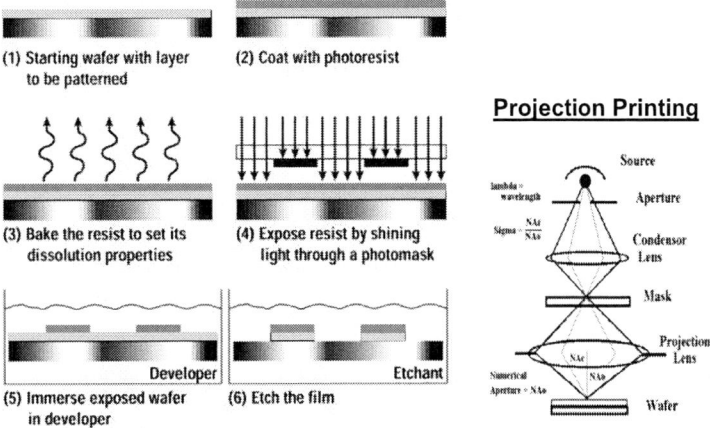

(1) Starting wafer with layer to be patterned

(2) Coat with photoresist

(3) Bake the resist to set its dissolution properties

(4) Expose resist by shining light through a photomask

(5) Immerse exposed wafer in developer

(6) Etch the film

Projection Printing

Courtesy EE143/Costas Spanos - UCB

Lithography Primer

- Scanner= Illumination equipment
  - Wafer is "stepped" under the mask to print, one field at a time
    - Scan the field by exposing a small "slit" at a time
    - Maximum field size limits maximum chip size (26mm x 33mm)
  - Just the lens weighs ~ 1000 pounds !
- The famous Raleigh Equation:
- Exposure = the amount of light or other radiant energy received per unit area of sensitized material.
- Depth of Focus (DOF) = a deviation from a defined reference plane wherein the required resolution for photolithography is still achievable. (affects 3D resist)
- Process Window = Exposure Latitude vs. DOF plot for given CD tolerance

$$Resolution = k_1 \frac{\lambda}{NA}$$

λ: Wavelength of the exposure system
NA: Numerical Aperture (measure of the size of the lens system)
k_1: process dependent adjustment factor

Diffraction at the Mask

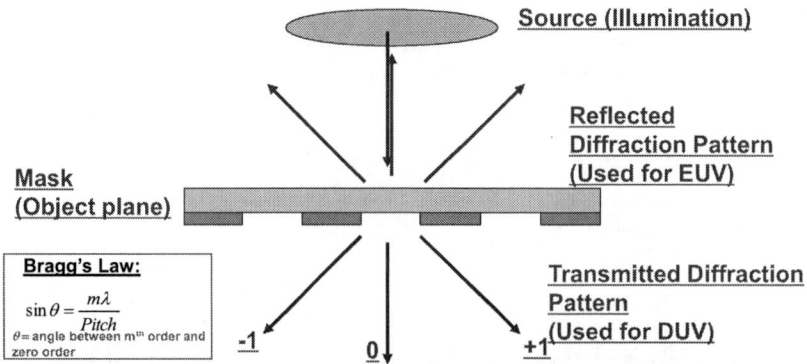

Bragg's Law:

$$\sin\theta = \frac{m\lambda}{Pitch}$$

θ = angle between m^{th} order and zero order

- Interaction between the illumination and the mask generates a diffraction pattern.
- For periodic structures, the diffraction pattern is discrete plane waves (or orders)
- Need at least two orders to pass through the lens to form the image

Courtesy Andrew Neureuther - UCB

Bigger Lenses: Nothing comes for free!

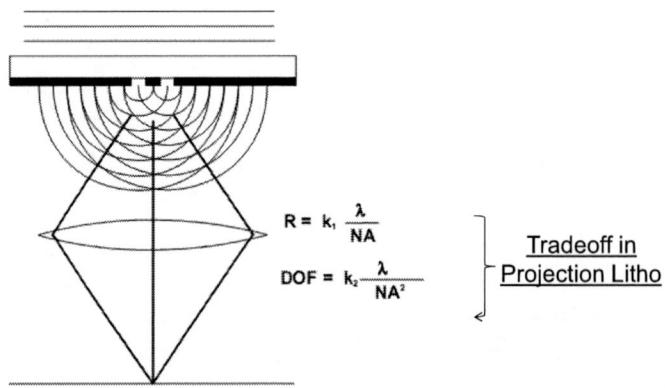

$$R = k_1 \frac{\lambda}{NA}$$

$$DOF = k_2 \frac{\lambda}{NA^2}$$

Tradeoff in Projection Litho

Courtesy Lars Liebmann, IBM

Resolution Enhancement (RET)

- The light interacting with the mask is a wave
- RET is wavefront engineering to enhance lithography by controlling fundamental wave properties

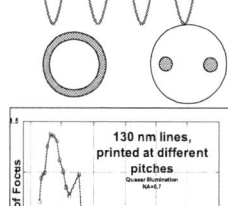

  - Wavelength (λ)
    - $\rightarrow$ tough (requires complete retooling)
  - Direction: Off-Axis Ilumination
    - "Shift" the diffraction orders to improve printing of certain pitches, while compromising others
    - "forbidden pitches" for layouts
  - Amplitude: Optical Proximity Correction (OPC)
    - Bias or add features to tweak the diffraction pattern
    - Mask no longer looks like the actual taped out design
    - Increases the cost of the mask ~10X
    - Model-based and OPC compute costs 1000s of CPU hours for typical modern designs
  - Phase: Attenuated or Alternating Phase Shift Mask

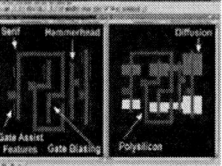

Courtesy F. Schellenberg, Mentor Graphics Corp.

ELECTRICAL MODELING OF IMPERFECT LITHOGRAPHIC PATTERNING

Lithographic WYSIWYG Breakdown

[VLSID'10]

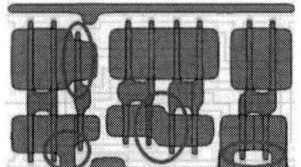

What designer sees   What silicon shows

- Existing compact device models (e.g., BSIM) do not handle non-rectangular geometries.
- Device models for shape imperfections :
  - Polysilicon gate shape contours
  - Diffusion rounding
  - Line-end shortening : gate not completely formed
  - Line-end rounding : "tapering", "necking" or "bulging"

Is Interconnect Modeling Important ?

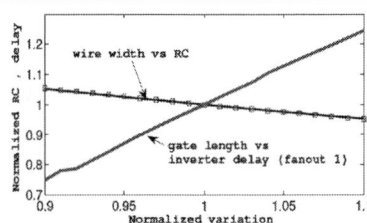

- Probably not..
  - Litho impacts wire width↑↓ (w)
    - w↑ → R_{wire}↓, C_g↑, C_c↑
  - Wires are *long* → averaging effects
  - Semi-global and global wiring (M3+) is wide and regular → patterning less of an issue
  - M1/M2 impact on power/performance is small
  - Caveat: contacts (and via) R variation may be non-negligible

Simulation at Chip-Level

- Delay and switching power <3%.
- Impact of wire variation is exaggerated as averaging effect is ignored. → *Let us concentrate on devices*

Interconnect layers (variation)	Δ delay (%)	Δ Switching power (%)
M2 (+10%)	0.89	1.46
M2 (-10%)	-0.75	-0.69
M3 (+10%)	1.90	2.83
M3 (-10%)	-1.62	-1.85
M4 (+10%)	0.77	1.64
M4 (-10%)	-0.65	-0.84
M5 (+10%)	0.08	0.50
M5 (-10%)	-0.07	0.13
M6 (+10%)	0.22	0.65
M6 (-10%)	-0.19	0.00

Total gates=43K Total area=$0.2mm^2$

FreePDK 45nm process

Polysilicon Rounding Model

[SPIE'06]

- Line-edge roughness and poly **rounding lead to NRG transistor**

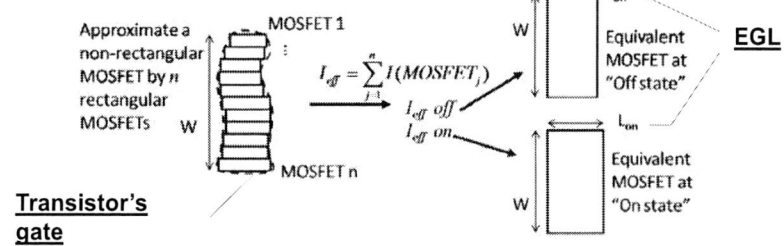

Transistor's gate

- Equivalent gate length (EGL) can be used to represent the current behavior of the transistor to communicate to SPICE
- Transistor "slice" currents can depend on distance from edge (location-dependent Vth)

Compact Model for Circuit Simulation

- EGLs depend on transistor working states
  - EGLs are extracted at $|V_{gs}| = 0$ and $|V_{gs}| = V_{dd}$ for leakage and timing analysis, respectively
- Alternatives :
  - Model a transistor by multiple smaller transistors connected in parallel
    - Accurate but number of transistors increases

    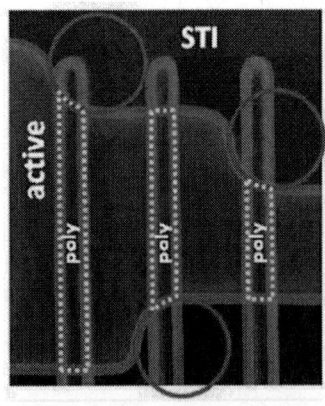

  - Fit one L_{eff} and V_{th} for I_{on} and I_{off}
  - Express gate length as a function of V_{gs} in device's model (e.g., BSIM)
  - Model the impact of gate length variation using voltage dependent current source

---

Its not only "L": Diffusion Rounding

[ASPDAC'08, VLSID'10]

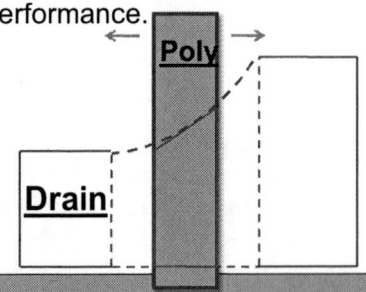

Victor Moroz, Munkang C. & Xi-Wei Lin SPIE 2009

- Diffusion rounding occurs due to printing imperfection.
  - **Diffusion routing**
  - Pwr/Gnd connections
- Modeled as trapezoid gate to investigate electrical performance.

Developing a Physical Diffusion+Poly Rounding Model

- To capture two dimensional E field, slice channel according to its distribution
  - For each slice, $L_{eff-i} = L_i$
- Effective width is derived using gradual channel approximation :

$$W_{eff-i} = \frac{(W_{s-i} - W_{d-i})}{\ln(W_{s-i}/W_{d-i})}$$

- V_{th} varies due to NWE and asymmetry between source and drain
- Using charge sharing model:

$$\Delta V_{th\text{-effective}} = \Delta V_{th-\text{Narrow width}} + \Delta V_{th-CS}$$

$$\Delta V_{th-CS} = \frac{qN_a W_c}{2LC_{ox}} \left[\frac{2(L_d W_d + L_s W_s)}{W_d + W_s} - (L_d + L_s) \right]$$

- Each slice is trapezoidal with equivalent (rectangular) L,W and V_{th}

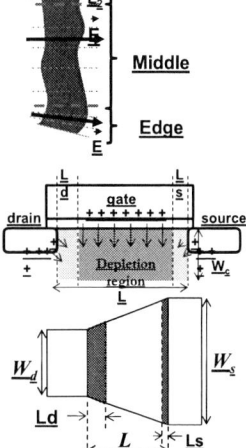

TCAD vs Model (Diffusion Rounding only)

- Asymmetrical I_{on}/I_{off} when rounding happens at Drain/Source terminals
  - ΔVth varies according to drain/source ratio
  - Behavior is *NOT* symmetric w.r.t source/drain

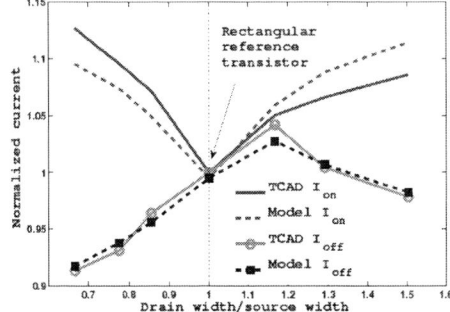

Application to Logic Cells

		NAND_X1		NOR_X1	
		Original	Spacing Reduced	Original	Spacing Reduced
Delay	nominal (no defocus)	1.00	1.00	1.00	0.99
	worst (100nm defocus)	1.05	1.04	1.05	1.05
Leakage	nominal (no defocus)	1.00	1.00	1.00	1.01
	worst (100nm defocus)	0.91	0.91	0.90	0.90
	area	1.00	0.95	1.00	0.95

- At 100nm defocus
  - Δ Delay   = 5%
  - Δ Leakage = 9%
- Design rule can be optimized.

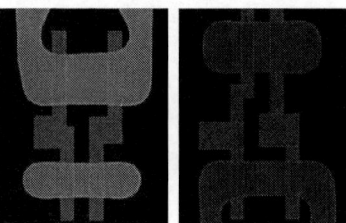

NAND2_X1   NOR2_X1

Electrical Impact of Line-End Imperfections

[JM3'10]

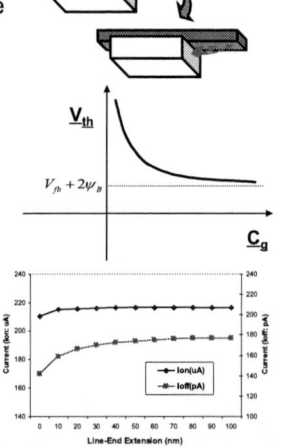

- **LEE vs. Capacitance**

Line-end extension increases C_g because there exists fringe capacitance between line-end extension and channel.

Increasing LEE

- **Capacitance vs. V_{th}**

C_g affects V_{th}, **narrow width effect**
  - C_g increases $\rightarrow$ V_{th} decreases
  - C_g decreases $\rightarrow$ V_{th} increases

$\underline{V_{th}}$

$V_{fb} + 2\psi_B$

$\underline{C_g}$

- **V_{th} vs. Current**

I_{on} and I_{off} are functions of V_{th}
  - V_{th} increases $\rightarrow$ I_{on}, I_{off} decrease
  - V_{th} decreases $\rightarrow$ I_{on}, I_{off} increase

Misalignment Model

- There exists misalignment error between polysilicon and active layer
- Overlap region (=actual channel) can vary according to misalignment error
  - Increased linewidth variation
- Misalignment has a probability, P(m)

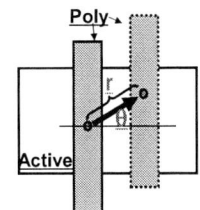

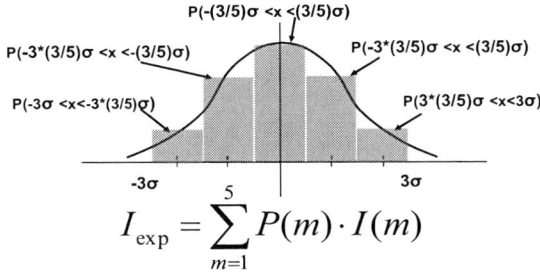

$P(-(3/5)\sigma < x < (3/5)\sigma)$

$P(-3*(3/5)\sigma < x < -(3/5)\sigma)$

$P(-3*(3/5)\sigma < x < (3/5)\sigma)$

$P(-3\sigma < x < -3*(3/5)\sigma)$

$P(3*(3/5)\sigma < x < 3\sigma)$

-3σ 3σ

$$I_{exp} = \sum_{m=1}^{5} P(m) \cdot I(m)$$

Optimizing Line-End of SRAM

<u>SRAM Bitcell Layout vs. Line-End Design Rule</u>

Width constraint graph:
- Longest path determines the width of a bitcell
- LEE(b) is common for all possible path

<u>(Line-End Length, Sharpness) vs. (Leakage, Area)</u>

Large *n* is better for leakage variation but it increases OPC and Mask costs.

<u>According to the taper shape, LEE design rule can be optimized to reduce bitcell size.</u>

Line-End Shortening (LES)

[DAC'07]

- Polysilicon does not cover active region completely
  - Sources: Misalignment and line-end pullback

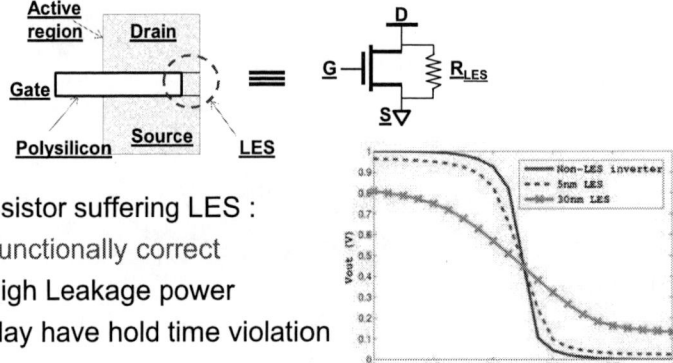

- Transistor suffering LES :
  - Functionally correct
  - High Leakage power
  - May have hold time violation

Design Flow Integration

- Full-custom/Analog designs
  - SPICE or SPICE-like analyses flows
  - Weq, Leq per transistor is sufficient
- Cell-based digital designs
  - Static analysis flows based on standard cell abstraction
    - One cell is 2-100 transistors
    - Timing/power views stored in pre-characterized ".lib" files
  - State of art 45nm logic designs have 10M+ cells and 50M+ transistors →Hierarchy preservation essential
    - One way: create multiple cell "variants" corresponding to different printed contours

DOUBLE PATTERNING AND DESIGN

NanoCAD Lab, UCLA

Double Patterning Lithography (DPL)

- The main form of double patterning requires two lithography steps with two different masks
  - Interleaved exposures → "doubled" resolution
- Will be used at 22nm and below.

MASK 2

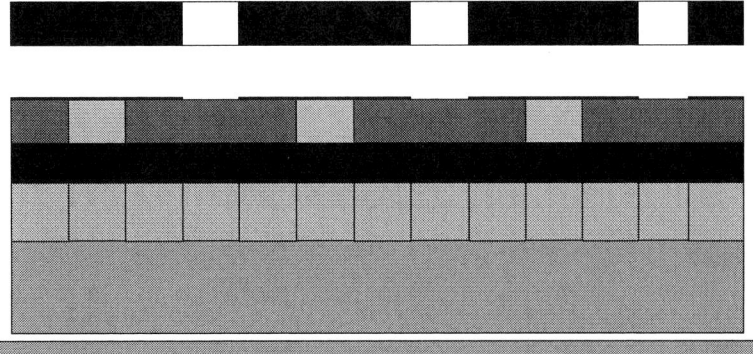

LELE Double Patterning

Layout	
Decomposed Layout (aka colored layout)	
Mask 1 (bright field)	
Resist Expose 1	
Hard-mask Etch 1	
Mask 2 (bright field)	
Resist Expose 2	
Hard-mask Etch 2	
Final Substrate Etch	

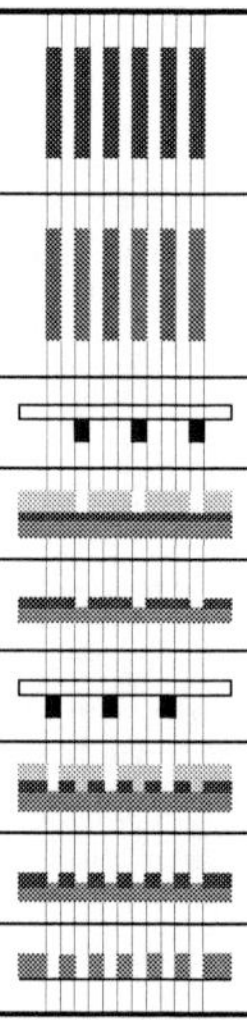

Challenges

Two independent exposures:

• overlay error affects space

• dimensional variation can cause bi-modal distribution

Courtesy Andres Torres

Self-Aligned Double Patterning (SADP)

Layout	
Decomposed Layout (aka colored layout)	
Mask 1 (mandrel mask)	
Resist Expose 1 (mandrel)	
Sidewall Spacer Deposition	
Mandrel Removal	
Mask 2 (block or trim mask)	
Resist Expose 2 (block)	
Final Substrate Etch	

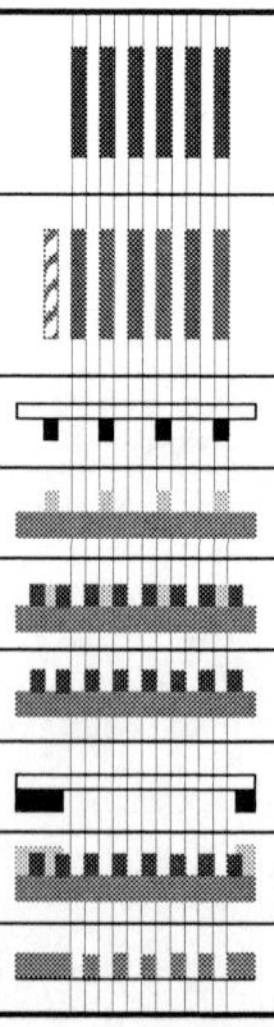

Challenges

• Sidewall spacer comes in one dimension only, either fixed space or fixed feature width

• Width variation on 'mandrel' causes 'pitch clustering'

Courtesy Andres Torres

Options Galore

- Litho-Etch-Litho-Etch (LELE) = Double patterning
- Spacer litho = Sidewall image transfer (SIT) = Self-aligned double patterning (SADP)
- All methods have "positive" and "negative" flavors
  - i.e., printing "lines" vs. "spaces"
- Option of using an extra trim or block mask
  - Required for SADP
- LELELE (Triple patterning) is already under serious consideration

Multi (Double)-Patterning Challenges

- Overlay control
- Layout decomposition into two patterns
  - Depends on what exposure system is used
- Two exposure + etch →decreased throughput
- Increased mask cost
- The bimodal problem
- Possibly more weird layout constraints

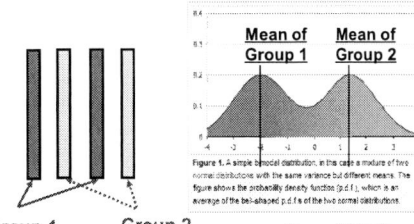

Group 1   Group 2

Source: Wikipedia

Figure 1. A simple bimodal distribution, in this case a mixture of two normal distributions with the same variance but different means. The figure shows the probability density function (p.d.f.), which is an average of the bell-shaped p.d.f.s of the two normal distributions.

DPL Layout Decomposition

- Two features assigned opposite colors if their spacing is less than the minimum coloring spacing

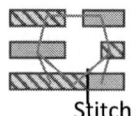

- If two features within minimum coloring spacing cannot be assigned different colors
  - THEN at least one feature must be split into two or more parts → stitch

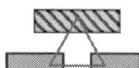

Stitch

- Stitches = possible locations for "pinching" due to rounding of line-end
  - No stitching in SADP
- Individual exposures themselves have to be manufacturable
- Many layouts are *not* decomposable
  - Odd cycle of conflict relationships

Ensuring DP Compliance

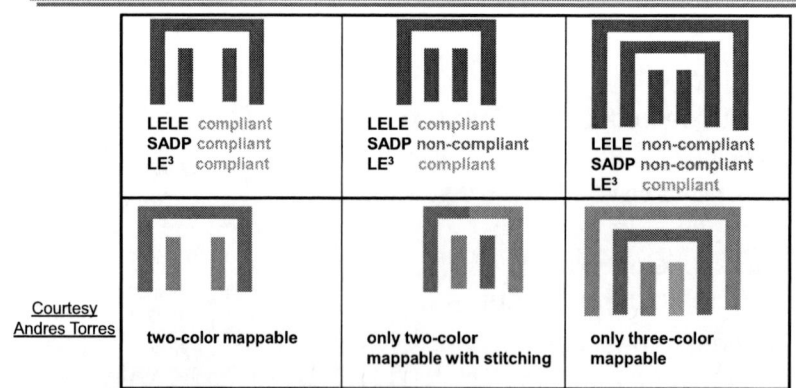

Courtesy Andres Torres

- Minimize conflicts during decomposition
- Conflict free standard cells during library migration
  - Area overhead within cells as well as between cells of the same row to resolve conflicts
  - Does not solve conflicts at boundary of cells from different rows
- Gridded layouts result in virtually no conflicts

DP Layout Decomposition

[ICCAD'11, TCAD'13]

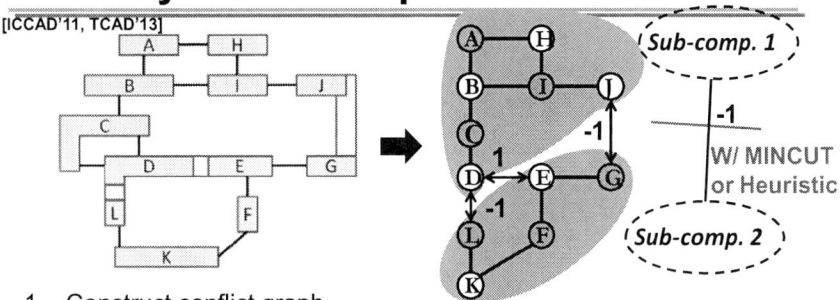

1. Construct conflict graph
2. Identify connected components & sub-components
3. Alternating coloring
4. Flip coloring of sub-components → minimize stitches
   - Optimal MINCUT partitioning or $O(n)$ heuristic
- Consider all candidate stitches → conflict-free solution if one exists

Compaction-Based Conflict Removal

[ICCAD'11, TCAD'13]

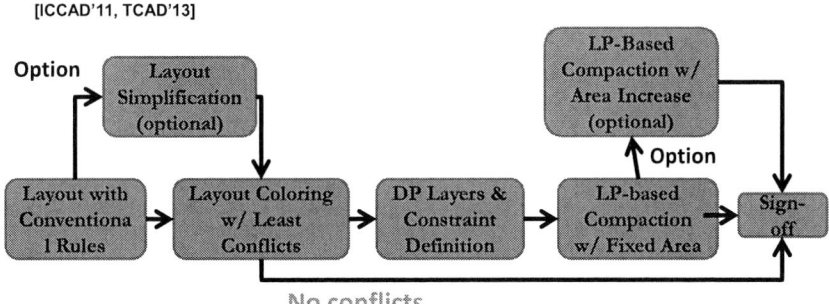

- Define individual DP layers – e.g., M1 → M1A, M1B
  - Individual layers to define DP constraints
  - Union (M1A, M1B) to define layer-to-layer constraints
- Rules & conflicts legalization **across all layers simultaneously**
  - Solving conflict on one layer does not create another elsewhere
  - No need for iterative loop of coloring + legalization

Sacrificing Layout Features When Necessary

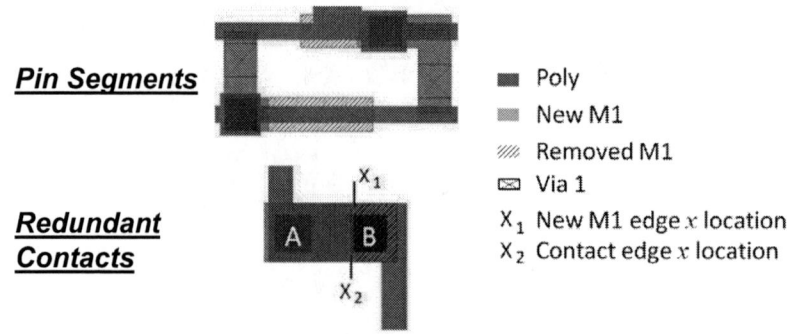

Pin Segments

Redundant Contacts

- Poly
- New M1
- Removed M1
- Via 1
- X_1 New M1 edge x location
- X_2 Contact edge x location

- Remove non-critical features *before* coloring to aid coloring
- **Add recommended constraints** to add the features back after legalization

Conflict Removal Results

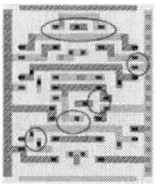 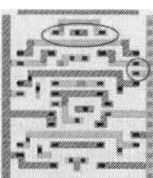 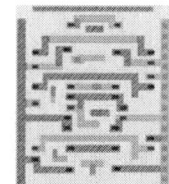

**Original
5 conflicts**

**Same area,
2 conflicts**

**No conflicts, 6% area
increase**

- Achieved **DP-compatible cells**
  - Simple cells → No area overhead
  - Complex cells/macros → Modest area overhead (at most 9%)

- Less than **1 min** in real time for largest macro (460 trans.)

Preferred Coloring & Layout Simplification Effects

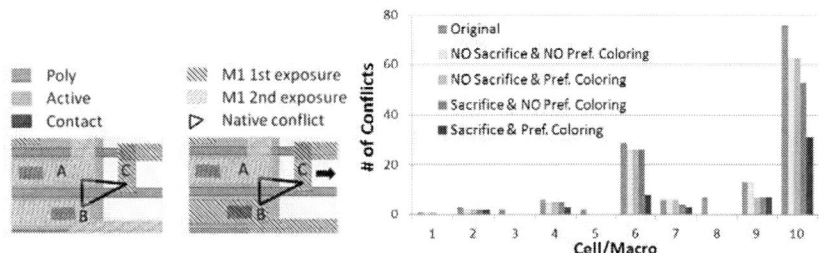

- Coloring of conflicts affects efficiency of conflict removal
- Give higher priority to vertical violations
- Efficient only when **both** methods are applied
  - 4X less conflicts for cells
  - 2X less conflicts for macros

Overlay Error Impact in DP

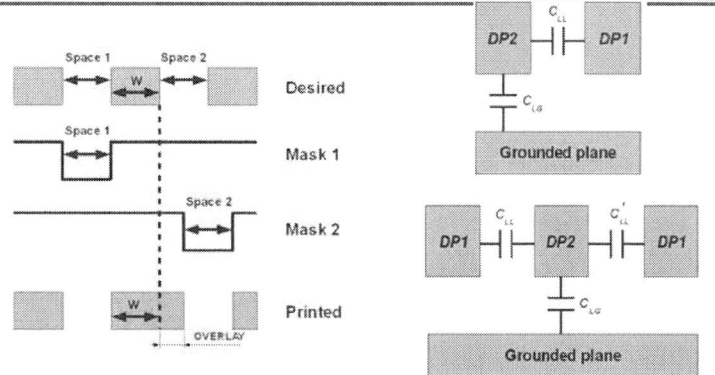

- Overlay translates into CD variation in DP
  - 3X tighter requirement
- Electrically evaluate overlay impact on design
  - Are things really bad ?

Overlay Impact in DP [SPIE'09, TSM'10]

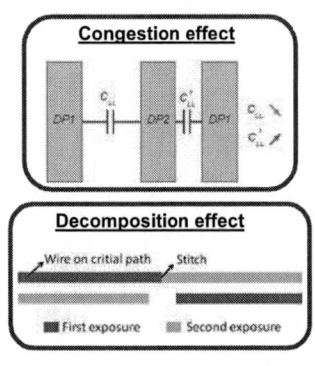

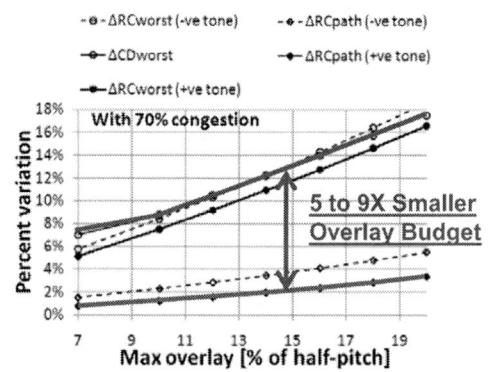

- Max worst-case ΔRC is only 6%, down to 3% with reduction effort

Bimodality Problem in DPL

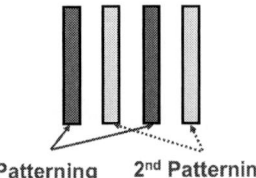

1st Patterning   2nd Patterning

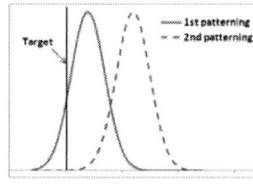

- Different exposure/etch steps → two CD populations
- Large CD/delay variability (e.g., 34% 3σ increase by ASML study)

$$3\sigma^2_{pooled} = \frac{3\sigma^2_{p1}}{2} + \frac{3\sigma^2_{p2}}{2} + \left(\frac{3}{2}\left|\mu_{p1} - \mu_{p2}\right|\right)^2$$

- Loss of spatial correlation
- Timing problems: clock skew and worse timing slack (e.g., 53ps and 46ps assuming 6nm CD difference [Jeong ASPDAC09])

WHAT LIES AHEAD

NanoCAD Lab, UCLA

Next Gen Litho Tool Requirements

- Looking into the future
  - Resolution < 14nm
  - CD uniformity < +/- 10%
  - Overlay ~ 20% (less for some flavors)
  - Throughput ~ 100 wafers/hour
- Strong NGL candidates
  - Multiple (2+) patterning
  - EUV Lithography
  - Ebeam direct write
  - Directed Self-Assembly

Extreme Ultraviolet Light Lithography

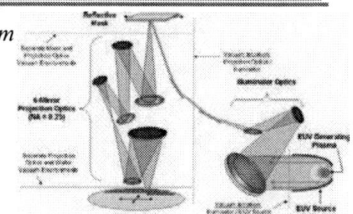

- Much shorter wavelength $\lambda = 11 - 13\ nm$
  - Promise of greatly improved resolution
- Expected to help with 10nm and beyond (always "n+1" tech.)
- Challenges
  - Creating defect-free mask blanks
    - Plan is to even inspect blank mask plates!
  - Generating EUV light
    - Very high powered EUV sources are needed for commercial production
  - Reflective optics problems
  - Need for improved resists

EUV uses *reflective* masks
Masks are 40 bilayers of MoSi

Massively Parallel E-Beam Direct Write

- EBDW: accurate printability and maskle
  - Can be a good option of prototyping/low
  - May need to use 100's of parallel beams to achieve acceptable throughput
- Challenges:
  - Data volumes
    - Every parallel write beam needs to have its own data channel
  - Increasing throughput
    - Issue with secondary electrons causing "blur" in the resist limit the energy of each beam

REBL from KLA-Tencor

Design Impact:
- Limited to resolution of the system
- Due to the scanning nature of the system better CD control along the scan direction.

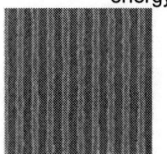

22nm L/S
Courtesy of CEA-LETI

22nm contacts

Courtesy Andres Torres and KLA-Tencor

Directed Self Assembly (DSA)

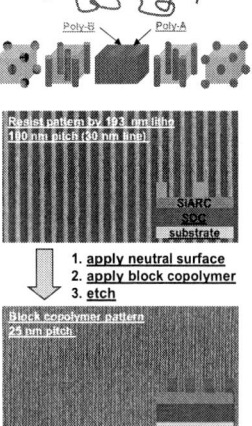

- Based on phase separation in block copolymers
- Benefits
  - Very high ultimate resolution << 10 nm
  - Can be implemented existing exposure tooling
  - Large multiples of frequency multiplication
- Remaining Challenges:
  - extensive materials and process development needed
  - CD control
  - Defects
- *Design Impact:*
  - Lines-Spaces: Uni-directional groupings
  - Contacts: Supports only single size circular contacts (no rectangles or any other form)

Conclusions

- Design and Lithography are co-developed

- Regularity (in all ways you can think of) is (almost) always helpful for patterning but can hurt density scaling

- Be prepared for increasingly unusual layout restrictions and electrical effects coming from lithography

NanoCAD Lab, UCLA

Low Power Chip & System Design for Biomedical Applications

Prof. Brian Otis
botis@uw.edu

University of Washington
Electrical Engineering
Seattle, WA

Goal of this talk...

o IC design and system building techniques that enable new wireless sensors

o A few case studies

o Challenge: thin-film integration of complex systems

Progression of impact

Research tools

Clinical/athletic

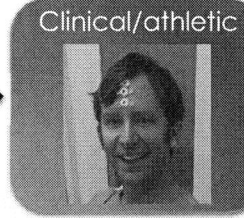

Mainstream

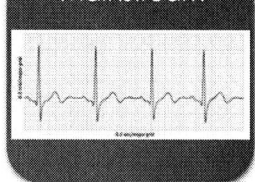

1. Enable new paradigm

1. Enable new paradigm
2. Regulatory (FDA, FCC)
3. Security

1. Enable new paradigm
2. Regulatory (FDA, FCC)
3. Security
4. Standardization
5. Reliability
6. Cost
7. User experience
8. Constant connectivity to web

Encounternet

Female Manakin, Costa Rica, 2011, Mennill et al.

- On-board storage of "encounter logs", periodically upload to basestations

- >10 years and millions of encounters logged

- ~1k on order from researchers

Encounternet: system anatomy

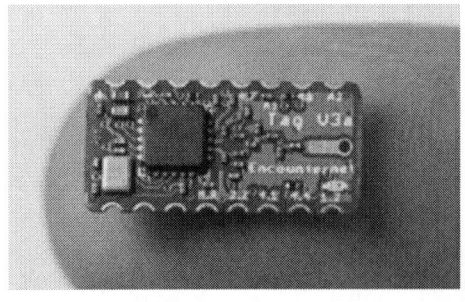

- Sub-gram Encounter logging
- Range: 20m
- Store & upload to basestation

Encounternet on Crows

- Goal: understand social network topologies

- Crow experiment (34 crows tagged)

- First week: 190,000 bird/basestation logs

- 28,000 unique encounters

Current Biology 22 (17), R669-R671.

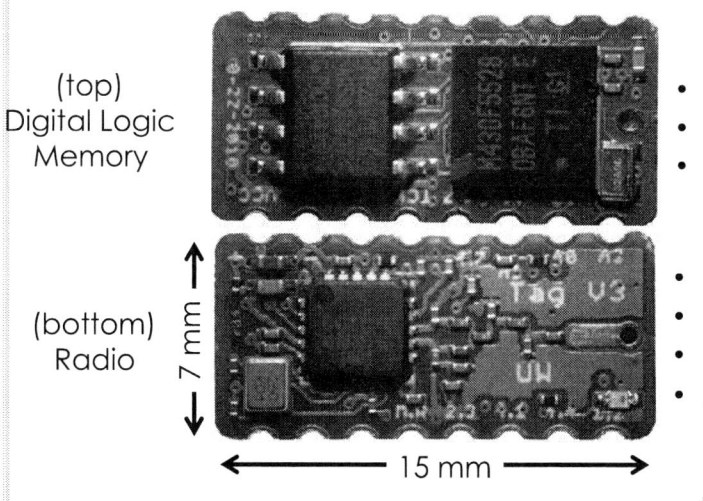

(top) Digital Logic Memory

(bottom) Radio

7 mm

15 mm

- Memory
- Microcontroller
- Watch Crystal

- RF Transceiver
- Matching network
- RF crystal
- Antenna connector

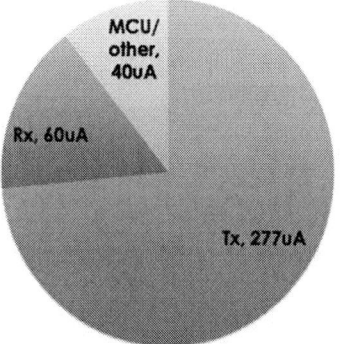

Animal work is a great
technology driver

It foreshadows really hard
emerging problems

Case study: wireless electrophysiology

Modality	Bandwidth	Amplitude	Spatial Resolution	Invasiveness
Single-Unit	.1-7 kHz	< 500 μV	0.2 mm	Invasive
Nerve cuff	< 5 kHz	< 10 μV	-	Mod. Inv.
LFP	< 200 Hz	< 5 mV	1 mm	Mod. Inv.
ECoG	.5-200 Hz	< 100 μV	5 mm	Mod. Inv.
EEG/evok	< 100 Hz	10-20 μV	30 mm	Non-Inv.
ECG	< 100 Hz	< 10 mV	-	Non-Inv.
EMG	< 1 kHz	< 10 mV	-	Non-Inv.
Power Line	50/60 Hz	10's mV	-	Pervasive

Tissue interface power consumption

PERFORMANCE COMPARISON OF BIOPOTENTIAL AMPLIFIERS

	BPA1	BPA2	BPA3	[8]	[25]	[12]	[13]
Vdd (V)	1	1	1	+/-2.5	1.8-3.3	2.8	0.8-1.5
I_{Amp} (μA)	12.5	0.8	12.1	16	1.2	2.7	0.33
NEF	4.5	1.9	2.9	4.0	4.9	2.67	3.8
NEF $^2 \cdot$ Vdd	20.3	3.6	8.4	80	43.2	20	11.6
Gain (dB)	40.5	36	40	39.5	45.5	30.8	40.2
1 dB comp.(@ Input) (mV)	3	1.7	4	—	—	—	—
$v_{ni,RMS}$ (μV)	3.2	3.6	2.2	2.2	0.93	3.06	2.7
PSRR (dB)	≥ 60	5.5	≥ 80	> 85	—	75	62-63
Bandwidth (Hz)	.4-8.5k	.3-4.7k	.05-10.5k	025-7.2k	.5-180	45-5.3k	3m-245
Area (mm^2)	.047	.046	.072	.16	—	.16	1
Technology (μm)	.13	.5	.13	1.5	0.8	.5	.35

<10uW for a general-purpose electrophysiological interface amplifier

ADC power consumption

Fundamental limit of sampled system

$$E_{min} = \frac{P_{min}}{f_{snyq}} = 8kT \cdot SNR = 8kT \cdot 10^{(6.02B+1.76)/10}$$

Target 10 bits at 10 kSamples-per-second
State-of-the-art power ~ 50 nW
Practical target for industry SoC applications ~ 500 nW

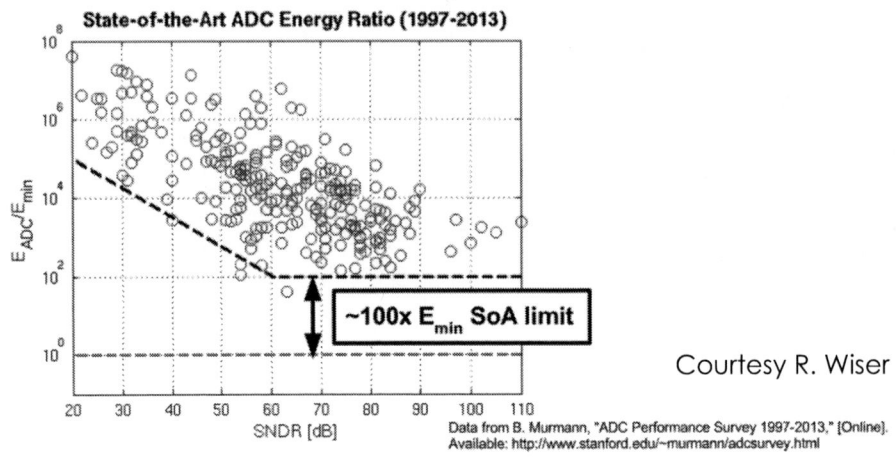

Courtesy R. Wiser

Need to work on wireless

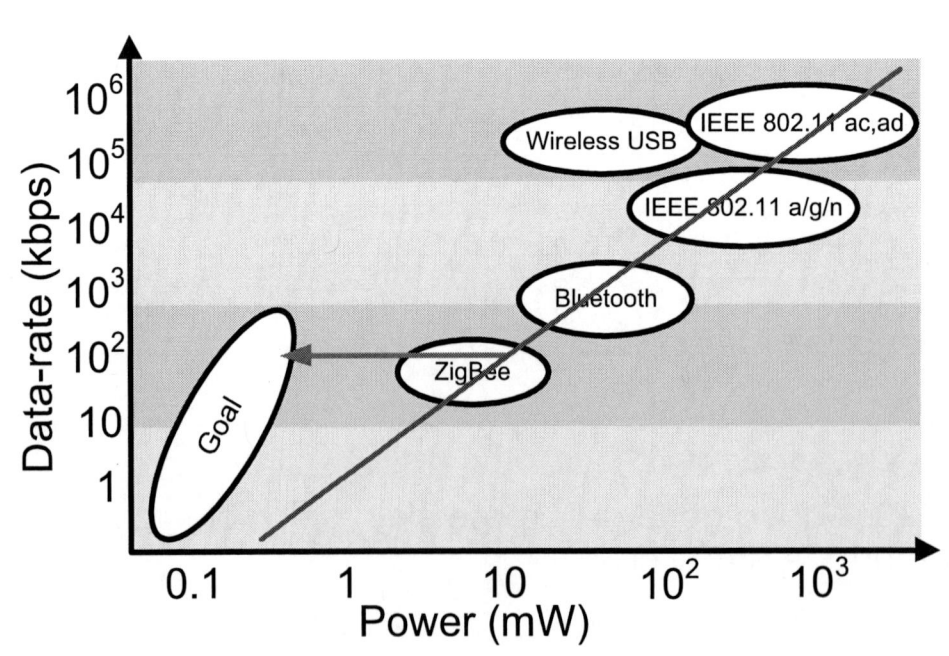

ISM/MICS-band transmitter (400 MHz)
1st gen: 500uW
2nd gen: 100uW

Link demonstration

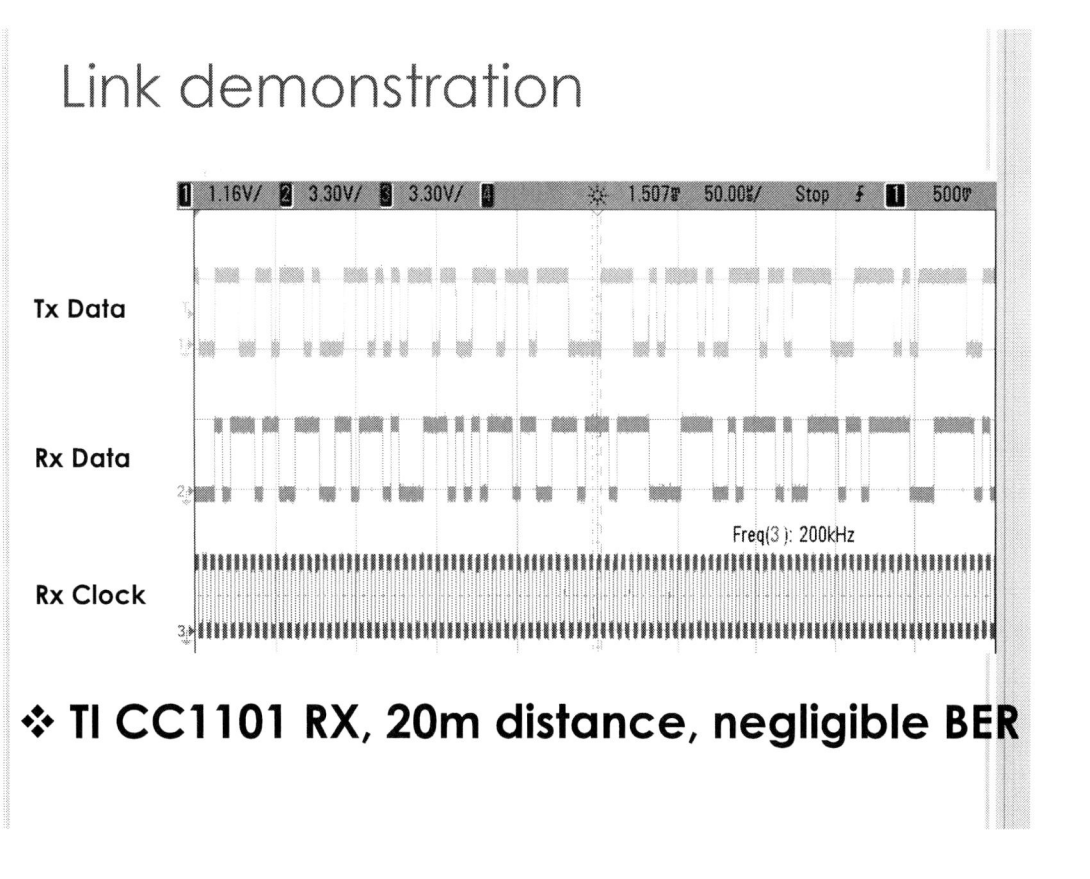

❖ **TI CC1101 RX, 20m distance, negligible BER**

The Bumblebee: 3rd gen

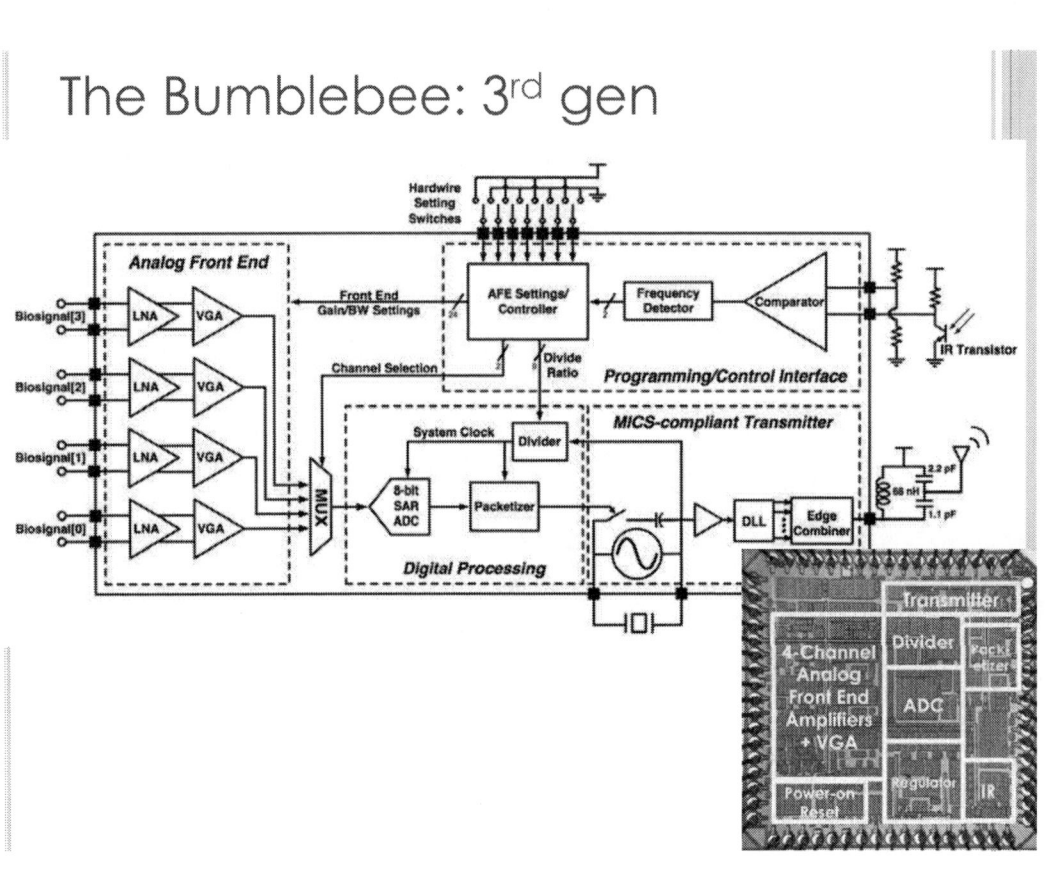

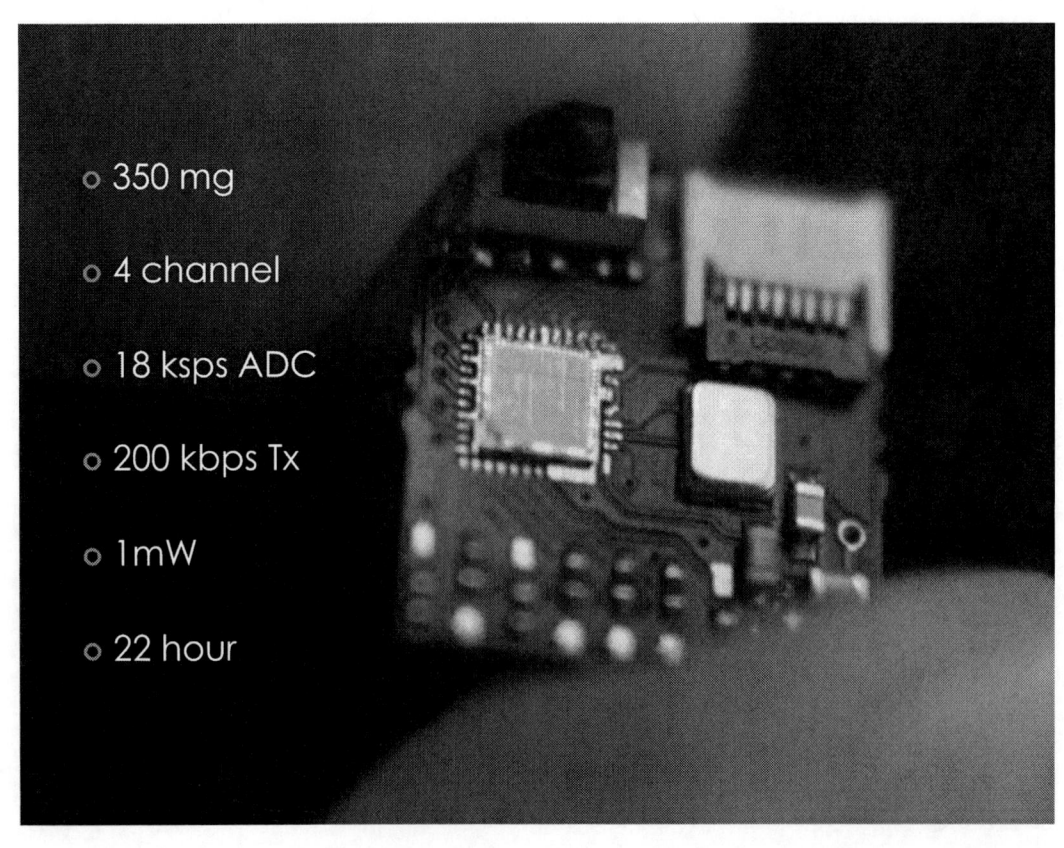

- 350 mg
- 4 channel
- 18 ksps ADC
- 200 kbps Tx
- 1mW
- 22 hour

Access point

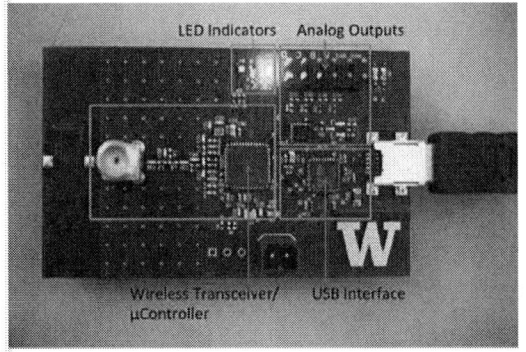

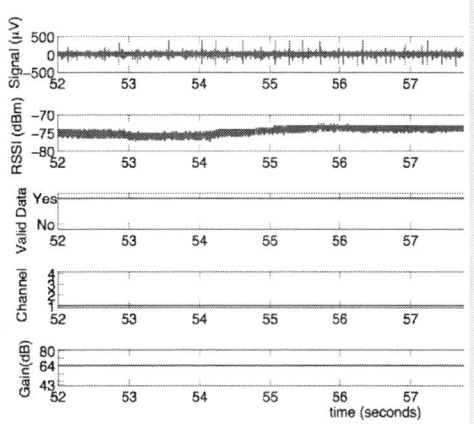

- USB/PC interface
- Analog out for audio spike observation
- LEDs indicate signal strength
- Indoor range ~24 meters

Human ECG experiment

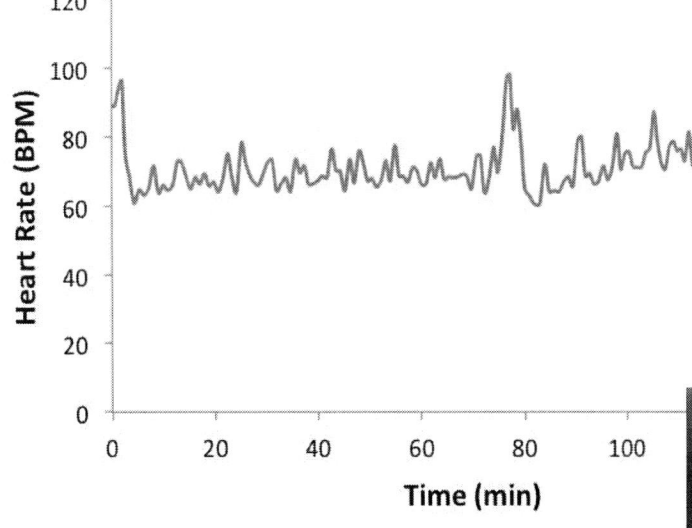

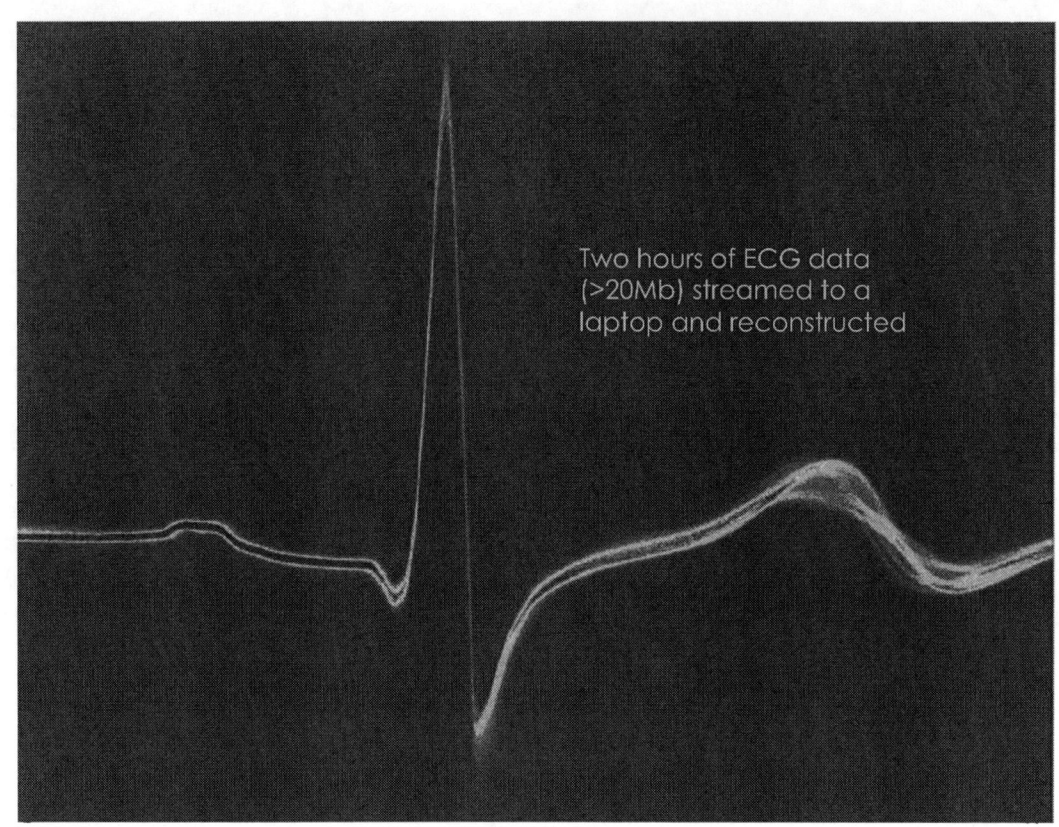

Two hours of ECG data (>20Mb) streamed to a laptop and reconstructed

Sensory-evoked spikes from auditory cortex
(collaboration with the Allen Institute for Brain Sciences)

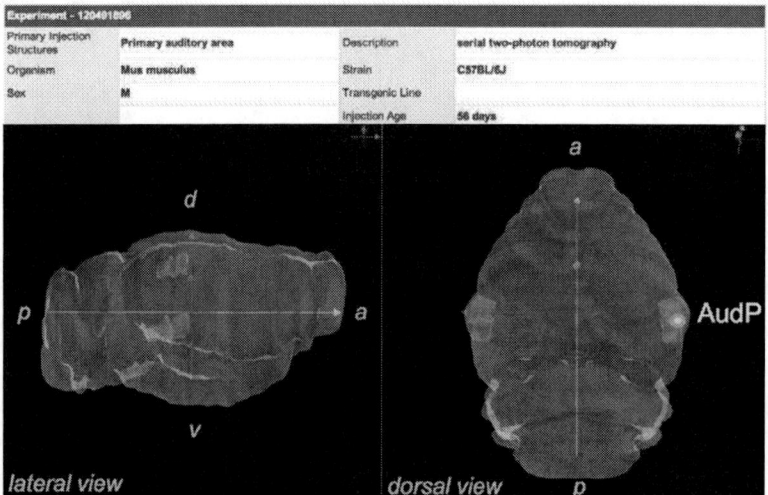

A tungsten wire electrode was implanted in layer IV/V of auditory cortex. To parse receptive fields, pure tones ranging in frequency and intensity were administered to an awake, behaving animal. Duration of tones was 35 ms (5 ms onset) with a 200 ms inter-tone interval, and the range was 1 to 32 kHz in 0.5 kHz increments, repeated 25 times.

http://connectivity.brain-map.org/projection/experiment/thumbnails/120491896

Sensory-evoked spikes from auditory cortex
(collaboration with the Allen Institute for Brain Sciences)

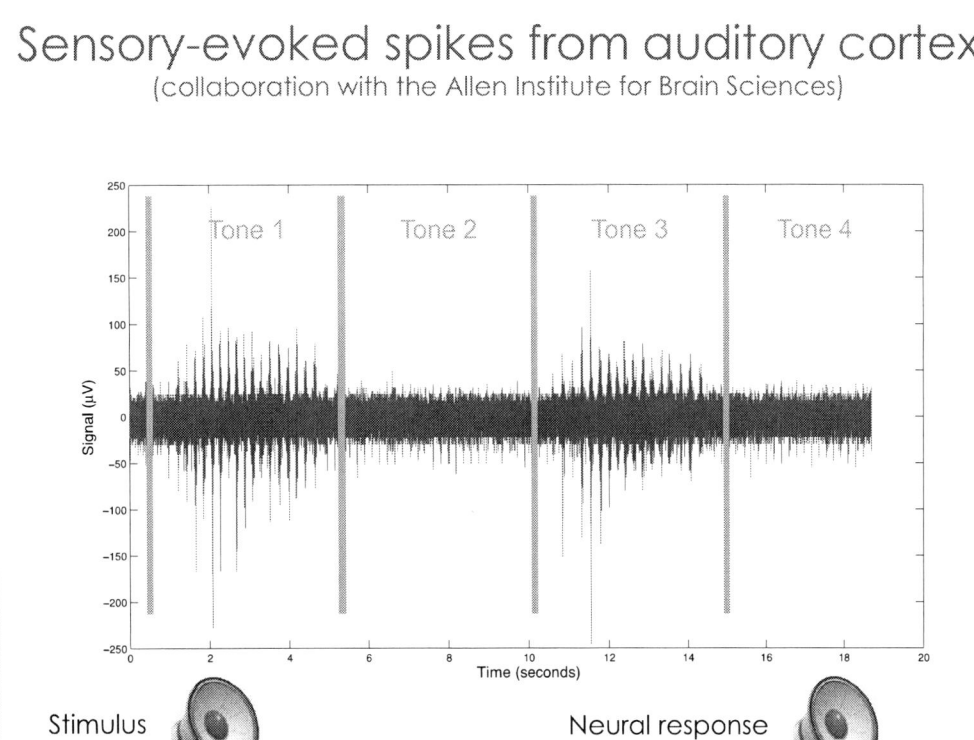

Stimulus   Neural response

Overnight recording session

o Long term *in vivo* recording performed overnight

o 11.5 hours of data collected

o Over 750 million samples received wirelessly (> 6 Gb of data)

o $1.4 * 10^{-7}$ % data loss due to dropped packets

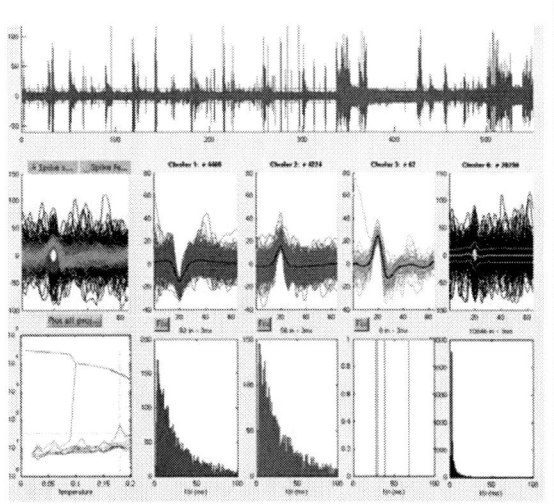

Still need a battery.

And a crystal.

Can we do better?

RFID to the rescue

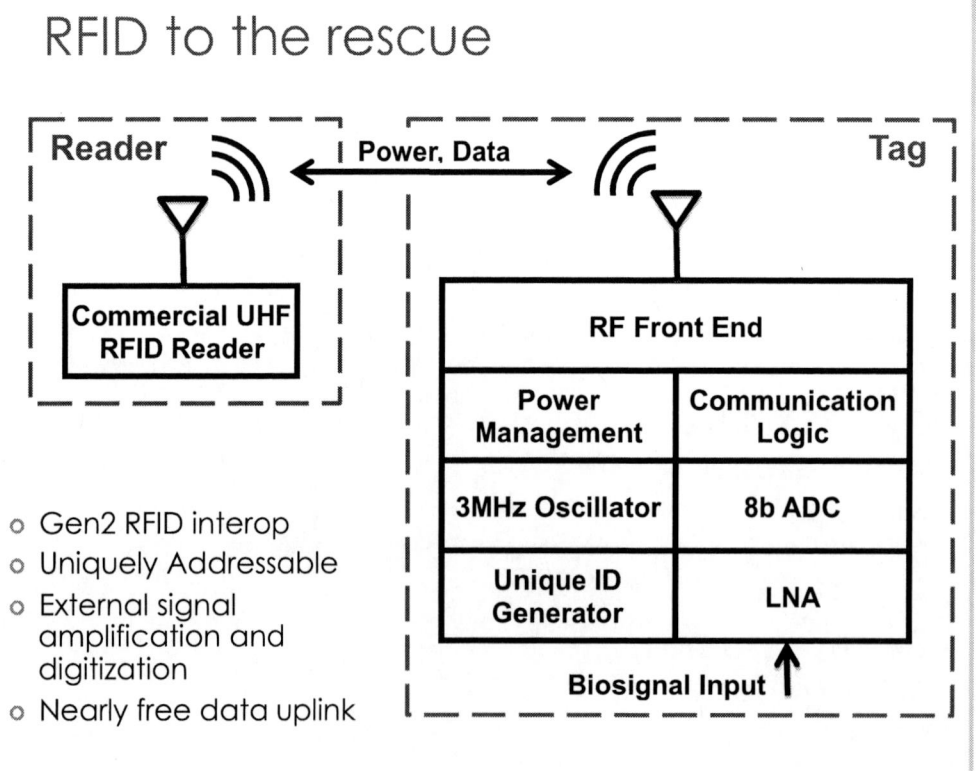

- Gen2 RFID interop
- Uniquely Addressable
- External signal amplification and digitization
- Nearly free data uplink

RFID-based wireless sensor

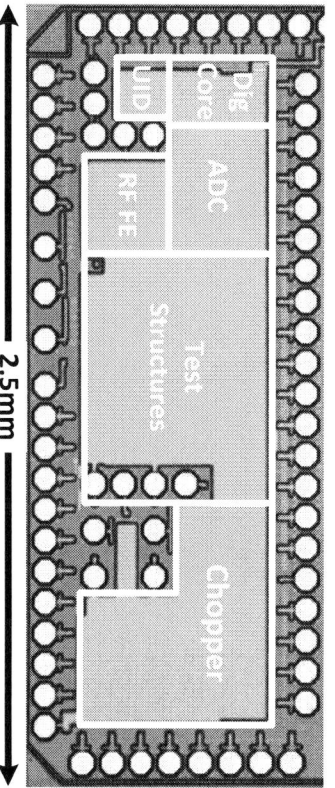

2.5mm

- -12 dBm sensitivity
  - 3m range

- RFID Gen2 Compatible
  - Talks to COTS Readers

- Zero trimming

- Demonstrates in-flight insect recording

Experimental setup

The in-flight temperature of the dominant flight muscles of Manduca (the dorsolongitudinal muscles: DLM).

The copper-constantan thermocouple was inserted into the DLM , approximately 3 mm below the dorsal aspect of the cuticle.

In-Flight Video

Courtesy Tom Daniel

Experimental results

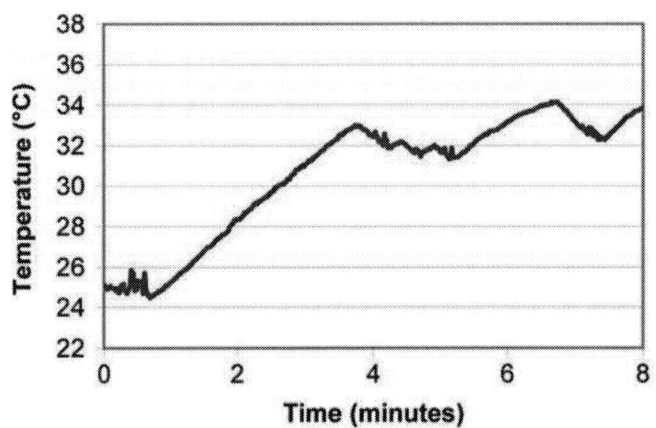

- Recorded from 1-2 m using an Impinj RFID reader

- Resting/cooling took place at 4m and 7m

Case study: glucose

1. 25.8M people in the US have diabetes (8.3% of population)

2. 79M adults in the are pre-diabetic (35% of population)

3. Immediate need for better glucose monitoring technologies for diabetics on insulin

4. Non-invasive continuous glucose monitoring would have a profound impact on diabetes prevention and treatment

Ref: Centers for Disease Control, http://www.cdc.gov/diabetes/pubs/pdf/ndfs_2011.pdf

Potential platform: the contact lens

- Maximum thickness <200μm

- Wireless power transfer

- Severe energy storage/ decap constraints

- Biocompatible integration of entire system

- Unobtrusive reader

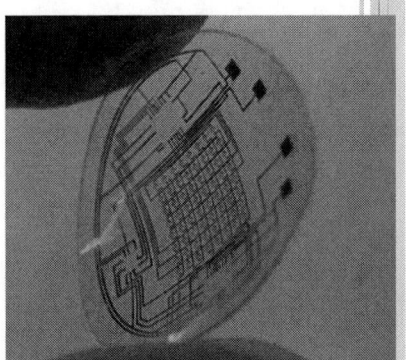

On-lens display driver chip

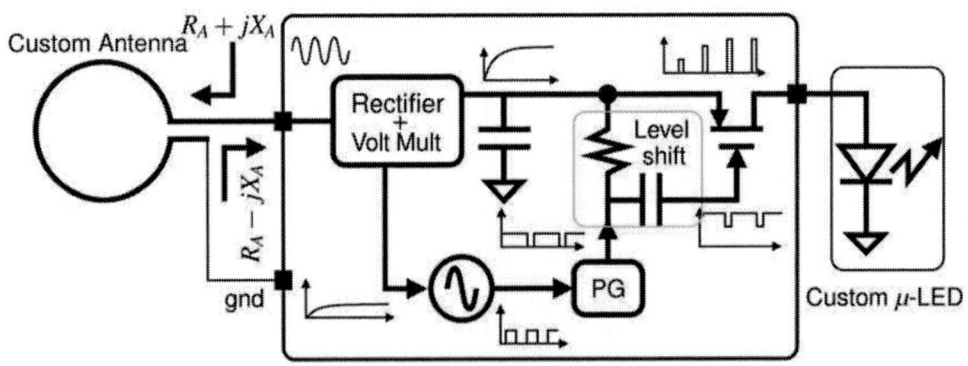

- Verify power transfer/comm feasibility
- Toward a display on-lens

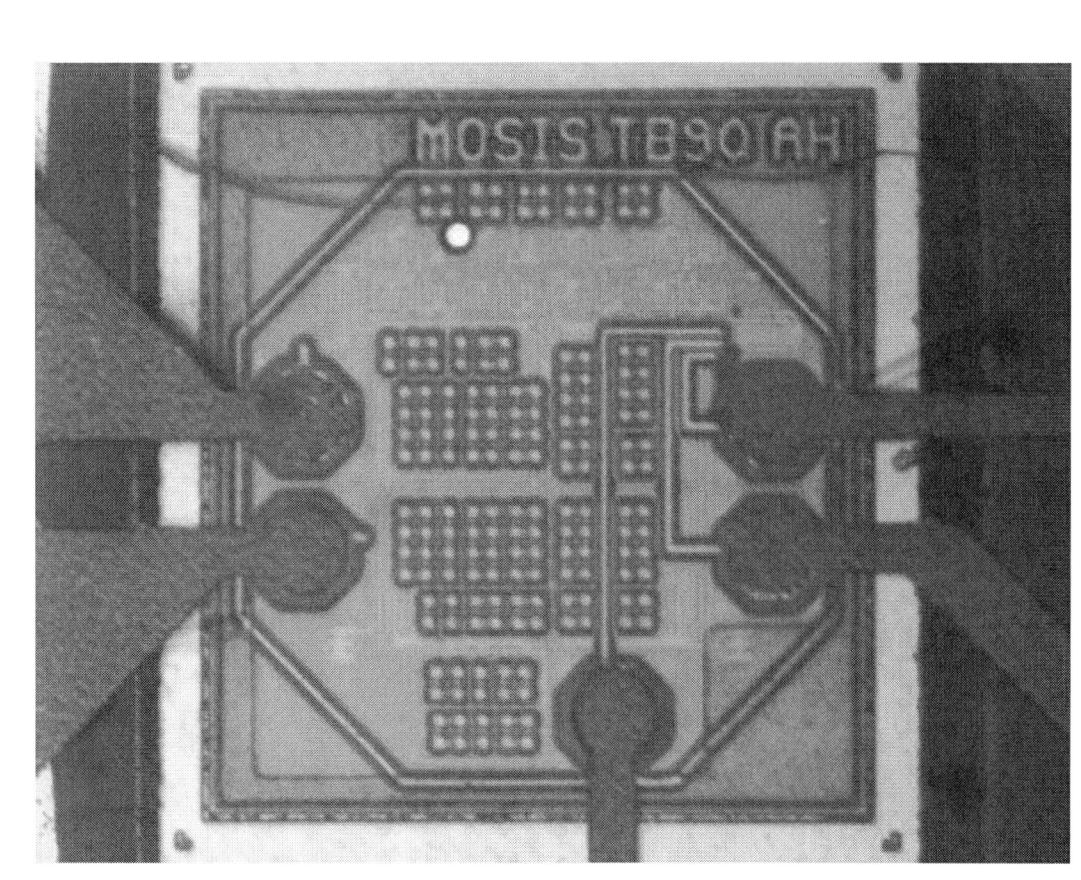

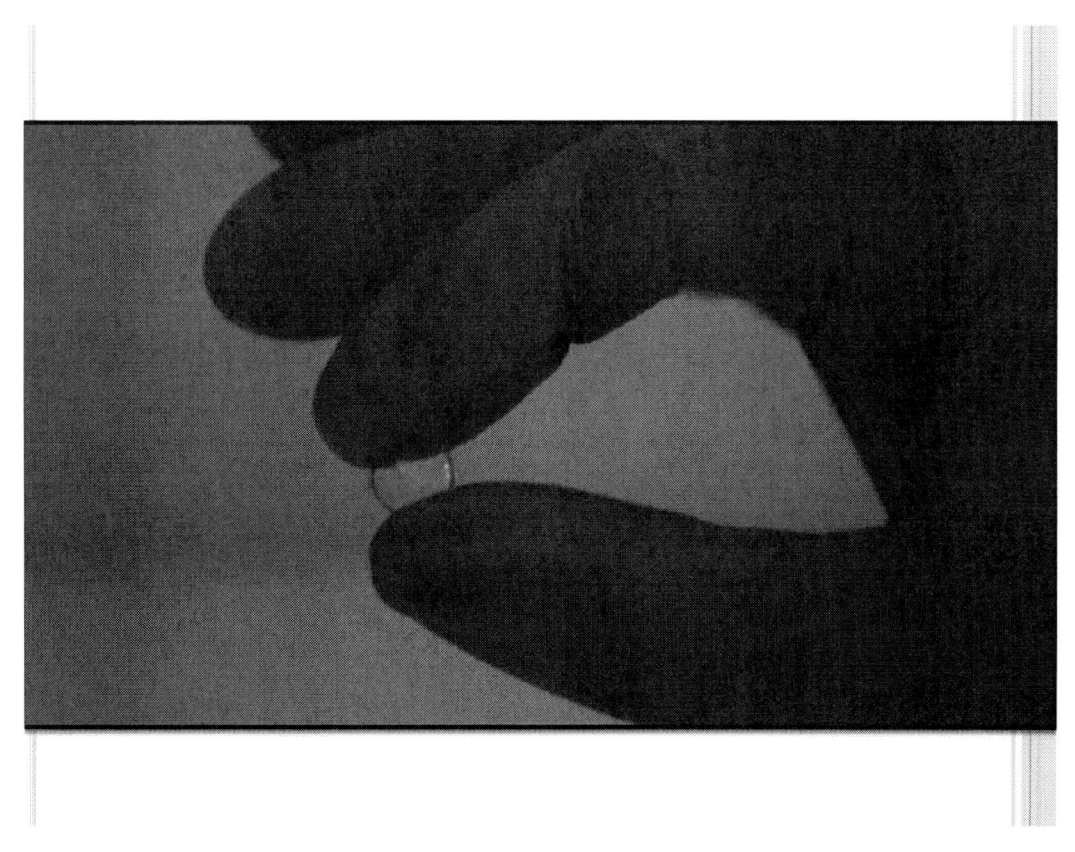

On-eye testing

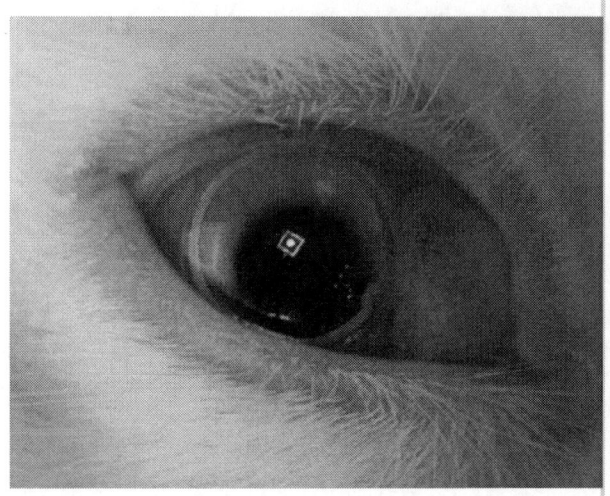

- Can integrate antenna, IC, LED on contact lens

- Can wirelessly transmit > 10μW

- What about sensing?

Courtesy Dr. Tueng Shen
U.W. Ophthalmology

Wireless glucose sensing on-lens

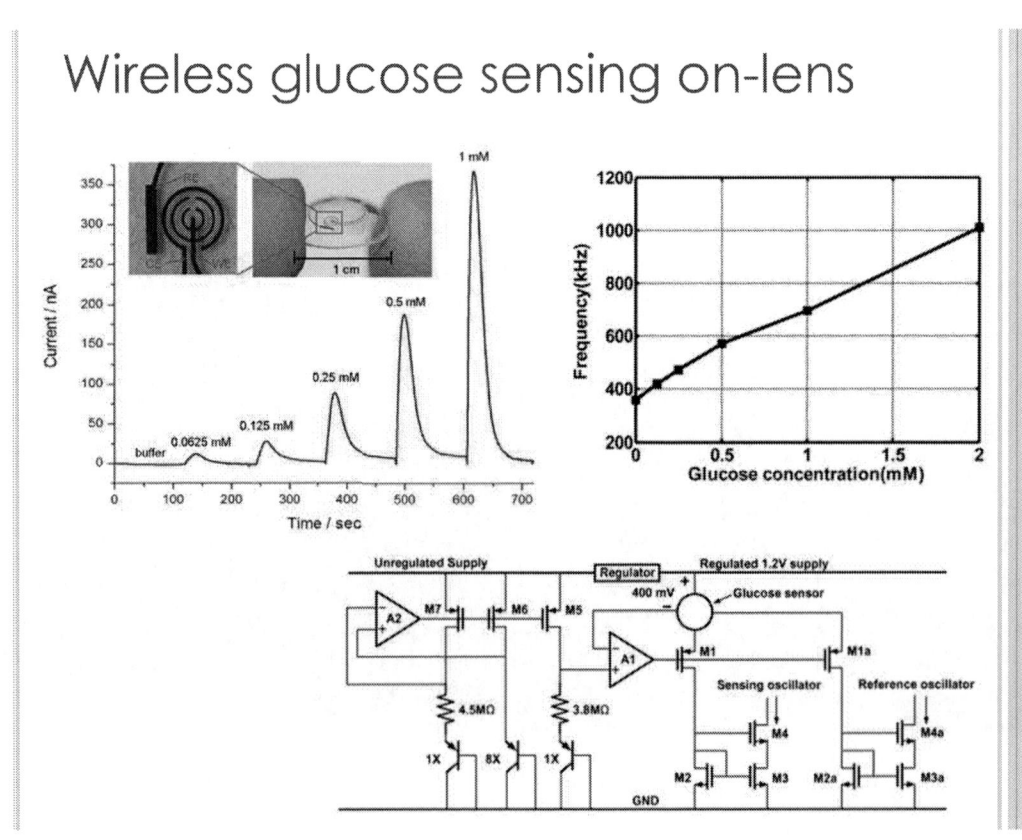

Wireless glucose sensing on-lens

Beyond glucose

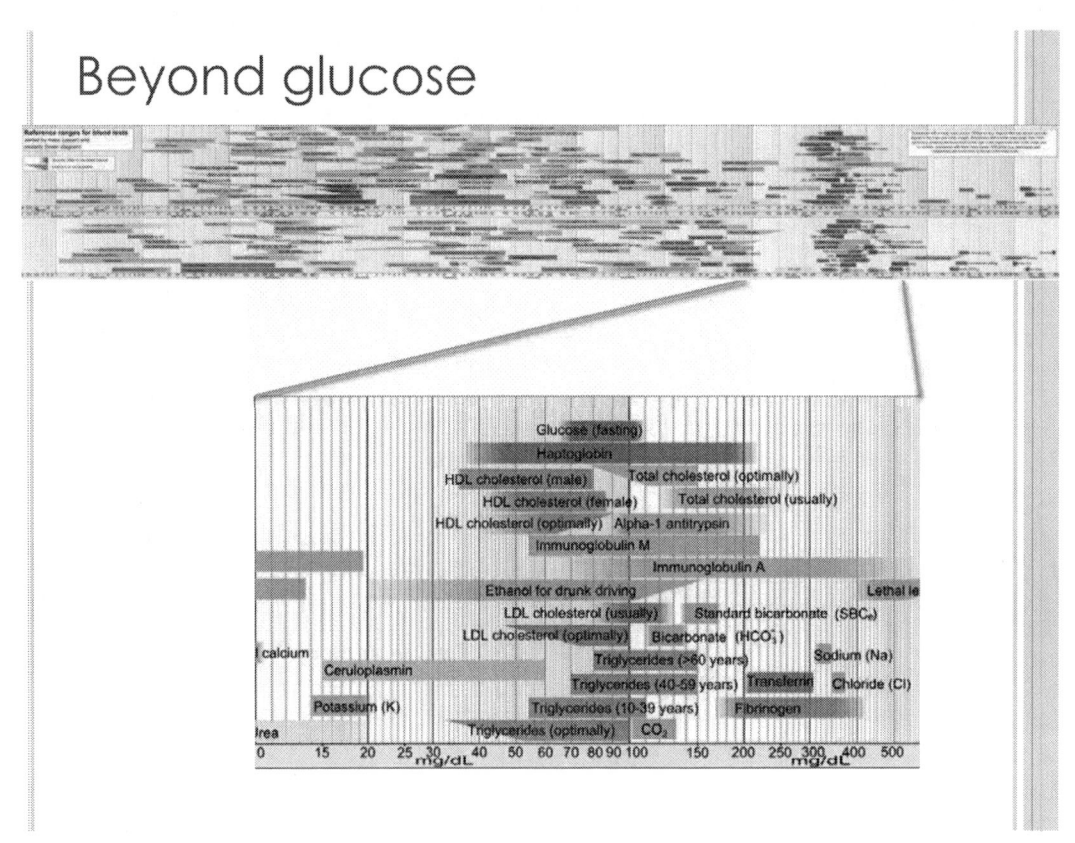

Further miniaturization

1. Need to shrink frequency references to allow true peer-to-peer communication

2. Need thin-film sensors (physical, chemical)

FBAR-based circuits

Collaboration with Avago Technologies

FMOS:
FBAR + CMOS

FBAR

Electronics
integrated into
hermetic lid

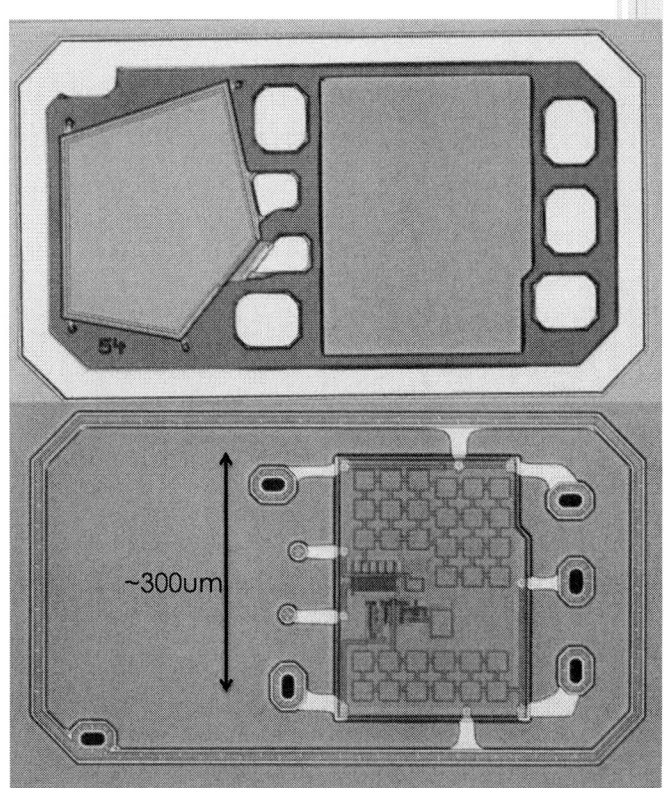

~300um

FMOS vs. Quartz

Frequency drift over temp/time

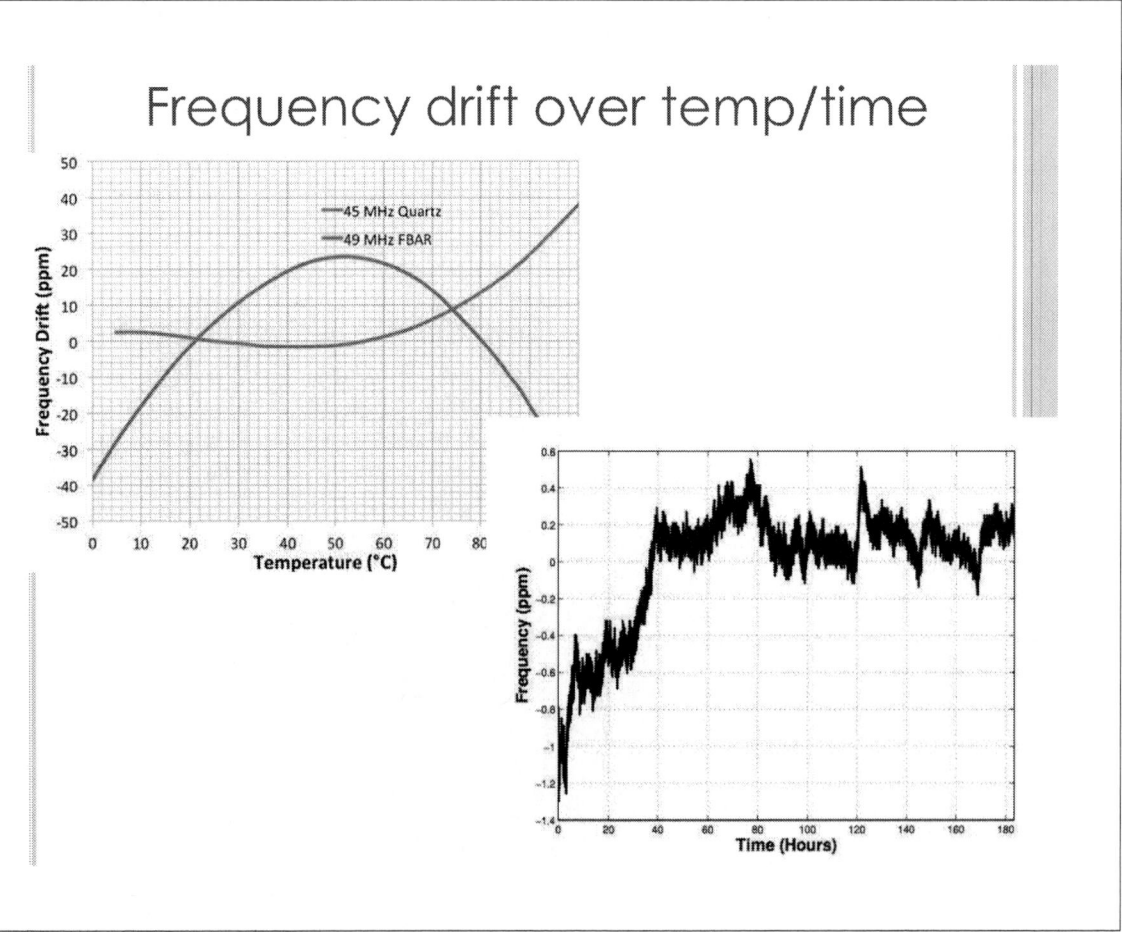

Sensing

1. The FBAR resonant frequency is extremely sensitive to mass (~1 ppm/picogram)
2. On-chip oscillators will faithfully track this resonant frequency
3. We have demonstrated on-chip electronics to resolve frequency down to roughly 0.5 Hz (10 ppb)

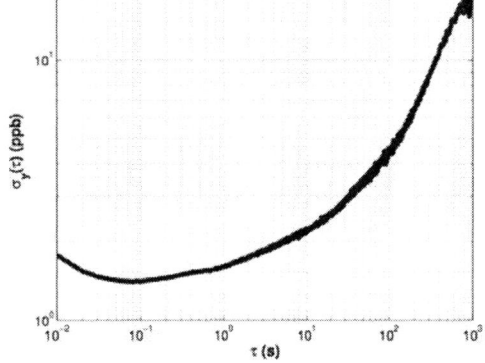

Post-processing

1. Goal: 80 um holes etched into the backside of the wafer

2. Etched using an inductively coupled plasma etcher with pulsed SF_6, C_4F_8

3. Depth was monitored using optical profilometry (etch depth ~ 220 um)

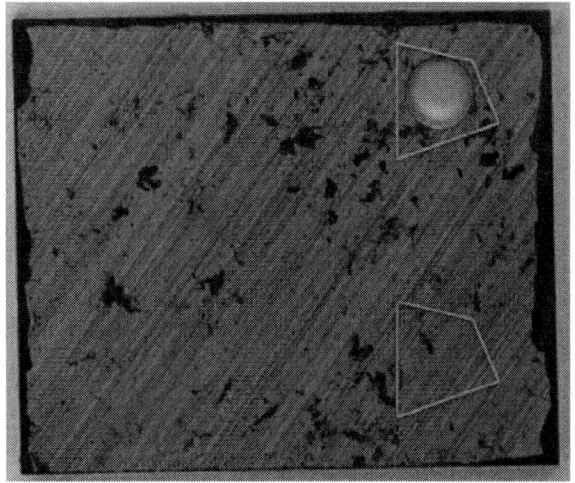

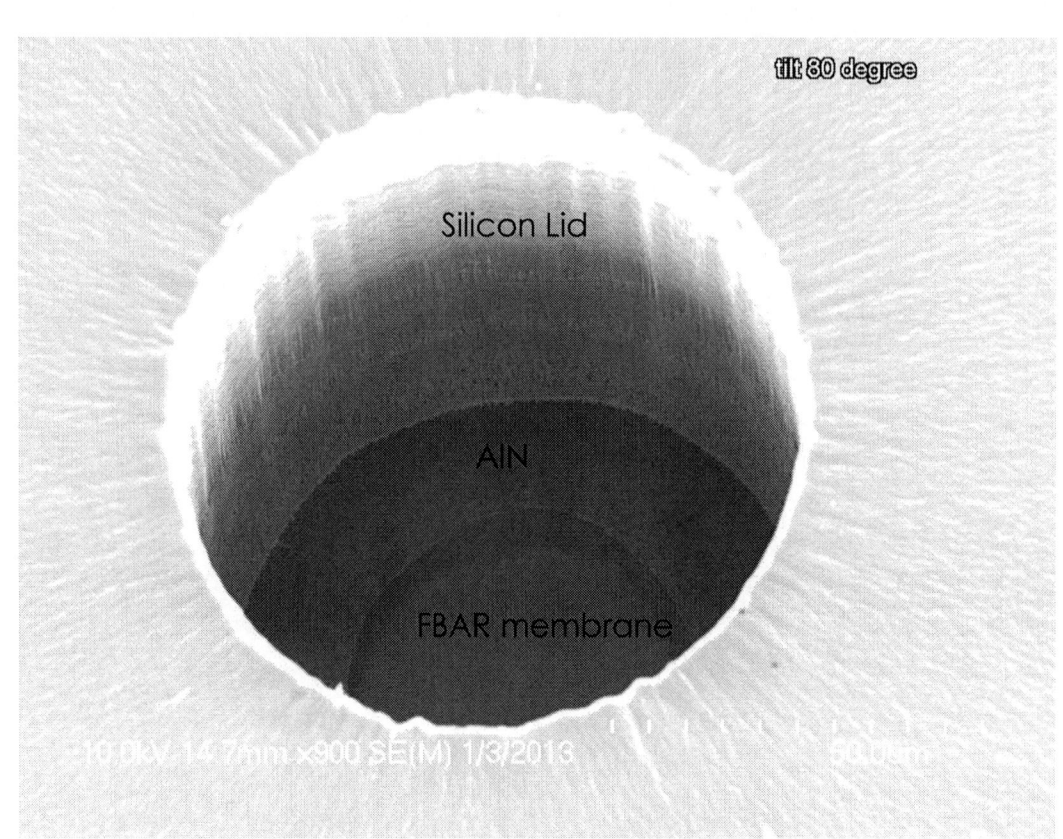

FBAR is not damaged from top side view

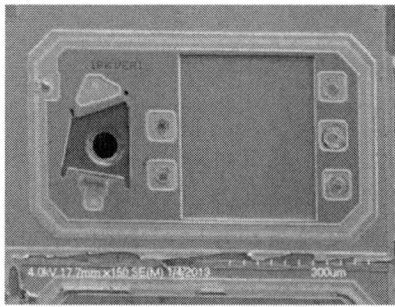

The lid after removing FBAR

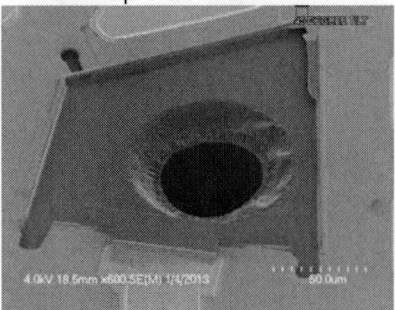

Show a successful etch-through hole

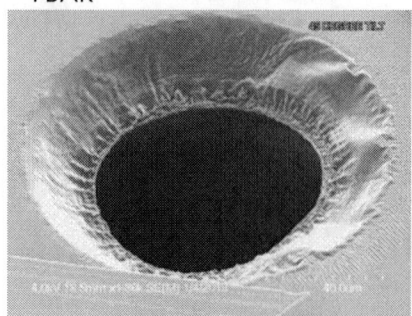

Reasonable undercut when etching through AlN

Initial results

1. Oscillators tested after etching: fnominal ~ 1.5 GHz

2. 20nm of PdAu sputtered. Oscillators re-tested: fmass1

3. 20nm of PdAu sputtered. Oscillators re-tested: fmass2

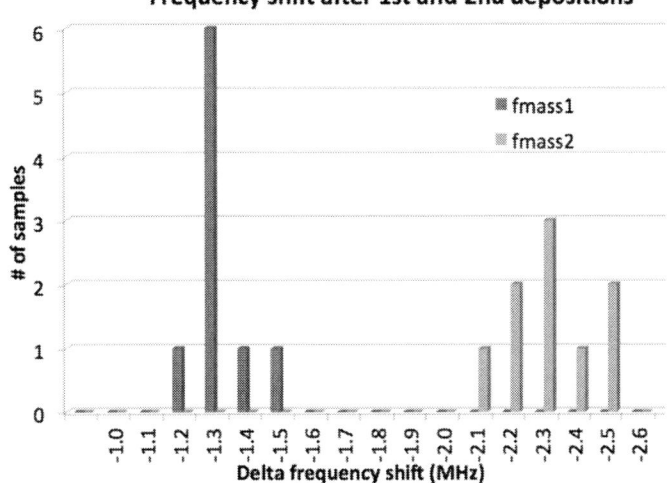

So what?

1. We can make extremely small battery-powered and battery-free wireless systems

2. Chip designers can have a profound influence on next generation biomedical sensing systems

3. Huge challenges and opportunities remain in the development of complex thin-film systems

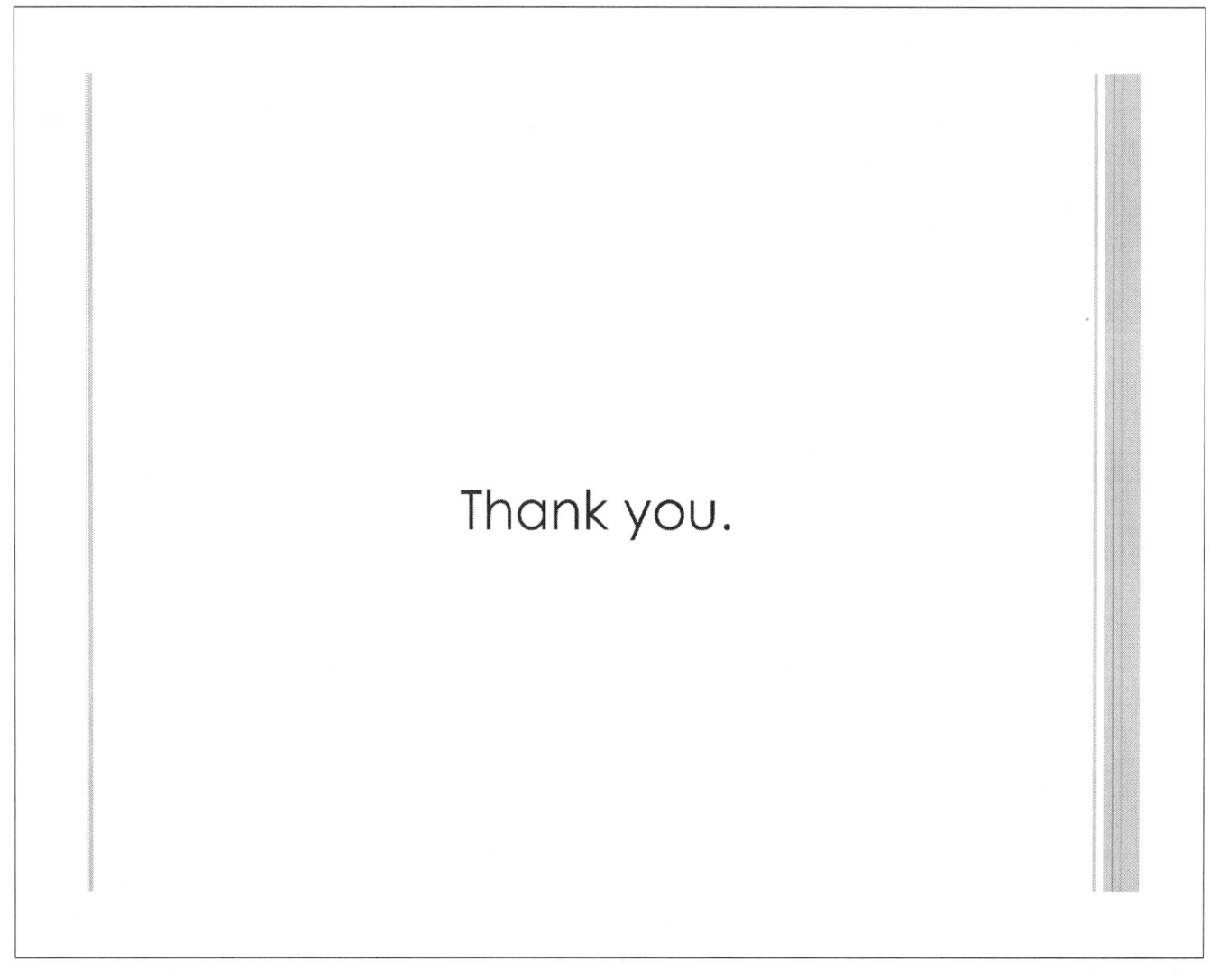

Thank you.

ST-MRAM Fundamentals, Challenges, and Applications

T. Andre, S.M. Alam, D. Gogl, C.K. Subramanian, H. Lin, W. Meadows, X. Zhang, N.D. Rizzo[2],
J. Janesky[2], D. Houssameddine[2], J.M. Slaughter[2]
Everspin Technologies, Austin, TX; Chandler, AZ [2]

Abstract— **Magnetoresistive Random Access Memory (MRAM) technology emerged from research and development into volume production within the last decade in the form of Toggle MRAM. The latest Magnetic Tunnel Junction (MTJ) based memory technology, Spin-Torque MRAM, has reached the level of customer sampling, offering higher density and bandwidth. Spin-Torque MRAM enables new applications, offers a wide range of features for use in embedded memory, and has the potential to extend to technology nodes beyond the capability of DRAM. This paper describes the devices, fundamental circuit challenges, and applications of this evolving MTJ based memory.**

I. INTRODUCTION.

Magnetic materials have been used in data storage elements since the early days of transistor based memory devices [1]. In the early 1990s, research into MRAM began using Giant Magnetoresistive (GMR) devices, with the potential of yielding a fast, unlimited endurance, non-volatile memory [2]. The pursuit of MRAM shifted in focus from GMR to Magnetic Tunnel Junction (MTJ) based devices in the late 1990s [3, 4]. These early MRAM approaches switch between two stable states using magnetic fields generated by currents flowing through conductors in close proximity to the storage elements (Fig. 1a). Of these field based MRAM approaches, the most robust and the basis for today's volume production devices is Toggle MRAM [5]. Toggle MRAM, available in densities up to 16Mb, has been adopted in a wide range of applications that previously required SRAM with either a battery backup supply or infrequent backup of the SRAM to conventional non-volatile devices to achieve fast non-volatile storage with unlimited endurance. Generating the field to switch between states in these MRAM devices, however, requires a significant level of current which does not reduce as the MTJ scales to smaller geometries. To address this challenge to scaling of field based MRAM devices, a new approach to switching the state of the MTJ has evolved called Spin-Torque MRAM (ST-MRAM) [6]. ST-MRAM switches the MTJ between two stable states by flowing current directly through the MTJ (Fig 1b). Development of ST-MRAM over the last several years has resulted in fully functional 64Mb DDR3 ST-MRAM devices [7, 8]. Scaling of ST-MRAM improves greatly over that of field based MRAM since the current required for switching reduces as the area of the MTJ

is scaled to smaller geometries. This lower switching current enables not only scaling to increased densities, but also enables sustained high-bandwidth write operations at DDR3 performance levels and beyond. Maintaining a required level of data retention, however, is a challenge for ST-MRAM as the MTJ is scaled. The next evolution of MTJ based memory uses perpendicular magnetic anisotropy (PMA) to achieve improved data retention at lower switching currents and has been considered a prime candidate for high density memory at nodes below 20nm [9].

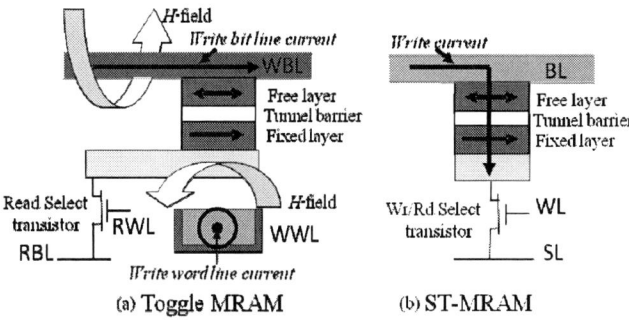

Fig. 1. (a) Field switching MRAM; (b) Spin-Torque switching MRAM

II. MAGNETIC TUNNEL JUNCTION STATES

A magnetic tunnel junction is constructed using a reference layer, typically with a fixed magnetic polarization, a storage layer that can be switched between two stable magnetic polarization states, and a tunnel barrier between the two magnetic layers. The resistance states of the MTJ are distinguished by the magnetoresistance effect, wherein the resistance across the tunnel barrier depends on the magnetic polarization of the storage layer relative to the reference layer. As shown in Fig. 2a, when the magnetic polarization of the storage layer is aligned with the magnetic polarization of the reference layer, the tunnel barrier exhibits a lower tunneling resistance and when the polarization of the storage layer is opposite that of the reference layer, the tunnel barrier exhibits a higher tunneling resistance.

The Magnetoresistance Ratio (MR) is defined as the percentage increase in resistance from the low resistance state to the high resistance state. MR, which is typically quoted using a very low bias measurement, decreases as the magnitude of the voltage across the MTJ increases (Fig. 1b). MR also decreases with increasing temperature.

978-1-4673-6145-3/13 $31.00 © 2013 IEEE

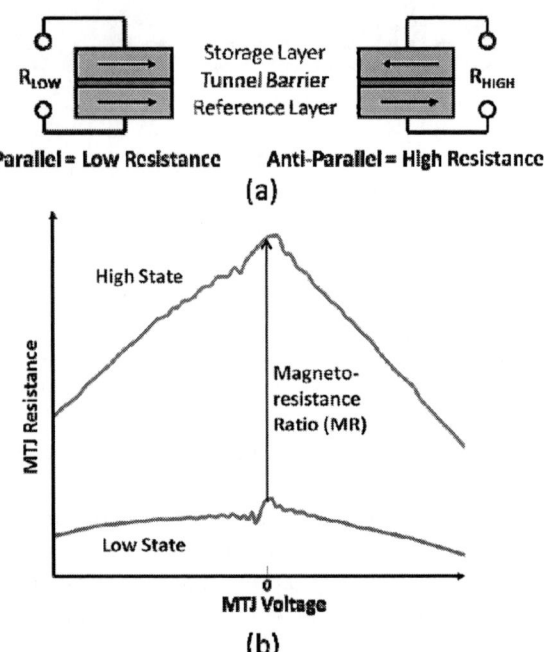

Fig. 2. (a) MTJ resistance states; (b) Resistance vs bias voltage

III. FIELD SWITCHING MRAM

MRAM products available in the market today use magnetic field to switch the storage layer between the two states. Initial approaches to switching MTJ devices apply a hard axis field along a row of MTJ elements and easy axis fields along columns of MTJ elements, switching the MTJ elements at the intersection of the row and columns when sufficient field is applied to each axis. With this initial approach, the state written to the MTJ depends on the polarity of the easy axis field. One significant challenge to this approach is that half selected devices along a row or column which are exposed to a single axis of applied field can be disturbed with excess current in one axis. Achieving a reliable window of switching functionality with this initial approach requires a very tight distribution of switching thresholds for the MTJ elements in an array.

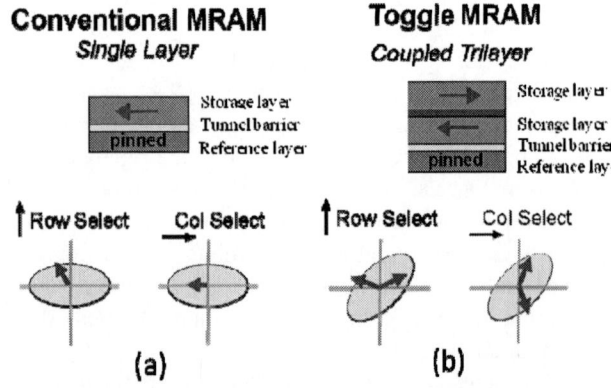

Fig. 3. (a) Basic MTJ and (b) Toggle MTJ, showing storage layer response to half selects on each axis

Toggle MRAM was developed to address this challenge using a different MTJ stack and an offset timing sequence on the programming lines. The Toggle MTJ, as seen in Fig. 3, adds a second storage layer to the stack, coupled to the first storage layer with a synthetic anti-ferromagnetic coupling layer that acts to keep the magnetic polarization of the two storage layers opposite to each other. The Toggle MTJ is oriented at an angle relative to the programming lines, which, with the stability added by the coupling layer, provides significant immunity to half select disturbs.

Switching the Toggle MTJ between states requires an offset timing sequence applied to the programming lines as shown in Fig. 4. The sequence applied to the program lines toggles the state from high to low resistance or from low to high resistance. Therefore, in order to set the Toggle MTJ to desired the state, the initial state of the MTJ must be read and a toggle sequence applied only if necessary.

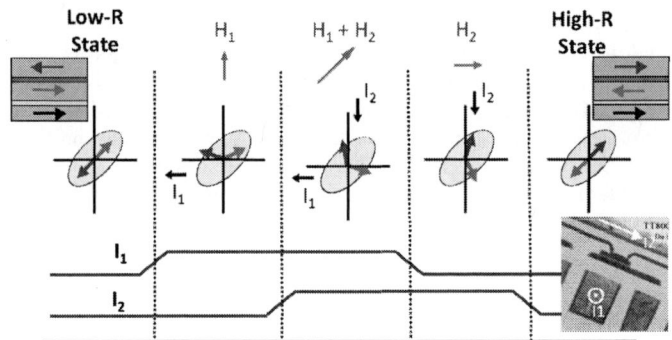

Fig. 4. Toggle MRAM switching sequence

The improved switching window of Toggle MRAM is demonstrated in Fig. 5. The initial conventional approach achieves only a very narrow window of operation with sufficient current applied to switch all the MTJ elements in an array without excess current on either axis causing half select disturbs. In contrast, Toggle MRAM is tolerant of significant overdrive on each axis beyond the point where all MTJ elements are switched. There is, however, an upper bound to the applied current for Toggle MRAM beyond which the coupling between the storage layers is saturated and switching is not guaranteed. This saturation level decreases with increasing temperature, reducing the operating window slightly.

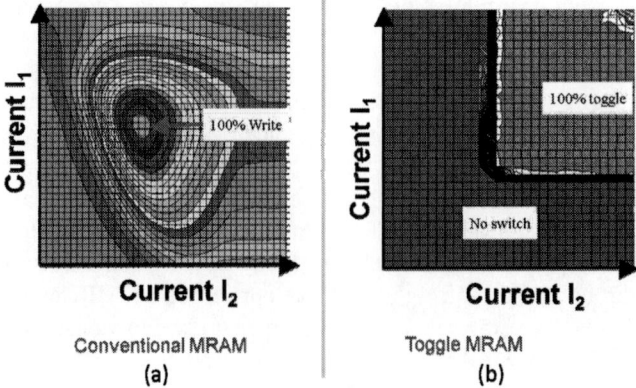

Fig. 5. Switching window for (a) initial conventional MRAM, and (b) Toggle MRAM

To avoid the saturation current level at high temperature

while applying sufficient current to switch the entire population of bits reliably across temperature, adjustment of the applied current level across temperature is typically required. Figure 5 shows an example of the average switching and saturation fields for each axis over temperature.

Toggle MRAM technology has enabled robust and reliable commercially available products, however, there are challenges in scaling the technology to support higher densities. The field magnitude required for field switching MRAM generally increases for smaller MTJ. The required program line current can be maintained by moving the metal lines closer to the MTJ, however, any significant reduction in current is difficult to achieve. Without significant reduction in switching current, which can be well over 10mA per axis, significant reductions are not likely in the supply voltage level required to sustain the IR drop along the program lines, the area of the circuitry controlling the switching current, or the power required to achieve an increased write bandwidth.

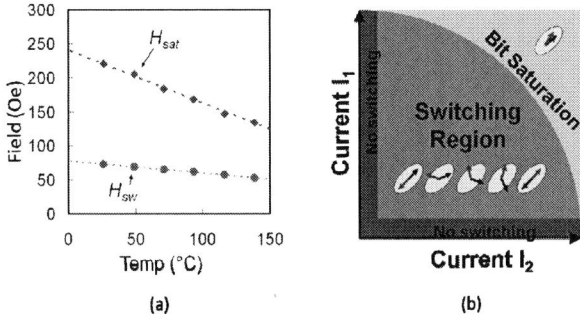

Fig. 6. (a) Average switching and saturation field; (b) Switching and saturation regions

IV. SPIN-TORQUE MRAM

Spin-Torque (Transfer) MRAM (ST-MRAM, or STT-MRAM) switches the state of the MTJ by flowing a sufficient density of current (Jc) across the tunnel barrier. When switching from the high resistance (anti-parallel) to the low resistance (parallel) state, current flows from the free storage layer to the fixed reference layer. As electrons move through the fixed layer, their spin is polarized to match the fixed layer. When a sufficient density of polarized electron flow is achieved, the free layer changes state to align with the fixed layer. Switching from the low resistance (parallel) to the high resistance (anti-parallel) state is accomplished by applying current in the opposite direction, with current flowing from the fixed reference layer to the free storage layer. In this switching direction, electrons with spin opposite the fixed layer are reflected back to the free layer causing the free layer to switch to the state opposite to the fixed layer.

One key component of the MTJ material stack for ST-MRAM is the use of Magnesium Oxide for the tunnel barrier rather than the Aluminum oxide used in Toggle MRAM today. As shown in Figure 7, a significantly lower resistance-area product (RA) can be achieved with Magnesium Oxide than typically achieved with Aluminum oxide. This lower RA is essential to achieving the required current density for Spin-Torque switching at voltages low enough for reliable operation.

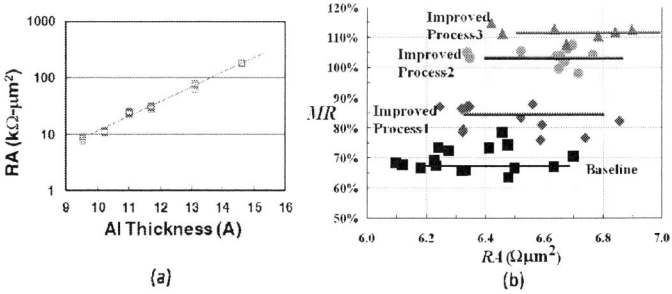

Fig. 7. (a) RA vs Al thickness for Aluminum Oxide; (b) MR vs. RA for Magnesium Oxide

Figure 8 shows an example of how ST-MRAM may scale relative to the current drive of a minimum feature size transistor, such as in a DRAM device [8]. The Ic(MTJ) curve represents the average current required to switch an MTJ and the Isw(array) curve shows that an increased level of current is required to switch the distribution of MTJ in an array.

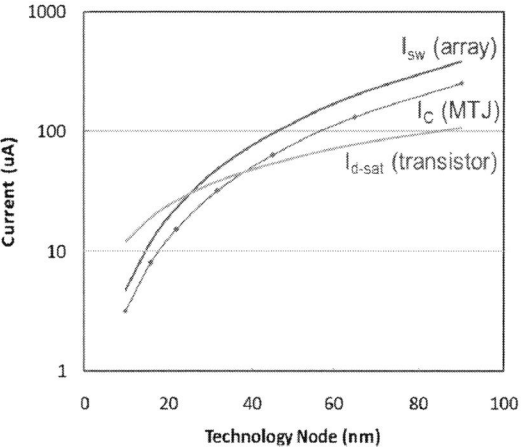

Fig. 8. Example scaling curve for ST-MRAM switching current

Compared to the >10mA of switching current on each axis for Toggle MRAM switching, a reduction to hundreds of microamps to switch an ST-MRAM bitcell enables 90nm products to achieve the DDR3-1600 write bandwidth of 1.6-3.2GB/s chip compared to the 30-60MB/s achieved by existing Toggle MRAM products consuming similar write power. Unlike Toggle MRAM, the required switching current for ST-MRAM is reduced proportional to the area of the MTJ, scaling faster at smaller technology nodes compared to the more linear scaling of the minimum sized transistor drive current. At the scaling rate described in Fig 8, ST-MRAM becomes a viable candidate for DRAM replacement at technology nodes below 20nm. Significant challenges exist to achieving this scaling path, one of which is maintaining a minimum level of data retention as the MTJ is scaled.

Data retention time for the MRAM device is an exponential function of the energy barrier (Eb) to reverse the state. Energy barrier is typically measured in units of Eb/kT, where k is Boltzmann's constant and T is temperature in degrees Kelvin. The average Eb/kT required to achieve 10 year data retention depends strongly on the Eb distribution (sigma) for each MTJ in the array, but can be in the order of 60-80. Figure 9 shows measurements of average Eb/kT for MTJ with different area and free layer thickness. Scaling of the MTJ to smaller area causes a reduction in data retention which can be compensated for with an increase in the thickness of the free layer. Increasing the free layer thickness, however, can increase the current density (Jc), and therefore voltage, required to switch the MTJ unless other improvements are made to the MTJ that reduce switching voltage without reducing Eb. MTJ using perpendicular magnetic anisotropy (PMA) have been shown to have an increased Eb for a given Jc compared to MTJ with in-plane magnetic anisotropy and therefore are the focus of research for ST-MRAM at advanced process nodes.

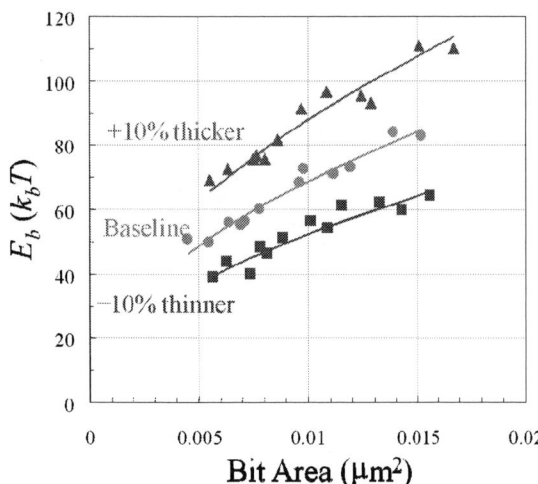

Fig. 9. Eb/kT vs. MTJ area for different relative free layer thicknesses

One key challenge in achieving a reliable ST-MRAM product is applying a sufficiently high write voltage to switch the distribution of MTJ in an array without overstressing the MTJ and limiting the device endurance. Fig. 10 shows an example switching distribution and example breakdown distributions for an array of MTJ [11]. The Time Dependent Dielectric Breakdown (TDDB) failure mechanism requires lower applied voltage levels to achieve higher levels of endurance, which is shown in the shift from the one cycle breakdown curve to the breakdown distribution for many cycles. Sustaining reliable operation over the product lifetime requires good control of the applied write voltage.

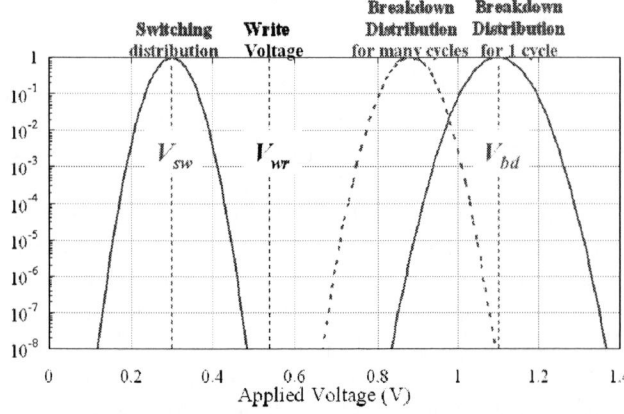

Fig. 10. ST-MRAM applied write voltage window vs. switching and breakdown distributions

V. MRAM ARRAY CONFIGURATIONS

Field Switching MRAM and ST-MRAM have both been pursued with a variety of array architectures. The highest density approach, which is also the most challenging, is the cross-point architecture [12]. The cross-point architecture requires no transistors in the MTJ array, but rather connects the MTJ directly to metal conductors organized in rows and columns with one MTJ at each intersection. The main challenge to achieving a functional design using the cross-point architecture is management of current sneak paths in the array. Where each MTJ is uniquely associated with one row conductor and one column conductor, each row conductor has an MTJ connected to each of the other column conductors, which similarly each have MTJs connected to each of the other row conductors. Therefore, a selected column conductor has multiple sneak paths to the selected row conductor through different numbers of series MTJs. Applying a bias to the unselected column and/or row conductors may help increase the percentage of the current flowing through the target MTJ from the selected conductors, however, this may be difficult to control. Even with high resistance MTJ and complex bias control, the cross-point architecture has proven to be an extremely challenging approach for either field based or Spin-Torque switching.

The most common approach to MRAM is the use of one transistor and one MTJ per bit (1T1MTJ). For Toggle MRAM, the MTJ is typically accessed through a bit line connected to the drain of the select transistor, which has a word line connected to its gate and the MTJ connected to its source. The second terminal of the MTJ is typically connected at the top to a conductor that is at a ground potential during read. Most MRAM use the write current conductor above the MTJ (write bit line) as the ground conductor, however, an alternate architecture using metal local interconnect (MLI) allows the MTJ to be isolated from the write conductor, reducing cross talk between the write and the read circuitry and providing more time for the write pulses to complete without disturbing a subsequent read operation.

For ST-MRAM, the 1T1MTJ architecture typically uses two column conductors, a source line connected to the select transistor and a bit line connected to the top of the MTJ. With the two column conductors, current can be applied down through the MTJ with a higher potential applied to the bit line and a lower potential applied to the source line, or up through the MTJ with a higher potential applied to the source line and a lower potential applied to the bit line. An alternate architecture uses a local source line along the row direction to connect a small group of MTJ and distributes the return current among the unselected bit lines.

Other architectures have been published demonstrating the potential trade-off between area and performance, such as a 2T2MTJ architecture [13] with increased performance for an increased area or a 1T4MTJ architecture [14], with reduced area and a more manageable signal than the cross-point architecture. Products to date have found the best option for performance and area to be the 1T1MTJ approach, however, the inherent flexibility offered by the MTJ in the architectures

978-1-4673-6145-3/13 $31.00 © 2013 IEEE

it supports as well as how the MTJ itself can be optimized to balance its properties may have great advantages when used in embedded applications.

VI. MRAM READ CIRCUIT CHALLENGES

Whether used in a field switching or a Spin-Torque switching approach, the read properties of the MTJ leave little margin for sense amplifier design. Figure 11 shows a measured resistance distribution for a Toggle MRAM device using an Aluminum Oxide tunnel barrier. Even with the very tight distribution shown, accounting for the further reduction in MR for high temp, the variation in the reference to distinguish the states, and part to part variation, the signal remaining for the sense amplifier to distinguish can be well below 1uA.

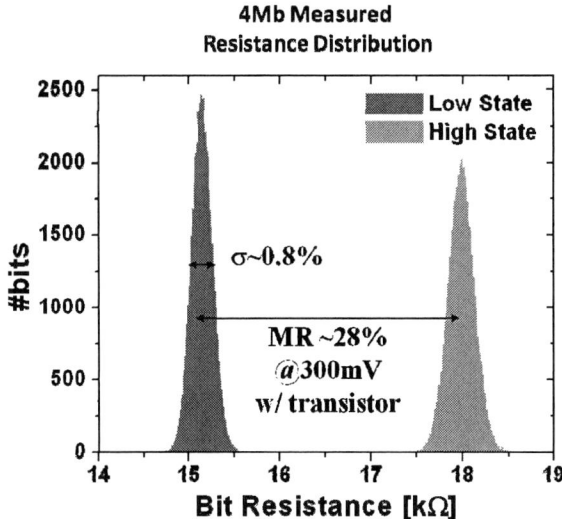

Fig. 11. Resistance distribution for a Toggle MRAM device

Achieving fast read speed while minimizing the mismatch of an MRAM sense amplifier is a challenge. Mismatch of any analog circuit is a function of several factors, which, after optimization of overdrive and other factors, typically become a function of the square root of the area of critical devices. As the area of the critical devices increases to reduce the mismatch, self biasing circuits using diode connected devices often suffer in speed due to the increased gate capacitance. One approach shown in Fig. 12a that has been proven to be robust in Toggle MRAM product uses a three input sense amplifier and a source side current averaging connection to eliminate the gate connection on the reference output node and minimize the output capacitance of the preamp [15]. This approach has been further optimized (Fig. 12b) to eliminate all self biasing gate connections using a static bias generated from a replica circuit to enable improved performance with reduced mismatch. Both options include the gain stages shown and latches, not shown. In practice, precharge and equalization devices are also required to maximize performance.

The read properties of ST-MRAM offer similar challenges to Toggle, but with some difference in the key parameters. The use of a Magnesium Oxide tunnel barrier exhibits higher MR than Aluminum Oxide, achieving over 100% as seen in Fig. 7b. However, as the MTJ is scaled to smaller geometries the resistance variation, or relative sigma, increases.

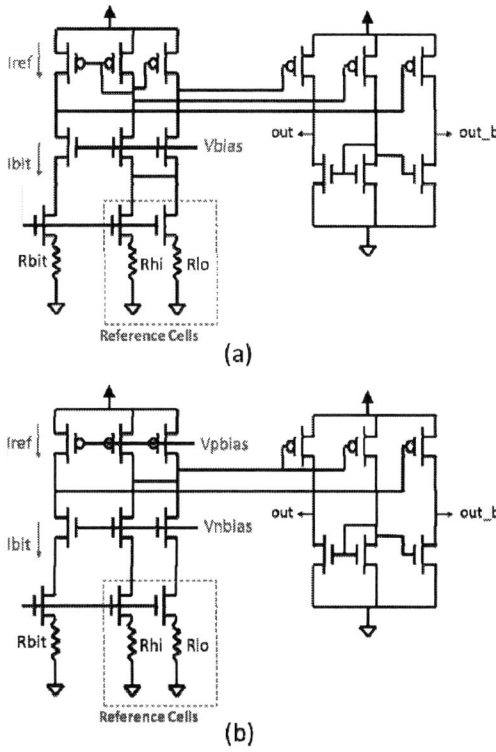

Fig. 12. Simplified schematics of sense amplifier preamp and gain stages used in Toggle MRAM device

As can be seen to a small degree in Fig. 11, the resistance relative sigma tends to be the same for the high resistance state as for the low resistance state, causing the absolute resistance sigma value to be higher for the high resistance state. This effect is more pronounced at higher MR, as shown in Fig. 13, causing the point where the distributions cross with the same probability to be much closer in resistance to the low resistance state than the high resistance state. Note that this ideal reference point corresponds to the average conductance of the two states and that the three input sense amplifiers shown in Fig. 12 achieve this ideal reference point.

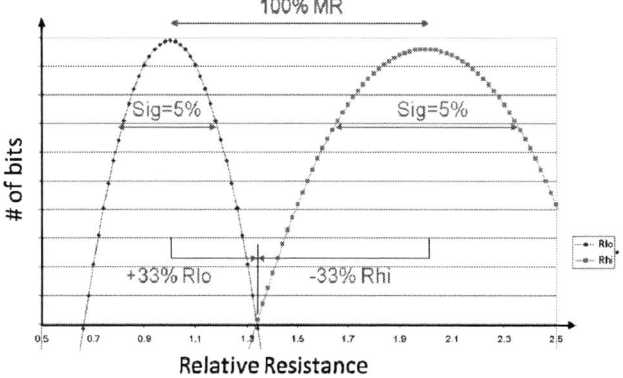

Fig. 13. Example distribution with 100% MR

The high MR and low RA properties of Magnesium Oxide

tunnel barriers provide the opportunity for higher differential read current, if the resistance sigma is sufficiently low. However, to avoid disturbing the state of the MTJ, a low bias voltage must be applied during reads. As shown in the example distributions in Fig. 10, the read voltage required to avoid disturbs is a function of the switching distribution and may require voltages below 100mV. In addition to the challenge of minimizing the mismatch in the sense amplifier, this low bias voltage adds to the challenge by requiring circuitry with increased tolerance to noise on the supplies and bias signals. Signal margin is also reduced by variation in the reference level, as well as reduction in MR due to bias or temperature. For all of these reasons, circuitry with maximum tolerance to variation of the MTJ signal is required to maintain read performance as the area of the MTJ is scaled.

VII. MRAM WRITE CIRCUIT CHALLENGES

The write circuit challenges for Toggle MRAM and ST-MRAM are different, however, share a common goal of applying the maximum voltage or current for a given circuit area while maintaining control of the applied level.

A. Toggle MRAM

For Toggle MRAM, the array efficiency is influenced greatly by the length of the programming lines used in the array. The shorter the lines, the more frequently the large current drivers must be repeated. Therefore, it is most efficient for a write current driver circuit to regulate the current with minimal voltage drop across the transistors and maximum voltage drop across the array programming lines. The load line analysis plot shown in Fig. 14 demonstrates how regulation near the linear region of the write driver device can cause variation in the applied current. This variation can be minimized using replica circuitry to control the bias applied to the driver to account for the average driver instance entering the linear region. However, due to the magnitude of the current driven by the driver, such a replica circuit, even scaled to a fraction of the final driver size, can require a significant level of product standby current.

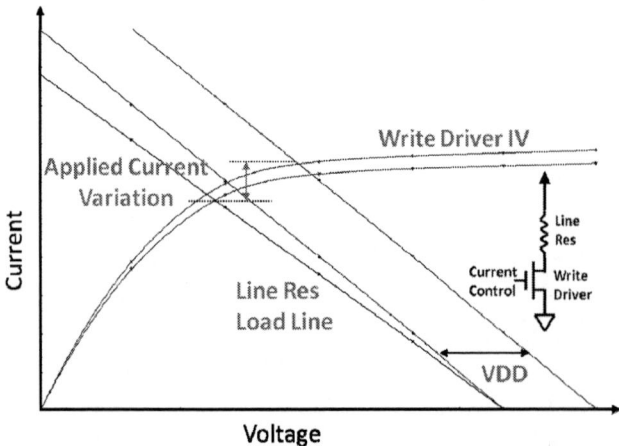

Fig. 14. Load line analysis plot of Toggle MRAM write driver

B. Spin-Torque MRAM

ST-MRAM design has a similar goal of minimizing the voltage drop across the series transistors to maximize the voltage regulated across the MTJ. When writing the state of the MTJ using Spin-Torque switching, the current must flow through the associated select transistor. The current drive capability of the select transistor is a function of the width of the device. With the scaling example shown in Fig. 8, the select transistor area required to drive the required current is likely to be larger than the MTJ area requirement for most technology nodes above 20nm. For this reason, optimizing the select transistor width has a direct impact on the memory cell size. At the optimal cell size, the resistance of the select transistor is similar to the resistance of the MTJ, providing about 0.5V across the MTJ when 1V is applied between the bit line and the source line. There is, however, an inherent asymmetry in the resistance of the select transistor for a given word line voltage, depending on the direction of the current flow. As shown in Fig. 15, when a positive bias is applied to the SL and a lower bias is applied to the BL, the voltage drop across the MTJ lifts the source of the select device and reduces the applied Vgs. This asymmetry adds challenge to achieving the necessary switching current in that direction. Techniques, such as using voltages above VDD on the word line, can help to reduce the select device impedance and compensate for the gate to source voltage lost due to source lifting.

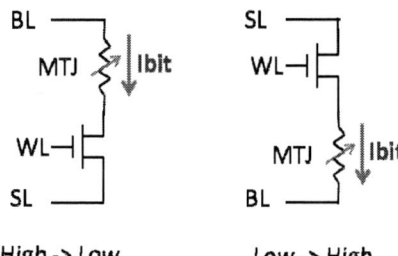

Fig. 15. Direction of current flow to write each state in ST-MRAM, assuming the fixed reference layer is the bottom layer of the MTJ

When the resistance of the select transistor is comparable to the resistance of the MTJ, a range of applied voltage for each MTJ in an array can exist due to variation in MTJ resistance and variation in transistor resistance. As shown in Fig. 16, regulating the voltage between the bit line and the source line applies a range of voltage across each MTJ due to the voltage divider between the select transistor and the MTJ. Applying a constant current eliminates the variation due to the select device resistance, however, the linear relationship between MTJ resistance and applied voltage may not be ideal to switch all of the lower resistance MTJ while avoiding poor endurance due to TDDB for the higher resistance MTJ. Providing sufficient drive capability at the low end of the resistance distribution is the key requirement defining the select transistor size. The overdrive at the high end of the resistance distribution can be reduced with further increase in the select device size by improving the ratio of the MTJ resistance in the voltage divider, however, the cost in cell area is not desired.

Some overdrive, however, may be acceptable depending on the breakdown distribution characteristics.

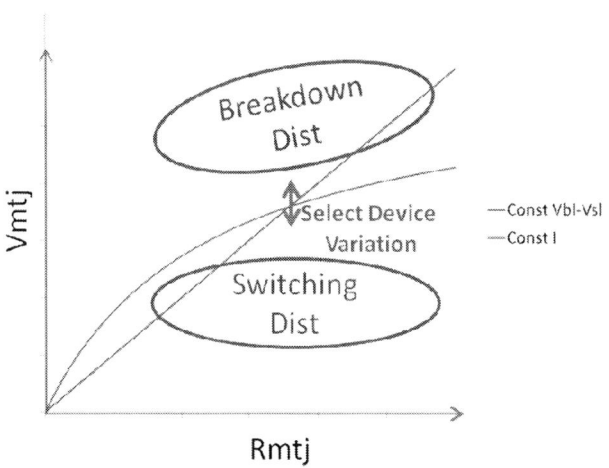

Fig. 16. Example applied voltage plots vs. switching and breakdown distributions for constant bit line to source line voltage and constant current

VIII. ST-MRAM SILICON RESULTS

A die photo of our 64Mb DDR3 ST-MRAM is shown in Fig. 17. The DDR3 interface highlights the write bandwidth achievable due to the lower switching current of ST-MRAM compared to field switching MRAM. The 64Mb DDR3 ST-MRAM is fabricated using a standard 90nm CMOS process and uses 4 Cu layers with one Al RDL layer. The MTJ layers are located between the third and fourth Cu layers. The device is packaged in a BGA package using JEDEC standard DDR3 ball locations.

An eight bank architecture is used to meet the JEDEC DDR3 standard. The banks are physically located side by side along the width of the die photo in Fig. 17. The wordline drivers can be seen as vertical stripes in the center of each bank. Column circuitry, including sense amplifiers and write driver circuitry for the source lines and bit lines, can be seen as horizontal stripes of circuitry.

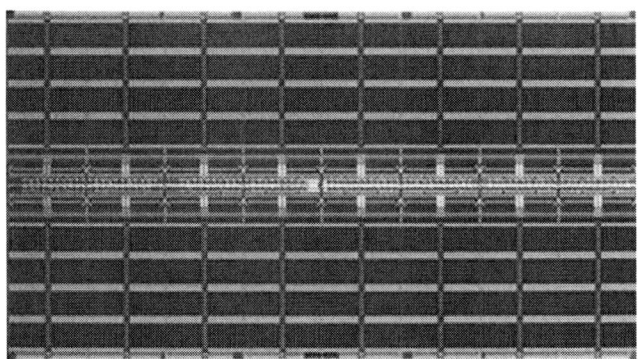

Fig. 17. Die photo of the Everspin 64Mb DDR3 ST-MRAM

Figure 18 shows a shmoo of the applied voltage in each switching direction. Measured after cycling each bit five hundred thousand times, the results demonstrate full functionality in the region where sufficient voltage is applied.

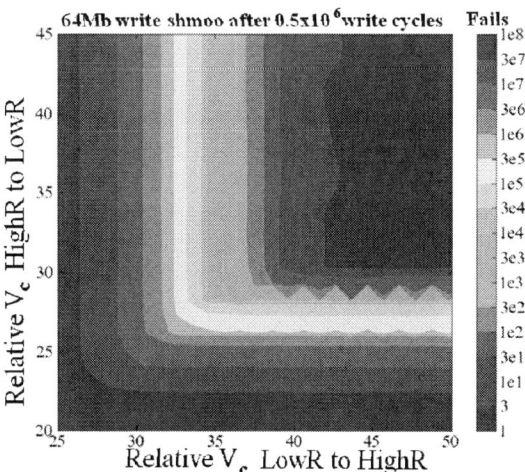

Fig. 18. Measured Shmoo plot of failing bits vs. applied write voltage for each switching direction

It was announced in November 2012 that this 64Mb ST-MRAM device has been sampled to key customers [7]. In addition to the memory test results demonstrating functionality up to 800MHz, the 64Mb devices have been validated with DIMMs on evaluation boards incorporating a variety of DDR3 controller IP.

IX. MRAM APPLICATIONS

Toggle MRAM currently in volume production is used in multiple applications where fast, unlimited endurance, nonvolatile writes are required. These applications include RAID storage systems, industrial computing and automation, automotive control units, and aerospace systems. With a 35ns cycle time specification and available in densities up to 16Mb, Toggle MRAM is ideal for applications that previously required battery backed SRAM, FeRAM, or NVSRAM products.

Designed into the RAID storage systems solutions provided by customers such as Dell and LSI, MRAM performs a write journal data log function that enables recovery of data in the event of a system failure [16].

In industrial computing applications, such as the Emerson Network Power NVME7100, MRAM stores critical data and program storage, enabling frequent updates without wear out of the memory or loss of data [17].

Automotive control systems, such as in the BMW Motorad Motosport engine control unit RSM5, leverage the extended temperature range (-40°C to 125°C) supported by Toggle MRAM [18].

Aerospace applications, such as the Angstrom Aerospace TAMU magnetometer subsystem used in the Japanese SpriteSat research satellite, benefit from the immunity of the Toggle MTJ to radiation induced soft errors [19].

Where Toggle MRAM offers a highly reliably solution in place of battery backed SRAM devices, Spin-Torque MRAM expands the density and bandwidth capability to offer the same benefit to applications currently using battery or capacitor backed DRAM [20].

Achieving the performance level of DDR3 DRAM enables

978-1-4673-6145-3/13 $31.00 © 2013 IEEE

using ST-MRAM to protect write data in storage systems not yet written to the drive. Many of these systems use batteries or super capacitors to preserve the data in DRAM until it can be stored elsewhere. Batteries and capacitors have a negative impact on system cost and reliability, providing an ideal opportunity for an improved solution using ST-MRAM. In addition to these applications, the higher density provided lowers the cost per bit for applications currently using high densities of Toggle MRAM.

Perhaps the application space with the widest range of use for ST-MRAM is embedded memory solutions. The relative ease of integration for MTJ devices, without any disruption to the CMOS devices, makes ST-MRAM a viable embedded nonvolatile technology at nodes where traditional nonvolatile technology, like Flash, have not yet or cannot easily be integrated. In addition, ST-MRAM technology offers RAM level of performance for embedded DRAM or embedded SRAM applications with the ability to have zero power data retention. With the standard architecture tradeoffs typical to RAM design and the additional optimization possible with the size or construction of the MTJ, ST-MRAM technology offers an extremely wide range of features for embedded applications.

The largest market potentially within reach of ST-MRAM technology is the high density memory market served by DRAMs. ST-MRAM, when scaled below 20nm to support multi-gigabit density devices, may offer an easier to implement solution than DRAM at the advanced process nodes as well as the ability to eliminate the performance overhead due to the frequent refresh requirement of DRAM.

X. Conclusion

MTJ based memory technology has proven to offer reliable solutions that have been adopted in a wide range of applications. Toggle MRAM solved the issue of half select disturbs and became the basis for the first commercial MRAM products, supporting densities up to 16Mb with a 35ns cycle time for both write and read operations. With the limited scaling capability of Toggle MRAM, the next generation of MTJ based memory, ST-MRAM, has been demonstrated at 64Mb density and offers much higher write bandwidth, achieving DDR3-1600 write and read bandwidth. The circuit design challenges for ST-MRAM include variation tolerant read solutions and write solutions that offer reasonable regulation of the applied voltage while minimizing the cell area. ST-MRAM expands the applications for MRAM products, and in addition, offers a wide range of features for embedded memory solutions.

Acknowledgement

The authors would like to acknowledge the many contributors to the material in this paper as well as support from across the entire Everspin team. It is great respect for the challenges faced by all of the departments at Everspin along with our combined persistence that delivered MRAM to the market and will continue to expand the available features.

This work was supported in part by NSF SBIR Phase II under Award 1058552.

References

[1] E. Leroy Younker, "A Transistor-Driven Magnetic-Core Memory," *IRE Trans. on Electronic Computers*, vol. EC-6, no. 1, pp. 14-20, Mar. 1957.

[2] K.T.M. Ranmuthu, *et al.*, "High Speed (10-20ns) non-volatile MRAM with folded storage elements," *IEEE Transactions on Magnetics*, vol. 28, no. 5 part 2, pp. 2359-2361, Sept 1992.

[3] W.J. Gallagher, *et al.*, "Microstructured Magnetic Tunnel Junctions," *Journal of Applied Physics*, vol. 81, no. 8, pp. 3741-3746, Apr. 1997.

[4] S. Tehrani, *et al.*, "Progress and outlook for MRAM Technology," *IEEE Transactions on Magnetics*, vol. 35, no. 5 part 1, pp. 2814-2819, Sept 1999.

[5] M. Durlam, *et al.*, "A 0.18um 4Mb Toggling MRAM," *IEEE International Electron Devices Meeting 2003*, pp. 34.6.1-34.6.3, Dec 2003.

[6] J. C. Slonczewski, "Current-driven excitation of magnetic multilayers," *J. Magn. Magn. Mater.*, vol. 159, pp. L1-L7,1996.

[7] "Everspin debuts first Spin-Torque MRAM for high performance storage systems" (Press Release), Nov 2012, Available: http://www.everspin.com/PDF/ST-MRAM_Press_Release.pdf

[8] J. Slaughter, *et al.*, "High density ST-MRAM technology (invited)," *IEEE International Electron Devices Meeting 2012*, pp. 29.3.1-29.3.4, Dec 2012

[9] W. Kim, *et al.*, "Extended scalability of perpendicular STT-MRAM towards sub-20nm MTJ node," *IEEE International Electron Devices Meeting 2011*, pp. 24.1.1-24.1.4, Dec. 2011

[10] M. Durlam, *et al.*, "Nonvolatile RAM based on magnetic tunnel junction elements," *IEEE International solid State Circuits Conference 2000*, pp. 130-131, Feb 2000.

[11] N.D. Rizzo, *et al.*, "A Fully Functional 64 Mb DDR3 ST-MRAM Built on 90 nm CMOS Technology," *IEEE Trans. Magn.*, vol. 49, NO. 7, July 2013.

[12] N. Sakimura, *et al.*, "A 512kb cross-point cell MRAM," *Solid-State Circuits Conference, 2003. Digest of Technical Papers. ISSCC. 2003 IEEE International* , vol., no., pp.278,279 vol.1, 2003

[13] H. Tanizaki, *et al.*, "A 1Mb High-Density Toggle-MRAM with Symmetrical Read/Write Operations," *Non-Volatile Semiconductor Memory Workshop, 2007 22nd IEEE* , vol., no., pp.63,65, 26-30 Aug. 2007

[14] T. Inaba, *et al.*, "Resistance ratio read (R^3) architecture for a burst operated 1.5V MRAM macro," *Custom Integrated Circuits Conference, 2003. Proceedings of the IEEE 2003* , vol., no., pp.399,402, 21-24 Sept. 2003

[15] T.W. Andre, *et al.*, "A 4-Mb 0.18-μm 1T1MTJ toggle MRAM with balanced three input sensing scheme and locally mirrored unidirectional write drivers," *Solid-State Circuits, IEEE Journal of* , vol.40, no.1, pp.301,309, Jan. 2005

[16] "Customer Case Studies: Storage Systems" [Online], Available: http://everspin.com/technology.php?qtype=9

[17] "Customer Case Studies: Industrial Computing" [Online], Available: http://everspin.com/technology.php?qtype=8

[18] "Customer Case Studies: Automotive" [Online], Available: http://everspin.com/technology.php?qtype=12

[19] "Customer Case Studies: Aerospace" [Online], Available: http://everspin.com/technology.php?qtype=11

[20] J. O'Hare, "MRAM: Disruptive technology for storage applications," [online] Available: http://www.electroiq.com/articles/sst/2013/03/mram--disruptive-technology-for-storage-applications.html

978-1-4673-6145-3/13 $31.00 © 2013 IEEE

Scaling Challenges of NAND Flash Memory and Hybrid Memory System with Storage Class Memory & NAND flash memory

Ken Takeuchi

Department of Electrical, Electronic and Communication Engineering, Chuo University
Tokyo, Japan, E-mail: takeuchi@takeuchi-lab.org

Abstract- **This paper summarizes the scaling challenges of the conventional 2D floating-gate cell NAND flash memories [1, 2]. The scaling trends and limits of the bulk and SOI NAND flash memories are investigated in terms of short channel effects and channel boosting leakage from 20nm to below 10nm generation using 3D-device simulation. In the bulk NAND cell, 13nm generation is the scaling limit for realizing both channel boosting during program-inhibit and SCE suppression. The SOI NAND cell scaling limit is decreased to 8nm generation. Then, scaling problems and device design for 3D-stackable NAND flash memory are investigated [3]. Control gate length (L_g) and spacing (L_{space}) are paid attention since they can be separately varied in 3D NAND and significantly affect the cell area of the 3D NAND as well as the electrical characteristics. L_g and L_{space} should be the same to cope with the tradeoff between memory window and disturbance. If the number of stacked layers is 18 with the layer pitch of 40nm, the effective cell size of the 3D NAND corresponds to that of 15nm planar NAND technology. Then, this paper discusses an error prediction (EP) low density parity check (LDPC) error correcting code (ECC) which realizes an over 10-times extended lifetime [4, 5]. As the design rule shrinks, the floating gate (FG)-FG capacitive coupling among neighboring memory cells seriously degrades the memory cell reliability. The EP-LDPC ECC calibrates the inter-cell coupling without access time penalty. Finally, this paper overviews a state-of-the-art hybrid memory solution with storage class memory (SCM) and NAND flash memory for the big data solid-state storage system [5, 6]. Data fragmentation of MLC NAND flash memory is suppressed and efficient MLC NAND flash usage is realized by storing small hot data to SCM. The 3D TSV hybrid SSD realizes 11 times performance increase, 6.9 times endurance enhancement and 93% write energy reduction.**

I. INTRODUCTION

SSDs and emerging storage class non-volatile memories such as PCRAM, ReRAM and MRAM have enabled innovations in various nano-scale VLSI memory systems for personal computers, smart phones, tablets and enterprise servers. There is a strong demand for continuous scaling in floating-gate (FG) type NAND flash memories below 20nm generation. However, serious program disturb errors due to the interferences from neighboring cells [7-10] become prominent. On the other hand, channel engineering and its scaling limitation in NAND flash memory cells are also concerns [11, 12]. Due to the large EOT (16~20nm) in NAND cells, short channel effect (SCE) is degraded in scaled generations. As a result, DIBL induced program disturb has been reported [13]. Also, operation margins may decrease in MLC technologies since large S-factor worsens on/off current ratio [14]. However, suppressing SCE by high channel doping

concentration leads to junction leakage during program-inhibit in bulk NAND flash memories [11].

A thin body, thin BOX fully depleted SOI NAND flash memory is one of the candidates for the future scaled NAND flash memory for the excellent SCE controllability [13, 15]. Moreover, the drawbacks of thin body SOI transistors such as high parasitic resistance, low V_{TH} controllability by channel doping and V_{TH} increase by quantum confinement are less critical in the NAND flash memory than in logic device because the NAND flash memory does not require high cell current during the read and precise initial V_{TH} control (V_{TH} is controlled by the amount of electrons in the FG).

II. SCALLING CHALLENGES OF 2D-NAND FLASH MEMORIES [1, 2]

In this section, the scaling trends of bulk and SOI NAND cells are newly investigated in terms of SCE and leakage during channel boosting from 20nm to below 10nm generation using 3D-device simulation [1, 2]. The SOI NAND pushes the scaling limit of the short channel effect (SCE) and channel boosting leakage from 13nm to 8nm generation.

Fig. 1 shows the program operation of a NAND flash memory. In a program-inhibit bit-line, channel voltage of the NAND string is boosted up to more than 8V to avoid V_{PGM} cell disturb. Thus, the junctions of the NAND cells must withstand high voltage stress. Otherwise, junction leakage occurs and program-inhibit fails because the channel voltage does not sufficiently increase. Hence, channel doping concentration cannot be increased in a bulk NAND cell for suppressing SCE. In SOI NAND flash cell, thinner BOX has better SCE characteristics while too thin BOX may cause BOX leakage during channel boosting. If there is no BOX leakage, channel leakage is greatly reduced because the junction area is very small.

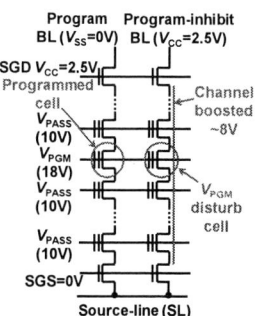

Fig. 1 Schematic of the program operation in a NAND flash memory. Channel voltage is boosted in program-inhibit bit-line and high drain-substrate voltage is applied to the cells

Considering above, the device design is discussed. Figs. 2(a) and 2(b) show the device structure of the bulk and SOI NAND flash memory cells at 15nm generation used in this simulation. Bulk source/drain junction depth X_j and SOI thickness T_{SOI} are fixed to 6nm. Punch-through stopper (PTS) layer, where the doping concentration is higher than the other channel region, is added to the bulk cell. T_S (distance between PTS layer and source/drain junction) is changed to control the SCE. Although SCE is better in smaller T_S, the junction leakage increases. For the SOI NAND cell, T_{BOX} (BOX thickness) is varied.

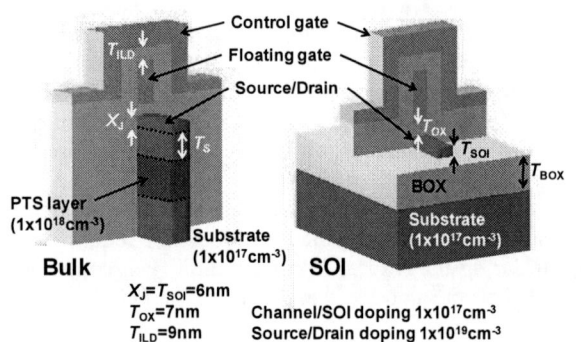

Fig. 2 Device structure of (a) bulk and (b) SOI NAND flash memory cells at 15nm generation for the 3D-simulation in this work. Punch-through stopper (PTS) layer is used in the bulk NAND cell to suppress SCE.

Channel leakage during channel boosting of the bulk and SOI NAND cells having the same SCE characteristics are compared. Junction electric field ($E_{junction}$) and BOX electric field (E_{BOX}) are evaluated for the channel leakage. Although the bulk and SOI NAND cells have the same V_{TH} roll-off characteristics, S-factor and DIBL in the SOI NAND cell are better.

Figs. 3(a) and 3(b) show the potential profile during the channel boosting simulation at 11nm generation for bulk and SOI NAND cells, respectively. SCE in both bulk and SOI NAND cells are made the same as the case in 15nm generation. Source and drain voltages are set to 8V. In the bulk NAND cell (Fig. 3(a)), $E_{junction}$ is found to be higher than 1MV/cm, which is the critical value for the junction leakage. Thus, channel boost in the bulk cell fails at 11nm generation. On the other hand, E_{BOX} in the SOI NAND cell is found to be below 10MV/cm where FN tunneling starts to occur. Note that BOX is made of thermal oxide and its quality is higher than that of the oxide (typically TEOS) filled between the FGs and CGs. Therefore, no BOX leakage occurs even at 11nm generation in the SOI cell.

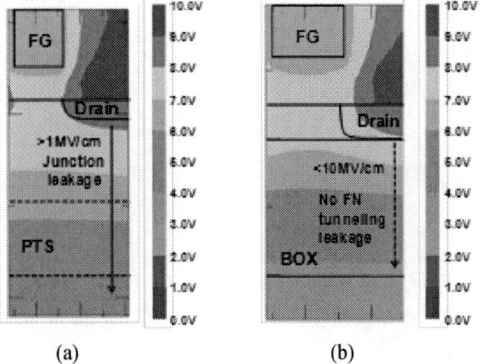

Fig. 3 Potential profile of the (a) bulk and (b) SOI NAND flash memory cells at 11nm generation during channel boosting.

Figs. 4(a) and 4(b) are the $E_{junction}$ and S-factor as a function of T_S in bulk technology, respectively, at 13nm and 11nm generation. Although $E_{junction}$ is below 1MV/cm except for T_S=0nm at 13nm generation, $E_{junction}$ exceeds 1MV/cm at 11nm generation for all T_S. S-factor for 11nm generation is also worse than 13nm generation.

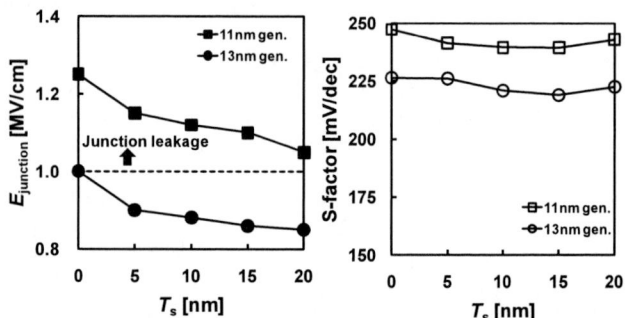

Fig. 4 (a) Electric field at the drain-substrate junction $E_{junction}$ in a bulk NAND cell as a function of T_s (distance between PTS layer and drain junction). (b) S-factor as a function of T_s. At 11nm generation, the bulk NAND cell fails to boost the channel by the junction leakage due to the high $E_{junction}$.

E_{BOX} and S-factor in the SOI NAND cell as a function of T_{BOX} are shown in Figs. 5(a) and 5(b). E_{BOX} is still below 10MV/cm at 11nm generation for all T_{BOX}. S-factor can be also slightly improved by reducing T_{BOX}.

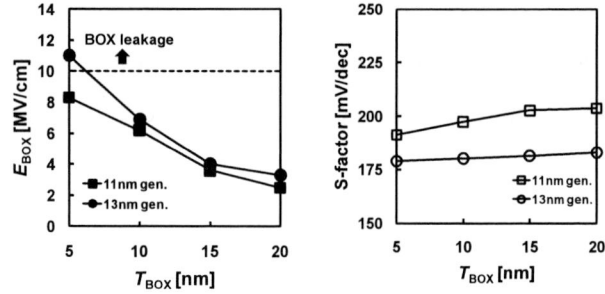

Fig. 5 (a) Electric field of the BOX layer at the drain E_{BOX} in a SOI NAND cell as a function of T_{BOX} (BOX layer thickness). (b) S-factor as a function of T_{BOX}. Above T_{BOX}=5nm, the SOI NAND cell satisfies both small E_{BOX} and S-factor even at 11nm generation.

The scaling trends for the $E_{junction}/E_{BOX}$ and S-factor of the bulk and SOI NAND cells are shown in Figs. 6 and 7. The scaling limit of the bulk NAND cell from the perspective of the junction leakage failure is 13nm generation. In the SOI NAND cell, the scaling limit decreases to 8nm.

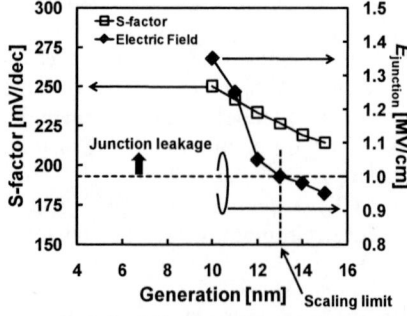

Fig. 6 Scaling trend of the S-factor and $E_{junction}$ in a bulk NAND cell. The scaling limit due to the junction leakage by high $E_{junction}$ is 13nm generation.

978-1-4673-6145-3/13 $31.00 © 2013 IEEE   808

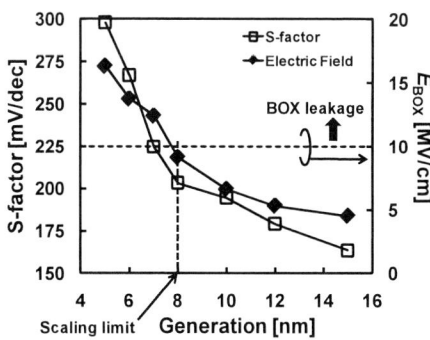

Fig. 7 Scaling trend of the S-factor and E_{BOX} in a SOI NAND cell. Compared with the bulk cell, the scaling limit of the SOI cell extends to 8nm generation.

III. 3D-STACKABLE NAND FLASH MEMORY DESIGN [3]

3D-stackable NAND flash memory (3D NAND) [15-18] has been attracting much attention to overcome the scaling limit of the planar NAND flash memory. In these devices, the number of stacked layers Nlayer is increased to reduce the bit cost instead of shrinking the size of cell area in planar direction using expensive lithography. Scaling and design methodologies for 3D NAND are needed because they are completely different from the planar NAND. Fig. 8(a) shows the simplified cross sectional view of the bit-cost scalable (BiCS) [16, 17] type 3D NAND. One of the problems of the 3D NAND is the decrease of cell density in the planar direction. The BiCS hole must be filled with O/N/O film (~20nm) and silicon channel. Since the O/N/O film is not aggressively scaled to maintain memory window and reliability, the diameter of the BiCS hole is not so scalable. Therefore, N_{layer} should be increased to compensate this drawback under finite taper angle θ in the BiCS hole (Fig. 8(a)). On the other hand, since the minimal line and space lithography pattern is not required for the control gate (CG) formation in 3D NAND, CG length L_g and spacing L_{space} can be separately chosen. This design flexibility is only allowed for 3D NAND. Thus suitable device design for 3D NAND can be explored in terms of L_g and L_{space}.

3D NAND scaling and design methodologies are investigated [17]. Comparing with the planar NAND, the requirements for the L_g and L_{space} of the 3D NAND are comprehensively studied from the cell size and electric characteristics..

The effective cell area (A_{eff}) of the 3D NAND is discussed. A_{eff} is A/N_{layer} where A is the cell area in each layer [19]. The A_{eff} of the 3D NAND in Fig. 8(a) is approximately given as (square layout is assumed [17]),

$$A_{eff} = \{2R_B + 2N_{layer}(L_g + L_{space})\tan\theta + 2t_{ONO} + F\}2/N_{layer} ,$$

where R_B, t_{ONO} and F are the bottom radius of the BiCS hole, total thickness of O/N/O layer and feature size (spacing between BiCS hole), respectively. Fig. 8(b) shows the effective cell area as a function of Nlayer with various layer pitches ($L_g + L_{space}$). eff for the planar NAND is also shown in the figure.

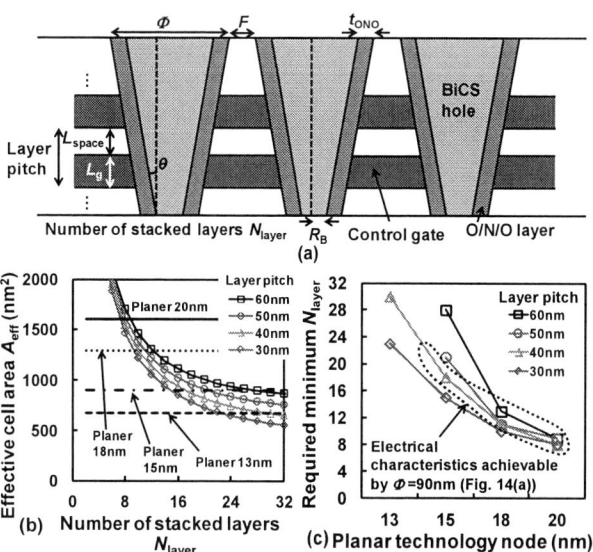

Fig. 8 (a) Cross sectional view of 3D-stackable NAND. (b) Effective cell area A_{eff} (A/N_{layer} : A is cell area in each layer) versus the number of stacked layers N_{layer}. t_{ONO}=20nm, R_B=10nm, F=40nm [5] and θ=1°. (c) Required minimum N_{layer} against technology node of planar NAND cell.

The A_{eff} of 3D NAND strongly depends on the cell pitch. Also, from Fig. 8(b), increasing Nlayer becomes less effective for A_{eff} reduction. This is because the BiCS hole pitch increases as the total height of the 3D NAND increases in the presence of the BiCS hole taper. The required minimum Nlayer to achieve A_{eff} smaller than the planar NAND cell is shown in Fig. 8(c). To achieve effective cell size of 15nm planar NAND technology, 40nm layer pitch with 18 layers is required.

In [3], L_g and L_{space} design window for 40nm layer pitch is explored and achieved by 3D device simulations. Figs. 9(a) and 9(b) show the design window of L_g and L_{space} for 3D and planar NAND. The criteria for unacceptable regions (shaded regions in Fig. 9) are assumed as follows; V_{th} roll-off < -3V, S.S. > 300mV/dec, V_{th} shift < 2V in the programmed cell, V_{th} shift > 0.6V by the neighboring cell and E_{ox_ngb}/E_{ox_pgm} > 0.6. From Fig. 9(a), L_g= L_{space}=20nm (layer pitch of 40nm), is achievable in 3D NAND in terms of the electrical characteristics. Same L_g and L_{space} are preferable to cope with the tradeoff between the large V_{th} shift for the programmed cell and the small V_{th} shift by the neighboring cell. For further improvement, the diameter of the BiCS hole should be decreased. According to Fig. 8(c), the evaluated device structure can realize 15nm planar technology with 18 layer stacks.

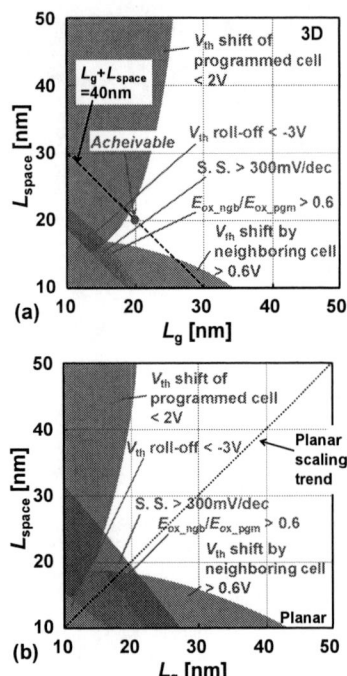

Fig. 9 L_g and L_{space} design window for (a) 3D NAND and (b) planar NAND structure.

Table 1 summarizes the comparison of the 3D and planar NAND. 3D NAND achieves very good I_{on}, S.S. and low V_{pgm} compared with planar NAND. Slight degradations in V_{th} roll-off and V_{th} shift by the stored electrons in the neighboring cell are observed only at the small L_g and L_{space} region.

Table 1 Summary of 3D and 2D NAND

	I_{on}	V_{th} roll-off	S.S.	V_{th} shift (programmed cell)	V_{th} shift (neighboring cell)	Tunnel oxide electric field (E_{ox_ngb}/E_{ox_pgm})	V_{pgm}
Planar NAND	Poor	Fair	Poor	Fair	Fair	Fair	20V
3D NAND	Very good	Poor at small L_g, L_{space}	Very good	Fair	Good at large L_g & L_{space}, Poor at small L_g & L_{space}	Fair	17V
Preferable scaling parameter	L_{space}	-	-	L_{space}	L_g	-	L_{space}

IV. ERROR PREDICTING (EP) LDPC ECC [4, 5]

As the design rule shrinks, the floating gate (FG)-FG capacitive coupling among neighboring memory cells seriously degrades the memory cell reliability. To enhance the error correction capability, an LDPC ECC is proposed for 1Xnm flash memories instead of the Bose-Chaudhuri-Hocquenghem (BCH) ECC. In the 2 bit/cell, 3 reference voltages (V_{ref}) are needed for the BCH. The conventional LDPC requires many, e.g. 21, V_{ref} to get accurate V_{TH} information. The inter-cell coupling is also considered to calibrate the interference. However, the increase in V_{ref} number requires more sequential read cycles. Assuming 50μs cell read time and 21 V_{ref}, the read access time is as much as 1050μs. In case of the 3bit/cell or 4bit/cell, the read access time increases by twice or five-times, which is unacceptably long. To realize both fast read and high reliability, the error

prediction LDPC (EP-LDPC) utilizing only 3 V_{ref} is proposed as show in Fig. 10 [4, 5].

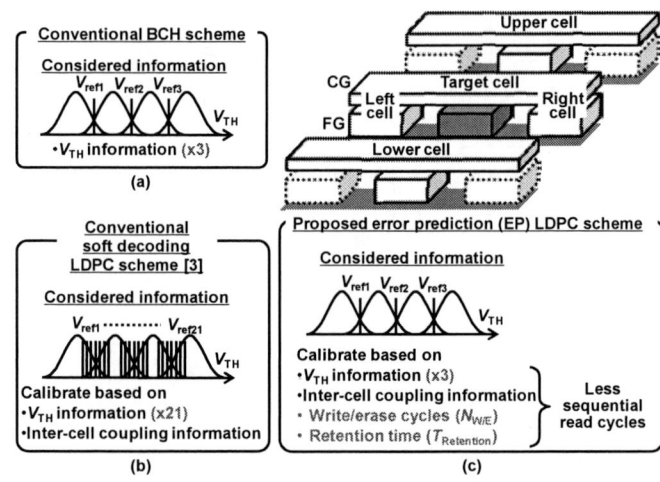

Fig. 10 Comparison of the conventional and the EP-LDPC ECC [4, 5].

The read is 7-times faster than the conventional LDPC. The EP-LDPC corrects errors most effectively because in addition to the V_{TH} and the inter-cell coupling, the write/erase cycles and the retention time are considered for the calibration. As a result, over 10-times extended lifetime is realized.

Fig. 11 shows the hardware architecture of the SSD with ED-LDPC. The error prediction sequence is realized with the simple logic gates in the NAND controller. The additional NAND controller circuit area to the conventional LDPC is negligibly small.

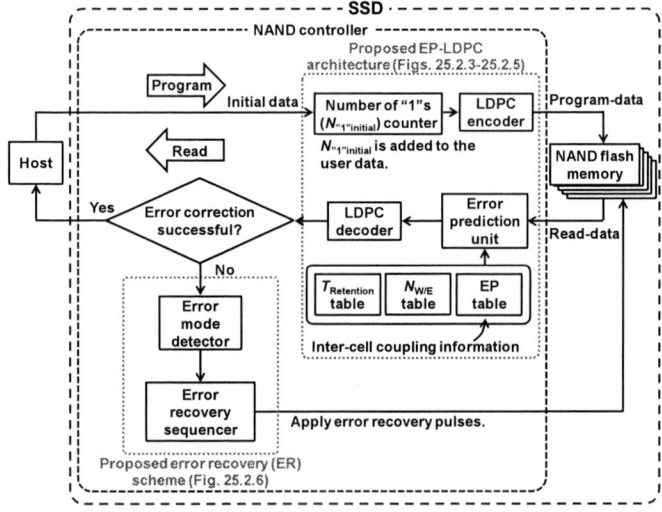

Fig. 11 Hardware architecture of SSD with EP-LDPC ECC [4, 5].

V. HYBRID MEMORY SOLUTION WITH STORAGE CLASS MEMORY & NAND FLASH MEMORY [5, 6]

There is a growing demand for a high performance, highly reliable and low power SSD. A 3D TSV-integrated SSD with hybrid memory configuration which uses storage class memories (SCMs) and NAND flash memories is a promising solution. Among various SCMs, ReRAM is the best candidate due to its high speed, low power operation and potentially high scalability [20, 21]. In [5, 6], the detailed specifications for the ReRAM and architecture for the hybrid SSD are proposed.

The block diagram of the hybrid SSD is shown in Fig. 12(a). The ReRAM uses NAND-like I/F. The polling (Ready/Busy status), which is used in NAND I/F, allows a variable access time. Fig. 12(b) shows the physical image of the hybrid SSD with TSVs.

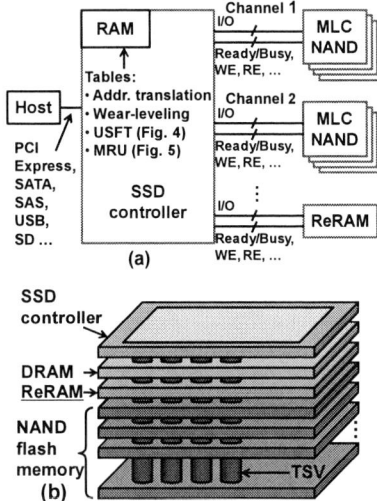

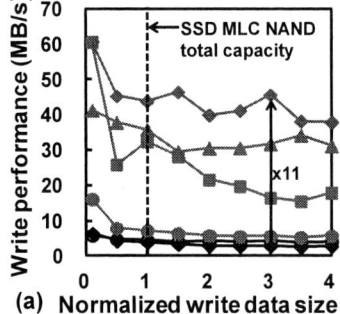

Fig. 12 (a) Block diagram of the proposed 3D TSV-integrated hybrid ReRAM/MLC NAND SSD. Proposed ReRAM uses NAND-like I/F. (b) Physical image of the proposed SSD.

In [5, 6], three data management algorithms are proposed for the 3D hybrid SSD. The key idea is to store hot fragmented data less than the page size to ReRAM and use MLC NAND for sequential data. To evaluate the hybrid 3D hybrid NAND SSD a TLM (transaction level modeling) -based SSD emulator that can comprehensively simulate performance, energy consumption and P/E cycles has been developed. The results for the write performance, write energy and average P/E cycles are shown in Fig. 13.

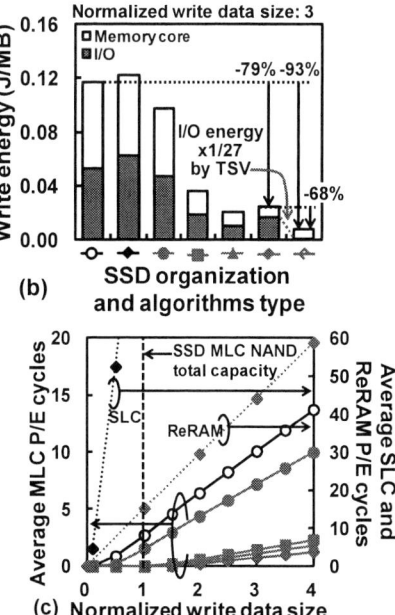

Fig. 13 (a) Write performance, (b) write energy and (c) average P/E cycles of the conventional and hybrid SSDs. The horizontal axis for (a) and (c) is the data size written to the SSD normalized by the SSD MLC NAND total capacity. 100ns/sector is assumed for the ReRAM write and read latency.

Compared with the conventional MLC NAND SSD, the hybrid SSD shows 11 times higher performance and 79% lower write energy (Figs. 13(a) and 13(b)). By using 3D TSV interconnects, the I/O energy is reduced by 27 times because the huge capacitance of the wire bonding is almost eliminated. As a result, the total SSD energy reduction reaches 93%. Furthermore, the slope of the average MLC NAND P/E cycles is decreased by 6.9 times in Fig. 13(c) by the hybrid SSD. This directly corresponds to a reduction in the replacement cost of a SSD storage system because the slope determines the aging speed of the SSD. In ReRAM, a data fragmentation does not occur because the partial overwrite is possible. As a result, the slope of the ReRAM P/E cycles is limited to 28 times of that of the MLC NAND in the hybrid SSD. Assuming MLC NAND endurance of 3×10^3, the required P/E cycles for ReRAM is less than 10^5, which is acceptable for the ReRAM device characteristics. Fig. 14 shows the valid page map of the conventional and hybrid SSD. The valid pages are scattered in the conventional SSD indicating that frequent overwrites have occurred to the MLC NAND. On the other hand, the hybrid SSD efficiently uses ReRAM and shows less fragmentation of MLC NAND because overwrites to MLC NAND are suppressed.

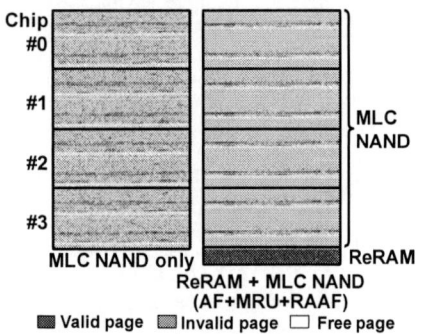

Fig. 14 Comparison of the SSD valid page location.

The required ReRAM latency to obtain sufficient improvements is also investigated in [5, 6]. Fig. 15 shows the SSD write performance and energy as a function of the ReRAM write latency. ReRAM read latency is also varied. From the figures, both ReRAM write and read latency should be less than 3us to maintain high performance and low power operation. Considering 50ns write pulse, the 3us access is achievable for ReRAM in write verify operation.

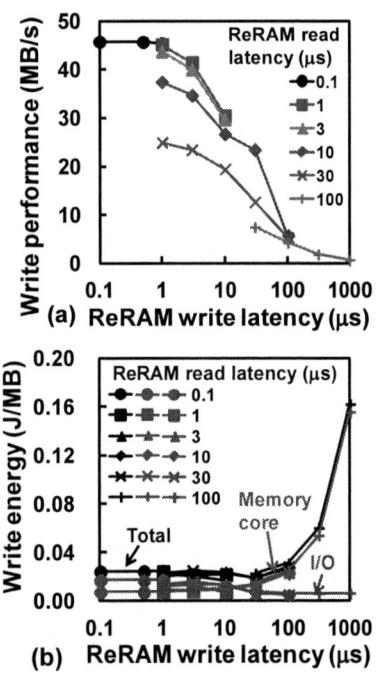

Fig. 15 (a) Write performance and (b) write energy of the hybrid SSD with various ReRAM write and read latency.

VI. CONCLUSION

Scaling challenges of 2D and 3D NAND flash memory and solution for the scaling blockage are discussed. The scaling trends and limits of the bulk and SOI NAND flash memories are investigated in terms of short channel effects and channel boosting leakage from 20nm to below 10nm generation using 3D-device simulation. In the bulk NAND cell, 13nm generation is the scaling limit for realizing both channel

boosting during program-inhibit and SCE suppression. The SOI NAND cell scaling limit is decreased to 8nm generation.

Then, scaling problems and device design for 3D-stackable NAND flash memory are investigated. If the number of stacked layers is 18 with the layer pitch of 40nm, the effective cell size of the 3D NAND corresponds to that of 15nm planar NAND technology.

An error prediction (EP) low density parity check (LDPC) error correcting code (ECC) which realizes an over 10-times extended lifetime is discussed. As the design rule shrinks, the floating gate (FG)-FG capacitive coupling among neighboring memory cells seriously degrades the memory cell reliability. The EP-LDPC ECC calibrates the inter-cell coupling without access time penalty.

Finally, the hybrid memory solution with storage class memory (SCM) and NAND flash memory are reviewed for the big data solid-state storage system. Data fragmentation of MLC NAND flash memory is suppressed and efficient MLC NAND flash usage is realized by storing small hot data to SCM. The 3D TSV hybrid SSD realizes 11 times performance increase, 6.9 times endurance enhancement and 93% write energy reduction.

ACKNOWLEDGMENTS

The author sincerely appreciates Kousuke Miyaji, Shuhei Tanakamaru, Hiroki Fujii, Koh Johguchi, Kazuhide Higuchi, Chao Sun, Yuki Yanagihara, Masafumi Doi and Chinglin Hung for their continuous hard work and efforts. This paper is the summary of their originally reported work about the NAND flash memory, SSD and hybrid memory systems [1-4, 6].

REFERENCES

[1] K. Miyaji et al., *SSDM*, pp. 128-129, 2011.
[2] K. Miyaji et al., *JJAP*, vol.51, no. 4, p. 04DD12, 2012.
[3] Y. Yanagihara et al., *IMW*, 2012.
[4] S. Tanakamaru et al., *ISSCC*, pp. 424-425, 2012.
[5] K. Takeuchi, *Inside Solid State Drives*, Chapter 7,13, 2012.
[6] H. Fujii et al., *Symp. VLSI Circuit*, pp. 134-135, 2012.
[7] K. Prall et al., *NVSMW*, pp. 5-10, 2007.
[8] M. Park et al., *IEEE EDL*, pp. 174-177, 2009.
[9] J. D. Lee et al., *NVSMW*, pp. 31-33, 2006.
[10] Y. S. Kim et al., *IRPS*, pp. 599-604, 2010.
[11] A. Torsi et al., *IEEE ED*, pp. 11-16, 2011.
[12] A. C. K. Chan et al., *IEEE ED*, pp. 2054-2060, 2004.
[13] D. Oh et al., *NVSMW*, pp. 5-7, 2008.
[14] L. Perniola et al., *Symp. VLSI Tech.*, pp. 42-43, 2007.
[15] H. Tanaka et al, *Symp. VLSI Tech.*, pp. 14-15, 2007.
[16] Y. Fukuzumi et al, *IEDM*, pp. 449-452, 2007.
[17] J. Jang et al, *Symp. VLSI Tech.*, pp. 192-193, 2009.
[18] J. Kim et al, *Symp. VLSI Tech.*, pp. 186-187, 2009.
[19] S. Aritome, *IMW short course*, 2011.
[20] Y. S. Chen et al., *IEDM*, pp.717-720, 2011.
[21] K. Higuchi et al., *SSDM*, pp.1011-1012, 2011.

A 28nm High Density 1R/1W 8T-SRAM Macro with Screening Circuitry against Read Disturb Failure

M. Yabuuchi, H. Fujiwara, Y. Tsukamoto, †M. Tanaka, S. Tanaka and K. Nii

Renesas Electronics Corporation, 5-20-1 Josuihon-cho Kodaira, Tokyo 187-8588, Japan

† Renesas Electronics Corporation, 4-1-3 Mizuhara Itami Hyogo 664-0005, Japan

Abstract— **We developed a high density 1R/1W SRAM macro based on 8T-SRAM with an effective scheme for Design for Testability. To achieve a smaller Macro area, a differential sense amplifier is introduced to read the data, where the reference voltage for reading 0/1 data is generated by unselected cell array. In addition, we proposed a screening test circuit for read disturb operation. A 512 kbit two port SRAM macro based upon 28nm process was designed, confirming experimentally that the worst minimum operation voltage (Vmin) can be reproduced by our test circuit. The bit density of 3.16 Mb/mm² was achieved, which is the highest among recent literatures.**

Keywords- 1R/1W SRAM, 8T-SRAM, High density, Screening test circuit, Disturb failure

I. INTRODUCTION

Rapid increase of embedded SRAM capacity can be read from many publications such as a mobile application processor with more than 100 Mbit embedded SRAM [1] and a network processor with huge capacity. This means the importance of minimizing SRAM bit density. In these processors, a two port SRAM (2P-SRAM) macro with a read and write port (1R/1W) is widely used and various techniques for its periphery have been reported [2-4]. Ref. [2] used a dual-port SRAM (DP-SRAM) bit cell to achieve fast access time. In general, DP-SRAM cell area is ~30% larger than 2P-SRAM cell, so that 2P-SRAM macro is preferable for the purpose of large capacity. Ref. [3] and [4] used 2P-SRAM with divided bit line scheme, which results in a large area overhead due to several local sense amplifiers. Therefore, it is necessary to develop 8T-SRAM macro without relying on divided bit line. Furthermore, in dealing with 2P-SRAM cell with large capacity, we need to consider degradation of operating margin or that of minimum operating voltage (Vmin) due to random dopant fluctuations [5]. In particular, 2P-SRAM has a specific degradation referred to as the "read disturb issue," which should be also treated to enhance variation margin. In this work, to realize a high density 2P-SRAM macro, we propose a new scheme for sense amplifier as well as a test circuit that can reproduce the read disturb condition. This enables us not only to achieve the highest density but also to simply screen out the read failure.

II. READ DISTURB ISSUES OF 1R/1W 8T-SRAM

Fig. 1 shows schematic of DP-SRAM and 2P-SRAM, respectively. As for Fig. 1 (a), since the internal nodes of DP-SRAM suffers from "dual" pre-charged BLs (ABT/B, BBT/B), the static noise margin (SNM) gets worse compared to the

standard 6T SRAM, resulting in a large cell area. In contrast, the read operation is executed via RBL in Fig. 1 (b), 2P-SRAM is free from SNM. In the 2P-SRAM, read stability of reading data "1" is different from that of reading data "0." In "0" read operation, read word line (RWL) is activated and the internal node MB is at VDD. Consequently, voltage level of read bit line (RBL) is pulled down by read current which flows through two transistors (RPG, RPD). Thus, "0" read stability is reduced by degradation of read current due to random variation in both RPG and RPD. On the other hand, in "1" read operation, RPG is turned off because MB is 0 V, which keeps the RBL voltage level at VDD. Thus, "1" read stability is affected by the amount of the leakage current in RPD, to which lower Vth for the RPD, lower Vth for PGR or higher Vth for the PDR can contribute.

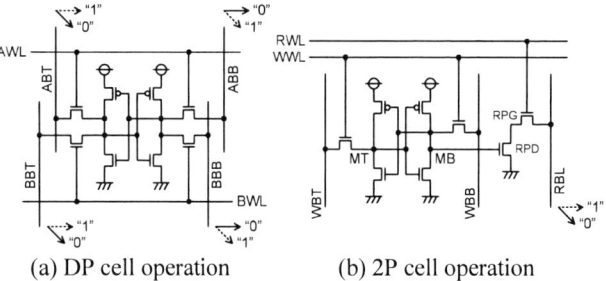

(a) DP cell operation   (b) 2P cell operation

Fig. 1 Schematic of DP SRAM vs 1R/1W 2P SRAM

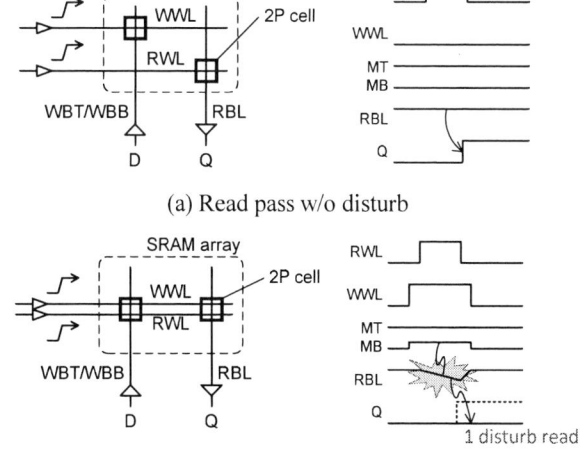

(a) Read pass w/o disturb

(b) Read failure by disturb

Fig. 2 Read disturb failure issue

Fig. 2 depicts a problem during "1" read operation. In Fig. 2 (a), if the read word line (RWL) and write word line (WWL)

978-1-4673-6145-3/13 $31.00 © 2013 IEEE   813

in a different row are activated, the read operation is successfully done. In Fig. 2 (b) where RWL and WWL in the same row are activated, as for the cell to be read, since both the write bit lines (WBT/WBB) are pre-charged at VDD, the MB node which should be kept at 0 V would be pulled-up slightly by WBB via PGR (half-selected). As a result, the RPD turns on weakly, which pulls down the RBL voltage. If the sense amplifier detects this voltage drop, then this is regarded as "0" read, resulting in the read failure (Output "Q" in Fig. 2 (b) stays at 0). This is called as a read disturb failure since the read operation is disturbed by the half-selected WWL. Fig. 3 is the read current distributions for "0" and "1" read operation. Note that the disturbed "1" read leakage current has a broader distribution compared to the "0" read current. This indicates that the larger the SRAM capacity gets in a product chip, the higher "1" read leakage current cell becomes comparable to "0" read current. Fig. 4 indicates Monte Carlo simulation results for the RBL swing with and without read disturb. Due to the fact that the 32 row cells have a smaller BL capacitance, "1" read operation with read disturb shows a large variation in the RBL swing, which causes the read failure.

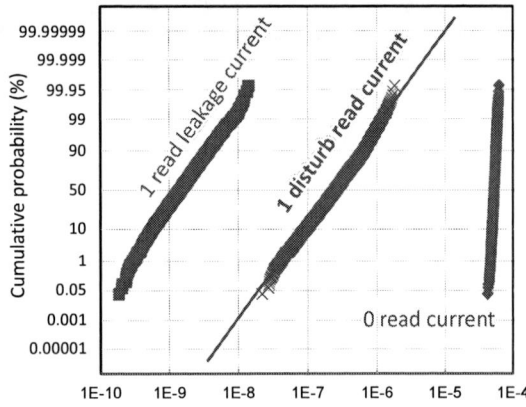

Fig. 3 Monte Carlo simulation of read current at 0.8V, FF, 25°C

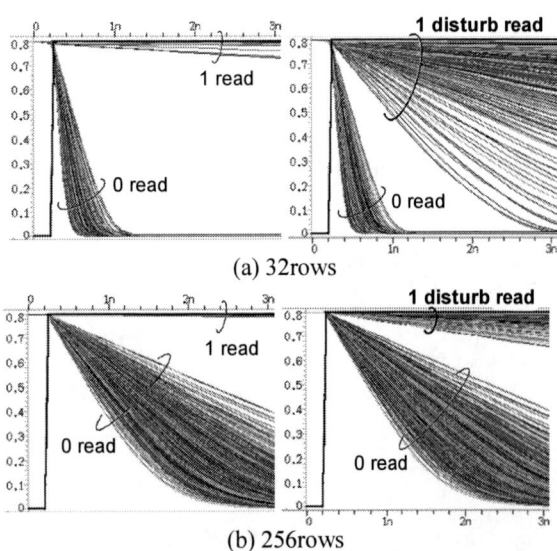

(a) 32rows

(b) 256rows

Fig. 4 Waveform of RBL swing at 0.8V, FF, 25°C

III. SCREENENG TECHNIQUE FOR READ-DISTURB FAILURES

The read-disturb failure is inevitable if the write port and read port are in asynchronous operation. In order to prevent the yield loss due to this failure mode in actual products, we have to reproduce the read-disturb mode effectively and adopt a new screening test before shipping. From the view point of turn-around-time for product testing, giving every possible WL skew (skew between RWL and WWL) for every chip is not a possible solution. Therefore, we propose a test scheme [6] so that the read disturb failure can be reproduced. Fig. 5 shows the basic timing chart for our proposal. SEL is a mode select signal to switch normal mode or test mode. In the normal mode (SEL=L), WWL and RWL are operated in asynchronous operation, and rise according to the activation of write-port clock (CLKA) and read-port one (CLKB), respectively. WWL pulse width is larger than RWL because WWL needs to be a long pulse not to degrade the write ability. In the test mode (SEL=H), TCLK is set to be a common clock supply both for write and read port. The pulse width of RWL should be larger than that of WWL in order to stabilize the SRAM internal node for disturbing (floated MB node in Fig. 2. (b)). This enables to keep "1" read current flowing during the activation of RWL. Fig. 6 depicts our proposed circuitry. Three selectors are introduced to the inputs of write/read clock generators as well as the output of read clock one, which realizes the operations explained in Fig. 5. The delay chains are included to realize the delayed activation of RWL compared to that of WWL.

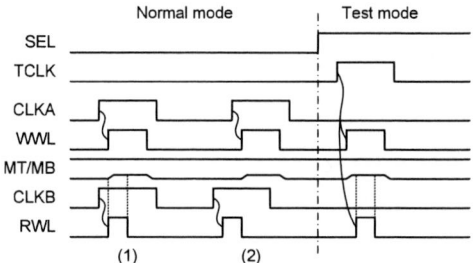

Fig. 5 Timing chart of screening test

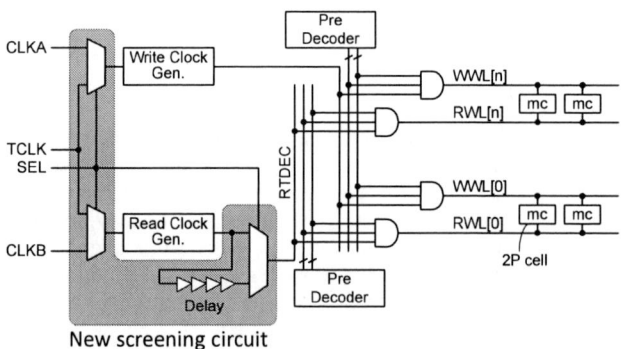

Fig. 6 Proposed clock line block chart

IV. HIGH DENSITY 8T-SRAM MACRO ARCHITECTURE

In conventional 6T SRAM, since a bit cell has a complementary bit line (BL) pair, a differential voltage sense amplifier (SA) can be used to read out a data from the bit cell,

978-1-4673-6145-3/13 $31.00 © 2013 IEEE

which enables a fast read operation without considering the speed degradation due to long BL capacitance. However, this SA is not applicable to the 2P-SRAM because the bit cell has a single-end BL for reading. To achieve a faster read operation, divided BL method is generally used [2,3], which inevitably accompanies with an area penalty due to the introduction of several local SAs. Therefore, we propose a new differential amplifier scheme by dividing memory array into two (upper and lower) MATs; differential voltage between RBL of the selected MAT and that of the unselected MAT is amplified.

Fig. 7 indicates the proposed schematic using differential SA. RTDEC is the internal clock for read (see Fig. 6), Y is the column select signal, and RDE is the read enable signal. In the read operation, column signal Y0 is selected so that upper read BL URB0 and lower one LRB0 connects to SA. Additionally, LRB0 is connected to GenRef circuit by UBYL signal. GenRef circuit is implemented in order to generate the reference voltage (Vref) for reading "0" or "1". Note that there is another method to generate Vref, such as using voltage divider, but we adopt a method of discharging current. This is because, in a compiled RAM with small row structure, it is difficult to control the sense timing because of a small margin between Disturb "1" read and "0" read operations. The margin is defined by the distribution of RML swing for "1" read and "0" read, which can be observed in Fig. 4. Assuming that the discharging current flowing in Genrel is I_{ref}, the voltage levels for LCR (V_{LCR}), UCR for 0 read (V_{UCR0}) and UCR for disturbed "1" read (V_{UCR1}) are expressed as follows:

$$V_{LCR} = VDD - \frac{T_{DWL} I_{ref}}{C_{LRB}} , V_{UCR0} = VDD - \frac{T_{RWL} I_{cell}}{C_{URB}} , V_{UCR1} = VDD - \frac{T_{RWL} I_{leak}}{C_{URB}}$$

where discharging time T_{DWL} and T_{RWL} are the same due to synchronous operation, while each capacitance C has the same value since the BL capacitance for upper and lower MATs are equivalent. Defining that ΔV_0 and ΔV_1 are the sensing voltage for V_{UCR0} and V_{UCR1} measured from V_{LCR}, which leads to,

$$\Delta V_0 = V_{UCR0} - V_{LCR} = \frac{T_{WL}}{C_{RB}}\left(I_{cell} - I_{ref}\right), \Delta V_1 = V_{LCR} - V_{UCR1} = \frac{T_{WL}}{C_{RB}}\left(I_{ref} - I_{leak}\right)$$

In order to gain the maximum sense margin, the condition of $\Delta V_0 = \Delta V_1$ is required. From this equation, I_{ref} should be expressed as below.

$$\Delta V_0 = \Delta V_1 \Leftrightarrow I_{cell} - I_{ref} = I_{ref} - I_{leak} \Leftrightarrow I_{ref} = \frac{I_{cell} + I_{leak}}{2}$$

Since I_{leak} is generally much smaller than I_{cell}, it is found that the I_{ref} should be half of I_{cell}. GenRef is designed to generate this reference current, where transistor sizes L/W are appropriately optimized not to be affected by the random dopant fluctuations. Fig. 8 represents the distributions for 1read disturb current, 0 read current and I_{ref}, confirming that reference current and voltage are between 0 and 1 read. This is how the voltage differential between UCR and LCR is amplified by the SA and read out to Q. Fig. 9 summarizes the simulated waveform for proposed circuit. In the read cycle, reference voltage (LCR) successfully comes between 0 read and 1read (UCR). Also note that in the disturb test cycle, the pulse width of RWL pulse is inside that of WWL, which reproduce the worst condition due to the read disturb.

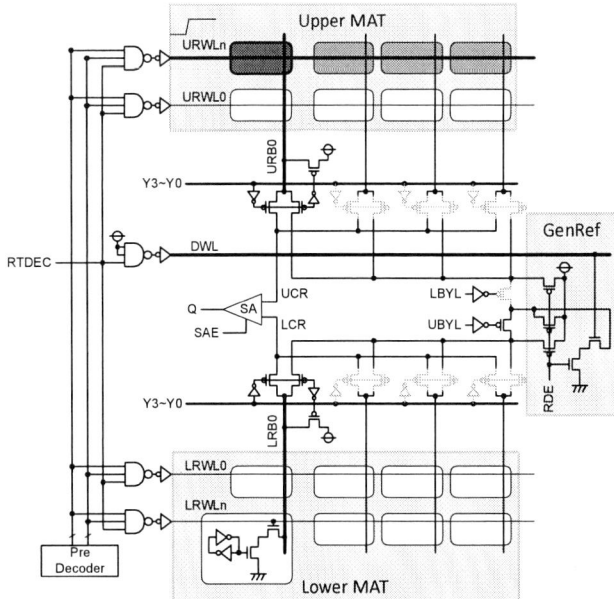

Fig. 7 Proposed deferential sense amplifier schematic

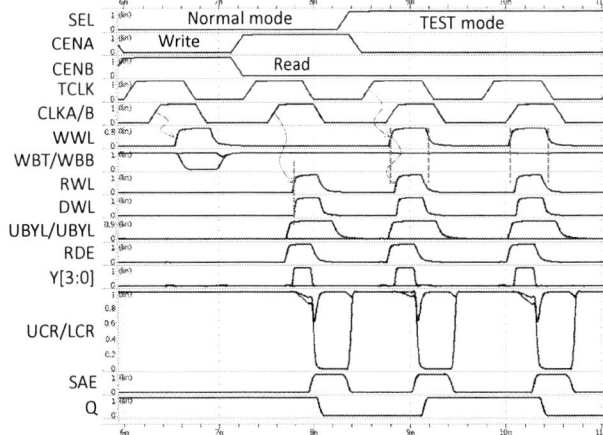

(a) Distribution of current   (b) RBL swing

Fig. 8 Monte Calro simulation of Iref and read current

Fig. 9 waveform of read, write and disturb test operation at 1.0V, TT, 25°C

978-1-4673-6145-3/13 $31.00 © 2013 IEEE   815

V. FABLICATION AND EVALUATION USING 28NM TECHNOLOGY

The test chip including the proposed 512 kbit 2P-SRAM macro was designed and fabricated using 28-nm high performance High-K and Metal Gate bulk technology. The designed 512 kbit macros are shown in Fig. 10. Table I is the features of proposed macro. Note that the bit density of 3.16 Mb/mm² is achieved, which is the highest density compared with other papers. The area overhead due to this test circuit is 0.01%. Fig. 11 plots the distribution of the measured access time, indicating the median is 593 ps at 1.0 V. Fig. 12 compares standby, active power and access time with the macro using conventional DP-SRAM, confirming lower power and higher speed simultaneously. Fig. 13 shows the "1" read Vmin in normal operation and screening test mode operation. Since our architecture can reproduce the read disturb situation which is specific for the 2P-SRAM, we succeeded in extracting the worst Vmin higher than Vmin in the normal operation.

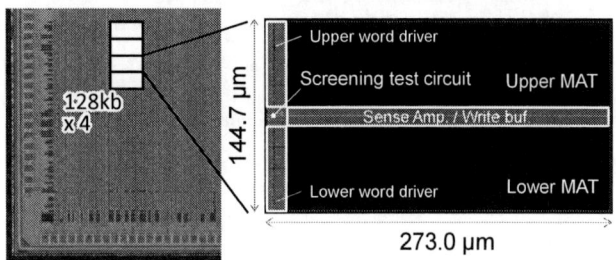

Fig. 10 Microphotograph of the test chip

TABLE I Features of fabricated SRAM macro

Process	28nm HK+MG CMOS bulk technology
Memory Capacity	Total : 512 kb
	Macro: 128 kb (2 kword x 32b) x 4
Physical size	144.7 μm x 273.0 μm
Bit density	3.16 Mb/mm²
Area penalty	0.01% @ Screening test circuit

TABLE II Comparison of SRAM bit density

	Ref.[2]	Ref.[6]	Ref.[7]	Proposed
Technology	28nm	28nm	45nm	28nm
Clock	Sync.	Async.	Async.	Async.
Cell type	DP	DP	DP	2P
Port	1R/1W	2RW	2RW	1R/1W
Bit density	1.57 Mb/mm²	2.02 Mb/mm²	2.09 Mb/mm² @ 28nm equivalent value	3.16 Mb/mm²

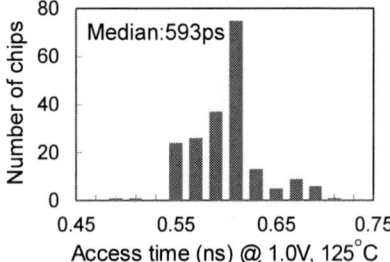

Fig. 11 Measured access time

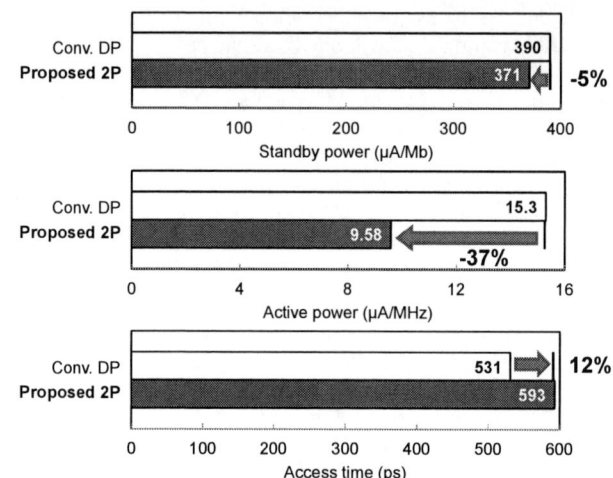

Fig. 12 Comparison of power and speed at 1.0V, 125°C, median

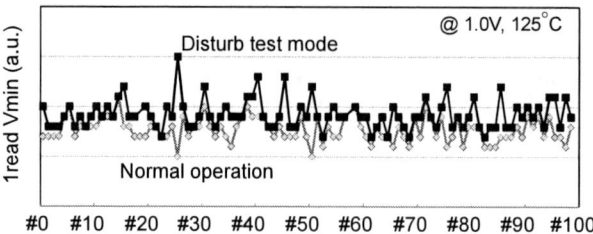

Fig. 13 Comparisons of Vmin between proposed test mode

VI. CONCUSION

In this study, we proposed a new screening test circuit and architecture for high density 2P-SRAM macro. The fabricated test chips using 28-nm process achieved the bit density 3.16 Mb/mm² and demonstrated 593 ps read access time at 1.0 V operation. In addition, our scheme enabled to extract the worst Vmin due to the read disturb test effectively.

REFERENCES

[1] M. Fujigaya et al., "A 28hm High-K Metal-Gate Single-Chip Communications Processer and LTE/HSPA+-Capable Baseband Processor," ISSCC Dig. Tech. Papers, pp.156-157, 2013.

[2] Y. Ishii et al., "A 28nm 360ps-access-time two-port SRAM with a time-sharing scheme to circumvent read disturbs," ISSCC Dig. Tech. Papers, pp.236-238, 2012.

[3] S. Ishikura et al.,"A 45 nm 2-port 8T-SRAM Using Hierarchical Replica Bitline Technique With Immunity From Simultaneous R/W Access Issues," JSSC pp.938-945, 2008.

[4] S. P. Park et al., "Column-Selection-Enabled 8T SRAM Array with ~1R/1W Multi-Port Operation for DVFS-Enabled Processors," ISLPED, pp.303-308, 2011

[5] Y. Tsukamoto et al., "Dynamic Stability in Minimum Operating Voltage Vmin for Single-port and Dual-port SRAMs," CICC, pp.1-4, 2011

[6] Y. Ishii et al., "A 28 nm Dual-Port SRAM Macro With Screening Circuitry Against Write-Read Disturb Failure Issues," JSSC, pp.2535-2544, 2011

[7] K. Nii et al., "A 45-nm single-port and dual-port SRAM family with robust read/write stabilizing circuitry under DVFS environment," VLSI cirucits, pp.212-213, 2008

A HKMG 28nm 1GHz Fully-Pipelined Tile-able 1MB Embedded SRAM IP with 1.39mm^2 per MB

Ming-Zhang Kuo, Osamu Takahashi, Ping-Lin Yang, Cheng-Chung Lin, Min-Jer Wang, Ping-Wei Wang, Sang-Hoo Dhong

Taiwan Semiconductor Manufacturing Company, Design Technology Platform, R&D, Hsinchu, Taiwan

Abstract-**A fully-pipelined tile-able 1MB SRAM IP with a 0.127um^2 cell in a HKMG 28nm bulk technology has an area of 1.39mm^2/MB with 79.2% array efficiency. It operates with 2-cycle latency up to 1GHz. The no-repair hardware has a circuit limited yield of 99.92 and 53% at 100 and 850MHz, respectively with 0.75V V$_{DD}$. A Data Retention Voltage of 0.42V has been measured. (Keyword: SRAM, 28nm, eDRAM)**

I. Introduction

A fully-pipelined tile-able 1MByte SRAM IP using a 0.127um^2 SRAM cell in a High-K Metal-Gate (HKMG) 28nm bulk technology [1] has been developed. The objectives of this IP are; (1) to provide a tile-able embedded SRAM IP of semi-custom quality as an alternative to existing memory compilers, and (2) to develop a highly scalable SRAM solution competitive in power, performance, area, and cost (PPAC) to an eDRAM [2,3] in a 28nm node and beyond.

The importance of memory has been growing rapidly in a SoC. The memory elements in a SoC are traditionally implemented by a memory compiler. A memory compiler can cover a wide range of the design space with a variable array size; however, its flexibility and productivity come at the cost of an inferior PPAC to that of a full custom design. A large memory subsystem, such as L2 and L3, worsens the situation.

We propose a tile-able semi-custom memory IP with a fixed array size as a mitigating solution. A size of 1MB is chosen to primarily satisfy L2 and L3 memory requirements. Currently an eDRAM is a preferred solution in this arena. However, scaling of a logic-technology compatible eDRAM cell has been more difficult than that of a SRAM cell, giving an increased cell size in F^2, shown in Fig. 1.

In Fig.2, we reduced further the difference in mm^2/MB between the two to 28%, employing a minimalistic circuit-design approach and technology-design co-development model. With a difference of 28%, the extra processing cost of an eDRAM multiplied by its typical relative area ratio to logic circuitry in a SoC tilts the PPAC in favor of a SRAM.

Previous eDRAM work [6] also reported an increased latency of a high density SRAM over that of an eDRAM from a longer wire delay. This issue was addressed by pipelining the memory fully. Full pipelining provides a much higher throughput, mitigating any potential performance loss due to an increased latency.

The total power consumption of the IP is kept competitive through a low DRV (<0.5V) in a data retention mode and a nominal V$_{DD}$ of 0.85V. An eDRAM requires a high voltage (~1.5V) for its word-lines constantly for refresh operation, giving much higher total power consumption.

The IP supports a 72-64 DEDSEC code identical to the previous work [6] for soft-error protection. An eDRAM has a lower SER than an SRAM; however, the ECC code in an eDRAM corrects additional variable-retention time (VRT) fails. VRT fails increase substantially in a finer feature technology. We believe that the overall bit-error rate (BER) for the eDRAM [6] and this IP are comparable when protected by an ECC code; albeit, any quantitative analysis of the overall BER difference was forgone for a lack of resource.

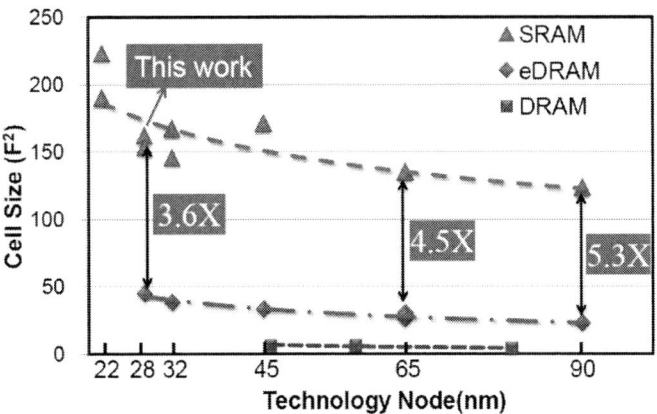

Fig. 1. Normalized cell area in F^2, scaling of eDRAM cell has been more difficult than that of a SRAM cell

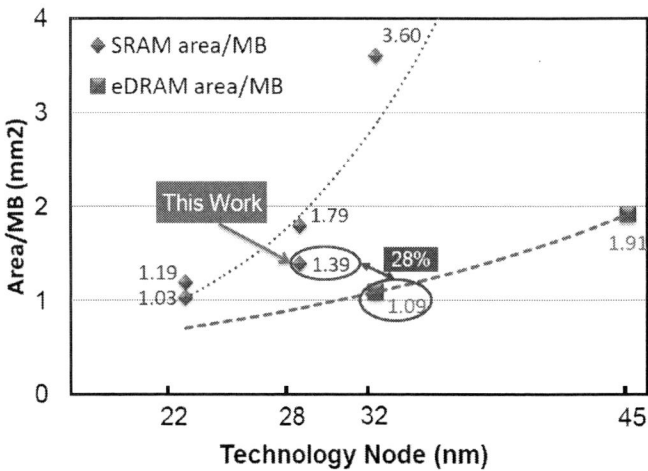

Fig. 2. Comparison of Area/MB (mm^2) .

978-1-4673-6145-3/13 $31.00 © 2013 IEEE

II. 1MB SRAM IP

A. Overview

The 1MB SRAM IP has an array efficiency of 79.2%, achieving 1.39mm²/MB (1.56mm²/MB not counting ECC bits as data bits). Its target frequency is 1GHz at a technology nominal power supply of 0.85V, 90°C, SS corner with a 4.5σ slow cell. The simulated average active (R&W) and static power numbers are 135 and 53mW, respectively, at 1GHz, 0.85V V_{DD}, 90°C, TT corner. The SRAM cell (0.127um²) was co-developed by technology and design teams. The minimalistic design style adopted also reduces technology learning cycles, which makes in turn technology-design co-optimization easier.

B. Array Architecture

A 1MB block (Fig. 3) consists of 64 sub-arrays. Each sub-array has 16kB (512 words & 148 bit-line pairs) with a 2:1 multiplexing for top/bottom and a 4:1 multiplexing among the interleaving columns. Four redundant and 16 ECC bits provide column reparability and support a 72-64 DEDSEC code.

The design has a latency of 2 cycles in a 1MB configuration. As shown in Fig. 4, address decoding is done in the first cycle. The result is stored in the word-line latches. The second cycle consists of word-line activation and latching of the output data or writing to a cell. Fig. 5 shows a 2kb column slice and the differential sense amplifier (SA) used. Activation of the SA is done by a delayed half-cycle clock. The slower clock delays the leading edge of SA activation signal and increases bit-line signal development time.

C. Larger Memory Subsystem Application

Using the 1MB IP and other supporting IPs, a larger memory subsystem can be built. Fig. 6 illustrates a possible 4MB configuration with IP blocks such as an eFuse, ECC generator, and a store queue to resolve any data contention issue at the same address. This operates in 4 cycles, having one cycle added before and after the 1MB IP cycles. The concept can be further extended to satisfy the requirements of SoC and 3D chips.

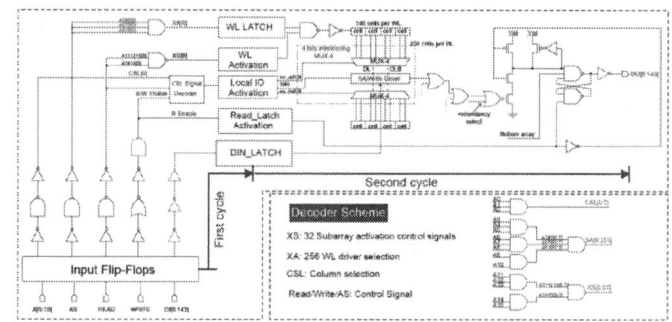

Fig. 4. Read and write path of 1MB SRAM IP

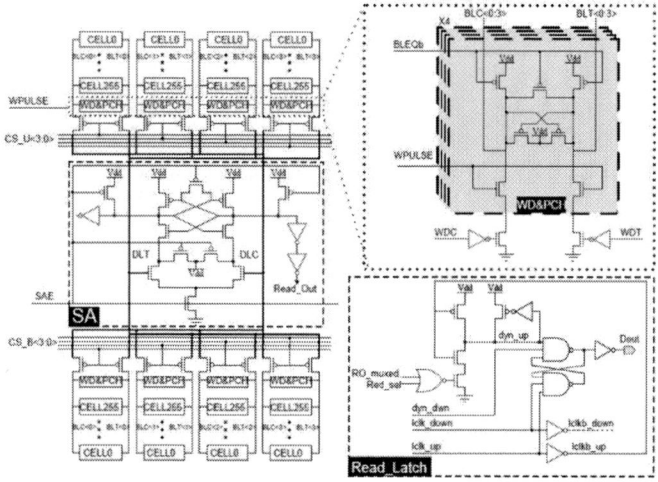

Fig. 5. 2kb column slice schematics including sense amplifier, write driver, and read out latch

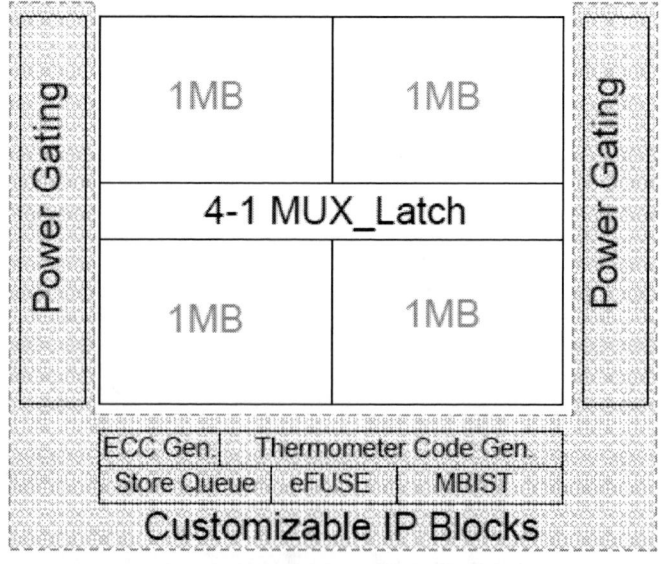

Fig. 6. A possible configuration of 4MB block

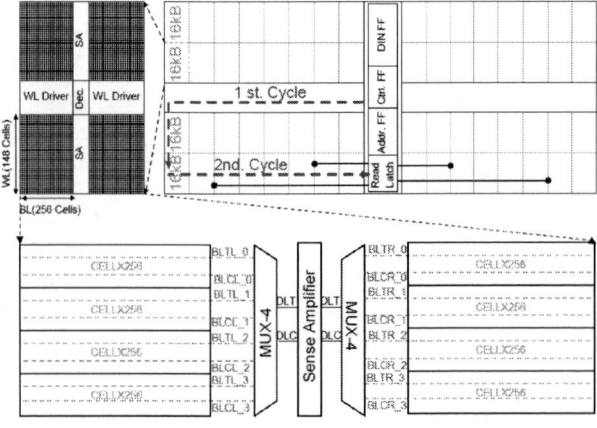

Fig. 3. 1MB IP block diagram with its cycle boundary

D. Cell Leakage Tolerance Simulation

A correct read operation occurs when (1) correct data were stored in previous write operations and (2) the SA offset voltage is smaller than the bit-line differential signal reduced by the total leakage current from the unselected cells. The SRAM cell was optimized for write operation. More detailed statistical analysis of cell leakage showed that guard-banding the cell leakage current by 10X or more should suffice for the required CLY without any R/W assistance circuitry. The optimum number of cells per bit-line (256cells/BL) was determined by simulations using a sub-array leakage model (Fig. 7). The selected cell was modeled as a 4.5σ slow cell and the total leakage current from unselected cells were modeled using nominal devices at the FF corner. This combination of the cell models was dictated from the leakage analysis mentioned previously. The simulations showed that the 256 cells/BL configuration can tolerate up to 11 times the total nominal leakage current of the unselected cells. This gives an adequate signal margin to the failing point of 14 times leakage, meeting the CLY requirement.

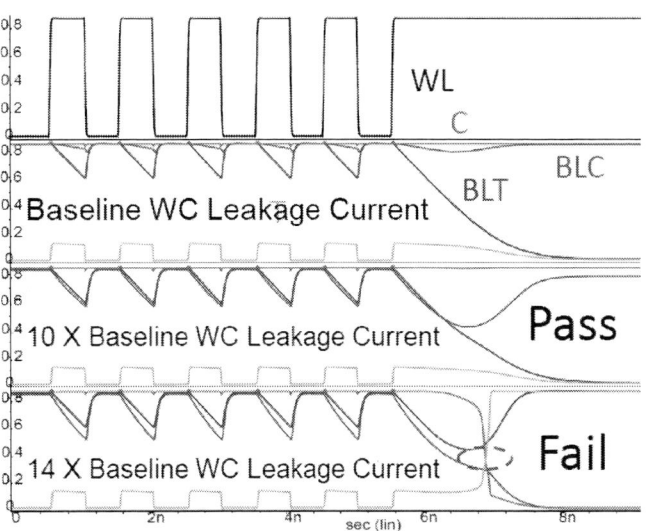

Fig. 8. Simulation results with 1X, 10X, and 14 X leakage currents at FF corner. Incorrect read occurs when leakage current is 14X of baseline.

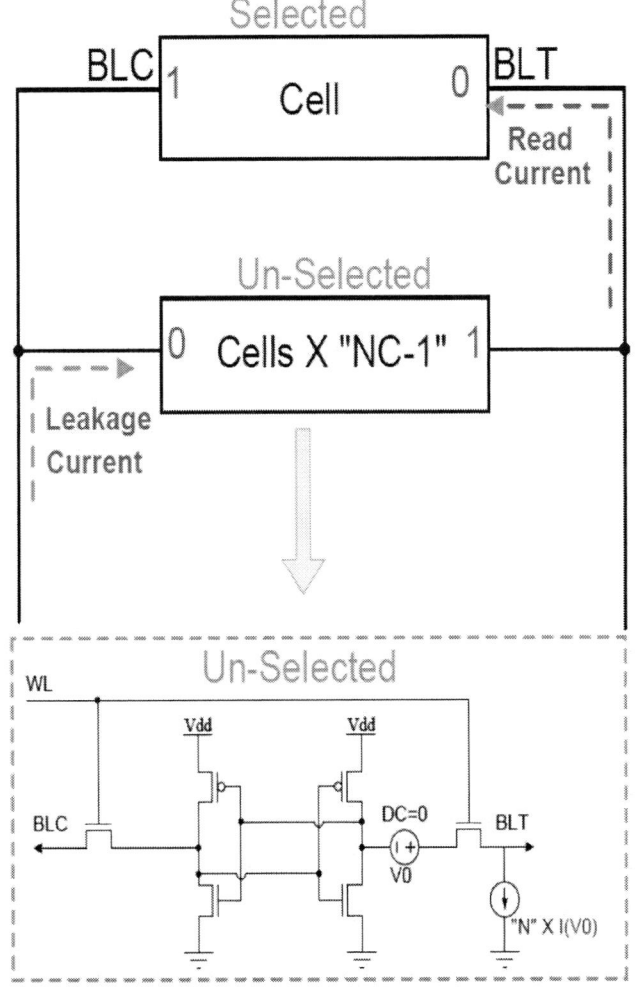

Fig.7. Cell leakage current tolerance simulation model. The selected cell is modeled by 4.5σ slow cell and the un-selected cells are modeled by nominal cell at FF corner

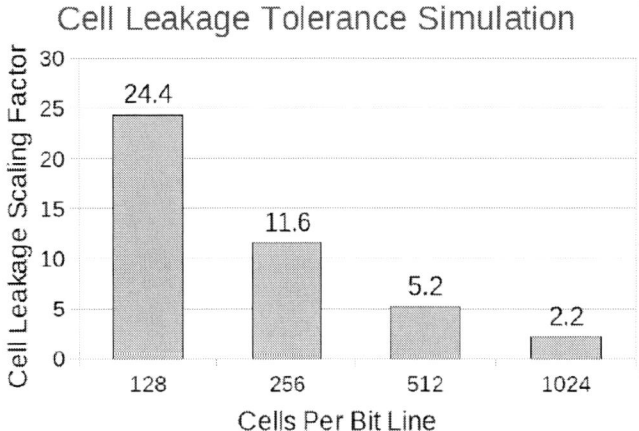

Fig. 9. Leakage tolerance analysis results with different array configuration. 256 cells/BL could tolerate 11.6X leakage current of baseline.

III.   Measurement Result

Fig. 10 shows the chip micrograph of 1MB SRAM IP which was fabricated in a HKMG 28nm technology with 6 metal layers. Fig. 11 shows the measured CLY of the 1MB IP without repair, achieving 99.92 and 53% for 100 and 850MHz, respectively, at 0.75V V_{DD}. A DRV of 0.42V (Fig. 12) was measured using a sequence of Write "1", retention period (RP) of 1sec, Read/Verify (R/V) "1", Write "0", RP 1sec, and R/V "0" for 256 addresses and 144 I/O. It agreed within 75mV to a simulation-predicted DRV of 350mV. The measured static current is 10.14mA at 0.85V V_{DD}, 60°C, SS corner as shown in Fig.12. The total current (Fig. 13) in an active mode is 23.25 and 50.86mA, respectively, at 100 and 600MHz, 60°C, SS corner, 0.9V.

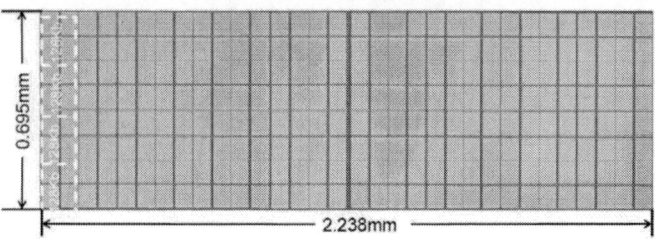

Fig. 10. Chip Micrograph of 28nm 1MB SRAM IP

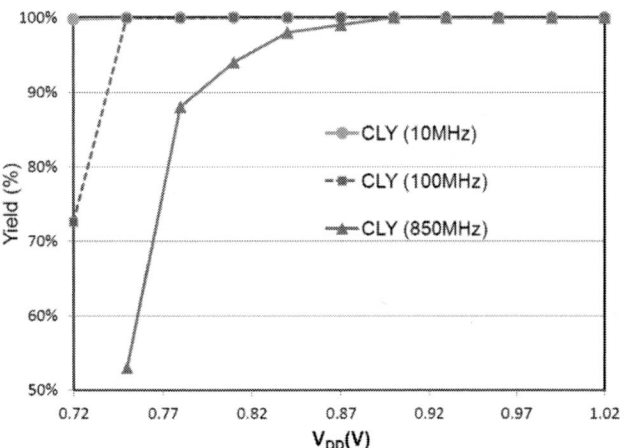

Fig. 11. 1MB IP without repair CLY distribution at different frequencies

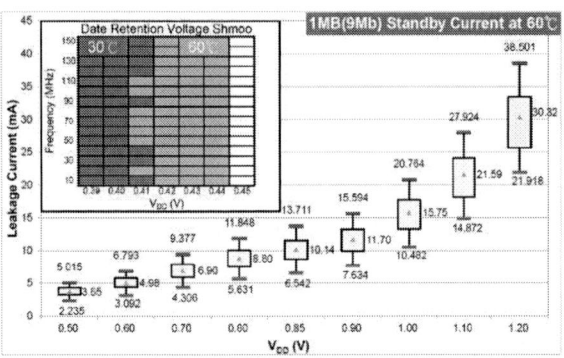

Fig. 12. DRV and box plots of leakage current measurements at 60°C, SS corner

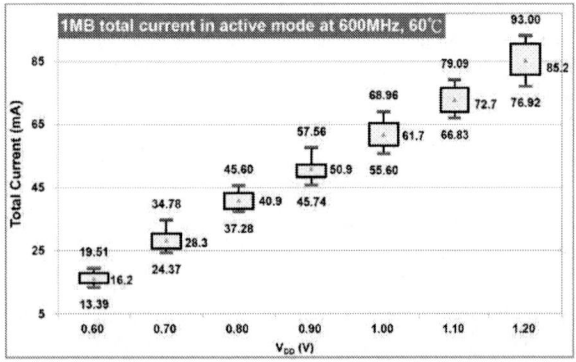

Fig. 13. Box plots of active current measurements at 600MHz, 60°C, SS corner

IV. Conclusions

A fully-pipelined 1MB SRAM IP in a HKMG 28nm bulk technology has been developed. The IP is tailored as an alternative to a memory-compiler generated large SRAM and a potential eDRAM substitute in a 28nm technology and beyond. It operates with 2 cycles at a frequency up to 1GHz, 0.85V, 90°C, SS corner. It is tile-able and easily expandable to a larger configuration. A delayed half-cycle clock scheme is chosen to ensure sufficient sense margins by slowing down the frequency when needed. The technology-design co-developed SRAM cell and a minimalistic design approach gave a highly competitive PPAC, $1.39mm^2$/MB with 6 metal layers. Extensive leakage simulations and analysis determined the array architecture, meeting the CLY requirement with a sufficient manufacturing guard band. Hardware results showed adequate design margins. The characteristics of the 1MB IP are summarized in the table I.

Table I
1MB IP Characteristics Summary

Technology	28nm Bulk CMOS
Technology Features	High-K metal gate, strained silicon
Metal Usage	6 metal layers
IP Block Size	1MB(9Mb)
Memory Cell Area	$0.127um^2$
IP Array Density	$1.39mm^2$/MB(8Mb)
Array Efficiency	79.2%
Access Cycles / Frequency	2 cycles / up to 1GHz
Nominal Supply Voltage	0.85V

References

[1] S. Y. Wu, et al., "A Highly Manufacturable 28nm CMOS Low Power Platform Technology with Fully Functional 64Mb SRAM Using Dual/Tripe Gate Oxide Process," Dig. Symp. VLSI Tech., pp. 210-211, June 2009.

[2] K. C. Huang, et al., "A High-Performance, High-Density 28nm eDRAM Technology with High-K/Metal-Gate," IEDM Dig. Tech. Papers, pp. 24.7.1-24.7.4, Dec. 2011.

[3] J. Golz, et al.,"3D Stackable 32nm Higk-K/Metal Gate SOI Embedded DRAM Prototype," Dig. Symp. VLSI Circuits, pp. 228-229, June 2011.

[4] D. Weiss, et al., "An 8MB Level-3 Cache in 32nm SOI with Column Select Aliasing," ISSCC Dig. Tech. Papers, pp. 258-259, Feb. 2011.

[5] E. Karl, et al., "A 4.6GHz 162Mb SRAM Design in 22nm Tri-Gate CMOS Technology with Integrated Active VMIN-Enhancing Assist Circuitry," ISSCC Dig. Tech. Papers, pp. 230-231, Feb. 2012.

[6] J. Barth, et al., "A 45nm SOI Embedded DRAM Macro for the POWER[TM] 32MByte On-Chip L3 Cache," IEEE J. Solid-State Circuits, vol. 46, pp. 64-75, Jan. 2011.

A Bit-by-Bit Re-Writable Eflash in a Generic Logic Process for Moderate-Density Embedded Non-Volatile Memory Applications

Seung-Hwan Song, Ki Chul Chun, and Chris H. Kim

Dept. of ECE, University of Minnesota, 200 Union Street SE, Minneapolis, MN 55455 USA

Abstract- **A bit-by-bit re-writable embedded flash memory is demonstrated in a generic 65nm logic process for moderate-density embedded non-volatile memory applications. The proposed 6T embedded flash memory cell improves the overall cell endurance by eliminating redundant program/erase cycles without disturbing cells in the unselected wordlines. A multi-story high voltage switch utilizes four boosted supply levels generated by a compact voltage doubler based on-chip negative charge pump.**

I. INTRODUCTION

Embedded Non-Volatile Memory (ENVM) technology has been employed in a number of applications such as post-silicon tuning, memory repair, on-line field test, and secure ID storage [1-6]. ENVM is also a critical component for self-healing applications where information regarding time dependent failure mechanisms such as circuit aging must be retained during system power off periods. Anti-Fuse One-Time-Programmable (AF-OTP) NVM memory [1, 2] has been extensively used for memory repair in standard logic processes; however, it does not allow the cell to be re-programmed after the anti-fuse has been programmed. Moreover, the Charge Pump (CP) has to provide a relatively large program current which incurs a significant area overhead. A single-poly embedded flash (eflash) memory, on the other hand, utilizes Fowler-Nordheim tunneling [3-6] which can support multiple program and erase operations (e.g., >1000 times) without a significant program current.

Table I compares the key features of the various single poly eflash memories. The 10T differential eflash cell [4] stores both the true and complementary values for high voltage margin at the expense of a large cell size and is capable of a bit-by-bit write; however, the high write voltage (VPP) and write protection voltage (hVPP) are applied to the BL's during the write operation which results in disturbance issues in the unselected WL cells. Another prior WL-by-WL erasable eflash [6] does not require boosted BL's, allowing disturbance free erase and program operations for multiple unselected WL's; however, it cannot erase the cell data on a bit-by-bit basis, as every cell on a selected WL is exposed to a high voltage stress and is erased simultaneously prior to the program operation. This inadvertently results in some '0' cells being cycled unnecessarily (Fig. 1). Since only 32 out of 128 cells are updated at a time, for a 32b data bus and a 128 cells/WL architecture, the '0' cells in the unselected columns and some '0' cells in the selected columns are unnecessarily erased to '1' state and then programmed again to '0' state. Thus, the overall cell endurance can be severely impacted

especially when only a small fraction of the stored data needs to be updated frequently.

TABLE I. SINGLE POLY EFLASH MEMORY OPTIONS

Logic Eflash	10T Eflash [4]	5T Eflash [6]	6T Eflash (This Work)
Unit Cell Schematic			
Process	0.18μm Logic	65nm Logic	65nm Logic
Tunnel Oxide	~7nm	~5nm	~5nm
Write Method	FN Tunneling	FN Tunneling	FN Tunneling
Write Voltage	10V	10V	-7.2V
Unsel. WL's	Disturbed	Not Disturbed	Not Disturbed
Bit-by-Bit Write	Supported	Not Supported	Supported
Capacity	192b	2kb	4kb
Cell Size	220μm²	8.62μm²	15.3μm²

Prior WL-by-WL erasable single-poly eflash [6]

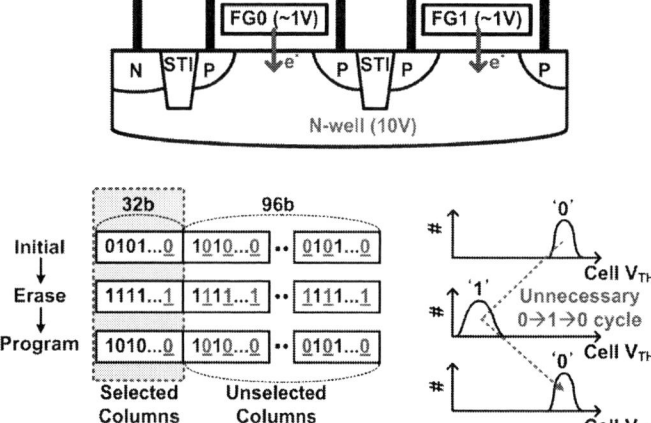

Fig. 1. In prior WL-by-WL erasable eflash memories, high voltage is applied to the tunnel oxide in the entire cells of the selected WL, causing an electron tunneling from the floating gate, which adds an unnecessarily erase/program cycle for the unchanged '0' cells in the selected and unselected columns.

To improve the overall cell endurance characteristic without disturbing multiple unselected WL's, we present a fully logic-compatible eflash memory capable of writing data on a bit-by-bit basis without utilizing boosted BL voltages by applying high voltages to the selected cell tunnel oxide in a

978-1-4673-6145-3/13 $31.00 © 2013 IEEE

bit-by-bit manner via selective FG boosting technique. It eliminates the aforementioned redundant cycling issue in the proposed 6T eflash cell which consists of a coupling transistor (M_1), a program transistor (M_2), a read transistor (M_3), and three select transistors (S_1-S_3). The proposed bit-by-bit re-writable eflash memory includes a 4kb eflash cell array, a multi-stage negative CP, multi-story High Voltage Switches (HVS), and low voltage Sense Amplifiers (SA) & BL drivers and was implemented using 1.2V core and 2.5V I/O transistors readily available in a standard CMOS process (Fig. 2).

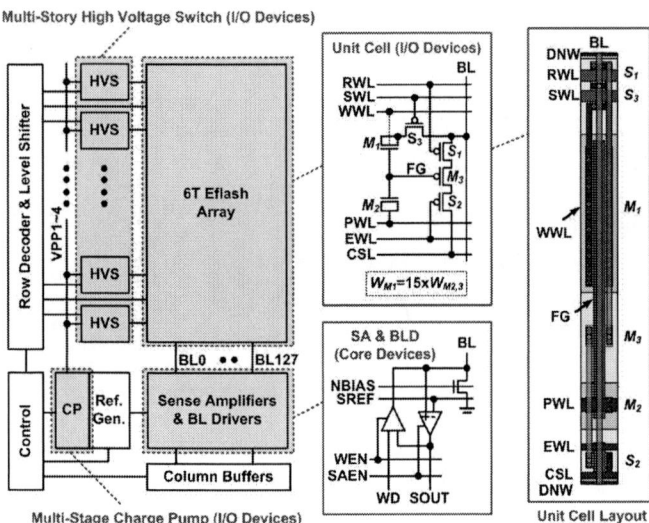

Fig. 2. Proposed bit-by-bit re-writable embedded flash memory including 6T cell array, charge pump, high voltage switch, sense amplifiers & BL drivers was implemented using core and I/O devices in a standard CMOS process.

II. Proposed Eflash Cell for Bit-by-Bit Write

To enable a bit-by-bit write, the proposed eflash cell boosts the floating gate (FG) of the each cell selectively via preferential coupling as illustrated in Fig. 3. Compared to the prior WL-by-WL erasable eflash (Fig. 1), the proposed cell does not share the source and drain node (SD) between adjacent cells in the WL direction, allowing it to have different voltage levels for each bit cell in the same WL. During write operations of the proposed eflash, WWL is switched to the negative boosted voltage (i.e. -7.2V), making FG node of '0' BL cell coupled down greater than the FG node of '1' BL cell. This is because the select TR (S_{3A} in Fig. 4) of '0' BL cells is turned off and the source and drain node of '0' BL cells (SD0) is floated, while the select TR (S_{3B} in Fig. 4) of '1' BL cells is turned on and the source and drain node of '1' BL cells (SD1) is tied to VDD (i.e. 1.2V). This FG node voltage difference can be utilized for the bit-by-bit write operations of the proposed 6T eflash cell.

The bias conditions for bit-by-bit write operations of the proposed 6T eflash cell are illustrated in Fig. 4. During write '0' phase (Fig. 4 top), PWL is switched to 1.6V and the large voltage difference between FG0 and PWL node enables only '0' BL cells to be selectively written to '0' states, making lose electrons from FG. During write '1' phase (Fig 4 bottom), PWL is switched to -7.2V and the large voltage difference

between FG1 and PWL nodes enables only '1' BL cells to be selectively written to '1' states, making gain electrons in FG.

Fig. 5 shows a read bias condition where the '1' cell flows higher current than '0' cell, raising BL voltage level above the SA reference (SREF). The read reference level (VRD) is provided through multi-story HVS in the WL driver (Fig. 8). The negative boosted writing voltage associated with NMOS coupling transistor (M_1) enables non-negative VRD levels, while not inverting the coupling transistor (M_1) with a floating channel during read operation. The overall bit-by-bit data update sequence of the selected WL can consist of the original eflash data read from the selected WL to column buffers (step 1), modify the column buffers to new data (step 2), and conduct write '0' and '1' phases (step 3).

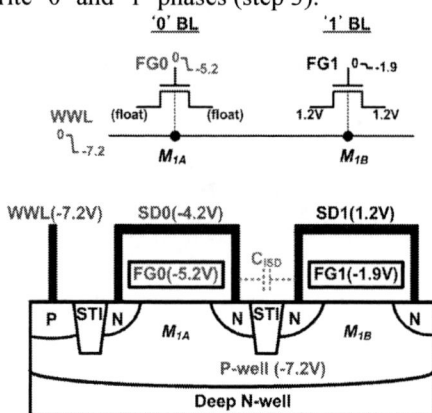

Fig. 3. Selective FG boosting is enabled by the preferential coupling for '0' BL cells where the source and drain node (SD0) of the coupling transistor (M_{1A}) is floated, while the corresponding node (SD1) for '1' BL cells is tied to VDD (1.2V). The differently boosted FG voltage levels for '0' and '1' BL's are utilized for the bit-by-bit electron tunneling occurred in the program transistor M_{2A}, M_{2B} (shown in Fig. 4) during write operations.

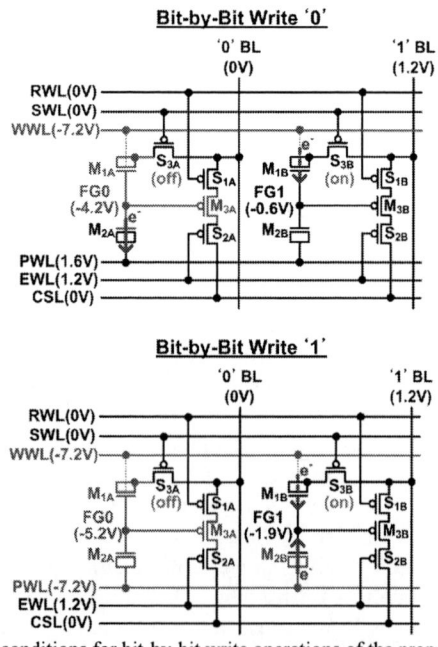

Fig. 4. Bias conditions for bit-by-bit write operations of the proposed 6T eflash cell. '0' BL cell loses electrons from FG during bit-by-bit write '0' operation, while '1' BL cell gains electrons in FG during bit-by-bit write '1' operation.

978-1-4673-6145-3/13 $31.00 © 2013 IEEE

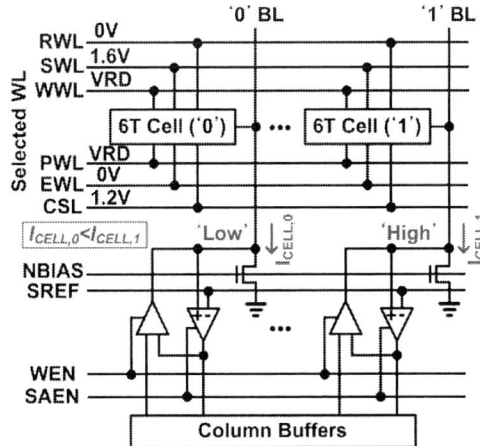

Fig. 5. Read bias condition of the proposed 6T eflash cell.

III. Negative Charge Pump and High Voltage Switch

The proposed 6T eflash cell requires a negative high voltage to be applied to WWL/PWL in the selected WL during write operations. In this work, a compact voltage doubler circuit [7] was cascaded to generate multiple negative supply levels VPP1-VPP4 (Fig. 6). Parasitic metal-to-metal capacitors (C_M) were utilized for the pumping capacitors. The write voltage level (VPP4) is regulated by comparing the resistively divided voltage level against a reference voltage (REF) and gating on or off the pumping clock. A deep n-well surrounds the VPP1-VPP4 p-wells for isolation purposes. All the devices are 1.2V core and 2.5V I/O devices provided in 65nm standard logic process. The measured CP output characteristic shows a reliable output voltage beyond the typical cell write current range (<1µA) (Fig. 7).

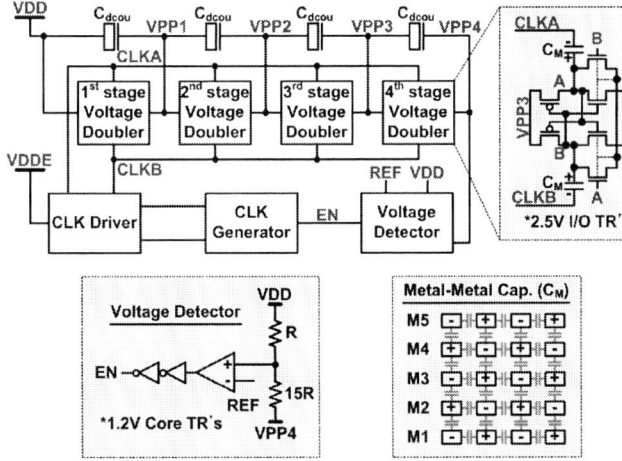

Fig. 6. Voltage doubler based negative charge pump generating VPP1-VPP4 levels is implemented in a 65nm standard logic process.

The proposed multi-story negative HVS, a refined version of the positive HVS presented in [6], utilizes the boosted negative supply levels VPP1-VPP4 (Fig. 8). All TR's in this HVS are implemented using 2.5V standard I/O devices. When SEL switches from VPP1 to VDD, nodes C and E are pulled-down to VPP3 and VPP2, and nodes A, B, D are pulled-up to VPP3, VPP2 and VPP1, respectively, making intermediate

node 'M' and output node WWL/PWL are connected to VPP3 and VPP4 levels, respectively. When SEL switches from VDD to VPP1, nodes C and E are pulled-up to VPP2 and VPP1, and nodes A, B, D are pulled-down to VPP4, VPP3 and VPP2, respectively. As a result, node 'M' and the output node WWL/PWL are driven to VPP1 and VRD/VSS. Similar to the previous HVS design [6], the pulse width and transistor sizes are optimized such that the intermediate latch states switch reliability while static power consumption kept small so as to minimize the current load to CP. The measured waveforms (Fig. 9) show VPP4 signal generated from the CP and WWL/PWL signals applied to the selected WL via the designed HVS for a bit-by-bit write '0' operation.

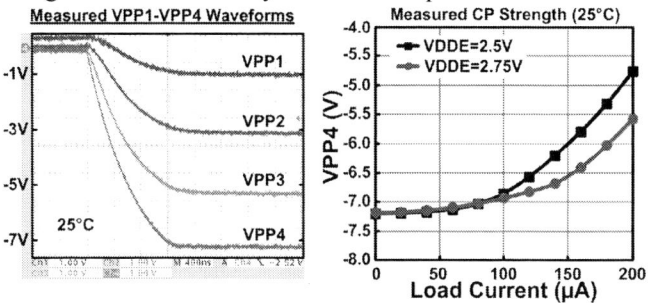

Fig. 7. Measured output characteristic of the voltage doubler based negative charge pump.

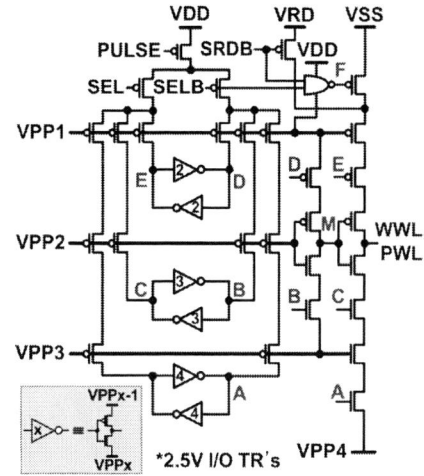

Fig. 8. Multi-story negative HVS implemented in a standard logic process.

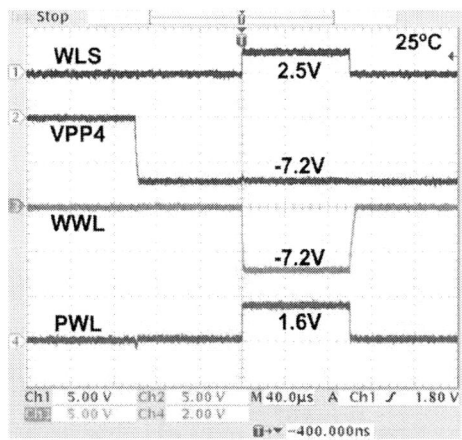

Fig. 9. Measured CP and HVS waveforms for a bit-by-bit write '0' pulse.

IV. Eflash Test Chip Results

A 4kb eflash test macro was implemented in a 65nm low power standard CMOS logic process. Fig. 10 shows the measured bit-by-bit write '0'/'1' phases and disturbance characteristic of the proposed 6T eflash cell. The inhibited '1' BL cells are disturbed increasing the cell V_{TH} and the sensing margin during bit-by-bit write '0' phase, while no apparent disturbance in the inhibited '0' BL cells is observed during bit-by-bit write '1' phase. Table II shows various write patterns tested in this work producing the different inter-cell SD capacitance (C_{ISD} in Fig. 3) values. The measurement result shows that (0101) pattern needs 10× more pulses than (0000) pattern to complete the write '0' phase, as higher C_{ISD} reduces the boosting effect of FG node of '0' BL cell, slowing down the write '0' speed. Fig. 11 shows the measured cell endurance and retention characteristic. The die photograph and feature summary of the fabricated test chip are shown in Fig. 12.

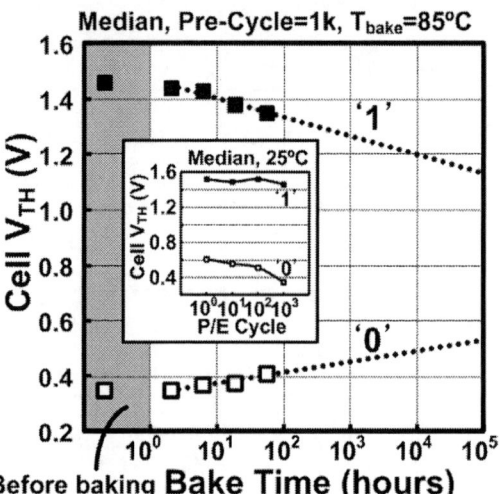

Fig. 11. Measured cell endurance and retention characteristic of the proposed 6T eflash cells.

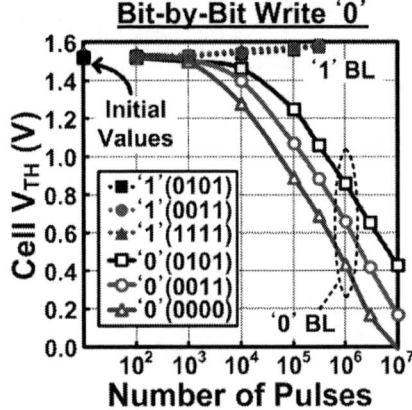

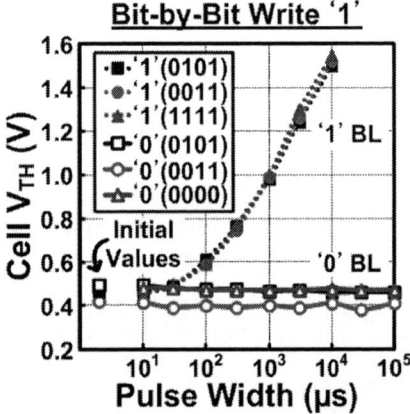

Fig. 10. Measured bit-by-bit write '0'/'1' phase and disturb results.

Table II. Test pattern dependency of the inter SD coupling (C_{ISD})

Pattern Name	BL # 0	1	2	3	•••	C_{ISD}
(1111)	1	1	1	1	•••	0
(0101)	0	1	0	1	•••	++
(0011)	0	0	1	1	•••	+
(0000)	0	0	0	0	•••	0

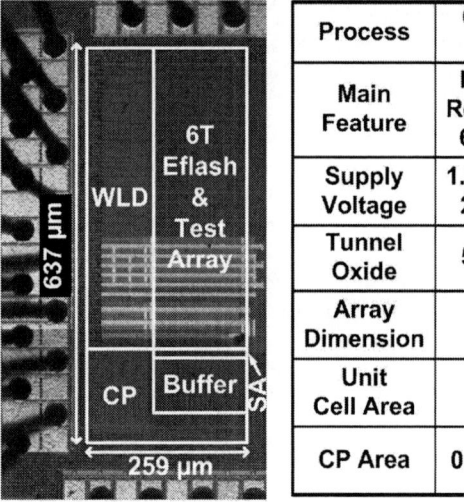

Process	65nm LP CMOS
Main Feature	Bit-by-bit Re-Writable 6T Eflash
Supply Voltage	1.2V (Core), 2.5V (I/O)
Tunnel Oxide	5nm (I/O)
Array Dimension	32 WL, 128 BL
Unit Cell Area	15.3μm²
CP Area	0.0214mm²

Fig. 12. Die photograph and chip summary.

References

[1] S. Kulkarni, S. Pae, Z. Chen, et al., "A 32nm High-k and Metal-Gate Anti-Fuse Array Featuring a 1.01μm2 1T1C Bit Cell," IEEE Symp. on VLSI Technology, pp. 79-80, 2012.

[2] K. Matsufuji, T. Namekawa, H. Nakano, et al., "A 65nm Pure CMOS One-Time Programmable Memory Using a Two-Port Antifuse Cell Implemented in Matrix Structure," IEEE Asian Solid-State Circuits Conf. (ASSCC), pp. 212-215, 2007.

[3] J. Raszka, M. Advani, V. Tiwari, et al., "Embedded Flash Memory for Security Applications in a 0.13μm CMOS Logic Process," IEEE Int. Solid-State Circuits Conf. (ISSCC), pp. 46-47, 2004.

[4] P. Feng, Y. Li, N. Wu, "An Ultra Low Power Non-volatile Memory in Standard CMOS Process for Passive RFID Tags," IEEE Custom Integrated Circuits Conf. (CICC), pp. 713-716, 2009.

[5] H. Chen, S. Wang, W. Ching, et al., "Single Polysilicon Layer Non-Volatile Memory and Operating Method Thereof," US Patent 8199578, June 12, 2012.

[6] S. Song, K. Chun, C. H. Kim, "A Logic-Compatible Embedded Flash Memory Featuring a Multi-Story High Voltage Switch and a Selective Refresh Scheme," IEEE Symp. On VLSI Circuits, pp. 130-131, 2012.

[7] P. Favrat, P. Deval., M. Declercq, "A High-Efficiency CMOS Voltage Doubler," IEEE J. Solid-State Circuits, vol. 33, no. 3, pp. 410-416, March 1998.

978-1-4673-6145-3/13 $31.00 © 2013 IEEE

Tail-Bit Tracking Circuit with Degraded VGS Bit-Cell Mimic Array for a 50% Search-Time and 200mV Vmin Improvement in a Ternary Content Addressable Memory

Igor Arsovski, Travis Hebig, John Goss, Paul Grzymkowski, Josh Patch

IBM Systems and Technology Group,
1000 River Road, Essex Junction, VT 05452 USA
Phone: 802-769-3347, e-mail: arsovski@us.ibm.com

Abstract— **A memory sense-amplifier timing circuit emulates the behavior of weak memory tail-bits to improve Tail-Bit Tracking (TBT) across Process, Voltage and Temperature. The TBT circuit is used to generate timing for a search operation in a 32nm Ternary Content Addressable Memory (TCAM) compiler resulting in 200mV Vmin improvement at a constant performance, and 50% improved search-time performance at a constant Vmin. This TBT circuit was implemented in 32nm High-K Metal Gate SOI process to achieve 0.60V operation and support up to 1G search/sec throughput on a 2048x640bit TCAM instance.**

I. INTRODUCTION

As CMOS device geometries scale into advanced technology nodes, large systematic and Random Device Variation (RDV) make memory voltage and performance scaling increasingly difficult. The key challenge is the modeling and tracking of highly skewed memory cells also known as tail-bits. Fig. 1 illustrates the problem by showing the race path between memory cell signal development (shown in red), and a delay circuit used to generate the sense amplifier (SA) timing (shown in blue). Although tail bit tracking is a challenge in all memories this paper focuses of the tracking of tail bits during a single-ended TCAM search operation. By examining the delay for adequate bit-cell signal development (top right of Fig.1), it is evident that RDV-induced tail-bits have a significantly larger low-voltage delay degradation than average bit-cells. Historically, memory SA timing was generated using logic delays, over-bounding these tail-bit delays across the PVT application space. This over-bounding approach achieves good yield, but results in loss of memory performance (Fmax) and increase in memory minimum operating voltage (Vmin). Recently, a number of SRAM SA timing circuits have been proposed to improve both average SRAM cell signal development [1], and weak-bit tracking [2,3], resulting in improvements in both Vmin and Fmax. This paper presents a programmable TBT circuit that uses only six additional transistors over the average-bit tracking circuit presented in [1] resulting in significantly lower area than what is presented in [2,3]. This area-optimized TBT circuit is used to generate SA timing for a self-referenced TCAM search operation [4] to achieve 50% search-time performance improvement at a constant Vmin, and a 200mV Vmin improvement at constant performance.

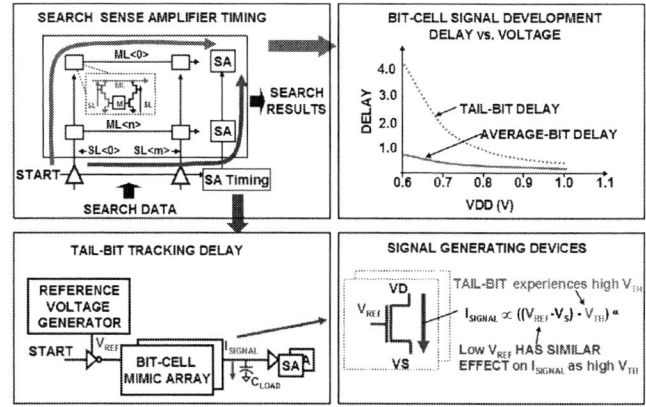

Fig. 1. Memory Sense Amp (SA) timing needs to account for weak tail-bit signal development. Tail-Bit-Tracking (TBT) circuit with degraded VGS bit-cell mimic array improves tail-bit tracking across process, voltage, and temperature (PVT).

II. TAIL-BIT TRACKING (TBT) CIRCUIT

The high level concept of this TBT circuit is shown in the lower portion of Fig.1. The circuit works by first generating a reduced reference voltage (VREF) that is then used as the reduced VGS for multiple parallel signal generating devices present in the bit-cell mimic array. When examining the bit-cell mimic array current equation (1) it is evident that a degradation in the VGS component has a similar effect as the increase in the VTH component present in weak tail-bits.

$$I_{SIGNAL} \propto (VGS - VTH)^{\alpha} \qquad (1)$$

Fig. 2 shows the circuit level implementation of the TBT. The delay starts by activating the reference voltage generation circuit which uses a six-transistor programmable source-follower to generate one of four different voltage values, starting from a full cell supply voltage (VDD), to progressively lower voltages (VDD-VLVT), (VDD-VRVT), and (VDD-VHVT) where the VxVT is the threshold voltage of the corresponding device type. The adjustable voltage drop generated by this source follower is used to emulate the VTH increase in skewed tail-bits. With 4-5σ skewed tail bits seeing roughly a VT shift, the circuit shown in Fig. 2 can achieve tail-bit delay emulation with only a 6 transistor area overhead. Once generated this VREF voltage is fed into the

978-1-4673-6145-3/13 $31.00 © 2013 IEEE

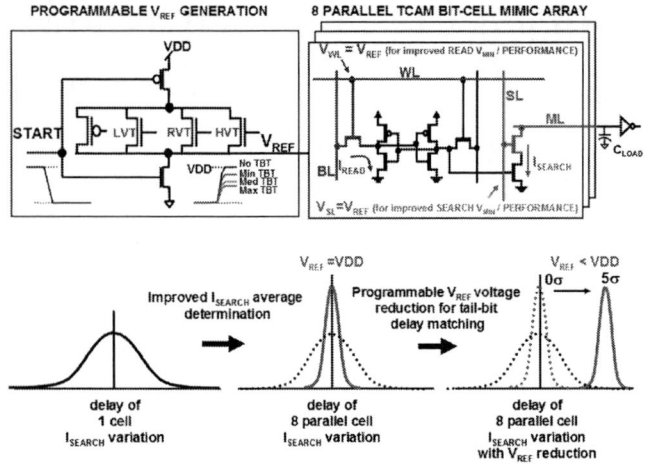

Fig. 2. Memory TBT circuit uses eight parallel bit-cells to extract the average-bit signal development delay. Reduced VREF degrades bit-cell mimic array VGS resulting in better tail-bit delay emulation.

critical signal generating devices of the bit-cell mimic array to emulate the current produced by the highly skewed tail-bits. This tail-bit signal current is then integrated on a load capacitor C_{LOAD} to produce the desired delay. Although the same approach could be used in a 6T SRAM cell timing delay (shown in blue), this work focuses on the performance-critical TCAM search operation (shown in red). As shown at the bottom of Fig. 2, the parallel activation of 8-bitcells extracts the average cell-current, reducing the timing circuit delay variation, while the reduced VGS=VREF voltage shifts the delay to cover the tails of the bit-cell delay. Although more advanced VREF generation circuits exist [2,3], the advantages of the source follower implementation are its good tracking with low area overhead.

III. SIMULATION RESULTS

Fig. 3 shows transient simulations of the TBT sense delay versus voltage in the 2048x640bit TCAM instance across four different TBT settings. The settings are post-silicon adjustable based on the compiled TCAM configuration and the amount of bit-cell skew present. The No TBT, Low TBT, Med TBT, and Max TBT settings are shown to track approximately 0σ, 3σ, 4σ, and 5σ bit-cell VT variation. As shown in Fig. 3 the reduced VGS=VREF results in a larger delay at low voltage, allowing better tracking of weak tail-bits, while adding much less delay at higher voltage , improving performance. By comparing the No TBT and Max TBT curves it is evident that for 1.8x degradation in sense delay at 0.95V, the TBT circuit generates a 3.5x improvement in signal development time at 0.60V. Thus for a set signal margin at 0.6V, the Max TBT setting would result in a 50% sense-time improvement at 0.95V. To reduce timing uncertainty with the source-follower voltage regulation the source-follower devices are implemented using multi-finger non-minimum size devices driving a small memory mimic cell array.

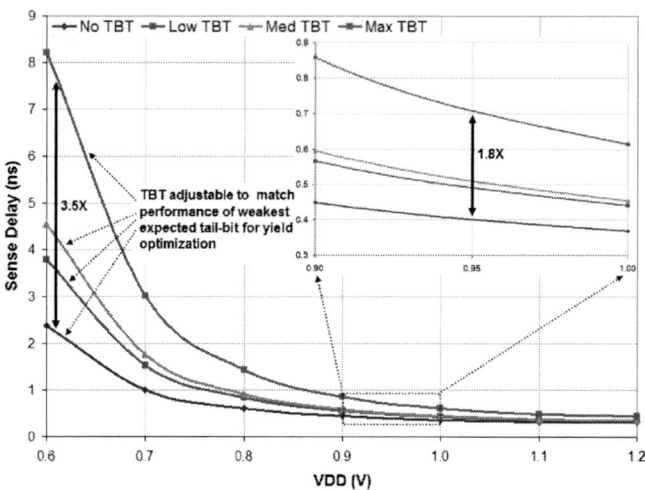

Fig. 3. TBT circuit simulation of delay across voltage for varying amounts of VREF reduction. When compared to average-bit tracking (No TBT), the tail-bit tracking circuit (Max TBT) shows 3.5X longer 0.60V low-voltage delay for only 1.8X performance degradation at the 0.95V performance corner.

Fig. 4 illustrates the quality of the TBT circuit across voltage through a simulated 5σ weak tail-bit signal margin development for both No TBT and Max TBT settings. For comparison, the No TBT signal margin is shown equalized to Max TBT at either 0.95V (solid blue) or 0.60V(broken blue). With Max TBT enabled (solid red), the signal margin at sense-time is maintained within +/-14% across the application voltage range. With No TBT enabled, the circuit behaves similar to [1], and shows a much wider signal spread of +/-42%. Thus, the Max TBT setting results in either 50% search sense-time reduction at 0.95V (at constant 0.60V signal margin), or 200mV Vmin improvement (at constant 0.95V signal margin) shown by Max TBT at 0.60V having the same signal margin as No TBT at 0.80V.

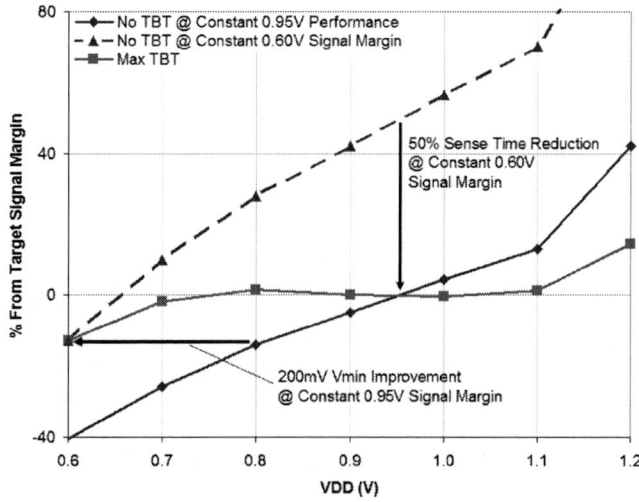

Fig. 4. Simulated 5σ bit-cell signal margin vs. voltage for No TBT and Max TBT. Normalized signal margin shows Max TBT provides 50% sense-time reduction at 0.95V (at constant 0.60V signal margin), and 200mV Vmin improvement (at constant 0.95V signal margin).

IV. HARDWARE RESULTS

Extensive functionality verification shows excellent yield throughout the complete TCAM compiler configuration space spanning a voltage range of 0.6V-1.1V across the full process window. Extensive Built-In Self-Test (BIST) verification has verified functionality across both nominal as well as worst-case noise-inducing test patterns ensuring reliable operation. Fig. 5 shows hardware measurements of fixable yield for a 1Mb TCAM instance for the four different TBT delay settings. Measurements were collected by running embedded BIST at many VDD_A/VDD_B corners on a sample of parts, then repeating testing with various TBT control register settings. BIST results for 1Mb TCAM instances are averaged together and compared among the TBT settings. The measurements were performed on characterization wafers with intentionally induced variations in VTH and Lpoly to capture a full manufacturing process range. It can be seen that by adjusting the amount of TBT, the TCAM achieves a 200mV Vmin improvement at a fixed yield.

Fig. 6 shows silicon measurements exceeding 1Gsearch/sec on TCAM hardware across a full process range. The measurements were taken using a McLeod loop [5] test structure in the silicon. This test circuit puts the memory in the critical path of a ring oscillator and additionally includes multiplexer to remove the memory from the critical path. The period of oscillation is measured with and without the memory in the critical path and therefore the performance of the memory can be determined. Leakage and dynamic power measurements of the same hardware support 0.76W power consumption at 1Gsearch/sec throughput on a 2048x640bit TCAM instance.

Fig. 7 shows a microphotograph of the largest compiled TCAM instance, 2048x640, implemented in 32nm High-K Metal Gate SOI process. The 16 bank x 8 field organization of this TCAM instance is noted and rows and columns of embedded Deep-Trench (DT) capacitors can be seen. Table 1 covers the specifications of the full TCAM compiler which supports a wide range of capacities, operating voltages and temperatures. To sustain worst case Ldi/dt power-supply noise voltage collapse, inherent to TCAMs, 5% of the TCAM area is covered with DT capacitors. DT capacitors enjoy 20-25X higher capacitive density than conventional planar capacitors [6] and allow high density TCAM implementation. By placing these highly efficient decoupling capacitors right next to the circuits with the highest current demand (i.e. SL drivers/ ML sense-amps) this TCAM design reduces the transient noise by 50% which in turn results in both tighter timing jitter, lower voltage and consequently lower power.

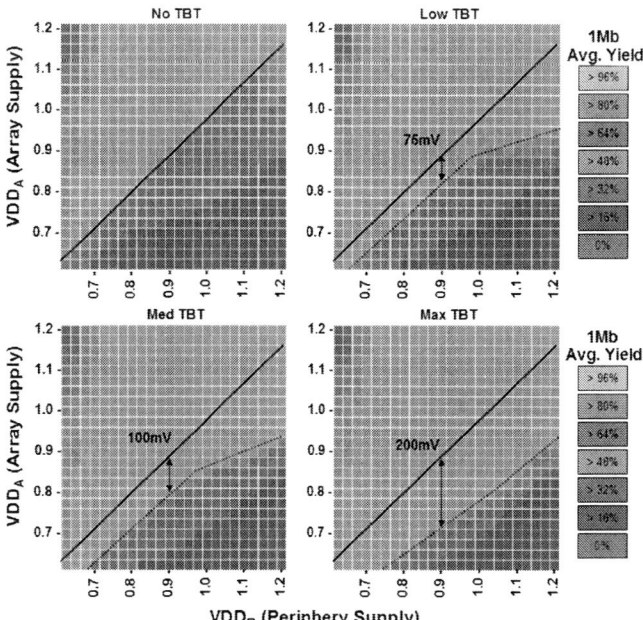

Fig. 5. Measured silicon hardware showing Vmin reduction through fixable yield of a 1Mb TCAM instance across four different TBT settings.

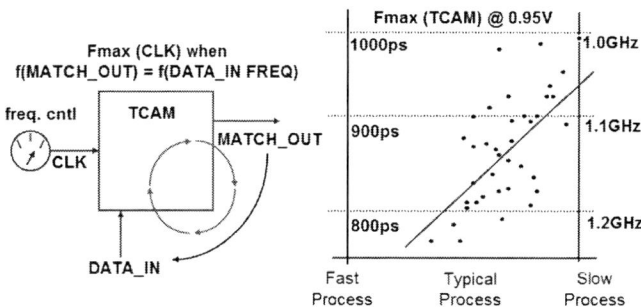

Fig. 6. Measured TCAM cycle-time showing 1GHz operation across the full process. Large scatter caused by VT and Lpoly striped characterization wafers.

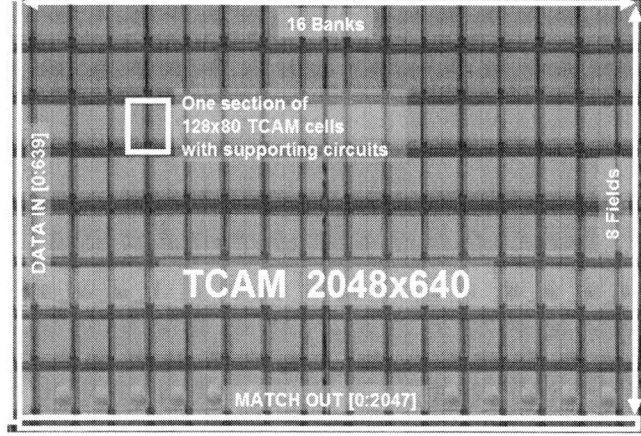

Fig. 7. TCAM microphotograph showing TCAM 2048x640 instance organization.

978-1-4673-6145-3/13 $31.00 © 2013 IEEE

TABLE I. TCAM COMPILER SPECIFICATIONS

Min Compiled Instance	64 entry X 12 bits
Max Compiled Instance	2048 entry X 640 bits
Max TCAM Organization	16 X 8 banks
Max Bank Organization	128 words x 128 bits
Performance (@ 0.95V)	1.0 GHz
Power (2048x640 @ 1GHz)	1.1 W
Density (2048x640)	0.84 Mb/mm2
Technology	32nm high-K SOI process With Embedded Deep-Trench DECAP
Power Saving Options	Selective Bank Activation Pre-search Activation
Noise Reduction	>12nF of Embedded DECAP
Functional Voltage Operating Temperature	0.60V – 1.2V -40°C – 125°C

V. CONCLUSIONS

The explosive growth of internet traffic has created a large demand for faster and lower-power network search engines, while the transition from IPv4 to IPv6 has nearly quadrupled the required network look-up table capacity. These developments have made the TCAM an increasingly important building block in today's network data processors. This paper described a tail-bit tracking-circuit for improved tail-bit delay tracking across PVT. The tail-bit tracking circuit is used to generate timing for a search operation in a 32nm TCAM resulting in 50% search-time performance improvement at a constant Vmin, and a 200mV Vmin improvement at constant performance. The TCAM compiler also employs self-referenced sensing [4] that enables high-performance sensing in at highly-variable sub-40nm technology and allows high-density architectures with up to 128 TCAM cells/sense-amplifier. The proposed TCAM compiler was implemented in 32nm High-K Metal Gate SOI process to achieve a record 1Gsearch/sec throughput on a 2048x640bit TCAM instance while consuming only 0.76W and 0.58-fJ/bit/search. To reduce the effect of power supply noise, 5% of the TCAM area is covered with embedded deep-trench capacitors which reduce power-supply voltage collapse by 53% and enable reliable operation.

ACKNOWLEDGMENTS

The authors thank Nitin Sharma, Sarah Braasch, and John Chickanosky for management support and the worldwide IBM Silicon Solutions Engineering memory organization for technical discussions.

REFERENCES

[1] Pilo, H.; Arsovski, I.; Batson, K.; Braceras, G.; Gabric, J.; Houle, R.; Lamphier, S.; Radens, C.; Seferagic, A., "A 64 Mb SRAM in 32 nm High-k Metal-Gate SOI Technology With 0.7 V Operation Enabled by Stability, Write-Ability and Read-Ability Enhancements," *Solid-State Circuits, IEEE Journal of*, vol.47, no.1, pp.97,106, Jan. 2012

[2] Kawasumi, A.; Takeyama, Y.; Hirabayashi, O.; Kushida, K.; Tachibana, F.; Niki, Y.; Sasaki, S.; Yabe, T., "A 47% access time reduction with a worst-case timing-generation scheme utilizing a statistical method for ultra low voltage SRAMs," *VLSI Circuits (VLSIC), 2012 Symposium on*, pp.100,101, 13-15 June 2012

[3] Kushida, K.; Hirabayashi, O.; Tachibana, F.; Hara, H.; Kawasumi, A.; Suzuki, A.; Takeyama, Y.; Fujimura, Y.; Niki, Y.; Shizuno, M.; Sasaki, S.; Yabe, T., "A trimless, 0.5V–1.0V wide voltage operation, high density SRAM macro utilizing dynamic cell stability monitor and multiple memory cell access," *Solid State Circuits Conference (A-SSCC), 2011 IEEE Asian*, pp.161,164, 14-16 Nov. 2011

[4] Arsovski, I.; Hebig, T.; Dobson, D.; Wistort, R., "1Gsearch/sec Ternary Content Addressable Memory compiler with silicon-aware Early-Predict Late-Correct single-ended sensing," *VLSI Circuits (VLSIC), 2012 Symposium on*, pp.116,117, 13-15 June 2012

[5] Wagner, O.; McLeod, M.H., "A New Method for Improved Delay Characterization of VLSI Logic," *Solid-State Circuits Conference, 1982. ESSCIRC '82. Eighth European*, pp.102,105, 22-24 Sept. 1982

[6] Wendel, D.; Kalla, R.; Cargoni, R.; Clables, J.; Friedrich, J.; Frech, R.; Kahle, J.; Sinharoy, B.; Starke, W.; Taylor, S.; Weitzel, S.; Chu, S.G.; Islam, S.; Zyuban, V., "The implementation of POWER7TM: A highly parallel and scalable multi-core high-end server processor," Solid-State Circuits Conference Digest of Technical Papers (ISSCC), 2010 IEEE International, pp.102,103, 7-11 Feb. 2010

978-1-4673-6145-3/13 $31.00 © 2013 IEEE

A 1.8mW 2MHz-BW 66.5dB-SNDR $\Delta\Sigma$ ADC Using VCO-Based Integrators with Intrinsic CLA

Kyoungtae Lee, Yeonam Yoon, and Nan Sun

Department of Electrical and Computer Engineering
The University of Texas at Austin, Austin, TX 78712, USA
Email: kyoungtae.lee@utexas.edu, irunjurun@utexas.edu, nansun@mail.utexas.edu

Abstract—**This paper presents a scaling-friendly continuous-time closed-loop VCO-based $\Delta\Sigma$ ADC. It uses the VCO as both quantizer and integrator, and thus, obviates the need for power-hungry scaling-unfriendly OTAs and precision comparators. It arranges two VCOs in a pseudo-differential manner, which cancels out even-order distortions. More importantly, it brings an *intrinsic* clocked averaging (CLA) capability that automatically addresses DAC mismatches. The prototype ADC in 130nm CMOS occupies a small area of 0.03mm² and achieves 66.5dB SNDR over 2MHz BW while sampling at 300MHz and consuming 1.8mW under a 1.2V power supply. It can also operate with a low analog supply of 0.7V and achieves 65.8dB SNDR while consuming 1.1mW. The corresponding figure-of-merits (FOMs) for the two cases are 0.25pJ/step and 0.17pJ/step respectively.**

I. INTRODUCTION

The conventional way to build continuous-time $\Delta\Sigma$ ADCs relies on the use of voltage comparators and integrators based on operational transconductance amplifiers (OTAs). This well-established design methodology encounters severe difficulties in advanced nanometer-scale CMOS processes due to reduced power supply voltages and transistor intrinsic gains. A small power supply leads to a small signal swing and decreased dynamic range. It also increases the precision requirement for the comparator. The reduced transistor gain makes it hard to design high-gain OTAs that are usually needed to ensure linearity. Thus, a new design scheme for $\Delta\Sigma$ ADCs that is compatible with CMOS scaling is highly desirable.

There have been emerging efforts in the research community to use ring voltage-controlled oscillators (VCOs) to construct $\Delta\Sigma$ ADCs [1]–[5]. VCOs can act as both quantizers and integrators. There are several advantages for using VCOs: 1) they contain only simple inverters, and thus, are easy to design; 2) they are area and power efficient; 3) they operate well under low power supply; and more importantly 4) they are scaling friendly. As technology scales, the transistor speed f_T increases, leading to a shorter inverter delay and a higher timing resolution. As a result, the performance of VCO-based $\Delta\Sigma$ ADCs improves naturally with CMOS scaling.

Despite the many advantages mentioned above, VCO-based $\Delta\Sigma$ ADCs suffer significantly from the VCO's nonlinear voltage-to-frequency conversion. Typically, the linearity of an open-loop VCO is limited to about 40 dB. To overcome this nonlinearity problem, many design techniques have been proposed. One way as shown in Fig. 1(a) is to use digital background calibration to correct the nonlinearity of the VCO

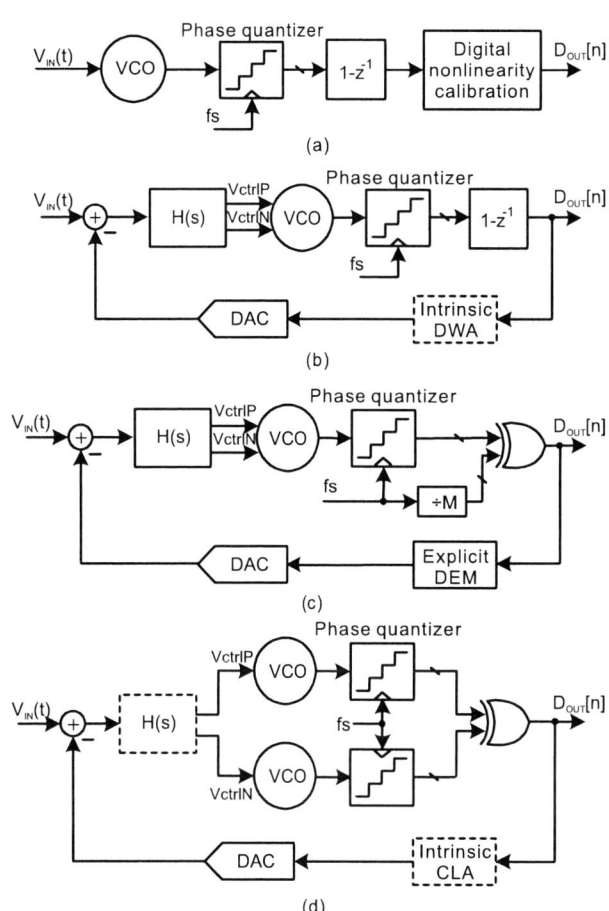

Fig. 1. (a) open-loop VCO-based $\Delta\Sigma$ ADC with nonlinearity correction; (b) closed-loop $\Delta\Sigma$ ADC using VCO-based quantizer; (c) closed-loop $\Delta\Sigma$ ADC using VCO-based integrator; (d) proposed closed-loop $\Delta\Sigma$ ADC using VCO-based integrator with intrinsic CLA.

[1]. The merit of this scheme is that it is mostly digital, but the drawback is that its calibration algorithm is complicated. In addition, its input signal swing is limited. It also requires an extra dithering DAC and a highly-linear V/I converter. The other way to solve the VCO nonlinearity problem is to put the VCO in a closed loop with an analog filter $H(s)$ preceding it [see Fig. 1(b)] [2]. The gain provided by $H(s)$ helps suppress the VCO nonlinearity. Moreover, this scheme has an intrinsic dynamic element matching (DEM) capability. Due to the digital differentiation at the VCO output, the elements in the feedback DAC are naturally selected in a

978-1-4673-6145-3/13 $31.00 © 2013 IEEE   829

barrel-shifted fashion. This implicitly implements dynamic weighted averaging (DWA). As a result, the DAC mismatches are automatically shaped to the first-order. However, the disadvantage of this scheme is that the analog filter $H(s)$ is scaling unfriendly and consumes large power and area. In addition, because of the digital differentiation, the VCO is only used as a quantizer. Its input still sees a large signal swing, which results in large distortions and limits the overall ADC linearity. To address this issue, Ref. [3] proposes to use the VCO phase, instead of its frequency, as the output [see Fig. 1(c)]. The VCO phase is measured by comparing it with a reference phase derived from the sampling clock. Since there is no differentiation in the loop, the VCO acts as an integrator, which increases the loop gain significantly. The VCO input swing is significantly reduced, and thus, its nonlinearity problem is solved. The price for using this technique is that the intrinsic DEM capability is lost, as the VCO phase output is thermometer coded. To address the DAC mismatch problem, an explicit DEM block has to be used, which leads to additional design complexity, chip area, and power. Furthermore, it increases the propagation delay of the feedback path, causing degradation in the quantization noise shaping and potentially destabilizing the loop.

To simultaneously use the VCO as an integrator and maintain an intrinsic DEM capability, we propose a novel VCO-based $\Delta\Sigma$ ADC architecture shown in Fig. 1(d). We arrange two VCOs in a pseudo-differential manner. The phase output of the dual VCO is measured by comparing the phase of one VCO with that of the other. The advantage of this scheme is that the feedback loop operates only on the difference between the two VCO phases and does not control the VCO center frequency. As a result, unlike [3], the VCO center frequency does not have to be locked to a fraction of the sampling frequency; it can be freely chosen. As a result, we can lower the VCO frequency to save power. More importantly, as will be explained in Sec. II, **the proposed scheme results in a natural rotation of the DAC selection pattern at twice the speed of the VCO center frequency**. This effectively realizes the data-independent DEM scheme of clocked averaging (CLA) [6]. In other words, the proposed ADC architecture has an intrinsic CLA capability. CLA modulates the DAC mismatch errors to twice the VCO center frequency and moves them out of the signal band. As a result, no explicit DEM technique is needed. Furthermore, because the dual VCO is used as an integrator, it provides a large loop gain to suppress the VCO nonlinearity. Its pseudo-differential structure also helps cancel out even-order distortions. Additionally, a large number (e.g., 25) of VCO stages can be used to increase the effective phase quantizer resolution and further suppress the signal swing at the VCO input. As a result, the analog filter $H(s)$ is no longer needed from the linearity point of view. Consequently, the proposed $\Delta\Sigma$ ADC architecture does not need any analog filter, external DEM, or calibration. It is only comprised of VCOs, phase quantizers, and DACs. Thus, it is highly scaling friendly; it occupies a small area and consumes low power, especially in advanced processes.

A prototype ADC implemented in 130nm CMOS occupies an active area of only 0.03mm^2. It achieves 66.5 dB SNDR over 2MHz bandwidth (BW) while consuming only 1.8mW from a 1.2V power supply. It can also work with a low analog supply of 0.7V and achieves 65.8 dB SNDR while consuming only 1.1mW. The corresponding figure-of-merits (FOMs) for the two cases are 0.25pJ/conversion-step and 0.18pJ/conversion-step, respectively, which compares favorably to the state-of-the-art $\Delta\Sigma$ ADCs.

The paper is organized as follows: Sec. II explains the intrinsic CLA technique; Sec. III presents the prototype ADC design; and Sec. IV shows the measurement results.

II. INTRINSIC CLA

DAC mismatches can severely affect the linearity of $\Delta\Sigma$ ADCs. For example, we simulated a first-order $\Delta\Sigma$ ADC using a 25-element thermometer-coded DAC. Without mismatch, the SNDR at the oversampling ratio (OSR) of 75 is 80 dB. However, with 1% element mismatch added, the SNDR is degraded to 54 dB [see Fig. 2(a)].

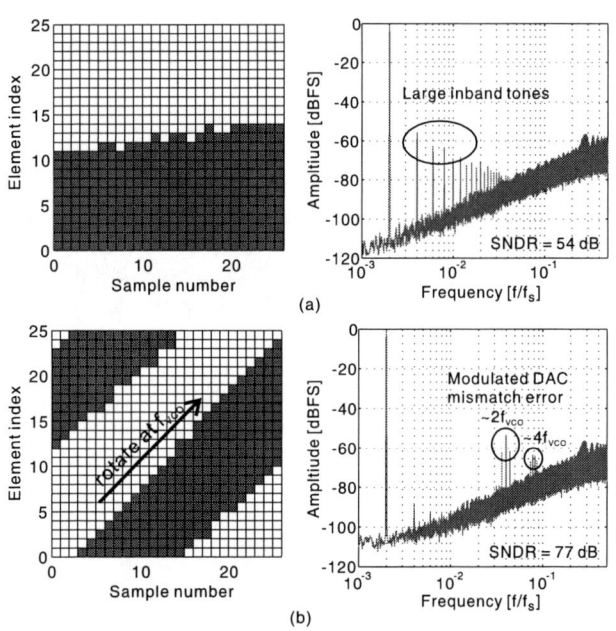

Fig. 2. DAC element selection pattern and ADC output spectrum with (a) thermometer coding and (b) intrinsic CLA with $f_{VCO} = f_s/50$. The solid box in the selection pattern means that its corresponding element is selected.

Our proposed $\Delta\Sigma$ ADC has an intrinsic CLA capability that effectively addresses the DAC mismatch issue without any hardware cost. The way to understand it is as follows. As discussed earlier, the ADC output is obtained by subtracting the phase of one VCO from that of the other. Consider the dual VCO as a differential circuit. The feedback loop only operates on the differential phase of the dual VCO; it does not control the common-mode phase. As a result, the common-mode phase simply accumulates and wraps around 2π at twice the speed of the VCO center frequency, $2f_{VCO}$. Consequently, the middle point of the DAC selection pattern also rotates at $2f_{VCO}$. This leads to the DAC selection pattern shown in Fig.

978-1-4673-6145-3/13 $31.00 © 2013 IEEE

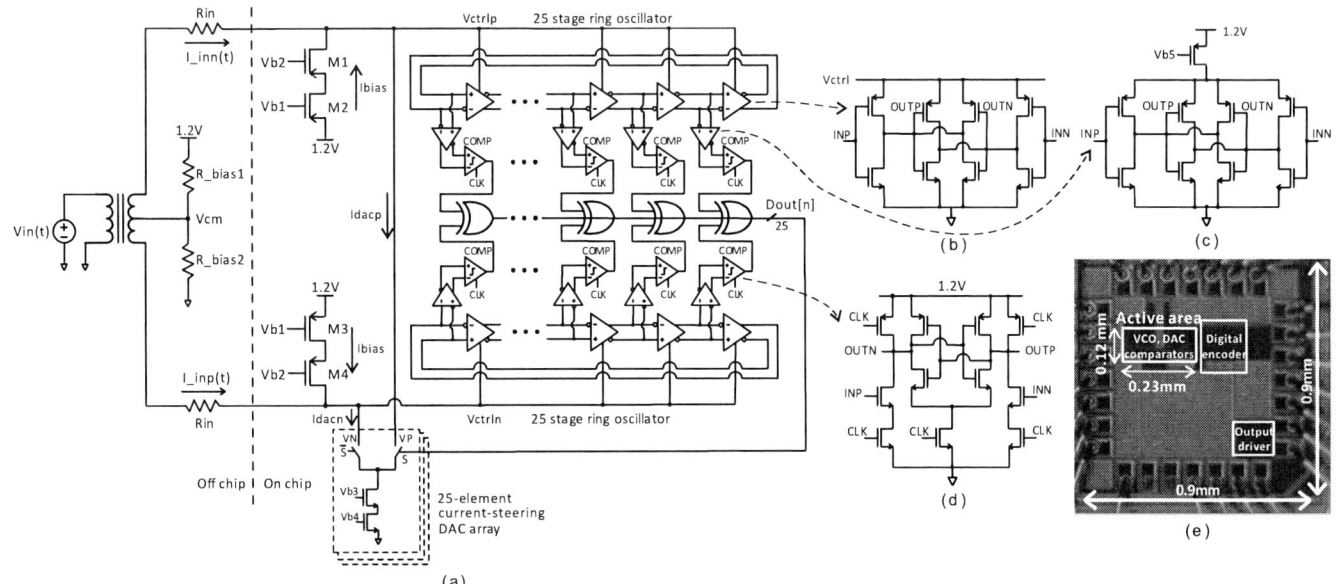

Fig. 3. Schematics of (a) ADC architecture; (b) delay cell; (c) VCO output buffer; (d) comparator; and (e) die photo

2(b), where we assume $f_{VCO} = f_s/50$. This selection pattern matches the result of the CLA technique [6]. Since the DAC elements are rotated at $2f_{VCO}$, the DAC mismatch errors are up-converted to VCO even-order harmonics and moved out of the signal band [see Fig. 2(b)]. Thus, the SNDR is significantly improved to 77 dB, and is only 3 dB lower than that of the ideal case. Hence, no external DEM technique is needed in the proposed VCO-based ADC. This intrinsic CLA allows the use of small DAC elements to reduce power and area.

III. PROTOTYPE ADC DESIGN

Fig. 3(a) shows the ADC architecture. The input is converted to a current by a resistor R_{in}, which is placed off-chip for tunability. R_{in} is chosen to be large enough (2.4 $k\Omega$) so that it allows a large input swing of 2.4V Vpp and contributes negligible noise current.

The VCO consists of 25 delay cells, whose schematic is shown in Fig. 3(b). The VCO works essentially in the current domain. The reason for using current to control frequency is that it is much more linear than using voltage, which is shown in the measured VCO tuning curves of Fig. 4(a) and (b). This is because current is directly related to how fast charge moves, which fundamentally sets the frequency. When choosing the VCO center frequency f_{VCO}, we want it to be small to minimize the power, but we also want it to be large enough so that: 1) the frequency tuning is linear; 2) the up-converted DAC mismatch errors are out of the signal band; and 3) the VCO output impedance, inversely proportional to f_{VCO} [see Fig. 4(c)], is low enough to receive the majority of the input and feedback current. After balancing these tradeoffs, we set f_{VCO} to be 4.5MHz, which leads to a low VCO current of 130uA, a low VCO voltage of 400mV, and a low VCO output impedance of 1kΩ.

Each VCO cell connects to a replica buffer [see Fig. 3(c)], which helps isolate the VCO from the kickback noise of the

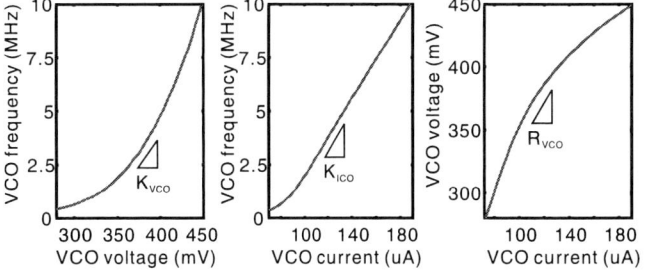

Fig. 4. Measured VCO tuning characteristics: (a) voltage vs. frequency; (b) current vs. frequency; and (c) current vs. voltage.

comparator [see Fig. 3(d)]. The comparator outputs from both sides go to an XOR gate, which then drives the feedback current-steering DAC.

Because the VCO only requires a low voltage of 0.4V, the proposed ADC also supports an ultra-low-power mode with a low analog supply of 0.7V. In this case, the 0.7V supply is connected to the common-mode voltage V_{cm} at the input transformer. The DC current required by the VCO and the DAC are solely supplied by the common-mode input current, and the bias current sources M1-M4 are turned off. This significantly reduces the power consumption. The price to pay is that the input swing is reduced 0.9V Vpp due to the reduction in R_{in}, which is required to supply enough current for the VCO and the DAC.

IV. MEASUREMENT RESULTS

A prototype ADC is implemented in 0.13um CMOS and occupies an active area of only 0.03mm^2 [see Fig. 3(e)]. It samples at 300 MS/s and has a signal bandwidth of 2 MHz. The ADC output waveform is shown in Fig. 5. The maximum output is 25, as the VCO has 25 stages.

The output spectrum for a -0.9 dBFS 661.5 kHz input is shown in Fig. 6. The 1st-order noise shaping is clearly seen.

978-1-4673-6145-3/13 $31.00 © 2013 IEEE   831

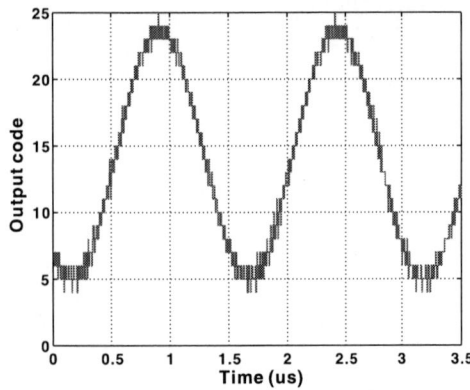

Fig. 5. Measured time-domain ADC output.

There are tones centered around the VCO harmonics, which is the effect of the intrinsic CLA as explained in Sec. II. The SNDR is 66.5 dB. The SFDR of 73 dB is limited by the 2nd-order distortion, which results from layout mismatches that cause imperfect cancelation of even order distortions. Fig. 7 shows the SNDR and SNR with varying input amplitudes.

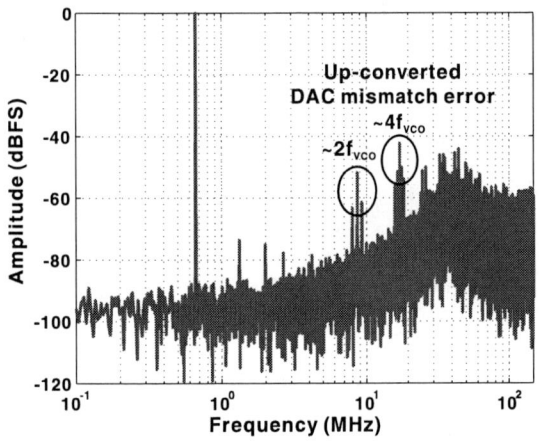

Fig. 6. Measured 65536 point fft plot with −0.9 dBFS and 661.5 kHz input.

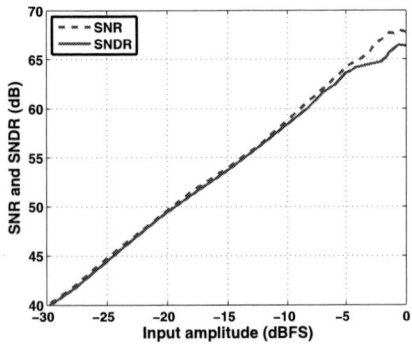

Fig. 7. Measured SNR and SNDR vs. input amplitude.

The measured analog power is 1.13mW, which includes the VCOs, the DAC, the replica buffers, and the comparators. The measured digital power is 0.62mW, which includes the XORs, flip-flops, and the thermometer-to-binary encoder. When operating in the ultra-low-power mode, the analog power drops

to 0.45mW and the SNDR is slightly lowered to 65.8 dB. The corresponding Walden figure-of-merits (FOMs) with 1.2V and 0.7V analog power supplies are 0.25 pJ/conversion-step and 0.17 pJ/conversion-step, respectively. Table I compares this work with state-of-the-art $\Delta\Sigma$ ADCs that have similar sampling rate, bandwidth, and SNDR. Note that its area is almost an order of magnitude smaller than others, and its FOM is less than half of that of the second best.

TABLE I
COMPARISON WITH STATE-OF-THE-ART $\Delta\Sigma$ ADCs

	Process (nm)	Area (mm^2)	f_s (MHz)	BW (MHz)	V_{ddA} (V)	SNDR (dB)	Power (mW)	FOM (pJ/step)
[5]	180	0.26	128	2	1.8	63.5	6	1.24
[7]	250	1.4	70.4	2.2	2.5	78.5	62.5	2.07
[8]	250	0.42	150	2	1.5	63.4	2.7	0.56
[9]	180	0.4	128	2	1.8	68	11	1.34
[10]	90	0.3	312	1.92	1.2	62.4	5	1.21
This	130	0.03	300	2	1.2	66.5	1.75	0.25
work					0.7	65.8	1.06	0.17

V. CONCLUSION

The paper presented a low-power small-area VCO-based $\Delta\Sigma$ ADC in 130nm CMOS. It is highly scaling friendly, as it only consists of VCOs, DACs, and comparators. It does not need any OTA. The DAC elements do not need to match well, as the ADC has an intrinsic CLA capability. It is envisioned that the performance of this ADC will be further significantly improved in advanced processes, such as 32nm or beyond.

VI. ACKNOWLEDGEMENT

We thank MOSIS MEP program for chip fabrication.

REFERENCES

[1] G. Taylor and I. Galton, "A mostly-digital variable-rate continuous-time delta-sigma modulator ADC," *IEEE J. Solid-State Circuits*, vol 45, pp. 2634-2646, Dec. 2010.

[2] M. Z. Straayer and M. H. Perrott, "A 12-bit, 10-MHz bandwidth, continuous-time $\Delta\Sigma$ ADC with a 5-bit, 950-MS/s VCO-based quantizer," *IEEE J. Solid-State Circuits*, vol 43, no 4, pp. 805-814, April 2008.

[3] M. Park and M. H. Perrott, "A 78 dB SNDR 87 mW 20 MHz bandwidth continuous-time $\Delta\Sigma$ ADC with VCO-Based integrator and quantizer implemented in 0.13 um CMOS," *IEEE J. Solid-State Circuits*, vol 45, no 12, pp. 3344-3358, December 2010.

[4] K. Reddy, S. Rao, R. Inti, B. Young, A. Elshazly, M. Talegaonkar, and P. K. Hanumolu, "A 16mW 78dB-SNDR 10MHz-BW CT-$\Delta\Sigma$ ADC using residue-cancelling VCO-based quantizer," *IEEE International Solid-State Circuits Conference*, Feb. 2012, pp. 152-154.

[5] J. Hamilton, S. Yan, and T. R. Viswanathan, "An uncalibrated 2MHz, 6mW, 63.5dB SNDR discrete-time input VCO-based $\Delta\Sigma$ ADC," *IEEE Custon Integrated Circuits Conference*, pp. 1-4, Sept. 2012.

[6] B. H. Leung and S. Sutarja, "Multibit AX AID converter incorporatin a novel class of dynamic element matching techniques," *IEEE Trans. Circuits Syst. II*, vol. 39, pp. 35-51, Jan. 1992.

[7] T. H. Chang, L. R. Dung, J. Y. Guo, and K. J. Yang, "A 2.5v 14-bit, 180-mW cascaded $\Delta\Sigma$ ADC for ADSL2+ application," *IEEE J. Solid-State Circuits*, vol. 42, pp. 2357-2368, Nov. 2007.

[8] T. Song, Z. Cao, and S. Yan, "A 2.7-mW 2-MHz continuous-time $\Delta\Sigma$ modulator with a hybrid active-passive loop filter," *IEEE J. Solid-State Circuits*, vol. 43, pp. 330-341, Feb. 2008.

[9] Y. Aiba, K. Tomioka, Y. Nakashima, K. Hamashita, and B. S. Song, "A fifth-order G_m-C continuous-time $\Delta\Sigma$ modulator with process-insensitive input linear range," *IEEE J. Solid-State Circuits*, vol. 44, pp. 2381-2391, Sep. 2009.

[10] M. Anderson and L. Sundstrom, "Design and measurement of a CT $\Delta\Sigma$ ADC with switched-capacitor switched-resistor feedback," *IEEE J. Solid-State Circuits*, vol. 44, pp. 473-483, Feb. 2009.

A 50MHz bandwidth, 10-b ENOB, 8.2mW VCO-based ADC enabled by filtered-dithering based linearization

Abhishek Ghosh, *Student Member, IEEE* and Sudhakar Pamarti, *Member, IEEE*
Department of Electrical Engineering, University of California, Los Angeles, CA 90095
Email:{abhishek,spamarti}@ee.ucla.edu

Abstract—**A dithering technique for linearization of VCO-based ADCs is proposed. The proposed technique conditions the signal to the VCO input to appear as white noise thereby eliminating spurious signal content arising out of the VCO non-linearity. The technique, thus obviates the need for power-hungry digital calibration techniques or expensive front-end loop-filters. A prototype implementation (in 65nm CMOS) based on the technique achieves 10-b ENOB in digitizing signals with 50MHz bandwidth consuming 8.2mW at an FoM of 90fJ/conv.step.**

I. INTRODUCTION

With both wireless communication and imaging applications pushing for higher data rates there is an ever-increasing demand for wideband, high resolution analog-to-digital converters (ADCs). Furthermore, such ADCs are often assembled in millions (imaging) or deployed in portable devices imposing strict power consumption limits on a single ADC. To this effect, oversampling noise-shaping converters have been a popular choice for digitizing wideband signals with a high resolution. In fact, continuous-time delta-sigma modulators (CTDSMs) have enjoyed a renewed interest in recent literature. However, the problem with such noise-shaping ADCs is their high power consumption which makes them unsuited for the above-mentioned applications.

Voltage-controlled oscillator (VCO)-based ADCs serve as a power-efficient, technology-scalable and simple alternative to conventional $\Sigma - \Delta$ modulators [1]–[4]. The fact that VCO-based architectures are mostly digital makes them amenable to a digital synthesis flow and hence highly attractive from a simplicity-of-design point of view.

A. VCO-based ADCs and prior-art

The basic concept of a VCO-based ADC is shown in Fig. 1. The input signal $V_{in}[n]$ is applied to the control voltage of a ring-VCO . The frequency of oscillation is proportional to $V_{in}[n]$ as shown in Fig. 1. Digital counters following the VCO count the number of rising/falling edges of the VCO output(s), $\Phi_i[n]$, during each sample period T_s. The counter outputs are accumulated to result in the final digital output, $y[n]$. In spite of the above-mentioned advantages, however, such ADCs are plagued by an inherently non-linear voltage vs. frequency curve (K_{VCO} tuning-curve nonlinearity), which limits the achievable spurious-free dynamic range (SFDR) to 6-7 bits. Several techniques have been employed to allay the

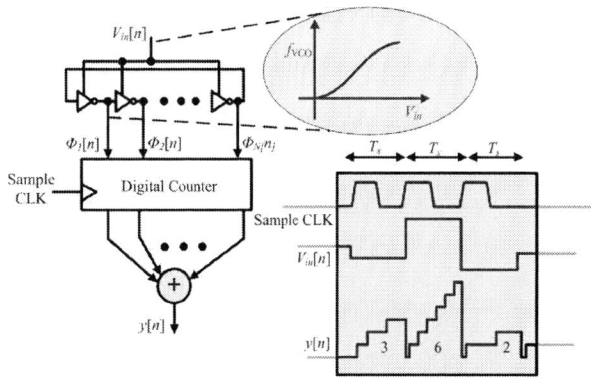

Fig. 1: VCO-based ADC: concept

impact of this nonlinearity on the ADC performance. In [1], the ring-ADC is used in a $\Sigma - \Delta$ feedback loop: the error signal (resulting from the negative feedback) which has a much lower dynamic range than the input is fed to the VCO, hence exercising a smaller region of the non-linear curve leading to lower distortion. However, the technique relies on 4th-order loop-filtering involving power-hungry op-amps in the signal path to achieve the desired performance; the op-amps increase power consumption, impose signal bandwidth constraints and negate several of the aforementioned scaling benefits offered by VCO-based ADCs. Recently, digital calibration has been proposed to linearize VCO-based ADCs. In [2], the VCO tuning curve is modeled as a Taylor series; a random calibration sequence and multiple digital correlators are used to estimate the Taylor series coefficients and the non-linear distortion terms are duly canceled by approximately inverting the Taylor series. However, this robust background technique suffers from increased digital power consumption due to the complex digital correction involved at the high sampling speed. The VCO-based ADC in [3] is used as a second stage in a 1-1 multi-rate MASH architecture where the first stage is a conventional $\Sigma - \Delta$ ADC. There have been previous efforts to use a $\Sigma - \Delta$ ADC as the second-stage in a 0-3 MASH architecture [5], mostly to reduce the dynamic range of the second stage. However, in [3], the input to the VCO-based ADC may be highly signal-dependent(quantization noise of a $\Sigma - \Delta$ ADC can have significant correlations with the input) and hence cause tones from the VCO-nonlinearity.

With a different approach to tackle the problem in a power-

978-1-4673-6145-3/13 $31.00 © 2013 IEEE

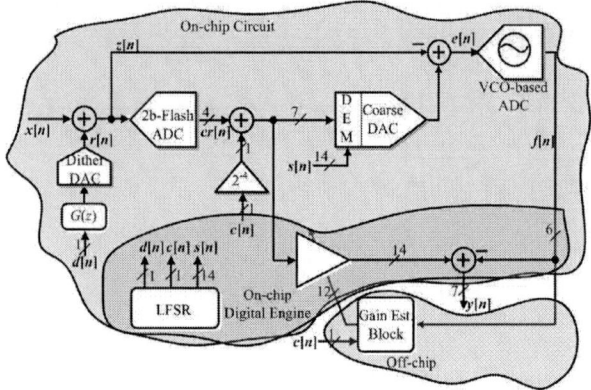

Fig. 2: Top-level architecture

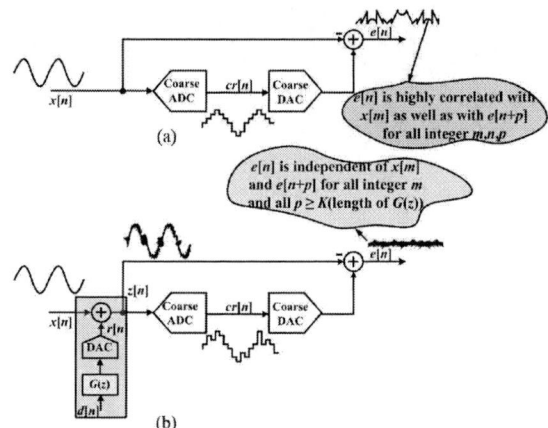

Fig. 3: Evolution of sequences for a sinudoidal input in coarse stage(a) without dither (b) with filtered dither

efficient way, this work employs shaped digital dither to essentially randomize the input to the VCO-based ADC in a 0-1 MASH architecture, thereby rendering it insensitive to VCO non-linearity.

II. PROPOSED ARCHITECTURE

The proposed architecture is presented in Fig. 2. The VCO-based ADC is embedded in a 0-1 MASH architecture [3], [5] where "0" represents an oversampled, very coarse ADC and "1" represents the *fine* VCO-based ADC. The outputs of the two ADCs are combined through gain calibration, using the gain estimation block, to account for inevitable gain mismatches between the two paths. The 0-1 architecture reduces the dynamic range of the VCO-based ADC input: a smaller portion of the VCO tuning curve is exercised thereby lessening the impact of non-linearity [3], [5]. The main contribution of this work lies, however, in demonstrating a signal conditioning algorithm to whiten the VCO input, as discussed next.

A 2-level, zero-mean random sequence, $d[n]$ (independent and identically distributed with $\mathbb{P}r(d[n] = -1) = \mathbb{P}r(d[n] = 1) = 0.5$), is digitally high-pass filtered ($G(z)$ in Fig. 2) and added to the input of the coarse-ADC using a digital-to-analog converter (DAC). The gain of the DAC is chosen such that the resultant additive dither is bounded within $[-\Delta/2, \Delta/2]$, where Δ is the quantization step-size of the coarse ADC.

A behavioral signal-processing illustration of the effects of the proposed technique is shown in Fig. 3. Fig. 3(a) shows time-domain waveforms for a sinusoidal input in the coarse ("0") stage without additive dither. It is evident that the coarse-stage residue (quantization error of the coarse ADC) is periodic for a low-resolution ADC and hence tonal. Furthermore, it can be proved that the quantization error $e[n]$ is highly correlated with the input $x[m]$ for all integer m, n [6], which when passed through a non-linearity (VCO) can produce tones. Increasing the resolution of the coarse stage reduces this correlation, but does not eliminate it; furthermore it increases power consumption.

Fig 3.(b) illustrates time-domain waveforms in the coarse stage in presence of the filtered dither. It can be proved that the error-sequence $e[n]$ is white, independent of the input signal and is uniformly distributed in $[-\Delta/2, \Delta/2]$ [6], [7]. This

ensures that the VCO nonlinearity acts on white noise from independent sources and hence loses potency to produce tones. Mathematically, if the VCO characteristics can be written as $f[n] = a_1 e[n] + h(e[n])$ where $h(u)$ is a memoryless polynomial nonlinear function, then from Fig. 2,

$$cr[n] \text{(Coarse-ADC output)} = x[n] + e[n] + r[n]$$
$$f[n] \text{(Fine-ADC output)} = a_1 e[n] + h(e[n]) + e_2[n]$$
$$y[n] \text{(Total output)} = a_1 cr[n] - f[n]$$
$$= a_1 x[n] + a_1 r[n] - h(e[n]) - e_2[n] \quad (1)$$

where $r[n], e[n]$ and $e_2[n]$ are the filtered dither, quantization error from the coarse-ADC and quantization error from the fine (VCO-based) ADC respectively.

As explained above, dither makes the VCO-input $e[n]$ white, which makes the resultant error ($h(e[n])$ in Eqn. 1) devoid of any spurious tones and spreads it in $[0, F_s/2]$, which causes negligible error due to oversampling. Furthermore, as an additional advantage the random nature of the VCO input ensures that inter-stage delay mismatches in the VCO do not degrade the SFDR.

Note that any arbitrary dither will *not* make $h(e[n])$ tone free. The dither filter, $G(z)$ is specially structured (each tap is a power of 2) with a finite impulse response (FIR) of length K. It has been proven in [6], [7], that the specially filtered additive dither decorrelates the quantization noise of the coarse-ADC $e[n]$ and the input signal $x[n]$ besides whitening $e[n]$. The filter $G(z)$ needs to cater to two purposes:

- $G(z)$ should conform to the structure for whitening the error-sequence $e[n]$ (power of 2 coefficient) [7]
- $G(z)$ should be aggressively high-pass shaped to minimize any in-band SNR degradation

The second condition is a consequence of Eqn. 1 wherein $y[n]$ has a residual $r[n]$ (the filtered dither) term.

III. IMPLEMENTATION DETAILS

A. Coarse Stage

A top-level circuit description is presented in Fig. 4. The incoming signal $x(t)$ is sampled on the bottom-plates of a bank

978-1-4673-6145-3/13 $31.00 © 2013 IEEE

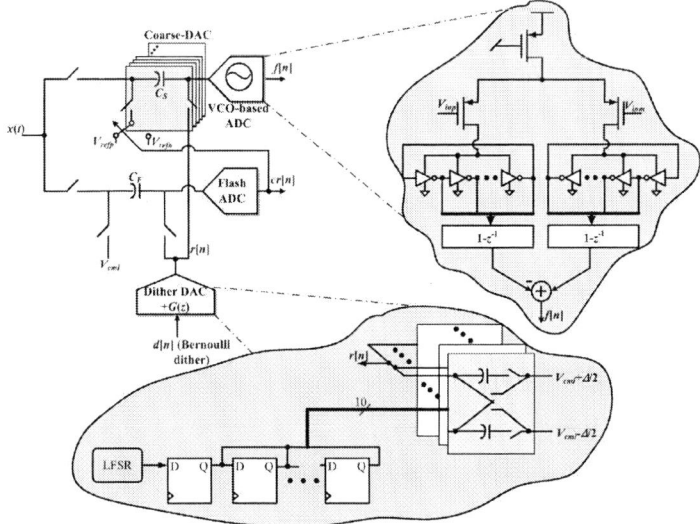

Fig. 4: Circuit Details

of capacitors ($C_s + C_F$)using bootstrapped switches while the top-plates are connected to the dither-DAC output $r[n]$, the latter acting as the common-mode for the sampling operation. The dither filter used in this work is

$$G(z) = 1 - 4z^{-1} + 4z^{-2} - z^{-3} + 2z^{-4} - 2z^{-5} + z^{-6} - 4z^{-7} + 4z^{-8} - z^{-9}$$

This filter choice is not unique and as discussed in Section II, out of the several filters that effectively whiten the VCO input [7], this filter ensured minimal SNDR degradation from the added dither due to its aggressive high-pass shape. The dither-DAC is implemented as a random input $d[n]$ being passed through a 10 bit shift register, with the outputs from the shift register switching a weighted capacitor array (based on the FIR filter weights). The sign inversions in $G(z)$ are effected through a differential swapping between the positive and the negative nodes as shown in Fig. 4. It should be borne in mind that errors in the dither-DAC (mismatches in the capacitors) introduce negligible errors, by virtue of the randomness of the dither input. The charge samples stored on C_F are used to resolve two bits using a flash ADC, serving as the coarse stage from the last subsection. The flash-ADC is a conventional 5-level structure with sufficient redundancy to tolerate comparator offsets. The comparators are implemented as a dynamic preamplifier followed by a dynamic latch [8]. C_s is used to compute the coarse-stage residue through simple charge sharing between the capacitors based on the decision by the flash ADC. Note that no residue amplifier is used to save static power. The resulting gain error caused due to parasitics is calibrated out as discussed next.

B. Fine Stage and Other Digital Conditioning

The VCO-based ADC is implemented as a 33-stage differential architecture as shown in Fig. 4 [1]. The differential input voltage is passed through a g_m-stage and the resulting signal currents control the oscillation frequency. The outputs of the flash-ADC and the VCO-ADC are combined using background gain calibration based on a correlation based gain estimator [9]. A random sequence $c[n]$ which follows the statistics $\mathbb{Pr}(c[n] = -\Delta/16) = \mathbb{Pr}(c[n] = \Delta/16) = 0.5)$ (referred to the coarse-DAC output) is added to the coarse-stage output $cr[n]$ as shown in Fig. 2. Consequently, $c[n]$ traverses the same path as the error signal $e[n]$ and hence when the fine-stage output $f[n]$ is correlated against $c[n]$, the gain through the coarse-stage DAC and the fine-stage can be estimated. Dynamic element matching (DEM) is employed to counter any static mismatches from the coarse DAC [10] using random sequences $s[n]$, as shown in Fig. 2. A segmented DEM architecture [10] is chosen since the addition of $c[n]$ to $cr[n]$ results in 7 effective number of bits going to the coarse DAC (Fig. 2).

IV. MEASUREMENT RESULTS

The ADC is realized in 65nm CMOS technology at an analog(coarse-stage, VCO, dither-DAC) supply voltage of 1V and a digital(gain calibration, combination) supply voltage of 0.8V (die photo in Fig. 6). The ADC sampling rate F_s is set at 1GHz. The total power consumed is about 8.2mW with a break-up of 5.5mW for the analog domain and 2.7mW for the digital domain. The low-speed correlation-based gain estimation is done off-the-chip using MATLAB while the on-chip digital engine multiplies $cr[n]$ with this estimate and adds it to $f[n]$ (Fig. 2)-the latter constitutes bulk of the digital power consumption. The measured power spectral density (PSD) plot for a tonal input at −4dBFS at 3.9MHz is shown with and without the dithering mode in Fig. 5, with the former showing almost 10dB improvement in measured SNDR. The measured SNDR over a 50MHz bandwidth is about 60dB resulting in an FoM of 90fJ/conv.step. The SNDR follows an expected almost-linear behavior with the input amplitude for different input frequencies with slight degradation near the full-scale as shown in Fig. 7(a). Fig. 7(b) plots the SNDR improvement for different input frequencies without and with the dithering technique (for a -4dBFS input), where the latter

978-1-4673-6145-3/13 $31.00 © 2013 IEEE

shows significant SNDR improvement. The reconfigurable nature of the ADC is highlighted in Fig. 8 which shows an almost uniform measured FoM over a considerable signal bandwidth. Table 1 puts the present work in perspective with recent competing works and shows that it advances state-of-the-art with a novel signal-processing application in circuits.

V. CONCLUSION

An open-loop signal-conditioning based technique for mitigating the effects of nonlinearity in VCO-based ADCs is presented. The technique relies on using the VCO-based ADC as the second stage in a 0-1 MASH architecture as well as applying a specially filtered dither to the input signal for whitening the VCO input. The prototype built based on the proposed technique is able to achieve the best FoM out of all published VCO-based ADCs.

REFERENCES

[1] Park, M.; Perrott, M.H., "A 78 dB SNDR 87 mW 20 MHz Bandwidth Continuous-Time $\Delta - \Sigma$ ADC With VCO-Based Integrator and Quantizer Implemented in $0.13\mu m$ CMOS," *Solid-State Circuits, IEEE Journal of* , vol.44, no.12, pp.3344,3358, Dec. 2009

[2] Taylor, G.; Galton, I., "A reconfigurable mostly-digital $\Delta\Sigma$ ADC with a worst-case FOM of 160dB," *VLSI Circuits (VLSIC), 2012 Symposium on* , vol., no., pp.166,167, 13-15 June 2012

[3] Asl, S.Z.; et. al"A 77dB SNDR, 4MHz MASH $\Delta\Sigma$ modulator with a second-stage multi-rate VCO-based quantizer," *Custom Integrated Circuits Conference (CICC), 2011 IEEE* , vol., no., pp.1,4, 19-21 Sept. 2011

[4] Reddy, K.; et. al "A 16mW 78dB-SNDR 10MHz-BW CT-$\Delta\Sigma$ ADC using residue-cancelling VCO-based quantizer," *Solid-State Circuits Conference Digest of Technical Papers (ISSCC), 2012 IEEE International* , vol., no., pp.152,154, 19-23 Feb. 2012

[5] Gharbiya, A.; Johns, D.A., "A 12-bit 3.125 MHz Bandwidth 0Ũ3 MASH Delta-Sigma Modulator," *Solid-State Circuits, IEEE Journal of* , vol.44, no.7, pp.2010,2018, July 2009

[6] Pamarti, S.; Delshadpour, S.; , "A Spur Elimination Technique for Phase Interpolation-Based Fractional- N PLLs," *Circuits and Systems I: Regular Papers, IEEE Transactions on* , vol.55, no.6, pp.1639-1647, July 2008

[7] Ghosh, A.; Pamarti, S. ; "Filtering of subtractive discrete dither in quantizers: some new results", to be published in *International Conference on Acoustics, Speech and Signal Processing*, 2013

[8] Wang, Y.T.; Razavi, B.; , "An 8-bit 150-MHz CMOS A/D Converter," *IEEE Journal of Solid-State Circuits*, vol. 35, pp. 308-317, March 2000.

[9] Siragusa, E.; Galton, I. ; "A digitally enhanced 1.8-V 15-bit 40-MSample/s CMOS pipelined ADC", *IEEE Journal of Solid State Circuits*, December 2004.

[10] Chan, K.L.; Zhu, J.; Galton, I ; "A 150MS/s 14-bit Segmented DEM DAC with Greater than 83dB of SFDR Across the Nyquist band", *2007 IEEE Symposium on VLSI Circuits*, pp. 200 Ũ 201, June 14, 2007.

Parameter	[1]	[2]	[3]	[4]	This work
Technology(nm)	130	65	130	90	65
F_s(MHz)	900	2400	1200	600	1000
Bandwidth(MHz)	20	37.5	4	10	30/50
SNDR(dB)	78.1	70	77	78	64/60
Power(mW)	87	39	13.8	16	8.2
FoM(fJ/conv.step)	330	201	298	125	87/94
Area(mm^2)	0.45	0.075	0.7	0.36	0.62

TABLE I: Comparison with state-of-the-art

Fig. 5: Die micrograph

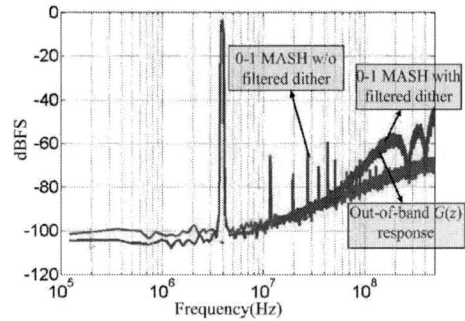

Fig. 6: Measured performance with and without technique

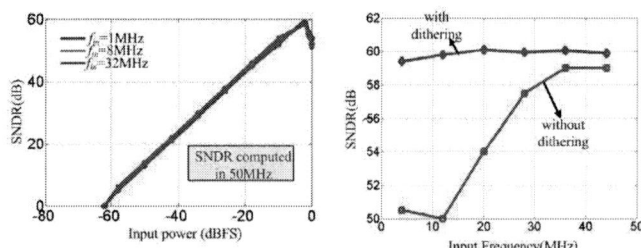

Fig. 7: Measured (a) SNDR vs input power (b) SNDR vs input frequency with and without dithering techniques

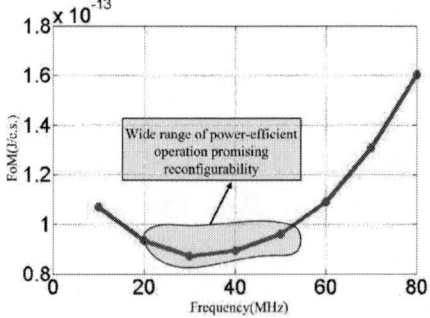

Fig. 8: Measured FoM with bandwidth

A Reconfigurable $\Delta\Sigma$ Modulator with up to 100 MHz Bandwidth using Flash Reference Shuffling

Trevor Caldwell, David Alldred, Zhao Li
Analog Devices, Inc.
Toronto, ON, Canada

Abstract—A reconfigurable 65 nm continuous-time low-pass $\Delta\Sigma$ modulator operates with a sampling frequency from 491 MHz to 1536 MHz, a signal bandwidth from 10 MHz to 100 MHz, and a dynamic range of 75.4 dB to 62.8 dB, respectively. Reference shuffling in the flash ADC is used to improve the linearity of the flash and DAC, while also increasing the highest sampling rate and bandwidth of the modulator.

I. INTRODUCTION

Wideband direct-conversion receivers offer compelling benefits for basestation transceiver implementations including component count reduction, board area reduction, simplified external filtering, and shorter development cycles for applications across multiple frequency bands. However, direct-conversion receiver architectures present several challenges towards meeting wideband multi-carrier system requirements including high bandwidth and linearity requirements, tolerance of out-of-band blocking signals, and low $1/f$ or flicker noise.

This paper describes a continuous-time $\Delta\Sigma$ modulator that achieves the required resolution and bandwidth for such applications. Section II discusses the $\Delta\Sigma$ leapfrog architecture while Section III describes the main circuits designed for the $\Delta\Sigma$ modulator. Section IV presents the flash ADC calibration and reference shuffling that improves the speed and linearity of the $\Delta\Sigma$ modulator. Section V summarizes the results from a 65 nm CMOS test chip and Section VI concludes the paper.

II. $\Delta\Sigma$ ARCHITECTURE

A. System

A direct-conversion receiver is shown in Fig. 1. For the target application, the $\Delta\Sigma$ ADC requires an equivalent single-tone linearity better than -76 dBc (-70 dBc with the TIA) so as not to reduce the noise figure of the overall system by more than a fraction of a dB. Since conversion down to DC is required, the flicker noise corner must be less than 120 kHz. An SNR of 68 dB (65 dB with the TIA) is required in the 50 MHz (and lower) mode, while 62 dB resolution is required in the 100 MHz mode.

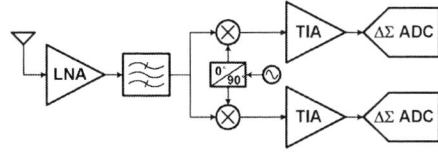

Fig. 1. Direct-conversion receiver architecture.

B. Architecture

The $\Delta\Sigma$ modulator that meets these targets is the 4th-order continuous-time leapfrog [1] with distributed feedback structure shown in Fig. 2. The feedback style was chosen to avoid out-of-band peaking from the *signal transfer function* (STF) that could be problematic in the presence of out-of-band blockers. These blockers could overload the modulator if there is insufficient filtering in the previous receiver stage. An additional direct feedback path through DAC_5 allows the internal 17-level flash ADC to output DAC signals one clock cycle late, thus relaxing the flash ADC speed requirement.

The leapfrog structure is different from a typical cascade-of-resonators with distributed feedback architecture since an additional feedback path is added from the output of the third amplifier to the input of the second amplifier (R_{23}). This adds more control to the zero locations in the *noise transfer function* (NTF); if the two resonators are designed with the same resonant frequency, the additional feedback path will split the NTF zeros so that they can still be placed optimally across the signal band in the NTF. Noise (flicker or thermal) from the third amplifier gets input-referred through the initial placement of the first resonator's NTF zero (i.e., before splitting), while the overall modulator maintains its optimally placed NTF zeros. This adds an extra degree of freedom when trying to allocate the input-referred noise contributions from the various stages of the modulator, giving the option to balance the high-frequency thermal noise with the low-frequency flicker noise.

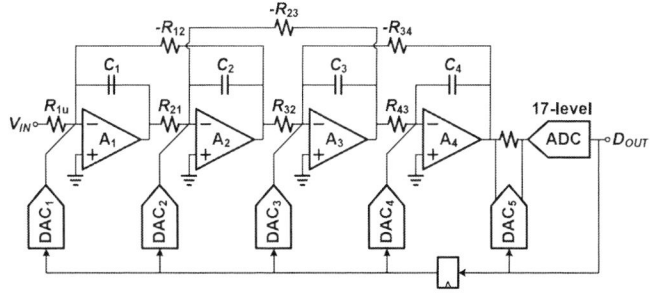

Fig. 2. Continuous-time leapfrog with distributed feedback $\Delta\Sigma$ architecture.

III. CIRCUITS

A. Amplifiers

Due to the high linearity and bandwidth requirements, the amplifiers need to maintain a gain higher than 40 dB up to the edge of the signal band. Assuming a first-order roll-off, this requires an amplifier with a unity-gain frequency greater than

978-1-4673-6145-3/13 $31.00 © 2013 IEEE

5 GHz, which would result in a power inefficient amplifier design because high-f_T biasing is needed for transistors handling the signal. Multi-stage multi-path feedforward amplifiers with a higher-order roll-off can obtain a high gain at high frequency with a lower unity-gain frequency [2]. The general architecture of the amplifier is shown in Fig. 3. The amplifier paths will contribute phase at different frequencies depending on their gain relative to the other paths at each frequency. Both lower order paths are designed to dominate the gain (and therefore phase) near the unity-gain frequency, keeping the amplifier conditionally stable as the phase is reduced.

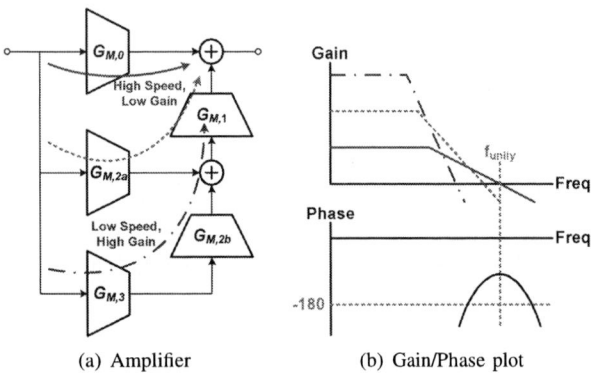

(a) Amplifier   (b) Gain/Phase plot

Fig. 3. Conceptual diagram of a multi-stage multi-path feedforward amplifier.

Another advantage of these amplifiers is their ability to decouple requirements such as noise, speed, swing and gain. The low-frequency in-band performance is determined by the third-order path through $G_{M,3}$, $G_{M,2b}$ and $G_{M,1}$; noise is dominated by $G_{M,3}$; output swing is determined by $G_{M,1}$; and stability and gain shape are determined by the interaction of the three paths.

The four amplifiers in the $\Delta\Sigma$ ADC are designed almost identically; $G_{M,3}$ is the only transconductance that changes. Amplifiers 3 and 4 use half the bias current of amplifiers 1 and 2 since their input-referred noise contribution is less. Also, since the ADC is designed for a direct-conversion receiver, the input pair of amplifiers 1 and 3 is 4x larger to reduce the flicker noise. This is not necessary in amplifiers 2 and 4 since their input-referred noise contribution is reduced at low frequency by an additional integrator stage.

B. Digital-to-Analog Converters

The linearity requirements in the current-mode DACs necessitate some form of dynamic element matching to reduce mismatch since sizing alone would be prohibitively large and add extra capacitance to the circuit. It would also only solve mismatch errors related to the current source, but not mismatch errors associated with the switches or their drivers.

A conventional *data-weighted averaging* (DWA) scheme reduces distortion caused by current-source mismatch but also increases second-order distortion from activity-dependent errors. This occurs because the pointer that determines which DAC elements to use is updated based on the previous DAC value, resulting in switching activity that is signal-dependent. Instead, a scheme where the pointer is updated by a fixed amount every clock cycle is employed. This results in a less aggressive mismatch-shaping scheme, but it uses all DAC elements over time, and it does not produce signal-dependent switching activity. This scheme has the potential to introduce mixing terms between out-of-band blocker signals and the rotating frequency (i.e., f_S/M where M is the number of quantizer levels), so an additional random component is added to the fixed pointer increment to spread out these tones (this pointer updating block is shown in the lower-left of Fig. 5 as part of the flash reference shuffling).

C. Flash ADC

The internal 17-level flash ADC requires 16 comparators running at a sample rate of up to 1536 MHz. The delay and bandwidth must be good enough to meet feedback timing and loop delay requirements while maintaining a low input capacitance and power dissipation. The direct feedback path through DAC_5 relaxes the timing requirements so that a clock cycle is available for the flash ADC to fully update its value and pass it to the DACs.

The comparator is shown in Fig. 4 [3]. A triode PMOS device M_1 acts as the load for the pre-amplifier, and M_3 disconnects the pre-amplifier from the comparator during regeneration. Transistor M_7 contributes additional current during pre-amplification for calibration purposes. The worst-case regeneration time constant of the tank is 10 ps.

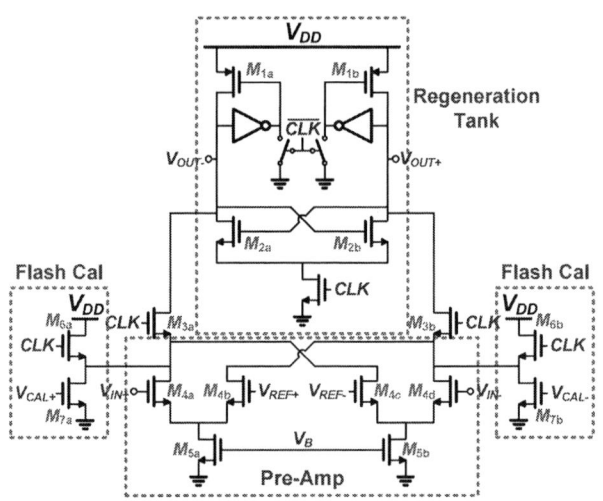

Fig. 4. Comparator schematic with flash calibration.

IV. FLASH CALIBRATION AND REFERENCE SHUFFLING

A. Calibration

The input transistors of the comparator are small to reduce capacitance and increase speed, but with a σ of 0.75 LSB the additional offsets in the comparator also reduce the SQNR of the modulator by up to 6 dB. Calibration is added to the comparator (see Fig. 4) where transistor M_7 adds current to the pre-amplifier branches to compensate for these offsets. A 3-bit calibration code selects an appropriate voltage for V_{CAL} for each of the 16 comparators. This can compensate for offset errors as large as 2 LSB.

The calibration code is determined by a foreground calibration scheme where a digital RMS meter measures the total noise output from the $\Delta\Sigma$ modulator by simply accumulating

978-1-4673-6145-3/13 $31.00 © 2013 IEEE   838

the squared time-domain outputs. The calibration codes are adjusted to minimize this total noise by intelligently searching the possible calibration codes with an on-chip algorithm.

B. Reference Shuffling

Mismatch shaping in the DACs typically tightens the constraints on the comparator regeneration time constant since a portion of the clock period is used to update the DAC pointer. Instead of trying to reduce the time constant on comparators that are already operating with a 1536 MHz clock rate, the DAC mismatch shaping is moved out of the feedback path to the reference voltages within the flash ADC [4], leaving a full clock period to update the DAC mismatch shaping and switch matrix, and to latch the comparators and drive the DACs. As shown in Fig. 5, the DAC mismatch shaping logic is still used to update the pointer in the flash reference switch matrix, but it is the reference voltages of the flash comparators that are switched while the individual comparators within the flash are hard wired to their respective DAC cells for maximum speed.

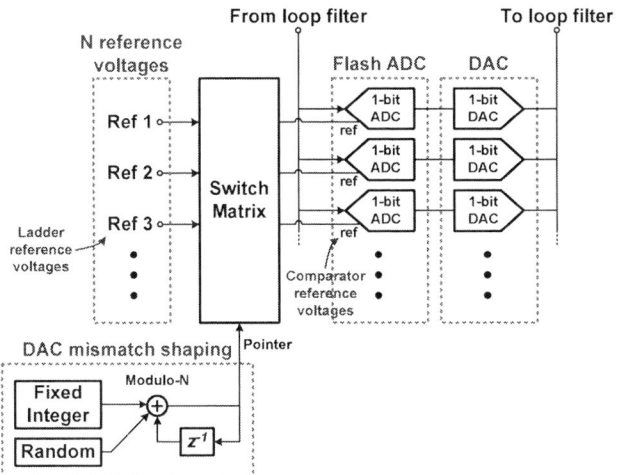

Fig. 5. Flash ADC architecture with shuffling for DAC mismatch shaping. Note that while only one DAC is shown, all DACs are attached to the comparators.

Moving the shuffling to the flash ADC also reduces the non-linearities associated with the flash ADC. While typically this is not a problem in a $\Delta\Sigma$ modulator where the OSR is high and the SQNR is much lower than other noise sources within the modulator, that is no longer the case in this modulator since it operates with a bandwidth of 100 MHz at a sampling frequency of 1536 MHz, which is an OSR of only 7.7.

One difficulty with using a switch matrix for the references in the flash ADC is the settling time of the comparator reference voltages; if the reference voltage changes by a large amount - from the top of the reference ladder to the bottom of the reference ladder - the transition can take too long to settle. As a result, a zipper rotation path for the reference switch matrix is used, as shown in Fig. 6. While typically the references would cycle through the values one by one with the top reference value rotating down to the bottom, the zipper rotation scheme has the references cycling through two references at a time, except at the top and bottom where they only move by one. It is clear that no reference voltage ever has to change by more than two reference levels when the pointer increments by one.

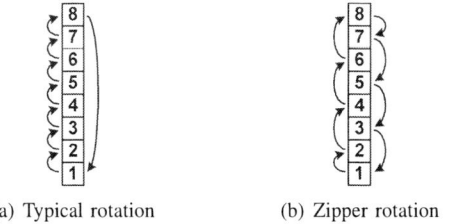

(a) Typical rotation   (b) Zipper rotation

Fig. 6. Switch matrix rotation scheme in the flash ADC. Note that while 16 reference voltages are used in the modulator, only 8 are shown here.

The circuit used to perform the reference shuffling is shown in Fig. 7. The input signal $sel[15:0]$ is a one-hot thermometer-coded signal that is boosted to a higher voltage, which then drives an NMOS switch matrix. There are 256 pairs of NMOS switches which are necessary to pass the 16 potential combinations (as determined by $sel\_bst[15:0]$) of the 16 reference voltages. The boosting circuit is necessary to increase the speed of the shuffler since only half a clock period is available to adjust the reference voltages, the τ of the transmission gate is 100 ps, and the settling requirement on the references is 4τ. The boosted gate has a τ of only 32 ps, and with a booster delay of 88 ps this results in a total delay of 296 ps on the worst-case corner which less than half of a 651 ps clock period.

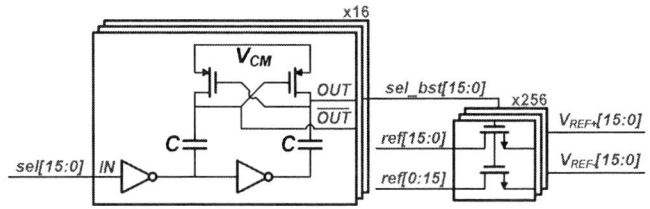

Fig. 7. Flash ADC reference shuffler.

V. EXPERIMENTAL RESULTS

The 4th-order continuous-time $\Delta\Sigma$ modulator was fabricated in 65 nm CMOS. A chip micrograph of the dual ADC for a direct-conversion receiver is shown in Fig. 8. The total area of an individual ADC core is 2000 μm by 400 μm. The ADC can only be tested with a signal when it is connected to the *transimpedance amplifier* (TIA), the preceding stage in the receiver. It can be tested in isolation by disconnecting it from the TIA, but with no signal present.

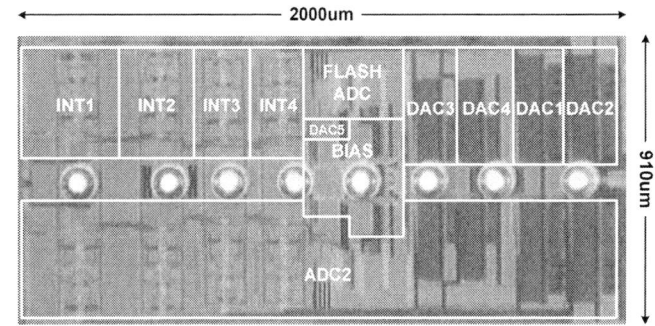

Fig. 8. Die photo of dual ADC.

978-1-4673-6145-3/13 $31.00 © 2013 IEEE

An output spectrum of the undecimated data is shown in Fig. 9 for a sampling frequency of 1229 MHz with a 15 MHz input and an SNDR of 62.9 dB. The reference shuffling keeps the 3rd-order distortion below -70 dBc (with the TIA). The low flicker noise necessary for direct-conversion receivers is evident with the low corner frequency of about 100 kHz. The STF is also shown and has a maximum variation of 0.24 dB within the signal band. Also, as expected in a feedback-style $\Delta\Sigma$ modulator, there is no out-of-band peaking.

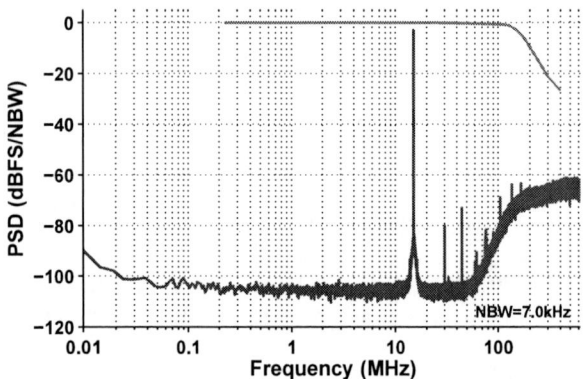

Fig. 9. Output spectrum for input at 15 MHz with a sampling frequency of 1229 MHz. The STF is also plotted above the signal.

SNDR vs input amplitude is plotted in Fig. 10. The peak SNDR is 62.9 dB and the dynamic range is 69.5 dB. The SNDR degrades at higher input signal amplitudes as the noise floor rises due to increased DAC noise. The quoted SNDR is 4.2 dB lower than expected due to the TIA which has roughly equal noise and distortion; with the TIA disconnected (but no signal present), the noise floor drops by 4.2 dB.

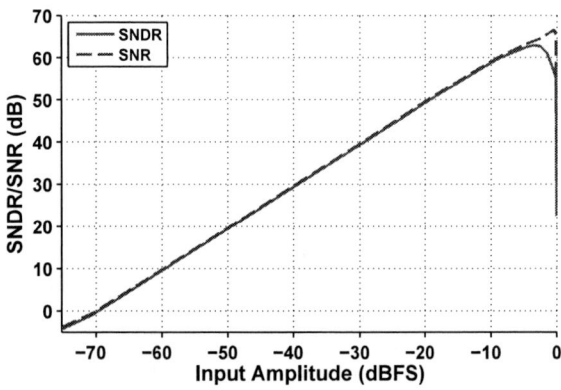

Fig. 10. SNDR vs input amplitude for an input at 15 MHz with a sampling frequency of 1229 MHz.

The four different modes of the reconfigurable $\Delta\Sigma$ are summarized in Fig. 11. The lower bandwidth modes (10 MHz and 20 MHz) have an OSR of 24.6, the 50 MHz mode has an OSR of 12.3, and the higher bandwidth mode (100 MHz) has an OSR of only 7.7. The SNDR drops in the higher bandwidth modes due to the lower OSR, and as expected the 100 MHz mode is the least power efficient since the SQNR is within a few dB of the SNR, a necessary trade-off when trying to increase the signal bandwidth at low OSRs. The experimental results in the 1229 MHz mode are summarized in Table I.

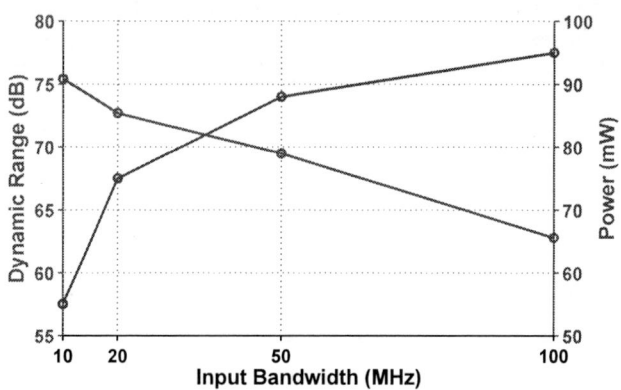

Fig. 11. Dynamic range and power for the four different ADC modes with signal bandwidths of 10 MHz, 20 MHz, 50 MHz and 100 MHz and OSRs of 24.6, 24.6, 12.3 and 7.7, respectively.

TABLE I. EXPERIMENTAL RESULTS

Parameter	Measurement
Sampling Frequency	1229 MHz
Signal Bandwidth	50 MHz
Input Signal	15 MHz
Peak SNDR[1]	62.9 dB
Peak SNR[1]	66.5 dB
Distortion (3rd)[1]	70.2 dBc
Dynamic Range[1]	69.5 dB
Power	88 mW
FOM$_S$[2]	157 dB
Technology	65 nm
Supply	1.3 V
Chip Area	0.8 mm^2

[1] Measurement includes TIA and ADC

[2] $\text{FOM}_S = \text{DR} + 10\log_{10}(\text{BW}/\text{P})$

VI. CONCLUSION

In this paper a reconfigurable continuous-time low-pass $\Delta\Sigma$ modulator for direct-conversion receivers was demonstrated. The speed and linearity of the flash ADC and DAC were improved with both flash calibration and reference shuffling. High signal bandwidth of up to 100 MHz was obtained with a power efficient feedback-style $\Delta\Sigma$ modulator which, while less efficient than a feedforward modulator, is far more tolerant of out-of-band blockers. Low flicker noise and high bandwidth were obtained with multi-stage multi-path feedforward amplifiers.

REFERENCES

[1] M. Van Valkenburg, *Analog Filter Design*. Oxford University Press, Inc., 1982.

[2] A. Thomsen, et al., "A Five Stage Chopper Stabilized Instrumentation Amplifier Using Feedforward Compensation," *Symposium on VLSI Circuits Digest of Technical Papers*, pp. 220-223, 1998.

[3] H. Shibata, et al., "A DC-to-1 GHz Tunable RF $\Delta\Sigma$ ADC Achieving DR=74 dB and BW=150 MHz at fo=450 MHz Using 550 mW," *IEEE Journal of Solid-State Circuits*, vol. 47, pp. 2888-2897, December 2012.

[4] W. Yang, et al., "A 100mW 10MHz-BW CT $\Delta\Sigma$ Modulator with 87dB DR and 91dBc IMD," *ISSCC Digest of Technical Papers*, pp. 498-499, Feb. 2006.

A 10-MHz Bandwidth 70-dB SNDR 640MS/s Continuous-Time ΣΔ ADC Using Gm-C Filter with Nonlinear Feedback DAC Calibration

Jiageng Huang, *Student Member, IEEE*, Shiliang Yang, Jie Yuan *Senior Member, IEEE*

Electronic & Computer Engineering Department, Hong Kong University of Science & Technology, Clear Water Bay, Hong Kong
(Tel: +852-2358-8844, Email: jghuang@ust.hk, Tel: +852-2358-8029, Email: eeyuan@ust.hk)

Abstract— So far almost all wide-band (>10MHz) continuous-time ΣΔ ADCs use active-RC filters. Although this type of filter possesses excellent linearity, it behaves as a resistive load for the preceding stage, which requires a signal buffer. For high speed and high resolution applications, a low noise and low distortion signal buffer consumes quite large power. Gm-C filters have small capacitive loads, which avoids the need of a signal buffer. As a result, the ADC could achieve better power efficiency. However, due to the poor linearity, previous ΣΔ ADCs with Gm-C filters have low SNDR. In this work, the first Gm cell's nonlinearity is calibrated by a nonlinear feedback DAC. The principle of the novel compensation technique is that transfer functions of the Gm cell and the feedback DAC match each other so that the DAC has the same nonlinearity as the first Gm cell. Therefore the distortion is removed at the output. The concept is implemented in a 640MS/s CT ΣΔ ADC for 10MHz signal bandwidth in a 0.18-μm CMOS process. After calibration, the modulator achieves 77/76/70dB DR/SNR/SNDR with 60mW power. The SNDR and power efficiency are much improved compared to previous ΣΔ ADCs with Gm-C filters. Without the power-hungry signal buffer, this novel on-chip calibration technique enables Gm-C-based ΣΔ ADC with similar SNDR and power efficiency as active-RC-based ΣΔ ADCs.

Index Terms— Analog-to-digital converter, continuous-time ΣΔ modulator, Gm-C filter, nonlinear feedback DAC, calibration.

I. INTRODUCTION

Active-RC filters are mostly used in wide-band (>10MHz) CT ΣΔ ADCs due to their excellent linearity at this moment. Very few wideband CT ΣΔ ADCs are designed with Gm-C filters [1], [2]. Limited by the nonlinearity, they have low SNDR (<70-dB) and power efficiency, which makes them unattractive compared to ΣΔ ADCs using active-RC filters.

However, CT ΣΔ ADCs with active-RC filters have a key drawback. Active-RC filters appear as resistive loads to the preceding stage so that signal buffers are required in the actual ADC as shown in Fig. 1. Due to the high gain of the loop filter, most of the signal current is subtracted by the feedback DAC1 and only small residual current steers into the integration capacitor (C1). If the signal buffer only passes signal current, which means k=1 in Fig. 1, the signal buffer would consume the same power as DAC1 which accounts for nearly 20% of the total ADC power according to [3]. In reality, a low distortion buffer requires the biasing current at least 4~5 times larger than the peak signal current I_m to ensure high linearity. Moreover, high supply voltage is usually needed to allow large input swing, which further increases the signal buffer power. Taking all these into consideration, signal buffers could consume as much

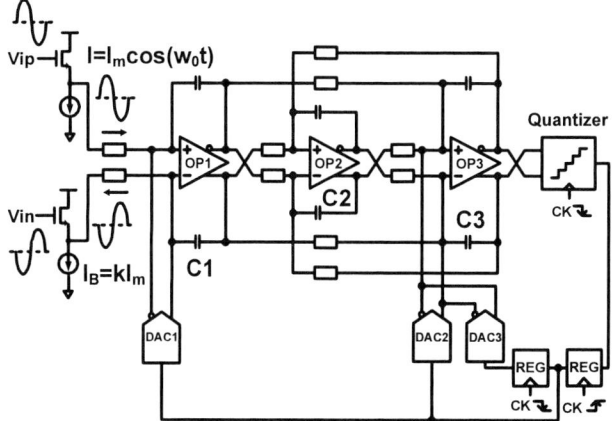

Fig.1. Signal buffer drives the resistive load of the active-RC filter.

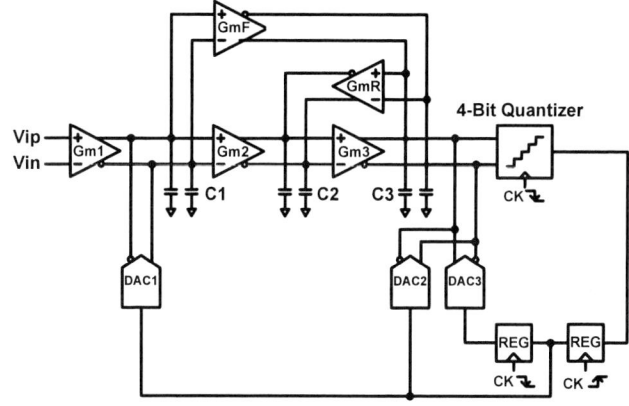

Fig.2. Our designed CT ΣΔ ADC based on Gm-C loop filter.

power as the total active-RC-based CT modulator. In publications, this driving power was provided by the testing equipment and was usually not included in FOM (figure-of-merit) calculations.

In this work, we designed a wide-band CT ΣΔ modulator with Gm-C filter. Its architecture is shown in Fig.2. With the small capacitive load at the input, no signal buffer is needed for this design. Gm-C filters are mainly limited by their poor linearity at the moment. In this work, we designed a novel on-chip calibration technique to compensate the nonlinearity of Gm1 so that high SNDR and power efficiency can be achieved for Gm-C-based wide-band CT ΣΔ ADC for the first time.

The following sections are organized as follows: Section II describes the proposed calibration technique. The circuit implementation is discussed in Section III. Section IV presents measurement results. Conclusions are drawn in Section V.

978-1-4673-6145-3/13 $31.00 © 2013 IEEE

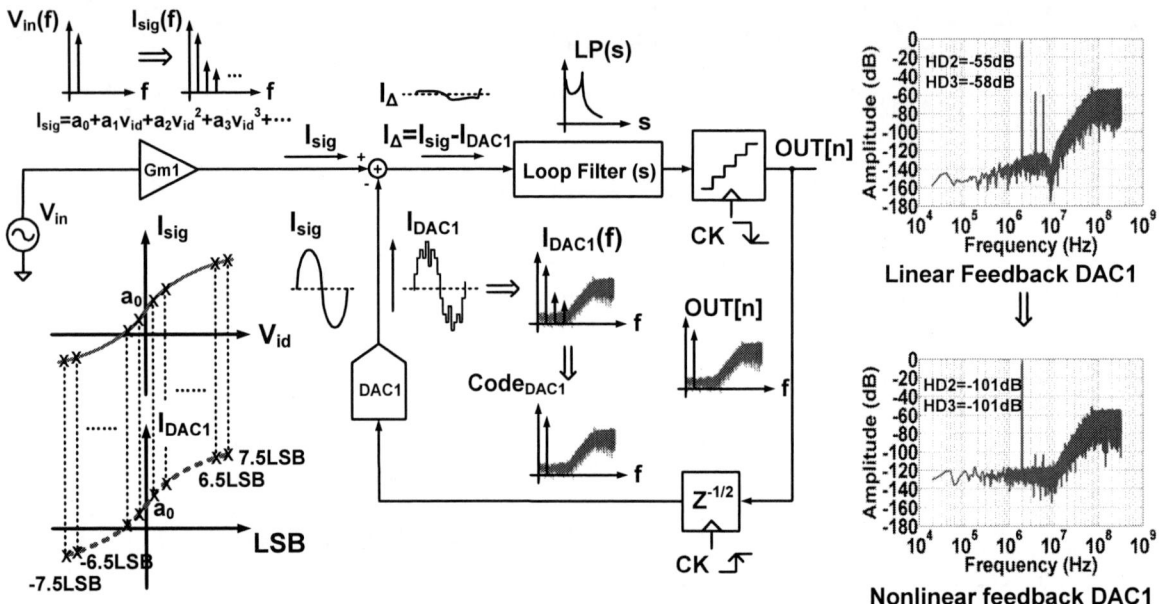

Fig.3. The new on-chip calibration scheme to compensate the nonlinearity of Gm1.

II. PROPOSED CALIBRATION TECHNIQUE

System-level simulations show that the nonlinearity of the modulator is mainly due to the first Gm1. Our calibration scheme illustrated in Fig.3 aims to compensate the nonlinearity of Gm1.

As shown in Fig.3, Gm1 converts the input voltage V_{in} to a signal current I_{sig} according to its V-I curve in red. This modulator has a 4-bit flash ADC with 16 quantization levels (from -7.5LSB to 7.5LSB). DAC1 converts this 4-bit code to a 16-level feedback current I_{DAC1}. In conventional designs, these 16 feedback current levels are designed to have very high linearity as its nonlinearity directly adds to the input signal. In most designs, real-time DEM (dynamic element matching) is needed to achieve this high linearity. Essentially, this highly linear DAC is required because of the highly linear V-I converter at the input.

In our current Gm-C modulator, Gm1 is highly nonlinear instead. Consequently, strong distortion exists in the spectrum of the input current I_{sig}. With this input, a highly linear DAC1 is not needed any longer. Instead, if the feedback current from DAC1 matches the output current of Gm1 at these 16 levels, the digital output code from the ADC will actually be linear to the input voltage V_{in}. In other words, the V-I transfer curve of the feedback DAC1 at the 16 discrete points should match Gm1's V-I transfer curve as shown in Fig.3. The blue dashed line is a fitting line of these 16 discrete points and it should match Gm1's red V-I transfer curve. As a result, the feedback DAC1 has the same nonlinearity as the first Gm cell.

The loop filter (LP(s)) provides high gain in the signal band. Similar to a high-gain OPAMP in feedback, most of the signal current I_{sig} is subtracted by the feedback current I_{DAC1}. Only a small current difference I_Δ flows into the loop filter. In other words, the feedback current I_{DAC1} consists of two parts, a signal current and a high-frequency residual current. This residual current is the quantization noise. The simulated I_{DAC1} spectrum from the nonlinear DAC1 is shown in Fig.3. The signal part of I_{DAC1} (f) has the same distortion as the spectrum of input signal current I_{sig} (f). Owing to the matched nonlinearity between DAC1 and Gm1, The distortion is eliminated at the ADC output.

System-level simulations are performed on this modulator. The spectra at the modulator output are illustrated in the right part of Fig.3. If a linear DAC1 is adopted, the output spectrum on the top indicates strong second-order and third-order distortion introduced by the nonlinear Gm1. With a matched nonlinear DAC1, the spectrum at the bottom shows much reduced distortion.

III. CIRCUIT IMPLEMENTATION

A. Gm-C Integrator

The schematic of the first Gm cell is shown in the left of Fig.4. A source degeneration resistor is adopted to linearize the Gm cell within the large input signal range. The linearity of the Gm cell is directly related to the swing ratio k, which is the peak single-ended signal current $1/2I_{max}$ over the bias current I_B. Smaller k results in higher linearity at the cost of higher bias current and power. In this design the swing ratio is set to 0.4 leading to an SFDR of 54dB from Gm1 alone. In terms of noise, simulation results show that, under the typical corner, Gm1's input referred in-band noise (from DC to 10MHz) accounts for 21% of the modulator's total in-band noise.

Similar source degenerated Gm cells are employed in the following stages. Their noise and linearity requirements are much relaxed since the noise and nonlinearity are shaped and suppressed by the gain of the preceding stages.

B. Calibration Circuits.

The calibration circuits are shown in Fig.4. The on-chip calibration overhead is a 16-level calibration signal generation DAC (CSG-DAC), a calibration DAC (Calib-DAC) in DAC1, a latch and digital logics for successive approximation (SAR) based comparison.

978-1-4673-6145-3/13 $31.00 © 2013 IEEE

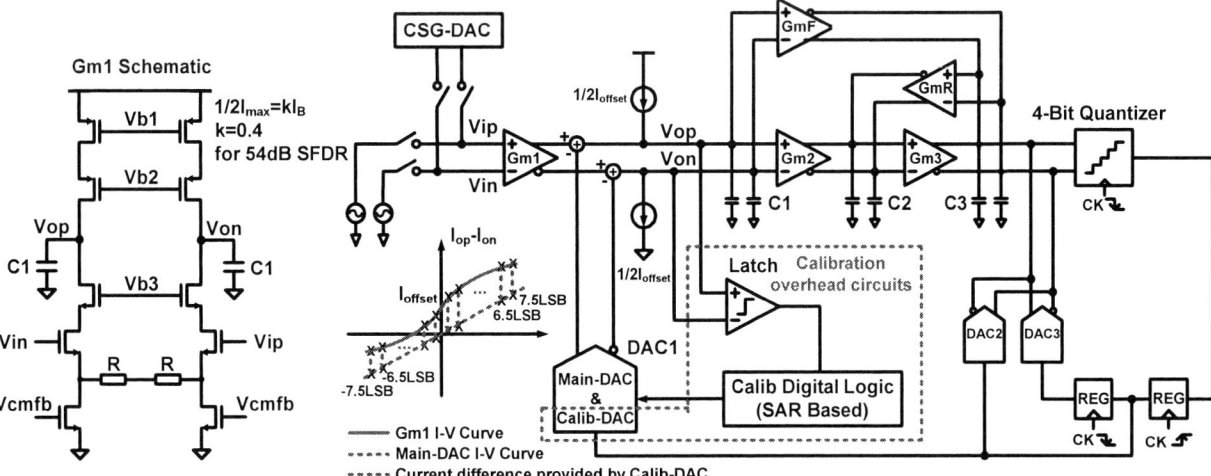

Fig.4. New Gm-C-based CT ΣΔ ADC including on-chip calibration circuits and Gm1 schematic.

During the calibration procedure, as shown in Fig.5, CSG-DAC generates 16-level voltage signals (from -7.5LSB to 7.5 LSB) sequentially to calibrate the 16-level feedback current from DAC1. With an input voltage, the Gm1 output current (red line in Fig.4) is subtracted by the feedback current from the Main-DAC (green dashed line in Fig.4). The current difference integrates on C1. A small offset current I_{offset} is deliberately added to Gm1's output. This small offset current will make the differential integration current always positive, which means V_{op} is always higher than V_{on} at the beginning. Then Calib-DAC only needs a pull-down current source at the V_{op} node. This current source is used to reduce this integration current through an SAR procedure. At the end of the SAR procedure, V_{on} and V_{op} are nearly equal. Consequently, the Gm1 current and the total DAC1 feedback current are equal at this input voltage. The 16-level feedback current from DAC1 is calibrated sequentially. When the calibration finishes, the Gm1 current and DAC1 feedback current match with each other at these 16 voltage levels. The Calib-DAC code is stored in a register. The Calib-DAC current is represented by dashed blue line in Fig.4. The V_{op} and V_{on} changes during the SAR procedure are plotted in Fig. 5.

A highly linear CSG-DAC is critical to this calibration scheme. This can be achieved using DEM on CSG-DAC during the calibration. Speed and noise are not issues for CSG-DAC. Hence, this DAC needs much less area and power than the highly linear feedback DAC1 in conventional designs, which requires real-time DEM run at the high speed clock frequency.

During normal data conversion, Calib-DAC provides a quite small current compared to Main-DAC. Hence, the power overhead of this calibration scheme is negligible.

IV. MEASUREMENT RESULTS

This prototype chip was fabricated in a 0.18-μm CMOS process. The die photo is shown in Fig.6. The calibration overhead circuits occupy about 15% of the overall modulator.

During the testing, the ADC operates at 640MS/s and consumes a power of 60mW. A 1.92MHz sinusoidal test signal is used because its 5-th order distortion is still in the signal band. Hp 8657B is used to generate the 640MHz clock of the modulator. Its RMS (root-mean-square) jitter is <300fs, which

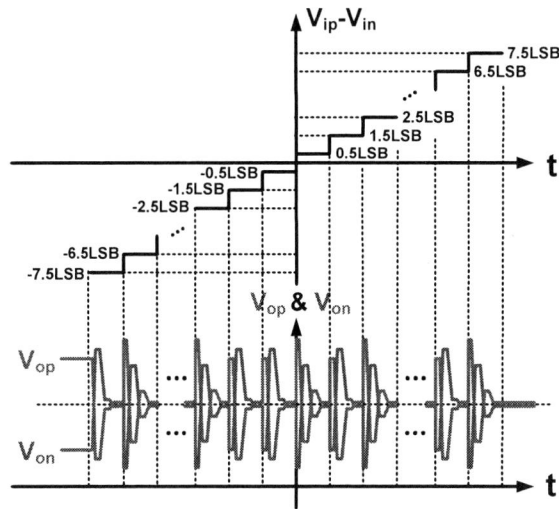

Fig.5 Gm1 output voltage during the calibration procedure.

amounts to <12μVrms in band jitter noise for the CT modulator. Silicon measurements show that the ADC achieves 76dB SNR over the 10-MHz signal bandwidth. After calibration, the SNDR can reach up to 70dB with -6.5dBFS input. The measured modulator spectrum is shown is Fig.7. This is equivalent to an ENOB of 11.3-bit. The dynamic range (DR) of the modulator is 77dB. Fig.8 shows the SNR/SNDR versus input power before and after calibration. With the calibration procedure enabled, the distortion with large input signal is much suppressed and the SNDR is increased by 10dB.

At this moment, due to the limitation of our testing equipment and the passive input signal filter, the measurable peak SNDR is limited by the distortion in the input signal (-67dBFS at maximum input) to the chip, which is clearly seen with spectrum analyzer. We are in the process of ordering a new band-pass filter with lower distortion. With lower input distortion, the calibration technique is expected to achieve higher SNDR.

Table I compares our design with other wideband CT ΣΔ ADCs presented recently. [1, 2] are the only two Gm-C-based wideband CT ΣΔ ADCs to authors' knowledge. Other ADC

978-1-4673-6145-3/13 $31.00 © 2013 IEEE   843

Fig.6 Die photo of the prototype ADC.

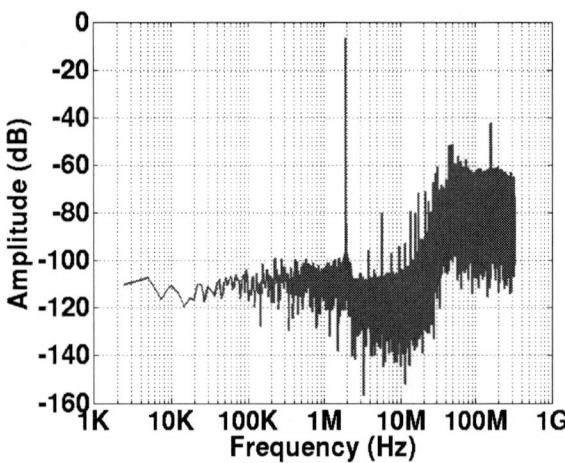

Fig.7. Spectrum for peak SNDR achieved with -6.5dBFS input, HD3=-73dB. The 3-rd harmonic comes from input signal. With a low distortion filter, the SNDR could be improved.

designs use active-RC filters. With the new calibration scheme, SNDR and FOM are much improved over previous Gm-C CT ΣΔ ADCs. Although designs with active-RC filter appear to have lower FOM, the signal buffer power is not included in these values. With this power included, the FOM could be doubled. With better testing equipment, our modulator should achieve higher SNDR with smaller FOM. Considering both effects, our Gm-C design is quite competitive with other CT ΣΔ ADCs in 0.18-μm CMOS process ([2][4][5][6]) and is comparable to CT ΣΔ ADCs in more advanced processes ([3][7][8][9]).

V. CONCLUSION

A new on-chip calibration technique for CT ΣΔ modulators with Gm-C filters is proposed. It uses a nonlinear feedback DAC to match the nonlinearity of the first Gm cell so as to remove its distortion at the modulator output. With the proposed calibration method, the Gm-C-based wideband CT ΣΔ modulator can achieve high SNDR for the first time. Its power efficiency is much improved and is comparable to

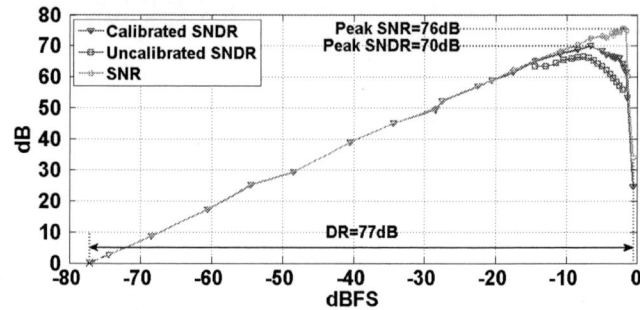

Fig.8 SNR/SNDR vs. input power before and after calibration.

Table I. Performance Summary
([1][2] use Gm-C filters and others use Active-RC filters)

Reference	BW/Fs (MHz)	SNDR (dB)	Power (mW)	Process (μm)	FOM (pJ/Conv)
[1]JSSC2006	10/320	54	30.4	0.25	3.76
[2]JSSC2010	18/360	63	230	0.18	5.86
[4]JSSC2012	16/800	65	47.6	0.18	1.02
[5]JSSC2011	5/850	56	6	0.18	1.15
[6]JSSC2010	25/400	68	48	0.18	0.48
[7]JSSC2011	20/640	64	58	0.13	1.13
[3]JSSC2009	20/900	78	87	0.13	0.33
[8]JSSC2011	20/2560	61	7	0.065	0.19
[9]JSSC2011	20/250	60	10.5	0.065	0.32
This work	**10/640**	**70**	**60**	**0.18**	**1.16**

active-RC-based wideband CT ΣΔ ADCs. The power-hungry signal buffers are no longer needed with this ADC.

REFERENCES

[1] Jesus Arias and Peter Kiss, "A 32-mW 320-MHz Continuous-Time Complex Delta-Sigma ADC for Multi-Mode Wireless-LAN Receivers" *IEEE J. Solid-State Circuits*, vol. 41, No.2, pp. 339—351, Feb. 2006.

[2] Yun-Shiang Shu and Junpei Kamiishi, "LMS-Based Noise Leakage Calibration of Cascaded Continuous-Time ΔΣ Modulators" *IEEE J. Solid-State Circuits*, vol. 45, No.2, pp. 368— 379, Feb. 2010.

[3] Matthew Park and Michael H. Perrott, "A 78 dB SNDR 87 mW 20 MHz Bandwidth Continuous-Time ΔΣ ADC With VCO-Based Integrator and Quantizer Implemented in 0.13μm CMOS" *IEEE J. Solid-State Circuits*, vol. 44, No.12, pp. 3344—3358, Dec. 2009.

[4] Vikas Singh, "A 16 MHz BW 75 dB DR CT ΔΣ ADC Compensated for More Than One Cycle Excess Loop Delay" *IEEE J. Solid-State Circuits*, vol.47, No.8, pp. 1884—1895, Aug. 2012.

[5] Bart De Vuyst, "A 5-MHz 11-Bit Self-Oscillating ΣΔ Modulator With a Delay-Based Phase Shifter in 0.025mm²" *IEEE J. Solid-State Circuits*, vol.46, No.8, pp. 1919—1927, Aug. 2011.

[6] Cho-Ying Lu, "A 25 MHz Bandwidth 5th-Order Continuous-Time Low-Pass Sigma-Delta Modulator With 67.7dB SNDR Using Time-Domain Quantization and Feedback" *IEEE J. Solid-State Circuits*, vol.45, No.9, pp. 1795—1808, Sep. 2010.

[7] Jun-Gi Jo, "A 20-MHz Bandwidth Continuous-Time Sigma-Delta Modulator With Jitter Immunity Improved Full Clock Period SCR (FSCR) DAC and High-Speed DWA" *IEEE J. Solid-State Circuits*, vol.46, No.11, pp. 2469—2477, Nov. 2011.

[8] Enrique Prefasi, "A 7 mW 20 MHz BW Time-Encoding Oversampling Converter Implemented in a 0.08mm² 65 nm CMOS Circuit" *IEEE J. Solid-State Circuits*, vol.46, No.7, pp. 1562—1574, Jul. 2011.

[9] Vijay Dhanasekaran, "A Continuous Time Multi-Bit ΔΣ ADC Using Time Domain Quantizer and Feedback Element" *IEEE J. Solid-State Circuits*, vol.46, No.3, pp. 639—650, Mar. 2011.

Algorithmic Nonlinear Macromodeling: Challenges, Solutions and Applications in Analog/Mixed-Signal Validation

Chenjie Gu
Intel Corporation
chenjie.gu@intel.com

Abstract—Analog/Mixed-Signal validation at the system level is becoming increasingly important as more electrical bugs are caused by the interaction among various circuit blocks. While hand-crafted behavioral models and linear models are still most widely used among designers, there is an increasing need for automatic behavioral modeling tools which capture low-level nonlinear behaviors in the circuit. This paper discusses challenges and difficulties of algorithmic nonlinear macromodeling, and reviews a series of recently developed techniques. In particular, we study the behavioral modeling problem from the perspective of projection in the state space defined by voltages and currents. We review a few nonlinear macromodeling techniques from the projection perspective, and demonstrate the model accuracy and computational efficiency compared to transistor-level models and linear models.

I. INTRODUCTION

System-level simulation and validation of analog/mixed-signal circuits are becoming increasingly important. While it is relatively easy to design individual circuits that meet their specifications, it is often hard to predict how the entire system behaves and how variations in individual blocks affect the system behavior. In fact, it has been witnessed that many bugs in mixed-signal circuits are exhibited after system-level integration[1]. Hence, it is crucial to be able to simulate the whole system, in order to validate its functional correctness as well as performance specifications.

However, even with the most advanced fast-SPICE techniques [2] and parallel computing capabilities, it is prohibitive to perform transistor-level simulations of many analog/mixed-signal circuits, such as high-speed I/O links, analog-to-digital converters, phase-locked loops, power amplifiers, *etc.*. For example, a typical multi-Gb/s I/O link [3] is composed of various circuit blocks, including channel, clock distribution, clock recovery and compensation loops. The high-level specification of an I/O link is the Bit Error Ratio (BER) which is typically of the order of 10^{-12} or less. This means that merely to get a very rough estimate of the BER, more than 10^{13} clock cycles need to be simulated. This is practically impossible to finish within reasonable amount of time.

To enable system-level simulation, often linear/nonlinear behavioral models are hand-crafted for every circuit block, based on designers' understanding of circuit dynamics. For examples, a transmission line can be modeled by a linear time-invariant (LTI) system, or equivalently characterized by its pulse response; a continuous-time linear equalizer can be approximated by a simple one-zero two-pole transfer function. However, such models could take weeks to build, and even so, they might show a large deviation from the transistor-level model behaviors. There are two major difficulties in the model building process. First, it is not straightforward to incorporate nonlinearities in the behavioral model – this usually requires good understanding of the circuit, and even so, seemingly insignificant nonlinearities can be easily ignored. However, it is known that many circuit blocks are inherently nonlinear and the nonlinear effect is crucial for circuit performances such as distortion. Second, it is not easy to automatically build behavioral models

from transistor-level circuit netlists. Such a bottom-up approach is important because it ensures that the behavioral model faithfully captures circuit characteristics in the transistor-level description.

Algorithmic macromodeling techniques (*a.k.a.*, behavioral modeling, model order reduction) provide an alternative of generating high-level behavioral models from transistor-level circuit netlists. In particular, many MOR algorithms for LTI systems have been proposed and extensively used [4], [5], [6], [7], [8]. In contrast, the development of such algorithms for nonlinear systems is not as mature, but there have been a few promising methods developed recently. As is explained in Sec. IV, the difficulty stems from intrinsic circuit nonlinearities which make both theory and analysis much harder than that of the linear circuits. However, since engineered circuits usually have regularized behaviors (as compared to arbitrary nonlinear dynamical systems), it is possible, with reasonable assumptions, to rigorously define and analyze specific nonlinear circuit responses, and subsequently develop algorithms for creating behavioral models that retain such responses.

In this paper, we review several recently developed nonlinear macromodeling algorithms, classified into two categories. First, we discuss methods that generate differential equation-based behavioral models that can be inserted into a SPICE/Verilog-AMS simulation flow. The key idea is to define a nonlinear manifold on which circuit dynamics evolve. In particular, we show that the model is able to reproduce nonlinear behaviors that small-signal linear models cannot predict. We also show that the oscillator phase macromodels, such as ISF and PPV, are special cases of projection-based methods if the manifold is defined to be the periodic steady state at the nominal condition.

Second, we review two techniques that build finite state machine (FSM) models and event-driven models, respectively. The FSM model is especially useful in modeling nonlinearities in digital-like circuits. The event-driven model is especially efficient in modeling nonlinear circuits with limited memory and limited input stimulus variability. An important feature of such behavioral models is that they are computationally extremely efficient in a system-level simulation flow – they have shown orders of magnitude of speedup in industrial examples.

Our intent, however, is not only to introduce a few recent nonlinear macromodeling techniques to the design community. More importantly, we demonstrate that the behavioral modeling problem can be treated from the perspective of projection in the state space, *i.e.*, by appropriately defining the projection of state variables, we can create behavioral models that capture linear/nonlinear circuit behaviors. The projection framework for behavioral modeling is a concept that might be less familiar to the design community, and we believe it is an important perspective that might inspire new behavioral modeling techniques to be developed.

The rest of the paper is organized as follows. In Sec. II, we review the analog/mixed-signal simulation flow, and formulate the behavioral modeling problem. We then briefly discuss a few typical macromod-

978-1-4673-6145-3/13 $31.00 © 2013 IEEE   845

eling techniques for linear and nonlinear circuits in Sec. III, and discuss the challenges and difficulties for macromodeling nonlinear circuits in Sec. IV. In Sec. V and Sec. VI, we present the details of a few recently developed nonlinear macromodeling algorithms, and we validate the algorithms by comparing the simulation results against full-SPICE and linear macromodels.

II. BACKGROUND AND PROBLEM FORMULATION

A. Analog/Mixed-Signal Simulation Flow

Pre-Silicon validation of analog/mixed-signal circuits consists of simulations at various levels of abstraction. Take high-speed I/O links as an example, we would like to validate high-level behaviors such as BER of the link or cross-talk across neighboring lanes, as well as low-level circuit characteristics such as the linearity of a phase interpolator and the phase noise of a PLL. While transistor-level SPICE simulations are typically done for each small circuit block, it is impossible to perform simulations of an entire I/O link at the transistor level, simply because the resulting system of differential equations is too large. Moreover, besides the circuits along the signal path, there are increasingly more compensation and digital calibration circuits that are necessary for countering various sources of variability (such as PVT). Such mixed-signal circuits involve a large number of digital signals in a feedback loop, and/or involve a training circuit (either by finite state machine, or software/BIOS) that has a lot of inputs and takes a long time to settle. These features make the simulation even slower to finish.

In a typical bottom-up validation flow, we perform SPICE simulation for each circuit block, and progressively build behavioral models for the simulation at the next level. For example, consider the validation of a voltage-controlled oscillator (VCO) and its impact to the BER of an I/O link. SPICE simulation of the VCO can be performed to validate its linear frequency-voltage relationship, as well as phase noise characteristics. Then a phase model could be built for the VCO, and be used in the simulation of a PLL to validate the PLL locking time and output jitter distribution. A clock jitter model can be further built for the PLL, and be fed into the transmitter/receiver of the high-speed I/O link in a BER simulation flow.

While this flow is widely used and has worked reasonably well in the past, it causes increasingly more problems that lead to unacceptable design and validation productivity. We believe that there are two main reasons for the low productivity in the traditional flow.

First, hand-written models take long time to build and are error-prone. Design knowledge is heavily used during the model building process. In fact, it usually takes an expert analog designer to write a good behavioral model. Even so, the model is vulnerable to human errors which could be very hard to be identified.

Second, hand-written models often ignore low-level behaviors (*e.g.*, nonlinearities) in the circuit, not only because these circuit blocks are complicated by themselves, but also that the circuit could have interactions with many surrounding circuits including the power grid, neighboring channels, *etc.*. It is well acknowledged that even if each individual analog circuit block is fully validated against its specifications, the system integration could lead to failures at a higher-level. Therefore, using an over-simplified behavioral model could easily lead to ignorance of "bad" behaviors in the circuit, therefore leaving bugs to post-Silicon which require much higher cost to fix.

To conclude the above discussion, it is desirable to have a tool that automatically creates behavioral models that faithfully reproduce original circuit behaviors. In the following, we formulate the algorithmic behavioral macromodeling problem.

B. Problem Formulation

Given a circuit netlist, modified nodal analysis [9] constructs a system of differential-algebraic equations (DAE) which describes the dynamics of the circuit. Typically, the DAE model can be written in the form of

$$\frac{d}{dt}q(x) + f(x) + Bu(t) = 0, \quad y = h(x), \quad (1)$$

where $x \in \mathbb{R}^N$ are the state variables (*e.g.*, representing node voltages); $q(x), f(x) : \mathbb{R}^N \rightarrow \mathbb{R}^N$ are Q-V and I-V functions; $h(x) : \mathbb{R}^N \rightarrow \mathbb{R}^{N_o}$; $B \in \mathbb{R}^{N \times N_i}$; $u \in \mathbb{R}^{N_i}$ are inputs, $y \in \mathbb{R}^{N_o}$ are outputs; N is the dimension of the state space (*e.g.*, number of nodes); N_i is the number of inputs; N_o is the number of outputs.

For simplicity, it is often more intuitive to think in terms of the ordinary differential equation (ODE) model which is a special case of Eqn. (1) by choosing $q(x) = -x$, *i.e.*,

$$\frac{d}{dt}x = f(x) + Bu(t), \quad y = h(x). \quad (2)$$

Loosely speaking, a behavioral model is a "simpler" model that captures "important" dynamics of the original full model Eqn. (1). To be more precise, one has to define

1) What is the model candidate? (*e.g.*, a nonlinear/linear differential equation model, a frequency response, a power-spectrum density at the output, a finite state machine, *etc.*)
2) What are the desired behaviors to be retained in the model? (*e.g.*, DC/AC response, harmonic distortion, phase noise, circuit delay, stability, passivity, *etc.*)

Depending on the model candidate and desired behaviors, the corresponding mathematical problem can be very different. In this paper, we classify the model candidates into DAE models and non-DAE models. The DAE models is separated out because that they are in the same form of SPICE-level models and therefore are compatible with SPICE simulation. In particular, the DAEs for the macromodel can be described by

$$\frac{d}{dt}q_r(x_r) + f_r(x_r) + B_r u(t) = 0, \quad y = h_r(x_r), \quad (3)$$

where $x_r \in \mathbb{R}^{N_r}$ are the state variables of the behavioral model, $N_r \ll N$; $q_r(x_r), f_r(x_r) : \mathbb{R}^{N_r} \rightarrow \mathbb{R}^{N_r}$, $h_r(x_r) : \mathbb{R}^{N_r} \rightarrow \mathbb{R}^{N_o}$ are nonlinear functions; $B \in \mathbb{R}^{N_r \times N_i}$. Note that Eqn. (3) and Eqn. (1) share the same inputs $u(t)$ and outputs $y(t)$, and therefore are used to describe the same circuit.

III. PREVIOUS WORK

In this section, we review a few representative algorithms that create macromodels for linear time-invariant (LTI) and nonlinear circuits in Sec. III-A and Sec. III-B, respectively. While they are not meant to be a complete list of all the available techniques, they cover the major ideas behind algorithmic behavioral macromodeling techniques. Due to page limit, without getting into too much detailed algorithms, we sketch out the key ideas of each method, and we summarize them in terms of the model candidate and retained behaviors, as mentioned in Sec. II-B.

A. LTI Circuit Macromodeling

For LTI circuits, the corresponding circuit DAEs Eqn. (1) take a special form where $q(x) = Cx$ and $f(x) = Gx$ are linear functions of x, *i.e.*

$$C\frac{d}{dt}x + Gx + Bu = 0, \quad y = D^T x, \quad (4)$$

978-1-4673-6145-3/13 $31.00 © 2013 IEEE   846

where $C \in \mathbb{R}^{N \times N}$, $G \in \mathbb{R}^{N \times N}$, $B \in \mathbb{R}^{N \times N_i}$, $D \in \mathbb{R}^{N \times N_o}$, N_i is the number of inputs and N_o is the number of outputs.

LTI reduced models are therefore in the form of

$$C_r \frac{d}{dt} x_r + G_r x_r + B_r u = 0, \quad y = D_r^T x_r, \quad (5)$$

where $C_r \in \mathbb{R}^{N_r \times N_r}$, $G_r \in \mathbb{R}^{N_r \times N_r}$, $B_r \in \mathbb{R}^{N_r \times N_i}$, $D_r \in \mathbb{R}^{N_r \times N_o}$, N_r is the order of the reduced model.

Since LTI models can be equivalently defined by their transfer functions $H(s)$, a straightforward modeling method is to fit a reduced transfer function $H_r(s)$ that gives a good approximation to $H(s)$.

A well-known approach is the Padè approximation[4], i.e., to find a rational reduced transfer function in the form of

$$H_r(s) = \frac{a_0 + a_1 s + a_2 s^2 + \cdots + a_m s^m}{b_0 + b_1 s + b_2 s^2 + \cdots + b_n s^n} \quad (6)$$

which satisfies

$$\begin{aligned} H_r(0) &= H(0), \\ \frac{d^k}{ds^k} H_r(0) &= \frac{d^k}{ds^k} H(0), \quad k = 1, 2, \cdots, m + n, \end{aligned} \quad (7)$$

where $\frac{d^k}{ds^k} H(0)$ is also referred to as the k-th moment of the transfer function. Therefore, any method that leads to a macromodel satisfying Eqn. (7) is also called a **moment-matching** method.

There have been a series of work developed based on this idea, among which two major ones are AWE[5] and PVL[6]. AWE[5] explicitly computes the moments and finds a Padè approximation $H_r(s)$ with $m = n - 1$ that matches these moments. PVL[6] avoids numerical inaccuracy in the explicit moment computation in AWE, and uses the Lanczos process to compute the Padè approximation of the transfer function.

However, the explicit moment calculation in the Padè approximation methods can have serious numerical inaccuracy, and this inaccuracy will be inherited to the reduced model. Furthermore, the transfer function model is a frequency-domain model, and is not directly applicable in time-domain simulations.

Linear projection based methods prove to be perfect for creating time-domain models. In such methods [7], [8], projection matrices V and W are constructed, and the corresponding reduced model is in the form of Eqn. (5) with the matrices defined by

$$C_r = W^T C V, \quad G_r = W^T G V, \quad B_r = W^T B, \quad D_r = V^T D. \quad (8)$$

The intuitive explanation for the choice of V and W is that the state variable x and the residual $\dot{x} - Gx - Bu$ approximately lie in the subspace defined by V and W. Therefore, by considering only the dynamics in $\text{span}(V)$ and $\text{span}(W)$ for x and the residual, respectively, the macromodel can still capture the major dynamics of the original circuit. Besides, a lot of study has been done to formally prove the properties of the macromodel for different choices of V and W.

For example, a popular method, named PRIMA[8], chooses $V = W$ that satisfies $\text{span}(V) = \mathcal{K}_{N_r}(G^{-1}C, B)$, where the Krylov subspace $\mathcal{K}_N(A, b)$ is defined by $\mathcal{K}_N(A, b) = \text{span}(Ab, A^2 b, \cdots, A^N b)$. It can be shown that such time-domain models also satisfy the moment-matching property Eqn. (7), and yet avoids the numerical issues in the transfer function fitting methods. Furthermore, with different choices of V and W, we might also guarantee other linear circuit properties, such as balanced controllability and observability [10], passivity[8], stability[7], etc..

B. Nonlinear Circuit Macromodeling

1) Linear Time-Varying Approximation and Macromodeling: Certain nonlinear circuits may be well-approximated by linear time-varying (LTV) systems, partly because these systems are engineered to work under time-varying operating points. For example, in clocked circuits and RF circuits such as mixers, the local clock input is fixed. Linearizing the nonlinear circuit around its periodic steady state (with only the clock input present) leads to a linear periodic time-varying system. Similarly, in switched-capacitor circuits, the clock input is fixed, and the circuit is approximately working periodically under two modes (clock high and clock low).

LTV Macromodeling techniques [11] extend the notion of transfer function for LTI systems to time-varying transfer functions. It can be shown that using a Padè-like approximation, or an appropriate projection, we can obtain macromodels whose time-varying transfer functions approximate the original one, in the sense of moment-matching.

2) Volterra-Based Methods: Generally, for a nonlinear system, we may derive a Volterra series approximation[12] which describes its local nonlinear responses. Therefore, instead of matching the moments of transfer functions, we may match the moments of Volterra kernels (higher-order transfer functions), and this leads to the Volterra-based methods such as NORM[13] as well as [14], [15]. In Volterra-based methods, we also choose projection matrices V and W that cover a series of Krylov subspaces so that the projected models match the specified moments of Volterra kernels.

3) Phase Macromodels: Yet another importance set of nonlinear macromodeling techniques are phase macromodeling techniques [16], [17], [18] which are particularly useful for oscillator circuits. For oscillator circuits, the key behavior of interest is the phase/frequency deviation at the output. Both the ISF model[16] and the PPV model[17] compute a periodic time-varying function which can be thought of as the sensitivity of the phase deviation with respect to the input perturbation/noise. Such macromodels can either be computed by performing a set of transient analysis, or be directly derived from the matrices involved in the Periodic Steady State (PSS) analysis such as harmonic balance and shooting.

4) Trajectory-based Methods: Trajectory-based methods rely on a set of *training* waveforms of the circuit, obtained from either simulations or measurements. Such methods try to match the circuit behaviors that are exhibited in the training waveforms. Three popular techniques in this category include system identification, proper orthogonal decomposition (POD) and trajectory piece-wise linear (TPWL) methods.

System identification methods [19], [20] view the circuit as a black/grey box. With certain assumptions of the model structure for the circuit, a parameterized circuit model is fitted according to the training waveforms. The fitting algorithm can be as simple as a least squares method, or a more complicated optimization such as steepest gradient search or simulated annealing.

POD builds macromodels by linear projection. The difference from other projection methods is that POD constructs the matrices V and W from the training waveforms so that the subspaces defined by V and W minimizes certain error metric between the projected reduced model and the full model.

TPWL is another method that is based on linear projection. The key idea is that at each point in the training trajectory, the circuit can be well approximated by its linearized model for which a linear macromodel can be built. Therefore, the subspaces (defined by the projection matrices V and W) of the macromodel should be the aggregated subspaces defined by all linearized models. Compared to

978-1-4673-6145-3/13 $31.00 © 2013 IEEE

POD methods, TPWL model not only matches the trajectories, but also ensures that the transfer functions of the linearized macromodels match those of the original model.

C. Summary of Algorithmic Macromodeling Techniques

We summarize a few existing macromodeling techniques (including the ones to be presented in this paper) in Table I, according to the model candidate and behaviors retained. It is worthwhile to mention that from Table I, there is no single method that fits all applications. Especially for nonlinear circuits, one has to identify the behavior of interested to be retained, and then choose the appropriate behavioral modeling technique.

IV. CHALLENGES IN ALGORITHMIC NONLINEAR MACROMODELING

While the methods mentioned in Sec. III have witnessed some success in building nonlinear behavioral macromodels, they do have at least two drawbacks that need to be treated carefully, as follows:

1) Limited guarantee of reduced order model behaviors. This is especially a problem for nonlinear macromodeling methods. In particular, in trajectory-based methods, we are only able to quantify the behavior of the reduced model locally or with respect to given training waveforms. This means that if certain behavior is not exhibited in the training waveforms, it might be missing in the macromodel.

2) Potentially limited computational efficiency. In methods that rely on linear projection, such as POD and TPWL, it might be the case that the subspace covered by the trajectories is as large as the full model – that is, no computational benefit will be gained by using the macromodel. Furthermore, unlike the linear macromodels, the nonlinear functions in nonlinear behavioral models might be expensive to compute, thus defeating the purpose of building behavioral models.

Rethinking about the nonlinear macromodeling problem, we believe there are three major difficulties that stand out as obstacles for further advancement of nonlinear macromodeling algorithms:

1) Lack of canonical representation of a nonlinear model. LTI models (linear differential equations) $C \frac{d}{dt}x + Gx + Bu = 0$ have a very special and compact matrix representation (C, G, B). There are also other equivalent canonical representations such as controllable/observable canonical forms [25]. However, there has been hardly any literature on canonical forms of general nonlinear systems.

2) Lack of characterization of system responses. LTI systems can be characterized by their transfer functions which are extensively used as a performance metric to preserve in linear model order reduction. However, for nonlinear systems, there is no simple way to derive analytical expressions of system responses such as harmonic distortion, oscillator frequency.

3) Difficulties intrinsic in the current projection framework. LTI macromodeling techniques are often formulated within a linear projection framework. However, linear projection may not be efficient and reasonable for creating nonlinear macromodels, as it leads to less computational speedup and lacks solid theoretical support.

V. NONLINEAR PROJECTION FRAMEWORK FOR NONLINEAR MACROMODELING

The key idea in linear projection-based methods is to construct an appropriate subspace, and build the macromodel by restricting the dynamics of the circuit within that subspace. Intuitively speaking, it

is equivalent to removing unimportant nodes in the circuit (*e.g.*, if a node is connected to ground via a very large capacitance, then it may be assumed that the voltage at that node almost stays constant, and therefore the circuit may be simplified), but in a more general way. Hence, the key to the success of such behavioral modeling methods is the validity of the assumption that the *circuit dynamics indeed stay in a restricted subspace in the state space* (*i.e.*, the space that is composed of node voltages and branch currents).

To gain insights into how this assumption holds for nonlinear circuits, we perform transient analysis of a CML buffer circuit and a ring oscillator circuit, and plot the simulation trajectories of these two circuits in Fig. 1a and Fig. 1b, respectively. In Fig. 1a, the three axes represent the input and output voltages. The red points are the DC operating points for different v_{in} values, and the blue points are the trajectories when a sinusoidal signal is applied at the input. Under this input stimulus, the output waveforms are distorted, and obviously, the trajectories are not restricted in a linear subspace. However, the trajectories lie approximately on a two-dimensional surface, which colored in yellow. In Fig. 1b, the three axes are the node voltages at the output of three inverters. In this case, the output waveforms are highly nonlinear (as opposed to sinusoidal in LC oscillators), and the trajectories are also not restricted in a linear subspace. However, even with perturbations in the circuit (*e.g.*, from power supply noise), the trajectories stay very close to the closed orbit in Fig. 1b, which is a one-dimensional manifold.

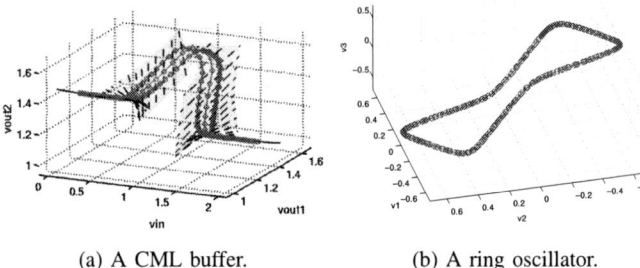

(a) A CML buffer. (b) A ring oscillator.

Fig. 1: Trajectories obtained from SPICE simulation.

A. Nonlinear Projection Framework

The examples above illustrate that circuit nonlinearity can lead to distortions in the waveforms, and the corresponding trajectories can cover a high-dimensional subspace in the state space, thus making it hard to build macromodels using linear projection. Despite this fact, however, it is likely that the trajectories stay close to a certain nonlinear manifold, depending on specific circuit characteristics. We try to exploit this intuition by constructing appropriate manifolds for building behavioral macromodels.

Our method, named **ManiMOR**, follows a nonlinear projection framework. The algorithm identifies two manifolds $\mathcal{V}$ and $\mathcal{W}$ (defined by nonlinear functions $v : \mathbb{R}^{N_r} \to \mathbb{R}^N$ and $w : \mathbb{R}^N \to \mathbb{R}^{N_r}$) that are sub-manifolds of $\mathbb{R}^N$, and builds the behavioral model Eqn. (3) by performing the projection, as shown in Algorithm 1.

Algorithm 1 Nonlinear Projection Framework

Inputs: $r(x) = \frac{d}{dt}q(x) + f(x) + b(t) = 0$

Outputs: $w(r(v(x_r))) = w\left(\frac{d}{dt}q(v(x_r)) + f(v(x_r)) + b(t)\right) = 0$

  1: Construct a mapping $v : \mathbb{R}^{N_r} \to \mathbb{R}^N$;

  2: Construct a mapping $w : \mathbb{R}^N \to \mathbb{R}^{N_r}$;

  3: Define the reduced model to be $w(r(v(x_r))) = 0$.

In the following, we define several different manifolds that capture

TABLE I: Summary of Algorithmic Macromodeling Techniques

Methods	Model Candidate	Behaviors Retained	Applications
AWE[5], PVL[6]	transfer function (rational function)	moments of the transfer functions	linear circuits
PRIMA[8]	linear differential equations	moments of the transfer functions	linear circuits
TBR[10]	linear differential equations	Hankel singular values	linear circuits
NORM[13]	polynomial differential equations	moments of Volterra kernels	weakly nonlinear circuits
TPWL[21]	piecewise-linear differential equations	moments of transfer functions of linearized models	nonlinear circuits
ISF[16]	time-varying phase sensitivity	oscillator phase sensitivity	oscillator circuits
PPV[17]	scalar differential equation	oscillator phase sensitivity	oscillator circuits
ManiMOR[22]	piecewise-linear differential equations	DC and AC responses	nonlinear circuits
NTIM[18]	scalar differential equation	phase response	nonlinear circuits
DAE2FSM[23]	finite state machine	discretized training I/O trajectories	nonlinear circuits
RAVEN[24]	event-driven model	training I/O trajectories	nonlinear circuits

different circuit characteristics, as well as numerical methods to compute them. For simplicity, we will focus on the definition of $v(\cdot)$, because $\mathcal{V}$ defines the manifold where the state variables evolve, and $w(\cdot)$ can often be chosen as $\frac{dv(\cdot)}{dx}(\cdot)$. Due to page limits, we do not cover all the details of theory and algorithms. More detailed derivation and discussion can be found in [22], [18], [26].

B. DC Manifold

DC response is one of the most important circuit responses to be retained in a behavioral model. A straightforward approach to capture the DC response is to explicitly include all the DC operating points in the model.

For a nonlinear system Eqn. (2), the DC operating points are the solution to

$$f(x) + Bu = 0, \quad u \in [u_{min}, u_{max}], \quad (9)$$

where u_{min} and u_{max} are min/max bounds of the input u according to physical constraints. If the nonlinear function $f(\cdot)$ is invertible, we have

$$x = f^{-1}(-Bu). \quad (10)$$

Eqn. (10) is the $v(\cdot)$ in Algorithm 1, and it defines a one-dimensional manifold (parameterized by u) consisting of all DC solutions. This constitutes the first dimension of the manifold, and we call it **DC manifold**.

C. AC Manifold

The AC response is defined in terms of the linearized models around the DC solutions. Hence, at each DC operating point, any LTI macromodeling technique described in Sec. III can be used to construct a local subspace that approximates the AC response, and the AC manifold can be defined by stitching all these local subspaces together.

Specifically, for each linearized model, the first dimension of the reduced subspace is determined by the DC manifold, i.e., the first Krylov vector, Krylov-subspace based methods appear to be a natural choice for computing the tangent spaces at each point on the manifold. Using Arnoldi-based method [8], for example, we have that at each point x, the local linear subspace (i.e., the tangent space of the manifold) is defined by the column span of a nonlinear projection matrix $V_{N_r}(x)$ that satisfies

$$\text{span}(V_{N_r}(x)) = \text{span}(G^{-1}(x)B, G^{-2}(x)B, \cdots, G^{-N_r}(x)B), \quad (11)$$

where $G(x) = \frac{\partial f}{\partial x}(x)$.

The manifold is stored by a collection of points on the manifold, and the corresponding tangent spaces. To identify points on the AC manifold, we notice that since the tangent space around a point x gives a good local approximation of the manifold, the points close to x is also on the manifold. Therefore, we may "grow" the manifold by integrating along the Krylov vectors to obtain the manifold. That is, by solving

$$\frac{\partial x}{\partial x_r} = V_{N_r}(x) \quad (12)$$

to obtain the projection $x = v(x_r)$. We call a manifold an **AC manifold** if at any point on the manifold, Eqn. (12) is satisfied – such manifolds lead to macromodels that match up to the N_r-th moment of the original linearized models, and therefore the AC responses are well retained in the macromodel.

D. Periodic Manifold

The **periodic manifold** is appropriate for circuits that operate around a closed orbit in the state space, similar to that in Fig. 1b. The manifold is parameterized by a single variable, such as the arc-length along the orbit.

For oscillator circuits, the periodic manifold can be naturally chosen as the oscillator periodic steady state. The natural parameterization of the manifold is time $t \in [0, T]$ where T is the period of the oscillator. It can be shown that by appropriately projecting the full model onto this manifold, we can construct the traditional oscillator phase models such as the ISF model [16] and the PPV model[17]. However, the derivation of the projection in this case is a bit involved, and readers can refer to [26] for more details.

This manifold is not only useful in building phase macromodels for oscillatory circuits, but also non-oscillatory circuits. Take an inverter chain circuit as an example, if the input signals are digital-like waveforms (i.e., square waveforms switching between 0 and Vdd with various slopes), the trajectories of the circuit are approximately the same. To the first order, only the delay and slope of the output waveform vary for different input waveforms. More generally, many digital-like circuits have the feature that for different input stimuli, the trajectories in the space stay on an orbit in the state space, but the rate of change along the orbit varies. This implies that a phase macromodel would be a perfect candidate for such circuits. By choosing a periodic manifold, and restricting the dynamics on the manifold, we may build a phase macromodel that captures timing responses [18].

E. Model Storage and Computation

The manifolds and the projection functions are stored in a non-parametric piece-wise linear fashion. For example, the function $x = v(x_r)$ can be defined as

$$x = \sum_{i=1}^{N_s} K_i(x_r) x_{r,i}, \quad (13)$$

where $x_{r,i}$ is the parameterized coordinates of the i-th point on the manifold, and $K_i(x_r)$ is a weighting (kernel) function that satisfies

$$K_i(x_r) \geq 0, \quad \sum_{i=1}^{N_s} K_i(x_r) = 1, \quad K_i(x_{r,i}) = 1. \quad (14)$$

For the nonlinear functions such as $f_r(\cdot)$ in Eqn. (3), we approximate them using a weighted linear combination of linear functions, *i.e.*

$$f_r(x_r) = \sum_i K_i(x_r)(f_{r_i} + G_{r_i}(x_r - x_{r_i})), \quad (15)$$

where f_{r_i} and G_{r_i} can be pre-computed and stored in a look-up table. Therefore, the computation of the model consists of a few table loop-up operations and an interpolation operation.

F. Connections to Trajectory-Based Methods

In trajectory-based methods, the trajectories are sampled to identify the region in the state space where circuit dynamics evolve. However, trajectory-based methods embed these trajectories in a linear subspace, instead of a manifold as is done in ManiMOR. This is the key difference between ManiMOR and trajectory-based methods.

Because of that difference, trajectory-based methods usually result in a model of much larger size than that of ManiMOR. Consider a nonlinear system which has different AC modes (corresponding to local subspaces) when operating under different DC equilibria. Suppose that the dynamics of each linearized system along the equilibria could be approximately embedded in a q-dimensional linear subspace, then for the entire system, a q-dimensional nonlinear manifold is adequate to embed all the q-dimensional linear subspaces, while a much larger linear subspace is needed to include all the q-dimensional linear subspaces. This explains why nonlinear projection leads to smaller behavioral macromodels.

G. Connections to TP-PPV

The oscillator phase macromodels [16], [17], while excellent in capturing the sensitivity to "small-signal" perturbations, are not accurate when "large-signal" inputs are applied. To address this problem, TP-PPV[27] has been proposed which combines several PPV macromodel at several different DC inputs.

From a projection perspective, it is in fact a simplified version of ManiMOR for oscillators. TP-PPV computes the limit cycles for the oscillator circuit at different DC bias, and stitches them together to obtain a cylinder-like manifold in the state space – this is an extension of the periodic manifold mentioned in Sec. V-D. In TP-PPV, the variable α and the DC input u parameterize this manifold, and the time-domain waveforms are restored by interpolation on this manifold given the phase deviation and input waveforms. Hence, TP-PPV can be thought of as a special case in the nonlinear projection framework, except that the projection is implicitly done by formulating the PPV equation[27].

H. Examples and Experimental Results

We illustrate the application of ManiMOR on a current-mode logic (CML) buffer chain [28], which has also been studied in previous literatures on nonlinear macromodeling. We compare the macromodel against the full-SPICE model as well as existing techniques (such as small-signal linear model and the TPWL model), and we show that the ManiMOR model is able to capture nonlinear effect that is exhibited in the full-SPICE model but is missing in other linear/nonlinear macromodels.

The circuit diagram of the CML buffer is shown in Fig. 2. In the experiments, we use the BSIM3 model for the MOSFETs in our

simulations, and the size of differential algebraic equations for the full circuit is 52.

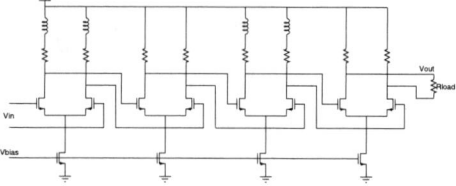

Fig. 2: Circuit diagram of a current-model logic (CML) buffer.

Applying ManiMOR, a size 5 model is generated for this circuit. The nonlinear manifold constructed by ManiMOR is visualized in Fig. 1a (only first two dimensions are plotted) by picking three node voltages as axes. Clearly, the manifold is nonlinear, and blue circled points, which are sampled on a transient trajectory, do stay close to this nonlinear manifold. This confirms that the manifold attracts system responses.

We first compare the ManiMOR model against the small-signal linear model. The small-signal model is created by linearizing the nonlinear SPICE model around its DC operating point. When a small-signal input is applied on the DC bias, the circuit still operates in the linear region, and therefore, the small-signal model tracks the full model. Since the ManiMOR model matches the AC response of the original circuit, not surprisingly, it also perfectly matches the full model, as is shown in Fig. 3a.

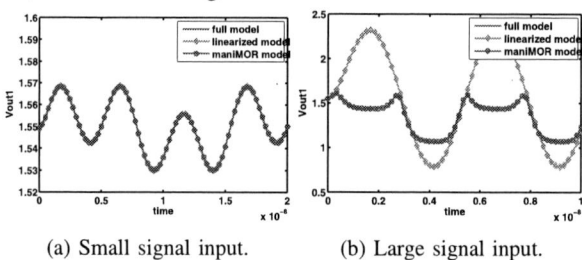

(a) Small signal input.   (b) Large signal input.

Fig. 3: Comparison of ManiMOR model with linearized reduced-order model. The waveforms of one output using different models are plotted.

However, when the input amplitude is large, the trajectory is no longer close to the DC solution, and exhibits nonlinear distortion in the waveform. In that case, the small signal model fails to match the waveforms of the full model. As shown in Fig. 3b, the output waveform is highly distorted. In contrast, the output waveform predicted by the ManiMOR model is almost overlapped with that of the full model – this is because the ManiMOR model uses different local models when the state variable travels to different regions in the state space. On the contrary, the small signal model, which makes the "linear" and "small-signal" assumption, still generates a sinusoidal output waveform, and therefore fails to capture the distortion effect.

We then compare the ManiMOR model to the TPWL model[21] by applying a step input as shown in Fig. 4a, which is chosen to traverse four DC operating points. As shown in Fig. 4b, without the heuristic to handle the DC solutions, the TPWL model of model size 10 fails to converge back to the correct DC operating points. After including the DC operating points into the projection matrix, the TPWL model of size 10 still leads to some error. Through a few trial-and-error iterations, we find that the model size must be at least 12 in order to match the original trajectory. This result is obtained by using a training input to be the same as the test input – potentially an even larger model will be needed.

978-1-4673-6145-3/13 $31.00 © 2013 IEEE

In comparison, ManiMOR model of size 5 renders an output waveform almost indistinguishable from that of the full model – it is less than half of the size of the TPWL model with the same accuracy. We conclude from this set of experiments that by performing nonlinear projection, the ManiMOR model is able to match the full model behaviors, and is superior to linearized models and previous linear-projection based TPWL model, in terms of model size and accuracy.

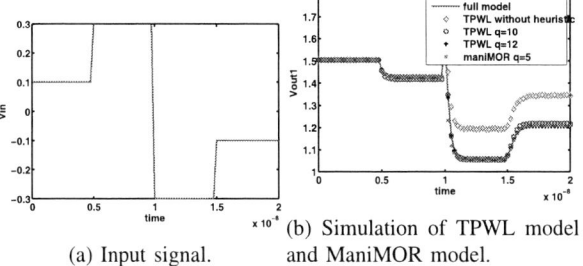

(a) Input signal.

(b) Simulation of TPWL model and ManiMOR model.

Fig. 4: Comparison of ManiMOR model with TPWL model. The waveforms of one output voltage using different models are plotted.

VI. NON-DAE MACROMODELING METHODS

While DAE macromodels are compatible with the SPICE simulation flow, they are commonly less used in higher-level simulation with large digital circuits (e.g., in system verilog). In this section, we briefly introduce two recently developed techniques that build non-DAE behavioral models which are more suitable for the simulation with large digital circuits.

A. DAE2FSM[23]

DAE2FSM creates a Mealy state machine model from circuit differential equations. The key idea of DAE2FSM is adapted from a finite automaton learning algorithm which iteratively queries input and output trajectories of the circuit, and uses the trajectories to incrementally build the FSM model.

The development of this technique is motivated by the fact that in mixed-signal circuits, digital circuit behaviors may not be accurately modeled by the ideal pre-specified finite state machines, due to parameter variations, crosstalk and coupling, environment changes, etc.. As a result, a bottom-up abstraction of the finite state machine from transistor-level differential equations is needed. While the initial application focuses on digital-like circuits, we discover that the algorithm can in fact work well for other nonlinear circuits as well. The FSM model implicitly characterizes the nonlinear behaviors in the circuit.

Without going into details of the algorithm[23], we illustrate the application of DAE2FSM to an erroneous latch circuit. One important erroneous behavior of latches is that when the input swing is reduced due to power optimization, the latch requires several consecutive "1"s or "0"s at the input in order to make a successful transition, as illustrated in Fig. 5.

Fig. 6 shows three finite state machine models built by DAE2FSM when inputs of three different voltage swing are applied. It can be verified that these models indeed capture the bit error mechanisms such as the one shown in Fig. 5, and they almost perfectly reproduce the SPICE simulation traces.

Each state in the FSM model has an interpretation in the original continuous state space. For example, Fig. 7 shows the state space partitioning which is constructed using Voronoi diagrams. The four points correspond to four states in Fig. 6b, and the (blue) trajectories in Fig. 7 are simulation traces that leads to the discrete points of

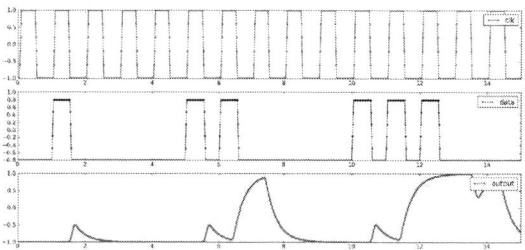

Fig. 5: Response of the latch when the input sequence is "10011000001001100000" and $V_{sw} = 1.6V$.

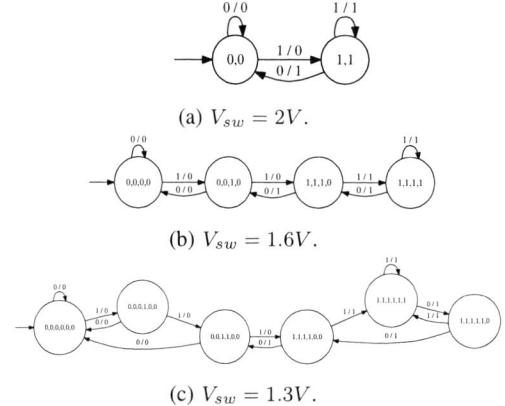

(a) $V_{sw} = 2V$.

(b) $V_{sw} = 1.6V$.

(c) $V_{sw} = 1.3V$.

Fig. 6: FSMs of a latch circuit under different input swing.

different states, and show typical transition dynamics of the latch. This partitioning of the state space also provides direct intuition on how the circuit behaves: the two states at two corners represent solid "0" and solid "1" states; the two states in the middle capture dynamics in the metastable region, and can be viewed as weak "0" and weak "1" states.

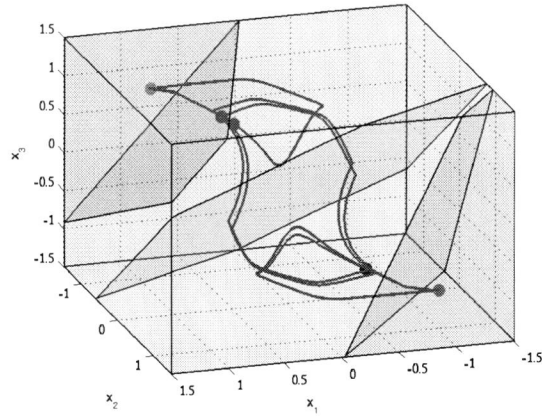

Fig. 7: Interpretation of states in the FSM ($V_{sw} = 1.6V$).

Fig. 7 shows another example that the behavioral modeling problem can be viewed from the perspective of state space projection. In the latch example, since the input is regular and the latch is designed to have two stable states, the trajectories of the circuit stay closely to the trajectories shown in Fig. 7. By constraining the dynamics of the circuit on these trajectories, and by discretizing the continuous states, we are able to create an FSM abstraction of the circuit.

It should be mentioned that even a simple FSM model such as the one in Fig. 6c takes more than a week to build by hand. While the FSM looks simple, it is not straightforward that a transistor-level netlist corresponds to such a finite-state machine. In contrast, the DAE2FSM algorithm created the model within a few minutes.

B. RAVEN[24]

RAVEN creates behavioral models (in SystemVerilog) that match a given set of input-output circuit responses. Given a set of input stimuli patterns (provided by designers), RAVEN performs circuit simulations to extract the corresponding output waveforms. The output waveforms are then parameterized and stored in terms of a few key parameters such as DC response, input-output delay, clock period, etc.. To accommodate mixed-signal validation, the RAVEN model is an event-driven one, i.e., the output changes only when an "event" is triggered at the input. The event, however, does not have to be a voltage change, but could be defined in a more general sense, such as the frequency/magnitude deviation of a sinusoidal stimulus.

Compared to the DAE behavioral macromodels, RAVEN model has the advantage of greater simulation speedup, but could be less efficient in modeling the memory in the circuit. RAVEN is similar to trajectory-based methods because it also relies on a training set to build the model. However, RAVEN incorporates more designers' knowledge in specifying input/output waveform types. In particular, RAVEN relies on the fact that the input stimuli during simulation is close to the ones in the training set, and that the outputs depend mainly on the inputs (as opposed to initial condition) – this is equivalent to say that RAVEN restricts the circuit dynamics on the training trajectories by matching the input-output response. As long as this assumption is satisfied, the RAVEN model is extremely efficient in terms of simulation speed and accuracy. It has shown orders of magnitude speedup in a few industrial examples, including phase-interpolator, VCO, thermal sensor, sigma-delta ADC, etc..

VII. CONCLUSION

In this paper, we reviewed the nonlinear behavioral macromodling problem, its challenges and existing solutions. We presented a nonlinear projection framework that unifies many existing techniques and provides a guideline for developing new nonlinear macromodeling algorithms, and we reviewed two classes of algorithms that create DAE and non-DAE models, respectively. With such tools becoming mature, the bottom-up simulation flow can be made much faster by automating the macromodeling process. In addition, we believe that the projection perspective of nonlinear macromodeling is a useful concept, and it may inspire new ideas for creating useful behavioral models of analog/mixed-signal circuits.

REFERENCES

[1] N. Hakim, A. Bhaduri, K. Donepudi, and S. Bodapati, "A Hybrid Electrical-Behavioral Modeling Approach for Pre- and Post-Silicon Electrical Validation," in *Custom Integrated Circuits Conference (CICC), 2012 IEEE*, Sept., pp. 1–5.

[2] M. Rewienski, *A Perspective on Fast-SPICE Simulation Technology*. Springer, 2011.

[3] G. Balamurugan, B. Casper, J. E. Jaussi, M. Mansuri, F. O'Mahony, and J. Kennedy, "Modeling and Analysis of High-Speed I/O Links," *Advanced Packaging, IEEE Transactions on*, vol. 32, no. 2, pp. 237–247, 2009.

[4] G. Baker and P. Graves-Morris, *Padé Approximants*. Cambridge Univ Pr, 1996.

[5] L. T. Pillage and R. A. Rohrer, "Asymptotic Waveform Evaluation for Timing Analysis," *Computer-Aided Design of Integrated Circuits and Systems, IEEE Transactions on*, vol. 9, no. 4, pp. 352–366, 1990.

[6] P. Feldmann and R. Freund, "Efficient Linear Circuit Analysis by Padé Approximation via the Lanczos Process," *Computer-Aided Design of Integrated Circuits and Systems, IEEE Transactions on*, vol. 14, no. 5, pp. 639–649, 1995.

[7] E. Grimme, "Krylov Projection Methods for Model Reduction," Ph.D. dissertation, University of Illinois, EE Dept, Urbana-Champaign, 1997.

[8] A. Odabasioglu, M. Celik, and L. T. Pileggi, "PRIMA: Passive Reduced-Order Interconnect Macromodeling Algorithm," *Computer-Aided Design of Integrated Circuits and Systems, IEEE Transactions on*, vol. 17, no. 8, pp. 645–654, 1998.

[9] L. Nagel, "SPICE2: a Computer Program to Simulate Semiconductor Circuits," Ph.D. dissertation, 1975, memorandum no. ERL-M520.

[10] B. Moore, "Principal Component Analysis in Linear Systems: Controllability, Observability, and Model Reduction," *IEEE Trans. Automatic Control*, vol. 26, pp. 17–32, Feb. 1981.

[11] J. Roychowdhury, "Reduced-order modeling of time-varying systems," *Circuits and Systems II: Analog and Digital Signal Processing, IEEE Transactions on*, vol. 46, no. 10, pp. 1273–1288, 1999.

[12] W. Rugh, *Nonlinear System Theory - The Volterra-Wiener Approach*. Johns Hopkins Univ Press, 1981.

[13] P. Li and L. Pileggi, "Compact Reduced-Order Modeling of Weakly Nonlinear Analog and RF Circuits," *Computer-Aided Design of Integrated Circuits and Systems, IEEE Transactions on*, vol. 24, no. 2, pp. 184–203, 2005.

[14] J. Phillips, "Projection-Based Approaches for Model Reduction of Weakly Nonlinear, Time-Varying Systems," *Computer-Aided Design of Integrated Circuits and Systems, IEEE Transactions on*, vol. 22, no. 2, pp. 171–187, 2003.

[15] C. Gu, "QLMOR: A Projection-Based Nonlinear Model Order Reduction Approach Using Quadratic-Linear Representation of Nonlinear Systems," *Computer-Aided Design of Integrated Circuits and Systems, IEEE Transactions on*, vol. 30, no. 9, pp. 1307 –1320, sept. 2011.

[16] A. Hajimiri and T. Lee, "A General Theory of Phase Noise in Electrical Oscillators," *Solid-State Circuits, IEEE Journal of*, vol. 33, no. 2, pp. 179–194, Feb.

[17] A. Demir, A. Mehrotra, and J. Roychowdhury, "Phase Noise in Oscillators: A Unifying Theory and Numerical Methods for Characterization," *Circuits and Systems I: Fundamental Theory and Applications, IEEE Transactions on*, vol. 47, no. 5, pp. 655–674, 2000.

[18] C. Gu and J. Roychowdhury, "Generalized Nonlinear Timing/Phase Macromodeling: Theory, Numerical Methods and Applications," in *Computer-Aided Design (ICCAD), 2010 IEEE/ACM International Conference on*, nov. 2010, pp. 284 –291.

[19] L. Ljung and E. Ljung, *System Identification: Theory for the User*. Prentice-Hall Upper Saddle River, NJ, 1987, vol. 280.

[20] R. Isermann and M. Mnchhof, *Identification of Dynamic Systems: An Introduction with Applications*. Springer, 2011.

[21] M. Rewienski and J. White, "A trajectory piecewise-linear approach to model order reduction and fast simulation of nonlinear circuits and micromachined devices," *Computer-Aided Design of Integrated Circuits and Systems, IEEE Transactions on*, vol. 22, no. 2, pp. 155–170, 2003.

[22] C. Gu and J. Roychowdhury, "Model Reduction via Projection onto Nonlinear Manifolds, with Applications to Analog Circuits and Bio-Chemical Systems," in *Computer-Aided Design, 2008. ICCAD 2008. IEEE/ACM International Conference on*, nov. 2008, pp. 85–92.

[23] ——, "FSM Model Abstraction for Analog/Mixed-Signal Circuits by Learning from I/O Trajectories," in *Design Automation Conference (ASP-DAC), 2011 16th Asia and South Pacific*, jan. 2011, pp. 7 –12.

[24] C. V. Kashyap and C. S. Amin, "Raven: A Tool for Automatic Generation of Analog Behavioral Models from Schematics," in *Frontiers in Analog Circuit (FAC) Synthesis and Verification*, 14-15 July 2011.

[25] S. Sastry, *Nonlinear Systems: Analysis, Stability, and Control*. Springer Verlag, 1999.

[26] C. Gu, "Model Order Reduction of Nonlinear Dynamical Systems," Ph.D. dissertation, University of California, Berkeley, 2011.

[27] X. Lai and J. Roychowdhury, "TP-PPV: Piecewise Nonlinear, Time-Shifted Oscillator Macromodel Extraction for Fast, Accurate PLL Simulation," in *Proceedings of the 2006 IEEE/ACM international conference on Computer-aided design*. ACM, 2006, pp. 269–274.

[28] J. Savoj and B. Razavi, "A 10-Gb/s CMOS Clock and Data Recovery Circuit with a Half-Rate Linear Phase Detector," *Solid-State Circuits, IEEE Journal of*, vol. 36, no. 5, pp. 761–768, 2001.

978-1-4673-6145-3/13 $31.00 © 2013 IEEE

Event-Driven Simulation of Volterra Series Models in SystemVerilog

Ji-Eun Jang, Si-Jung Yang, Jaeha Kim

School of Electrical Engineering and Computer Science, Inter-university Semiconductor Research Center
Seoul National University, Seoul, Korea

Abstract- **This paper presents a method to simulate the behavior of a weakly-nonlinear analog circuit in a digital logic simulator like SystemVerilog. To our knowledge, this is the first report on a true event-driven simulator that can compute the nonlinear circuit response modeled with a Volterra series. The core idea is to express a continuous-time signal as a linear combination of basis functions, $t^m exp(at) \cdot u(t)$, which was previously demonstrated for linear systems. To extend this approach to nonlinear systems, a Volterra series model is reformulated into a set of linear differential equations each corresponding to a different-order component of the nonlinear system response and describing the contribution of the initial condition explicitly. The presented simulator is demonstrated on the cases of estimating the spectral regrowth of a RF power amplifier and distortion-induced eye-opening reduction in a continuous-time linear equalizer. Compared to a commercial SPICE simulator, the proposed simulator achieves 300~1000× speed-up in performing time-domain simulations with the same level of accuracy.**

I. INTRODUCTION

Today's complex analog/mixed-signal (AMS) systems demand a fast behavioral simulator that can handle large digital systems and model continuous analog signals. While the existing mixed-signal simulators including Verilog-AMS aim to achieve this by combining a fast event-driven logic simulator with a SPICE-like analog simulator, they tend to exhibit slow simulation speeds due to the use of an ordinary differential equation (ODE) solver. A more promising approach is to model both the digital and analog circuit behaviors on a common event-driven simulation platform like SystemVerilog [1],[2], which was previously demonstrated for systems with linear analog components, such as phase-locked loops, decision feedback equalizers, and switching regulators.

The objective of this paper is to extend the approach in [1] to simulate weakly nonlinear circuit behaviors, such as spectral regrowth in RF transmitters due to a nonlinear power amplifier [3] and eye-opening reduction in wireline data receivers due to equalizer nonlinearity [4]. To date, the use of a numerical ODE solver has been a predominant way to simulate the time-domain responses of such nonlinear systems, however its speed and accuracy have competing dependencies on the simulation time step. In contrast, the proposed method computes and updates the corresponding output event only when a new input event arrives and stays idle thereafter. To the best of our knowledge, this is the first report on a truly event-driven simulator for nonlinear Volterra-series models.

However, such an extension is not straightforward when the system is not linear. First, one can no longer rely on the superposition principle to compute the output response contributed by all the past input events. Second, computing the distortion effects such as inter- or cross-modulation requires multiplication of signals.

To address these issues, this paper employs a perturbational form of a Volterra series model with an explicit notion on the initial conditions and introduces an s-domain operator for time-domain signal multiplication. Then, the behavior of linear and nonlinear analog circuits as well as digital circuits can be simulated on a single event-driven simulation platform of SystemVerilog, without involving numerical ODE integration. This extended simulator retains all the useful properties of the previous one in [1], whose simulation speed and accuracy do not hinge on fine simulation time steps.

The paper is organized as follows. Section II briefly reviews the previous event-driven method for linear systems in [1]. Section III and IV then describe how it can be extended to nonlinear systems and implemented in SystemVerilog. Section V discusses the cases of modeling a RF power amplifier and a high-speed data receiver with a continuous-time linear equalizer (CTLE), and Section VI concludes the paper.

II. TRULY EVENT-DRIVEN SIMULATION OF LINEAR SYSTEMS

This section summarizes our previously-published event-driven simulation method for linear systems [1]. The key idea is to express a continuous-time analog signal as a series of analog events, each of which describes the signal in a functional form of Eq. (1).

$$x(t) = \sum_i c_i t^{m_i-1} e^{-a_i t} u(t) \xrightarrow{\mathcal{L}} X(s) = \sum_i \frac{b_i}{(s+a_i)^{m_i}} \quad (1)$$

For instance, Fig. 1 illustrates how to express a sinusoidal signal whose phase is modulated by a binary data. While it would take a large number of samples to express this kind of signal using time-value pairs, it takes only four events at $t_1 \sim t_4$ using this functional form. That is, when the signal is expressed as a sum of two exponential functions, $c_1 e^{j\omega(t-ti)} u(t-t_i)$ and $c_2 e^{-j\omega(t-ti)} u(t-t_i)$, it is sufficient to update the coefficients c_1 and c_2 at the instant of each event, $t_i = t_1 \sim t_4$.

The key benefit of this signal expression is that it enables an algebraic computation of the output event upon the arrival of a new input event without involving numerical integration of ODEs. It is best explained using s-domain expressions after Laplace transforms. As given in Eq. (1), our signal expression can be converted to a sum of partial fractions in s-domain. If a linear dynamical system model is expressed as an s-domain transfer function, the output signal can be computed simply as a product between these two s-domain expressions, which again takes the same form after partial fraction decomposition. This operation is illustrated for a first-order RC low-pass filter with a transfer function $H(s)=1/(s/\omega_p+1)$ shown in Fig. 2. If an

978-1-4673-6145-3/13 $31.00 © 2013 IEEE

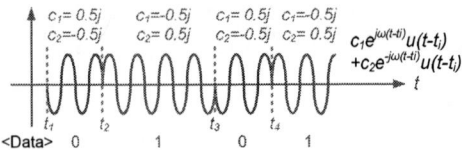

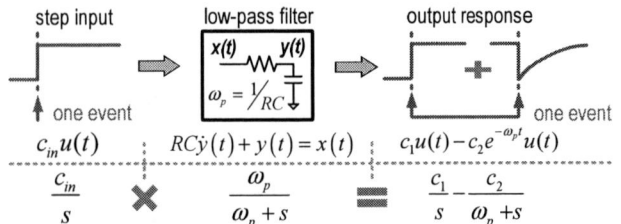

Fig. 1. An analog signal is expressed as a series of analog events in a functional form.

Fig. 2. A simple analog filter example to demonstrate the event-driven simulation method for linear circuits.

input event arrives as a step function, $c_{in}u(t)$, equivalent to c_{in}/s in s domain, the resulting output event is $c_{in}/s(s/\omega_p+1)$, which can be decomposed to a sum of a step $c_1u(t)$ and an exponential $c_2e^{-\omega_p t}u(t)$. Therefore, it is sufficient to update the output coefficients c_1 and c_2 only once whenever the input coefficient c_{in} changes, enabling a true event-driven operation.

III. EVENT-DRIVEN SIMULATION OF A WEAKLY NONLINEAR CIRCUIT BEHAVIOR

A Volterra series expresses a nonlinear system response $y(t)$ to an input $x(t)$ as a sum of partial responses $y_i(t)$:

$$y(t) = y_0 + \sum_{i=1}^{N} y_i(t) = y_0 + y_1 + y_2 + \cdots + y_N \qquad (2)$$

where $y_i(t)$ is computed as i-times repeated convolution with the i-th order Volterra kernel $h_i(\cdot)$ [5]:

$$y_i(t) = \int_{-\infty}^{+\infty} \cdots \int_{-\infty}^{+\infty} h_i(t_1, t_2, ..., t_i) \\ \times x(t-t_1)x(t-t_2)\cdots x(t-t_i)\, dt_1 dt_2 \cdots dt_i \qquad (3)$$

For a weakly nonlinear system, which exhibits only minor deviations from the linear response such as inter- and cross-modulation and gain compression, the output response can be fully described only with the first few orders.

To extend our previous approach for linear systems, it is more convenient to compute $y_i(t)$'s by solving a set of linearized differential equations than by directly computing Eq. (3). Such a set of differential equations can be obtained using a perturbation method [6]. While it is generally applicable to a broad class of weakly nonlinear systems, we will demonstrate the procedure for the case of decomposing a nonlinear single-input-single-output (SISO) system equation for the sake of brevity. Let's consider a nonlinear system with inter-modulation as well as cross-modulation between the input $x(t)$ and output $y(t)$:

$$\frac{d}{dt}\left(C_1 y(t) + C_2 y^2(t)\right) + \left(G_1 y(t) + G_2 x(t)y(t) + G_3 y^2(t)\right) \\ = U_1 x(t) + U_2 x^2(t) \qquad (4)$$

The procedure applies a small perturbation to the input, $x(t) = x_0 + \epsilon \Delta x(t)$, where ϵ is an arbitrarily small scalar value. According to Eq. (2) and (3), the output response $y(t)$ should take the form of:

$$y(t) = y_0 + \epsilon y_1(t) + \epsilon^2 y_2(t) + \cdots + \epsilon^n y_n(t). \qquad (5)$$

As Eq. (5) should satisfy the system equation, Eq. (4), for an arbitrary value of ϵ, we can obtain a set of n differential equations by equating each of the ϵ^i-coefficients to zero. For instance, listing only the first three equations:

$$\epsilon \;:\; \left(C_1\frac{d}{dt}+G_1\right)y_1 = U_1 x \qquad (6)$$

$$\epsilon^2 \;:\; \left(C_1\frac{d}{dt}+G_1\right)y_2 = \left(-C_2\frac{d}{dt}-G_3\right)y_1^2 - G_2 xy_1 + U_2 x^2 \qquad (7)$$

$$\epsilon^3 \;:\; \left(C_1\frac{d}{dt}+G_1\right)y_3 = \left(-C_2\frac{d}{dt}-G_3\right)y_1 y_2 - G_2 xy_2 \qquad (8)$$

These are the governing equations for the partial responses $y_1(t)$, $y_2(t)$, and $y_3(t)$ of the Volterra series. Among them, $y_1(t)$ is the response of the linearized system at (x_0, y_0). Note that $y_2(t)$ and $y_3(t)$ are the responses of linearized systems whose inputs include the inter- or cross-product terms between the lower-order partial responses or the input.

As stated in the introduction, to simulate the responses of the linear differential equations in Eq. (6)-(8) in an event-driven manner *à la* [1], we need to address two issues. First, we can no longer rely on the superposition principle to derive the overall system response to all the input events arrived in the past. In [1], the response was computed simply by super-imposing the individual output events each corresponding to an input event. Second, the decomposed linear system equations as shown in Eq. (6)-(8) may contain the inter- or cross-products between the input or output partial responses, requiring a nonlinear operation: *multiplication*.

The first issue is addressed by re-writing the set of partial response equations with an explicit notion on their initial conditions. The system response to a new input event can be computed as that of the system restarting at the corresponding time instant with its initial condition equal to the current system state [2]. Hence, the final system response can be generally computed without relying on the superposition principle. For instance, the set of revised equations for Eq. (6)-(8) considering the initial conditions is:

$$\left(C_1\frac{d}{dt}+G_1\right)y_1 - C_1 y_1(0) = U_1 x \qquad (9)$$

$$\left(C_1\frac{d}{dt}+G_1\right)y_2 - C_1 y_2(0) = \left(-C_2\frac{d}{dt}-G_3\right)y_1^2 + C_2 y_1^2(0) - G_2 xy_1 + U_2 x^2$$

$$\left(C_1\frac{d}{dt}+G_1\right)y_3 - C_1 y_3(0) = \left(-C_2\frac{d}{dt}-G_3\right)y_1 y_2 + C_2 y_1(0)y_2(0) - G_2 xy_2$$

where $y_1(0)$, $y_2(0)$, and $y_3(0)$ denote initial states of the system.

To address the second issue, we noted that multiplication between two signals expressed in the functional form of Eq. (1) can be more easily handled in time domain, rather than in s domain. Since any inter- or cross-product between the terms $c_i t^{m_i-1}e^{-a_i t}$ can be expressed in the same form, it follows that the resulting signal after the multiplication can also be expressed as Eq. (1). The computation of its coefficients $\{c_i$'s, m_i's, a_i's$\}$ is straightforward involving only simple arithmetic operations like addition and multiplication. Since all of our other computations for simulating a system response are carried out in s domain, it is convenient to define an s-domain operator A_2 that performs a time-domain multiplication

978-1-4673-6145-3/13 $31.00 © 2013 IEEE   854

between the two signals, $x_1(t) = \sum c_i t^{m_i-1} e^{-a_i t} u(t)$ and $x_2(t) = \sum c_j' t^{m_j'-1} e^{-a_j' t} u(t)$:

$$A_2[X_1(s), X_2(s)] = \sum_{i,j} \frac{(m_i + m_j' - 2)!}{(m_i - 1)!(m_j' - 1)!} \cdot \frac{b_i b_j'}{(s + a_i + a_j')^{m_i + m_j' - 1}} \quad (10)$$

Using the operator A_2, the partial responses of the output signal $y_1(t)$, $y_2(t)$, and $y_3(t)$ in Eq. (9) can be algebraically computed in s domain as follows:

$$Y_1 = \{C_1 y_1(0) + U_1 X\}/(sC_1 + G_1) \quad (11)$$

$$\begin{aligned} Y_2 = \{&C_1 y_2(0) - (C_2 s + G_3)A_2[Y_1, Y_1] + C_2 y_1^2(0) \\ &- G_2 A_2[X, Y_1] + U_2 A_2[X, X]\}/(sC_1 + G_1) \end{aligned} \quad (12)$$

$$\begin{aligned} Y_3 = \{&C_1 y_3(0) - (C_2 s + G_3)A_2[Y_1, Y_2] + C_2 y_1(0)y_2(0) \\ &- G_2 A_2[X, Y_2]\}/(sC_1 + G_1) \end{aligned} \quad (13)$$

Then the procedure to compute the output event when a new input event arrives is as follows. First, the present states of the system $y_1(0)$, $y_2(0)$, $y_3(0)$ are computed by evaluating $y(t)$ at the current time instant (assumed t=0 here for notational convenience). Second, the output partial responses $y_1(t)$, $y_2(t)$, and $y_3(t)$ are computed in s domain by evaluating Eq. (11)-(13) in sequence. Finally, summing the resulting Y_1, Y_2, and Y_3 (actually, just concatenating their terms) yields the expression for the updated final output $y(t)$ which is again in the functional form of Eq. (1).

IV. IMPLEMENTATION IN SYSTEMVERILOG

The described algorithm is implemented in SystemVerilog in order to realize a unified, event-driven simulation platform for both digital and analog circuits. In addition, SystemVerilog provides useful language extensions such as a composite data type *struct* and support for fast C extension calls (direct programming interface; DPI) that enable aesthetically clean model interfaces and efficient computation.

In particular, we adopt a *struct*-type namcd XREAL to bundle multiple variables required to express a continuous-time analog signal as a single variable [1]. As listed in Fig. 3(a), an XREAL signal has three member variables: *param_set*, *t_offset*, and *change_flag*. *param_set* is a C-pointer handle to a linked list containing a dynamic set of parameters for expressing the signal according to Eq. (1). *t_offset* is a *real*-type variable that indicates an actual time instant of the event within a time interval quantized by a simulation time step and *change_flag* is an event variable that indicates whether *param_set* has been updated.

Fig. 3(b) lists a SystemVerilog pseudo code for the SISO

```
(a) typedef struct {
        chandle param_set;
        real t_offset;
        event change_flag;
    } XREAL;

(b) module nonlinear_SISO(input XREAL x, output XREAL y);
        always @(x.change_flag) begin
            1. sample y₁(0), y₂(0), and y₃(0)
            2. update param_set of y₁, y₂, and y₃ using Eq. (11)-(13)
            3. compute y.param_set as the sum of y₁, y₂, y₃'s param_sets
            4. set y.t_offset equal to x.t_offset
            5. trigger y.change_flag
        end
    endmodule
```

Fig. 3. (a) A composite-type XREAL and (b) a pseudo-code for a nonlinear SISO system model in SystemVerilog.

nonlinear system model discussed in Section III. The true event-driven operation is exemplified by the *always* statement of which content is executed only once whenever a new input event arrives. It first samples the present states of the partial responses y_1, y_2, and y_3 and then updates the partial responses via Eq. (11)-(13) using sampled values as the initial conditions. The final output y is then computed as the sum of the resulting partial responses. Its *t_offset* is set identical to that of the input and its *change_event* is triggered to notify the subsequent blocks consuming y as their inputs. Many of the operations on XREAL-typed variables are performed in C using custom DPI function calls.

V. EXPERIMENTAL RESULTS

A. Class-A RF Power amplifier

An RF transmitter employing a phase-shift keying (PSK) modulation and a class-A power amplifier, shown in Fig. 4, is modeled in the Volterra series and simulated using the described method. The nonlinear circuit equation governing the response of the power amplifier is:

$$C_L \frac{dv_o(t)}{dt} + \frac{V_{CM} + v_o(t)}{R_L} + K\left(V_{OV} + v_i(t)\right)^2 (1 + V_o(t)) = I_{SS} \quad (14)$$

where the circuit parameters R_L, C_L, K, V_{OV}, and λ are assumed to be 50Ω, 100fF, 1.0A/V^2, 0.1V, and 1/50, respectively. This equation models the distortion due to the square-law dependence and channel-length modulation effect of the MOSFET device.

Time-domain simulation is performed in SystemVerilog to assess a third-order intercept point (IP3) and spectral re-growth property of the transmitter. Fig. 5(a) plots the main signal power, the third-order inter-modulation power (IM3), and the extrapolated IP3 when a two-tone sinusoidal input at 0.99-GHz and 1.01-GHz is applied. On the other hand, Fig. 5(b) shows the spectra of the output signal when a 1-GHz carrier signal is PSK-modulated with 1-Mbps data. For both cases, good agreement with the Cadence Spectre results is achieved with a third-order Volterra series model.

Thanks to the event-driven operation of the proposed simulator, the substantial speed-up of ~300× was observed compared to Cadence Spectre. For a 512-bit data transmission (corresponding to 512-μsec), it took only 0.38 seconds for our

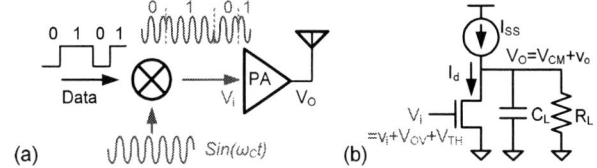

Fig. 4. (a) A RF transmitter employing a phase-shift keying modulation scheme and (b) a class-A power amplifier.

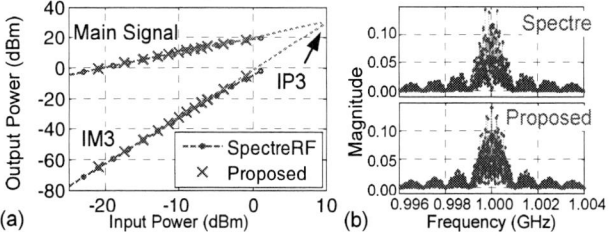

Fig. 5. (a) A third-order intercept point (IP3) and (b) output spectra of the power amplifier with data-modulated inputs compared to Spectre.

simulator while it took 112.5 seconds for Spectre. The speed-up is largely due to the fact that the functional form in Eq. (1) is particularly efficient in expressing a stiff signal such as a PSK modulated signal. For instance, it takes only 512 events for our proposed simulator to express the signal without any accuracy loss, while it takes one million samples for Spectre to Nyquist-sample the 1-GHz carrier signal.

B. Multi-level PAM Receiver with Continuous-Time Equalizer

The second example is a multi-level, high-speed data receiver, employing a continuous-time linear equalizer (CTLE) stage and a four-level pulse-amplitude modulation (4PAM) scheme as shown in Fig. 6 [4]. The multi-level receiver is particularly sensitive to the distortion, as it experiences non-uniform filtering/amplification depending on the signal level.

The CTLE in Fig. 6(b) is designed to compensate an 8-dB loss at 2GHz with a zero introduced by a source-degeneration capacitor (C_S) and resistor (R_S). The governing circuit equations with input v_{in} and output v_{out} are:

$$2I_{SS} = K(V_{OV} + v_{in} - v_{s+})^2 + K(V_{OV} - v_{in} - v_{s-})^2$$

$$\left(C_S\frac{d}{dt} + \frac{1}{R_S}\right)(v_{s+} - v_{s-}) = g_m(v_{in} - v_{s+}) + K(v_{in} - v_{s+})^2$$

$$\left(C_L\frac{d}{dt} + \frac{1}{R_L}\right)v_{out} = K(V_{OV} + v_{in} - v_{s+})^2 \quad (15)$$

where the circuit parameters g_m, K, C_S, R_S, C_L and R_L have values of 10mS, 0.1A/V², 800fF, 400Ω, 100fF and 200Ω, respectively.

Fig. 7 compares the eye-diagrams of the 2-Gbps 4-PAM signals before and after the CTLE stage for the transmit swing

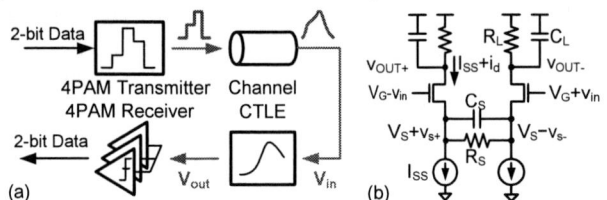

Fig. 6. (a) A 4-PAM high-speed I/O interface example and (b) circuit schematics of a CTLE.

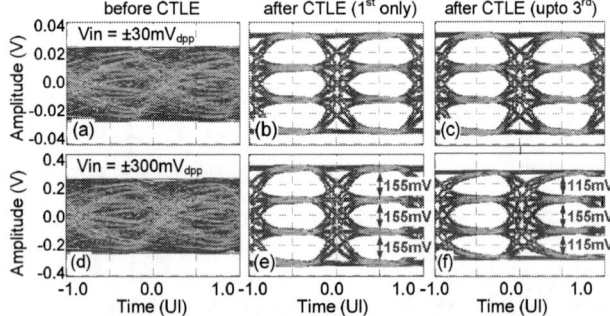

Fig. 7. Simulated eye-diagrams (a,d) before the CTLE and after (b,e) the CTLE without and (c,f) with the third-order distortion included for two swing levels: ±30mV$_{dpp}$ and ±300mV$_{dpp}$.

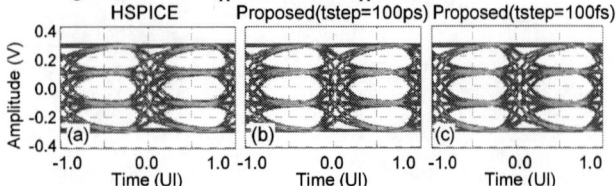

Fig. 8. Eye-diagrams of the CTLE output simulated by (a) HSPICE, and the proposed method with a time step of (b) 100ps and (c) 100fs.

of ±30mV$_{dpp}$ and ±300mV$_{dpp}$. Since the channel is modeled as a linear system with 7 poles, there is no difference in its output eye shapes between Fig. 7(a) and (d). The same is true when only the first-order response of the CTLE stage is modeled (Fig. 7(b) and (e)). However, when the third-order distortion response is included, the output signal exhibits different amount of distortion depending on the signal swing as shown in Fig. 7(c) and (f). In particular, for the input swing of ±300mV$_{dpp}$, the top-most and bottom-most eye openings are smaller (115mV) than that of the middle eye (155mV).

Fig. 8 compares the eye diagrams obtained with Synopsys HSPICE and the proposed simulator with different time steps of 100ps and 100fs. Note that the eye diagram obtained is identical to the one with HSPICE even with a coarse time step of 100ps (0.2 UI). In addition, simulating 2,000 bits of data pattern takes 26 seconds for our simulator on a Linux machine with AMD Phenom II X4 945 processor while HSPICE simulation takes 25,720 seconds (~990× speed-up). For the proposed simulator, the simulation time remains the same for the time steps of 100ps and 100fs.

VI. CONCLUSION

This paper presented a truly event-driven way to simulate a nonlinear analog circuit behavior modeled as a Volterra series on a SystemVerilog simulator. In order to extend the previous method demonstrated for linear circuits in [1], a Volterra-series model is reformulated into a set of linearized differential equations with an explicit notion on initial conditions. Then it follows that if the input signals are linear combinations of our basis functions, $c \cdot t^m exp(at) \cdot u(t)$, so is the output signal and it enables an event-driven simulation by updating only the coefficients $\{c\}$'s. The practical circuit examples demonstrated that the effects of nonlinear distortion can be accurately simulated at a 300~1000× faster simulation speed compared to SPICE-like ODE solvers. More importantly, the speed and accuracy of the simulation depend very weakly on the time step, confirming a truly event-driven simulation.

ACKNOWLEDGMENTS

This research was supported by the KCC (Korea Communications Commission), Korea, under the R&D program supervised by the KCA (Korea Communications Agency) (KCA-2013-(12-911-01-102)). CAD tool licenses are supported by the IC Design Education Center (IDEC) in Korea.

REFERENCES

[1] J.-E. Jang, et al., "True Event-driven Simulation of Analog/Mixed-signal Behaviors in SystemVerilog: a Decision-Feedback Equalizing (DFE) Receiver Example," in Proc. IEEE Custom Integrated Circuits Conf. (CICC), pp.1–4, Sep. 2012.

[2] J.-E. Jang, et al., "An Event-Driven Simulation Methodology for Integrated Switching Power Supplies in SystemVerilog," in Proc. ACM/IEEE Design Automation Conf. (DAC), to be published.

[3] A. Zhu, et al., "An Efficient Volterra-Based Behavioral Model for Wideband RF Power Amplifiers," in IEEE MTT-S Int'l Microwave Symposium Dig., pp. 787-790, Jun. 2003.

[4] T. Toifl, et al., "A 22-Gb/s PAM-4 Receiver in 90-nm CMOS SOI Technology," IEEE J. Solid-State Circuits, pp. 954-965, Apr. 2006.

[5] S. A. Maas, Nonlinear Microwave and RF Circuits, 2nd. Ed., Norwood, MA: Artech House, 2003, ch. 4.

[6] J. Roychowdhury, "Reduced-Order Modeling of Time-Varying Systems," IEEE Trans. Circuit and Systems II, pp. 1273-1288, Oct. 1999.

A Verilog Piecewise-Linear Analog Behavior Model for Mixed-Signal Validation

Sabrina Liao, Mark Horowitz

Department of Electrical Engineering, Stanford University, Stanford, CA 94305

Abstract-**Full chip mixed-signal validation requires simulating the entire design through a large number of test vectors, which makes fast, event-based Verilog models of analog circuits essential. We describe an extensible approach to creating these models that maps continuous signals into piecewise linear waveforms by creating analog events which contain a value and slope. By breaking analog circuits into sub-blocks with mostly unidirectional ports, we avoid explicit time integration, thus fitting well into an event-driven digital framework. The result is Verilog analog functional models that are pin-accurate, fast to simulate and capture the key dynamics in analog circuits. A 2.5V-1.8V buck converter and 1GHz PLL models are demonstrated.**

I. INTRODUCTION

Mixed-signal systems in which digital and analog circuits communicate across a tight interface are commonplace today. This tight coupling compounded with the large complexity of the system, and the large number of test vectors that need to be run at the system level requires efficient system-level simulation models. The only practical means of performing this validation is through efficient HDL simulator [8]. Thus it is essential to have analog functional models that would fit seamlessly into the digital validation framework.

Given the importance of full system simulation, there has been considerable work in this area. Verilog-AMS based method is one example [2]. These behavioral models use differential equations and time integration, and hence require solvers similar to those found in circuit simulators. Matlab/Simulink [6] is another common route. While these models are faster than the Verilog-AMS models, they are run in Matlab's event-driven simulator and have little resemblance to the physical circuit thereby limiting their value in validation effort. Analog behavioral modeling has also been attempted in digital Verilog and these models are typically constant time-step based. In [8], FIR filters and adjustable circuit parameters such as multiple biases were implemented. In [5], the constant time-step approach is supplemented with additional data such as the actual crossing time of a clock transition. There is also a large body of related research in macromodeling ([1],[4],[7]), which attempts to create extremely accurate models to replace SPICE simulation entirely.

II. MODELING APPROACH

Our goal is to create analog functional models that would benefit system-level validation in an event-driven, digital simulator. Therefore, these models must parallel standard digital models in certain aspects. First, they must be pin-accurate in order to check for equivalence with their schematics [3]. To provide confidence in system level validation, they should capture salient features of real circuits' behavior, but like digital standard cells, the analog blocks still

need to be verified using SPICE before being abstracted into a functional model. Lastly, the analog functional models must not slow down digital simulation.

A. Representing Continuous-Time Signals

The first challenge is representing analog signals in a digital simulator. Typical real number modeling [6][8] samples the continuous-time signals above Nyquist (see Fig. 1 a). Problems arise with this approach when an event occurs between sample points (eg. since V[2] has not arrived yet, we can't interpolate the correct sampled value). This type of issue is critical if we want to model the effect of clock jitter.

To address this issue, instead of a single value at each time sample, we represent the signal as an initial voltage and a slope. The signal continues on the same slope until the next event in the sequence. Pin-accuracy can be maintained by defining an analog signal as a structure in SystemVerilog (Fig. 1 b). Each module generates events to create a piecewise linear (PWL) model of the waveform with small error, generally using a spacing proportional to the time constant of the circuit.

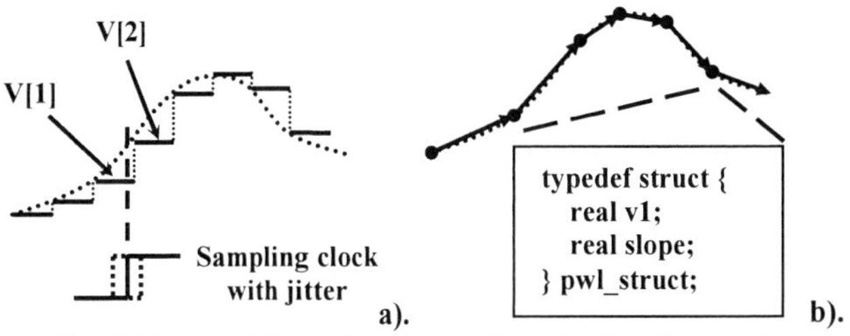

Fig. 1. Representations of continuous-time signals - a). piecewise constant b). piecewise linear

B. Avoiding Time Integration

To be computationally efficient, we want the analog cells to generate as few events as possible, and to be quick to evaluate. These requirements rule out building models based on nodal differential equations. Our strategy is to partition large analog circuits into unidirectional blocks so that closed-form equations can be derived for their outputs given some stimulus – this could be in some domain other than voltage/current, for example, time/phase domain for a VCO model.

Fig. 2 shows an example of the partitioning needed in the frontend of a single-slope ADC. Combining the sample-and-hold, capacitor, and ramp current source creates a circuit that can be modeled unidirectionally and analytically. We treat the combination as nearly linear systems with two modes – tracking and ramping – controlled by the sampling clock. Since these are nearly linear systems, they have transfer functions which lead to closed-form equations that describe the output. For example, the tracking phase response to a linear input is:

$$V_{out} = \beta t + \beta RC \left(e^{-\frac{t}{RC}} - 1\right) + \alpha \left(1 - e^{-\frac{t}{RC}}\right) + V_0 e^{-\frac{t}{RC}} \quad (1)$$

α is the starting voltage of the input; β is the slope of the input; and V_0 is the initial voltage on the capacitor. This equation is universal to any single pole system. The generalization to more complex systems can be formulated as the problem of fitting coefficients or state space matrices of nearly linear systems based on SPICE results.

Circuits where the transfer function can be controlled through other inputs, like AGC amplifiers, are easily handled by this framework. Here the output equations become a function of the controlling input. This approach is valid as long as the bandwidth of the control inputs is much less than the signal bandwidth which is nearly always true.

Weakly non-linear behavior can be included in a similar way by partitioning the overall behavior into several linear approximations (using methods in [7] for instance) and the module simply chooses the correct transfer function to use according to region of operation.

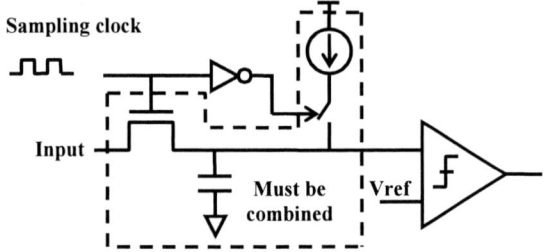

Fig. 2. Proper partitioning of analog blocks

C. Model Output Evaluation

Given the general concept of PWL representation of analog signals and the practice of partitioning analog circuits into unidirectional modules, it is straightforward to arrive at the output waveform of a model. For every input update, we must evaluate the output according to the time-domain response of a linear system to a ramp (solid circle to solid circle mapping in Fig.3). If the output waveform is not a linear ramp, then an internal reevaluation is scheduled to produce the next linear segment (hollow circle to solid circle in Fig.3).

Fig. 3. Model output evaluation

There are also input events that do not generate output events because they don't cause significant changes to the output. It is important to filter these events if the system modeled uses feedback (when events are not filtered, each output event generates a new input event, and the simulator will hang). Our solution is to create a standard gatekeeper on each model's output that only allows a new event on the output if it differs significantly from the old one. This is illustrated in Fig. 3 in which the last hollow circle on the input does not generate an output event.

III. MODELING FRAMEWORK

Commonly encountered model components can be grouped into one of four different variants, depending on whether the inputs and outputs are analog or digital.

A. Digital In & Out

Circuits with digital inputs and outputs are easily modeled in standard Verilog. The only caution is that when used in analog designs, these circuits are often analog in delay/phase (eg. in a PLL), and in these cases one needs to use accurate delay models for these gates.

B. Analog In & Out

These circuits are filters or linear systems completely described by their poles and zeros. Amplifiers, filters, and sample and holds all fall into this category. To construct these models we start with the transfer function of the system from every input to the output. From this transfer function and the waveform accuracy required, we set T, the delay this model will use before recalculating the output to determine if a new output segment is needed.

The Verilog model consists of an *always* block followed by a gatekeeper. The *always* block is the main evaluation loop that's sensitive to input updates and the internal evaluation signal that indicates the end of T. When triggered, it computes the current state of the system by projecting from the previous state. The equation used follows from the transfer function of the system. Then it calculates the state of the system in T seconds, generating a potential new output event. Lastly it cancels any pending internal evaluations of this block, since it just evaluated, and then schedules a new internal evaluation point T seconds in the future. These operations are illustrated in Fig. 4 a) and b). The dotted lines in the output row are the actual trajectories from state equations, while the solid lines are the PWL approximations provided to downstream circuits.

The gatekeeper module filters output events as previously described. Its operation is illustrated in Fig. 4 c). The continuation of the solid line from the previous evaluation cycle is very close to the actual response (dotted line), hence no new event occurs on the output – ie. no solid dot.

C. Digital ↔ Analog

The last two classes of circuits convert between digital and analog waveforms. These models can be easily constructed by an idealized converter and a filter. Digital to analog blocks (eg. DAC's) can be represented by a converter generating a PWL segment when its digital input changes, followed by a filter that represents the dynamics of the circuit. Analog to digital blocks such as comparators and voltage controlled oscillators, can be model by a filter that captures the dynamics of the "sampler", followed by an ideal slicer. In Verilog, the ideal slicer creates an event to change the digital output when the output of the filter reaches a certain threshold.

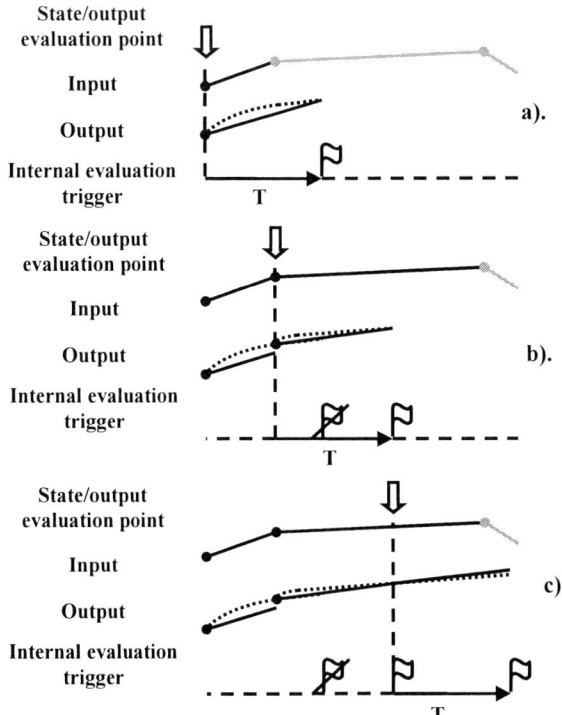

Fig. 4. Analog in and out model operation

IV. EXAMPLE MODELS

The buck converter is a 2.5V to 1.8V constant-offtime switching regulator with a synthetic ripple-based sensing mechanism (see Fig. 5 a). The 1GHz ring-based PLL block diagram is shown in Fig 5 b.

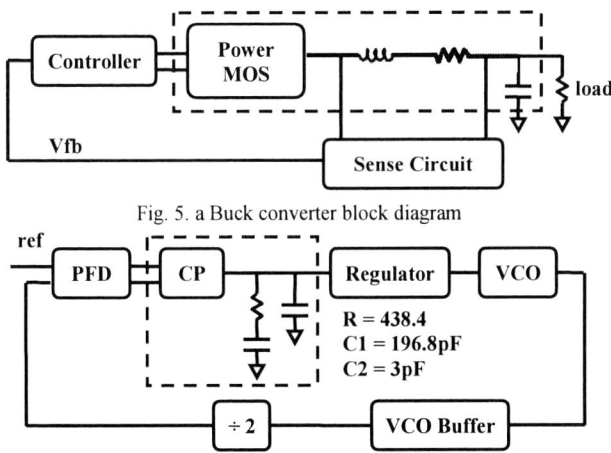

Fig. 5. a Buck converter block diagram

Fig. 5. b 1GHz PLL block diagram

A. Buck Converter Model

The power MOS and inductor/capacitor network is combined to form a single analog input and output module with its transfer function being a function of the load resistance and controlled by the gating signals on the power MOS. The sense circuit is a separate analog in/out module. The controller is mostly digital gates with a clocked comparator modeled as a filter, followed by a slicer and ideal delay (ie. an analog to digital block).

B. PLL Model

The phase frequency detector, VCO buffer and divider are regular digital logic, with added features of delay as a function of the supply, and jitter.

The regulator is a block with analog input and output. SPICE simulations showed that the input to output transfer function is not affected very much when the supply is varied between +/- 10%. Therefore, we treat Vdd as just another input and model the output as the sum of two transfer functions: one from the input and the other from Vdd. If the two transfer functions affected each other, the model template can be easily changed to select between versions of one transfer function according to the parameter that affects it. Noise performance is also extracted from SPICE and modeled.

The charge pump and the low pass filter must be grouped together to create a unidirectional model (a digital to analog block). The charge pump section is the converter that converts digital up/dn signals into PWL current pulses. We include effects such as supply dependent delay, up and down path mismatch, charge injection, as well as supply and drain voltage dependent current output. The low pass filter is similar to the regulator model with its output being a sum of two transfer functions: the filter impedance and the noise transfer function of the resistor noise source.

The VCO is an analog to digital block that operates nearly linearly in the phase domain. The filter portion of the VCO model is an integrator with phase as its state variable and

$$\Delta\phi = \int f_{VCO} dt \qquad (2)$$

as the equation that governs the evolution of this state variable. The slicer determines when the phase crosses π and schedules the output to be toggled at that time. We include substrate noise by fitting the VCO frequency as a function of control voltage and substrate voltage. We also include jitter by adding the phase noise accrued since the last evaluation point to the incremental phase resulting from the state equation alone.

C. Experimental Results

The models written here are pin-accurate once proper partitioning is done. Table 1 lists sample simulation speeds on a Dual-Core AMD Opteron Processor 2216.

TABLE 1
VERILOG MODEL AND SPECTRE SIMULATION SPEEDS

Circuit	Transient Time	Spectre	Verilog
Buck Converter	3us	25h 15min	0.46s
PLL	3us	33min 48s	4.4s

Fig. 6 shows the startup behavior of the buck converter with a load of 20Ω and Fig. 7 compares the model and Spectre responses to a load change at 1.5us from 20Ω to 10Ω. The locking behavior of the PLL is compared in Fig. 8. A static offset of 33ps is measured in Verilog and 34.5ps in Spectre. Next, we turn on the noise/jitter properties of the PLL sub-blocks and list their contributions to output jitter in Table 2. Adding all these yields an expected total jitter of 7.9ps. Then, with all the noise sources turned on in the model, we arrive at the jitter histogram shown in Fig. 9 using 8000 clock edges. The 8ps of total rms jitter achieved matches well with the

978-1-4673-6145-3/13 $31.00 © 2013 IEEE

expected value from Table 2 (jitter calculated from Spectre phase noise simulation is 7.78ps). We also compute the phase noise plot (Fig. 9) using 430k output edges and Welch's method with an FFT size of 2048 and overlap of 1024. As expected, we see that it peaks around the PLL bandwidth of 13MHz. To verify supply effects, voltage steps are applied to the PLL supply and the transient output clock edge offset from the reference vs. time is plotted in Fig. 10.

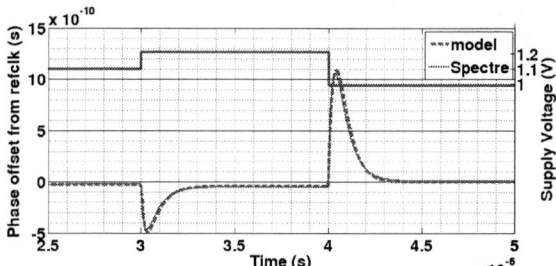

Fig. 10. PLL output phase response to supply disturbances

V. CONCLUSION

Creating pin-accurate, fast functional Verilog models of analog circuits has become essential for mixed-signal validation. To generate these models, we use a piecewise linear representation of the analog signals, which is compatible with event-driven simulation. Next we partition the circuits into unidirectional blocks and transform the inputs and outputs to a domain where the circuit is nearly linear, allowing us to create closed form solutions in response to PWL inputs. The result is a general approach for creating analog functional models. We demonstrated the utility of this approach by applying these techniques to a buck converter and 1GHz PLL. The resulting models have sufficient speed to be used in system level validation, and sufficient fidelity to correctly model real circuit behavior. The underlying principles are general, so we expect these same techniques will be useful in modeling most mixed signal systems.

ACKNOWLEDGEMENT

The authors would like to thank Sakshi Arora and Professor Bruce Wooley for the buck converter design and discussions.

REFERENCES

[1] W. Daems, G. Gielen and W. Sansen, "Simulation-based automatic generation of signomial and posynomial performance models for analog integrated circuit sizing," in *Proceedings of ICCAD*, pp. 70-74, 2001.

[2] K.S. Kundert, "Predicting the phase noise and jitter of PLL-based frequency synthesizers," in *Phase-Locking in High-Performance Systems: From Devices to Architectures*. IEEE Press, 2003.

[3] B.C. Lim, J. Kim and M. Horowitz, "An efficient test vector generation for checking analog/mixed-signal functional models," in *Proceedings of DAC*, pp. 767-772, June 2010.

[4] H. Liu, A. Singhee, R. Rutenbar and L. Carley, "Remembrance of circuits past: macromodeling by data mining in large analog design spaces," in *Proceedings of DAC*, pp. 437-442, 2002.

[5] M-J. Park, H. Kim, M. Lee and J. Kim, "Fast and accurate event-driven simulation of mixed-signal systems with data supplementation," in *Proceedings of CICC*, pp. 1-4, Sept. 2011.

[6] M. Van Ierssel, H. Yamaguchi and A. Sheikholeslami, "Event-driven modeling of CDR jitter induced by power supply noise, finite decision-circuit bandwidth and channel ISI," *IEEE Trans. Circuits and Systems I.* 55, 5, pp. 1306-1315, May 2008.

[7] J. Wang, X. Li and L. Pileggi, "Parameterized Macromodeling for Analog System-Level Design Exploration," in *Proceedings of DAC*, pp. 940-943, 2007.

[8] C. Werner et al., "Modeling, simulation and design of a multi-mode 2-10Gb/sec fully adaptive serial link system," in *Proceedings of CICC*, pp. 709-716, Sept. 2005.

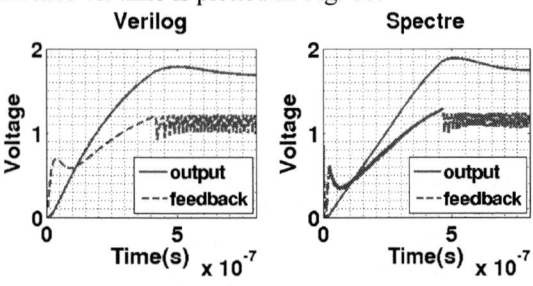

Fig. 6. Buck converter startup behavior

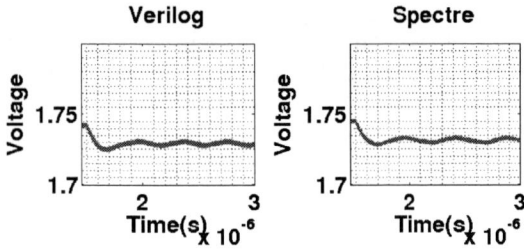

Fig. 7. Buck converter load response (20 to 10 ohm)

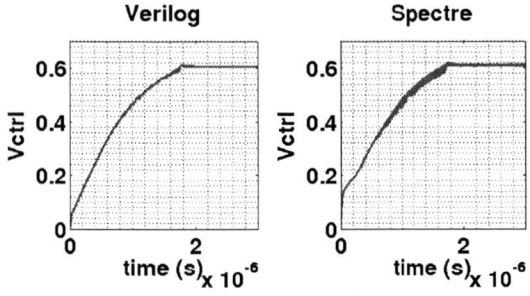

Fig. 8. PLL locking behavior

TABLE 2

SUB-BLOCK NOISE CONTRIBUTION TO PLL OUTPUT JITTER

Noise Source	PLL Output Jitter
PFD	0.556ps
CP	0.126ps
LPF	0.124ps
Regulator	0.612ps
VCO	7.85ps
VCO Buffer	0.394ps
Divider	0.126ps

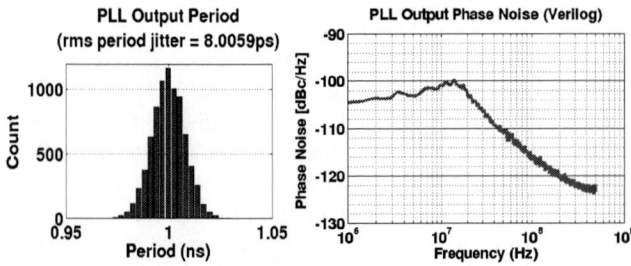

Fig. 9. PLL output jitter histogram and phase noise (Verilog simulation)

Advancements in High-Speed Link Modeling and Simulation
(An Invited Paper for CICC 2013)

Mike Peng Li, Masashi Shimanouchi, and Hsinho Wu

Altera Corporation

101 Innovation Road

San Jose, CA 95134

*Abstract-*As the high-speed I/O (HSIO) and serial link data rate keeps increasing, the requirements for accuracy and advanced capabilities of its modeling and simulation techniques get more stringent. Emerging requirements such as comprehending process, voltage, and temperature (PVT) variations at deep sub-micron process nodes or smaller, fully accounting for all the circuit blocks of the link, gap closing between modeling and measurements, have become critical and important, yet the conventional modeling and simulation methods cannot meet most or all of those requirements. In this paper, we will start with reviewing the status of techniques/methods used in recent HSIO simulation and modeling for signaling, integrated circuits (ICs), board circuits, such as behavioral statistical, SPICE and IBIS-AMI, outlining areas where they fall short, in comparison with those emerging requirements. We then will discuss the new methods and techniques that can meet and comprehend these emerging requirements and how they enhance and advance the accuracy and capability for the HSIO link modeling and simulation. We will give simulation and experimental results to demonstrate and quantify how to meet emerging requirements, as well as needed accuracy and capability advancements, with the new techniques.

I. INTRODUCTION

HSIO and link speed keeps increasing at a pace of doubling every 2-3 years on average [1], to meet the ever increasing bandwidth demands for network and computer systems. In addition to the bandwidth benefits, the speed doubling also enables density and throughput doubling for a given device or board. Using the PCI Express as an example, its per lance speed has increased from 2.5 Gbps of Gen 1, to 5.0 Gbps of Gen 2, to 8 Gbps of Gen 3, and to 16 Gbps of Gen 4. Another example of OIF CEI has its Gen1 speed at 6.5 Gbps, Gen 2 at 11.3 Gbps, Gen 3 at 28 Gbps, and Gen 4 at 56 Gbps. At 28 Gbps, the unit interval (UI) is only 35.7 ps for a NRZ modulated signal, and this timing budget has to be shared by the transmitter (TX), receiver (RX), and channel (CH), with TX and channel gets ~30% of UI each, and RX gets ~40% of the UI for their respective timing budgets, for most of the standards [2]. When all the timing and jitter impairments are considered, meeting a UI of 35.7 ps at 28 Gbps is a challenge task, and at 17.9 ps UI/56 Gbps, this task becomes even more challenge. Accurate and capable modeling and simulation techniques are critical for designing these HSIO circuits, devices, and links.

As the data rate increases, the inter-symbol interference (ISI) gets worse due to the lossy characteristics of the copper channel. To mitigate its ISI at higher data rates, various equalization circuits have been developed, including TX feed-forward equalizer (FFE), RX continuous linear equalizer (CTLE), and RX decision-feedback equalizer (DFE). Figure 1 shows a typical high-speed link and its subcomponents at 10 Gbps and above for a backplane (BP, typically has insertion loss (IL) > 25 dB or higher, and consists of PCBs, connectors, and vias). In addition to equalization circuit blocks, clock generation (CG) via a phase-locked loop (PLL) for TX, and clock recovery (CR) via PLL for RX are also shown. 3-4 tap TX FFE, 4th-8th order CTLE, and 5-taps or higher DFE are commonly needed for high-speed BP links (see, e.g., [3]), implying a complex equalization solution space.

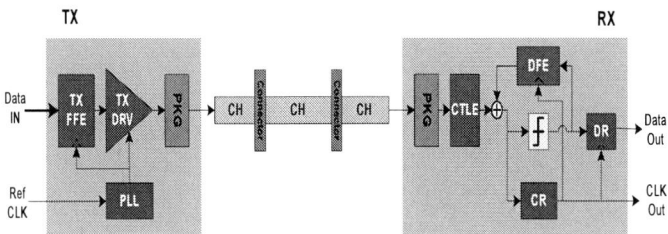

Fig. 1. Illustration of a High-Speed Link System

Transistor level based circuit simulator such as SPICE or HSPICE has not been effective for high-speed link as the simulation time would be too long to be practical/effective. For example, simulation time for a SPICE may take many hours even for a primitive link (i.e., no equalization or clock recovery involved) consisting only TX driver, RX buffer, and copper channel for a few hundred bits. The SPICE simulation time for a high-speed link would be highly prohibitive if various equalizations, clock generation and recovery are comprehended. Performance merit of the high-speed link is commonly defined by its bit error rate (BER) which is set at 1e-12 or 1e-15 by most standards [2]. The performance gap between what a practical SPICE simulator can deliver and standard requirement is huge.

To overcome the limitations of SPICE in simulating a high-speed link, an analytical worst case channel eye estimation method called peak-distortion analysis (PDA) was developed in early 2000s [4]. PDA treats a copper channel as a linear-time invariant (LTI) system, and the analysis was based on single-bit-response (SBR) and associated sampling cursors. While PDA method helps alleviating the long simulation time challenge of SPICE, it tends to give overly pessimistic results compared with reality, and as such statistical link simulation methods have been developed in the mid of 2000s (e.g., [5],[6]). The statistical methods build eye diagram by super-positioning time shifted (time delay or advance in integer multiple of UI) SBRs in a probabilistic manner that is equivalent to convolution of SBR cursor probability density functions (PDFs) statistically. Statistical methods can simulate a high-speed link with relatively fast throughput. IBIS-AMI standard [7] adopted LTI and statistical methods for high-speed link simulation for this reason.

Perhaps a major limitation of the statistical method is its limited capability in handling jitter and noise for TX and RX circuits, as well as jitter and noise interactions with TX, channel and RX. Progresses have been made to incorporate jitter and noise coverage in statistical methods (e.g., [8], [9]). However, those approaches often involve assumptions of noise to jitter conversion that does not apply when jitter and noise are independent. While LTI and statistical methods are appropriate to copper channels, it introduces inaccuracy or even error due to the non-linearities of TX driver and FFE, RX buffer and DFE, that are often overlooked. In recent years, as the high-speed links are widely adopted and deployed in network and computer systems, system designers and signal integrity engineers often demand a good correlation proof between simulation and measurement before they can adopt a simulator or simulation methodology for their link design validation and large sample pre-production simulations. In order to address emerging requirements of correlation and volume pre-production simulation, a high-speed link simulator needs to comprehend the device and channel PVT variations. Moreover, the number of possible equalization parameters for a link with FFE, CTLE, and DFE can be a few millions, yet, a time efficient and consistent optimization method is required or expected. However, correlation, PVT, and equalization optimization are neither addressed nor lighted touched-upon in most of the statistical methods published.

In this paper, we intend to highlight the emerging requirements and challenges in high-speed link simulation and discuss the advancements in solving those challenges. To enable a smooth content flow, we will give a high-level review of the statistical link simulation methods in section II. In section III, we will discuss two new simulation methods of time-domain full-waveform and hybrid simulation methods and associated CG, CR, data recovery (DR) modeling methods that are not comprehended in statistical methods. In addition, fast and computationally efficient optimization methods will be discussed in this section. Section IV will focus on PVT variations for the link simulation. Novel method invoking behavioral model coupling with the measurement based jitter and noise look-up table (LUT) to account for PVT and operating conditions, will be discussed. In section V, we will focus on correlation method, and the procedures needed to better refine the modeling and simulation methods, as well as to ultimately improve the link simulation accuracy in an iterative manner. Accuracy/correlation improvement and validation examples and data will be presented in this section. We will summarize and conclude in section VI.

II. OVERVIEW OF STATISTICAL LINK SIMULATION METHOD

In general, statistical methods treat each circuit block with equivalent or approximated higher-level behavior model that can be represented mathematically or algorithmically. Figure 2 shows a behavioral block diagrams and associated high-level and applicable mathematical representations corresponding to the functional block diagrams of Figure 1, for a statistical method.

LTI is often used for cascading link component blocks and calculating frequency-domain transfer function (TF) or time-domain impulse response (IR) from point-to-point. The overall IR from TX FFE, to RX CTLE, with TX driver, TX package, channel, RX package, and RX buffer in between, can be calculated via convolution chain process shown in Eq. (1):

$$h_{s1}(t) = h_{ffe}(t) * h_{dr}(t) * h_{txp}(t) * h_{ch}(t) * h_{rxp}(t) * h_{ctle}(t) \quad (1)$$

where * denotes convolution. Note that for simplicity, Eq. (1) is intended to be a generic formula for estimating overall IR from TX to RX, and can be represented in both continuous and discrete time-domain. In the case when FFE is implemented digitally, $h_{fir}(t)$ needs to be handled in discrete time-domain.

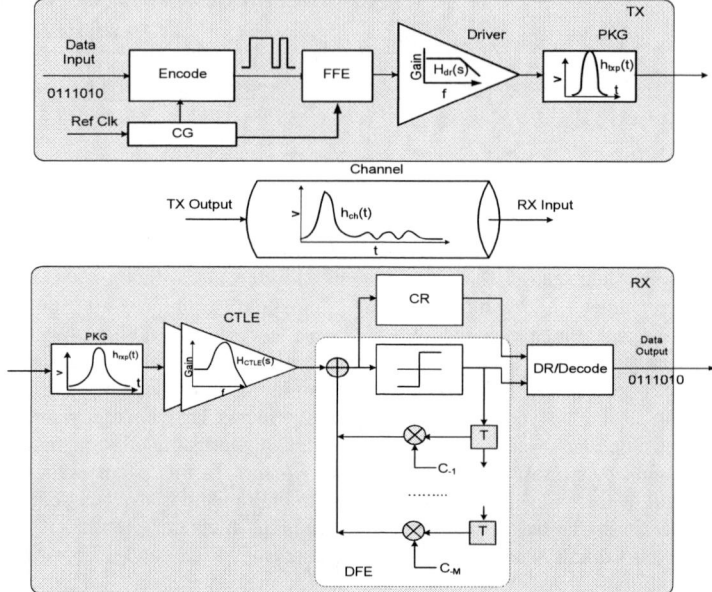

Fig. 2. An Illustration of a Statistical Modeling for a High-Speed Link

SBR can be calculated via convolving the IR with a single bit ideal square wave S(t) of the following

$$\Pi_{s1}(t) = S(t) * h_{s1}(t) \quad (2)$$

SBR Π_{s1} obtained in Eq. 2 does not include the DFE as it does not follow LTI rule. However, DFE may be modeled approximately by subtracting DFE tap weights from the SBR Π_{s1} as the follows:

$$\Pi_{s2}(t) \approx \Pi_{s1}(t) - \sum_{i=1}^{N} c_i comb(t - iT) \quad (3)$$

where the c_i are the DFE tap coefficients, T is the UI, and comb(t) is the comb function. With SBR Π_{s2}, statistical eye can be constructed by time-shifting (delay or advance of 1 to multiple UIs) Π_{s2} and probabilistic superposition (see, e.g., [4],[5]), and we denote the associated statistical eye PDF as $p_0(t, v)$. BER cumulative density function (CDF) and 'bath-tub' curves, can be derived from $p_0(t, v)$ [10].

Statistical methods may be extended to comprehend the jitter and/or noise effects, and this is often achieved by convolving the assumed jitter and/or noise PDFs with the statistical eye PDF $p_0(t,v)$.

In figure 3, we show an illustrative example of statistical simulation results for a PCIe3.0 backplane link running at 8.0 Gbps, with the TX having an FFE, and RX having both CTLE and DFE equalizers. The simulation takes 1.8 min to finish, based on an Intel Core i7 2.7GHz processor with 8GB RAM, and 32 bit Win 7 operating system.

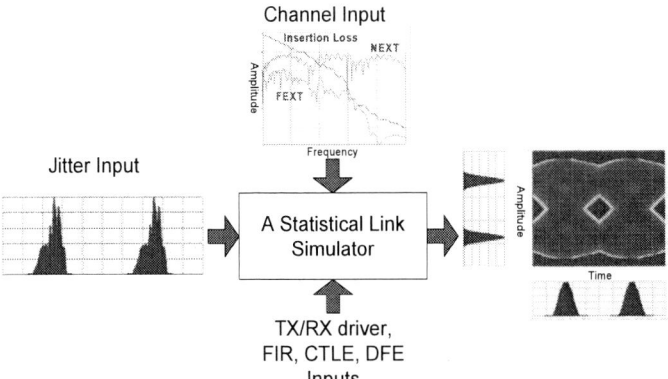

Fig. 3. An Illustration of a Statistical Simulation Example for PCIe3.0.

While statistical link simulation method offers a computationally effective way to estimate the link statistical eye, it is also subject to limitations, and we will list a few. First, equalization adaptation is hard to built-in, limiting its optimal solution search capability; second, interactions of various jitter and noise components from the TX with channel and RX are difficult to be comprehended, limiting its accuracy and coverage; third, CG, CR, and DR cannot be accounted for due to their time-domain operation nature; also limiting its coverage and accuracy. Apparently, new and advancements are needed for high-speed simulations beyond the statistical methods, and we will discuss them in section III.

III. FULL-WAVEFORM AND HYBRID SIMULATION METHODS

The basic concept for a full-waveform method is to simulate the waveform in time-domain from TX, to channel, and to RX, emulating signal follow and propagation path in an actual link. In this section, we will first discuss the full-waveform method, followed by the hybrid method, and then CG, CR, and DR modeling.

A. Full-waveform method

Let's denote the digital bit sequence or data pattern as $d_i(t) = d_i(iT)$, where d_i is either 1 or 0 bit for NRZ modulation, and i is the bit sequence index. The waveform at the TX output can be calculated as:

$$V_{TX0}(t) = V_0 d_i(t) * h_{ffe}(t) * h_{dr}(t) * h_{txp}(t) \qquad (4)$$

where V_o is the output voltage and we call $V_{TX0}(t)$ as TX deterministic waveform. Since the simulation is done in time-domain, jitter can be introduced via phase modulation, and noise can be introduced via amplitude modulation. Let's denote $\Delta t_{TX}(t)$ as TX jitter, and $\Delta V_{TX}(t)$ as TX voltage noise, TX waveform comprehending both jitter and voltage noise can be expressed as:

$$V_{TX}(t) = (V_0 + \Delta V_{TX}(t))d_i(t) * h_{ffe}(t) * h_{dr}(t + \Delta t_{TX}(t)) * h_{txp}(t) \quad (5)$$

with the obtained $V_{TX}(t)$ waveform comprehending TX deterministic behavior, jitter, and noise, its characteristics such as eye-diagram, rise/fall times, BER contour can be estimated subsequently[10].

The waveform at the channel output can be further calculated with the following Eq. (6)

$$V_{CH}(t) = V_{TX}(t) * h_{ch}(t) \qquad (6)$$

Similarly, the waveform at the receiver CTLE output can be calculated as:

$$V_{CTLE}(t) = (V_{CH}(t) + \Delta V_{RX}(t)) * h_{rxp}(t) * h_{ctle}(t) \qquad (7)$$

Here RX voltage noise $\Delta V_{RX}(t)$ is introduced and amplitude modulated in time-domain. As we had mentioned, DFE is a nonlinear system and LTI does not apply well. DFE output needs to be modeled in a mixed-signal manner and is given by the following:

$$V_{DFE}(t) = V_{CTLE}(t) + \sum_{j=1}^{N} c_{-j} V_{SL}(t - jT + \Delta t_{RX}) \qquad (8)$$

where V_{SL} is the voltage after the DFE slicer, and $\Delta t_{RX}(t)$ receiver jitter associated with recovered clock.

Eq. (1)-(8) outline the basics and theoretical frameworks for the full-waveform method. With them, waveforms at various observing points within a link can be modeled and estimated, along with associated characteristics such as eye-diagrams, rise/fall times, jitter and noise PDFs and BER contour CDFs. In practice, full-waveform method may take long time to finish if long pattern bit sequence is to be simulated. In this case, extrapolation from simulated probability level to a target probability level (e.g., 1e-12) may be carried-out [11] to alleviate the very long/non-practical simulation time limitation.

B. Hybrid method

While the full-waveform method enables the accurate and complete time-domain simulation of a high-speed link, it also faces the challenges of long simulation time when bit sequence is long (e.g, 1e12 bit) and to comprehend small probability (e.g., 1e-12) jitter and noise. To overcome this limitation, hybrid method was developed. A hybrid method maintains the full-waveform theoretical foundation and framework, but limits the time-domain jitter and noise to bounded or high-probability types (e.g., periodic jitter (PJ), bounded-uncorrelated jitter (BUJ), duty-cycle distortion (DCD), and ISI). For the unbounded jitter and noise (e.g., random jitter (RJ), random noise (RN)), they are modeled in statistical-domain. In essence, hybrid method invokes both time and statistical domains, and its name reflects this nature. Let's denote the bounded waveform corresponded PDF as $p_b(t, v)$, then the complete PDF comprehending both RJ and RN will be given by:

$$p(t,v) = p_b(t,v) * p_g(t) * p_g(v) \qquad (9)$$

where p_g represents a Gaussian distribution.

A hybrid method reduces the simulation time compared with a full-waveform method as its time-domain simulation potion only needs to cover bounded processes/effects that are commonly contained within 1 million bits or less in practice.

C. Clock generation and recovery

As we had mentioned in section I, jitter may be accounted for in a statistical method via convolution and by assuming a jitter distribution, rather than model the jitter from its root-causes. For a high-speed link, TX and RX clock jitter is determined by CG and CR circuits respectively and those circuits are commonly implemented with a PLL circuit. A PLL circuit operates in time-domain, therefore, clock jitter can be modeled in full-waveform or hybrid methods, rather than being assumed as in the case of statistical method.

978-1-4673-6145-3/13 $31.00 © 2013 IEEE

PLL performance and jitter modeling is a well-studied subject, and we will only give a high-level overview how this subject is handled in high-speed link simulation, and leave further reading to [10], [12].

Figure 4 shows a diagram illustrating PLL functional blocks, corresponding math model in complex S-domain, and jitter/phase noise (PN) injection and processes. Note that time and S-domain is related uniquely via Laplace (or complex Fourier)/inverse Laplace transformation, and they should be considered as equivalent.

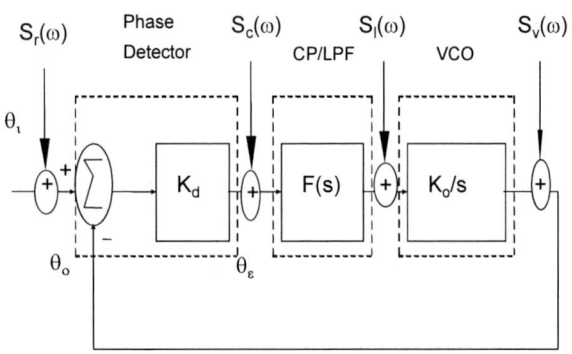

Fig. 4. PLL Components, Associated Transfer Functions, and Jitter/Noise Process Illustration.

The phase jitter/PN power-spectrum density (PSD) at the PLL output is determined by the following Eq.:

$$
S_o(\omega) = \begin{aligned} &S_r(\omega)\left|\frac{K_d K_o F(s)}{s + K_d K_o F(s)}\right|^2 + S_c(\omega)\left|\frac{K_o F(s)}{s + K_d K_o F(s)}\right|^2 \\ &+ S_l(\omega)\left|\frac{K_o}{s + K_d K_o F(s)}\right|^2 + S_v(\omega)\left|\frac{s}{s + K_d K_o F(s)}\right|^2 \end{aligned} \tag{10}
$$

where $S_r(\omega)$, $S_c(\omega)$, $S_l(\omega)$, $S_v(\omega)$ are jitter/PN PSD associated with reference clock input, charge-bump, low-pass filter (LPF), and voltage control oscillator (VCO), respectively, with $s=j\omega$. Jitter/PN variance can then be estimated by the following via Inverse Fourier Transformation (i.e., $\mathfrak{I}^{-1}$)

$$
\sigma_t^2(t) = 2(\sigma_0^2 - \mathfrak{I}^{-1}(S_0(\omega))) \tag{11}
$$

where σ_t^2 is the variance at time t, σ_0^2 is the total variance of the underline jitter/PN process[10].

The σ_t is then used for the jitter phase modulation in full-waveform or hybrid method for TX CG and RX CR. In the case of TX CG, σ_t comprehends uncorrelated reference clock jitter and intrinsic PLL jitter. Most recent CR circuits use dual-loop design [2] where a reference clock is not used during the mission operation phase, as such the recovered clock jitter only consists of PLL intrinsic jitter, and there will be no need to include reference clock jitter in σ_t. As for the correlated DCD, it is comprehended by controlling the rise/falling edge time transition of the digital bit sequence $d_i(t)$.

D. *Data recovery*

The DR is the last analog stage for a high-speed link where data after CTLE and DFE finally gets recovered by a data latch or sampler as show on the left side of figure 5. The clock input to the DR or data latch is the recovered clock that tends to move in-phase with data, or equivalently track the jitter on the data. The equivalent behavioral model for the DR is a difference function in time or phase domain, as shown on right side of figure 5.

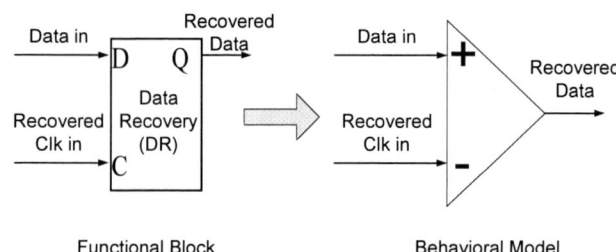

Fig. 5. Data Recovery Functional Block Diagram (left), and Corresponding Behavioral Model (right)

E. *Automatic Adaptation*

In both full-waveform and hybrid methods, automatic equalization and clocking optimization becomes possible, unlike the case of statistical method. Automatic adaptation and optimization is a key requirement for practical link design and operation for efficiency and productivity considerations. There are two basic optimization methods in practice, the first is TX first, RX second. In this method, a set or sets of TX equalization and clock optimal parameters are found by minimizing the error between the simulated waveform at the channel output vs the expected using methods such as Least-Mean-Square (LMS). Then the best set or sets of TX and RX equalization and clocking parameters are found by minimizing a figure of merit (FOM) such as eye-width (EW), eye-height (EH), eye-area (EA), or signal to noise ratio (SNR). Note that different FOMs will yield different optimal solutions. The second method is RX first, TX second. In the second method, a set or sets of RX equalization and recovered clock parameters are found, then each set or sets of parameters will cycle through all the possible TX equalization and clocking parameters to find the best set or sets of TX and RX equalization and clocking parameters via a FOM. The second method needs a back-channel to facilitate the adaptation. There could be other adaptation methods than those two mentioned. Further, there could be multiple sets of optimized equalization, clocking, and recovered clock parameters possible, depending on the optimization algorithm and FOM used.

F. *Simulation examples and comparisons*

Figure 6 shows an example of a full-waveform method, for the same PCIe3.0 channel shown in figure 3, but with a prbs2^23 data pattern (8 million bits). DCD, CG PLL and associated BUJ, PJ, RJ, and reference clock jitter, deterministic noise (DN), and RN for TX, insertion loss (IL), crosstalk (xtalk), and return loss (RL) for channel, and DCD, CR PLL and associated BUJ, PJ, RJ, DN, and RN for RX, as well as TX jitter and noise interactions with channel, are all comprehended. This is not possible in a statistical method. The simulation took 3.5 hours to finish, with the same computer system used in section II for the statistical method. Note that the BER CDF at probability level of 1e-6 or lower is extrapolated in Q-space using dual-Dirac jitter model [10]. The EW and EH are 0.47 UI, and 69 mv at BER of 1e-12.

978-1-4673-6145-3/13 $31.00 © 2013 IEEE

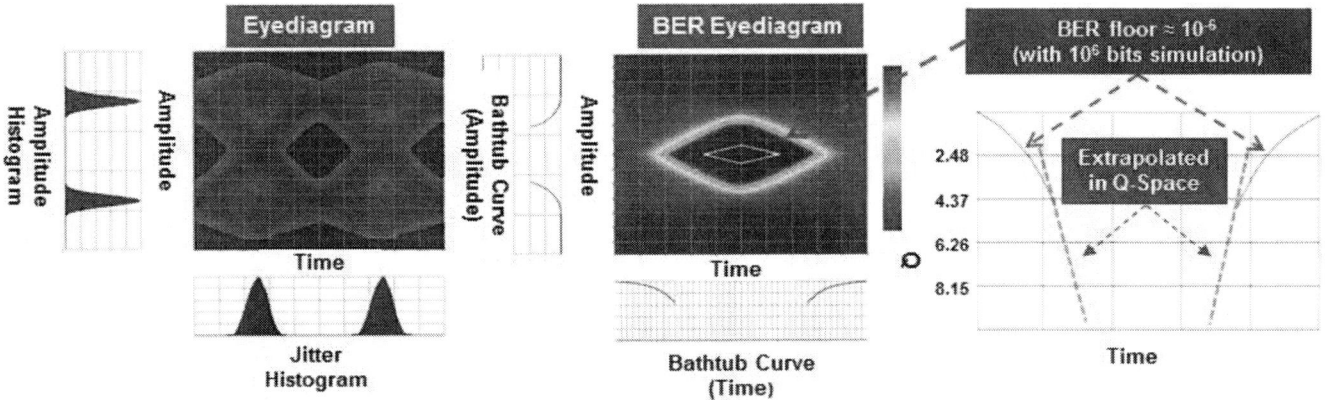

Fig. 6. Eye-Diagram and Associated Jitter/Noise PDFs, BER Eye and Associated Jitter/Noise CDFs, and a Zoom-In CDF in Q-Space.

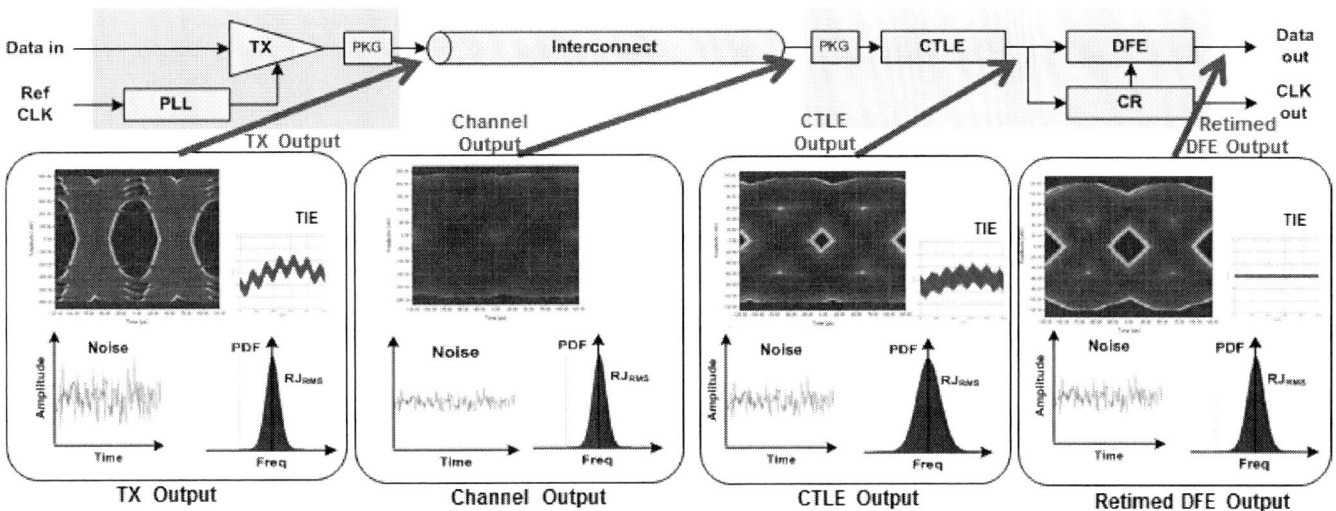

Fig. 7. Eye-Diagram, Noise vs Time, TIE, RJ PDF at Various Points with the Link.

Figure 7 shows a simulation example of hybrid method, for the same PCIe3.0 channel. 0.065 million bits are simulated and the worst ISI edge transition is covered, with the same device parameters as those used for figure 6. The simulation time is reduced to 3.5 minutes. Eye-diagrams, amplitude noise vs time function, time-interval error (TIE) vs time function, RJ PDF, at various observing points within a link are shown. At TX output, the eye-diagram is pre-emphasized/de-emphasized and low-frequency sinusoidal jitter can be seen in the TIE vs time plot. At channel output, the eye is closed due to the lossy channel and associated significant amount of ISI, even with TX FFE turned on. TIE is not shown and has no meaning as eye is closed. Amplitude noise is also reduced due to the lossy channel, while the RJ PDF width is also slightly reduced. At the CTLE output, eye is opened by the CTLE significantly, while the jitter vs time function and RJ PDF width get larger due to the CTLE amplification, compared with those at channel output. At the DFE output, eye gets further opened, and amplitude noise is reduced, due to the DFE signal only boost. Meanwhile, RJ PDF width, and TIE are significantly reduced due to CR low-frequency jitter tracking. The EW and EH are 0.45 UI and 66 mv respectively at BER 1e-12. In comparison with those obtained with full-waveform method, the difference are -0.02 UI (-4.3%) and -3 mv (-4.3%). Note that the hybrid shows slight pessimistic results compared with those of full-waveform, and this is due the fact that in

full-waveform, 8 million bits were simulated, corresponding to a "deeper" jitter tracking down to ~KHz, while for the hybrid, only 0.065 million bits were simulated, corresponding to a not so "deeper" jitter tracking down to ~100 KHz. On the other hand, statistical method cannot show any of the time-domain capabilities and jitter/noise channel interactions. We did not compare results from full-waveform and hybrid method with those from statistical method as they would not be apple-to-apple, due to the fact that statistical method cannot comprehends CG, CR, DR, and TX jitter/noise interactions with channel and RX, and automatic adaptation.

With the discussions of the statistical, full-waveform, and hybrid methods, we now can give a relative characteristic comparison in Table I, as absolute comparison requires specific simulation conditions/assumptions (as we have shown) and the results would be hard to generalize.

978-1-4673-6145-3/13 $31.00 © 2013 IEEE   865

Table I. Comparison between Different Simulation Methods

	Statistical	Hybrid	Full-Waveform
Eye-Diagram PDF/BER CDF	Y	Y	Y
Rise/Fall Time	N	Y	Y
Waveform/Transient	N	Y	Y
Jitter/Noise Components	Partial	Y	Y
TX Jitter/Noise Interaction with Channel and RX	N	Y	Y
CG, CR, DR	N	Y	Y
Automatic Adaptation	N	Y	Y
Overall Accuracy (Relative)	1	Better	Best
Overall Throughput (Relative)	1	Longer	Longest

IV. COMPREHENDING PROCESS, VOLTAGE, AND TERMPERATURE (PVT) VARIATIONS

If behavioral TX/RX models are based on generic LTI math representation (e.g., [5]), then they cannot comprehend the TX/RX PVT variations, a practical subject that a link design and simulation must address. If, however, the behavioral TX/RX models are developed based on SPICE, they are commonly based on typical PVT and its variations are not accounted for either.

To comprehend the device PVT variations, behavioral TX/RX models must be derived from the SPICE or equivalent models that are capable of comprehending transistor and IC PVT variations. This may be achieved by creating a bank of TX/RX models covering PVT variation space and corners. However, in practice, SPICE models can only comprehend deterministic behavior such as deterministic waveform and associated ISI, offering little or none knowledge on the IC/device non-deterministic characteristics such as jitter and noise and their PVT variations. Furthermore, jitter and noise of the TX/RX depend on IC/device operating conditions and are often affected by neighbor circuit activities, and this is illustrated in figure 8.

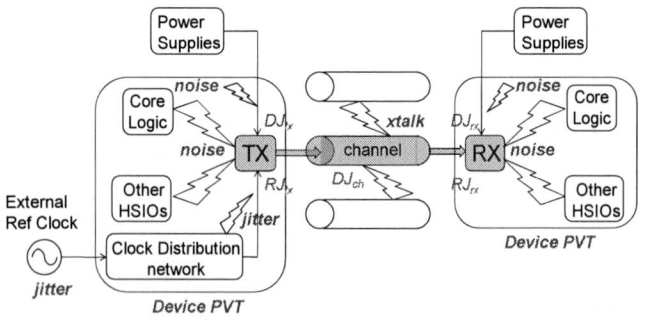

Fig. 8. Illustration of Sources of Variability for a High-Speed Link.

To account for the jitter and noise variability over operating conditions and PVT, a novel measurement based method has been developed recently. Early work in this area covers jitter only [13], and in this paper we extend the coverage to both jitter and noise. In this method, jitter and noise components of deterministic jitter (DJ), RJ,

deterministic noise (DN), and RN are measured over a wide range of TX and RX operating conditions, over various devices and channels, and over PVTs. The common operating condition/parameters for TX/RX may include, but not limited to: data pattern (pat), data rate (f_d), logic core and neighboring channel activity level. The specific operating conditions may include various output voltage (V_o), PLL bandwidth (f_{pll}), reference clock frequency (f_{osc}) for TX, and input voltage (V_i), clock recovery PLL bandwidth (f_{cr}) for RX. Multiple-dimensional jitter and noise component LUTs are created after the comprehensive measurements are made, which can be conceptually represented by the following equations:

$$(DJ, RJ, DN, RN)_{TX} = LUT_{TX}(pat, f_d, V_o, f_{pll}, f_{osc}, PVT) \quad (12)$$

for TX, and

$$(DJ, RJ, DN, RN)_{RX} = LUT_{RX}(pat, f_d, V_{id}, f_{cr}, PVT) \quad (13)$$

for RX. To manage the large amount of possible permutations and corresponding measurement time for the TX and RX LUTs, only carefully selected discrete sets of permutation for the measurements are carried out, and a continuous coverage of LUTs is achieved via interpolation and extrapolation techniques. Influence of environmental, neighboring activities, device-to-device, channel-to-channel variations are combined and degenerated to three classes of best, typical and worst.

Figure 9 is a snapshot example for a sub-dimensional TX RJ LUT.

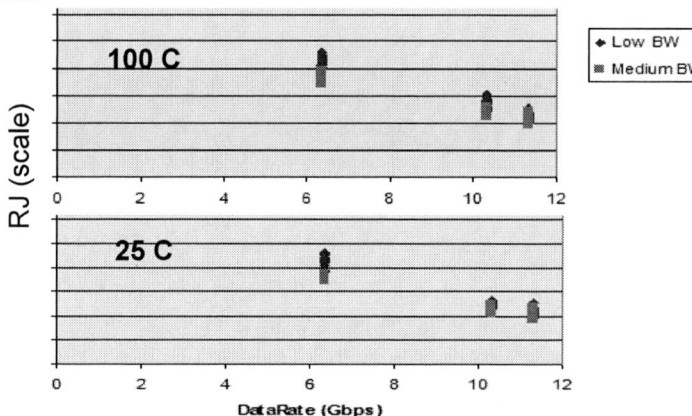

Fig. 9. A Sub-Dimension Example of TX RJ LUT

The accuracy of DJ, RJ, DN, and RN in the LUTs are largely determined by the laboratory instruments, and that is ~150 fs for timing, and ~1 mv for voltage, for the leading edge ones.

With PVT and operating condition variability coverage capable LUTs, deterministic behavioral models, and channel S-parameters all incorporated with the full-waveform, hybrid, or statistical method, we have achieved the goal of comprehending those variabilities for the high-speed link simulation.

V. ACCURACY AND CORRELATION

An importance performance merit for a simulator is its accuracy. For a high-speed simulator, the ultimate golden reference used to determine the accuracy/correlation is the actual measurements. There are two aspects associated with this subject, one is accuracy improvement, another is the accuracy validation, both are related and will be discussed in this section.

We will discuss accuracy improvement by case study as generic accuracy improvement for the entire link and all the circuit blocks is a big topic and beyond the scope of this paper. We choose TX driver and FFE equalization model accuracy improvement as the case study. In a behavioral modeling, the basic TX driver and FFE modeling consists of two steps. Step 1 is to model its edge shaping linear filter without FFE. Step-2 is to model FFE behavior, namely FFE voltage levels setting vs. tap-coefficients. When FFE is enabled, the behavior of TX driver plus FFE will have non-linear effects such as peak current limitation of the IC [2] that needs to be accounted for beyond the LTI. We call this "static" correction (over LTI) comprehending this non-linearity effect. Figure 10 shows an example illustrating the procedure. In figure 10a, we show two waveform zoom-ins from behavioral model vs from HSPICE model without the FFE turned-on, and we get a good correlation between those two, indicating LTI works well in modeling the driver. In figure 10b, we show three waveform zoom-ins when FFE is turned-on with 1st post-tap. Noticed that a really good correlation between behavioral model with non-linearity accounted for and HSPICE model, and not so good correlation between behavioral model without non-linearity accounted for and HSPICE model.

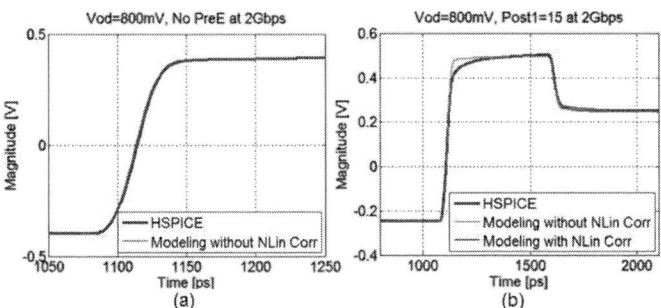

Fig. 10. (a) TX Waveforms from LTI Behavior (red) vs HSPICE, (b) TX Waveforms from LTI Behavior (red), vs Behavior accounts for Non-Linearity (green), vs HSPICE.

In figure 10 case study, the accuracy/correlation improvement via non-linearity comprehending was done based on lower data rate of 2 Gbps. It turns out that this non-linearity effect is data rate dependent, and simple, "static" non-linearity comprehending does not work well at higher data rate. What needed is data rate dependent or "dynamic" non-linearity comprehending. Figure 11 (a) and (b) show an example of how "dynamic" non-linearity comprehending improves the accuracy/correlation of the behavioral model waveform prediction vs the reference obtained via measurement by an oscilloscope using average mode, at a higher data rate of 14.1 Gbps. Noticed that the "static" non-linearity comprehending developed from lower data rate does not extend well to higher data rate.

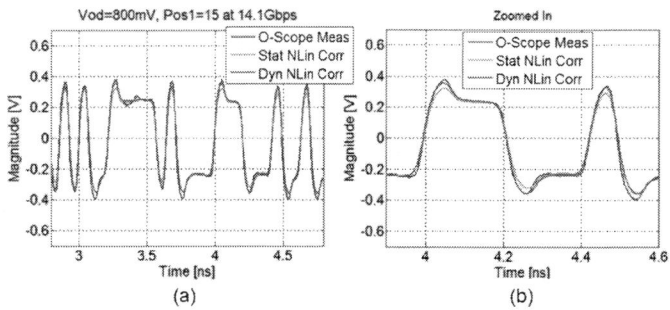

Fig. 11. (a) Waveform Predictions with Behavior Model Using "Static" Non-Linearity Comprehending (green), vs "Dynamic" Non-Linearity Comprehending (red), vs Measurement; (b) Zoom-in Comparison.

Figures 10 and 11 are two case-study examples demonstrating the accuracy/correlation improvement via developing advanced behavioral models for TX driver and FFE circuit blocks and their comparisons with HSPICE or measurements. This procedure needs to be done in an iterative manner. Obviously, this technique and other new techniques beyond LTI need to apply to other circuit blocks of the link in order to achieve a good accuracy and correlation. We summarize the general behavioral model accuracy/correlation improvement and validation procedure in figure 12.

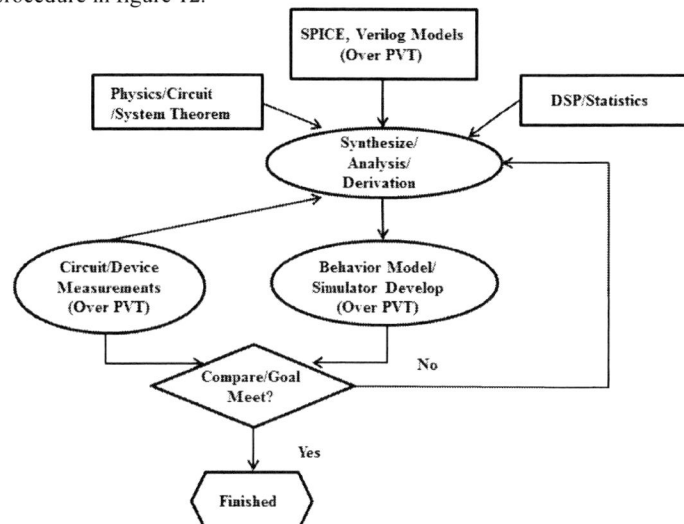

Fig. 12. A Flow-Chart for Accurate/Correlated Behavioral Model/Simulator Development

With a thorough practice following the flow-chart procedure in figure 12, a good accuracy and correlation can be achieved. Figure 13 shows two measured eye-diagrams vs two simulated ones using hybrid method overlaying on each other, where TX FFE is enabled, with jitter, noise, non-linearity effects all accounted for. The comparison data clearly shows a good match, indicating a good accuracy/correlation.

978-1-4673-6145-3/13 $31.00 © 2013 IEEE

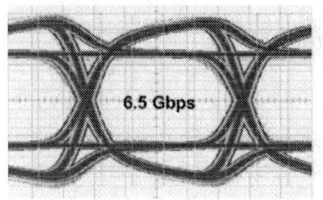 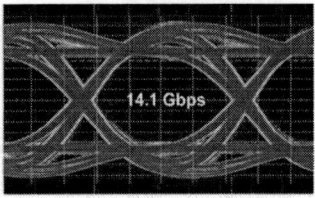

········ Measurement
——— Simulation

Fig. 13. Measured Eye-Diagrams (yellow, back), Overlayed with Simulated Ones (red, front) for 6.5 Gbps and 14.1 Gbps When TX FFE is Enabled.

VI. SUMMARY AND CONCLUSIONS

We have reviewed the high-speed I/O and link trend and technology roadmap that doubles its speed at a pace of 2-3 years. We then reviewed the challenges and progress paths for the corresponding simulation technologies. We further dove-in the details of the statistical method, reviewed its capabilities and limitations and challenges faced in meeting some of the recent emerging requirements, as the link speed and complexity both grow in recent years.

We introduced two new methods of full-waveform and hybrid, with full-waveform in time-domain, capable of comprehending most of the circuit blocks within a link, and hybrid in both time and statistical domains, striking a good balance of maintaining the advantages of both time and statistical domains, while overcoming their limitations. We identified and demonstrated the new capabilities beyond the statistical that both full-waveform and hybrid can offer and challenges they can solve. As every technique has advantages and limitations, we gave a summary comparison between those three simulation methods.

We discussed a practical and important, yet relatively under-studied/investigated subject of PVT and operating condition variation comprehending for high-speed simulation. We discussed a new modeling method that comprehends the deterministic IC/device PVT variations based on a bank of SPICE models, and non-deterministic jitter and noise PVT variations based on LUTs constructed from actual measurements under various operating conditions.

With simulation technology advancements via full-waveform and hybrid methods and PVT comprehending via SPICE and measurement based jitter and noise LUT, we discussed the accuracy and correlation for high-speed simulation. Like the PVT, this is also a practical and important, yet relatively under-studied subject. Methodology for accuracy/correlation improvement and validation are discussed, along with case study examples. Good correlation data for a hybrid method with PVT coverage is demonstrated.

As the industry starts to deploy Nx25G high-speed links today, and Nx50G links in the future, new simulation requirements and challenges (e.g., modulation scheme beyond NRZ such as PAMn, electro-optically integrated SERDES) will emerge. Continue advancements and innovations are required and expected for high-speed link simulation technology.

REFERENCES

[1] ITRS roadmap, *ITRS*, 2011 Edition:
www.itrs.net/Links/2011ITRS/Home2011.htm

[2] M. Li and S. Shumarayev, "Emerging Standards at ~ 10 Gbps for Wireline Communications and Associated Integrated Circuit Design and Validation", *IEEE Custom Integrated Circuits Conf. Procedings*, pp. 105-112, 2009

[3] "A 28 Gbps 4-Tap FFE/15-Tap DFE Serial Link Transceiver in 32 nm SOI CMOS technology", *IEEE Int. Solid-State Circuits Conference Digest*, pp. 324-325, 2012.

[4] B. Casper, M. Haycock, and R. Mooney, "An Accurate and Efficient Analysis Method for Multi-Gb/s Chip-to-Chip Signaling Schemes", *IEEE Very Large Scale (VLSI) Circuits Symp*, pp. 54–57, 2002.

[5] V. Stojanovic and M. Horowitz, "Modeling and Analysis of High Speed Links", *IEEE Custom Integrated Circuits Conf. Proceding*, pp. 589–594, 2003.

[6] A. Sanders, M. Resso, and J. D'Ambrosia, "Channel Compliance Testing Utilizing Novel Statistical Eye Methodology," *DesignCon* 2004.

[7] IBIS-AMI standard web site, http://www.vhdl.org/ibis

[8] B. Casper *et al.*, "Future Microprocessor Interfaces: Analysis, Design and Optimization", *IEEE Custom Integrated Circuits Conf. Procedings*, pp. 479–486., 2007.

[9] G. Balamurugan, B. Casper, J. Jaussi, M. Mansuri, F. O'Mahony, and J. Kennedy, "Modeling and analysis of high-speed I/O links," *IEEE Transactions on Advanced Packaging*, vol. 32, no. 2, 237–247., 2009.

[10] M. Li, "Jitter, Noise, and Signal Integrity at High-Speed", *Prentice Hall*, 2007.

[11] M. Shimanouchi, M. Li, H. Wu, "Comparison of Two Statistical Methods for High Speed Serial Link Simulation", *Designcon*, 2013.

[12] F. M. Gardner, "Phaselock Techniques", *John Wiley & Sons*, 2nd Edition, 1979.

[13] M. Li, and M. Shimanouchi, "New Hybrid Simulation Method for Jitter and BER in High-Speed Links", *Designcon*, 2011.

Structure-Aware High-Dimensional Performance Modeling for Analog and Mixed-Signal Circuits

Shupeng Sun[1], Xin Li[1] and Chenjie Gu[2]

[1]Electrical & Computer Engineering Department, Carnegie Mellon University, Pittsburgh, PA, USA, 15213
[2]Intel Strategic CAD Labs, Hillsboro, OR, USA, 97124

Abstract—Efficient high-dimensional performance modeling of nanoscale analog and mixed signal (AMS) circuits is extremely challenging. In this paper, we propose a novel structure-aware modeling (SAM) technique. The key idea of SAM is to accurately solve the model coefficients by applying an efficient statistical algorithm to exploit the underlying structure of AMS circuits. As a result, SAM dramatically reduces the required number of sampling points and, hence, the computational cost for performance modeling. Several circuit examples designed in a commercial 32nm CMOS process demonstrate that SAM achieves more than 2× runtime speedup over the traditional sparse regression technique without surrendering any accuracy.

I. INTRODUCTION

The aggressive technology scaling of integrated circuits (IC) leads to large-scale process variations [1], significantly impacting the parametric yield of analog and mixed-signal (AMS) circuits. As an important tool for variability analysis, performance modeling aims to approximate the circuit-level performance (e.g., frequency of ring oscillator) as an analytical (e.g., linear, quadratic, etc) function of device-level variations (e.g., ΔV_{th}, ΔL, etc) [2]-[3]. Once such performance models are available, they can be applied to several important applications such as yield estimation [4], worst-case corner extraction [5], design optimization [6], etc.

Although performance modeling was extensively studied in the past, several new technical challenges arise due to the recent evolution of nanoscale IC technologies. Today, a large number of random variables must be employed to accurately capture device-level variations. For example, more than 40 random variables are used to model the random mismatches for a single transistor in a commercial 32nm CMOS process. Even if we consider a small AMS circuit with 100 transistors only, there exist more than 4000 random variables for device-level mismatch modeling. In addition, it is almost impossible to pre-select a subset of these random variables for variability analysis, since the impact of device mismatches is circuit- and performance-dependent. It, in turn, requires us to fit a high-dimensional performance model based on a large number of sampling points that must be generated by expensive transistor-level simulations [7]. A major challenge here is how to minimize the required number of sampling points so that the modeling cost can be substantially reduced.

Recently, it has been found that although there are a large number of unknown coefficients associated with a high-dimensional performance model for a given performance of interest (PoI), most of these coefficients are close to zero [7]. By exploiting this unique property, sparse regression technique has been developed for high-dimensional performance modeling [7]. The key idea is to automatically identify the important (i.e., non-zero) model coefficients by applying a statistical algorithm. As such, a high-dimensional performance model can be accurately fitted with a small number of sampling points.

In this paper, we further improve the traditional sparse regression method and propose a novel *structure-aware modeling* (SAM) technique for AMS circuits. Our proposed method is motivated by the observation that many coefficients of a performance model often share similar magnitude. For instance, consider a ring oscillator that is composed of identically-sized inverters. In this example, the threshold mismatches (i.e., ΔV_{th}'s) of all NMOS (or PMOS) transistors have similar contribution to the performance variability due to its symmetric circuit topology. Therefore, the model coefficients associated with these ΔV_{th}'s should be similar.

Based on this observation, SAM attempts to group the "similar" model coefficients and solve them together. Towards this goal, a new algorithm, referred to as *simultaneous orthogonal matching pursuit* (S-OMP), is adopted from the statistics community [8] and applied to our high-dimensional performance modeling problem. In addition, a number of important heuristic methodologies are proposed in order to make S-OMP of great efficiency, as will be discussed in detail in Section II. Several circuit examples in Section III demonstrate that SAM achieves more than 2× runtime speedup over the traditional sparse regression technique without surrendering any accuracy.

The remainder of this paper is organized as follows. In Section II, we describe SAM and the S-OMP algorithm integrated with our proposed heuristics. The efficacy of SAM is demonstrated by several circuit examples in Section III. Finally, we conclude in Section IV.

II. METHODOLOGY

Without loss of generality, we assume that there are N device-level random variables $\boldsymbol{\varepsilon} = [\varepsilon_1 \ \varepsilon_2 \ ... \ \varepsilon_N]^T$ modeling process variations. Given a PoI denoted as y, performance modeling aims to construct an analytical function to capture the relation between y and $\boldsymbol{\varepsilon}$. In this paper, we use the following linear performance model as an example to illustrate our proposed SAM methodology:

$$y = \alpha_0 + \sum_{n=1}^{N} \alpha_n \cdot \varepsilon_n = \boldsymbol{\alpha}^T \cdot \mathbf{x}, \quad (1)$$

where $\boldsymbol{\alpha} = [\alpha_0 \ \alpha_1 \ ... \ \alpha_N]^T$ represents the model coefficients and $\mathbf{x} = [1 \ \varepsilon_1 \ \varepsilon_2 \ ... \ \varepsilon_N]^T$ contains the constant term 1 and the random variables $\boldsymbol{\varepsilon}$. It should be noted that SAM can be further extended to other nonlinear performance models, even though the details of nonlinear performance modeling are not discussed in this paper due to the page limit.

The unknown model coefficients $\boldsymbol{\alpha}$ in (1) are often solved from a set of sampling points. In particular, if M sampling points are collected, the following linear equation is formulated to solve $\boldsymbol{\alpha}$:

$$\mathbf{X} \cdot \boldsymbol{\alpha} = \mathbf{y}, \qquad (2)$$

where

$$\mathbf{X} = \begin{bmatrix} 1 & \varepsilon_1^{(1)} & \cdots & \varepsilon_N^{(1)} \\ 1 & \varepsilon_1^{(2)} & \cdots & \varepsilon_N^{(2)} \\ \vdots & \vdots & \vdots & \vdots \\ 1 & \varepsilon_1^{(M)} & \cdots & \varepsilon_N^{(M)} \end{bmatrix} \quad \text{and} \quad \mathbf{y} = \begin{bmatrix} y^{(1)} \\ y^{(2)} \\ \vdots \\ y^{(M)} \end{bmatrix}. \qquad (3)$$

In (3), $\varepsilon_n^{(m)}$ and $y^{(m)}$ denote the values of ε_n and y for the m-th sampling point, respectively.

For today's nanoscale AMS circuits, the total number of unknown model coefficients can be extremely large (e.g., $10^3 \sim 10^4$). However, most of these coefficients are close to zero, as is demonstrated in the literature [7]. In addition, a number of the model coefficients often share similar magnitude. For instance, consider the latch-based comparator in Fig. 1 as an example. In this compactor, each transistor contains 10 multipliers. If we consider the input offset as PoI, the model coefficients associated with the threshold mismatches (i.e., ΔV_{th}'s) of all multipliers of the two input transistors M_1 and M_2 should have similar magnitude, since these threshold mismatches have similar contribution to the variability of PoI. While such structure-based information defined by circuit topology is completely ignored by the traditional performance modeling techniques [7], it will be exploited in this paper to improve the modeling accuracy and/or reduce the modeling cost.

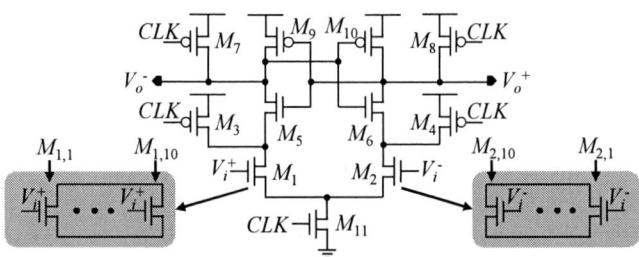

Fig. 1. Simplified circuit schematic is shown for a latch-based comparator where each transistor has 10 multipliers. For illustration purpose, the multipliers of the two input transistors M_1 and M_2 are highlighted in grey color.

In particular, we propose to "group" the model coefficients that are expected to have similar magnitude. These groups are formed purely based on circuit topology and design knowledge without running any transistor-level simulation. The aforementioned grouping strategy would help us to accurately identify the important (i.e., non-zero) model coefficients based on very few sampling points. Our proposed structure-aware performance modeling technique consists of two major steps: (i) model coefficient grouping, and (ii) model coefficient fitting. In what follows, we will describe the technical details of these two steps and highlight the novelty.

A. Model Coefficient Grouping

Our objective is to partition all model coefficients into several groups where the coefficients within the same group share similar magnitude. If a model coefficient has different magnitude from all other coefficients, it should form a separate group containing a single coefficient only. In this paper, the grouping step is performed by taking the design knowledge from a user as the input. The following shows two examples how coefficient groups can be formed.

- **Example I**: The same mismatch variables (e.g., ΔV_{th}'s) from all multipliers of the same transistor can be grouped together. Taking the comparator in Fig. 1 as an example, the threshold variations of all multipliers for each device (e.g., $\{\Delta V_{th1,k}; k = 1, 2, ..., 10\}$ for the transistor M_1) can be grouped together. Here, we assume that all multipliers of the same transistor are placed close to each other and there is no significant systematic difference between these multipliers. Hence, the mismatch variables of all multipliers will have a similar impact on the variability of PoI. Note that once the layout of a circuit is given, setting up the coefficient groups in this example is straightforward and it is almost independent of the circuit functionality.

- **Example II**: The same mismatch variables (e.g., ΔV_{th}'s) from different transistors can be grouped together, if our prior knowledge reveals that these mismatch variables have similar contribution to the variability of PoI. Again, taking the comparator in Fig. 1 as an example, it is easy to note that the mismatch variables of the two input transistors M_1 and M_2 have a similar impact on the variability of input offset. Hence, the mismatch variables $\{\Delta V_{th1,k}; k = 1, 2, ..., 10\}$ for M_1 and $\{\Delta V_{th2,k}; k = 1, 2, ..., 10\}$ for M_2 can be grouped together, if the input offset is considered as PoI. Compared to Example I, the grouping strategy in this example is circuit- and performance-dependent. It is based upon a deep understanding of the circuit operation.

Once the coefficient groups are identified, a statistical algorithm should be further applied to select the important groups containing non-zero model coefficients and then solve these coefficient values based on a small number of sampling points. In the next sub-section, a modified S-OMP algorithm will be introduced to address this coefficient fitting problem.

B. Model Coefficient Fitting

The key idea of S-OMP [8] is to identify the non-zero model coefficients based on the "correlation" between each group and the PoI. To illustrate the mathematical formulation of S-OMP, we first re-order all columns of the matrix $\mathbf{X}$ in (3) based on the coefficient groups. Namely, the matrix $\mathbf{X}$ is re-arranged as:

$$\mathbf{X} = \begin{bmatrix} \mathbf{x}_{1,1} & \cdots & \mathbf{x}_{1,N_1} & \cdots & \mathbf{x}_{K,1} & \cdots & \mathbf{x}_{K,N_K} \end{bmatrix}, \qquad (4)$$

where the vector $\mathbf{x}_{k,n}$ represents one column of the matrix $\mathbf{X}$ and all vectors $\{\mathbf{x}_{k,n}; n = 1, 2, ..., N_k\}$ belonging to the k-th group are adjacent to each other in the re-ordered matrix $\mathbf{X}$. After re-ordering, the matrix $\mathbf{X}$ in (4) shows a unique structure that contains K groups in total and N_k vectors associated with the k-th group.

Next, S-OMP selects the most important group based on the following criterion:

$$\arg\max_k \quad g_k = \frac{1}{N_k} \cdot \sum_{n=1}^{N_k} \left| \left\langle \mathbf{x}_{k,n}, \mathbf{y} \right\rangle \right|, \qquad (5)$$

where the operator $\langle \bullet, \bullet \rangle$ calculates the inner product between two vectors, and g_k stands for the "correlation score" for the k-th group. Intuitively, if the score g_k is large, it implies that all terms (either the constant term or the device-level random variables) within the k-th group are strongly correlated with the PoI $\mathbf{y}$. Hence, the corresponding model coefficients are likely to be non-zero, and the k-th group should be identified as an important group.

The correlation score g_k in (5) is derived based upon the assumption that we do not know the sign of the correlation (i.e., either positive or negative) between $\mathbf{x}_{k,n}$ and $\mathbf{y}$. In this case, the "average" correlation is calculated for the absolute value of the individual correlation. In practice, if we know the "relative sign" of the correlation (i.e., whether the correlation between $\mathbf{x}_{k,n}$ and $\mathbf{y}$ has the same sign for any $n \in \{1, 2, ..., N_k\}$), such sign information can be explicitly incorporated into the formulation in (5), resulting in the following criterion to calculate the correlation score:

$$\arg\max_k \quad g_k = \frac{1}{N_k} \cdot \left| \sum_{n=1}^{N_k} s_{k,n} \cdot \left\langle \mathbf{x}_{k,n}, \mathbf{y} \right\rangle \right|, \qquad (6)$$

where $s_{k,n} \in \{-1, 1\}$ represents the relative sign of the correlation between $\mathbf{x}_{k,n}$ and $\mathbf{y}$. For instance, if the correlation between $\mathbf{x}_{k,1}$ and $\mathbf{y}$ and the correlation between $\mathbf{x}_{k,2}$ and $\mathbf{y}$ share the same sign, $s_{k,1}$ and $s_{k,2}$ should be set to the same value. Otherwise, they should be set to two opposite values.

Compared to (5), Eq. (6) first calculates the average of the individual correlation, and then takes the absolute value of the average correlation. As such, the accuracy of the estimated correlation score g_k can be substantially improved. To understand the reason, we consider a simple example where the correlation between $\mathbf{x}_{k,n}$ and $\mathbf{y}$ is zero for any $n \in \{1, 2, ..., N_k\}$. In this case, the inner product $\langle \mathbf{x}_{k,n}, \mathbf{y} \rangle$ may not be exactly zero, since it is calculated from a small number of sampling points. Hence, if the correlation score g_k is estimated by (5), the average of the absolute value of the individual correlation is not equal to zero. On the other hand, the correlation score g_k calculated by (6) can be much closer to zero (i.e., the true correlation value), since it attempts to average out the error of the individual correlation. The difference between (5) and (6) will be further discussed in Section III, along with our numerical examples.

Once an important group is identified by S-OMP, all model coefficients associated with this group are considered to be non-zero. The performance model in (1) is fitted with all other model coefficients set to zero. Next, the modeling error is calculated and an additional group is identified by S-OMP to maximally reduce the modeling error. The aforementioned iteration steps are repeatedly performed until the modeling error is sufficiently small. Algorithm 1 summarizes the major steps of the proposed SAM technique based on S-OMP. Due

to the page limit, more details about S-OMP are not included in this paper, but they can be found in [8].

Algorithm 1: Structure-Aware Modeling (SAM)
1. Start from a set of sampling points and formulate the linear equation in (2).
2. Group the model coefficients based on circuit structure and design knowledge.
3. Re-order the matrix $\mathbf{X}$ to the form of (4) where the column vectors within the same group are adjacent to each other.
4. Set the residual $\mathbf{r} = \mathbf{y}$ and the set $\Omega = \{\}$.
5. If the sign information is unknown for the correlation between $\mathbf{x}_{k,n}$ and $\mathbf{y}$, select the most important group (say, the k-th group) by using:

$$\arg\max_k \quad g_k = \frac{1}{N_k} \cdot \sum_{n=1}^{N_k} \left| \left\langle \mathbf{x}_{k,n}, \mathbf{r} \right\rangle \right|. \qquad (7)$$

Otherwise, if the sign information is known, select the most important group (say, the k-th group) by using:

$$\arg\max_k \quad g_k = \frac{1}{N_k} \cdot \left| \sum_{n=1}^{N_k} s_{k,n} \cdot \left\langle \mathbf{x}_{k,n}, \mathbf{r} \right\rangle \right|. \qquad (8)$$

6. Update the set $\Omega = \Omega \cup \{(k, 1), (k, 2), ..., (k, N_k)\}$.
7. Determine the model coefficients defined by the set Ω by solving the following least-squares fitting problem:

$$\arg\min_{\alpha_{k,n}; (k,n) \in \Omega} \left\| \mathbf{y} - \sum_{(k,n) \in \Omega} \alpha_{k,n} \cdot \mathbf{x}_{k,n} \right\|_2^2. \qquad (9)$$

8. Update the residual $\mathbf{r}$:

$$\mathbf{r} = \mathbf{y} - \sum_{(k,n) \in \Omega} \alpha_{k,n} \cdot \mathbf{x}_{k,n}. \qquad (10)$$

9. If the residual is sufficiently small, stop iteration and set $\alpha_{k,n} = 0$ for any $(k, n) \notin \Omega$. Otherwise, go to Step 5.

III. EXPERIMENTAL RESULT

In this section, two circuit examples designed in a commercial 32nm CMOS process are used to demonstrate the efficacy of the proposed SAM method. Our objective is to build linear performance models to study the circuit-level performance variability with respect to device mismatches. Such a performance modeling problem is non-trivial, because device mismatches must be modeled by a large number of independent random variables, rendering a high-dimensional variation space. For testing and comparison purposes, two different techniques are implemented: (i) the traditional sparse regression method based on OMP [7], and (ii) the proposed SAM method (i.e., Algorithm 1). For each method, we measure the modeling error from 50 repeated runs where the sampling points are independently generated for each run. All numerical experiments are performed on a 2.5GHz Linux server with 64GB memory.

A. Ring Oscillator

Shown in Fig. 2 is the simplified circuit schematic of a ring oscillator consisting of 21 identically-sized inverters. There are 2835 independent random variables defined by the process design kit to model device mismatches. The oscillation frequency is considered as the PoI in this example.

Since all inverters in Fig. 2 are identically sized, the random mismatches of all NMOS (or PMOS) transistors should have a similar impact on the frequency variability. Hence, the mismatch variables associated with all NMOS (or PMOS) transistors are grouped together. Such a grouping strategy is similarly applied to the load capacitors that are connected to the output of each inverter.

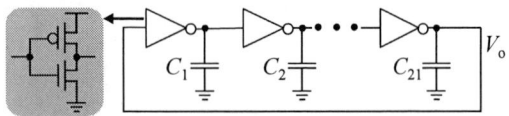

Fig. 2. Simplified circuit schematic is shown for a ring oscillator consisting of 21 identically-sized inverters.

In this example, the sign information can be easily defined. For all random variables within the same group, they play a similar role on the PoI and, hence, the corresponding model coefficients should have the same sign. Even though the sign information is available in this example, two versions of Algorithm 1, with and without the sign information respectively, are implemented and tested for comparison purposes.

Fig. 3(a) shows the modeling error as a function of the number of sampling points. Note that, given the same number of sampling points, SAM achieves superior modeling accuracy over OMP. On the other hand, to achieve the same modeling accuracy, SAM offers about 2× runtime speedup over OMP, as shown in Table I. It is also observed from Fig. 3(a) that if the sign information is incorporated into Algorithm 1, SAM is able to further reduce the modeling error, especially when the number of sampling points is small. This observation is consistent with our discussion in Section II.

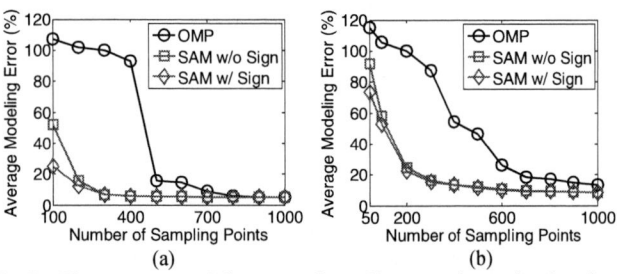

Fig. 3. The average modeling error from 50 repeated runs is plotted as a function of the number of sampling points: (a) ring oscillator, and (b) latch-based comparator.

TABLE I
COMPUTATIONAL COST FOR PERFORMANCE MODELING

	Ring Oscillator		Comparator	
	OMP	SAM	OMP	SAM
# of Sampling Points	800	400	1000	400
Modeling Error	5.72%	5.80%	13.33%	13.17%
Simulation Cost (Hour)	3.60	1.80	5.50	2.20
Fitting Cost (Sec.)	111.90	2.97	141.85	5.79
Total Cost (Hour)	3.63	1.80	5.54	2.20

B. Latch-based Comparator

As a second example, we consider the latched-based comparator shown in Fig. 1. In this example, there are 4950 independent random variables modeling device mismatches and the input offset is considered as our PoI. As previously mentioned in Section II, each transistor in Fig. 1 is composed

of 10 multipliers. The mismatch variables from all multipliers of the same transistor are grouped together. In addition, by considering the symmetry of the circuit topology, we group the mismatch variables corresponding to the following transistor pairs: $\{M_1, M_2\}$, $\{M_3, M_4\}$, $\{M_5, M_6\}$, $\{M_7, M_8\}$ and $\{M_9, M_{10}\}$. Based on the aforementioned grouping strategy, the sign information within each group can be derived according to the structure of the comparator.

Fig. 3(b) shows the modeling error as a function of the number of sampling points. Similar to the ring oscillator example, SAM requires substantially less sampling points than OMP to achieve the same modeling accuracy. Compared to OMP, SAM offers about 2.5× runtime speedup, as shown in Table I. However, unlike the ring oscillator example, the sign information does not make a significant difference in modeling accuracy for the comparator example. This observation is made, because the error posed by other steps in Algorithm 1 (e.g., the regression error of the least-squares fitting in Step 7 of Algorithm 1) dominates the overall modeling error, when the number of sampling points is extremely small. For this reason, even though the sign information helps to accurately identify the important groups, it does not substantially reduce the modeling error.

IV. CONCLUSION

In this paper, a novel SAM technique is proposed for structure-aware high-dimensional performance modeling of AMS circuits with consideration of process variations. The key idea of SAM is to group the model coefficients that have similar magnitude. As such, only an extremely small number of sampling points are required to fit a high-dimensional performance model, thereby substantially reducing the computational cost of performance modeling. As is demonstrated by our circuit examples designed in a commercial 32nm CMOS process, SAM achieves more than 2× runtime speedup over the traditional sparse regression technique.

V. REFERENCES

[1] Semiconductor Industry Associate, *International Technology Roadmap for Semiconductors*, 2011.

[2] X. Li, J. Le and L. Pileggi, *Statistical Performance Modeling and Optimization*, Now Publishers, 2007.

[3] T. McConaghy, "High-dimensional statistical modeling and analysis of custom integrated circuits," *IEEE CICC*, 2011.

[4] X. Li, J. Le, P. Gopalakrishnan and L. Pileggi, "Asymptotic probability extraction for nonnormal performance distributions," *IEEE Trans. on CAD*, vol. 26, no. 1, pp. 16-37, Jan. 2007.

[5] M. Sengupta, S. Saxena, L. Daldoss, G. Kramer, S. Minehane and J. Cheng, "Application-specific worst case corners using response surfaces and statistical models," *IEEE Trans. on CAD*, vol. 24, no. 9, pp. 1372-1380, 2005.

[6] A. Dharchoudhury and S. Kang, "Worse-case analysis and optimization of VLSI circuit performance," *IEEE Trans. on CAD*, vol. 14, no. 4, pp. 481-492, Apr. 1995.

[7] X. Li, "Finding deterministic solution from underdetermined equation: large-scale performance modeling of analog/RF circuits," *IEEE Trans. on CAD*, vol. 29, no. 11, pp. 1661-1668, Nov. 2010.

[8] J. A. Tropp, A. C. Gilbert, and M. J. Strauss, "Algorithms for simultaneous sparse approximation. Part I: Greedy pursuit," *J. Signal Process.*, vol. 86, pp. 572-588, Apr. 2006.

A $148fs_{rms}$ Integrated Noise 4MHz Bandwidth All-Digital Second-Order $\Delta\Sigma$ Time-to-Digital Converter Using Gated Switched-Ring Oscillator

Wonsik Yu, KwangSeok Kim, and SeongHwan Cho

Department of EE, KAIST

Daejon, Republic of Korea

Email: wonsik.yu@gmail.com, kskim99@gmail.com, chosta@ee.kaist.ac.kr

Abstract— This paper presents an all-digital second-order $\Delta\Sigma$ time-to-digital converter (TDC) by using switched-ring oscillator (SRO) and gated switched-ring oscillator (GSRO). Unlike conventional multi-stage noise-shaping (MASH) TDC using the SRO, the proposed TDC does not require complex calibration to compensate for the error from frequency difference between the SROs. The prototype TDC achieves $148fs_{rms}$ integrated noise and 80.4dB dynamic range in 4MHz signal bandwidth at 400MS/s while consuming 6.55mW in a 65nm CMOS process.

I. INTRODUCTION

High resolution time-to-digital converters (TDCs) are employed in digital PLLs, time-domain ADCs, jitter measurement and time-of-flight detection [1]–[3]. For low-bandwidth applications, $\Delta\Sigma$ TDCs exploiting noise-shaping property have been proposed to achieve high-resolution, high linearity. In [4], a GRO-TDC has been introduced as a first-order $\Delta\Sigma$ TDC, achieving picosecond time-resolution with 90dB dynamic range. Unfortunately, over-sampling ratio (OSR) of a GRO-TDC is limited by the input pulse rate (f_i) since sampling frequency (f_S) must be the same as f_i. Thus, the time-resolution is limited if the input pulse rate is low. To achieve high OSR for low input pulse rates, an SRO-TDC [5] was proposed, where f_S can be higher than f_i, resulting in a finer time-resolution. To further improve the time-resolution and signal bandwidth, a second-order MASH TDC using SROs has been proposed [8] since the OSR cannot be increased indefinitely. Unfortunately, a high-order $\Delta\Sigma$ TDC using SRO requires complex calibration to compensate for the error from frequency difference between the SROs. As a result, it consumes additional power and area as well as a long settling time [8].

In this paper, we propose a gated switched-ring oscillator (GSRO) which not only removes the need for complex calibration in high-order $\Delta\Sigma$ TDC but also allows for obtaining high OSR. Using the proposed GSRO, a novel second-order $\Delta\Sigma$ TDC is implemented, achieving low integrated noise and wide dynamic range with low complexity.

II. PROPOSED 1-1 MASH $\Delta\Sigma$ TDC USING GSRO

A 1-1 MASH TDC from [8] is shown in Fig. 1(a). The first stage SRO-TDC that performs first-order noise shaping is followed by a quantization error generator (QEGen) that produces

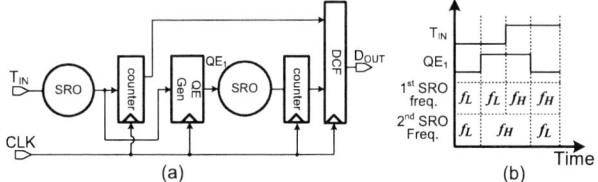

Fig. 1. Conventional implementation of the 1-1 MASH TDC using SRO. (a) Block diagram. (b) Timing diagram. f_L represents the minimum frequency and f_H represents the maximum frequency of the oscillator.

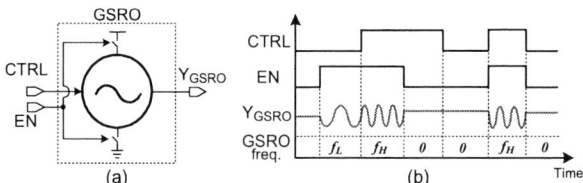

Fig. 2. (a) Block diagram and (b) Timing diagram of the GSRO.

a quantization error pulse. The first stage quantization error (QE_1) of width T_{QE1} is fed to the second stage SRO-TDC. Thus, it can be expected that this architecture will achieve second-order noise shaping. Unfortunately, there is frequency difference between the first and second stage oscillators during T_{QE1} as shown Fig .1(b), since frequency change of the first stage SRO cannot be tracked by the second stage SRO. Hence, an undesired gain is multiplied to QE_1, which destroys the second-order noise shaping [8]. Although the frequency difference between the first and second stage SROs can result from layout mismatch and manufacturing imperfections, frequency tracking error dominates the degradation of noise shaping. In [8], this problem is solved by using an off-chip calibration based on an LMS filter.

In this paper, we overcome this problem by proposing a GSRO whose operation principle is shown in Fig. 2. As can be seen, GSRO is basically an SRO with phase-holding gates added at supply and ground. Hence, GSRO acts as an SRO when the gates are closed and holds the phase like a GRO when the gates are open. Note that GSRO can also be con-

978-1-4673-6145-3/13 $31.00 © 2013 IEEE

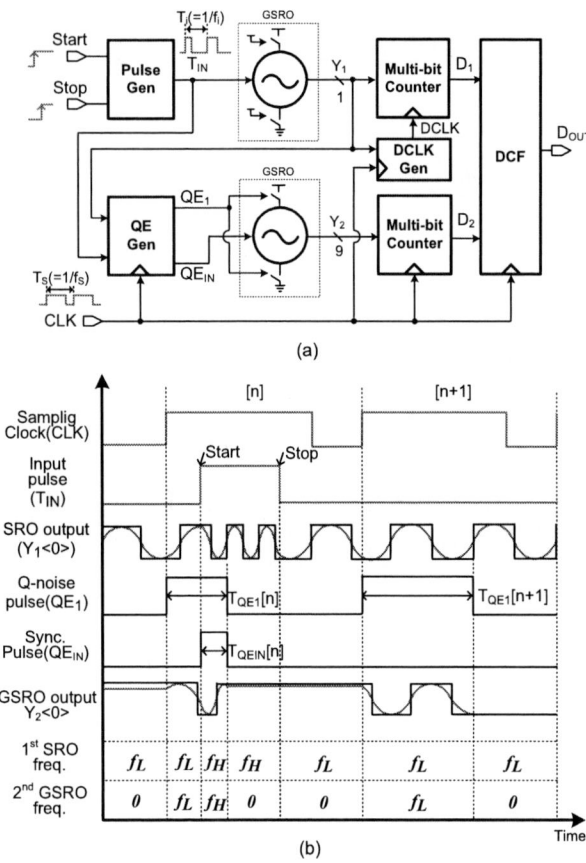

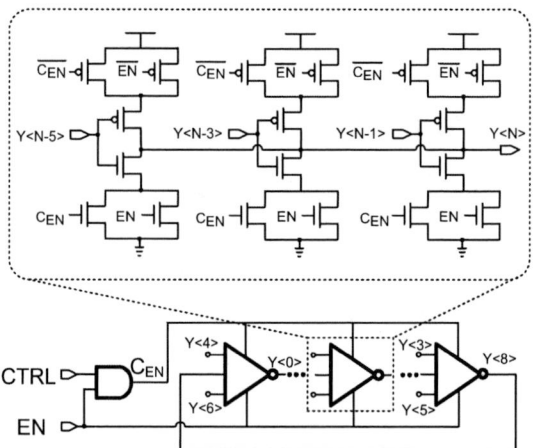

Fig. 4. Proposed 9-stage GSRO using multi-path structure.

Fig. 3. (a) Block diagram and (b) Timing diagram of the proposed $\Delta\Sigma$ TDC. Y_1 and Y_2 represent the output of the GSRO in the first and second stage, respectively.

sidered as frequency controllable GRO. The schematic of the GSRO is shown in Fig. 4, which is basically a gated delay-line with frequency control via CTRL. Multi-path structure [4] is applied to reduce the gating skew error due to leakage current. The block diagram and timing diagram of the proposed TDC using GSRO are shown in Fig. 3. In the first stage, GSRO is configured as an SRO by closing the EN gates. In the second stage, QEGen shown in Fig. 5(a) generates quantization error pulse QE_1 and a frequency sync pulse QE_{IN}. An offset is added to QE_1 to avoid narrow pulse width which leads to a deadzone problem. The offset is easily subtracted in the digital cancellation filter (DCF). Since QE_1 controls the gates of the GSRO and QE_{IN} controls the frequency of the GSRO, oscillation frequencies of the first stage SRO and second stage GSRO are the same during T_{QE1}, as shown in Fig. 3(b). Therefore, gain calibration is not needed and second-order noise shaping can be achieved by a simple DCF used in a typical MASH modulator as shown in Fig. 5(b).

Although the first stage SRO-TDC looks similar to the circuit shown in [5], there is a couple of key differences that are noteworthy. First, the first stage counter counts only one of the multi-phase outputs of the SRO instead of counting all the phases. As quantization error of the first stage is removed after the DCF, there is no need to minimize T_{QE1} by counting all the phases. As a result, complexity and power consumption of the first stage SRO-TDC is significantly reduced. Second, the frequency of the SRO is designed to be higher than f_S so that there exists at least one rising edge during a sampling period. This is because a residue pulse must be generated every cycle to complete the second-order noise-shaping. This is in contrast to [5] where the frequency of the SRO can be smaller than f_S and hence a residue pulse may not be produced every cycle. Therefore, we employ a multi-bit counter shown in Fig. 6(a) to generate a residue pulse every cycle instead of a one-bit counter. One drawback of the multi-bit counter is that meta-stability may cause large error. In order to reduce the effect of meta-stability, a delayed clock generator (DCLKGen) is proposed by sampling the counter output (CNT_{output}) with a delayed clock as shown in Fig. 6(b) and (c).

Similar to MASH ADCs, some non-idealities of the proposed MASH TDC benefits from noise shaping. For example, $1/f$ noise, phase noise, and gating skew error of the second stage GSRO is reduced as they are first-order noise-shaped and filtered. In addition, the proposed structure is immune to mismatch between the delay-cells of the GSRO, since only the second stage uses multi-phase outputs and the effect of mismatch is second-order noise-shaped.

III. EXPERIMENTAL RESULTS

A prototype of the proposed $\Delta\Sigma$ TDC was fabricated in a 65nm CMOS and it occupies an active area of $250 \times 210\mu m$ (0.053 mm^2) as shown in Fig. 7. The proposed TDC operates at 400 MS/s with 200 MHz input pulse rate. To verify the performance of the proposed TDC, power supply of an off-chip delay-line was modulated. The measured output spectrum of a 390 kHz, 19 ps peak-to-peak sinusoidal input with a time offset of approximately 1 ns is shown in Fig. 8, where it can

978-1-4673-6145-3/13 $31.00 © 2013 IEEE   874

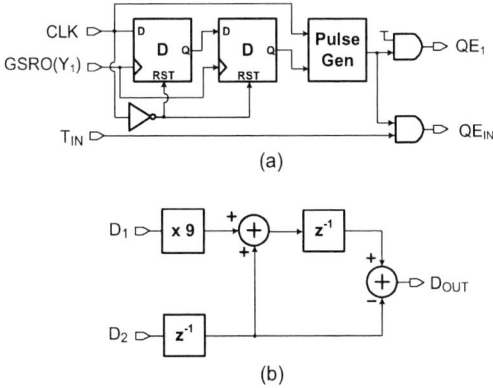

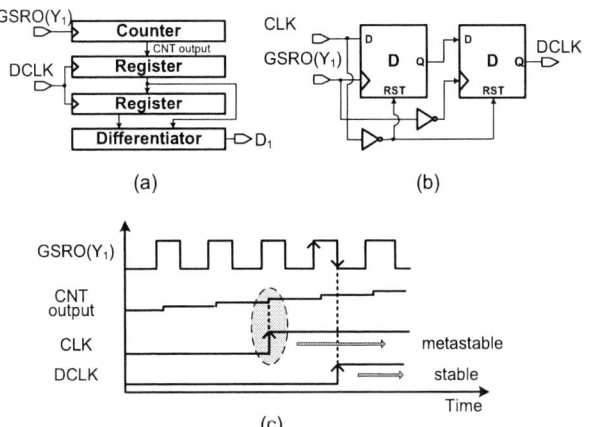

Fig. 5. (a) Schematic of quantization error generator (QEGen) (b) Block diagram of digital cancellation filter (DCF).

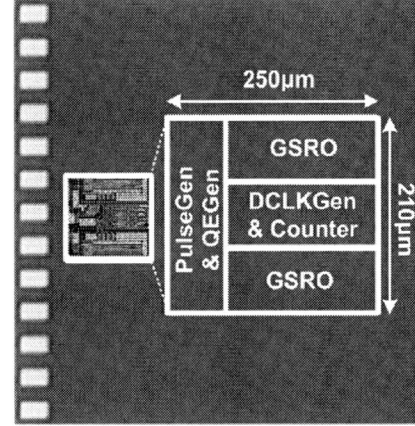

Fig. 7. Chip micrograph.

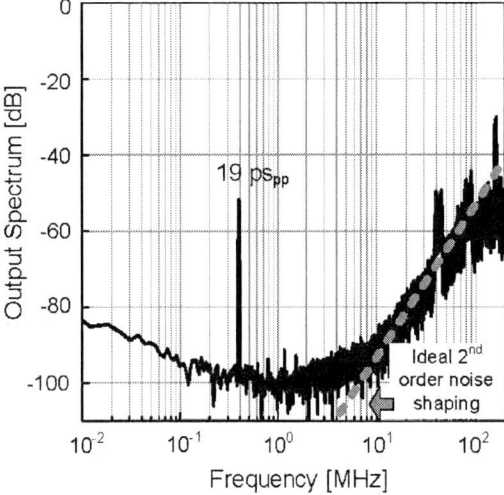

Fig. 6. (a) Schematic of the multi-bit counter (b) Schematic of the delayed clock generator (c) Timing diagram of the delayed clock generator.

Fig. 8. Measured output spectrum. 65,536pt FFT is performed with a Hanning window.

be seen that second-order noise-shaping is achieved with $1/f$ noise dominating at low frequencies. The measured integrated noise ($T_{int,rms}$) from 10 kHz to 4 MHz is -74.6 dB, which translates to 148 fs_{rms} at 200 MHz input pulse rate. The measured integrated noise for different OSRs is shown in Fig. 9, where the proposed TDC achieves better performance than a conventional SRO-TDC under the same OSR. Note that the integrated noise of the proposed TDC is limited by thermal and $1/f$ noise at higher OSRs (>100). The power consumption of the proposed TDC depends on the input pulse width, T_{IN}. The upper limit is 6.55 mW, which is when the input is always high. When the average input pulse width is 1 ns, the power consumption is 5.35 mW. The performance of the proposed TDC is summarized and compared with the recent state-of-the-art $\Delta\Sigma$ TDCs in Table I and Fig. 10. The proposed TDC achieves the widest bandwidth while achieving good FoM. Note that FoM1 used for $\Delta\Sigma$ converters is a better indication of the performance than FoM2 defined for Nyquist converters.

IV. CONCLUSION

In this paper, a novel all-digital second-order $\Delta\Sigma$ TDC has been proposed that achieves wide dynamic range, wide bandwidth and low integrated noise. Using the proposed GSRO, the order of noise-shaping has been increased by cascading GSRO-TDC without any calibration. Since time-resolution of the GSRO depends on the speed of logic gate, the proposed architecture is expected to gain higher performance with the continued scaling of the CMOS process.

V. ACKNOWLEDGMENTS

This research was supported by the Basic Science Research Program through the National Research Foundation of Korea (NRF) grant funded by the Korea government (MEST) (2012-0000701) and IDEC of KAIST. The authors would like to thank Dr. Hayun Chung for helpful discussions.

978-1-4673-6145-3/13 $31.00 © 2013 IEEE

TABLE I

PERFORMANCE SUMMARY AND COMPARISON WITH OTHER STATE-OF-THE ART $\Delta\Sigma$ TDCs.

	VLSI '09 [4]	CICC '10 [9]	JSSC '12 [6]	ISSCC '12 [5]	VLSI '12 [7]	VLSI '12 [8]*	This work
Process	130-nm	90-nm	130-nm	90-nm	90-nm	65-nm	65-nm
Scheme	GRO	Phase-domain $\Delta\Sigma$	Relaxation Osc.	SRO	Charge Pump**	FSO	GSRO
fs	50 MS/s	156.25 MS/s	50 MS/s	500 MS/s	90 MS/s	16 MS/s	400 MS/s
fi	50 MHz	156.25 MHz	50 MHz	80 MHz	90 MHz	16 MHz	200 MHz
Bandwidth	1 MHz	1 MHz	1 MHz	1 MHz	2.8 MHz	0.5 MHz	4 MHz
$T_{int,rms}$	80 fs	–	–	315 fs	–	–	148 fs
Resolution	1 ps	2.4 ps	5.6 ps	–	3 ps	35 ps	1.4 ps
Full scale range	12 ns	3.2 ns	20 ns	12.5 ns	5.55 ns	31.25 ns	4 ns
Power consumption	21 mW	2.1 mW	1.7 mW	2.1 mW	2.81 mW	0.28 mW	6.55 mW
Active die area	0.04 mm^2	0.12 mm^2	0.11 mm^2	0.02 mm^2	0.43 mm^2	0.0007 mm^2	0.05 mm^2
FoM1[dB]***	171	156	157	169	162	154	168
FoM2[fJ/step]****	228	399	293	92	159	305	96
Calibration	No	No	Background	No	No	Background	No

* Does not include power and area consumed by off-chip calibration.
** Charge-pump with noise-shaping single slope quantizer.
*** $FoM1 = DR + 10log_{10}(Bandwidth/Power)$ [dB], where $DR = 20log_{10}(T_{range}/T_{int,rms})$.
**** $FoM2 = Power/2^{Bit}/2/Bandwidth$, where Bit=number of bit : $Bit=(DR-1.76)/6.02$.

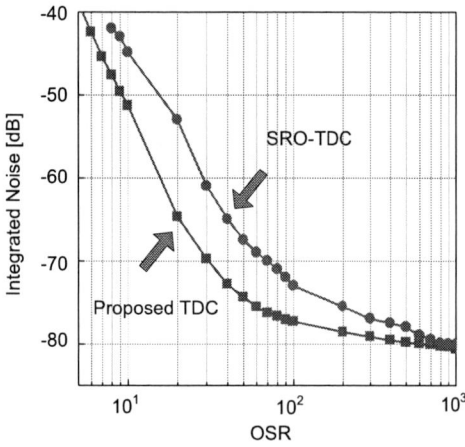

Fig. 9.   Measured integrated noise for different OSRs. The integrated noise is varied by changing signal bandwidth.

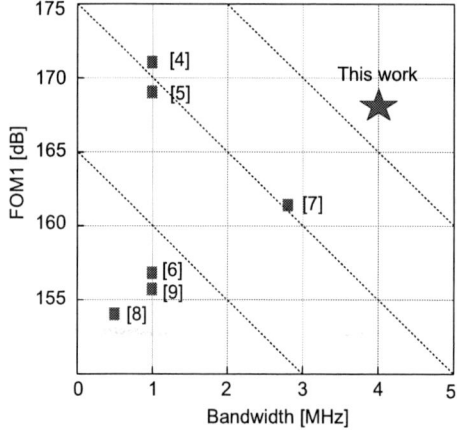

Fig. 10.   FoM comparison with recent reported $\Delta\Sigma$ TDCs.

REFERENCES

[1] K.-S.Kim *et al.*, "A 7 bit, 3.75 ps Resolution Two-Step Time-to-Digital Converter in 65 nm CMOS Using Pulse-Train Time Amplifie," *IEEE J. Solid-State Circuits*, vol. 48, no. 4, pp. 1009–1017, Apr. 2013.

[2] D.-W. Jee *et al.*, "A 2 GHz Fractional-N Digital PLL with 1b Noise Shaping $\Delta\Sigma$ TDC," *IEEE J. Solid-State Circuits*, vol. 47, no. 4, pp. 875–883, Dec. 2010.

[3] S. Naraghi *et al.*, "A 9-bit, 14 μW and 0.06 mm^2 Pulse Position Modulation ADC in 90 nm Digital CMOS," *IEEE J. Solid-State Circuits*, vol. 45, no. 9, pp. 1870–1880, Sep. 2010.

[4] M.Z. Straayer *et al.*, "A Multi-Path Gated Ring Oscillator TDC With First-Order Noise Shaping," *IEEE J. Solid-State Circuits*, vol. 44, no. 4, pp. 1089–1098, Apr. 2009.

[5] A. Elshazly *et al.*, "A 13b 315fsrms 2mW 500MS/s 1MHz Bandwidth

Highly Digital Time-to-Digital Converter Using Switched Ring Oscillators," in *IEEE Int. Solid-State Circuits Conf. Dig. Tech. Papers*, Feb. 2012, pp. 464–466.

[6] Y. Cao *et al.*, "1-1-1 MASH $\Delta\Sigma$ Time-to-Digital Converter With 6ps Resolution and Third-Order Noise-Shaping," *IEEE J. Solid-State Circuits*, vol. 47, no. 9, pp. 2093–2106, Sep. 2012.

[7] M. Gande *et al.*, "A 71dB Dynamic Range Third-Order $\Delta\Sigma$ TDC using Charge-Pump," in *IEEE Symp. VLSI Circuits Dig. Tech. Papers*, Jun. 2012, pp. 168–169.

[8] T. Konish *et al.*, "A 61-dB SNDR 700 μm^2 Second-Order All-Digital TDC with Low-Jitter Frequency Shift Oscillator and Dynamic Flipflops," in *IEEE Symp. VLSI Circuits Dig. Tech. Papers*, Jun. 2012, pp. 190–191.

[9] B. Young *et al.*, "A 2.4ps resolution 2.1mW Second-Order Noise-Shaped Time-to-Digital Converter With 3.2ns Range in 1MHz Bandwidth," in *Proc. IEEE CICC*, Sep. 2010, pp. 1–4.

A 0.84ps-LSB 2.47mW Time-to-Digital Converter Using Charge Pump and SAR-ADC

Zule Xu, Seungjong Lee, Masaya Miyahara, and Akira Matsuzawa

Department of Physical Electronics, Tokyo Institute of Technology
2-12-1 S3-27, Ookayama, Meguro-ku, Tokyo 152-8552 Japan
E-mail: xuzule@ssc.pe.titech.ac.jp

Abstract—We propose a time-to-digital converter (TDC) using a charge pump and a SAR-ADC. With this architecture, high time resolution is attainable by increasing the charging current or reducing the sampling capacitance. Thus, the resolution limitation in a delay-chain TDC does not exist. We propose to use a SAR-ADC attributed to its characteristics of compact structure, scalability, low power consumption, and small area. The prototype chip was fabricated in 65nm CMOS, achieving 0.84ps LSB, 2.47mW power consumption, and 0.06mm^2 area occupation. With 8-bit outputs, the DNL and INL are -0.7/1.0 LSB and -2.7/1.7 LSB, respectively.

I. INTRODUCTION

Having been functioning in a time-of-flight measurement system for decades, time-to-digital converters (TDC) are increasingly demanded since the advent of all-digital PLLs (ADPLL). Technology scaling shortens the delay of logic gates, endowing delay-chain TDCs' (Fig.1 (a)) applications in more and more systems. However, the time resolution with a logic gate delay fails to satisfy low-jitter ADPLLs that typically require pico-second resolution of a TDC. Thus, advanced techniques are demanded to break this limitation while keeping low power consumption and small area.

Various techniques have been proposed to enhance the time resolution. A Vernier TDC shrinks the resolution to the difference of two logic gate delays but it is vulnerable to PVT variations. Although a DLL calibration loop can be used for global calibrations, local mismatches still cause jitters that degrade the effective resolution and the linearity [1]. In a pipeline TDC, time amplifiers (TAs) are utilized to amplify the time residue for further quantization in following stages but the nonlinearity and mismatches of TAs require much calibration effort [2]. Against jitter and nonlinearity issues, oversampling and noise-shaping TDCs are proposed. However, the input signal bandwidth is inherently low, or on the other hand, the resolution is limited by the oversampling ratio (OSR) [3]. A statistic TDC utilizes the process variations of timing arbiters to accumulate a high-gain transfer function. Due to this characteristic, however, the detectable range is short and the performance is heavily dependent on the technology used [4]. All these solutions address the time

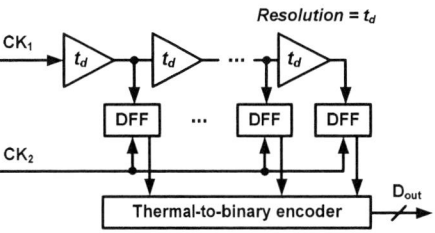

(a)   Conceptual block diagram of delay-chain TDC

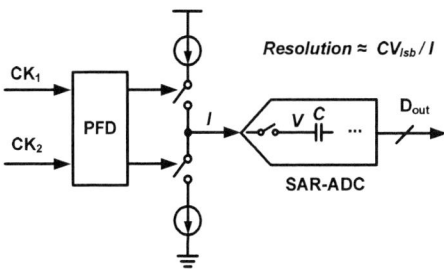

(b)   Conceptual block diagram of proposed TDC

Fig. 1. (a) Dealy-chain TDC and its resolution limitation, and (b) proposed TDC and its potential of high resolution.

quantization in the time domain which is still "rough" with technologies commonly used.

In this paper, we propose a solution exploiting the charge domain for high time resolution. Although some analog circuits are employed, they are kept simplified and compact for the balance between performance and power/area overhead. Meanwhile, our proposal still benefits from technology scaling, which will be described in the following sections.

II. PROPOSED SOLUTION

A. Concept and Conventional Issues

The conceptual block diagram of our proposal is shown in Fig. 1(b), where time interval is translated to charges on a capacitor and then the charges are quantized by an ADC. From a simplified equation: $t_{res} = CV_{lsb}/I$, the time resolution can be boosted by increasing the current, reducing the capacitance, or enhancing the

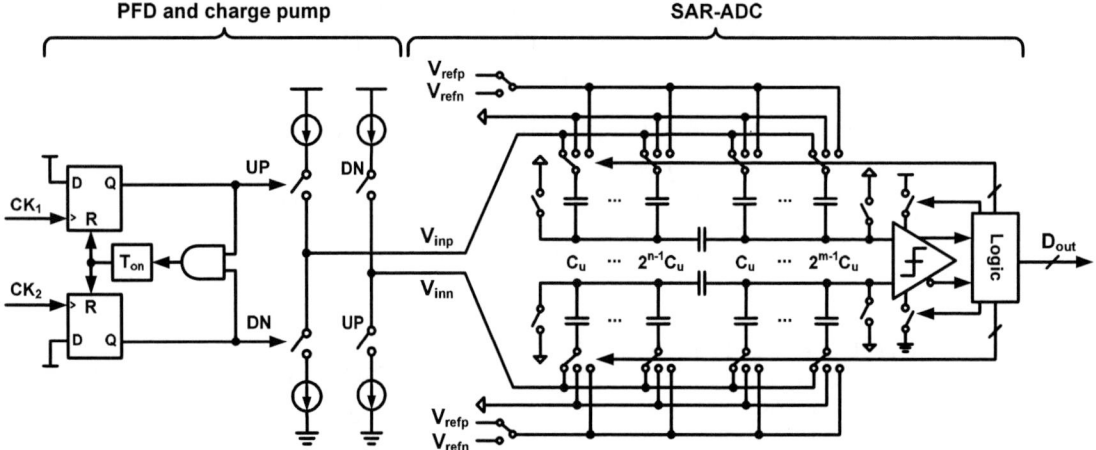

Fig. 2. Architecture of proposed TDC.

ADC's resolution. In the old days, however, the latter two approaches were impractical when implementing on an integrated circuit (IC) due to the low density of on-chip capacitors, and the unreasonably large power and area consumptions of an ADC. Consequently, on-chip solutions used a dual-slope-counter [5] or "semi-time-semi-voltage-domain" delta-sigma architectures where the signal bandwidths are limited [3][6].

B. Selection of the ADC and the Proposed Architecture

The type of the ADC affects the practicability of this proposal. A flash ADC is capable of extremely high speed with low resolution. To increase its dynamic range, the comparators increase exponentially so that significant power, area and calibration are involved. A pipeline ADC features high resolution and high speed. However, the amplifiers consume much power, and designing a high gain op-amp becomes tough with recent technologies. A delta-sigma ADC faces the similar situation due to its integrators, although it achieves extremely high resolution.

We propose to use a SAR-ADC attributed to its several advantages for the proposed TDC [7]. First, unlike other types of ADCs, it only contains one CDAC for both sampling and quantization, as well as one comparator. Thus, a compact structure is available. By using metal-oxide-metal (MOM) capacitors and a dynamic comparator, it squeezes power and area consumptions to be low with enough sampling rate for an ADPLL. Moreover, the binary-weighted CDAC suggests easier scalability of the resolution. The extension of the resolution is generally done by adding the capacitors other than the comparators or the op-amps that are required in other types of ADCs.

Fig.2 illustrates the architecture of our proposed TDC. A differential topology is used against the common-mode surges. An issue is that the voltage noise is also integrated during the sampling period. We will analyze this effect in the next section.

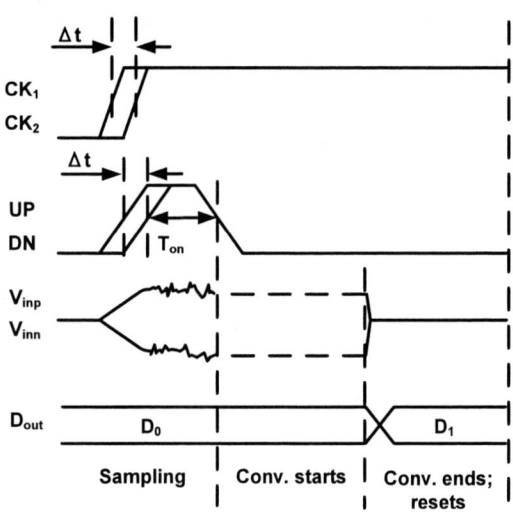

Fig.3 Timing diagram and the noise accumulation.

C. Timing, Noise, and Design Perspective

The major timing diagram of the proposed TDC is shown in Fig. 3. The rising edges of two input signals, CK_1 and CK_2 trigger PFD's outputs UP and DN to turn on the switches of the charge pump. It is known that the feedback delay, T_{on}, is used to remove the dead-zone effect but the voltage noise are accumulated on the capacitors during this period.

The accumulated average noise power can be calculated using (1), where C is the total sampling capacitance seen from the output of the charge pump, I_{cp} is the bias current of the current source, and g_{mn} and g_{mp} are the transconductances of NMOS and PMOS current sources, respectively.

978-1-4673-6145-3/13 $31.00 © 2013 IEEE   878

$$\sigma_{vn}{}^2 = \frac{4kT\gamma}{2C^2} \cdot \frac{(g_{mn} + g_{mp})}{I_{cp}} \cdot I_{cp} \cdot T_{on} \qquad (1)$$

Considering the pseudo-differential topology, using $t=CV/(2I)$, we derive (2), where V_{lsb} is the LSB voltage resolved by the ADC, and $\sigma_{vn,diff}$ is the sum of noise power of two outputs.

$$\sigma_{vn,diff}{}^2 = 2 \cdot kT\gamma \cdot \frac{T_{on}}{C_{lsb}} \cdot \frac{V_{lsb}}{t_{res}} \cdot \frac{(g_{mn} + g_{mp})}{I_{cp}} \qquad (2)$$

Equation (2) gives a perspective of designing the proposed TDC, where g_m/I_{cp} is a design parameter for transistor sizing of the current sources. Generally, small values of g_m/I_{cp} and C are preferred for low noise and small area, which are fortunately the outcomes of the technology scaling. Small C requires short T_{on} to stop the noise accumulation, which is also practical thanks to the shrunk switch sizes.

III. CIRCUITS DESIGN

A. Charge Pump and Switched-Capacitor Replica Biasing

A cascode pseudo-differential charge pump is designed as shown in Fig. 4. Dummy branches are used to reduce charge sharing, and also serve as the replica biasing during reset period to equal currents of PMOS and NMOS transistors. It typically requires an amplifier in a feedback loop, consuming some power.

For a low power design, we propose to use switched-capacitor (SC) feedback circuitry to generate the bias voltage. The idea is same as SC common mode feedback, well applied in fully-differential amplifiers, except that only one node, V_m, is sensed. For simplicity, output common mode is not regulated since the actual charging time is short enough and the outputs are reset cyclically. Although it is vulnerable to common-mode surges during either UP or DN is turned on, the feedback provides some power supply rejection because either dummy branch is also on at the same time. Hence, the surge from the power supply is sensed from V_m, and V_{bp_fb} is regulated.

B. SAR-ADC

The topology of the SAR-ADC is shown in Fig. 2. MOM capacitors and a dynamic comparator are designed. We use 8-bit output from a 12-bit topology since a 12-bit SAR-ADC is being developed in another project. If an actual 8-bit topology is used, the size of the CDAC can be shrunk to 1/16, as well as its power consumption. Fig. 5 shows the image layout of 12-bit, 10-bit, and 8-bit CDAC, implying this area shrinking. The increase of the noise may be concerned. However, if we rewrite (2) into (3), we find that the noise to minimum signal ratio does not change if only the ADC's resolution is scaled by shrinking the CDAC. For example, from 12-bit to 8-bit, C is reduced to 1/16, and V_{lsb} is increased to 16 times as well, i.e. the required charges do not change, if the reference voltage is unchanged. Therefore, the time

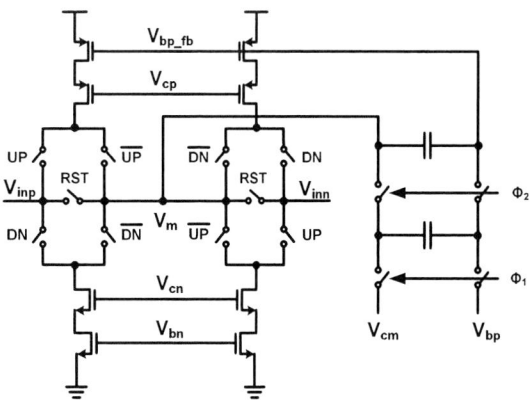

Fig. 4. Charge-pump with switched-capacitor replica biasing

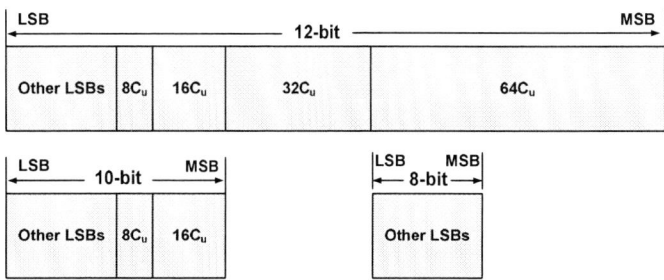

Fig. 5. Image layout of 12-bit, 10-bit, and 8-bit CDAC.

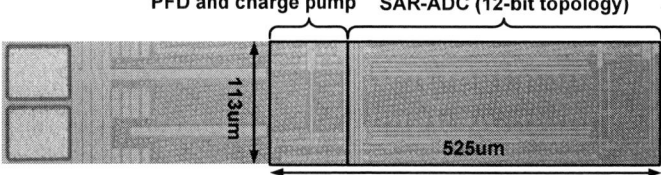

Fig. 6 Chip photo

resolution is not affected when scaling the SAR-ADC. To decrease this ratio, short T_{on} is important which is available with technology scaling as stated in section II.C.

$$\frac{\sigma_{vn,diff}{}^2}{V_{lsb}{}^2} = 2 \cdot kT\gamma \cdot \frac{1}{CV_{lsb}} \cdot \frac{T_{on}}{t_{res}} \cdot \frac{(g_{mn} + g_{mp})}{I_{cp}} \qquad (3)$$

IV. MEASUREMENT RESULTS

A prototype IC has been fabricated in 65nm CMOS. The chip photo is shown in Fig. 6, with 0.06mm² core area. With 40MHz input clocks and 1.0V power supply, the total power consumption is 2.47mW, where about 0.9mW is consumed by the charge pump.

To measure the DNL and INL, two frequencies with nominal 5Hz difference were input to the TDC, creating time ramps. With histogram method, DNL and INL are calculated as -0.7/1.0 LSB and -2.7/1.7 LSB, respectively, with 0.84ps per LSB, as shown in

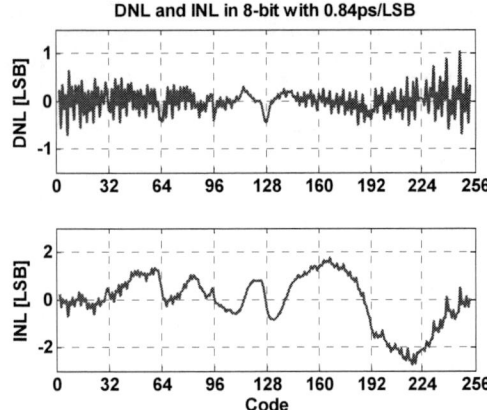

Fig. 7 Measured DNL and INL

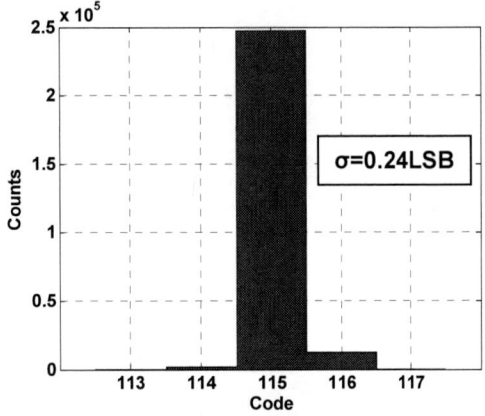

Fig. 8 Measured single-shot precision

Fig. 7. To measure the single-shot precision, one output from Agilent E4861B is split into two and fed to the TDC. The measured standard deviation is 0.24LSB as shown in Fig. 8, suggesting the TDC's low intrinsic noise.

The performance comparison is listed in Table I, manifesting the best balance of our work. Reference [2] has higher resolution but its power and area overheads are 4 and 5 times larger than ours, respectively. Although [1] achieves smallest area and lowest power among others, its resolution is 5 times lower than ours. Since the SAR-ADC has a 12-bit topology, if it is optimized to 10-bit, we expect 2.5 times higher sampling rate, 1/2 conversion energy, and 1/4 area occupation in the next work, as listed in Table I. Again, this is reasonable according to (3) and the corresponding design methodology of the SAR-ADC.

V. CONCLUSION AND FUTURE WORK

We have proposed a high resolution TDC using a charge pump and a SAR-ADC, proving that quantizing time in charge domain is practical to achieve high resolution with low power consumption and small area. We have given equations of the

TABLE I. PERFORMANCE COMPARISON

	[1]	[2]	[3]	[4]	This work	Next target
Type	Vernier	Pipeline	Delta-sigma	Stochastic	SAR-ADC	SAR-ADC
CMOS [nm]	65	130	130	65	65	65
Supply [V]	1.2	1.3	1.2	1.2	1.0	1.0
Resolution [ps]	4.8	0.63	3	3	**0.84**	0.84
Range [bits]	7	11	11	4	8	10
DNL [LSB]	<1	0.5	N/A	1.4	-0.7/1.0	N/A
INL [LSB]	3.3	2	N/A	1.5	-2.7/1.7	N/A
Frequency [MHz]	50	65	90 (OSR:16)	40	40	100
Power [mW]	1.7	10.5	3.2	8	**2.47**	4
Area [mm²]	0.02	0.32	0.43	0.04	0.06	0.015

involved voltage noise, showing a method of the parameter decision, the scalability of the SAR-ADC, and the benefits from the technology scaling. The prototype has achieved 0.84ps LSB, 8-bit range, 2.47mW power consumption, and 0.06mm² area occupation. The performance can be further optimized by designing the SAR-ADC properly.

In the future, we will implement this TDC into an ADPLL where low in-band phase noise can be expected using this high resolution TDC with low power consumption and small area.

ACKNOWLEDGMENT

This work was partially supported by NEDO, MIC, CREST in JST, STARC, HUAWEI, Berkeley Design Automation for the use of the Analog FastSPICE(AFS) Platform, and VDEC in collaboration with Cadence Design Systems, Inc.

REFERENCE

[1] L.Vercesi, A. Liscidini, R. Castello, "Two-Dimensions Vernier Time-to-Digital Converter," *IEEE J. Solid-State Circuits*, vol.45, no.8, pp.1504-512, 2010.

[2] Y. H. Seo, J. S. Kim, H. J. Park, J. Y. Sim, "A 0.63ps resolution, 11b pipeline TDC in 0.13μm CMOS," *Symp. VLSI Circuit 2011*, pp.152-153, Jun. 2011.

[3] M. Gande, N. Maghari, T. Oh, U. K. Moon, "A 71dB dynamic range third-order ΔΣ TDC using charge-pump," *Symp. VLSI Circuits 2012*, pp.168-169, Jun. 2012.

[4] M. Zanuso, S. Levantino, A. Puggelli, C. Samori, A.L. Lacaita, "Time-to-digital converter with 3-ps resolution and digital linearization algorithm," *Proc. ESSCIRC 2010*, pp.262-265, Sep. 2010.

[5] E. R. Ruotsalainen, T. Rahkonen, J. Kostamovaara, "An integrated time-to-digital converter with 30-ps single-shot precision," *IEEE J. Solid-State Circuits*, vol.35, no.10, pp.1507-1510, 2000.

[6] B. Young, S. Kwon, A. Elshazly, P. K. Hanumolu, "A 2.4ps resolution 2.1mW second-order noise-shaped time-to-digital converter with 3.2ns range in 1MHz bandwidth," *IEEE CICC 2010*, pp.1-4, Sep. 2010.

[7] A. Matsuzawa, Invited, "Analog and RF circuits design and future devices interaction," *IEEE IEDM 2012*, pp.14.3.1-14.3.4, Dec. 2012.

978-1-4673-6145-3/13 $31.00 © 2013 IEEE

A double-sampling cross noise-coupled Sigma Delta modulator with a reduced amount of opamps

Maarten De Bock, Pieter Rombouts

Ghent University (UGent), Dept. ELIS, Sint-Pietersnieuwstraat 41, 9000 Ghent, Belgium
Email: maarten.debock,rombouts@elis.UGent.be

Abstract—**This paper presents the design of a second order double-sampling split path Sigma Delta modulator with cross noise-coupling. The power budget for the double-sampling is reduced by using bilinear integrators, while cross noise-coupling between the two modulator loops increases the noise shaping to third order. The implementation of the noise-coupling is incorporated into the second integrator using a novel delaying feed-forward circuit. The complete modulator is integrated in a 130 nm CMOS technology and operates at a 120 MHz clock frequency. It achieves 77.8 dB dynamic range and 71.4 dB SNDR over a 5 MHz bandwidth.**

I. Introduction

Sigma Delta ($\Sigma\Delta$) modulation is widely used for low power high accuracy analog-to-digital (A/D) conversion. The conceptual diagram of a $\Sigma\Delta$-modulator is shown in fig. 1(a). A quantizer is embedded in a control loop with loop filter H. The digital output D is fed back toward the input of the modulator with a digital-to-analog converter (DAC). Assuming the DAC has a linear gain and modeling the quantizer as a linear gain with added white noise Q, we find the linear model of a $\Sigma\Delta$-modulator in fig. 1(b).

While recent research has focused more on modulators with a continuous time loop filter [1], discrete time loop filters remain popular because they are less sensitive to parameter variation and clock jitter. For a discrete time loop filter, we can write the output of the modulator as:

$$D(z) \approx V_{\text{in}}(z) + \underbrace{\frac{1}{1 + H(z)}}_{NTF(z)} Q(z)$$

The contribution of the quantization noise Q at the output is described by the noise transfer function $NTF(z)$. The ratio of the sample frequency to the signal bandwidth is called the oversampling ratio (OSR). Typically, the loop filter consists of a cascade of integrators and as such has a high gain within the signal bandwidth. This way, the contribution of the quantization noise is spectrally shaped outside the signal band and the signal to quantization noise ratio (SQNR) within the signal bandwidth is increased. The number of integrators in general determines the order of the loop filter and the order of quantization noise shaping. But also the contribution of the quantization noise at the higher frequencies must be controlled to prevent modulator overloading. A common strategy to design the loop filter is to place the poles of the loop filter in a Butterworth configuration [2].

A discrete time $\Sigma\Delta$-modulator is commonly implemented with switched-capacitor (SC) circuits. The power efficiency of

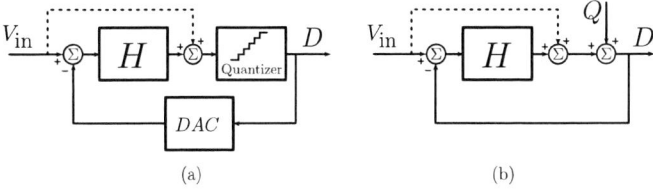

Fig. 1. (a) The $\Sigma\Delta$-modulator and (b) the linear model.

switched capacitor circuits can easily be doubled using double-sampling. However, special care must be taken with regard to quantization noise folding when using double-sampling [3]. This problem is tackled in section II.

In [4], [5] an enhanced split-architecture $\Sigma\Delta$ ADC is presented. Through the use of noise-coupling the effective noise shaping order is increased. This can lead to a significant reduction in power consumption for low-order modulators [6]. For the implementation of noise-coupling, an active adder is needed at the quantizer input. In section III, a novel delaying feed-forward circuit is presented which eliminates the need for an active adder by incorporating the noise-coupling in the last integrator stage.

Section IV then highlights the circuit design of the modulator and the measurement results for the integrated prototype are discussed in section V.

II. Double-sampling

When using double-sampling, all signals are updated during both clock phases and the sampling frequency is twice the clock frequency [7]. However, a straight-forward implementation of double-sampling using conventional SC-integrators suffers from quantization noise folding due to mismatch between the different capacitors used during the different clock phases [7]. The result of this quantization noise folding is that signals around $f_s/2$ (mainly the shaped quantization noise) are spectrally folded back into the low pass signal band. This severely degrades the SNR of the modulator. This problem can be tackled using the fully-floating bilinear integrator shown in fig. 2 [8]. The transfer function for this circuit can be written as:

$$\frac{V_{\text{out}}(z)}{V_{\text{in}}(z)} = \frac{1 + z^{-1}}{1 - z^{-1}} \frac{C_{\text{A}}}{C_{\text{FB}}}$$

For this fully floating bilinear integrator, it can be shown that mismatch between C_{A} and C_{B} does not lead to quantization noise folding. Moreover, due to the bilinear factor, the input capacitors (C_{A} and C_{B}) can be halved, reducing the power consumption of the opamp. However, the bilinear factor

978-1-4673-6145-3/13 $31.00 © 2013 IEEE
881

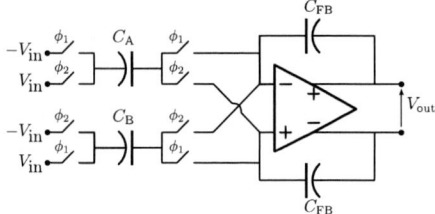

Fig. 2. The fully-floating bilinear integrator.

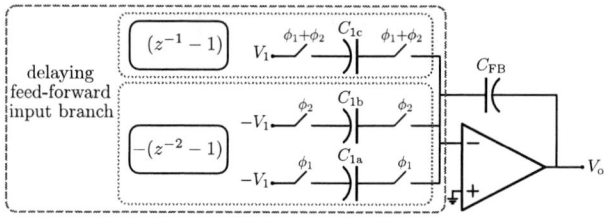

Fig. 4. A delaying feed-forward circuit.

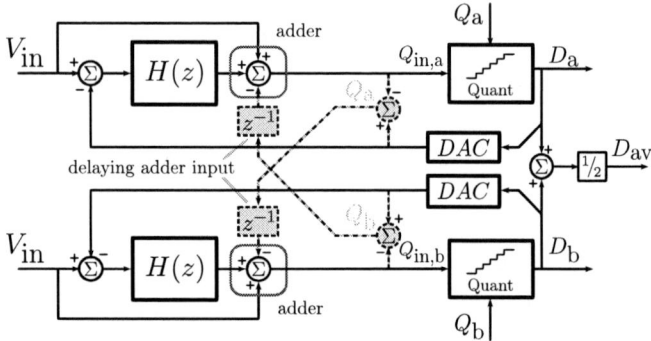

Fig. 3. Cross noise-coupling in a split path $\Sigma\Delta$-modulator.

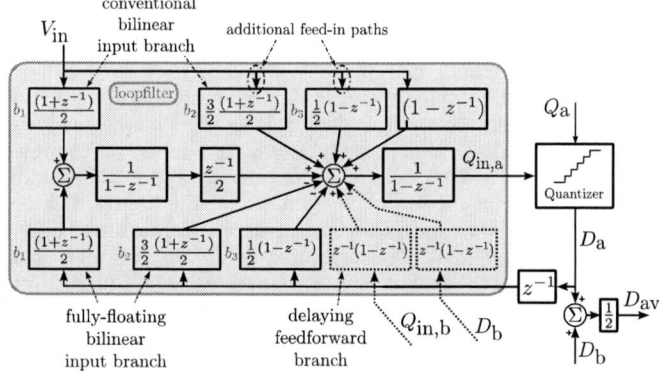

Fig. 5. The architecture of a single modulator loop.

$(1+z^{-1})$ modifies the loop filter and increases it's order. This problem can be tackled by using a slightly modified loop filter architecture as proposed in [3]. With this modified architecture the conventional loop filter design strategy of [2] can still be used.

III. NOISE-COUPLING

In a split-path $\Sigma\Delta$-modulator, two modulator loops are used in parallel and the average output of the two loops is then taken as the overall output. In [4], [5] it is proposed to use quantization noise-coupling between the two modulator loops, as shown in fig. 3. Here, the quantization noise from one modulator loop is injected into the other loop. For the other loop, the injected quantization noise is just a dither signal, but for the average output D_{av} the order of the quantization noise shaping is increased by one [4].

To implement the cross-noise coupling, the quantization noise of each modulator must be determined during each clock phase and subtracted from the quantizer input of the other modulator during the next clock phase, as shown in fig. 3. The quantization noise in a $\Sigma\Delta$-modulator can easily be determined by calculating the difference between the quantizer output and input. To reduce the power consumption and the required die area, the noise-coupling should be incorporated into the second integrator stage using feed-forward input branches. However, the signals that need to be added, the quantizer input and output, need an additional delay of one clock cycle (z^{-1}) before summation. This is almost inherent to the digital output of the quantizer, but as the quantizer input is an analog signal, this would require an additional sample-and-hold circuit to implement this delay. For this reason, we propose a new delaying feed-forward branch for which a single ended implementation is shown in fig. 4. It is a combination of three branches with capacitors C_{1a}, C_{1b} and C_{1c}, all equal in size. The input branches with capacitors C_{1a} and C_{1b} operate

in tandem and are alternately switched to the negative input node of the opamp during one clock phase, while they remain floating during the other clock phase. The input for these branches is the inverse of V_1, which is readily available in this fully differential implementation. The third input branch (C_{1c}) is similar to a conventional feed-forward branch as it is connected to the negative input node of the opamp during both clock phases. For the combination of the 3 input branches in fig. 4 we can write:

$$\frac{V_o(z)}{V_1(z)} = \underbrace{\frac{(z^{-2}-1)}{(z^{-1}-1)}\frac{C_1}{C_{FB}}}_{C_{1a}\ \&\ C_{1b}} - \underbrace{\frac{C_1}{C_{FB}}}_{C_{1c}} = z^{-1}\frac{(z^{-1}-1)}{(z^{-1}-1)}\frac{C_1}{C_{FB}}$$

We see that V_1 is summed at the opamp output with an additional delay of one clock phase.

IV. CIRCUIT LEVEL IMPLEMENTATION

The architecture of a single modulator loop is shown in fig. 5. The complete modulator then consists of 2 loops with outputs D_a and D_b of which the average output D_{av} is taken. The loop filter consists of a cascade of two integrator stages (2nd order noise shaping for single loop, 3rd order noise shaping for average output). The NTF is designed similar to [9], with an OSR of 24. The distributed feedback paths use the fully-floating bilinear integrator, while the input signal is sampled with a conventional bilinear integrator. The quantization noise coupling is incorporated into the second integrator stage using a conventional feed-forward input branch for the digital quantizer output D and the proposed delaying feed-forward input branch (fig. 4) for the analog quantizer input.

The modulator is integrated in UMC's 130 nm technology. This technology offers 2 flavors of 1.2 V transistors, high-speed (HS) and low-leakage (LL) with the latter having a

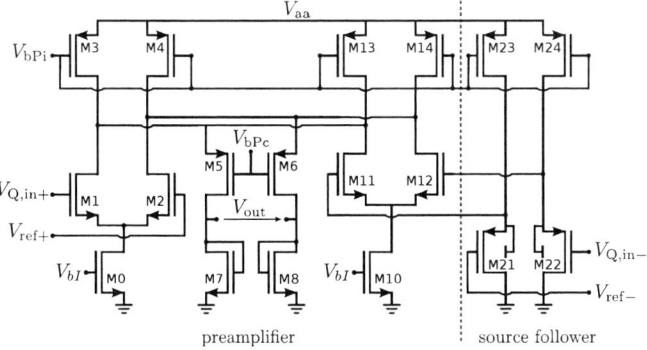

Fig. 6. The preamplifier determines the difference between the input signal ($V_{Q,in+} - V_{Q,in-}$) and the quantization level reference ($V_{ref+} - V_{ref-}$).

lower leakage current, but also a higher threshold voltage. To make the design independent of the technology choice, the more standard HS-devices were chosen. As the linearity of a standard CMOS transmission gate was insufficient, the bootstrapped switch is used [10]. The circuits are designed for a clock frequency of 120 MHz, which gives a 5 MHz signal bandwidth for an OSR of 24.

Because the output swing of the first integrator stage is reduced by applying an additional feedin path for in the input signal, a folded cascode OTA can be used. The operational amplifier for the second integrator stage needs to provide the full signal swing. For this reason, a two-stage miller compensated opamp is used.

The quantizer consists of 8 comparators to implement the 9 quantization levels for a differential full-scale of ± 900 mV. Each comparator consists of a preamplifier of which the output is fed into two D-latches (one for each clock phase), followed by two SR-latches. To implement the different quantization levels, the reference voltage of each quantization level is subtracted from the quantizer input signal at the preamplifier of each comparator, as shown in fig. 6. The preamplifier consists of two NMOS differential pairs (M1-M2 and M11-M12), of which the output is fed via current follower (M5-M6) into a diode connected NMOS (M7-M8) to provide an overal gain of $\approx g_{m,diff}/g_{m,diode}$. To limit the influence of large signal mismatch between the two differential pairs, ($V_{Q,in+} - V_{ref+}$) is compared to ($V_{ref-} - V_{Q,in-}$) rather than comparing ($V_{Q,in+} - V_{Q,in-}$) to ($V_{ref+} - V_{ref-}$). To reduce the systematic offset due to mismatch in bias current between the two differential pairs, large channels lengths are chosen for the bias transistors (M0, M10) to increase their output impedance. To implement the full signal swing, also PMOS source followers (M21-M22) are added at the extreme quantization levels, also shown in fig. 6.

The microscope picture of the integrated circuit is shown in fig. 7. The active area measures 0.625×0.525 mm^2. The two modulator loops are located on the left and right side of this figure, with an analog signal bus in between. A digital signal bus is located on the outside of the loops, and enters in between the first and second integrator stage to drive the feedback capacitor switches. To reduce the digital signal path length from one modulator loop to the other for the quantization noise-coupling, the digital signal bus crosses the analog signal bus. To prevent signal degradation by parasitic

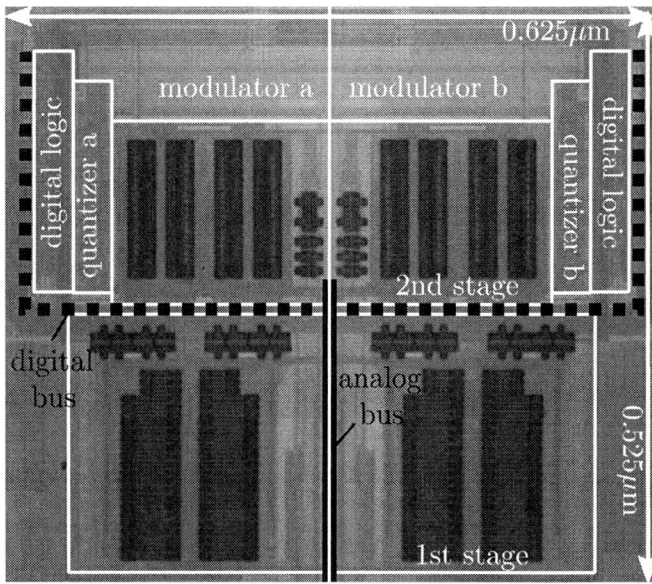

Fig. 7. Microscope photograph of the integrated circuit.

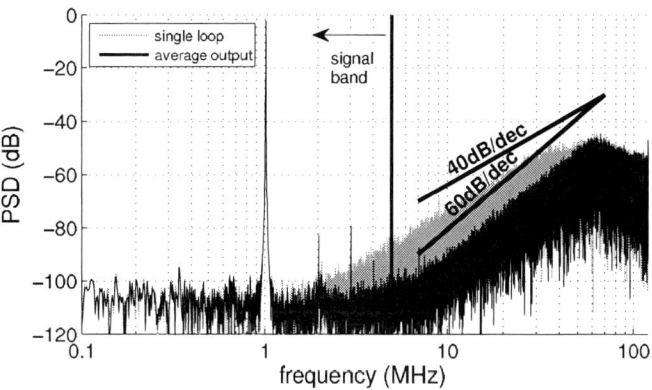

Fig. 8. Measured output spectrum for a single modulator loop and the average output (after calibration).

capacitive coupling between the analog and digital signals, a double shield to the analog and digital ground was used.

V. MEASUREMENT RESULTS

Initial measurements are done on a test die in a JLCC44 package which is mounted in a socket on a test PCB. For this test setup, the performance severely degrades for a clock frequency above 50 MHz. For this reason, the die is wire bonded directly on a new test PCB and the reported measurements are performed on this test PCB. Fig. 8 shows the measured output spectrum for a single modulator loop and the average output of both modulator loops for a -2 dBFS input tone at 1 MHz. This clearly shows that the noise-coupling increases the slope of quantization noise outside the signal bandwidth from 40 dB/decade to 60 dB/decade corresponding to 3rd order noise shaping.

The overall non-linearity of the modulator is limited by the mismatch in the DAC feedback capacitors. To eliminate the effect of this mismatch, an off-line calibration method similar to [11], [9] is used. The DAC-level error ε_i are determined using an offline calibration cycle and are then stored in a

978-1-4673-6145-3/13 $31.00 © 2013 IEEE

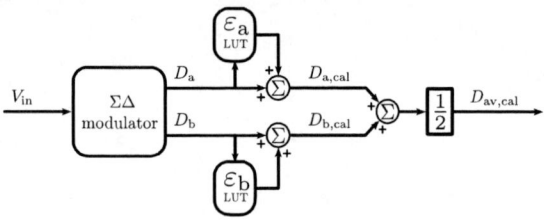

Process	130 nm 1P8M CMOS (UMC)		
Active Area	$625\,\mu m \times 525\,\mu m$ $(0.33\,\mu m^2)$		
Power consumption	12.1 mW (9.25 mW analog core)		
Supply voltage	1.2 V	V_{ref}	0.9 V
Signal bandwidth	5 MHz	DR	78.4
Clock frequency	120 MHz	peak SNR	73.4
Sample frequency	240 MHz	peak SNDR	71.4
OSR	24	FOM	157.5 dB

TABLE I. PERFORMANCE SUMMARY.

Fig. 9. The calibration scheme for the $\Sigma\Delta$-modulator.

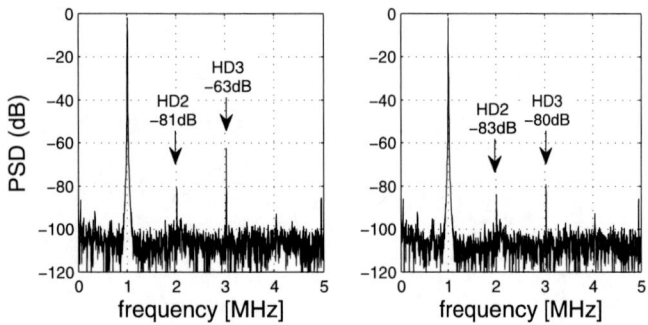

Fig. 10. Comparison of the signal band output spectrum (D_{av}) without (left) and with (right) calibration.

look-up table. Fig. 9 then shows the modulator during normal operation. The original modulator loop output is corrected using the look-up table, giving D_{cal}. Fig. 10 compares the power spectrum of the averaged output without and with this calibration procedure for the same $-2\,dBFS$ input tone at 1 MHz. The third harmonic is reduced from $-63\,dBFS$ to $-80\,dBFS$. As a result, the SNDR for this case increases from 60.2 dB to 71.4 dB.

Fig. 11 shows the measured SNR and SNDR vs input amplitude. Without calibration, both SNR and SNDR are limited by the DAC-nonlinearity, and the corresponding peaks are at 72.6 dB and 67.6 dB respectively. After calibration, the peak SNR and SNDR increase to 73.4 dB and 71.4 dB both at $-2\,dBFS$. The dynamic range (DR) is 78.4 dB which corresponds to 12.7 effective number of bits. The corresponding Schreier figure of merit (FOM $= SNDR_{dB} + 10\log\frac{BW}{P}$ [1]) is 157.5 dB. The performance details of the implemented prototype are summarized in table I.

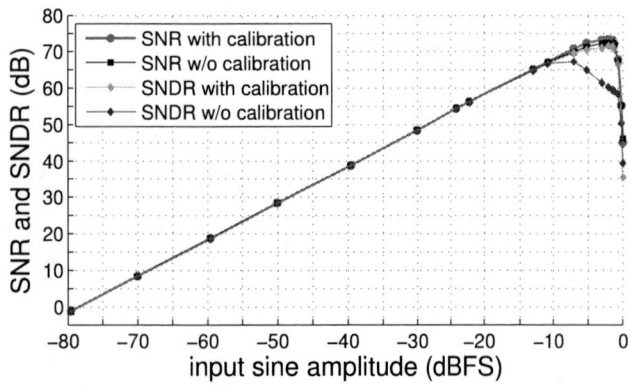

Fig. 11. SNR and SNDR vs input amplitude for a 1 MHz input tone (with and without calibration).

VI. CONCLUSION

This paper presents the design of a 2nd order double-sampling split path $\Sigma\Delta$-modulator with a 9-level quantizer. The power budget for the double-sampling is further reduced by using bilinear integrators. Cross noise-coupling between the 2 modulator loops is used to increase the noise shaping to 3rd order. The implementation of the noise-coupling is incorporated into the second integrator using a novel delaying feed-forward circuit. The complete modulator is integrated in a 130 nm CMOS technology and operates at a 120 MHz clock frequency. It achieves achieves 78.4 dB DR and 71.4 dB peak SNDR over a 5 MHz signal bandwidth for a power budget of 12.1 mW.

ACKNOWLEDGMENT

This work was supported by the Special Research Fund (BOF) of Ghent University.

REFERENCES

[1] B. Murmann, "ADC Performance Survey 1997-2013," *[Online]. Available: http://www.stanford.edu/%7Emurmann/adcsurvey.html.*

[2] R. Schreier, "An empirical study of high-order single-bit delta-sigma modulators," *IEEE Trans. Circuits Syst.-II*, vol. 40, no. 8, pp. 461–466, Aug. 1993.

[3] P. Rombouts, J. De Maeyer, and L. Weyten, "Design of double-sampling Sigma Delta modulation A/D converters with bilinear integrators," *IEEE Trans. Circuits Syst.-I*, vol. 52, no. 4, pp. 715–722, Apr. 2005.

[4] K. Lee and G. C. Temes, "Enhanced split-architecture Delta-Sigma ADC," *Electron. Lett.*, vol. 42, no. 13, pp. 737–739, Jun. 2006.

[5] K. Lee, J. Chae, M. Aniya, K. Hamashita, K. Takasuka, S. Takeuchi, and G. C. Temes, "A Noise-Coupled Time-Interleaved Delta-Sigma ADC With 4.2 MHz Bandwidth,-98 dB THD, and 79 dB SNDR," *IEEE J. Solid-State Circuits*, vol. 43, no. 12, pp. 2601–2612, Dec. 2008.

[6] F. Maloberti and E. Bilhan, "A Wideband Sigma-Delta Modulator With Cross-Coupled Two-Paths," *IEEE Trans. Circuits Syst.-I*, vol. 56, no. 5, pp. 886–893, May 2009.

[7] P. Rombouts, J. Raman, and L. Weyten, "An approach to tackle quantization noise folding in double-sampling Sigma Delta modulation A/D converters," *IEEE Trans. Circuits Syst.-II*, vol. 50, no. 4, pp. 157–163, Apr. 2003.

[8] D. Senderowicz, G. Nicollini, S. Pernici, A. Nagari, P. Confalonieri, and C. Dallavalle, "Low-voltage double-sampled Sigma Delta converters," *IEEE J. Solid-State Circuits*, vol. 32, no. 12, pp. 1907–1919, Dec. 1997.

[9] M. De Bock and P. Rombouts, "A 8 mW 72 dB Sigma Delta-modulator ADC with 2.4 MHz BW in 130 nm CMOS," *Analog Integr. Circuits and Signal Process.*, vol. 72, no. 3, SI, pp. 541–548, Sep. 2012.

[10] M. Dessouky and A. Kaiser, "Very low-voltage digital-audio Delta Sigma modulator with 88-dB dynamic range using local switch boot-strapping," *IEEE J. Solid-State Circuits*, vol. 36, no. 3, pp. 349 –355, Mar. 2001.

[11] X. Xing, M. De Bock, P. Rombouts, and G. Gielen, "A 40MHz 12bit 84.2dB-SFDR continuous-time delta-sigma modulator in 90nm CMOS," in *IEEE Asian Solid State Circuits Conference*, Nov. 2011, pp. 249 – 252.

978-1-4673-6145-3/13 $31.00 © 2013 IEEE

A Novel OTA-Based Fast Lock PLL

Mezyad Amourah, Sandeep Krishnegowda, Morgan Whately

Cypress Semiconductor, San Jose, CA, 95134

Email:mzar@cypress.com

Abstract- **This paper describes a novel fast lock scheme for phase-locked loops (PLLs). The proposed scheme uses a simple operational transconductance amplifier (OTA) to achieve significant reduction in PLL lock acquisition time without affecting PLL noise performance. The new scheme allows short starting time and fast dynamic power cycling for various sub-systems on SOC's. Multiple PLLs utilizing the new fast lock schemes were implemented in multi-port SRAM chip to provide frequencies from 400MHz to 1.6GHz, The chip was fabricated using 65nm CMOS process. Silicon measurements across corner lots show significant reduction in PLL lock time, by a factor of 6.5X, over device operating conditions.**

I. INTRODUCTION

Modern products use phase locked loops (PLLs) extensively to satisfy clocking requirements. For several cases PLL lock time is an important parameter that we may need to bring down especially for products involving multi-core Micro-processors, large system-on-chip (SOC's) with sub-systems turned ON and OFF, and applications where cascaded PLLs are turned ON sequentially [1]. PLL lock time is dependent on PLL loop Bandwidth (BW) or PLL natural frequency (ω_n) [2]. As shown in (1), PLL natural frequency suggests that PLL lock time is inversely proportional to the update rate (1/N), charge pump current (I_{CP}) and is proportional to loop filter caps sizes C_Z and C_P. PLL block diagram and LPF schematics are shown in Fig. 1.

$$\omega_n^2 = \frac{I_{CP}K_{VCO}}{2\pi N(C_Z + C_P)} \qquad (1).$$

Most of the PLLs used in the industry have charge pump current in the order of tens of micro-amps while loop filter cap sizes are in the order of hundreds of picofarads. PLL lock time is lengthened mainly by the time required to charge the PLL loop filter main capacitor, C_Z, to the required VCO control voltage. To speed up this process previous work has focused on pre-charging the main cap C_Z to a predetermined voltage before enabling the PLL, pumping up the charge pump current or using a second charge pump circuit, modified PFD and extra control logic to be able to increase the PLL bandwidth (BW) when the PLL is acquiring lock and bring the PLL charge pump current, I_{CP}, to its natural value once PLL is locked [3,4,5]. Increasing PLL loop BW by methods described above are not simple neither low cost. Increasing charge pump current do disturb the phase margin and worsen PLL noise performance, the use of a large charge pump current via second charge-pump circuit results in having a large switching current which couples to the supply and substrate affecting noise performance all over the chip. In this work we propose a new fast lock circuit which uses a simple, low

area, low power operational transconductance amplifier (OTA) that is much smaller than previous solutions, the extra current used to speedup lock process is a continuous one that is linearly dependent on how far we are from lock. The speedup current decreases gradually until turns off automatically as we attain lock without the need to extra control or logic circuits. Furthermore the use of the OTA is compatible with loop filter pre-charging and other loop BW increase schemes.

II. PROPOSED OTA BASED FAST LOCK SCHEME

If we observe PLL loop dynamics and follow on voltages and currents in the PLL low pass filter "LPF" as shown in Fig. 2. The low pass filter has two nodes: the input/output node called (V_{cntl}) because it is the provider of the control voltage to the voltage controlled oscillator (VCO), and the internal node called low ripple voltage node because the voltage there is a filtered version of V_{cntl} and we call it here (V_{lrpl}). During PLL operation we notice that when there is a difference between the phase-frequency-detector (PFD) output pulses, named UP/DN pulses, we charge or discharge the small cap, C_p, during that time. The slope of the LPF input/output voltage V_{cntl} will be proportional to I_{CP} and inversely proportional to capacitor value, C_P. When both UP and DN pulses are low we redistribute the charge from C_P to C_Z with a time constant shown in Fig. 3 remembering that $C_Z \gg C_P$. When the PLL is far from lock we notice that the voltage difference between the LPF input/output terminal V_{cntl} and the LPF internal low-ripple node V_{lrpl} is big and unidirectional as shown in Fig. 3, and once the loop gets close to lock that difference between V_{cntl} and V_{lrpl} drops down gradually to a very small value, ideally zero volts. This behavior happens because when we are far from lock we have a relatively large amount of control charge moved to C_P, representing the large skew between reference and feedback clock edges. This charge sampled at C_P is then redistributed to C_Z as described above, and the redistribution time is small compared to the reference clock period. When the PLL is close to lock then the correction charge sampled to C_P will be small and the redistribution time is longer which allows full charge redistribution which results in small ripple on V_{cntl} and very small difference between V_{cntl} and V_{lrpl}. Another important observation is that if the PLL is speeding up then V_{cntl} will be higher than V_{lrpl} and inversely when the PLL is slowing down then V_{cntl} will be lower than V_{lrpl} with the difference decreasing almost to zero when the PLL attains lock.

978-1-4673-6145-3/13 $31.00 © 2013 IEEE

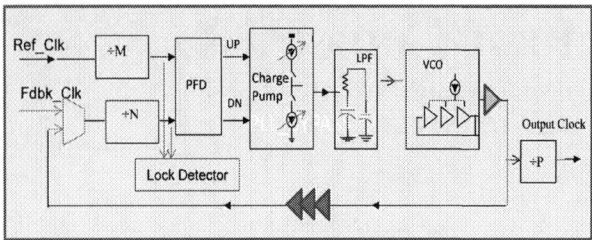

Fig 1: A general PLL block diagram

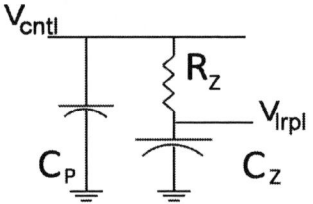

Fig. 2: A conventional PLL LPF schematics

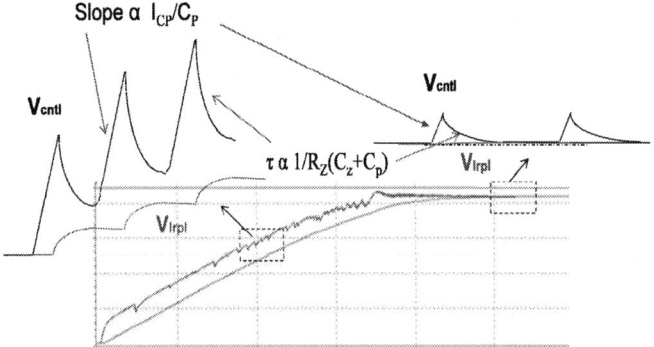

Fig. 3: PLL control voltage (V_{cntl}) and low ripple internal voltage (V_{lrpl}) relationship.

Previous lock time enhancement techniques where dependent on a large switching current that is turned ON when the PLL is far from lock. Those techniques result in PLL loop phase margin degradation and generating a lot of noise during that period [3,4].

The new fast lock technique depends on sensing the difference between V_{cntl} and V_{lrpl} voltages and generating a continuous time but variable current to charge/discharge the LPF large cap C_Z as shown in Fig. 3. This method is implemented by adding a simple wide operating range transconductance "G_m" amplifier (OTA) with inputs are V_{cntl} and V_{lrpl}. The output goes to V_{lrpl} to charge the capacitor C_Z directly. The amount of current is linearly dependent on the difference between V_{cntl} and V_{lrpl} as shown in (2) and Fig. 4.

$$I_{out} = G_m * (V_{cntl} - V_{lrpl}) \qquad (2)$$

The fact that the OTA is in a closed loop form is good because it will limit the effects of offset and gain error. For very low jitter implementations and to limit the OTA effect on the PLL phase noise performance once the PLL attains frequency lock we power it down. A PLL simulation

showing V_{cntl} and V_{lrpl} relation for different cases are shown in Fig. 5. The same figure shows the smaller difference between the two voltages when the PLL is close to lock or already locked. One of the major advantages of the OTA based fast lock scheme is that it is independent on the VCO, PFD, charge-pump circuit types which is not the case for most of the previously implemented ideas. Finally the OTA used is a traditional dual input folded cascaded amplifier. The OTA has NMOS and PMOS differential pairs at the input to have a wide operating range as the PLL frequency range is wide.

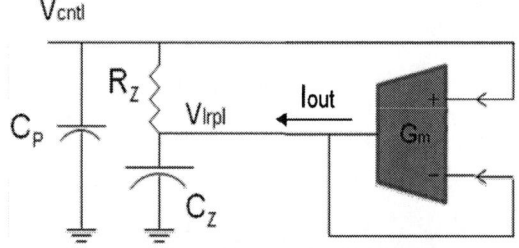

Fig. 4: OTA based Fast Lock implementation

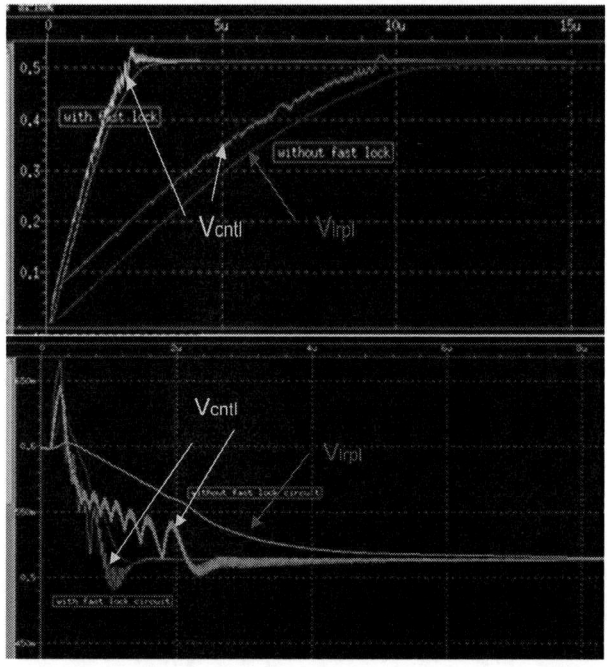

Fig. 5: relationship between V_{cntl} and V_{lrpl} for both normal case and fast lock, and for cases where the PLL is speeding up (top) or slowing down (bottom).

The new fast lock implementation faces two non idealities. OTA offset and OTA gain error. To overcome the offset we can implement a switched capacitor circuit to perfectly cancel the offset. Gain error is not a big issue for such an implementation because fixing gain error will not improve lock time due to the nature of the circuit and the large sizes of the capacitors in the loop filter besides we power down the OTA when frequency lock is attained.

A simple method to overcome the offset of the OTA is to add a series resistor between the output of the

OTA and its negative input which generates an offset that is variable with the OTA output current in the direction that cancels the negative effect of the OTA intrinsic offset and gain error as shown in Fig. 6..

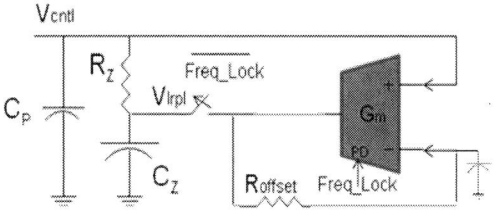

Fig. 6: Final implementation of the fast lock circuit

III.   MEASUREMENT RESULTS

The proposed fast lock scheme was implemented into several PLLs in a multi-port SRAM chip and fabricated using 65nm CMOS process. Die photo showing two PLLs in the center of the chip is shown in Fig. 7. On the chip during normal operation two PLL's are cascaded, while during BIST we cascade 3 PLL's. The PLLs operate at input frequency range of (20MHz-1.6GHz) and VCO range of (300MHz to 1.6GHz). The new fast lock technique is combatable with pre-initialized PLL startup and with regular loop BW increase. In Fig. 8 we show a frequency versus time capture for one of the PLLs tested with and without fast lock mode driven by a 100MHz reference clock and the VCO running at 1.6GHz.

The mean of the measured lock time for the same PLL obtained with 100MHz reference clock and VCO frequency of both 800MHz and 1.6GHz cases are shown in table 1. There are 5 corner lots, 3 units per corner lot, supply is in the range of 1.1V up to 1.35V and temperature values are in the range of (-40C, 50C, 140C). Lock time statistics are shown in Fig. 9. Finally PLL's silicon area is about 360umX360um while the OTA layout area is 60umX60um. The PLL consumes a total 10mA average current. The OTA consumes a maximum of 200uA total current but because we turn the OTA off after PLL locks then it consumes only leakage current. Moreover because the OTA output current decreases to the nanoamp range as the PLL approaches lock there will be no glitch on the output clock frequency when the OTA is powered down as shown in Fig. 8.

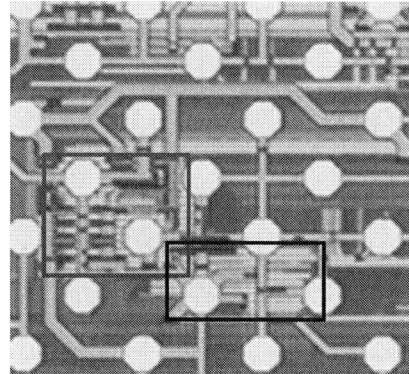

Fig. 7: Die photo showing two PLLs with fast lock option.

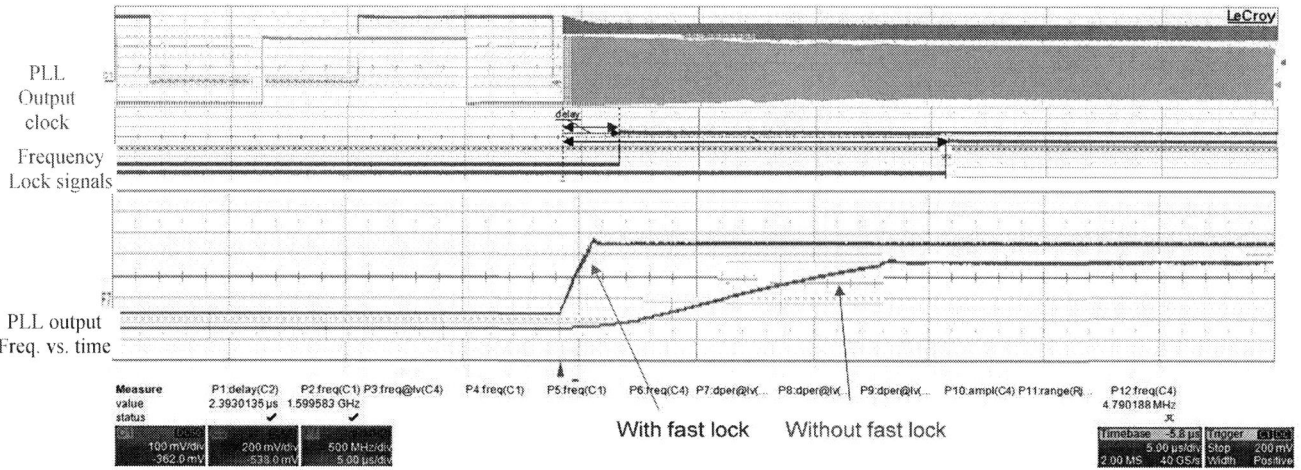

Fig. 8: Silicon measurement for PLL lock time w/wo fast lock with Fref=100MHz and FVCO=1.6GHz

TABLE I
LOCK TIME (AVERAGED OVER 90 MEASUREMENTS)

Silicon Corner	Lock time Normal/800MHz	Lock time Fast Lock/800MHz	Lock time Normal/1.6GHz	Lock time Fast Lock/1.6GHz	Lock time Fast Lock/800MHz/pre-initialized	Lock time Fast Lock/1.6GHz/ pre-initialized
TT	10.38uS	2.0uS	15.65uS	2.35uS	0.53uS	0.82uS
SS	10.75uS	2.07uS	16.3uS	2.56uS	0.575uS	0.9uS
FF	10.51uS	1.76uS	15.55uS	2.45uS	0.56uS	0.88uS
SF	11.39uS	2.2uS	16.89uS	2.73uS	0.59uS	0.94uS
FS	10uS	1.9uS	15.43uS	2.62uS	0.52uS	0.79uS

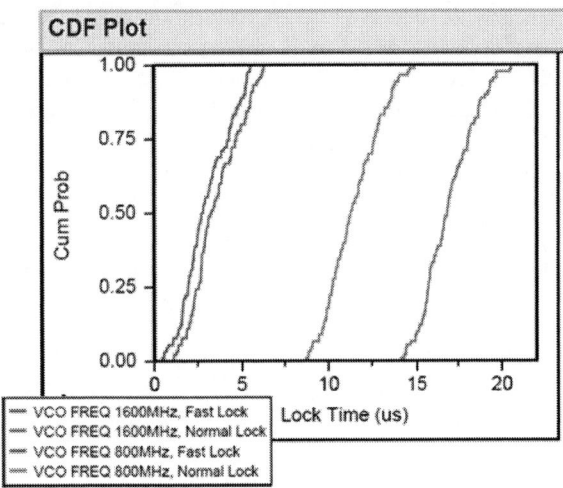

Fig. 9: Statistics of PLL lock time w/wo fast lock for fref=100MHz.

IV. CONCLUSION

A new fast lock technique has been implemented and tested showing a significant reduction in PLL acquiring lock time. The proposed scheme is very efficient, low cost, and easy to integrate. Silicon data shows up to 80% reduction in the maximum PLL lock times with greatly improved distributions over process, voltage and temperature. The new scheme is independent on the VCO, PFD, and charge-pump architectures.

REFERENCES

[1] N. Kurd, J. Douglas, P. Mosalikanti, R. Kumar,"Next Generation Intel Micro-architecture (Nehalem) Clocking Architecture," *IEEE Symposium on VLSI Circuits Digest of Technical papers*, pp. 62-63, 2008
[2] F. Gardner, "Charge-pump phase-lock loops," *IEEE Trans. Commun.*, vol. 28, no. 11, pp. 1849-1858, Nov. 1980.
[3] J. Maneatis, "Low-Jitter Process-Independent DLL and PLL Based on Self-Biased Techniques," *IEEE J. Solid-State Circuits*, vol. 31, no. 11, pp. 1723-32, Nov. 1996
[4] K. L. Wong, et al, "Cascaded PLL Design for a 90nm CMOS High Performance Microprocessor," *IEEE International Solid-State Circuits Conference*, 2003
[5] A. Bashir, et al, "Fast Lock Scheme for Phase-Locked Loops," *IEEE Custom Integrated Circuits Conference (CICC)*, 2009

UCSB

Trends, Possibilities and Limitations of Silicon Photonic Integrated Circuits and Devices

A Tutorial at the IEEE Custom Integrated Circuits Conference 2013

John Bowers

Director, Institute for Energy Efficiency
Kavli Professor of Nanotechnology
Departments of Materials and Electrical and Computer Engineering
University of California, Santa Barbara
bowers@ece.ucsb.edu
http://optoelectronics.ece.ucsb.edu/

Research at UCSB supported by Jag Shah and Josh Conway at DARPA MTO, Intel, Aurrion and HP

UCSB # Acknowledgements

Slides

Lionel Kimmerling (MIT), Tom Koch (Lehigh), David Miller (Stanford), Justin Rattner (Intel), Radha Nagarajan (Infinera), Garry Epps (Cisco), Donn Lee (Facebook), Keren Bergman (Columbia), Jeff Kash (IBM), Lorenzo Pavesi (Univ. of Trento), Haisheng Rong (Intel), Roel Baets (Ghent), Connie Chang Hasnain (UC Berkeley), Dal Negro (Boston University), Pallab Battacharya (Michigan), Meint Smit (IMEC)

Collaborators

UCSB: Sid Jain, Sudha Srinivasan, Jock Bovington, Daoxin Dai, Martijn Heck, Geza Kurzveil, Jon Peters, Jason Tien, Yongbo Tang

Intel : Richard Jones, Yimin Kang, Mario Paniccia, Hyundai Park, Matt Sysak

Aurrion Collaboration: Alex Fang, Greg Fish, Eric Hall, Brian Koch

Hewlett Packard: Di Liang, Marco Fiorentino, Ray Beausoleil

UCSB   Outline

- Motivation: Why Photonics? Why Silicon Photonics?
- Silicon photonic devices
  - Passive Devices
  - Lasers
  - Amplifiers
  - Modulators
  - Switches
  - Photodetectors
  - Isolators
- Silicon Photonic Integrated Circuits

UCSB   Why Photonics?

- Loss is low (0.2 dB/km at 1310 nm O-band and 1550 nm C-band for fiber compared to 1000 dB/km for coax at 10 GHz)
- Capacity is large (50,000 GHz for fiber compared to 20 GHz for coax)

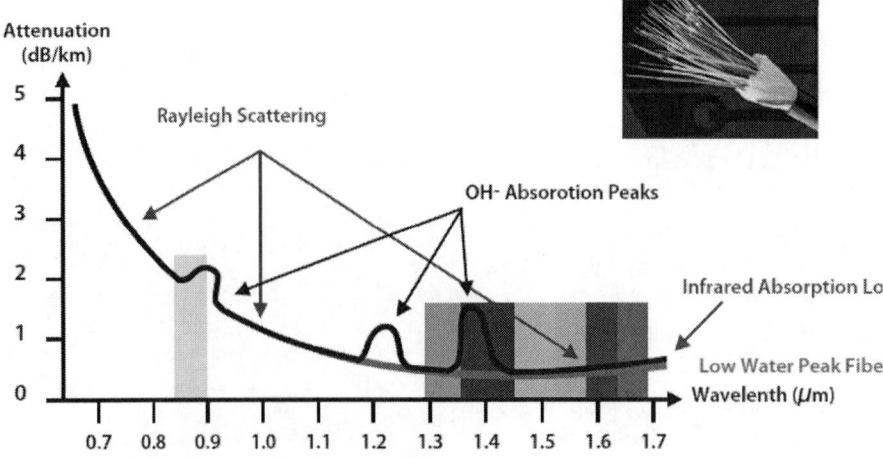

UCSB
HIGH-CAPACITY WDM SYSTEMS
COMMERCIAL CAPACITY SATURATION AT ~ 50 TB/S

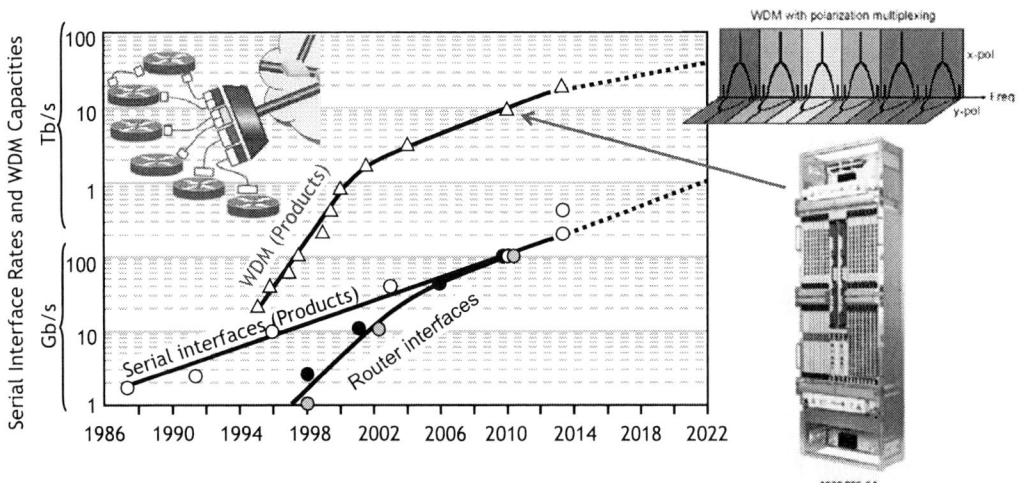

- WDM systems available up to ~10 to 20 Tb/s
- WDM capacity scaling has slowed from ~100%/year to ~20%/year in 2000

[P. J. Winzer, IEEE Comm. Mag. 26-30 (2010)]

UCSB
Why NOT Photonics?

- No significant integration (typically 2 or 3 devices/PIC)
- Cost is high
- Size is large
- Power is large (1 W/Gbps)

192 wavelength
transmission system

UCSB

Why not Integrate?

100 Gb/s Transmit

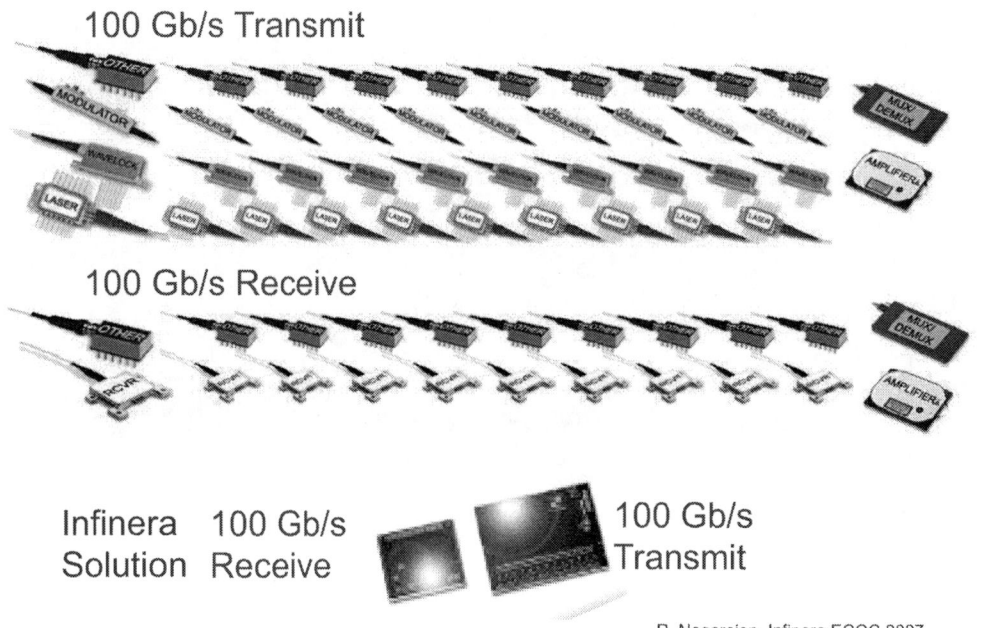

100 Gb/s Receive

Infinera 100 Gb/s
Solution Receive

100 Gb/s
Transmit

R. Nagarajan, Infinera ECOC 2007

UCSB

Why Silicon Photonics?

- Integrate photonics with electronics
  - Same wafer
  - Bump bonding of silicon PIC with silicon IC
    - Same coefficient of thermal expansion
  - 3D stacking

Cross-sectional view of an IBM Silicon Nanophotonics chip combining optical and electrical circuits
Vlasov et al. IEDM postdeadline

- Reduce cost by going to larger diameter wafers
  - InP limited by wafer breakage to 100 mm diameter
- Reduce cost by sharing VLSI facility with electronics
- Improve yield by taking advantage of silicon process development
- Volume driver: Solve IC interconnect bottleneck (from 4 Tbps to 1 Pbps). Optical transmitters/receivers on processors, memories, switches.

978-1-4673-6145-3/13 $31.00 © 2013 IEEE   892

UCSB — The Solution: Optical Interconnects

- 3D layer stacking will be prevalent in the 22nm timeframe
- Intra-chip optics can take

BUT: Silicon is reciprocal. How to make an isolator?

BUT: SiO2 is thermally resistive. So, power dissipation of active devices is a problem, particularly for rings and DWDM

integrated with high

BUT: Silicon is centrosymmetric (not electro-optic)! So, how to integrate modulators?

optimized for performance

BUT: Silicon has an indirect gap and is a poor absorber (not 1550 nm)! So, what about photodetectors?

BUT: Silicon has an indirect gap and doesn't emit light! So, how to integrate laser sources?

UCSB — Bringing Si Manufacturing to the Laser

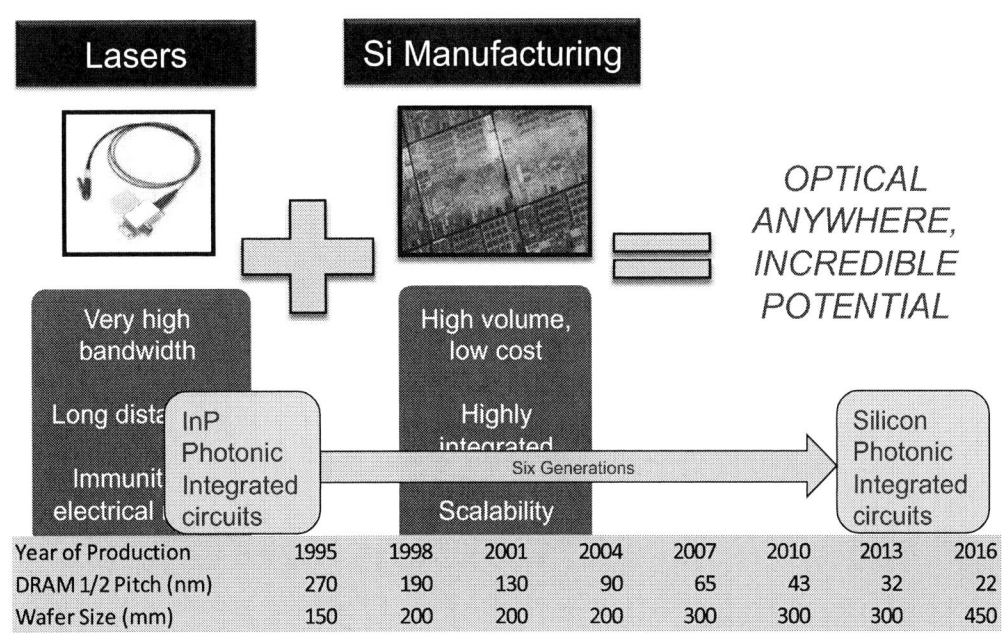

Year of Production	1995	1998	2001	2004	2007	2010	2013	2016
DRAM 1/2 Pitch (nm)	270	190	130	90	65	43	32	22
Wafer Size (mm)	150	200	200	200	300	300	300	450

UCSB Why Photonic Interconnects?

- Electrons are charged particles. They are fermions. Electronic crosstalk is inherent.

- Photons are bosons. They don't interact (you have to work hard to do so). Crosstalk is minimal. 50 Tbps on a waveguide is possible.

- For short lengths, power to drive an electrical connection is proportional to its length.

- Power to drive an optical connection is the same for 1 micron to 100 km.

UCSB Optical and Electrical Power Requirements

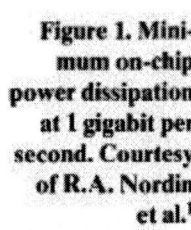

Figure 1. Minimum on-chip power dissipation at 1 gigabit per second. Courtesy of R.A. Nordin et al.[1]

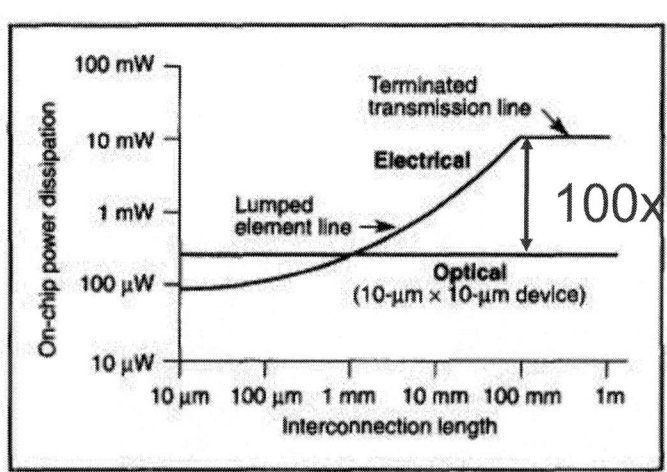

100x less power/line!

UCSB — Moving to Interconnects

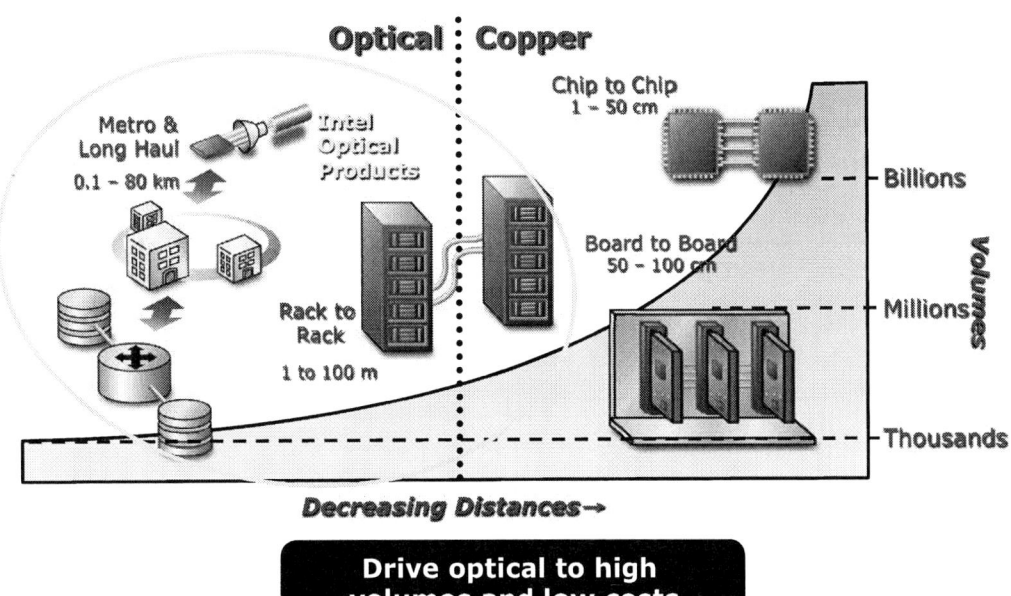

Drive optical to high volumes and low costs

UCSB — Multimode versus Single Mode Fiber

Pros: Cons:

- Multimode fiber

  - Low cost sources (VCSELs)   - Higher fiber cost
  - Cheaper packages   - Higher loss
  - Cheaper connectors   - Lower bandwidth

- Single mode fiber

  - Longer reach (>100 km)   - Higher cost transceivers (not
  - Higher bandwidth (>100 Gbit/s)   true with silicon photonics?
  - Lower fiber cost   - Higher cost connectors

Evolution of Optical Modules

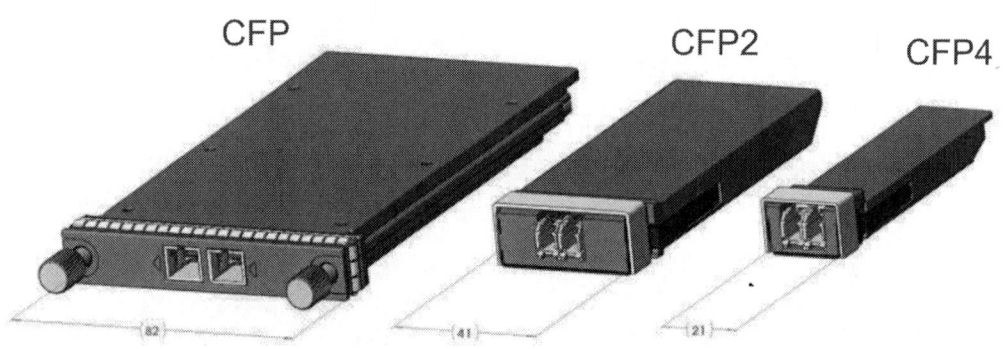

CFP   CFP2   CFP4

Chris Cole, Finisar, ECOC 2012

16 September 2012   10   *Finisar*

Evolution of Optical Transceivers

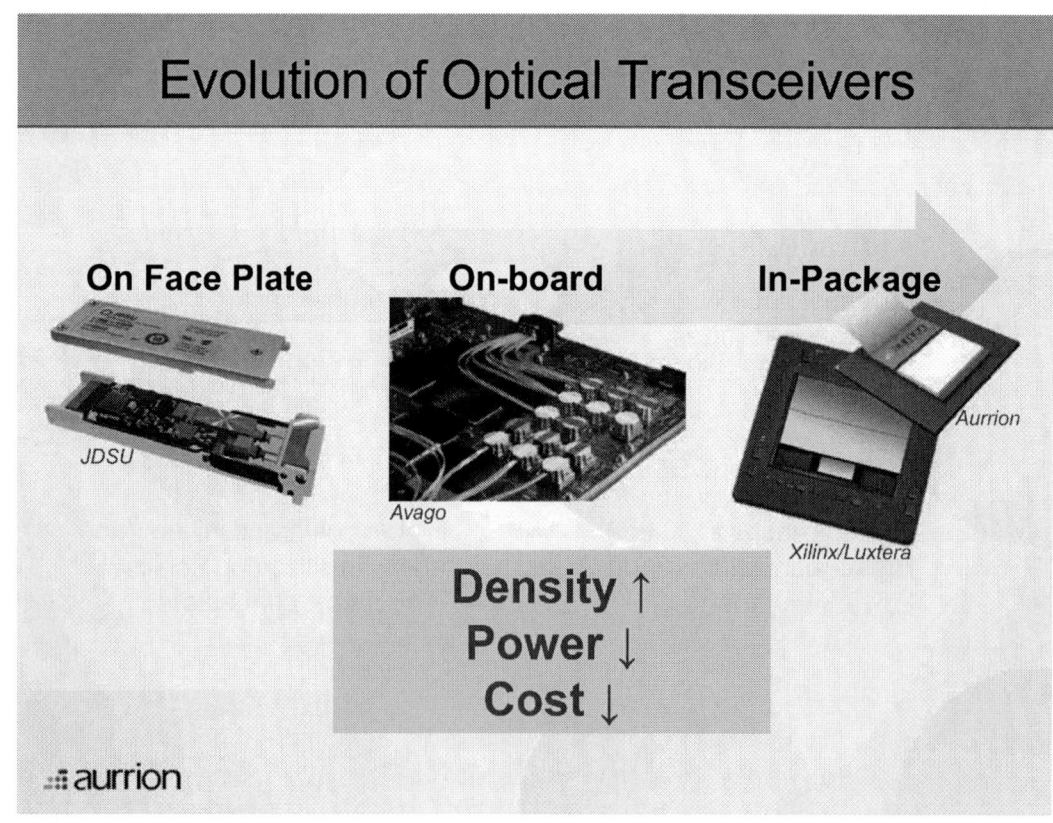

On Face Plate   **On-board**   **In-Package**

JDSU

Avago

Aurrion

Xilinx/Luxtera

Density ↑
Power ↓
Cost ↓

aurrion

In Package: Required Technology

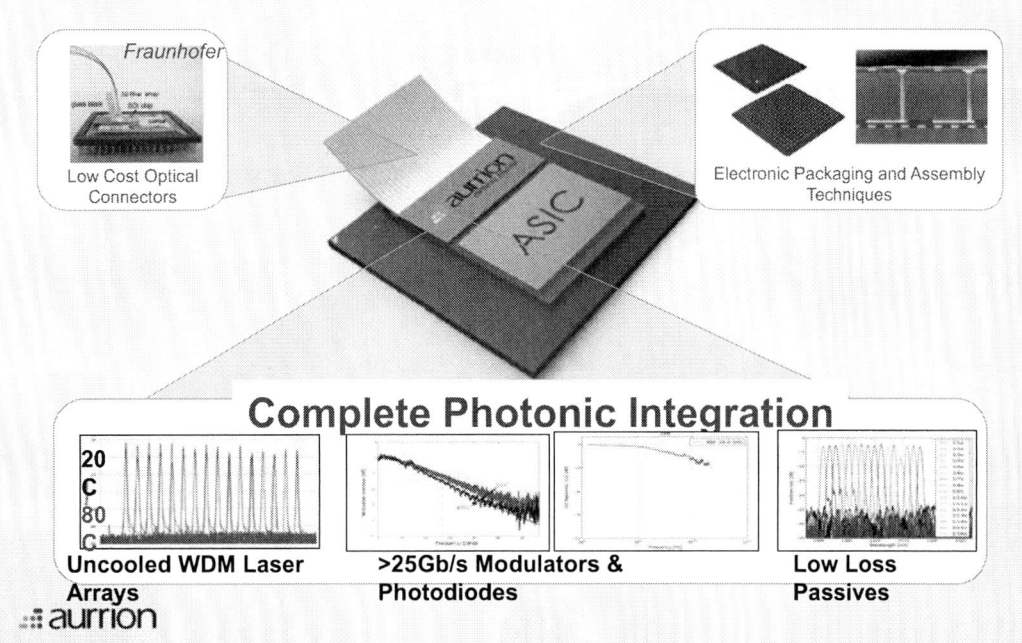

Fraunhofer

Low Cost Optical Connectors

Electronic Packaging and Assembly Techniques

Complete Photonic Integration

Uncooled WDM Laser Arrays

>25Gb/s Modulators & Photodiodes

Low Loss Passives

.: aurrion

UCSB   # Silicon Photonics

- Passive devices
  - Waveguides
    - Si
    - Silicon nitrides
  - Wavelength combiners
    - Echelle
    - Arrayed Waveguide Gratings (AWGs)
- Active devices
  - Lasers
  - Modulators
  - Photodetectors
- Photonic Integrated Circuits

UCSB Silicon Waveguides: SOI Substrates

- Low loss
- High index contrast: $n_{Si}=3.5$, $n_{SiO2}=1.5$
  - Small waveguides, sharp bends

Strip "Silicon wire"	**Quasi-Planar Ridge**	**Large-Area Rib**

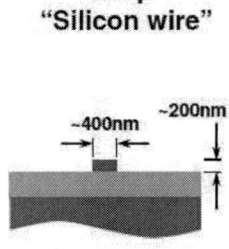

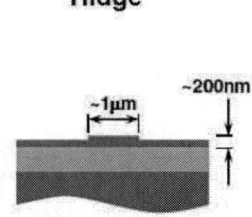

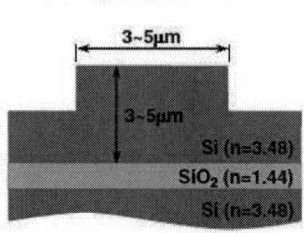

- Small mode-area
- Excellent bending performance (~μm range)
- Higher propagation loss

- Tightest vertical field concentrations for overlap with active phenomena
- Provides means for electrical access for engineering active devices
- Lower propagation loss

- Efficient coupling to SM fiber without taper
- Low propagation loss (~0.1 dB/cm)
- Poor bending radius

Koch, OFC Tutorial 2013

UCSB Passive Si Photonic Devices

- Low loss delay lines, Bends, Filters

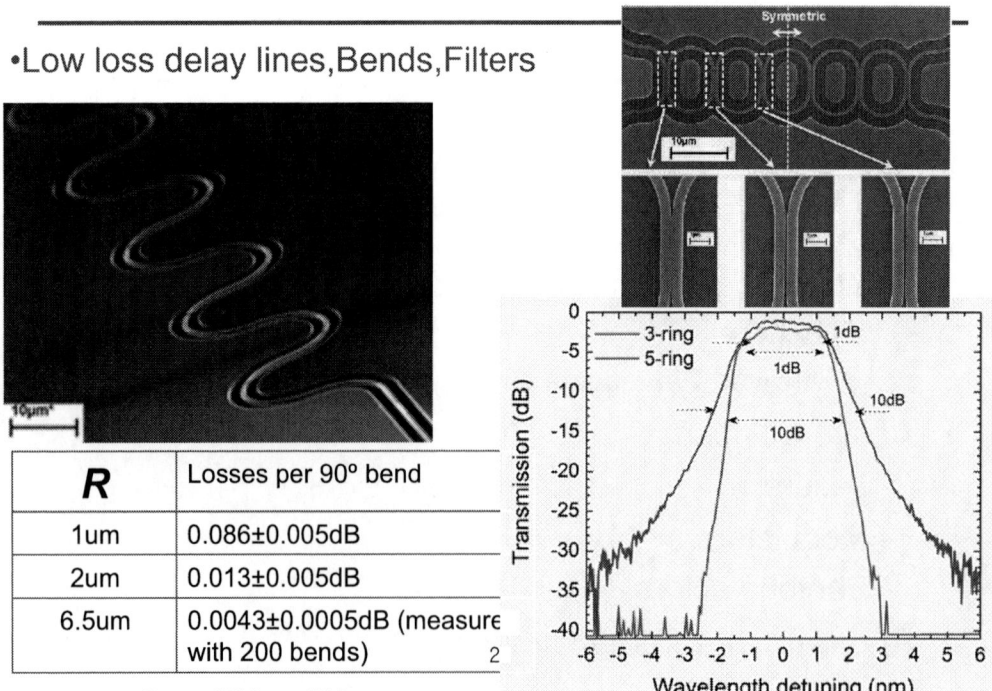

R	Losses per 90° bend
1um	0.086±0.005dB
2um	0.013±0.005dB
6.5um	0.0043±0.0005dB (measure with 200 bends)

Source: Y. Vlasov, IBM

UCSB

Si: High index difference
Low bending loss

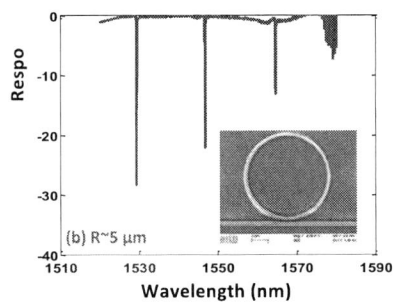 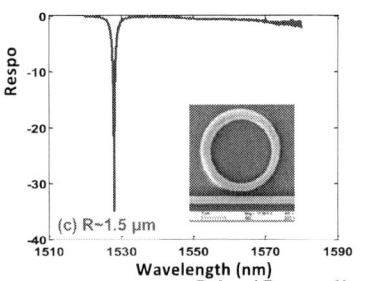

Dai and Bowers, Nanophotonics (2013)

- High coupling efficiency to optical fiber

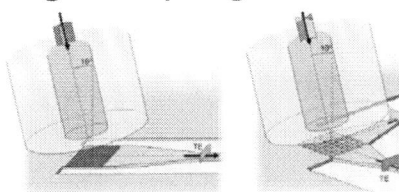 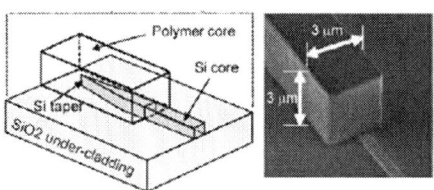

Grating coupler: 3~7dB/facet

Inverse taper: 0.5dB/facet
IEEE JSTQE. 11:232, 2005

W. Bogaerts, et al. Opt. Express 15, 1567 (2007)
Y. Tang, Opt. Lett. 35, 1290 (2010)
x. Chen. IEEE Photon. J. 1(3): 184-190, 2009

UCSB

Ultralow loss Silicon nitride/Silica
Waveguides (0.05 dB/m)

- Low loss requires high temperature processing.
- Not possible on InP or GaAs.
- Add InGaAsP at back end of process.

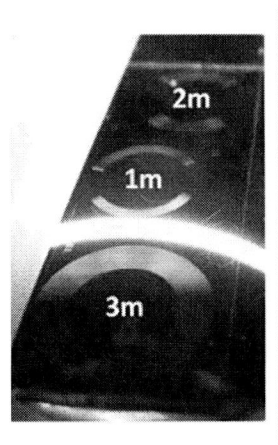 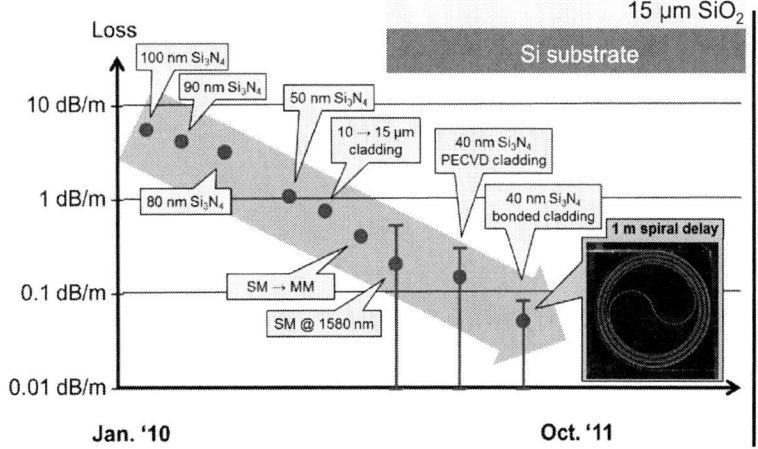

J.F. Bauters, et al., "Planar waveguides with less than 0.1 dB/m propagation loss fabricated with wafer bonding," Optics Express , 19, 24090, 2011

UCSB — Waveguide Loss Comparison

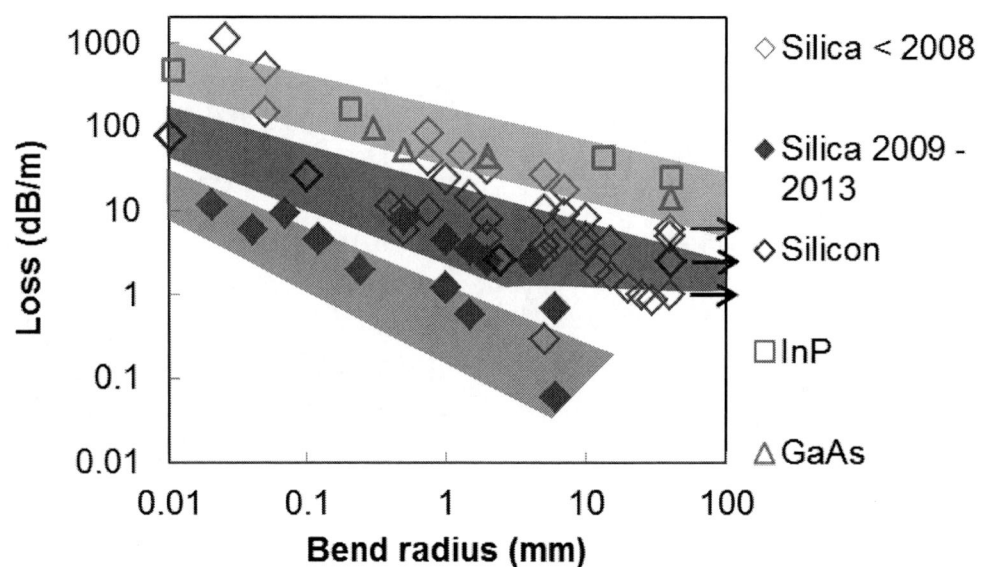

UCSB — Si$_3$N$_4$ Arrayed Waveguide Gratings

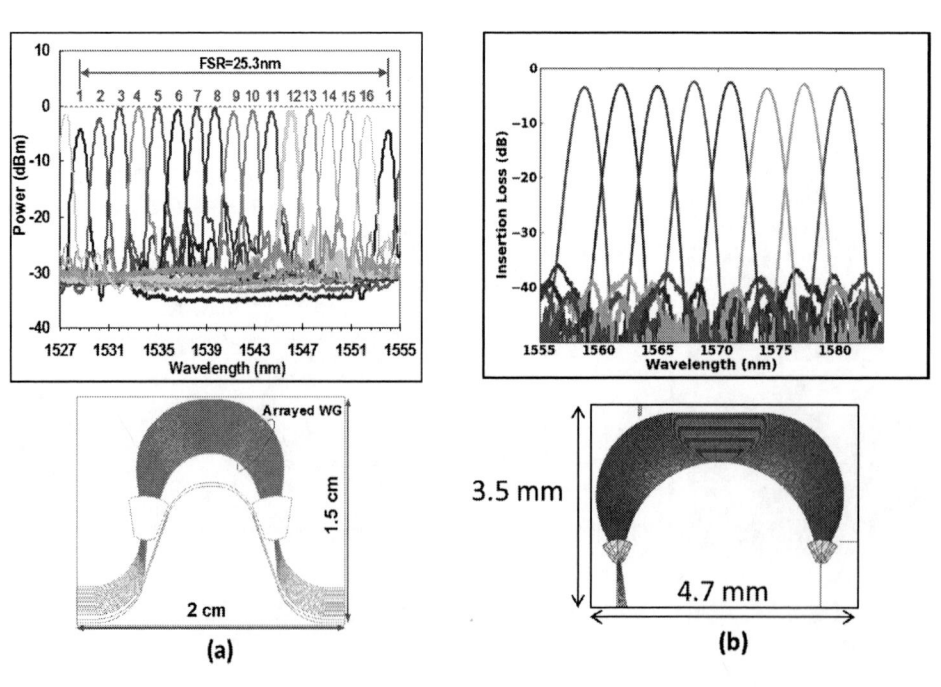

UCSB

Active Optoelectronics

Lasers, Modulators, Photodetectors

UCSB Silicon: Indirect Bandgap

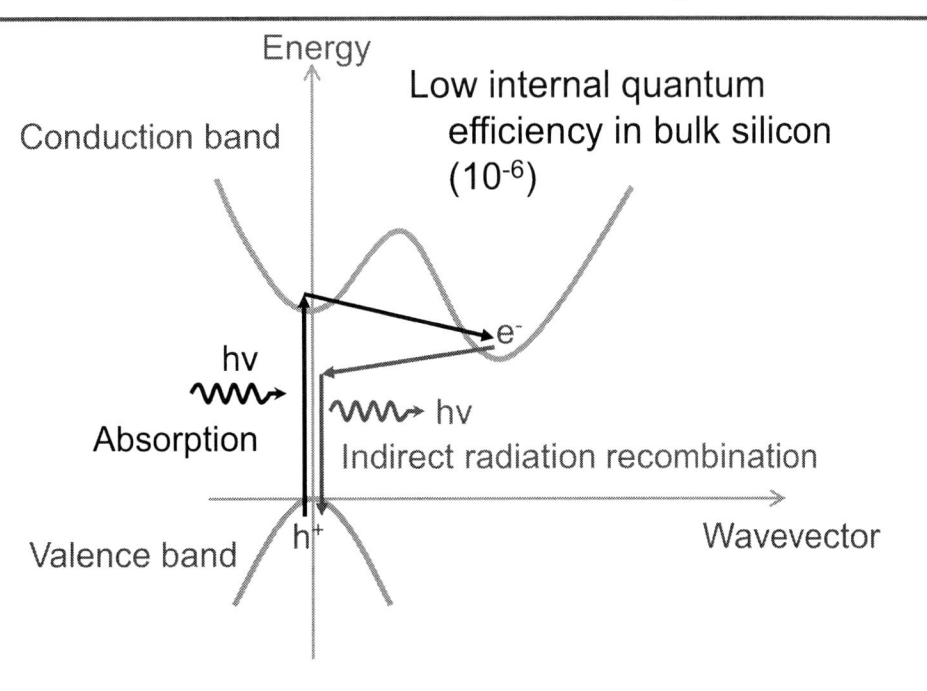

978-1-4673-6145-3/13 $31.00 © 2013 IEEE

UCSB Silicon light emission – How?

- Bulk silicon
- Low dimension Silicon
  - Silicon nanocrystal (Pavesi, …)
  - Periodic nanopatterned crystalline silicon (Jimmy Xu)
- Er dopants (Dal Negro,…)
- Raman laser (UCLA/Intel)
- Another material for gain (hybrid approach)
  - Epitaxial
    - Ge
    - GeSn
    - Quantum Dot
    - Pillars
  - Bonding
    - Die level
    - Wafer level (BCB or Molecular)

> Light emission but not lasing

> Lasing, but optically pumped

> Electrically pumped Lasing

UCSB Er Doped SiN Waveguides
Dal Negro (Boston Univ)

- EL centered at 1535 nm (Er $^4I_{13/2}$)
  - Small peak at 980 nm (Er $^4I_{11/2}$)

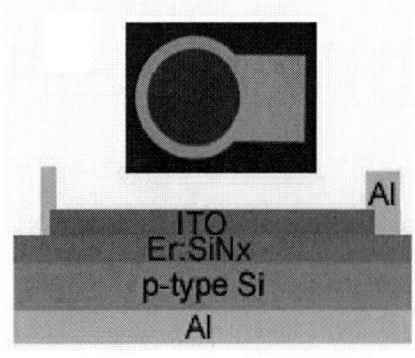
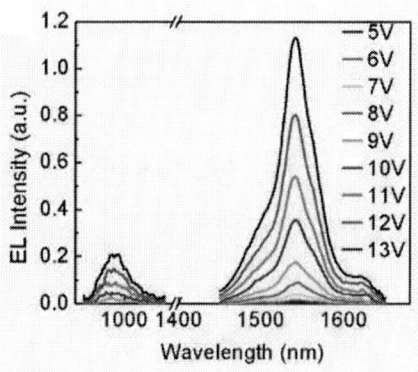

S. Yerci et al., IPR 2010

UCSB

Optically Pumped Er-doped microdisk laser on a silicon chip

- SiO_2 microdisk resonator on silicon;
- whispering gallery-type mode;
- $^4I_{15/2} \rightarrow {}^4I_{13/2}$ transition @ 1450 nm, lasing @ 1550 nm
- pump threshold 43 µW

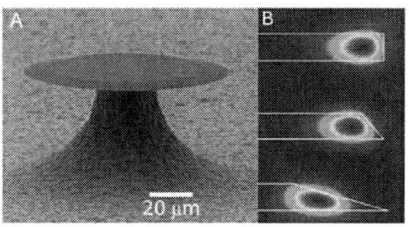

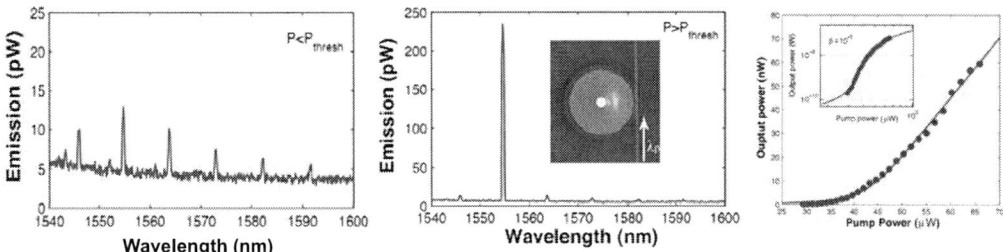

Kippenberg et al., Phys. rev. A, 2006

UCSB

CW Ring Raman silicon laser

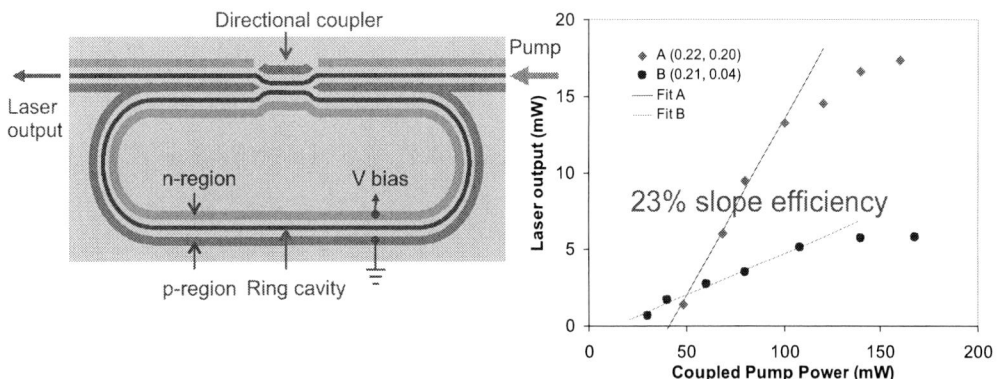

23% slope efficiency

A pump laser is still needed

H. Rong, et al., Opt. Express **14**, 6705-6712 (2006)

UCSB

Hybrid Approaches:
Use another material for gain

- Epitaxial growth on Si substrate
  - Strained Germanium (MIT)
  - SiGeSn (AFRL, U. Mass., Arizona)
  - Quantum Dot (Michigan)
  - MOCVD nanopillars on Si (UC Berkeley)
- Strained Ge Nanomembranes transferred to polyimide (Boston Univ., Univ Wisconsin)
- Bonded III-V layers on Si substrate
  - BCB (Ghent)
  - Molecular (UCSB, Intel, HP, Caltech, U. Tokyo, TIT, Ghent)

UCSB

Direct Gap Transition of Ge by Tensile Strain and n-type doping

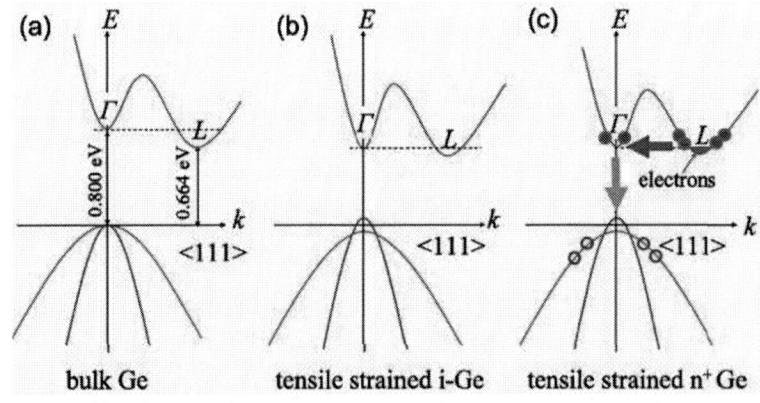

Liu et al, Opt. Express. 15, 11272 (2007)

- Efficient emission at 1550-1620 nm: <u>0.2-0.3% tensile strain</u> **plus**
- <u>n-type doping</u> to compensate energy difference between Γ and L valleys.

Slide courtesy of L. Kimmerling, MIT

UCSB Ge-on-Si: in-plane tensile strain

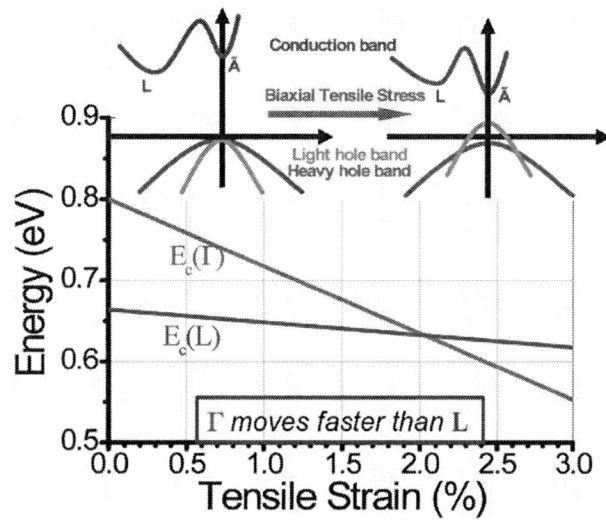

- Ge is pseudo-direct gap and compatible with CMOS.
- Tensile strain drives Ge towards direct gap behavior.

Slide courtesy of L. Kimmerling, MIT

UCSB Electrically pumped Ge-on-Si laser

- Pulsed operation at 15 C
- Threshold 280 kA/cm²
- 1.7 V at threshold
- Threshold current: 0.6 A*?
- Threshold power: 1 W*?
- Output: 1 mW at 370* kA/cm²
- Efficiency ~10⁻⁴

*Not stated in paper. Estimated.

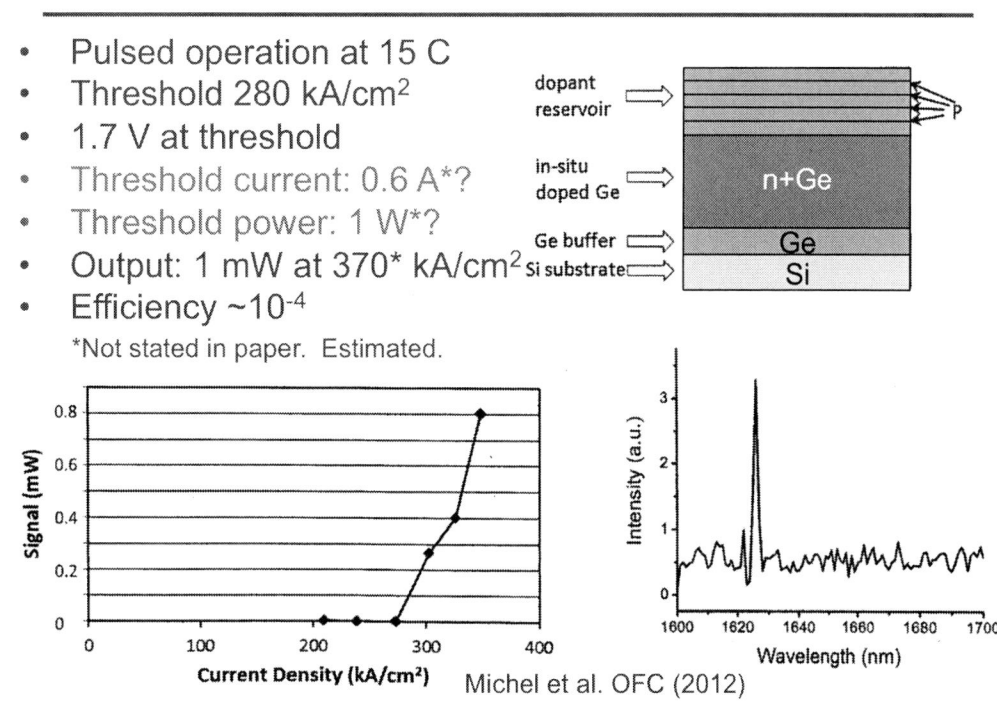

Michel et al. OFC (2012)

UCSB
Integrated Laser/Modulator on Silicon
Bhattacharya, University of Michigan

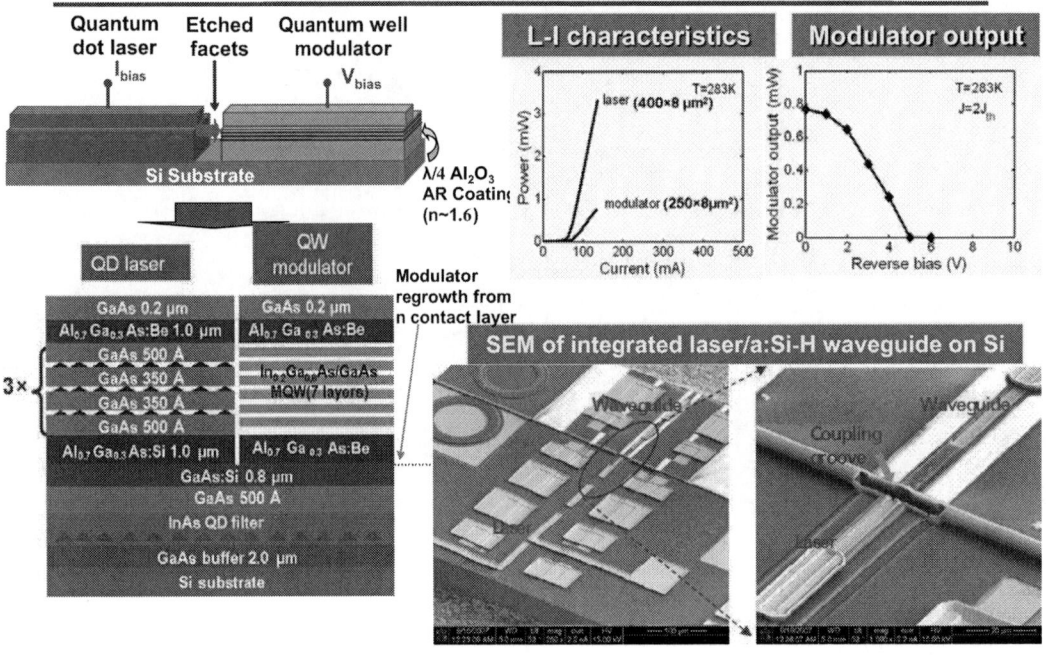

MBE re-growth, PECVD deposition of a:Si-H waveguides.   J. Yang, P. Bhattacharya,, *Optics Exp* 16, 5136 (2008)

UCSB
Epitaxial Quantum Dots
Seeds et al. UCL

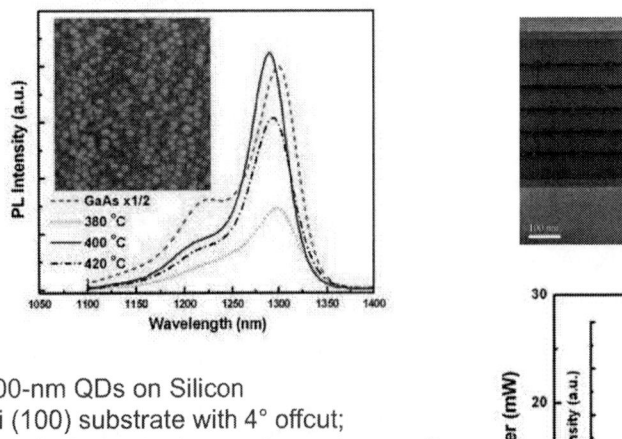

1300-nm QDs on Silicon
• Si (100) substrate with 4° offcut;
• InAs/InGaAs dot-in-a-well structure on Ge
• QD density of 4.3 Å~ 1010 cm-2;
•Pulsed RT Lasing at 1302 nm
Threshold current density of 725 A/cm2,
• Output power of 26 mW at RT;
• Lasing up to 42 oC

Liu et al., IPRM (2012), Opt. Express 19, 11381 (2011)

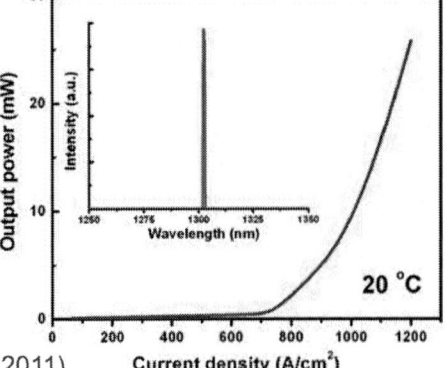

UCSB Bond III-V Lasers

- Another material for gain (hybrid approach)
  - Bonding: Die level
    - Flip-chip bonding

UCSB Die bond lasers one at a time (Luxtera)

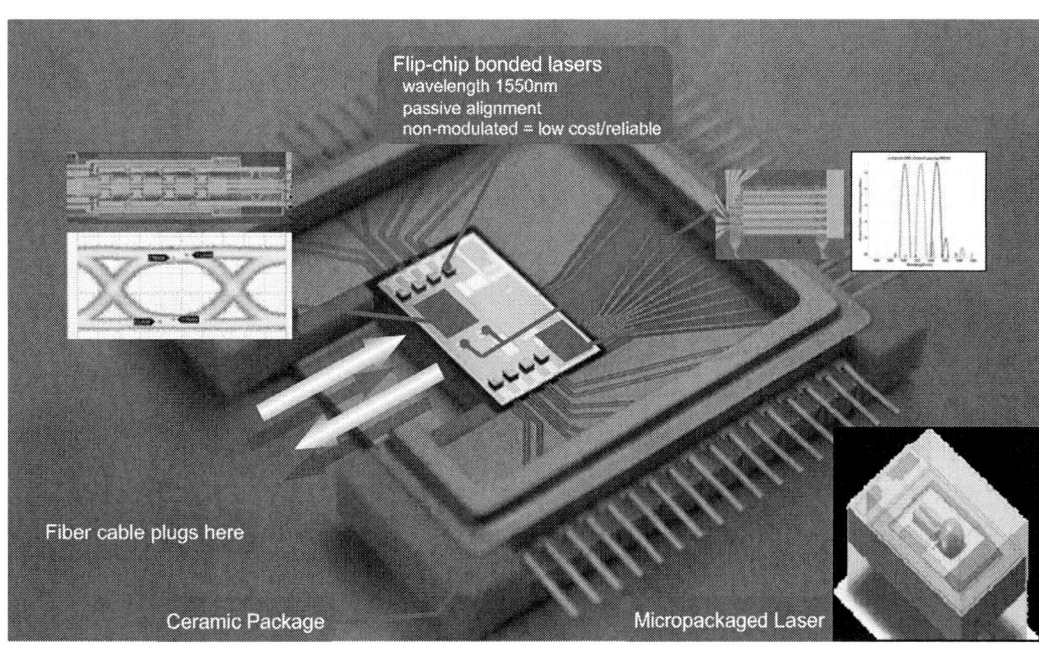

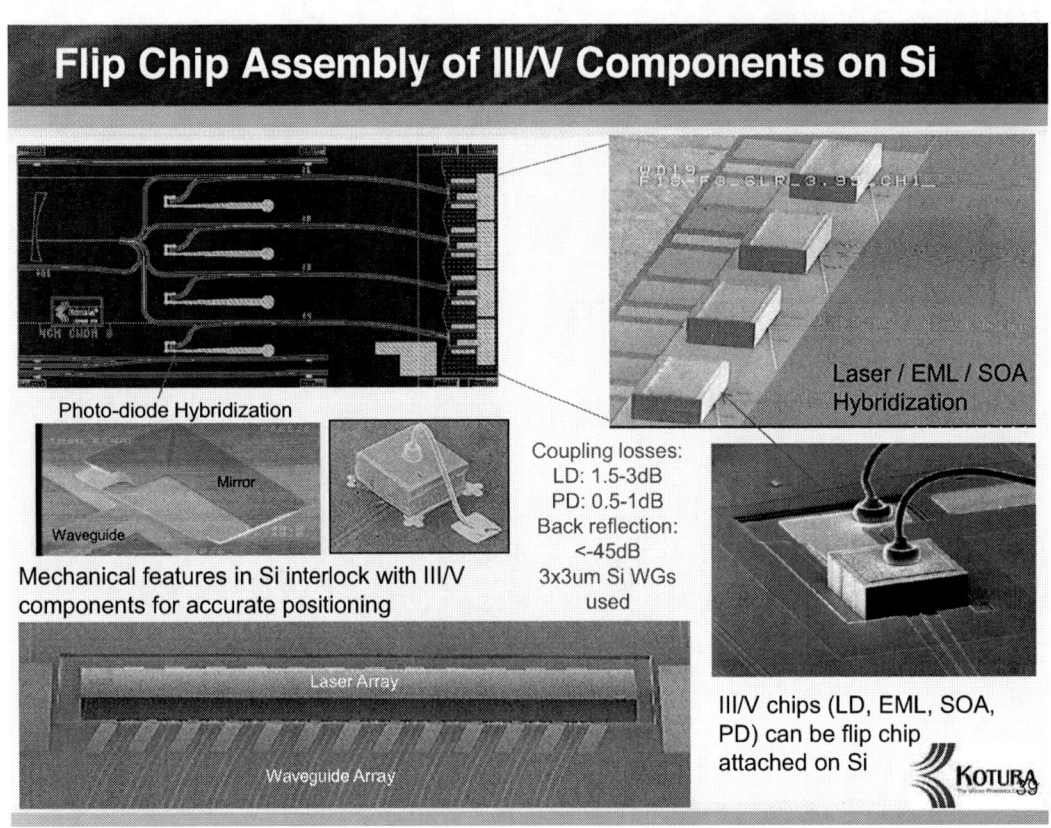

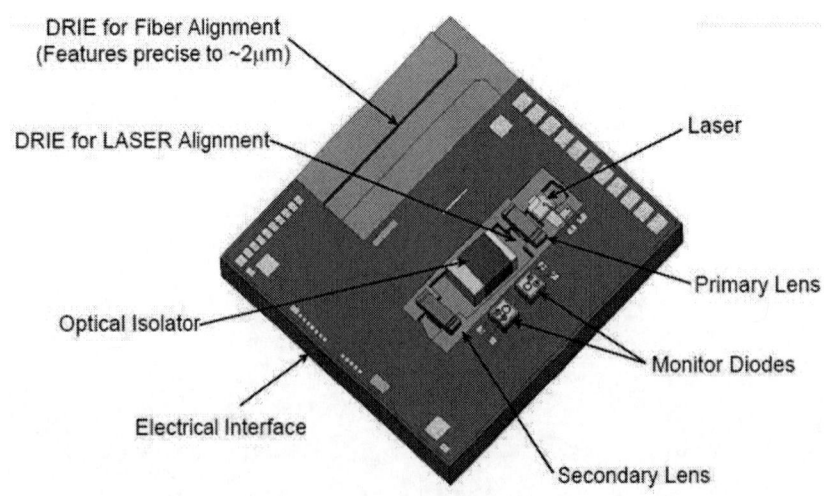

UCSB
Heterogeneous integration
UCSB, Intel, HP, Ghent, TIT, Caltech

Step 1: Bond InP-dies on SOI waveguide

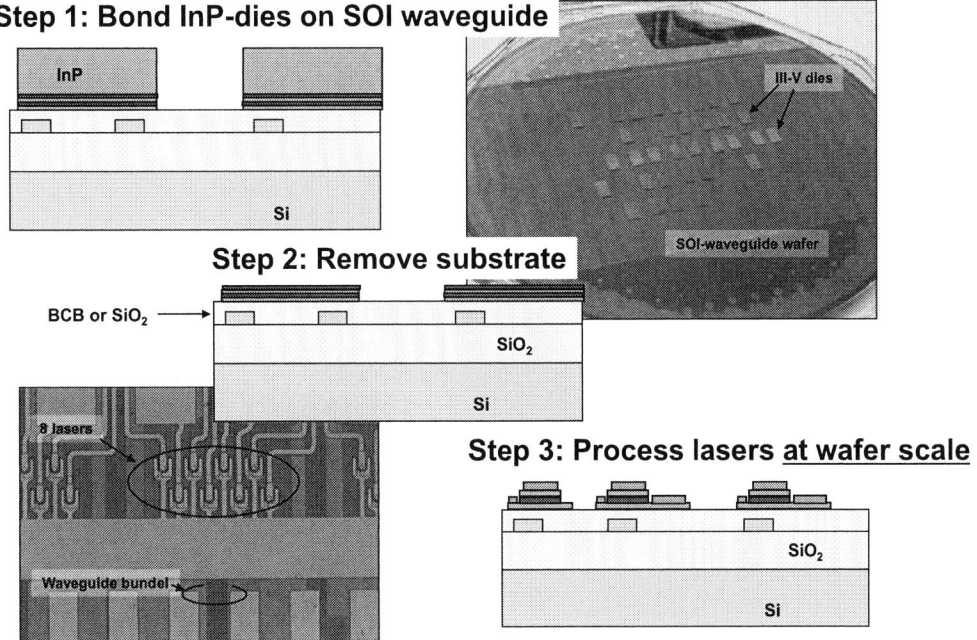

Step 2: Remove substrate

Step 3: Process lasers <u>at wafer scale</u>

UCSB
Heterogeneous integration

Two alternatives for the die-to-wafer bonding process

- Adhesive layer bonding
  - Planarization and bonding in single step (IMEC-Ghent University)
  - Ultra-thin bonding layers (sub 200nm) [1]

- Molecular bonding
  - InP on SOI-waveguides (UCSB, Intel, CEA-LETI, TRACIT) [2,3]

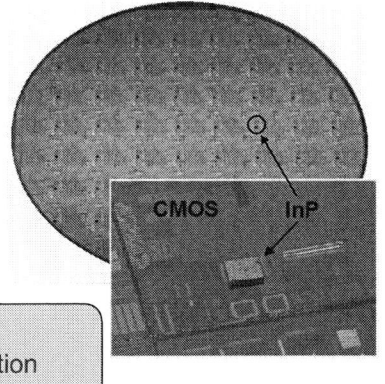

Direct bonding has the lowest thermal resistance, highest cw temperature operation (105C) and highest power (45 mW single mode)

[1] G. Roelkens et al., "Adhesive Bonding of InP/InGaAsP Dies to Processed Silicon-On-Insulator Wafers using DVS-bis-Benzocyclobutene", J. Electrochem. Soc., Volume 153, Issue 12, pp. G1015-G1019 (2006)

[2] D. Liang469. D. Liang, G. Roelkens, R. Baets, J. E. Bowers , "Hybrid Integrated Platforms for Silicon Photonics," Materials , 3 (3), 1782-1802 , March 12 , 2010

[3] M. Kostrzewa et al., 'InP dies transferred onto silicon substrate for optical interconnects application ', Sensors & Actuators A 125 (2006) 411-414

UCSB Scaling of Direct Bonded Wafers

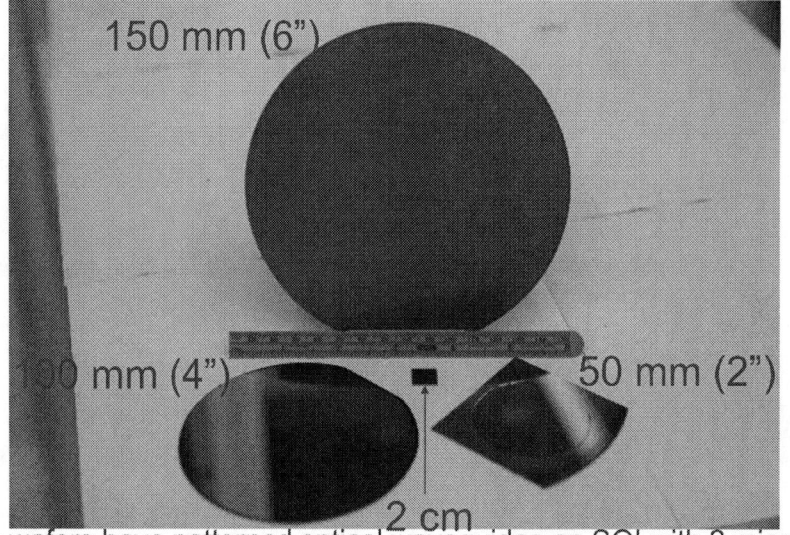

These wafers have patterned optical waveguides on SOI with 2 micron GaInAsP layer on top.

Oxygen plasma enhanced bonding: 300 C, 30 minutes

•D. Liang, G. Roelkens, R. Baets, J. E. Bowers , "Hybrid Integrated Platforms for Silicon Photonics," *Materials* , **3** (3), 1782-1802 . March 12 . 2010

UCSB   Hybrid Silicon Photonics

Direct Gap III-V
InGaAlAs

Silicon

- Optical gain from III-V Material
- Efficient coupling to silicon passive photonic devices
- No bonding **alignment** necessary: suitable for high volume CMOS
- All back end processing low temperature (<350 C)
- CW lasing to 105 C

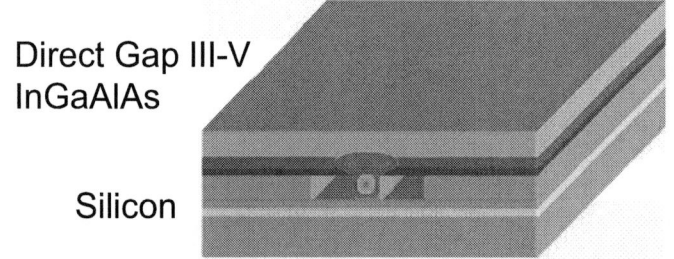

Liang and Bowers, Nature Photonics, **4**, 511, Aug. 2010.

UCSB   Device fabrication

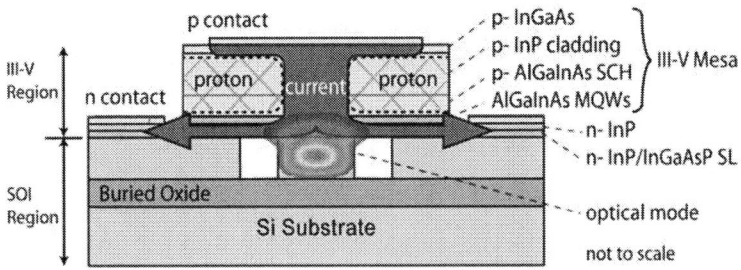

- Current injection:
  top-down; spreading via
  n-contact layer
- Proton implanted mesa:
  lateral current confinement

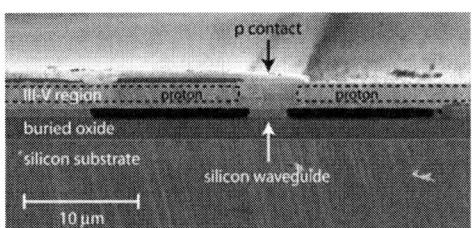

UCSB   **105 C** CW 1310 nm laser

- 1310 nm important for
  FTTH and data
  communications.
- Max fiber coupled output
  power: 5.5 mW
- Max operation temperature:
  105 °C
- T_0: 80 °C
- Injection efficiency: 52 %

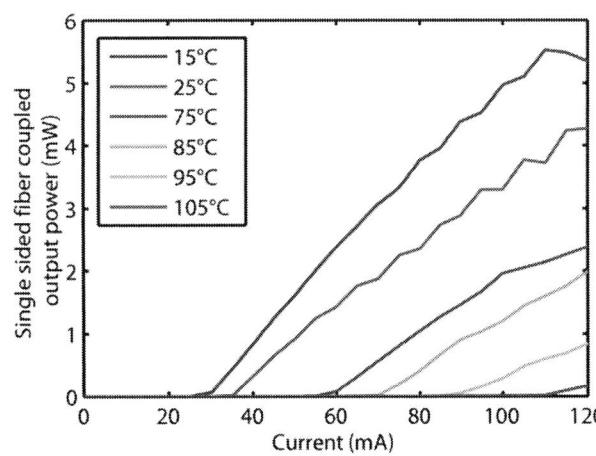

Chang et al., Optics Express 15(18), 11466, August (2007).

UCSB
Microring Lasers

- Fang et al., UCSB
- Liang et al., Hewlett Packard, UCSB
- Van Campenhout et al., Ghent

UCSB
Hybrid Silicon Racetrack Ring Laser with Integrated Photodetectors

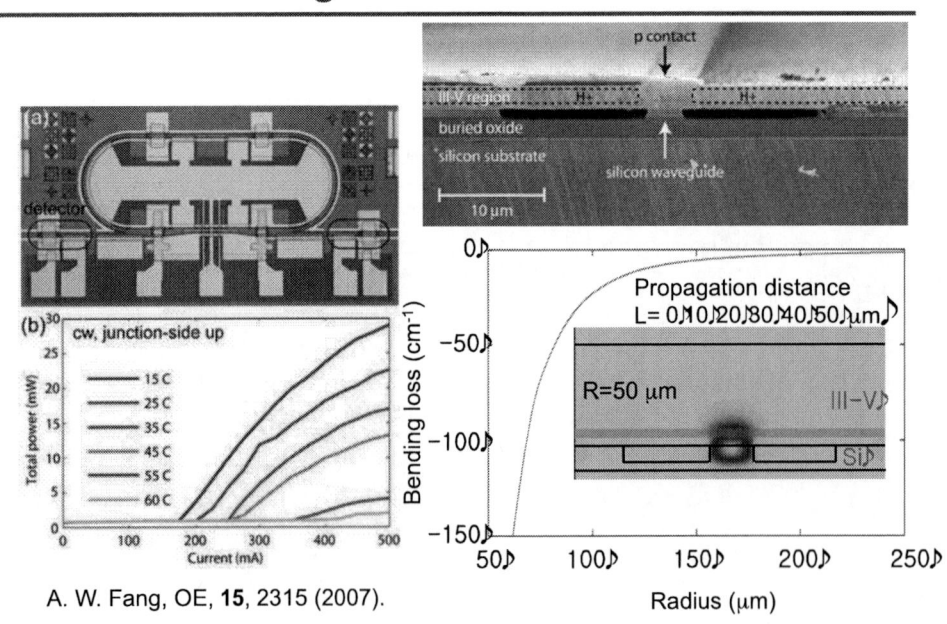

A. W. Fang, OE, **15**, 2315 (2007).

Hybrid Silicon Microring Laser

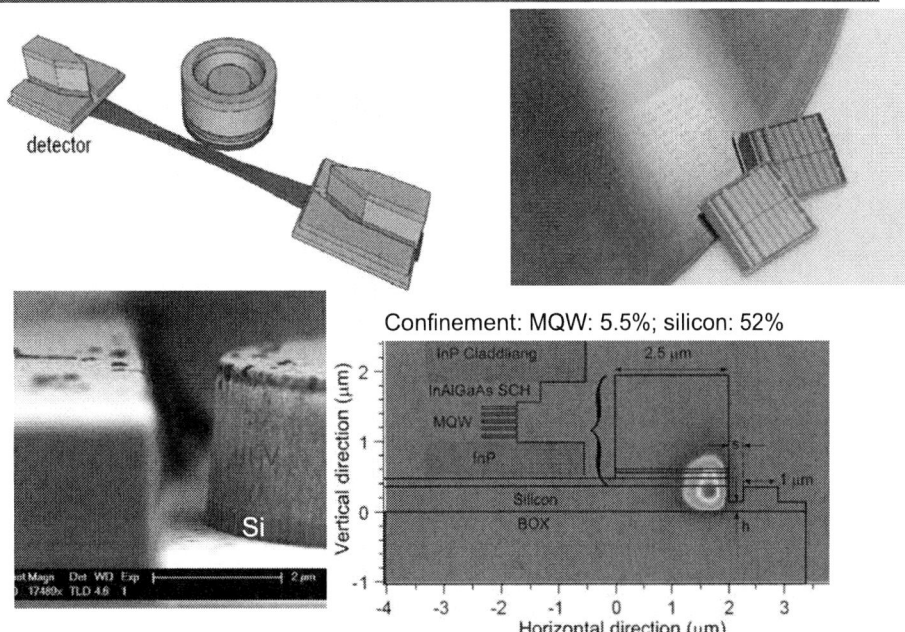

Confinement: MQW: 5.5%; silicon: 52%

D. Liang, et al. *Optics Express*, **17** (22), 20355-20364 , October 23 , 2009

Threshold Improvement

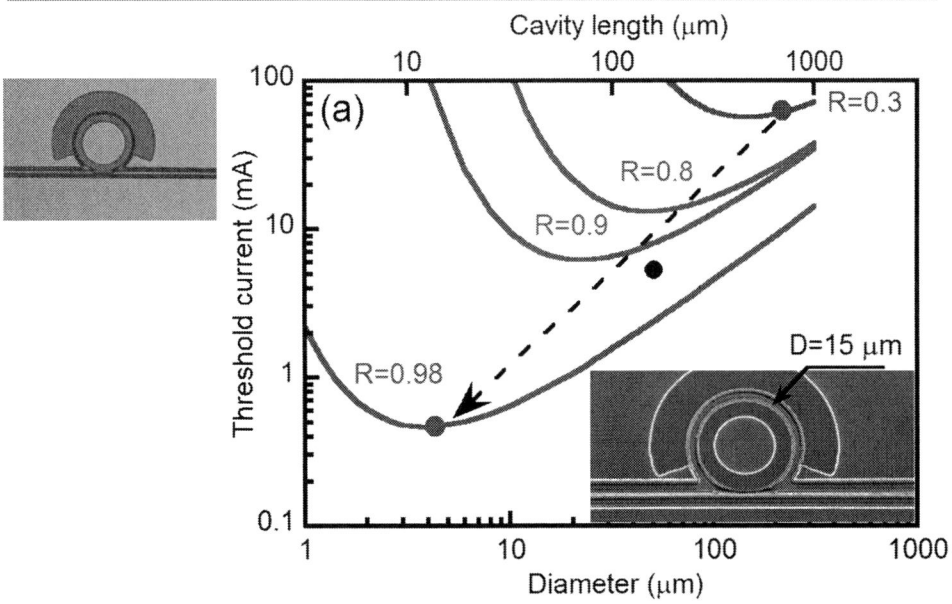

D. Liang et al., Group IV Photonics 2009

UCSB Scaling bottleneck: device heating

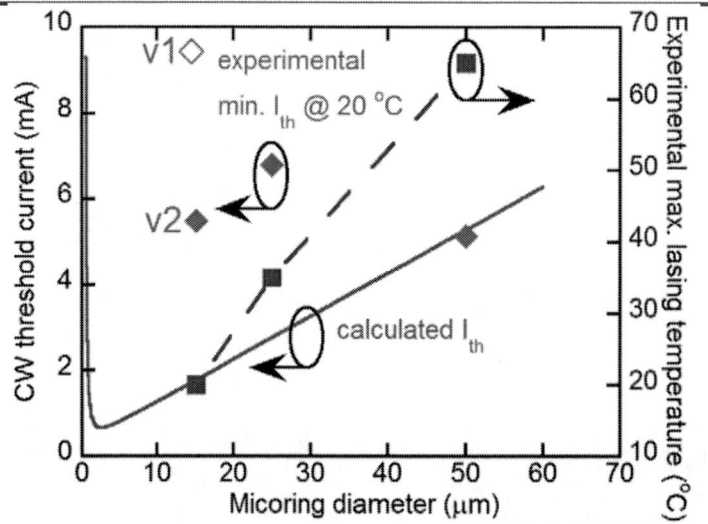

- Reducing ring diameter increases resistance → More heating → Lar...

Thermal shunts through SiO2 developed at Intel and HP to improve thermal resistance

UCSB HP photonics technologies

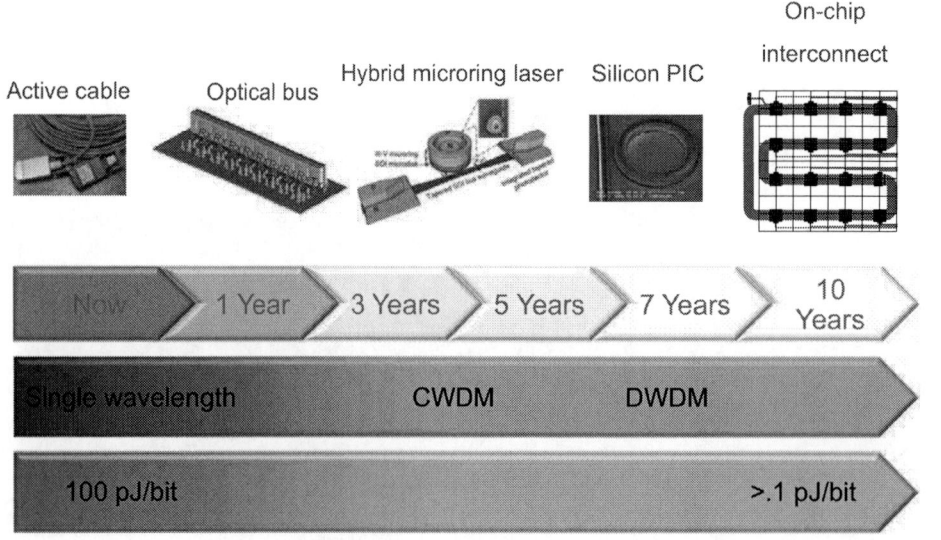

UCSB DWDM Ring Array (Ghent)

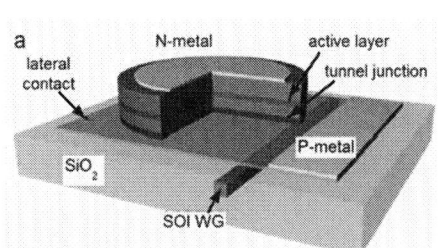

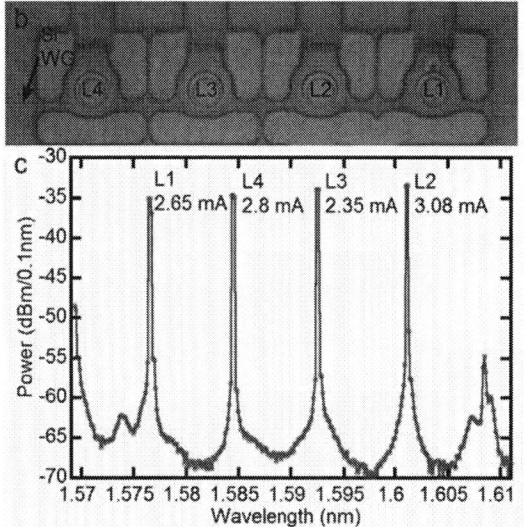

Van Campenhout, J. *et al.*, A Compact SOI-Integrated Multiwavelength Laser Source Based on Cascaded InP Microdisks. *IEEE Photonics Technology Letters* 20 (16), 1345-1347 (2008).

UCSB

Tunable Distributed Bragg Reflector (DBR) Laser Technology
Duan (Ghent) ECOC 2012
Sysak (UCSB/Intel): SGDBR

Distributed Feedback (DFB) Laser

Yariv (Caltech) IPC 2012, ISLC 2012
Srinivasan (UCSB) ISLC 2012
Sid Jain (UCSB) ISLC 2012

UCSB — Multiple bandgaps: Quantum Well Intermixing

Hybrid Silicon Sampled Grating DBR Tunable Laser

- Six electrically isolated regions
- Three III-V material bandgaps
  - Gain/backside absorber ●
  - Electroabsorption modulator ●
  - Front/rear mirrors and mode transformer ③
- Multiple bandgap functionality
  - Carrier injection for wavelength tuning
  - Shift mode transformers outside laser cavity

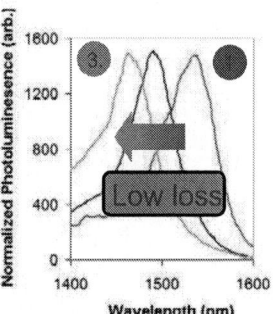

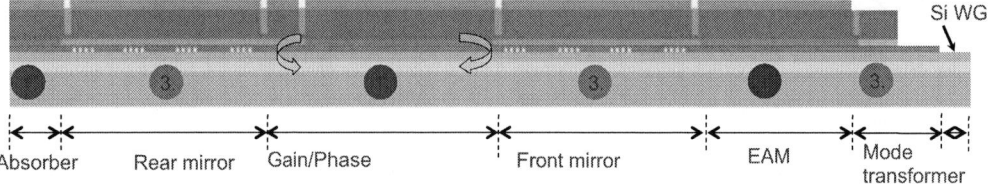

Absorber | Rear mirror | Gain/Phase | Front mirror | EAM | Mode transformer

M. Sysak, J.O. Anthes, J.E. Bowers, O. Raday, R. Jones , *Optics Express* , **16** (17), 12479-12486 , August 18 , 2008

UCSB — III-V/Si Tunable laser (Ghent)

- 8nm tuning range, thermo-optic tuning of silicon ring resonato

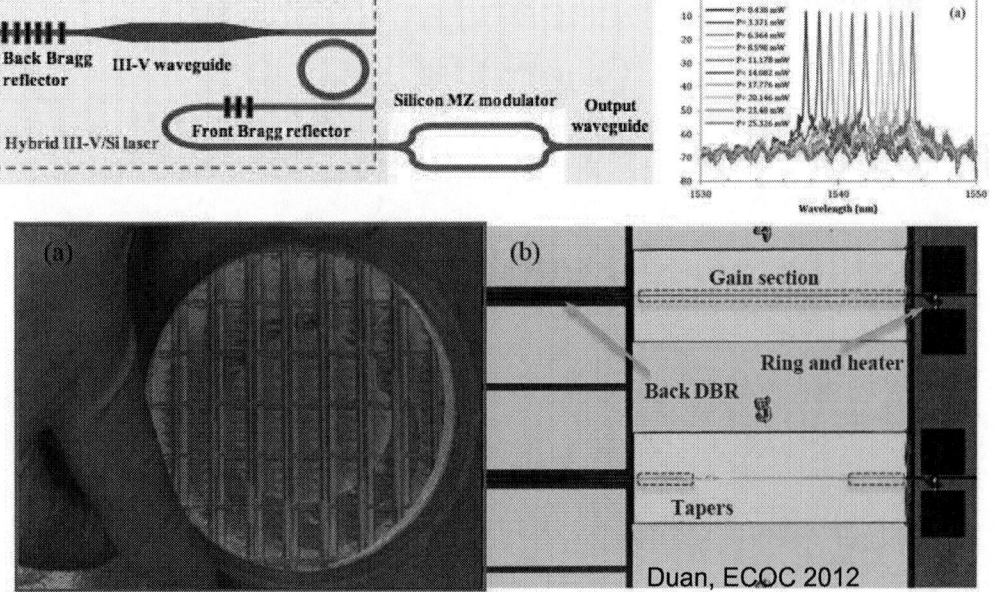

Duan, ECOC 2012

UCSB The future of the DFB laser (Yariv, 2012)

Hybrid Si/III-V semiconductor laser

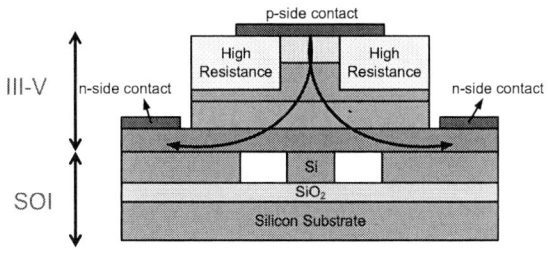

Cross section

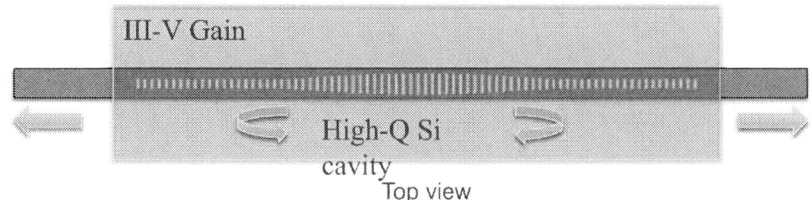

Top view

- Combines strengths of both worlds: high-Q resonator in Si, high gain in III-V
- Less sensitive to carrier fluctuations in the active region, lower α parameter

---

UCSB High-Q/Slow-light Si resonator: Structure

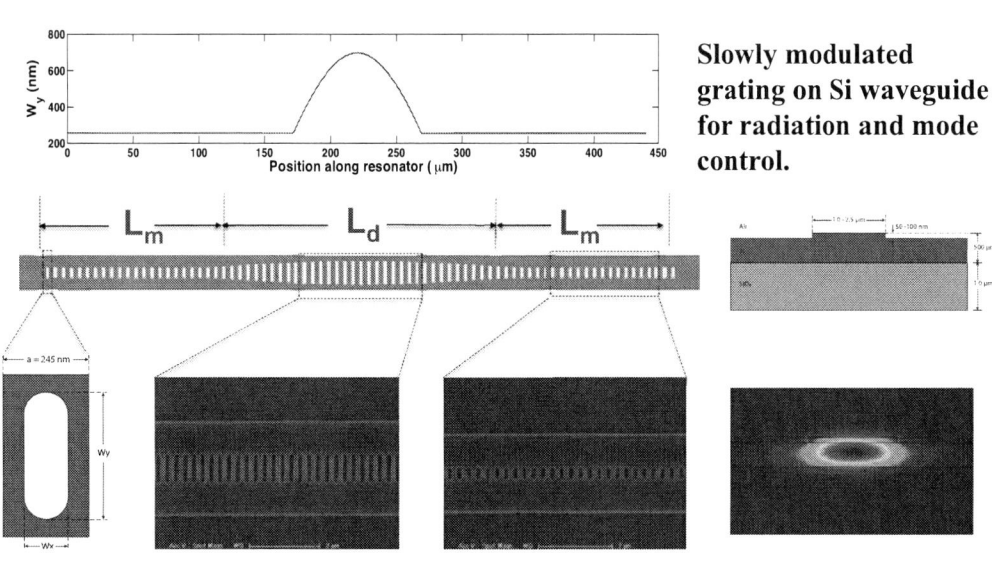

Slowly modulated grating on Si waveguide for radiation and mode control.

High-Q/Slow-light Si resonator: Experiment

UCSB

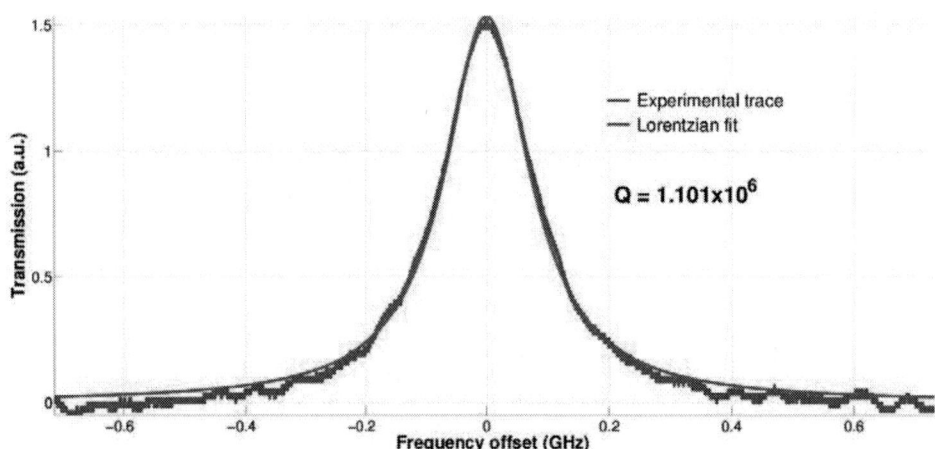

$Q = 1.101 \times 10^6$

Highest-Q grating-based resonator on Si or other material on substrate (not suspended) to date.

UCSB

DFB Performance
Hybrid Silicon Reliability and Yield

Srinivasan et al. (UCSB) ISLC 2012
Sid Jain et al. (UCSB) ISLC 2012

UCSB

16 DFB Arrays

- Quantum well intermixing to create 4 bandgaps on one wafer.
- Use for a broad spectrum of DFB/EAM integration

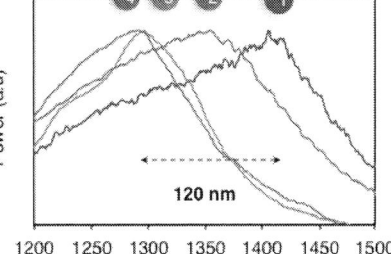

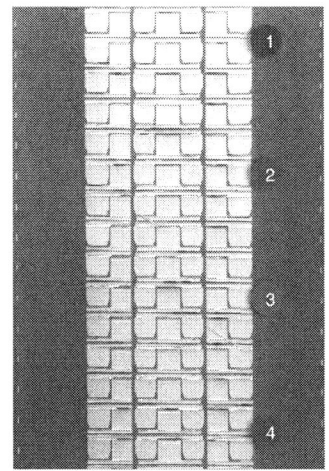

DFB   PD   DFB

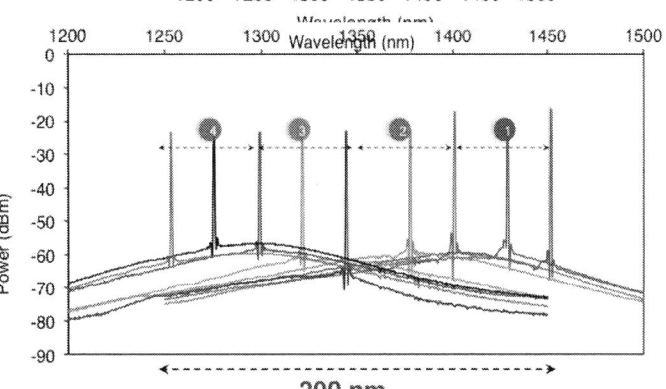

Sid Jain et al. (UCSB) ISLC 2012

Chip layout

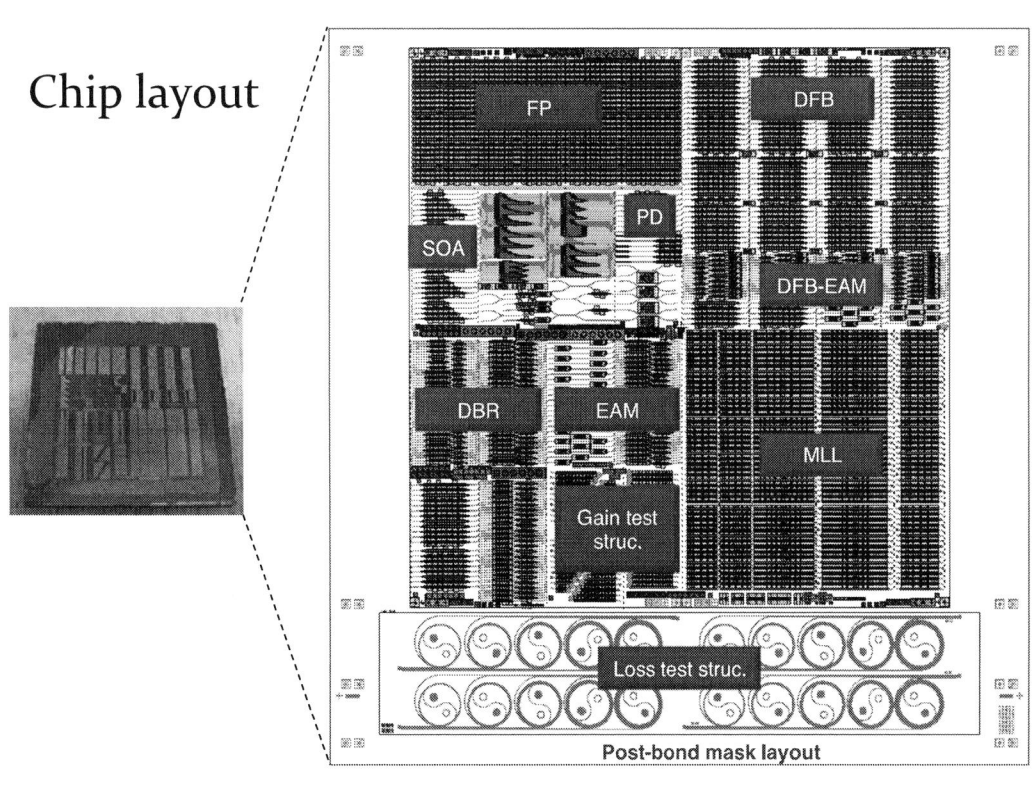

978-1-4673-6145-3/13 $31.00 © 2013 IEEE   919

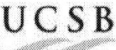

Distributed feedback laser structure

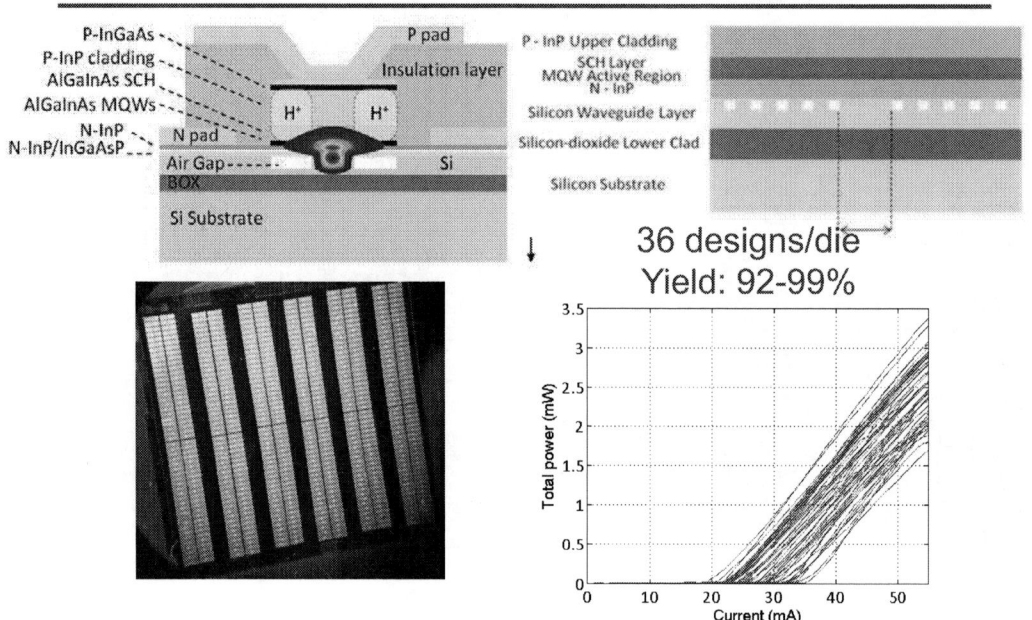

36 designs/die
Yield: 92-99%

S. Srinivasan and J.E. Bowers , "Reliability of hybrid III-V on Si distributed feedback lasers," International Semiconductor Laser Conference (ISLC) 2012 , San Diego, 2012

Temperature Dependence of DFB Aging

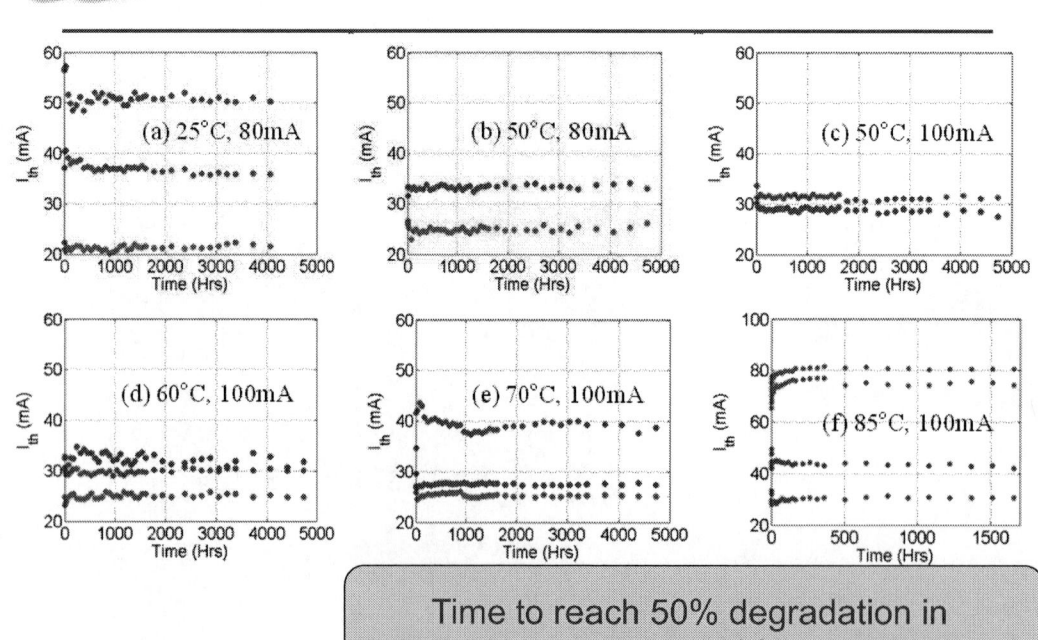

Time to reach 50% degradation in threshold current at 70°C is >40,000hrs

S. Srinivasan and J.E. Bowe[...] lasers," International Semiconductor Laser Conference (ISLC) 2012 , San Diego, 2012

UCSB

Bonded III-V QD Laser on Si
Arakawa, Univ. Tokyo

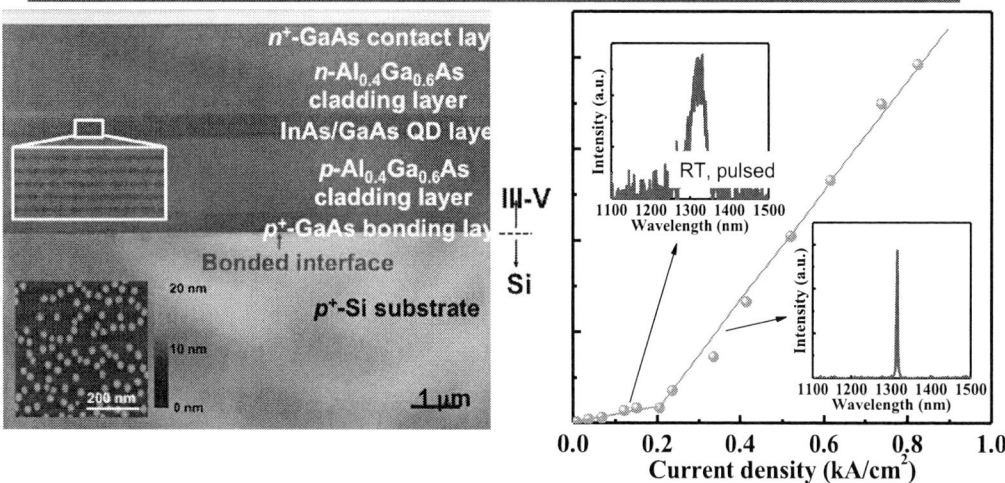

- Room temperature lasing at 1.3 μm O-band, ground state transition of InAs QDs

- Threshold current density 205 A/cm² (20.5 A/cm² per QD layer)

K. Tanabe et al, Scientific Reports 2, 349 (2012)

UCSB

Hybrid Silicon Evanescent
Optical Amplifiers

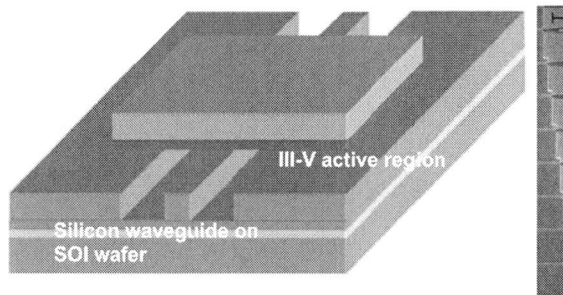

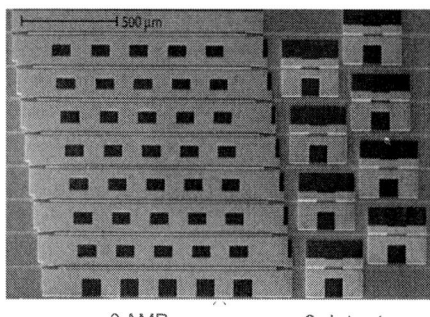

8 AMPs   8 detectors

- **Impact**
  - Electrically pumped amplifiers (unlike Raman or Erbium amplifiers)
  - Wider wavelength range than erbium amps: 1310 nm, S, C, L band operation

- **Issues:**
  - Minimize reflections at transitions for spectrally flat gain, and high gain.

H. Park et al., PTL, 19(4), 230, February (2007).

UCSB

Silicon Modulator

UCSB

EAM

Input   absorber   Output

$$T = \exp\left(-\alpha_0 L - \underbrace{\alpha(V)}_{\text{voltage-controlled loss}} L\right)$$

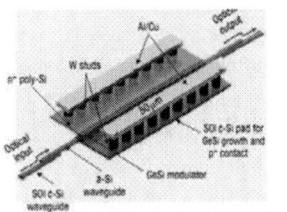

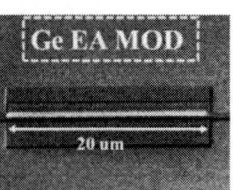

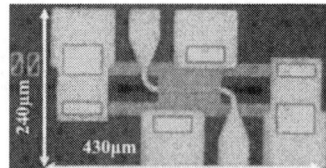

Liu, et al., Nat. Photonics, 2, 433-437, 2008

Lim, S*STAR, OFC OWQ2, 2011

Tang, et al., OE, 19(7), 5811-5816 (2011)

Group	f_{3dBe} [GHz]	Len [μm]	ER [dB|V_{pp}|Gb/s]	Type	Note
Liu, MIT, 2008	1.2	50	--	GeSi	Frank-Keldysh
Rong, Stanford, 2010	13	30	0.53|2.5|3.125	GeSi/Si	QCSE, λ=1408nm
Tang, UCSB, 2011	74	100	9.8|2|50	AlGaInAs	QCSE, hybrid
Lim, A*STAR, 2011	--	100	--|--|1.25	Ge	FK, λ=1600nm

UCSB UCSB Hybrid Silicon EAM Modulator

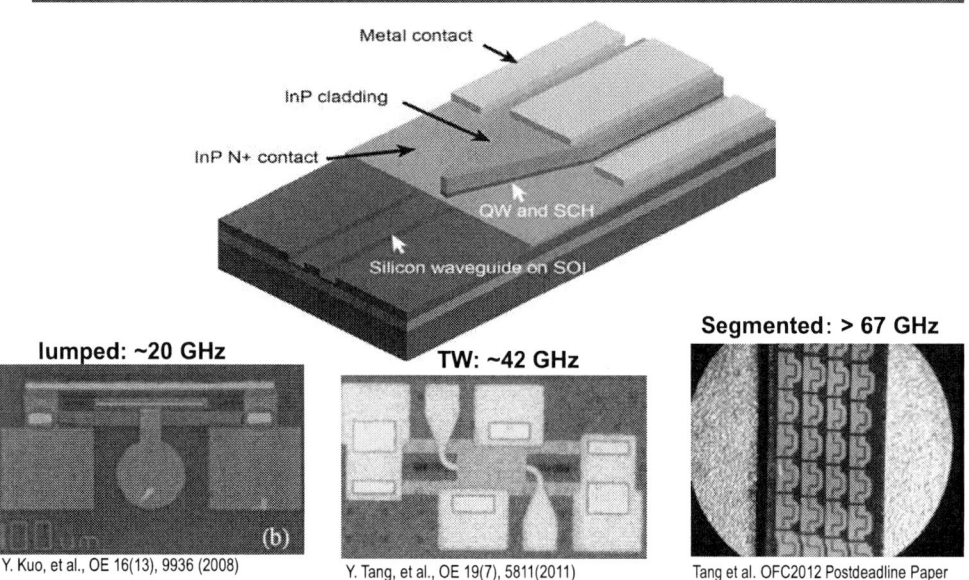

lumped: ~20 GHz

Y. Kuo, et al., OE 16(13), 9936 (2008)

TW: ~42 GHz

Y. Tang, et al., OE 19(7), 5811(2011)

Segmented: > 67 GHz

Tang et al. OFC2012 Postdeadline Paper

UCSB UCSB Hybrid Silicon HSEAM

➢Segmented electrode
➢Length: 100 µm
➢>67 GHz Bandwidth

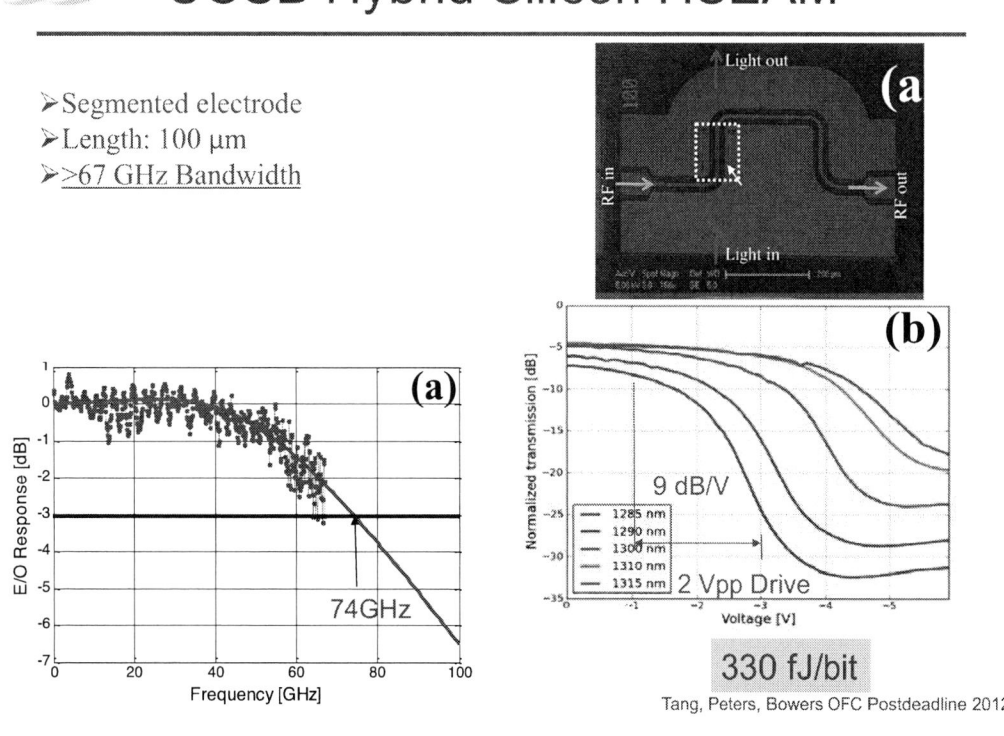

74GHz

9 dB/V

2 Vpp Drive

330 fJ/bit

Tang, Peters, Bowers OFC Postdeadline 2012

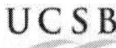

Large Signal Modulation

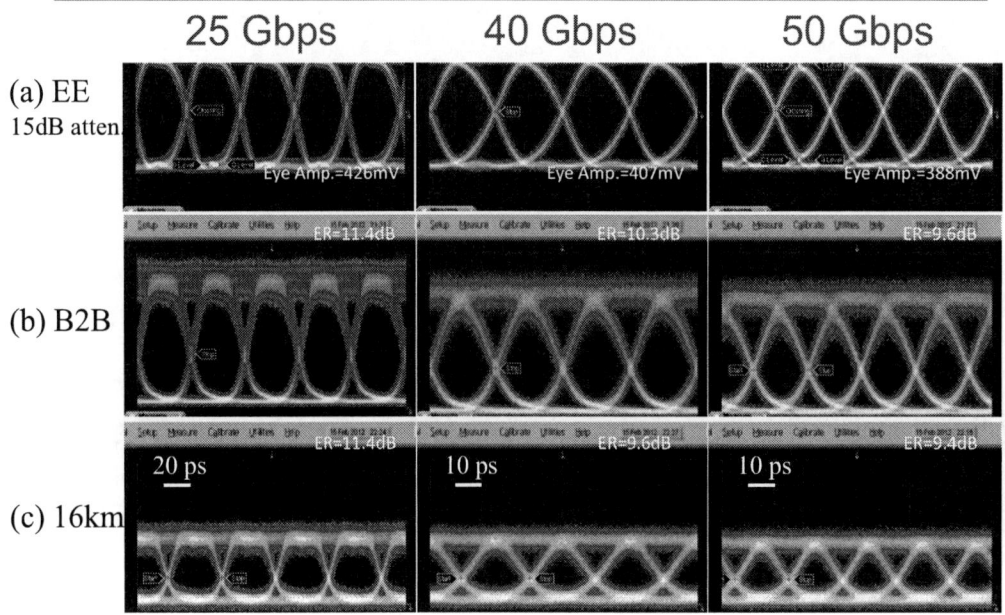

Index Change by Carrier Depletion

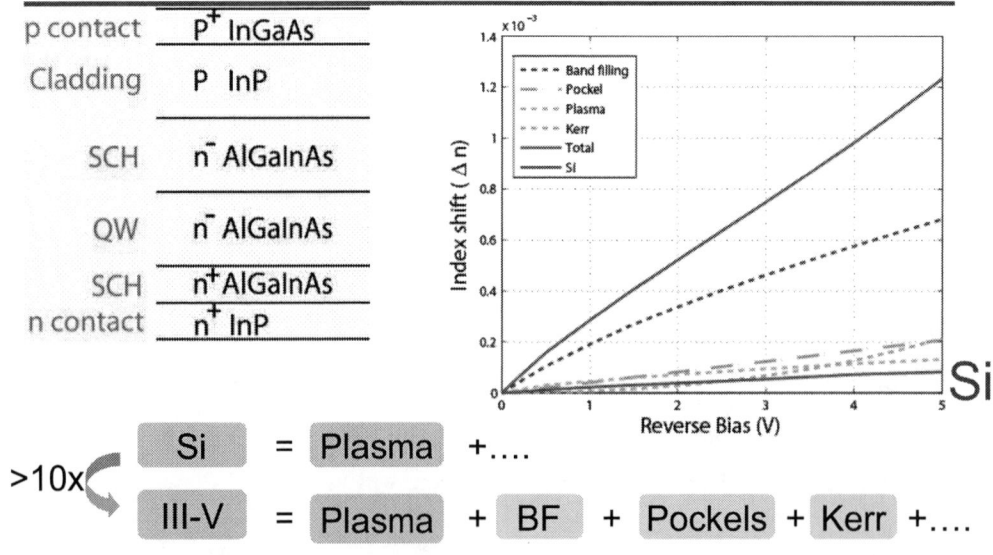

H.-W. Chen et al., "25Gbps Hybrid silicon switch using a capacitively loaded traveling wave electrode," Opt. Exp. 18, 1070 (2010).

UCSB # Silicon optical modulators

(1) MZ silicon modulator

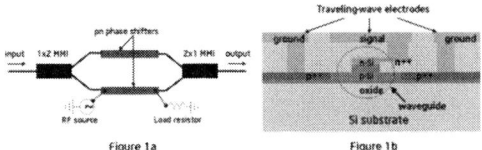
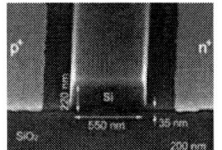

Intel. A. Liu. 40Gbps   IBM. W. Green. 10Gbps

(2) Microring/disk modulator

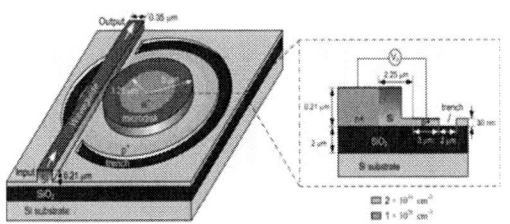
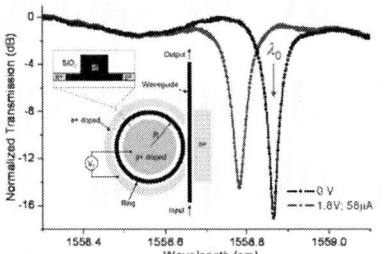

HKUST. A. Poon. 0.5 Gb/s   Cornel Univ. M. Lipson. 12.5Gpbs

UCSB # Mach Zehnder Modulators

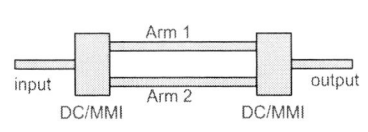

$$T = \frac{1}{2}\left[1 + \frac{2E_1 E_2}{E_1^2 + E_2^2}\cos(\Delta\phi)\right]$$

$$\Delta\phi = \underbrace{\frac{2\pi}{\lambda}\Delta n_{eff} L}_{dynamic} + \underbrace{\frac{2\pi}{\lambda} n_{eff}\Delta L + \Delta\phi_{coupler}}_{\Delta\phi_{bias}}$$

Group	f_{3dBe} [GHz]	Len [μm]	VπL [V·mm]	ER [dB|V_{pp}|Gb/s]	Type	Note
Basak, Intel, 2008	10	3450	33	3.8|1.1|10	MOS	
Fujikata, NEC, 2010	25?	120	5	3|3.5|12.5	MOS	
Green, IBM, 2006	--	200	0.36	--|1.2|10	PIN	Pre-emphasis, 7Vpp
Gu, UT Austin,2007	--	80	0.16	--|--|1	PIN	PhC waveguide
Basak, Intel, 2009	33	1000	40	1|1|6.2|40	PN	TW, 14Ω
Park, ERTI, 2009	7	1500	18	3|4|12.5	PN	
Feng, Kotura, 2010	12	1000	14	7|8|12.5	PN	
Marris, PSUD, 2008	10	4000	50	--|--|--	PININ	TW, 50
Liow, A*STAR, 2010	--	2000	26	6|5|10	PN	
Chen, UCSB, 2011	23	500	2	10|4|40	Hybrid	CLTWE, 25
Baehr-Jones, UW, 2010	3	1000	8	--|--|--	slot	polymer

UCSB — Ring Modulator

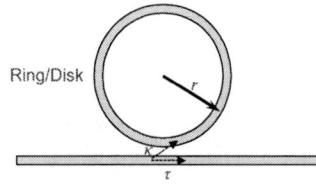

Ring/Disk

$$T = \left| \frac{\tau - \exp(-i\phi)}{1 - \tau \exp(-i\phi)} \right|$$

$$\phi = \underbrace{(\beta - i\alpha)2\pi r}_{\text{static phase shift}} + \underbrace{(\Delta\beta - i\Delta\alpha)l_a}_{\text{dynamic phase change}}$$

Group	f_{3dBe} [GHz]	len [µm]	Q	ER [dB|V_pp|Gb/s]	Type	Note
Xu, Cornell, 2007	--	30	39350	3.8|1.1|10	PIN	
You, ETRI, 2008	8	260	9482	1.2|4|12.5	PN	
Dong, Kotura, 2009	11	90	14500	6.5|2|10	PN	
Zheng, Sun, 2010	15	90	14500	3.1|2|5	PN	Driver integrated
Zhou, HKUST, 2006	0.51	30	16900		PIN	Microdisk
Liu, Ghent, 2008	--	45	--	--|1.1|2.73	PIN	Microdisk, gain-loss

UCSB — Modulator Survey

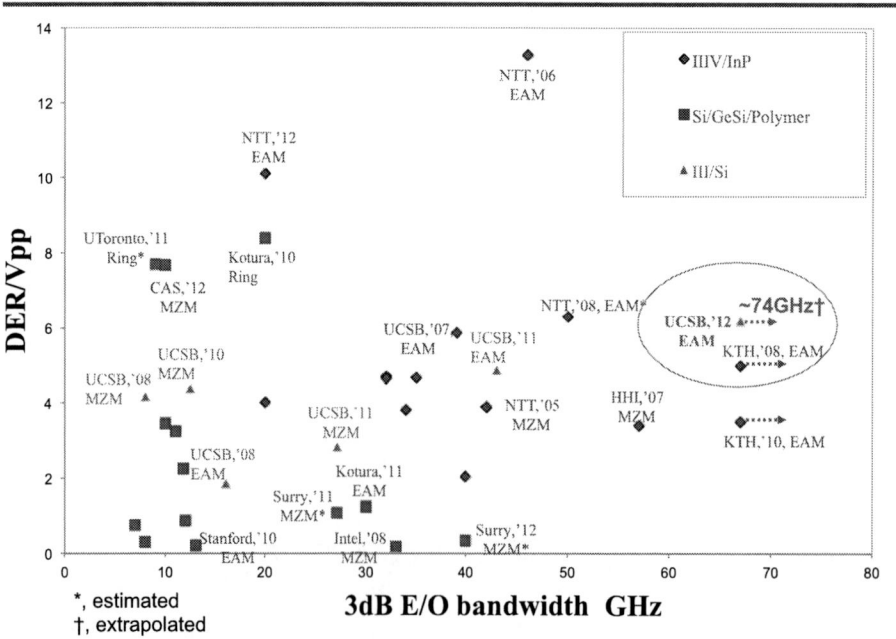

*, estimated
†, extrapolated

UCSB

Photonic Integrated Circuits

UCSB **Electronic-Photonic Integration**

- Numerous examples from IBM

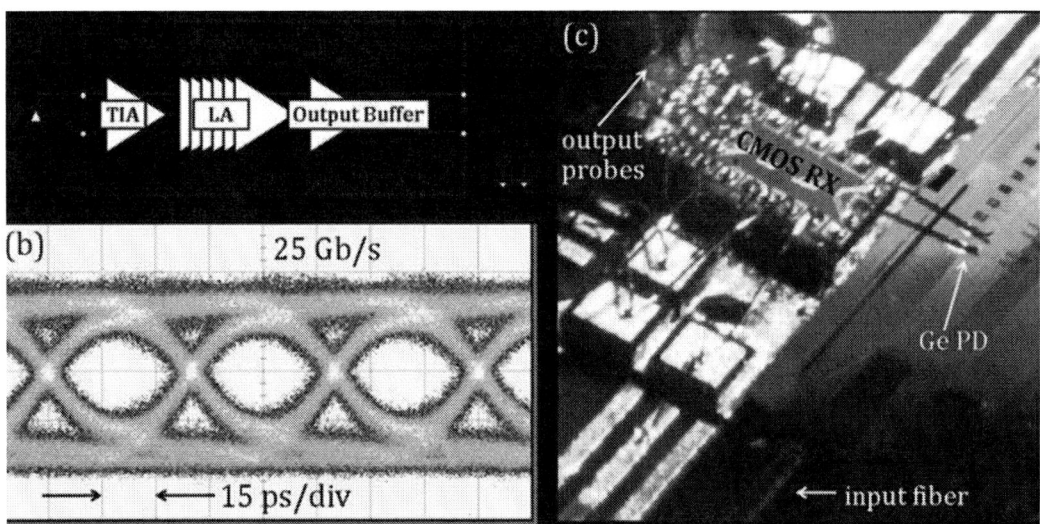

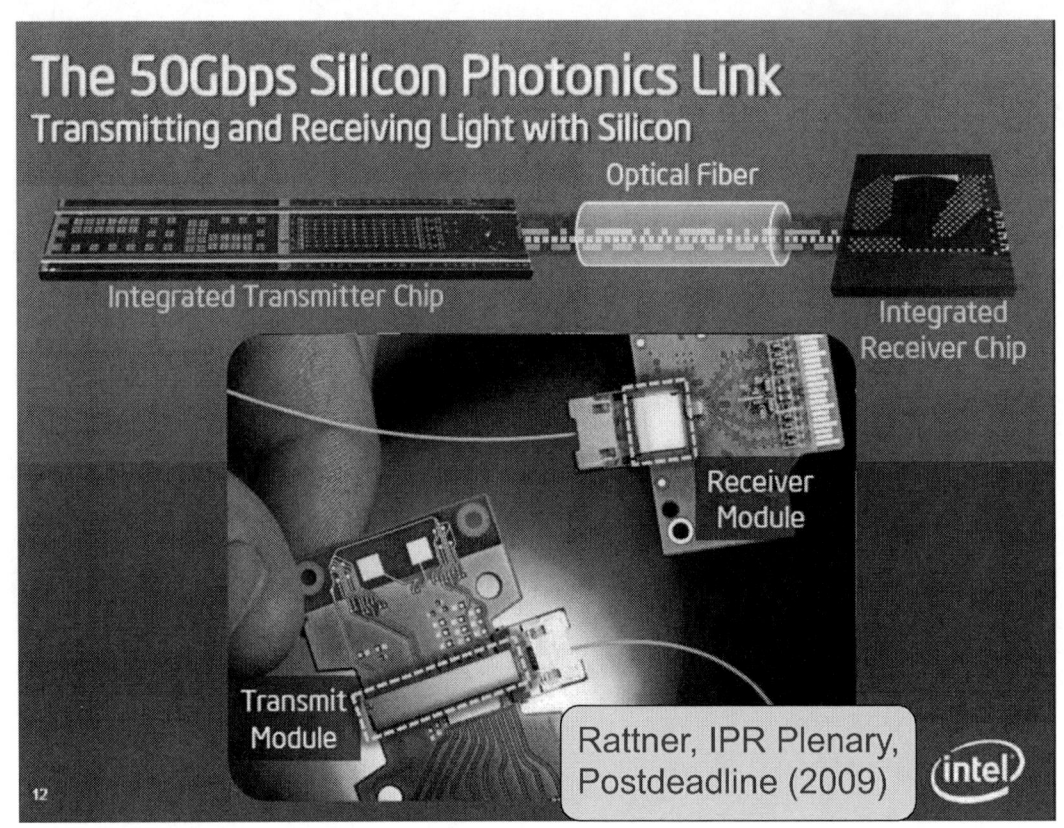

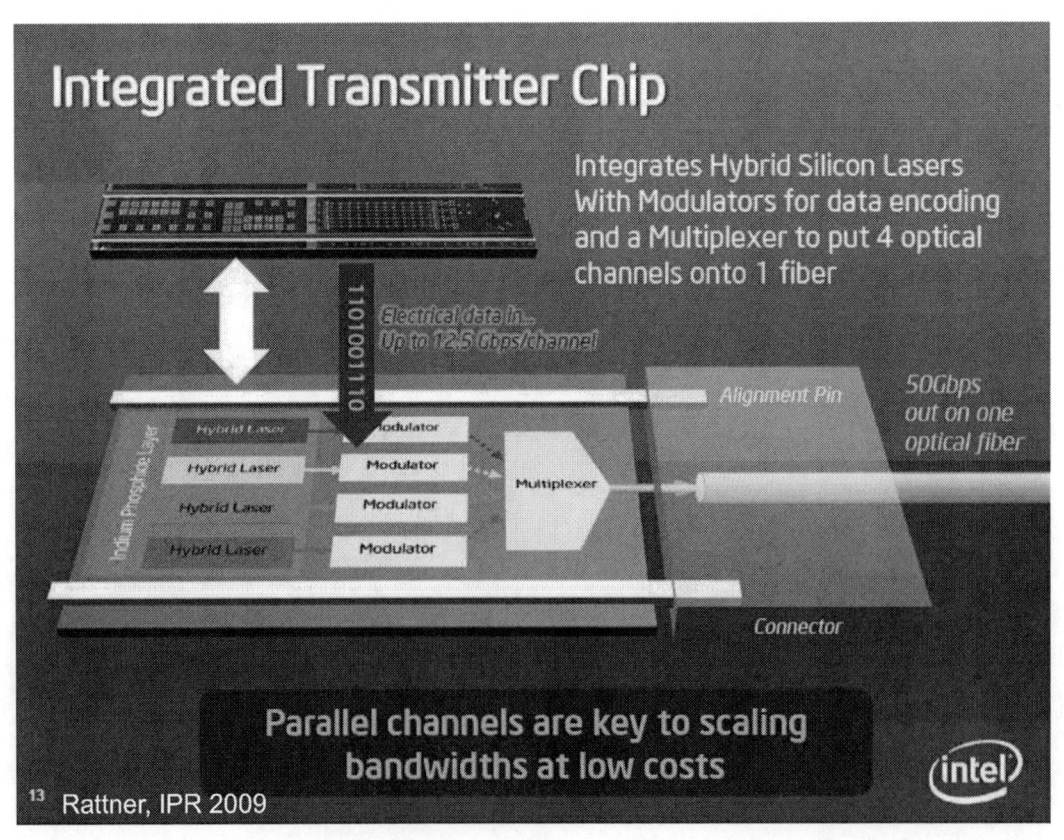

978-1-4673-6145-3/13 $31.00 © 2013 IEEE

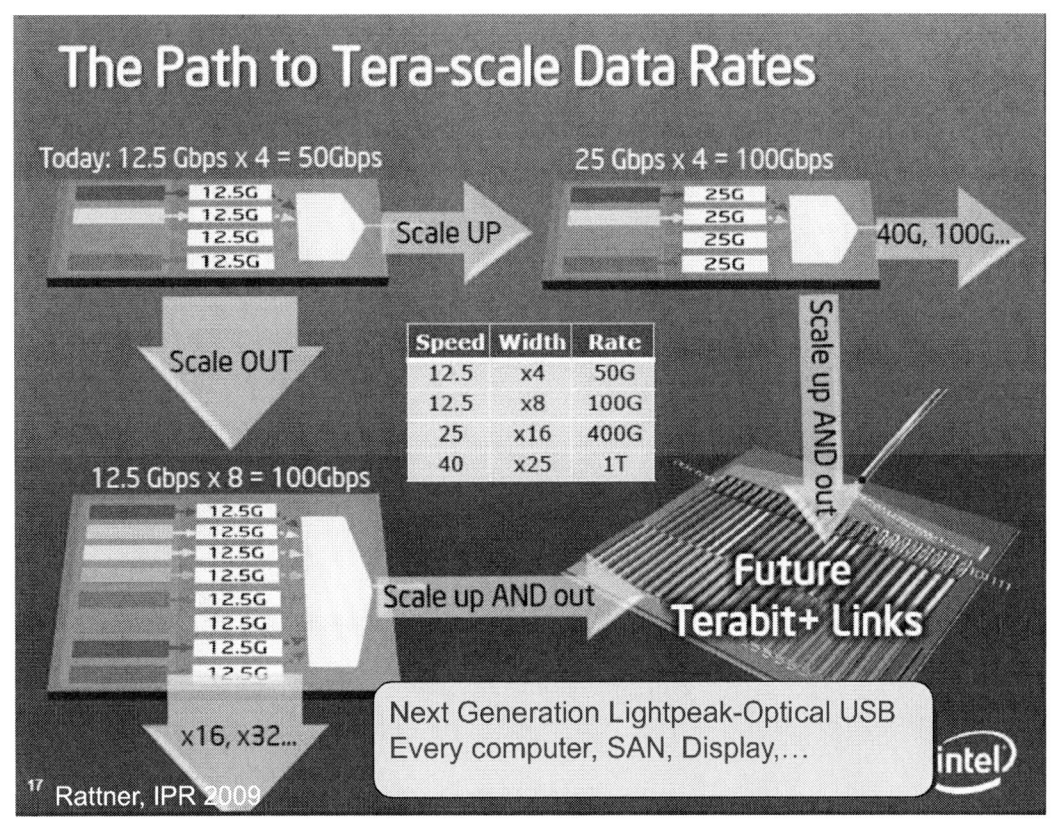

Hybrid Silicon Triplexers

UCSB

Laser, 1310/1500 nm MUX MMI, MZI 1550/1490 MUX, PDs integrated

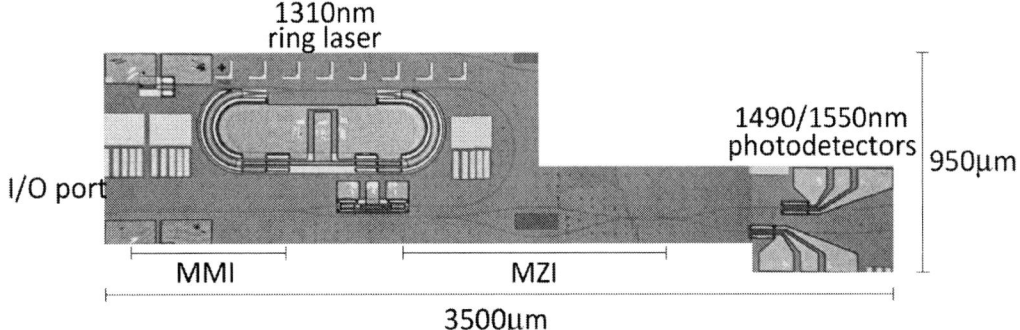

Chang et al. OFC 2010

Aurrion PIC Integration

Brian Koch *et al.*, OFC Postdeadline 2013

UCSB

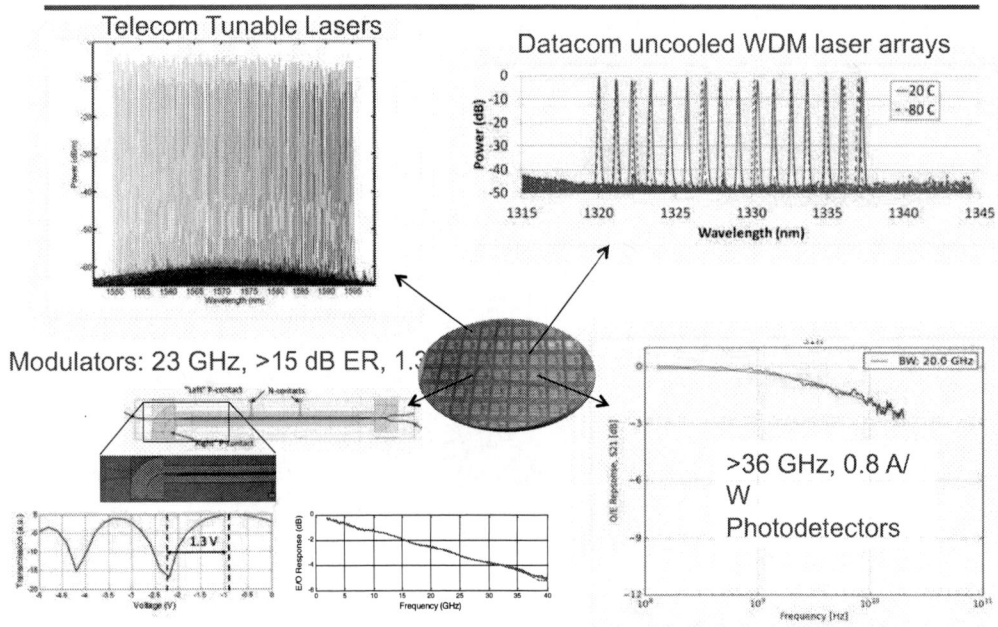

Telecom Tunable Lasers

Datacom uncooled WDM laser arrays

Modulators: 23 GHz, >15 dB ER, 1.3

>36 GHz, 0.8 A/W Photodetectors

Separate wafer: 45 mW CW Lasers

IBM Aurrion Hybrid Silicon 60 Gbps Receiver: CLEO Postdeadline 2013

IBM

60-Gb/s Receiver Employing Heterogeneously Integrated Silicon Waveguide Coupled Photodetector

Receiver Performance – BER Measurements

- Good sensitivities at speeds ≤ 40 Gb/s (< -11 dBm)
  - Degradation at higher speeds due to limitations in reference TX and RX
  - Power referenced to light coupled in silicon waveguide (~ 7dB coupling loss)
- Open eye margin at speeds ≤ 50 Gb/s (> 0.3 UI)
  - Equipment limitation (back-to-back eye margin) = _____ UI

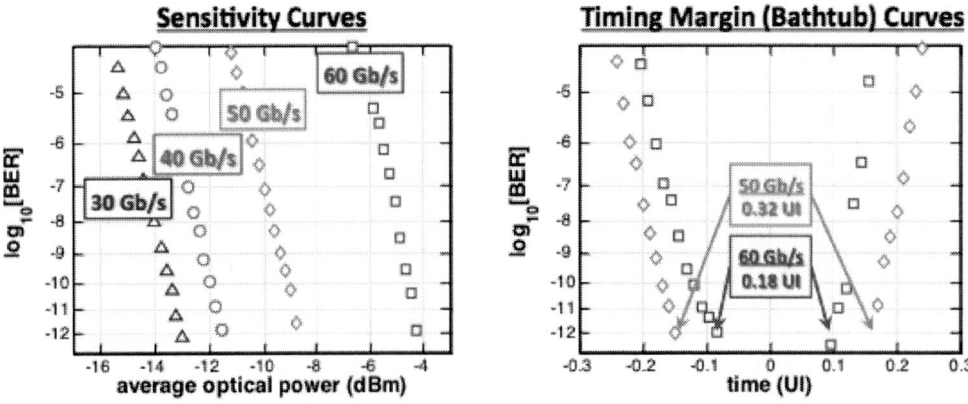

Approved for Public Release, Distribution Unlimited

CMOS Integration in Photonic IC

Chen et al. OFC 2013

- ❑ "Smart Photonics" – Integrated electronic w/ photonic ICs
- ❑ Avoid driving 50Ω terminations
- ❑ Self-calibration
- ❑ Active feedback control

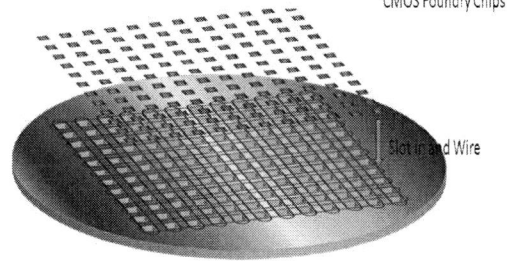

CMOS Foundry Chips

Slot in and Wire

Hybrid III-V Photonics Silicon Wafer

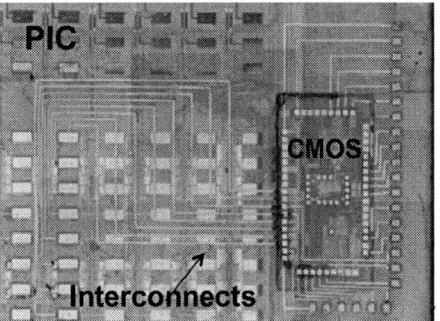

PIC

CMOS

Interconnects

Integration on InP or Si?

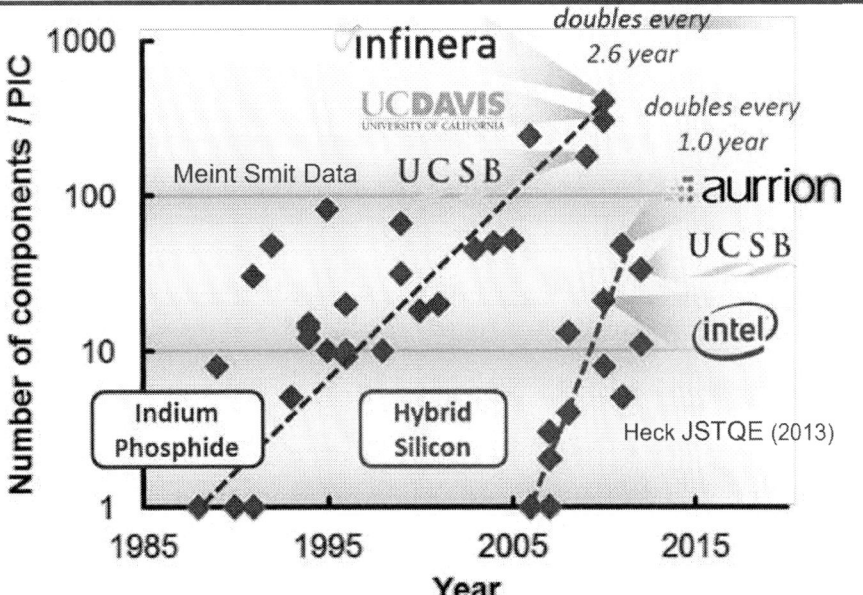

D. Liang, J. E. Bowers, "Photonic Integration: Si or InP Substrates"
Electronics Letters, **45** (12), June, 2009

UCSB — Integration on InP or Si?

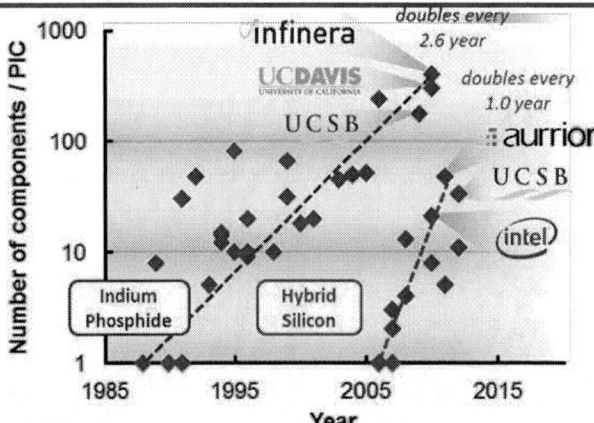

Why is the slope higher?
- Better process control
- More advanced process technology.
- More uniform materials
  Si vs $In_xGa_{1-x}As_yP_{1-y}$ for core
  SiO_2 vs $In_xGa_{1-x}As_yP_{1-y}$ for cladding

Which will go higher? (Higher level of Integration)
Silicon because
- Wafer size is larger
- Process equipment can handle larger wafers
- Economic driver is integration on Silicon ICs (market is >100x larger)
  - PCB/Backplane/Interrack/Data Center/ Teleco

UCSB — Remaining Challenges

- Light emission-Lasers and amplifiers
  - Lower threshold (0.1 mA). Higher gain (1 dB/mA)
  - Higher power
  - Higher efficiency
  - Higher yield (99.?%)
  - Better wavelength stability
- Integrated Isolators
- Thermal resistance (DWDM on chip)
- Athermal characteristics (no temperature stabilization in WDM systems)
- CMOS Compatibility
  - Widespread photonics in CMOS process lines

UCSB

Summary

- CMOS compatible silicon photonics combines two great inventions:
  - The transistor ⟹ microprocessor
  - The laser ⟹ optical communications

  and enable >100x lower power and lower power density.

- A suite of hybrid silicon devices demonstrated
  - Fabry Perot, DFB, Ring and Mode Locked Lasers
  - Optical amplifiers and integrated preamplifier arrays
  - 50 Gbps modulators demonstrated

- Low cost, high volume CMOS compatible hybrid silicon PICs will soon be commercially available.

A/D Converter Circuit and Architecture Design for High-Speed Data Communication

Boris Murmann

murmann@Stanford.edu

September 2013

978-1-4673-6146-0/13/$31.00 ©2013 IEEE

Abstract

As modern electrical and optical communication systems transition toward advanced modulation schemes, there exists a pressing need for power efficient A/D converters operating at tens of gigasamples per second. Within this context, this tutorial will cover relevant circuit- and architecture-level design techniques for high-speed CMOS A/D converters. At the circuit level, we will discuss fundamental challenges in the design of track-and-hold circuits and voltage comparators, which will also include a review of clock jitter and metastability. At the architecture level, we consider tradeoffs in the design of time-interleaved SAR and flash converters as well as techniques for the estimation, system-level budgeting and calibration of circuit imperfections.

Motivation – Increasing Need for Bandwidth

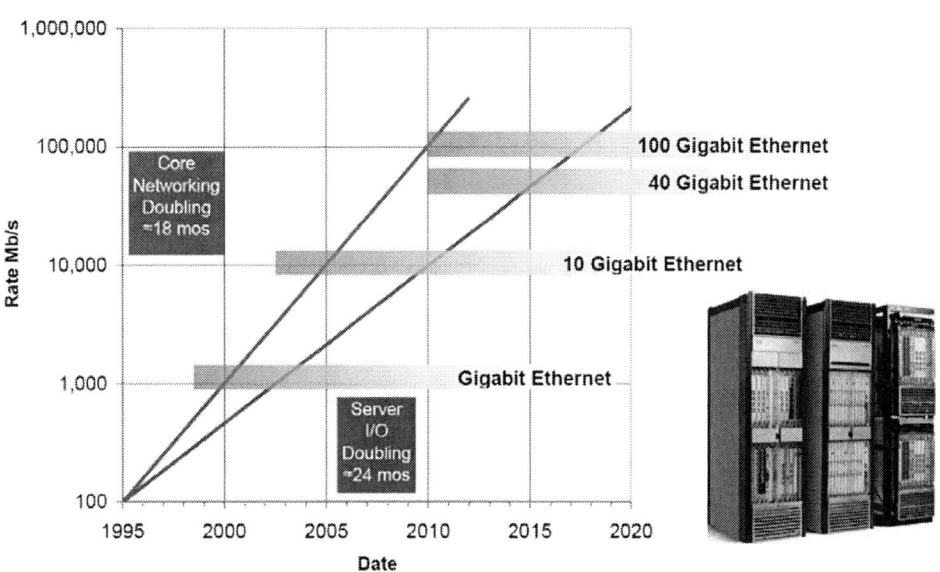

Motivation – Increasingly Complex Modulation

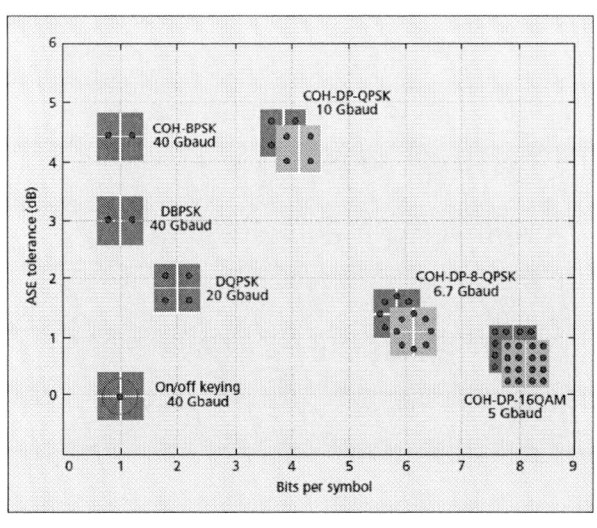

[K. Roberts et al., IEEE Communications Magazine, July 2010]

- Both electrical and optical systems are trending away from simple NRZ signaling

- Requires linear front-end circuits and very high-speed A/D converters

High-Speed ADC for Optical Communications

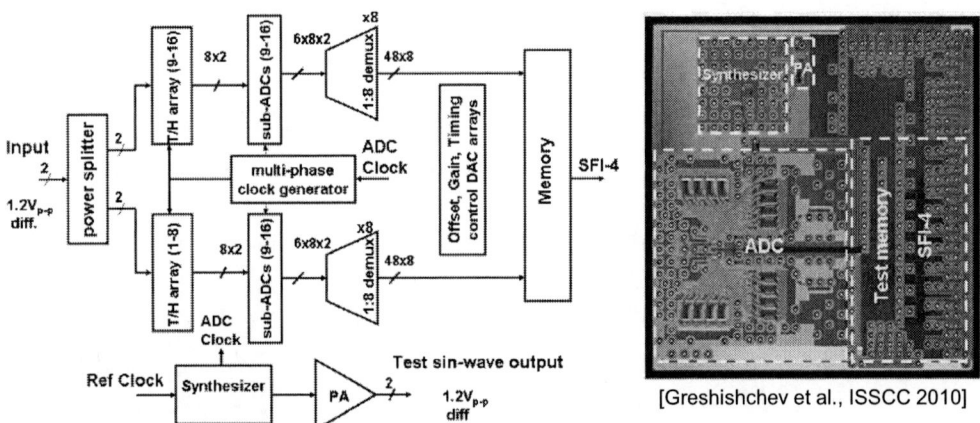

[Greshishchev et al., ISSCC 2010]

- 16x10 SAR ADCs interleaved to resolve 6 bits at 40GS/s

Tutorial Objective

- Covering everything there is to know about high-speed ADC design in 90 minutes is impossible

- We will focus on a few important "nuggets" that play a significant role in arriving at a viable design

- The discussion will span various abstraction levels that are intricately connected

Outline

- System level
  - Budgeting of nonidealities

- Architecture level
  - Time interleaving
  - Calibration
  - Flash ADCs
  - SAR ADCs

- Circuit level
  - Signal distribution
  - Voltage comparators

ADC Model

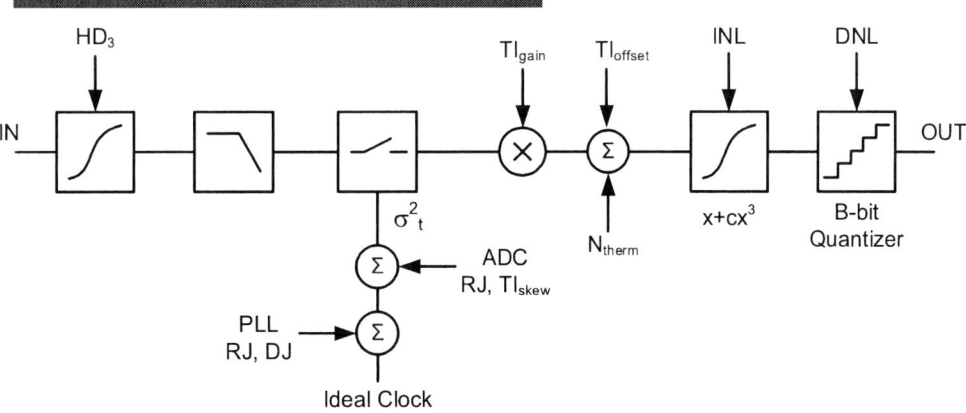

- σ^2_t lumps all timing artifacts (random jitter, TI skew, etc.)

- For a discussion on how to think about the various types of jitter (cycle-to-cycle, accumulated, etc.) see Da Dalt, TCAS1, 9/2002

Simplified ADC Model

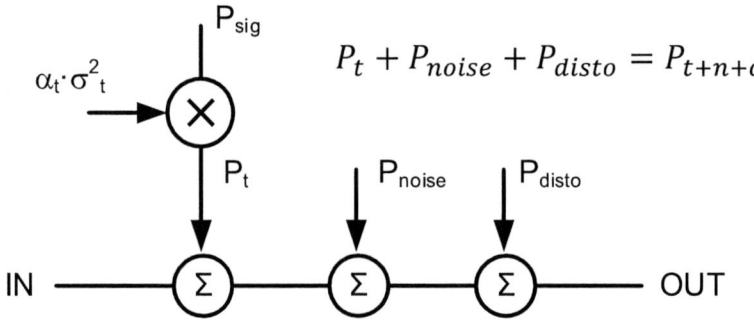

$$P_t + P_{noise} + P_{disto} = P_{t+n+d}$$

- The parameter α_t serves to convert timing errors into amplitude noise

Budgeting of Nonidealities

- Generally a complex task that requires iteration

- Suggested flow
  - Determine the total noise and distortion budget from the slicer SNR and peak-to-average power ratio of the signal
  - Estimate α_t for the given channel and RX, TX frequency response estimates
  - Distribute the noise and distortion budget
    - Timing errors, quantization noise, DNL noise, thermal noise, time interleaving errors, nonlinearities
  - Iterate!

Noise and Distortion Budget

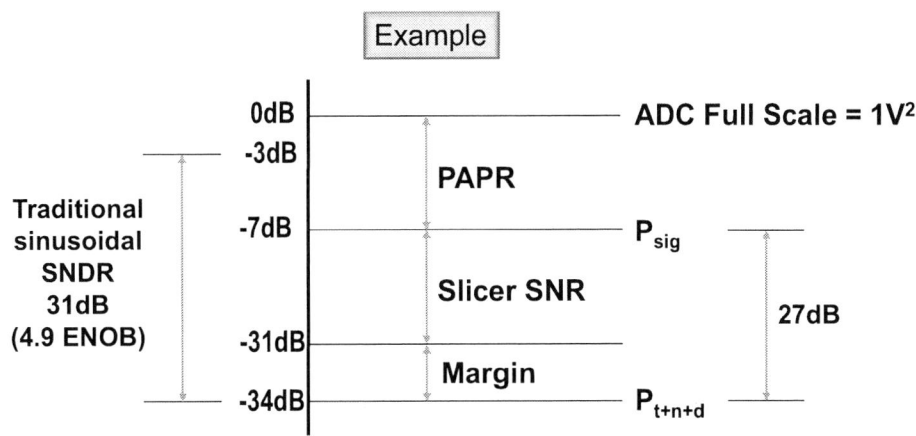

Example

- Start by assuming a 6-bit quantizer in the first iteration

Estimation of Timing Errors (1)

- How to find α_t?

- Textbook analysis typically assumes a sinusoidal input, which is far too pessimistic for a communication link

$$SNR_t = \frac{P_{sig}}{P_t} = \frac{\frac{1}{2}A^2}{\frac{1}{2}A^2\omega_{sin}^2 \cdot \sigma_t^2} = \frac{1}{\omega_{sin}^2 \cdot \sigma_t^2} \qquad \alpha_t = \omega_{sin}^2$$

Example: 10GHz sinusoid, 1ps$_{rms}$ jitter

$$\alpha_t \cdot \sigma_t^2 = (2\pi 10GHz)^2 1ps^2 = \frac{3.9 \cdot 10^{-3}}{ps^2} \cdot 1ps^2 = -24dB$$
$$\text{(ouch!)}$$

Estimation of Timing Errors (2)

- Da Dalt showed (TCAS1, 9/2002) that α_t follows from the curvature of the autocorrelation function (for arbitrary signals with second order differentiable autocorrelation)

$$\alpha_t = -\frac{R''(0)}{R(0)}$$

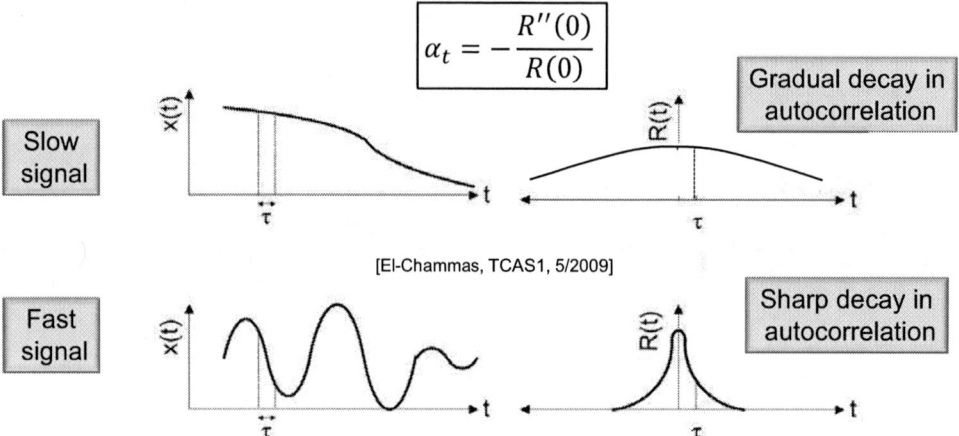

[El-Chammas, TCAS1, 5/2009]

Example (1)

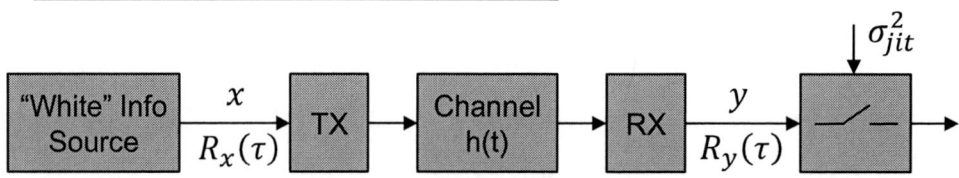

- Neglecting the frequency response of TX and RX for simplicity

$$R_y(\tau) = R_x(\tau) * h(t) * h(-t)$$

$$R_y(\tau) = \delta(\tau)\sigma_x^2 * h(t) * h(-t) = \sigma_x^2 \cdot \underbrace{[h(t) * h(-t)]}_{w(\tau)}$$

$$\alpha_t = -\frac{R_y''(0)}{R_y(0)} = -\frac{w''(0)}{w(0)}$$

Example (2)

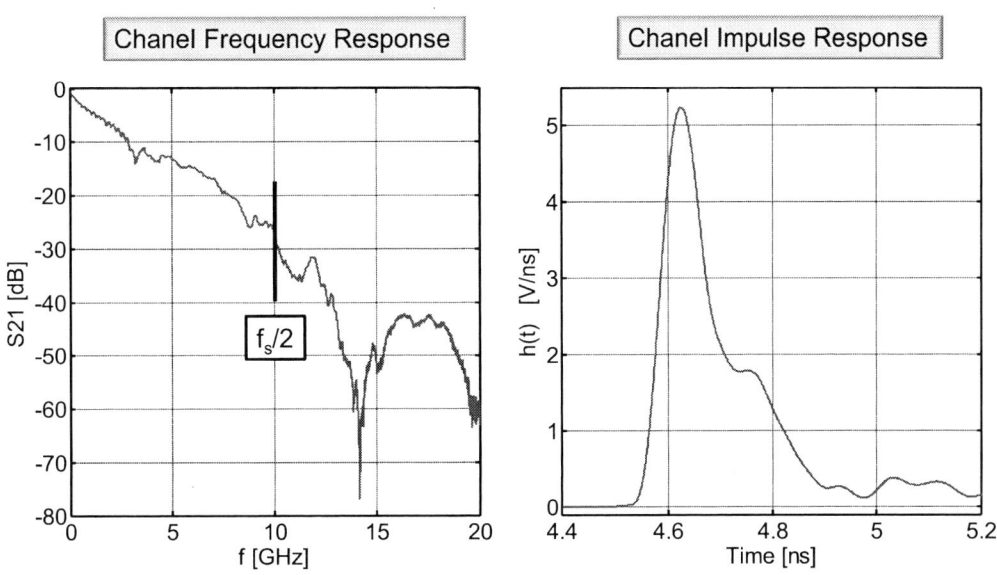

Example (3)

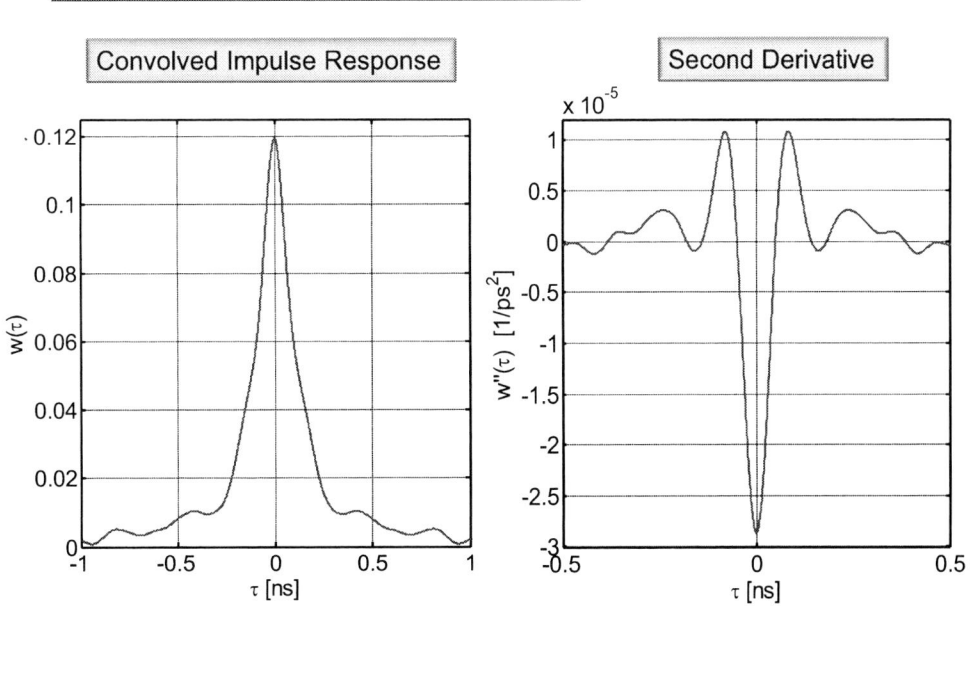

Example (4)

$$\alpha_t = -\frac{w''(0)}{w(0)} = \frac{2.8 \cdot 10^{-5}}{0.12}\frac{1}{ps^2} = 2.3 \cdot 10^{-4}\frac{1}{ps^2}$$

For $1ps_{rms}$ jitter:　　$\alpha_t \cdot \sigma_t^2 = -36.4dB$　　(OK!)

- Another way to think about this is to compute the sinusoid frequency that causes the same jitter

$$f_{sin,eq} = \frac{1}{2\pi}\sqrt{\alpha_t} = 2.4GHz$$

- Bottom line: the fact that the signal is wideband and filtered by the channel helps, but the jitter spec will still be "non-trivial"

Distribution of Noise and Distortion (1)

Total budget

$$\sqrt{P_{t+n+d}} = 1V \cdot 10^{-\frac{34}{20}} = 20mV_{rms}$$

Timing noise
(Assuming $\sigma^2_t = 4ps^2$)

$$\sqrt{P_t} = \sqrt{1V^2 \cdot 10^{-\frac{7}{10}} \cdot 2.3 \cdot 10^{-4} \cdot 4} = 13.5mV_{rms}$$

Quantization Noise

$$\sqrt{P_q} = \sqrt{\frac{1}{12}\frac{1V}{2^6}} = 4.5mV_{rms}$$

DNL Noise
(assuming 0.5 LSB uniformly distributed)

$$\sqrt{P_{DNL}} = \sqrt{P_q} = 4.5mV_{rms}$$

Remaining budget

$$13.3mV_{rms}$$

978-1-4673-6145-3/13 $31.00 © 2013 IEEE

Distribution of Noise and Distortion (2)

- Assuming that we split the remainder equally between the other significant nonidealities, the distribution looks as shown below
  - Here, TI artifacts model residual gain and offset errors (more later)
- This is just a starting point, iterate and optimize from here...

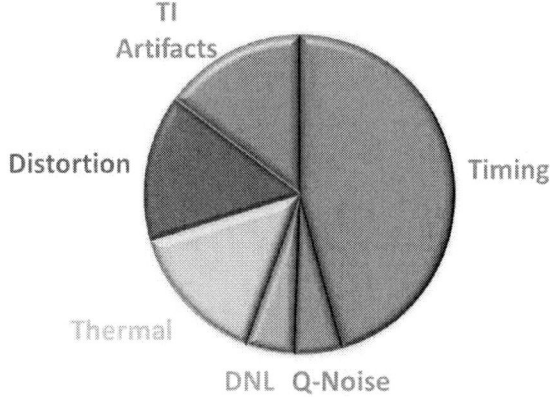

Outline

- System level
  - Budgeting of nonidealities

- Architecture level
  - Time interleaving
  - Calibration
  - Flash ADCs
  - SAR ADCs

- Circuit level
  - Signal distribution
  - Voltage comparators

Time-Interleaved ADC

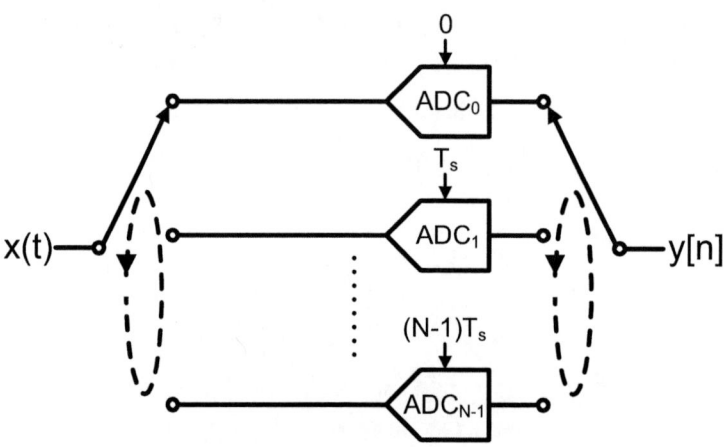

[W. Black and D. Hodges, JSSC, Dec. 1980]

Why Time Interleaving?

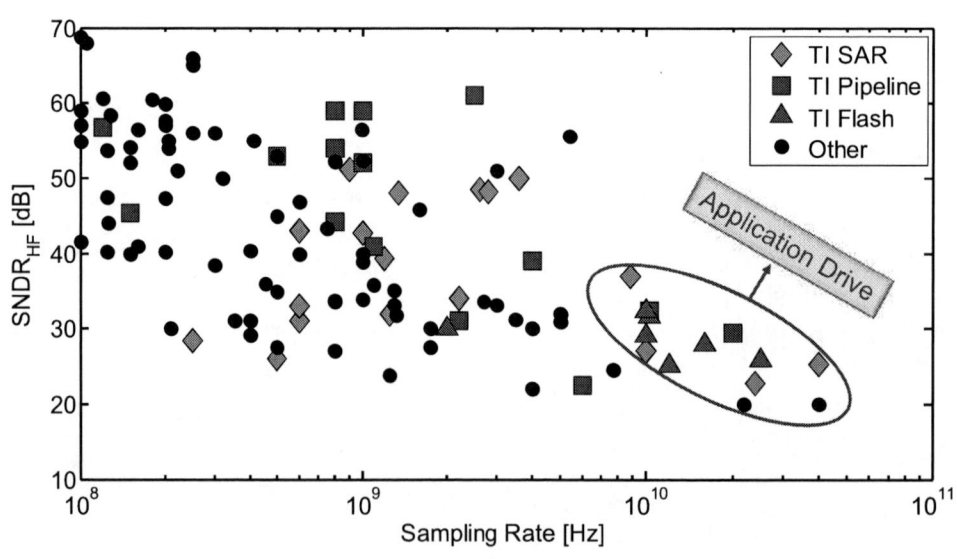

B. Murmann, "ADC Performance Survey 1997-2013," [Online]. Available: http://www.stanford.edu/~murmann/adcsurvey.html

How Many Channels?

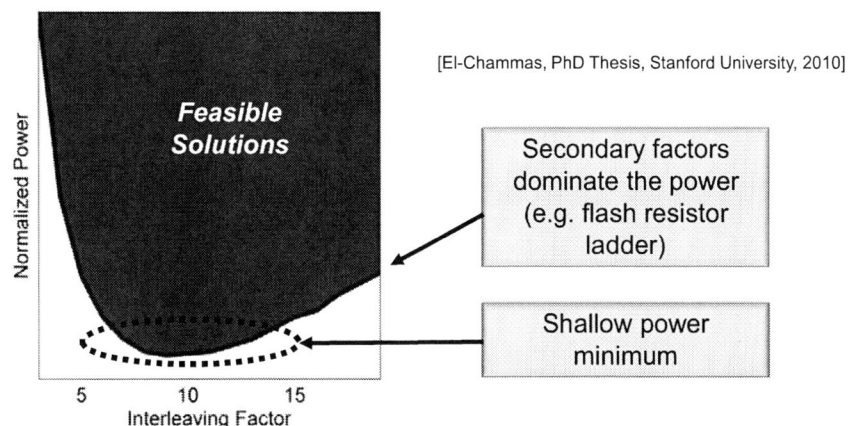

[El-Chammas, PhD Thesis, Stanford University, 2010]

- A complex optimization problem!
  - □ Luckily, first order analyses show that the optimum is shallow

Time Interleaving Errors

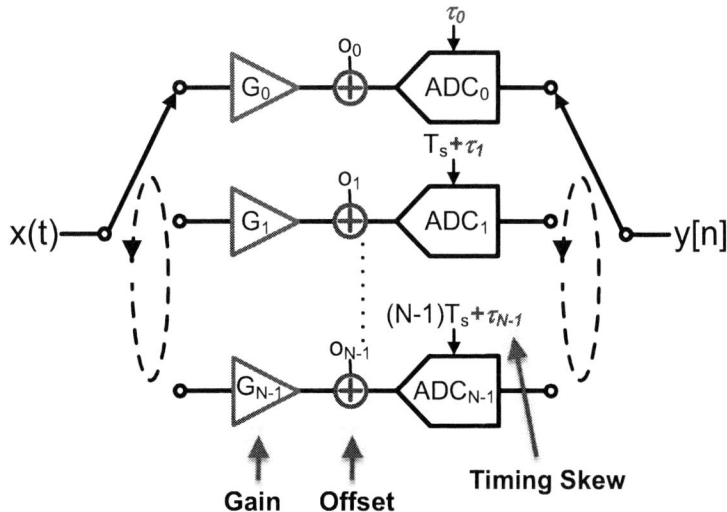

Compensation of Interleaving Errors

- Gain and offset
  - Several relatively "simple" techniques exist
  - Can compensate in analog and/or digital domain

- Timing skew
  - Much more difficult to handle
  - Popular solutions
    - Measure errors in digital domain, compensate via adjustable delay lines → typically incur a jitter penalty
    - Measure errors in digital domain, compensate by skewing equalizer taps
  - See overview paper by B. Razavi, CICC 2012

Fully Digital Gain Error Compensation

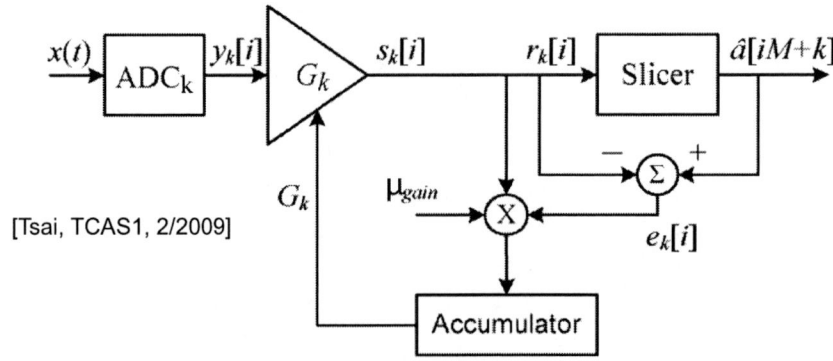

[Tsai, TCAS1, 2/2009]

- A similar approach can be used for digital offset compensation

Gain, Offset and Nonlinearity Cal via Trim-DACs

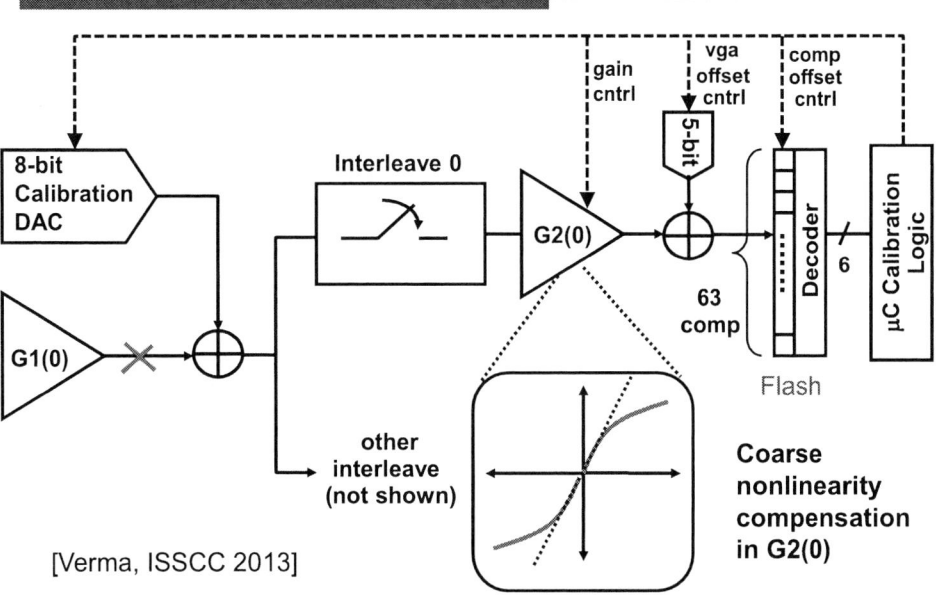

[Verma, ISSCC 2013]

Coarse nonlinearity compensation in G2(0)

Sources of Timing Skew

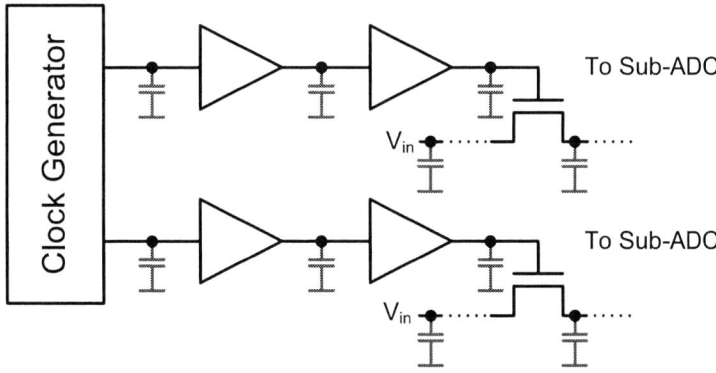

■ All traces and all transistors have mismatch, which can easily result in ~10ps total timing skew for a complex clock tree

[A. Agarwal, CICC 2008]

Local Re-Timing & Fine Tuning

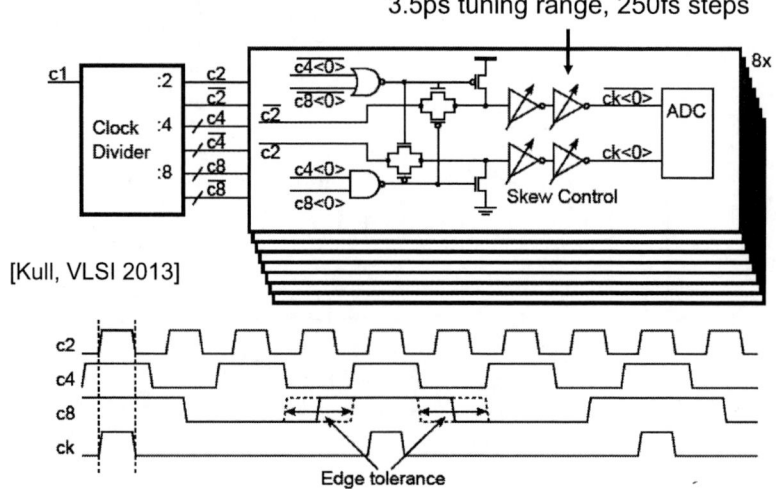

3.5ps tuning range, 250fs steps

[Kull, VLSI 2013]

- How to measure the residual skew?

Example

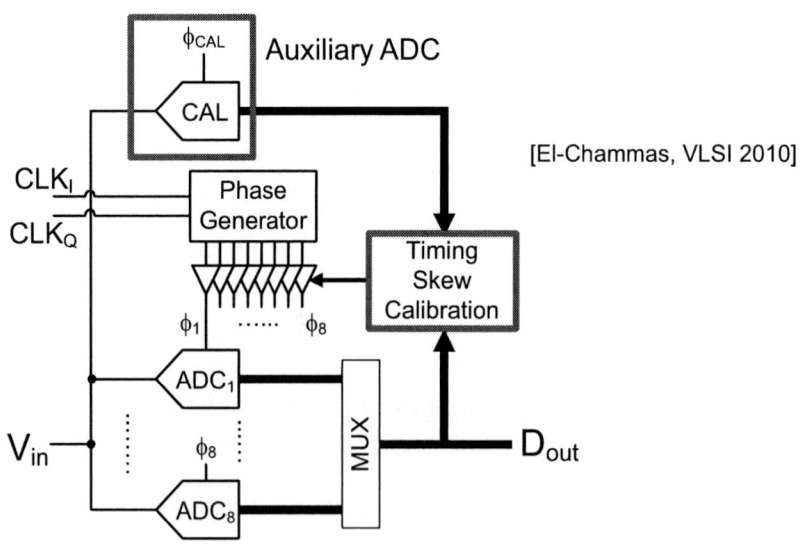

[El-Chammas, VLSI 2010]

Finding Skew via Correlation

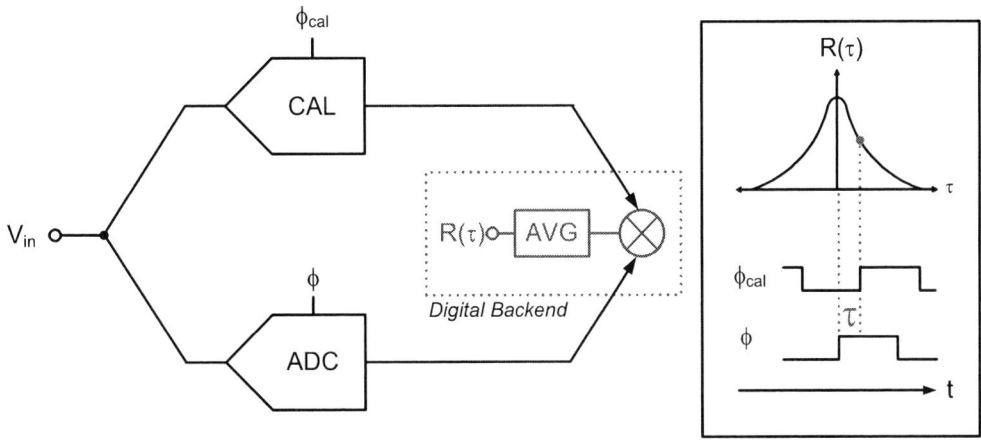

Adjust the Sampling Edge

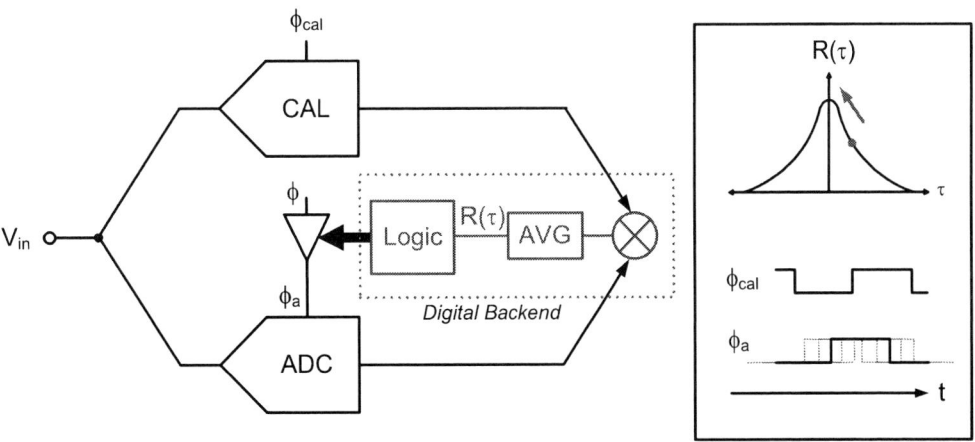

Use a "1-bit ADC"

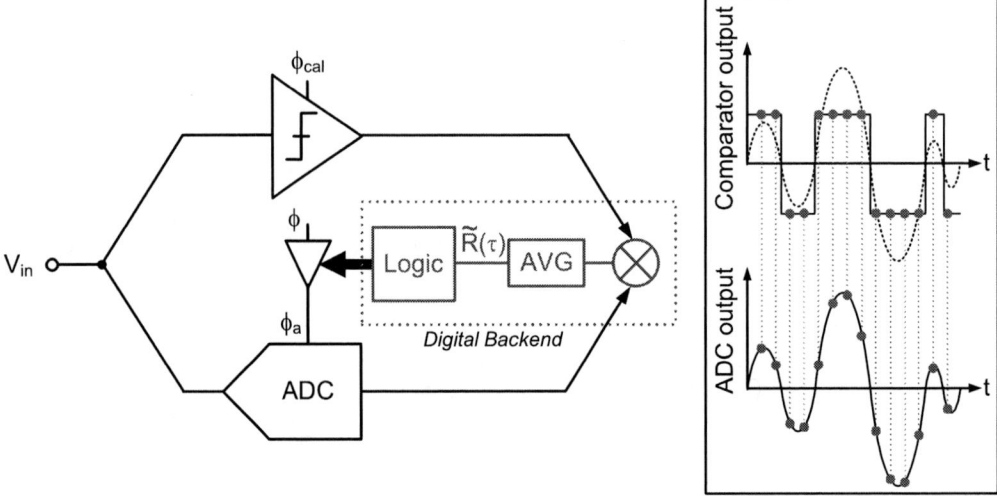

Calibrate All the Sub-ADCs

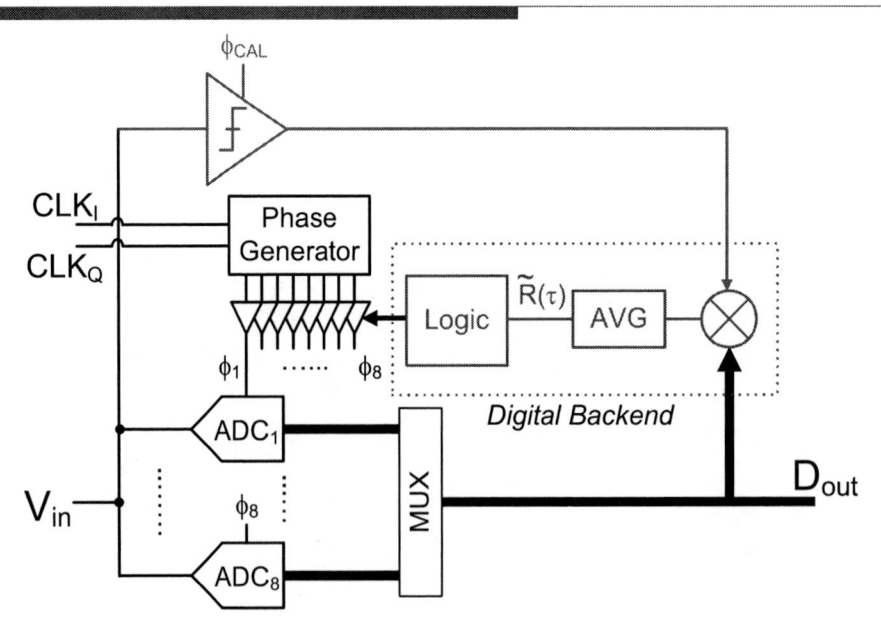

Clocking the Calibration Comparator

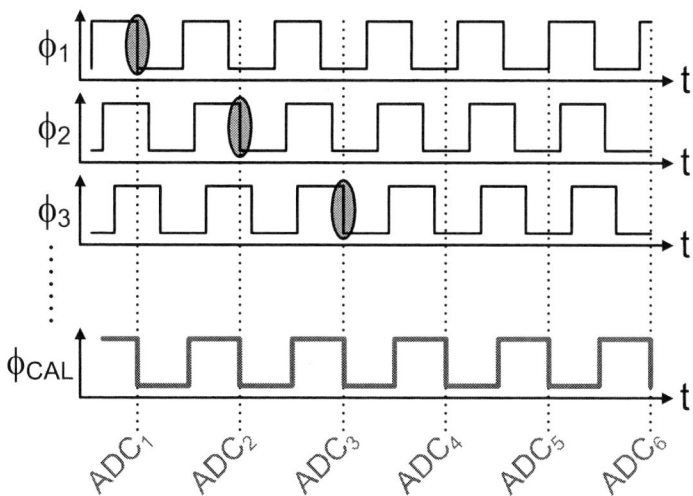

Period of ϕ_{CAL} is 9/8 period of $\phi_1 - \phi_8$

Clocking the Calibration Comparator

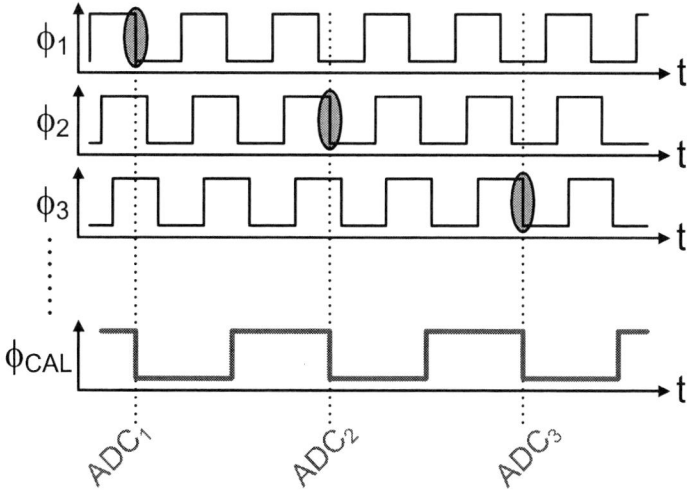

Period of ϕ_{CAL} is 17/8 period of $\phi_1 - \phi_8$

Calibration Clock Generation – Option 1

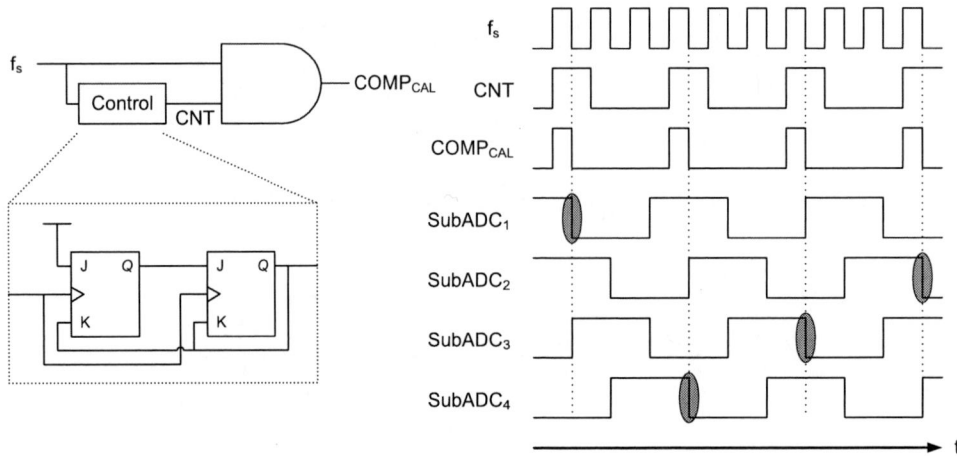

Calibration Clock Generation – Option 2

- If full-rate clock is not available, then an integer-N or a fractional-N PLL will work

- If interleaving factor = 4, then fsubADC = $f_s/4$. If reference clock is f_{subADC}, then one solution is f_{cal} = 4/M f_{subADC} (odd M)

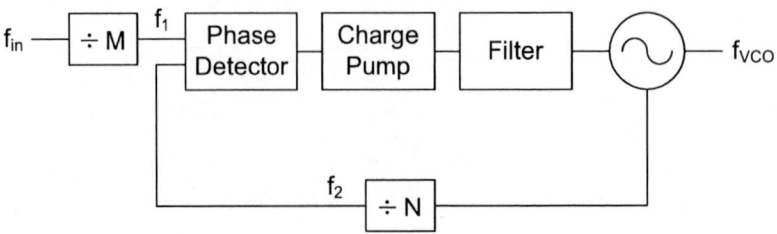

Measured Convergence Results

Average 500k samples → 8 ms / calibration cycle

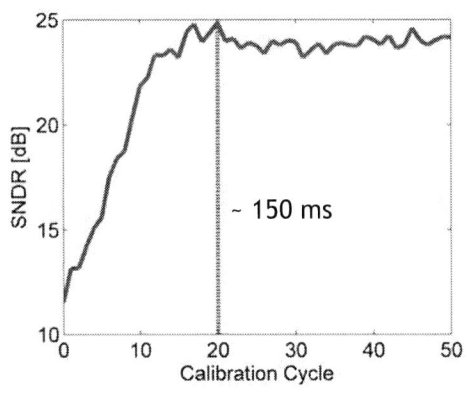 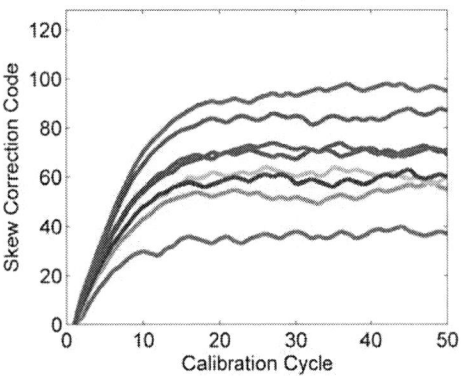

f_s = 12 GS/s, f_{in} = 8 GHz, f_{cal} = 480 MHz

Measured Input Frequency Sweep

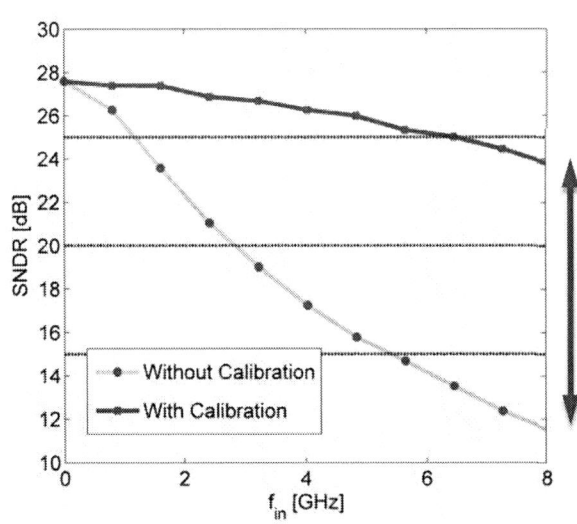

Residual Skew ~ 0.4 ps
Estimated Jitter ~ 0.6 p_{srms}

SNDR improves by 12 dB
for an 8 GHz sinusoidal input

Interesting Alternative

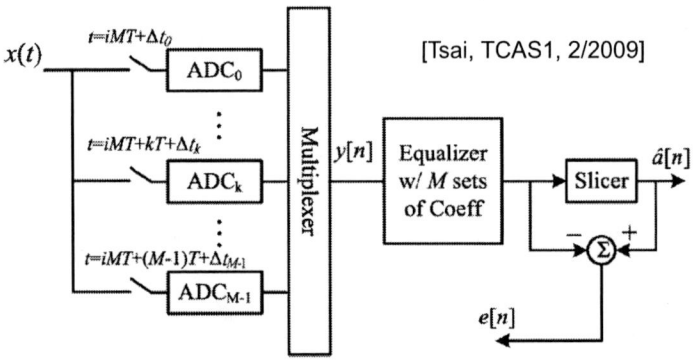

[Tsai, TCAS1, 2/2009]

- Split equalizer into M paths and adapt coefficients separately
- For complex equalizers typically found in ADC-based links, one must carefully evaluate if the power overhead is justifiable

Outline

- System level
  - Budgeting of nonidealities

- Architecture level
  - Time interleaving
  - Calibration
  - Flash ADCs
  - SAR ADCs

- Circuit level
  - Signal distribution
  - Voltage comparators

Flash ADC Design

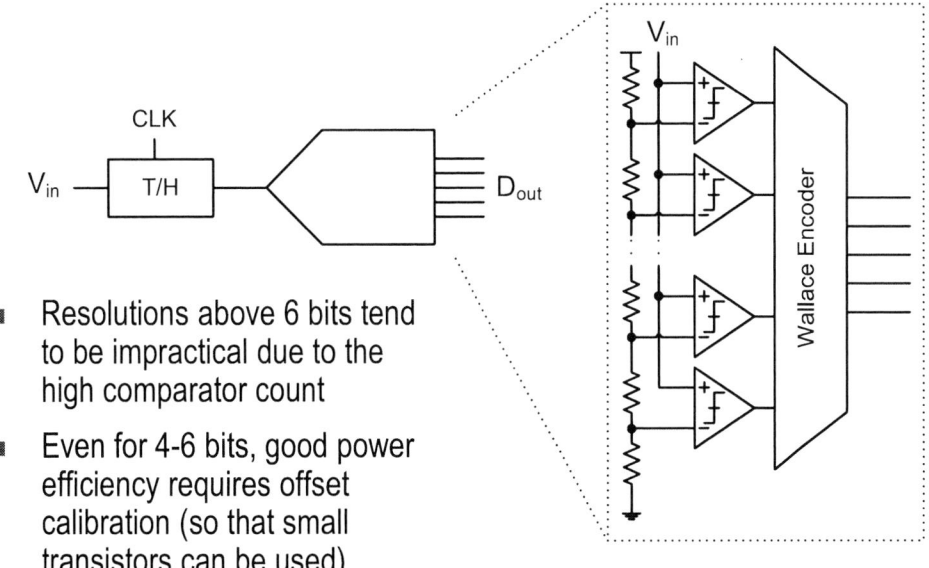

- Resolutions above 6 bits tend to be impractical due to the high comparator count

- Even for 4-6 bits, good power efficiency requires offset calibration (so that small transistors can be used)

Comparator with 5-bit Offset DAC

(Reset switches not shown)

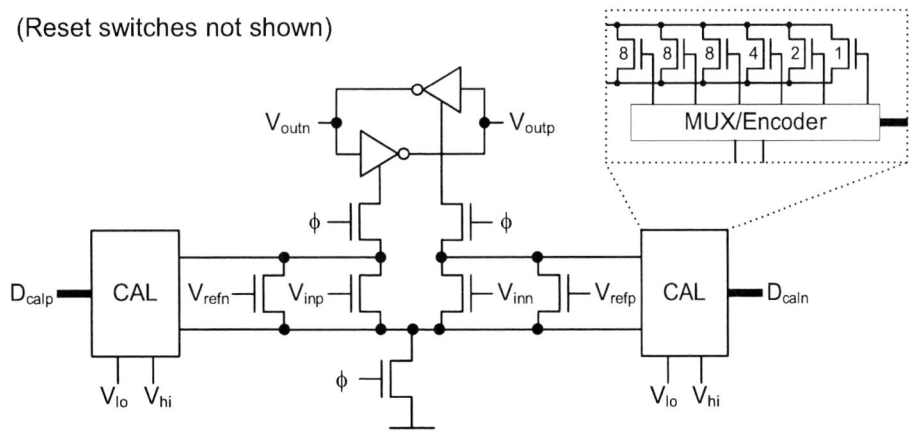

[M. El-Chammas, VLSI 2010]

Offset Calibration at Start-Up

- Output oscillates between one and zero as the input-referred offset converges to zero

- Start-up calibration can be OK at flash resolutions

- The correction circuit shown on the previous slide has a relatively low temperature dependence

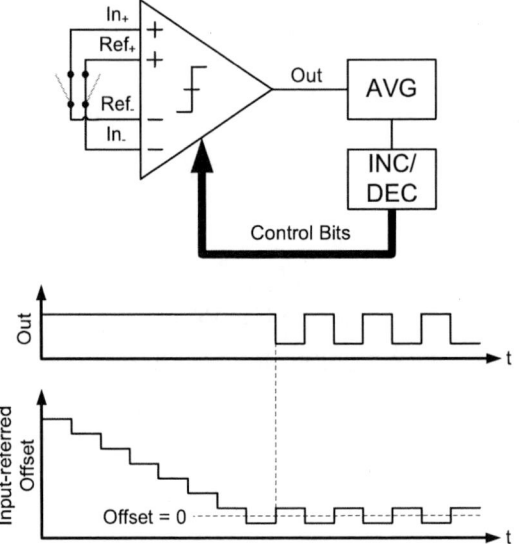

Tuning Range Issue (1)

- Ideally, the trim-DAC must bring the offset well within one LSB of the flash ADC

- The required trim-DAC resolution can become unacceptably large if near minimum-size transistors (with large V_t mismatch) in the latest technology are used

- A clever workaround is to utilize the error correction capability of the Wallace tree encoder (ones counter)

Tuning Range Issue (2)

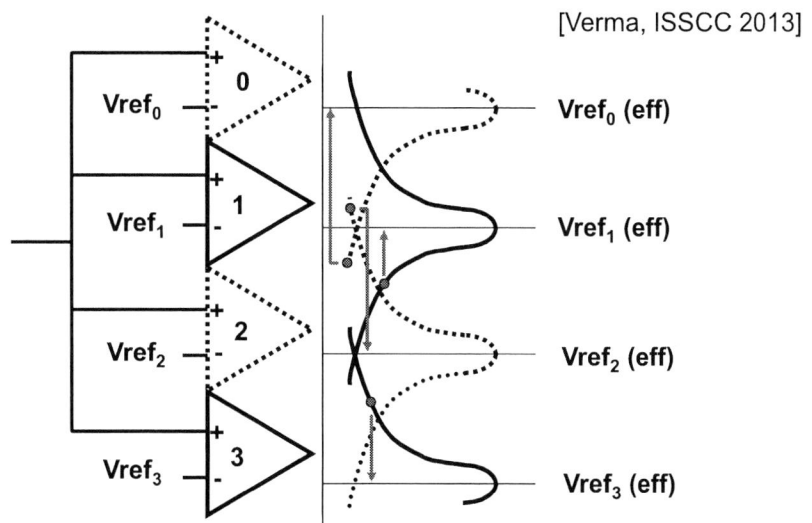

[Verma, ISSCC 2013]

Ones Counting is Equivalent to Reordering

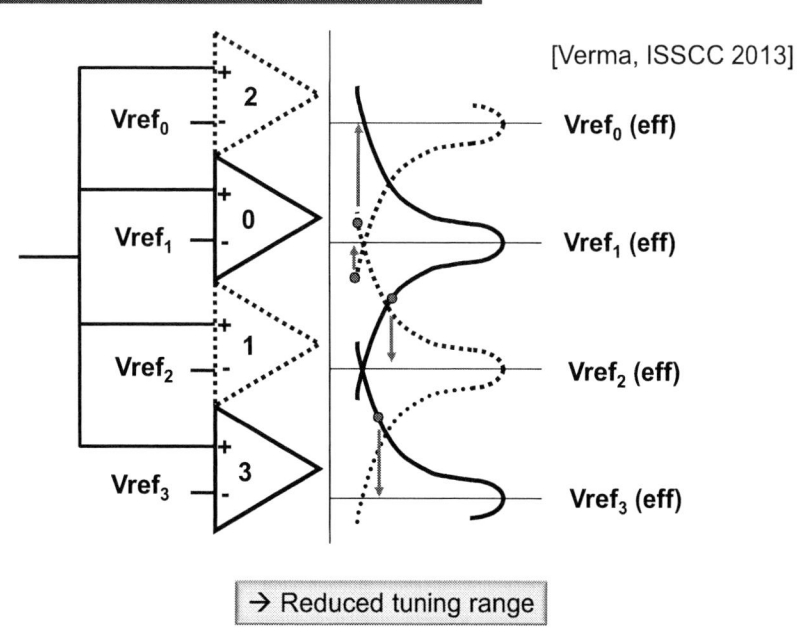

[Verma, ISSCC 2013]

→ Reduced tuning range

SAR ADC Design

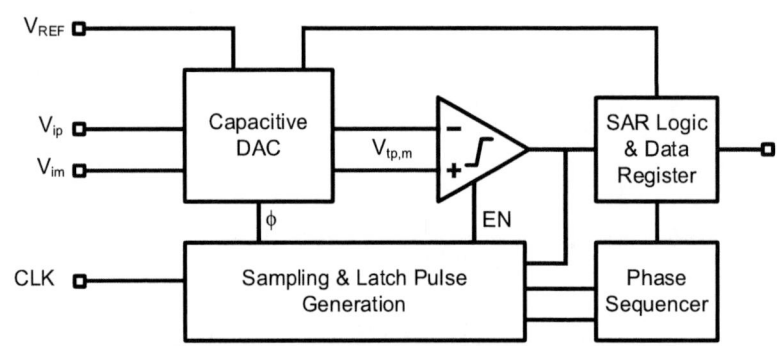

- Generally want to use asynchronous timing for improved speed and metastability rate [M. Chen, ISSCC 2006]

- In the latest CMOS, it is possible to resolve 8 bits at > 1GS/s
  - This helps reduce the interleaving factor and power to drive the array

State of the Art (8x Interleaved SAR Array)

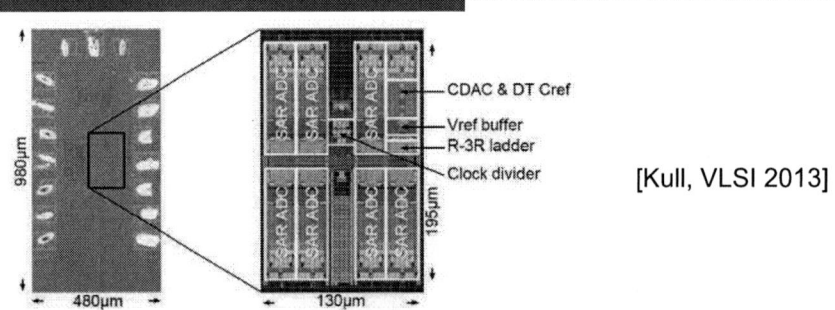

[Kull, VLSI 2013]

Specifications	[3]	[4]	[5]	This work	
Architecture	Flash	Ti-Pipeline	Ti-Flash	Ti-SAR	
CMOS Technology (nm)	65	90	65	32	
Resolution (bits)	6	6	6	8	
Supply Voltage (V)	1.3	-	1.5	1.0	1.1
SNDR (dB)	32	36.6	30.8	38.5	38.5
Sampling Speed (GHz)	5	10.3	16	8.8	10
Number of Channels	1	8	8	8	
Power (mW)	320	1600	435	35	49
FOM (fJ/conv. step)	1970	2790	2600	58	71
Area (mm²)	0.3	-	1.47	0.025	

DAC Settling Errors

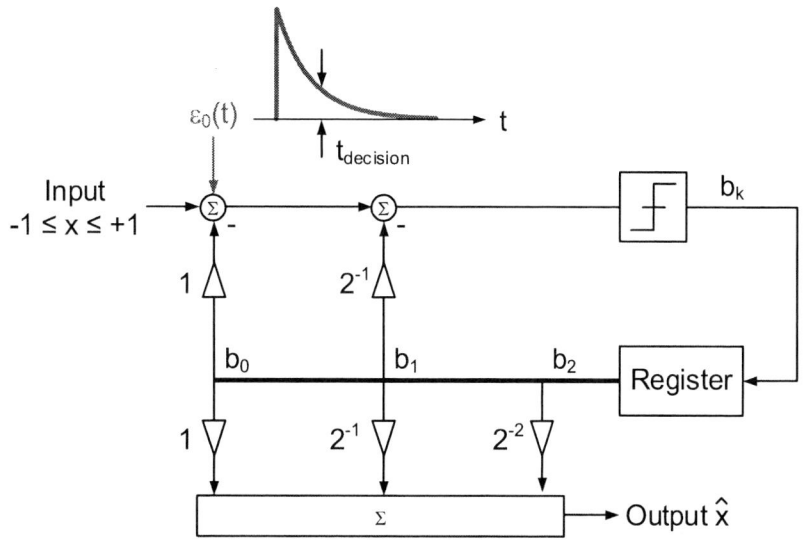

Error in First Decision

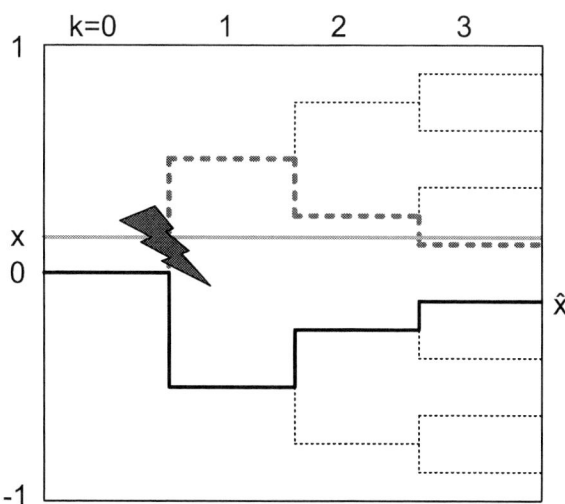

Redundancy (Radix < 2)

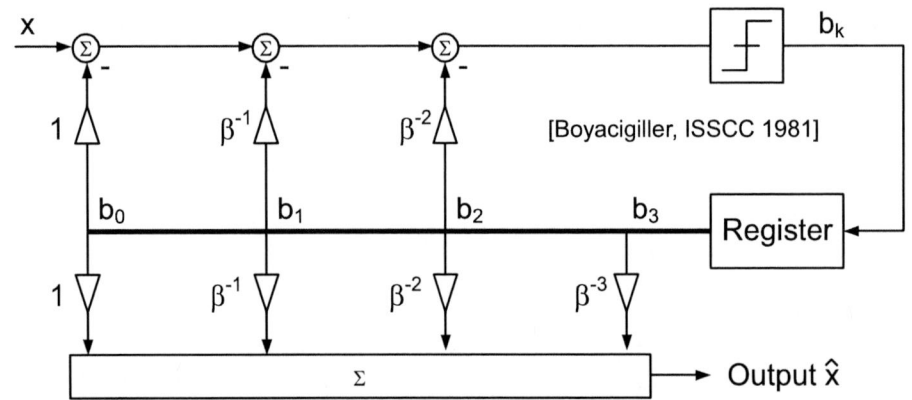

[Boyacigiller, ISSCC 1981]

$$\beta < 2 \qquad \sum_{i=1}^{B-1} \beta^{-i} > 1 \qquad \text{(Key property)}$$

Error in First Decision ($\beta = 2^{2/3}$)

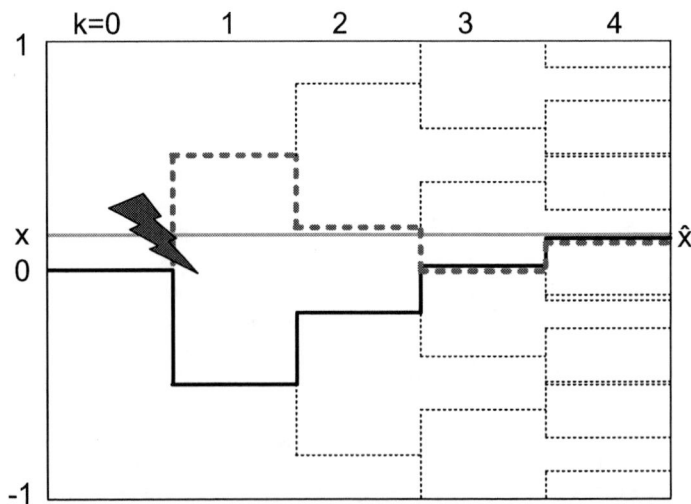

SAR ADC with Redundant Step

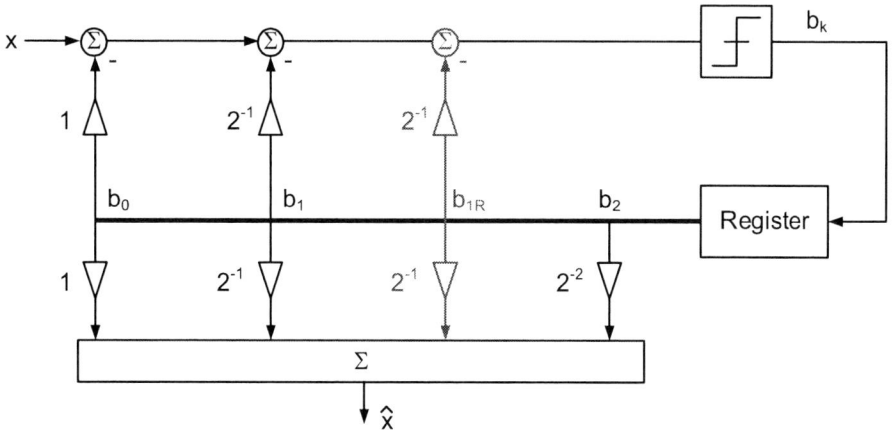

[C.-C. Liu, ISSCC 2010]

Extra Levels

■ The addition of the redundant step provides extra levels that help counter errors in previous decisions

b_2	$b_{1R} + b_2$
	+0.25+0.125 = 0.375
+0.125	+0.25-0.125 = +0.125
-0.125	-0.25+0.125 = -0.125
	-0.25-0.125 = -0.375

Error in First Decision

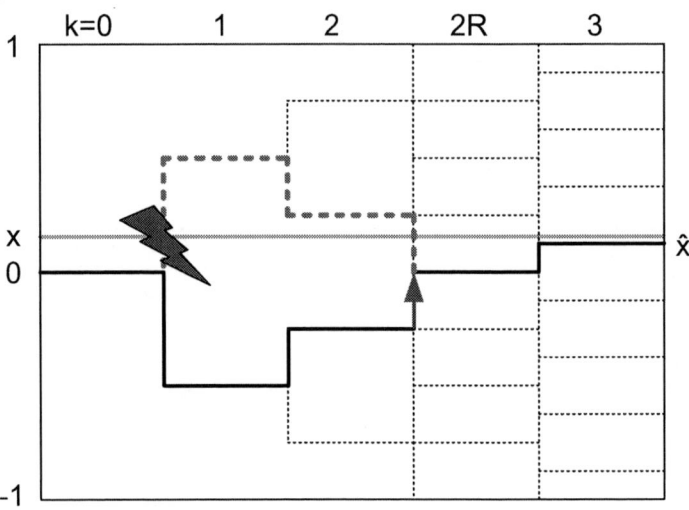

Potential Speed Benefit

Binary search algorithm   4bit

Step1	Step2	Step3	Step4

Exact DAC settling → Long time

A/D conversion time

Non-binary search algorithm

Step 1	Step 2	Step 3	Step 4	Step 5	Step 6

Correction of incomplete-settling error.

Incomplete DAC settling → Short time

Algorithm	Binary	Prior	Generalized
Time slot for each step	6.3τ	2.2τ	1.9τ
Number of steps	10	12	12
Total conversion time	56.7τ	24.2τ	20.9τ

T. Ogawa et al., IEICE Trans. on Fundamentals of Electronics, Communications and Computer Sciences, Feb. 2010.

- Benefit is clearly significant at high resolution
  - However, one must carefully evaluate if redundancy helps significantly at low resolutions (e.g. 6-8 bits)

Outline

- System level
  - Budgeting of nonidealities
- Architecture level
  - Time interleaving
  - Calibration
  - Flash ADCs
  - SAR ADCs
- **Circuit level**
  - Signal distribution
  - Voltage comparators

T/H and Buffering

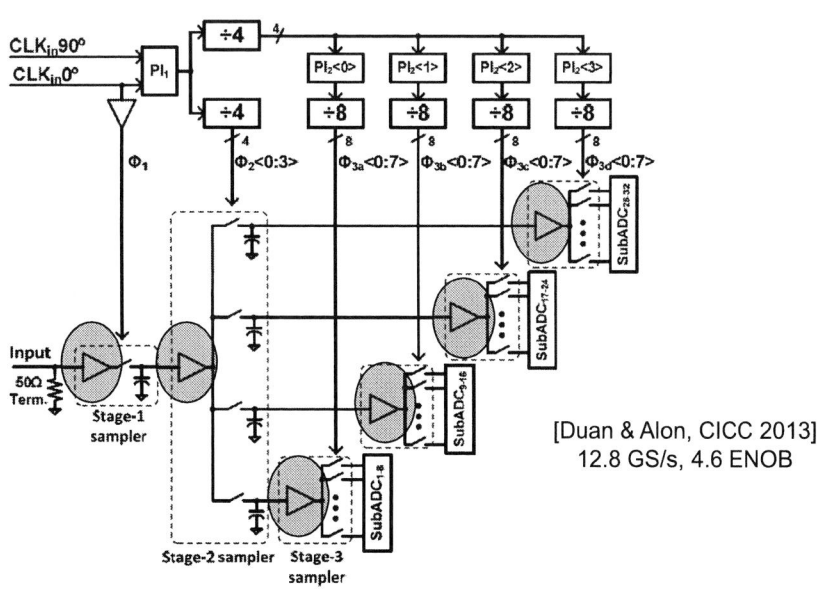

[Duan & Alon, CICC 2013]
12.8 GS/s, 4.6 ENOB

Innovative Solutions (1)

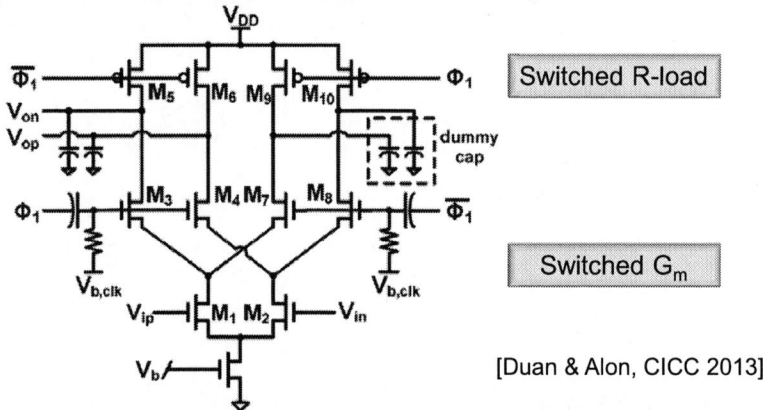

Switched R-load

Switched G_m

[Duan & Alon, CICC 2013]

- Analysis shows that this circuit can achieve higher bandwidth than a simple switch + capacitor
- However, one must carefully manage the distortion from the differential pair

Minimum Power of a Differential Pair

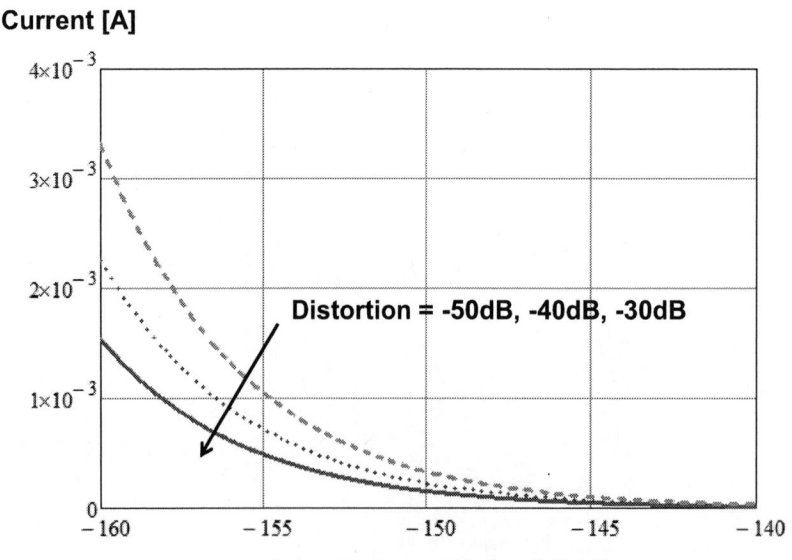

Distortion = -50dB, -40dB, -30dB

The Cost of Signal Distribution

- The SAR ADCs in a massively interleaved array have become quite efficient
    - But the power required to drive them is still relatively high
    - For example, in Duan's work (~4.6 ENOB), the buffer circuits consume about as much power as the ADC array (~32mW)

- Determining the "optimum" allocation of buffers and overall interleaving structure is still an open research problem
    - The problem can be analyzed similar to logic gate fan-out, except that in this case noise and distortion must also be considered

Innovative Solutions (2)

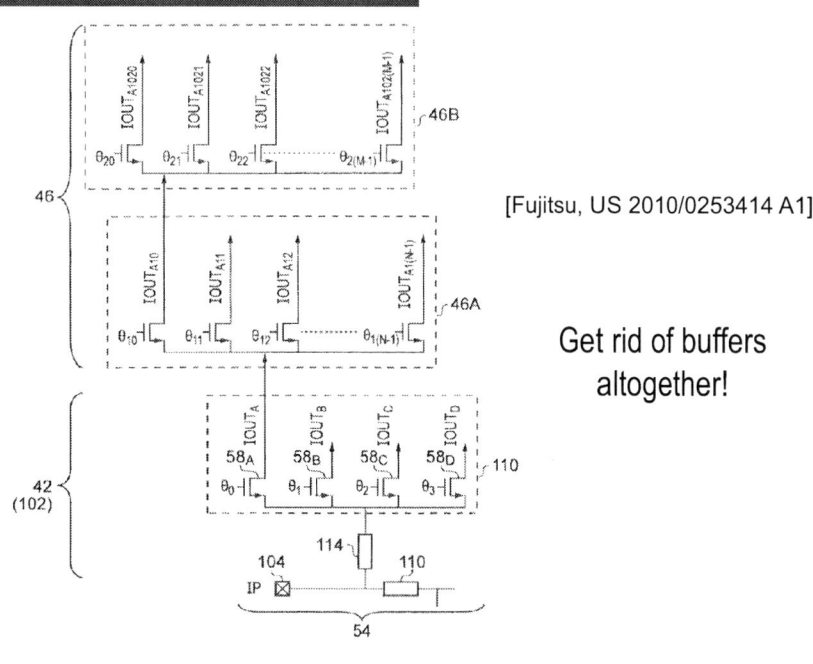

[Fujitsu, US 2010/0253414 A1]

Get rid of buffers altogether!

Comparator Design

- **Flash ADC**
  - Main issue is offset (as discussed previously)
  - Metastability issues are alleviated by fault tolerant encoder
  - In addition, one can employ latch pipelining, see e.g. [Portmann, JSSC 8/1996]

- **SAR ADC**
  - Metastability can be a <u>serious</u> showstopper
  - First need to determine system requirements
    - System with forward error correction (FEC) → ~10-6
    - System without FEC → ~10^{-15} (or better)

Typical Comparator Architecture

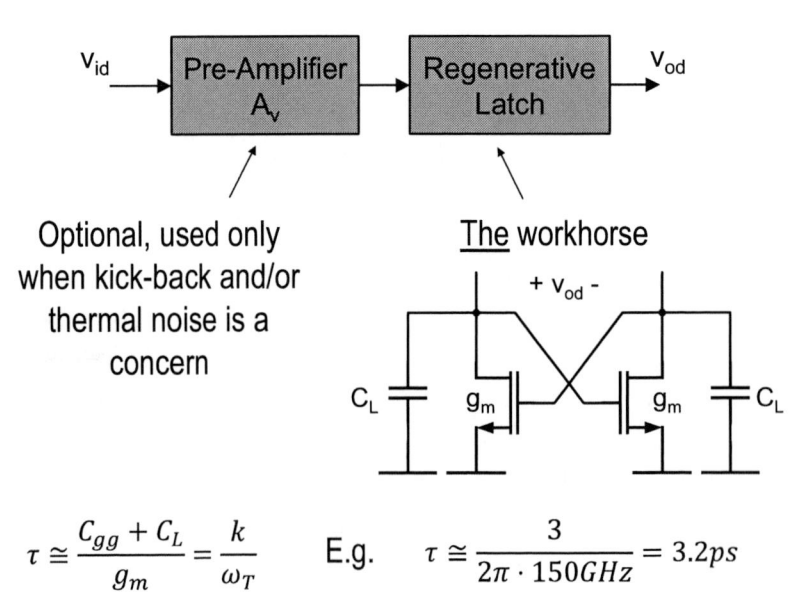

$$\tau \cong \frac{C_{gg} + C_L}{g_m} = \frac{k}{\omega_T} \qquad \text{E.g.} \qquad \tau \cong \frac{3}{2\pi \cdot 150GHz} = 3.2ps$$

Example without Pre-Amplifier

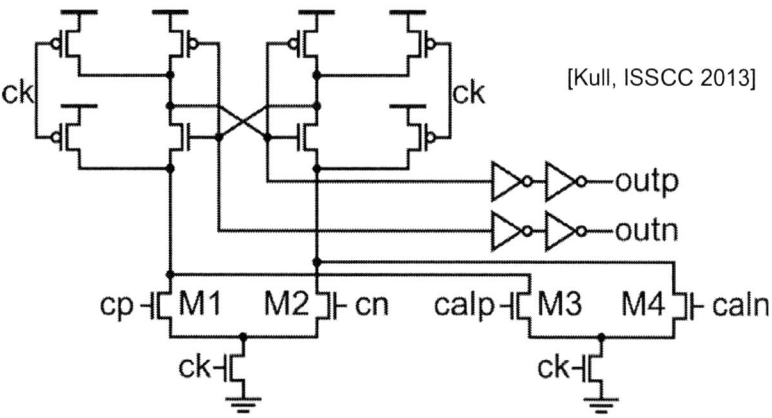

[Kull, ISSCC 2013]

- Input referred noise 1.3 mV$_{rms}$
  - □ Need to a add a pre-amp if this is too large for your application

Latch Regeneration Time

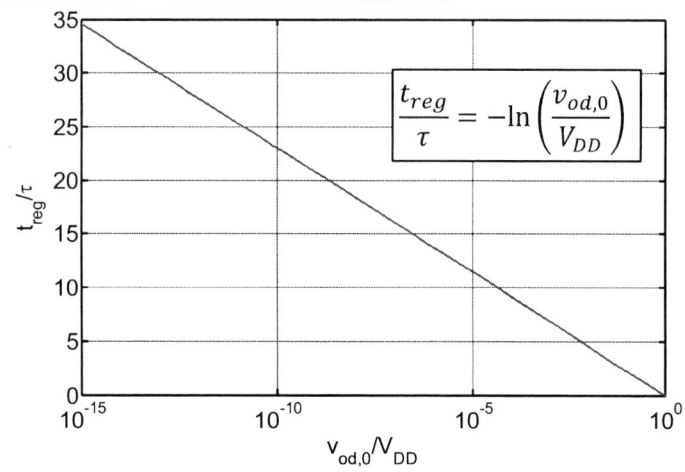

$$\frac{t_{reg}}{\tau} = -\ln\left(\frac{v_{od,0}}{V_{DD}}\right)$$

- Very small inputs lead to long decision time → metastable output
- Let's take a closer look at metastability in asynchronous SAR ADCs

Asynchronous SAR Timing

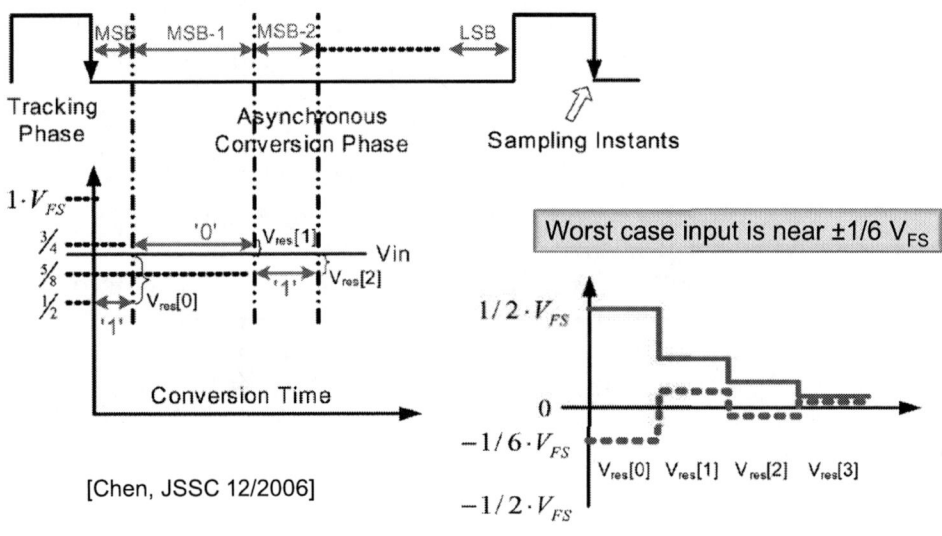

[Chen, JSSC 12/2006]

Worst case input is near ±1/6 V_{FS}

Example: 8-bit SAR (1)

- There are seven "easy" decisions, and one "hard" decision

- It can be shown that the sum of the easy decisions will take anywhere between 25-33τ, depending on the input

- For the hard decision, the input is within ±1LSB, and we can assume that it is uniformly distributed

$$P_{meta} = \frac{v_{od,0,min}}{LSB}$$

$$\frac{t_{hard}}{\tau} = \ln\left(\frac{V_{DD}}{v_{od,0,min}}\right) = \ln\left(\frac{V_{DD}}{P_{meta} \cdot LSB}\right) \cong \ln\left(\frac{2^B}{P_{meta}}\right)$$

Example: 8-bit SAR (2)

Number of time constants to be allocated for the hard decision, given a desired metastability rate (B=8)

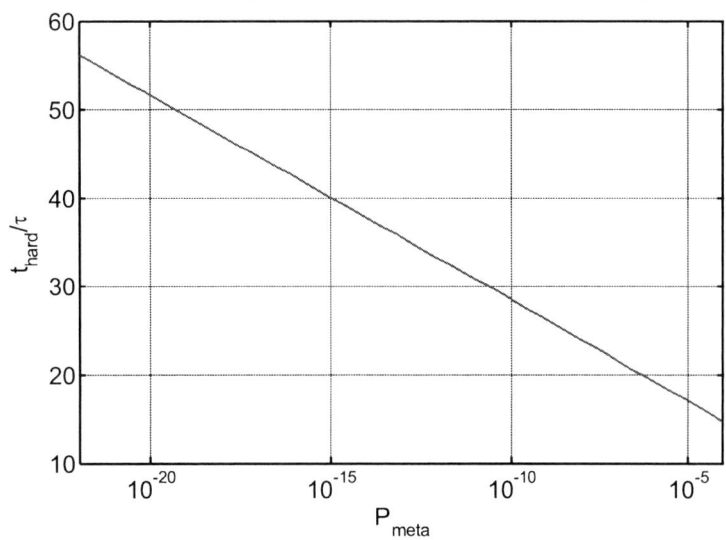

Example: 8-bit SAR (3)

- Continue using the following assumptions (design and technology dependent)
  - Total cycle time T_s = 1ns (1GS/s)
  - Sampling window = $T_s/8$
  - Sum of DAC settling times = $T_s/3$
  - Sum of logic delays = $T_s/6$
- The time available for regeneration is $t_{reg,tot}$ = (3/8)T_s = 375ps
- The metastability rate estimate as a function of τ is

$$P_{meta} \cong 2^B e^{-\frac{t_{hard}}{\tau}} = 2^B e^{-\frac{t_{reg,tot}-t_{easy}}{\tau}} \cong 2^B e^{-\frac{t_{reg,tot}-(25...33)\tau}{\tau}}$$

Example: 8-bit SAR (4)

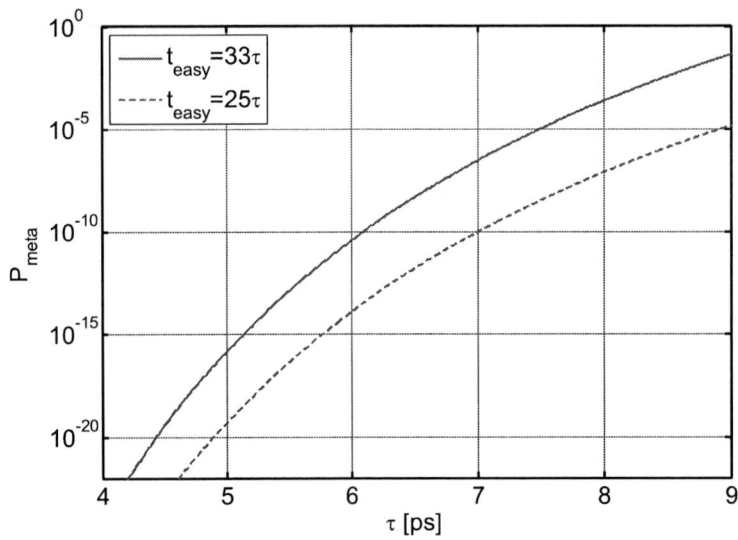

- Incredibly steep tradeoff!

Metastability Detectors

- Some people believe in "metastability detectors" — I don't
  - Why would I want to introduce any extra loading or logic delay to "detect" metastability? Just look at the plot on the previous slide.

- More fundamentally, most (if not all) ideas on metastability detection introduce new metastable points
  - Which may be harder to find/analyze than the original ones, and are hence almost always conveniently ignored
  - Interesting reference: R. Ginosar, "Fourteen ways to fool your synchronizer," Symposium on Asynchronous Circuits and Systems, May 2003

- We need to get our act together as a community and stop advertising metastability detectors without proper analysis and <u>measurements</u>!

Comparator Reset (1)

- Fast reset is just as as important as fast regeneration
- The design below ping-pongs between two comparators to increase the available reset time

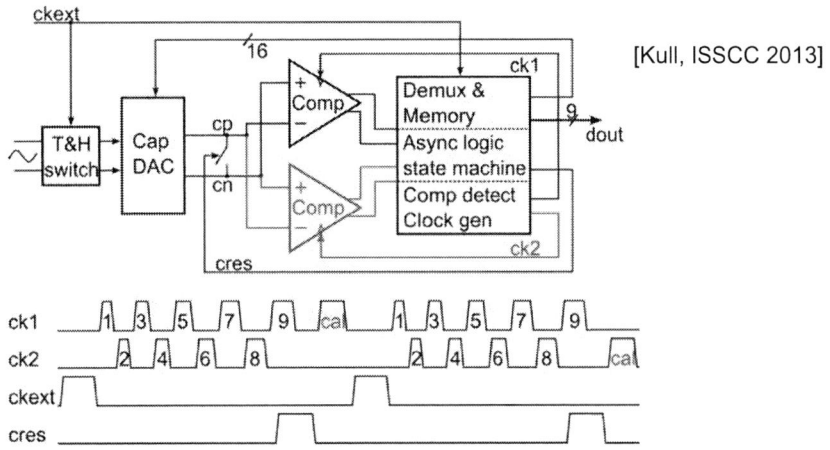

[Kull, ISSCC 2013]

Comparator Reset (2)

- Alternatively, one can try to optimize the reset time against the regeneration time

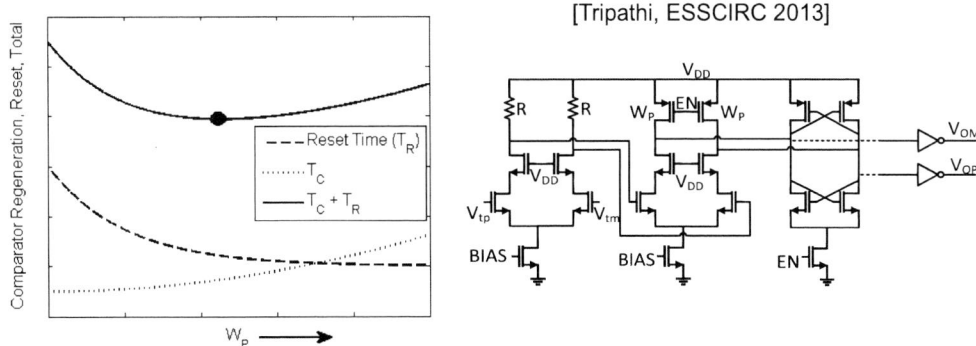

[Tripathi, ESSCIRC 2013]

Summary

- High-speed, low-to-moderate resolution ADCs running at tens of gigasamples per second are presently enjoying a strong application push

- Deriving proper circuit design specs requires close collaboration with the system folks
  - E.g. knowledge of channel, tolerable metastability rate, etc.

- We have come a long way in understanding and improving high-speed time interleaved ADCs
  - But much more work is needed to get to the power levels that upcoming applications will be happy with
  - Both technology scaling and improved design will play a critical role going forward

The Challenge Ahead

$$FOM_W = \frac{P}{f_s \cdot 2^{ENOB}}\Bigg|_{f_{in} \cong \frac{f_s}{2}}$$

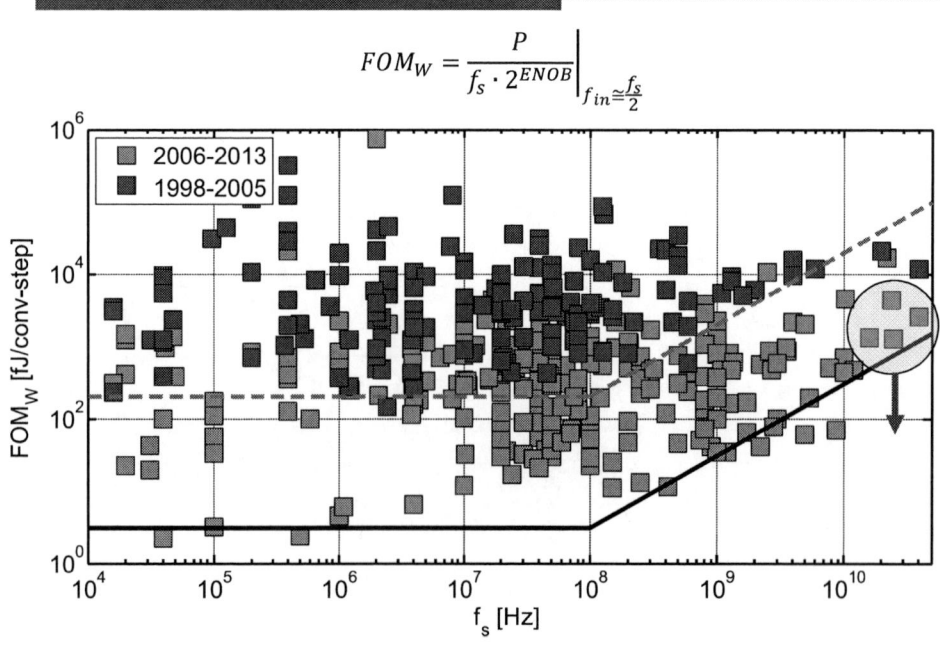

AUTHOR INDEX

Abdelfattah, K. 204
Abdelhalem, S. 49
Abdul-Latif, M. 37
Abugharbieh, K. 299
Agrawal, V. 236
Ahmadi, M. 485
Ahmed, K. 113
Aitken, R. 61
Akré, J. 712
Alam, S. 799
Ali, T. 37
Alioto, M. 263
Alldred, D. 837
Alon, E. 457
Amourah, M. 885
Anceau, F. 712
Andre, T. 799
Aouini, S. 636
Arora, S. 101
Arsovski, I. 825
Asada, K. 188, 212, 660
Atalla, E. 624
Bakhshiani, M. 1
Bakishev, T. 236
Balasubramanian, S. 684
Balsara, P. 624
Banijamali, B. 299
Bawa, G. 620
Becker, B. 640
Bellaouar, A. 624
Belostotski, L. 580
Ben-Hamida, N. 636
Besten, G. 311
Bhagavatula, S. 117
Billoint, O. 712
Bingert, R. 152
Blaauw, D. 13, 252, 263
Bo, Y. 716
Boling, E. 244
Borna, A. 279
Borremans, J. 156
Bourdoux, A. 343
Bousquet, J. 636
Boutros, K. 287
Bowers, J. 889
Braswell, B. 676
Brebels, S. 343
Brooks, D. 109, 708
Brooks, T. 204
Buckwalter, J. 576
Bull, D. 692
Bulzacchelli, J. 315
Burbach, G. 684
Burg, A. 256
Cai, S. 704

Caldwell, T. 837
Cao, P. 700
Casper, B. 323
Catli, B. 37
Cha, S. 156
Chan, M. 105
Chandra, V. 61
Chao, Y. 359
Chen, A. 204
Chen, B. 616
Chen, C. 664
Chen, D. 192, 664
Chen, F. 267
Chen, G. 224
Chen, J. 271
Chen, M. 176, 477, 584
Chen, S. 148, 656
Chen, W. 29, 200
Chen, Y. 29, 263
Chen, Z. 168, 208, 248
Chi, B. 168
Chi, T. 652
Chiang, P. 224, 656
Chiong, D. 477
Chippa, V. 696
Chiu, Y. 133
Cho, S. 873
Choi, J. 640
Chou, M. 200
Chu, R. 287
Chun, K. 821
Chung, Y. 481
Ciftcioglu, B. 592
Cimaz, L. 81
Clark, L. 236
Cline, B. 61
Cogal, Ö. 256
Cojbasic, R. 256
Colinet, E. 712
Colinge, J. 291
Craninckx, J. 156
Darabi, H. 160
Das, S. 692
Dasika, G. 692
De Bock, M. 881
Deng, C. 700, 704
Dhong, S. 291, 817
Ding, H. 734
Do, K. 640
Dong, Z. 53, 144
Driesen, J. 248
Duan, J. 192
Duan, Y. 457
Dudek, P. 632
Elad, D. 275

AUTHOR INDEX

Ema, T. 236
Emami-Neyestanak, A. 5
Esumi, A. 260
Fallahi, S. 37
Farahabadi, P. 339
Fei, W. 355
Feng, P. 271
Ferriss, M. 129
Fick, D. 252
Fischer, P. 628
Fojtik, M. 252
Friedman, D. 129, 275
Fu, Y. 176
Fujiwara, H. 813
Fukuoka, K. 453
Furtner, W. 81
Galal, S. 204
Galayko, D. 712
Gande, M. 720, 730
Gao, L. 208
Garrity, D. 676
Garverick, S. 596
Garzia, F. 248
Gerber, D. 93
Ghosh, A. 833
Ghosh, S. 680
Giannini, V. 343
Giridhar, B. 252
Gogl, D. 799
Goldbach, M. 684
Goss, J. 825
Groeseneken, G. 148
Grzymkowski, P. 825
Gu, C. 845, 869
Gudem, P. 49
Guerber, J. 730
Guo, Q. 17
Gupta, P. 746
Gupta, R. 311
Hamashita, K. 720
Han, J. 716
Hanson, H. 311
Hanumolu, P. 307
Hashemi, S. 469
Hashimoto, T. 453
Haslett, J. 580
Hebig, T. 825
Hellings, G. 148
Hernes, B. 196
Hershberg, B. 720
Hill, B. 608
Hiseh, Y. 176
Holdø, C. 196
Homayoun, A. 172
Hong, Z. 224

Hori, M. 236
Horowitz, M. 857
Houssameddine, D. 799
Hsiao, C. 137
Hsiao, S. 176
Hsu, W. 176
Hu, J. 592
Hu, W. 656
Huang, A. 620
Huang, J. 841
Huang, Z. 208
Hughes, B. 287
Hull, C. 279, 335
Hung, C. 176
Iizuka, T. 188, 660
Ishizone, Y. 660
Ismail, Y. 220
Ito, M. 260, 453
Jalali, M. 331
Janesky, J. 799
Jang, J. 184, 853
Jariwala, D. 628
Jaussi, J. 323
Javidan, M. 712
Jayakumar, H. 696
Jeng, M. 137
Jeong, S. 13
Jiang, X. 204
Joseph, A. 53
Joshi, V. 684
Jou, S. 29
Juillard, J. 712
Jung, B. 117
Jung, D. 572
Jung, M. 628
Jung, S. 121, 572
Kaald, R. 196
Kam, J. 283
Kapusta, R. 724
Karmazin, R. 608
Kawaguchi, H. 612
Kepler, N. 236
Khalaf, K. 343
Kibune, M. 331
Kida, T. 453
Kidd, D. 236
Kim, B. 588
Kim, C. 220, 588, 821
Kim, D. 267
Kim, H. 640
Kim, J. 33, 37, 121, 267, 853
Kim, K. 121, 640, 873
Kim, T. 121
Kim, W. 109
Kim, Y. 640

AUTHOR INDEX

Kinget, P. 93
Kline, M. 93
Komatsu, S. 212
Koo, J. 648
Korniienko, A. 712
Kotani, K. 240
Kousai, S. 644
Krishnan, G. 236
Krishnapura, N. 672
Krishnegowda, S. 885
Kumamoto, T. 232
Kuo, A. 244
Kuo, M. 817
Lam, H. 604
Larson, L. 49
Le, C. 93
Leblebici, Y. 228, 256
Lee, C. 29, 600
Lee, E. 9
Lee, H. 89, 97, 730
Lee, J. 121
Lee, K. 829
Lee, M. 176
Lee, R. 664
Lee, S. 152, 244, 473, 877
Lee, W. 572
Lee, Y. 33, 200, 263, 628
Lei, P. 664
Leinonen, P. 81
Lepkowski, W. 283
Leshner, S. 236
Levantino, S. 41, 524
Li, A. 351
Li, H. 152, 481, 616, 656
Li, J. 576
Li, K. 260
Li, M. 343, 861
Li, W. 208
Li, X. 129, 869
Li, Y. 267, 716
Li, Z. 837
Liao, S. 857
Libois, M. 343
Liempd, B. 156
Lim, I. 676
Lim, S. 628
Lin, C. 137, 817
Lin, D. 267
Lin, H. 799
Lin, J. 473
Linten, D. 148
Littow, M. 81
Liu, C. 604
Liu, J. 461
Liu, L. 700, 704

Liu, R. 224
Liu, W. 271
Liu, Y. 37, 200, 275
Liu, Z. 97
Loh, M. 5
Løkken, I. 196
Long, J. 343
Lopich, A. 632
Lu, F. 53, 144
Lu, M. 604
Lu, N. 152
Luo, X. 351
Luong, H. 351, 359
Lyden, C. 724
Ma, K. 176
Ma, R. 53, 144
Ma, S. 355
Macpherson, A. 580
Maeder, T. 256
Majerus, S. 596
Majidzadeh, V. 228
Mangraviti, G. 343
Manohar, R. 608
Mansuri, M. 323
Martens, E. 156
Mastrangelo, C. 17
Matsuo, M. 576
Matsuzawa, A. 473, 877
McAndrew, C. 676
Meadows, W. 799
Mehr, I. 204
Meinerzhagen, P. 256
Mikhemar, M. 160
Mirzaei, A. 160
Mitani, J. 236
Mitra, S. 640
Miura, S. 660
Miura, Y. 232
Miyahara, M. 473, 877
Miyano, S. 612
Moez, K. 339
Mohapatra, D. 628, 696
Mohseni, P. 1
Mok, P. 105, 489
Moon, U. 720, 730
Mori, R. 453
Morita, S. 453
Moriwaki, S. 236
Morrow, P. 628
Muhammad, K. 176
Mukai, H. 576
Mukhopadhyay, S. 113
Murakami, Y. 660
Murmann, B. 688, 934
Murphy, D. 160

AUTHOR INDEX

Nakura, T. 188
Nam, J. 477
Namgoong, W. 485
Nandwana, R. 307
Natarajan, A. 129
Nazemi, A. 37
Nii, K. 453, 813
Niknejad, A. 279, 335
Nomura, T. 453
Ochiai, T. 453
Oh, S. 180
Oh, T. 720
O'Mahony, F. 323
Otero, C. 608
Otis, B. 648, 772
Otsuga, K. 453
Paik, D. 473
Pamarti, S. 833
Pan, L. 152
Pao, C. 616
Papadopoulos, N. 600
Park, D. 89
Park, J. 572, 644, 652
Parker, B. 129, 275
Parlak, M. 576
Parvais, B. 343
Patch, J. 825
Pavan, S. 57
Payne, R. 461
Piazza, G. 648
Pietromonaco, D. 61
Pileggi, L. 129
Pitchumani, V. 628
Pivonka, D. 184
Plouchart, J. 129
Poon, A. 184
Putnam, C. 152
Qi, N. 168
Qin, Y. 224
Raczkowski, K. 343
Raghavan, P. 343
Raghunathan, A. 696
Rajesh, N. 57
Ramalingam, S. 299
Ranade, P. 236
Randall, M. 152
Rascoe, J. 734
Razavi, B. 172, 465, 469, 738
Ren, J. 355
Reynolds, S. 275
Rhee, W. 267
Rizzo, N. 799
Roberts, N. 180
Rogenmoser, R. 236, 244
Rombouts, P. 881

Roy, K. 696
Roy, R. 236
Ryu, K. 572
Ryu, S. 33
S, R. 672
Sachdev, M. 600
Sadhu, B. 129
Sahoo, B. 465
Sakakibara, K. 232
Samori, C. 41
Sanders, S. 93
Sanduleanu, M. 129, 275
Sauer, M. 640
Saxena, S. 307
Schächer, S. 81
Schenker, R. 77
Schmid, A. 228
Scorletti, G. 712
Segovai-Fernandez, J. 648
Senning, C. 256
Seomun, J. 640
Shahidi, G. 69
Shan, C. 712
Sharma, P. 584
Sheikholeslami, A. 21, 331
Sheinman, B. 275
Shekhar, S. 323
Shi, Z. 144
Shimanouchi, M. 861
Shinohara, H. 612
Shivashankar, K. 692
Shivnaraine, R. 331
Shlafman, S. 275
Sholz, M. 148
Sim, J. 13, 263
Singh, V. 77
Sinha, S. 61
Slater, C. 256
Slaughter, J. 799
Sobue, K. 720
Soens, C. 343
Someya, T. 188
Son, S. 33
Song, D. 121
Song, S. 821
Song, T. 628
Song, Z. 168
Spagnolo, A. 343
Springer, S. 152
Su, D. 101
Su, K. 137
Su, M. 29
Subramanian, C. 799
Sun, L. 53
Sun, N. 829

AUTHOR INDEX

Sun, S.	129, 869	Wang, S.	716
Surapaneni, R.	17	Wang, X.	53, 53, 144
Suster, M.	1, 17	Wang, Y.	248, 279
Suys, H.	156	Wang, Z.	168, 267
Sylvester, D.	13, 252, 263	Webb, C.	628
Szortyka, V.	343	Wei, G.	109, 708
Takahashi, O.	817	Wei, H.	465
Takamiya, M.	612	Wei, S.	700, 704
Takayanagi, K.	453	Wentzloff, D.	180
Takeuchi, K.	807	Whately, M.	885
Tamura, H.	21, 331	Wilk, S.	283
Tan, C.	604	Wojko, M.	236
Tanaka, H.	453	Wong, W.	600
Tanaka, M.	813	Woods, W.	734
Tanaka, S.	813	Wooley, B.	101
Tandon, J.	212	Wu, E.	299
Tanimoto, S.	232	Wu, H.	592, 861
Taylor, G.	628	Wu, M.	481
Tazzoli, A.	648	Wu, N.	271
Tekin, A.	204	Wu, P.	29, 299
Telstø, F.	196	Wyland, C.	299
Thornton, T.	283	Xie, J.	168
Thyagarajan, S.	335	Xu, J.	216, 592
Tian, Y.	604	Xu, W.	588
Tierno, J.	129	Xu, Y.	168, 168
Tong, T.	109, 708	Xu, Z.	877
Townsend, K.	192	Yabuuchi, M.	813
Tripathi, V.	688	Yakovlev, A.	184
Tsai, C.	616	Yaldiz, S.	129
Tsai, T.	200	Yamada, J.	453
Tse, J.	608	Yamada, T.	236
Tsukamoto, Y.	813	Yamaguchi, T.	212
Tsuruta, T.	236	Yang, C.	220
Tuinhout, H.	397	Yang, H.	208
Vaesen, K.	343	Yang, L.	656
Valdes-Garcia, A.	129, 275, 734	Yang, M.	600
VanBentum, R.	684	Yang, P.	817
Venkatram, H.	720, 730	Yang, S.	841, 853
Venugopalan, S.	684	Yang, X.	461
Vidojkovic, V.	343	Yang, Z.	216, 656
Wachnik, R.	152	Yeo, H.	33
Wakayama, S.	236	Yeric, G.	61
Wambacq, P.	343	Yin, J.	351
Wan, D.	664	Yin, S.	700, 704
Wan, L.	224	Yoon, Y.	829
Wan, Y.	628	Yoshimoto, M.	612
Wang, A.	53, 144, 363	Yoshimoto, S.	612
Wang, D.	53, 700	Youn, S.	121
Wang, F.	129	Young, D.	17
Wang, G.	133	Yu, H.	355
Wang, H.	271, 279, 628, 644, 652	Yu, W.	873
Wang, K.	176	Yuan, J.	841
Wang, L.	53, 144	Yuan, M.	200
Wang, M.	817	Yue, C.	53
Wang, P.	817	Zamudio, L.	684

AUTHOR INDEX

Zeng, T. 192
Zeng, X. 716
Zhang, C. 144
Zhang, D. 208
Zhang, F. 624
Zhang, J. 604
Zhang, P. 465
Zhang, W. 704
Zhang, X. 109, 708, 799
Zhao, D. 236
Zhao, H. 144
Zheng, S. 351
Zhou, C. 604
Zhu, H. 724
Zhu, M. 700
Zhu, W. 208
Zhuo, H. 267
Zianbetov, E. 712